AF528496

Instrumentation in Analysis of Food and Dairy Products

The Author

Dr M. K. Srivastava has done M.Sc (Organic Chemistry) and D.Phil (Micro Analysis of Organic Compounds) from Allahabad University, Allahabad. He has got one year training in Quality Analysis of Milk and Milk Products from National Dairy Development Board (NDDB), Anand and different trainings in analysis of Milk products and Minerals from National Dairy Research Institute (NDRI), Karnal.

He has more than 35 years working experience in Dairy industry and well versed with the Quality and Food Safety Management for Food and Dairy Products and Lab accreditation under NABL. He has good experience in dealing with National (BIS, Agmark, FSSA-2006, NABL etc) and International (Codex, IDF etc.) Quality as well as Accreditation agencies.

He is working as Senior Chemist in Pradeshik Co-operative Dairy Federation Ltd and worked as HOD (Q.A. and R&D) Incharge, Central Lab., Lucknow and undertaken various projects of Council of Science and Technology, U.P. UPDASP etc. He is also academic counsellor in IGNOU for DDT programme and visiting faculty for M.Sc. (Food Processing), Lucknow University. He has published 20 research papers and two books. He is an approved Food Safety Lead Auditor.

Instrumentation in Analysis of Food and Dairy Products

– Author –

Dr M. K. Srivastava

M.Sc., D.Phil (A.U.)

Senior Dairy Chemistry and Food Safety Lead Auditor
Department of Quality Assurance
Pradeshik Cooperative Dairy Federation Ltd.
29, Park Road, Lucknow
Uttar Pradesh

2019

Daya Publishing House®

A Division of

Astral International Pvt. Ltd.

New Delhi – 110 002

ISBN: 978-93-88173-52-0 (HB)

Published by : **Daya Publishing House®**
A Division of
Astral International Pvt. Ltd.
– ISO 9001:2015 Certified Company –
4736/23, Ansari Road, Darya Ganj
New Delhi-110 002
Ph. 011-43549197, 23278134
E-mail: info@astralint.com
Website: www.astralint.com

Digitally Printed at : **Replika Press Pvt. Ltd.**

This copy is for sale in Indian Sub-Continent.

Dedicated to my Respected

Father
Late Sri Raj Kishore Srivastava

Mother
Late Smt Shanti Srivastava

Eldest Brother
Mr S.K. Srivastava

Bhabhiji
Mrs Chanda Srivastava

Acknowledgements

The author is indebted to Managing Director, Pradeshik Co-operative Dairy Federation Ltd., Milk Commissioner (U.P.) and Principal Secretary (Dairy Development) Govt of U.P. for providing me opportunity to write this book.

The author expresses his gratitude and thanks to all Dairy chemists, Dairy/ Food Scientists, Dairy Institutions, Websites and Manufacturers of Lab Instruments under references for providing technical information.

The author is also thankful to Chief General Manager, PCDF, Lucknow and Managers (QA) of PCDF, especially Mrs S. Chopra and Mr H.S.Verma. for inspiring him for the same.

The author is genuinely indebted to his Mother, Father, Brothers, sisters, wife, kids and other family members for providing continuous moral support without which, it is impossible for author to complete the task.

Last but not least, author acknowledges his thanks to all his friends and all staff of Central Quality Assurance Lab, PCDF Lucknow.

Dr M. K. Srivastava

Preface

In India, dairy industries are mostly based on the buffalo milk and hence there is significant seasonal variation in the milk production. The aforesaid seasonal variation in milk production and low productivity of milch animals causes remarkable gap in "Demand and Supply" which encourages the antisocial element to adulterate the milk or to produce so called "Synthetic milk" (Milk likes liquid produced by addition of cheaper fat and chemicals) which ultimately deteriorate the quality of milk products.

Quality of milk and milk products play a very important role as far as consumer acceptability is concerned and if dairy industry is not giving emphasis on quality then it is very difficult to survive in present competitive national and international market.

The above cited national and international pressures on the quality of milk/milk products, compel the dairy industries to equip the laboratories with latest analytical equipments, skilled man power and technical literatures related to the analysis of quality of milk/milk products because the aforesaid requirements can only make the dairy industries able to evaluate the quality of milk and milk products properly and accurately. Realising the importance of compilation of technical literatures related to analysis of milk and milk products, author has made successful attempt to produce a compiled and completed advanced technical information related to the analysis of quality of milk and milk products in the form of **"Handbook on Analysis of Milk" (Chemical and Microbiological Analysis of Liquid Milk) and "Handbook on Analysis of Milk Products" (Chemical and Microbiological Analysis of Milk Products)**.

The Traditional analytical methods although giving accurate test results but these methods are generally time consuming. The Advanced and Instrumental analytical methods being accurate, quick, fast and reliable have established upper hand over traditional analytical methods. To provide a compiled and completed advanced technical information related to the instrumental analysis of milk, milk products, other food products *etc.*, author has now made an attempt to write a book entitled **"Instrumentation in Analysis of Food and Dairy Products" (Part 1 and 2).** The present **Part-1 publication** deals with Analysis of Body/Texture and Taste/ Flavour of Milk Products by Body Texture Analyser and Electronic Nose, Chemical analysis by Spectroscopic Methods (UV-Vis. Spectroscopy, IR Spectroscopy-NIR (Near Infrared), MIR (Mid Infrared), FTIR, Nuclear Magnetic Resonance-NMR Spectroscopy, Mass Spectroscopy, Microwave Spectroscopy, Atomic Spectroscopy, Ultrasound Spectroscopy), Chromatographic Methods (Paper Chromatography, Thin Layer Chromatography, Gas Chromatography, High Pressure Liquid Chromatography), Polarimetry, and Rapid Microbiological analysis where as **Part- 2 publication** deals with Immunoassay techniques Biosensor techniques, Electrophoresis, Polarography, Voltammetry and Microbiological analysis by PCR)/ Real Time PCR and Flow Cytometry. Author expects that the said publications are able to meet the requirements of dairy men as well as Dairy Educationists.

Dr M. K. Srivastava

Contents

— Part 2 —

Section A: Instrumentation in Analysis of Taste of Food/Milk Products

Section B: Instrumentation in Chemical Analysis

Section C: Instrumentation in Food Safety and Microbiological Analysis

Part 1

Section A

Instrumentation in Analysis of Body/Texture and Flavour of Milk Products/Foods

Chapter 1

Analysis of Body/Texture and Flavour of Milk Products

Sensory quality of food refers to a set of characteristics such as appearance, colour, taste, smell, aroma, texture and auditory properties. Each of the above characteristics can be viewed separately, but together they represent an important food quality determinant (1-3).Food acceptance among the consumer is governed by Appearance, Flavour, Texture and Nutrition. The organoleptic properties such as taste, texture, odour, mouth feel and appearance are of considerable importance in differentiating products in the market and can ultimately determine the success of the product. Generally in dairy industry, Body and Texture, Taste and Flavour and colour of milk products are being judged organoleptically and there is system which comprises organization of panel of Experts/Judges and evaluation of organoleptic quality as per scorecard *etc.* Assessment by human test panel increases time and money. The skill of judging the quality of milk/milk products/food organoleptically *i.e.* appearance, colour, odour and flavour can be acquired by some practice. An experienced person can find out bad samples of foods with fair degree of accuracy. In dairy industry, milk is received at village level Dairy cooperative societies/collection centers after performing aforesaid organoleptic tests. Realising the importance of aforesaid tests, scorecards are made available for milk/milk products on the basis of results obtained after performing these tests. The national (BIS) and international quality agencies have prescribed scorecards for quality of milk/milk products on organoleptic parameters to grade and to have good consumer appeal/acceptability. Some of which are given below:

Sl.No.	*Characteristics*	*Milk*	*Butter (Points)*		*Ghee*	*Ice cream*
			Min.	*Max.*		
1	Flavour and Odour	40	40	50	50	45
2	Sediment	10			15 (Free from Suspending Impurities)	-
3	Body and Texture	-	15	20	20	30
			at 15-16° C			
4	Colour	-	15	20	10	15
5	Appearance and Finish	-	5	10	-	-
6	Required scores for excellent Quality	40 and Above	95 and Above		91 and Above	-

Taste is an important organoleptic property governing acceptance of products for administration through mouth. Taste assessment is important quality control parameter for evaluating taste of foods. The primary method for the taste measurement of Food/Milk Products is by human panelists. The use of sensory panelists is very difficult and problematic in industry and this is due to the problems in recruiting experienced taste panelists, motivation and panel maintenance are significantly difficult. Therefore, analytical taste-sensing multichannel sensory system called as electronic tongue (e-tongue or artificial tongue) system could potentially assist, or even replace, a sensory panel in certain type of routine analysis in development and production. It gives precise measurement of taste. Electronic tongue measurements can be performed as often as analytically needed without the regulatory hurdles or expense of human testing. The exposure and risk of using employee to test products is greatly reduced. Thus, e-tongue includes benefits like reducing reliance on human panel.

Odour/Flavour testing seems mysterious and odour data mythical to most practitioners. In millennium past the "practice of odour" was in the hands of wizards and experts. For years engineers and operators have relied on "odour experts" to interpret odor testing results. Today odour, odour control *etc.* are understandable subjects for plant operators, facility managers, engineering practitioners, and consumers (citizens). Human olfaction supplies 80 per cent of flavor sensations during eating. Despite the importance of the olfactory sense to mankind, the sense of smell in man is often considered the least refined of the human senses, far less sensitive than that of other animals. For example, the human nose possesses only about one million aroma receptors that work in tandem to process olfactory stimuli whereas dogs have about 100 million receptors that distinguish scents at least 100 times more effectively than the average human [4,5].

Odor is measurable and quantifiable using standard practices as published by the **American Society of Testing and Materials (ASTM E679 and E544)** and by the

European Union. In 2000 the proposed European Normalization Standard, prEN 13725, will be implemented and become the de facto "International Standard" for odor/odour testing [6].

The Body and Texture of some of the Foods and Dairy Products is not only an important characteristic but also it has to be maintained to the required degree for **consumer acceptability.** Texture measurement may be carried out by either **Sensory analysis or Instrumental analysis**. Sensory methods permit more complex attributes to be evaluated by the trained panelists and thus cost more to administer. Instrumental methods, on the other hand, are reliable within their limitations and are more economical to use. Now Texture analyzers and some other instruments based on other techniques are also being used.

The information given in following table help us in understanding positive and negative points of the Organoleptic Sensory Analysis and Instrumental Texture Analysis [7].

Organoleptic Sensory Analysis	*Instrumental Texture Analysis*
It includes use of the senses of smell, taste, sound and touch. Evaluation of texture of food *etc.* by touch includes the use of the fingers, as well as the lips, tongue, palate and teeth in the mouth. Sensory methods of analysis are subject to wide variability, are labour intensive and therefore expensive.	Instrumental methods of assessing texture can be carried out under more strictly defined and controlled conditions. Problems of experimental variability are more likely to be caused by sample heterogeneity than by instrumental imprecision.
The changes in ingredient levels cause several simultaneous changes in product characteristics. Some of these changes are difficult to mask and thus tend to make sensory analysis difficult.	It is not the case with Instrumental Texture analysis. The change in products characteristics due to change in ingredients level, can be easily observed by this method.

The main goal of many texture studies is to devise one or more mechanical tests with the capacity to replace human sensory evaluation as a tool to evaluate a wide range of textural properties.

Numerous works have shown that electronic nose and electronic tongue pose as excellent non-destructive methods for the determination of volatile and non-volatile food compounds. **Although human olfactory and taste systems and sensory assessment of food cannot be replaced by any device.**

But presently, to overcome the problems raised during organoleptic evaluation, the instrumental techniques *e.g.* Body Texture Analyser, Electronic Nose, Electronic Tongue *etc.* are now being used to judge the Body and Texture and Taste and Flavour of Dairy/Food Products.

Source

1. *Piggot, J.R. (1994): Understanding flavour quality: difficult or impossible? Food Quality and Preference 5 (1-2), 167-171. doi: dx.doi.org/10.1016/0950-3293 (94)90024-8.*

2. *O'Riordan, J.P., Delahunty, C.M. (2003): Characterisation of commercial Cheddar cheese flavour. 1: traditional and electronic nose approach to quality assessment and market classification. International Dairy Journal 13, 355-370*

3.. *Biolatto, A., Grigioni, G., Irurueta, M., Sancho, A.M., Taverna, M., Pensel, N. (2007): Sea-sonal variation in the odour characteristics of whole milk powder. Food Chemistry 103, 960-967. doi: dx.doi.org/10.1016/j.foodchem.2006.09.050,*

4. *Ouellette, J. Electronic noses sniff our new markets. Indust. Physic. 1999, 5, 26-29.*

5. *Alphus D.Wilson and Manuela Baietto, Applications and Advances in Electronic-Nose Technologies, Sensors 2009, 9,5099-5148.*

6. *Charles M. McGinley, Michael A. McGinley, Donna L. McGinley, Odor Basics-Understanding and Using Odor Testing, presented at The 22nd Annual Hawaii Water Environment Association Conference Honolulu, Hawaii: 6-7 June 2000.*

7. *Operating Manual of Stable Micro Systems.*

1. Analysis of Rheological Properties of Foods [1,2,3]

In general, rheological properties are referred to as the "texture" of the food. Rheology is the science of flow and deformation of matter and describes the interrelation between force, deformation and time. Foods being diverse and complex materials in nature exhibit a wide range of different rheological properties. To characterize them, different instruments have been developed. The application of instruments vary according to the type of sample as well as the property to be measured.

On the basis of rheological classification, foods can be divided into following two groups:

(i) **Solid**: Solid foods are studied for their textural attributes using Texture analyser.

(ii) **Liquid:** The liquid foods are characterized for the flow properties using viscometer.

Semisolid foods, therefore, are studied using both instruments. Viscometer and Texture analyser (Universal) are two major such instruments which, when combined, can cover whole spectrum of rheological analysis. The sub-classification of Liquid and solids foods on rheological behavior are tabulated on next page.

Source

1. *Sanket Borad and Ashish Kumar Singh, Instrumental Measurement of Rheological Properties of Dairy Foods, Compendium on Advanced Techniques and Novel Approaches for Quality and Safety Evaluation of Dairy Foods-2015, National Dairy Research Institute, Karnal.*

2. *C. Wilkinson et al. (2000), Trends in Food Science and Technology, 11: 442-450.*

3. *K. Kohyama (2009), Chemistry and Biology, 47: 133-137.*

A. Rheological Analysis of Solid Food by Body Texture Analyser

The principle of a texture measurement system is to deform the sample in predefined controlled conditions and simultaneous measurement its response. The force experienced by the load cell equals to the stress generated in the food.

Sl. No.		Types of Food										
		Liquid Foods							Solid Foods			
1	Rheological Property	Flow							Deformation			
2	Sub classification	Viscous					Plastic		Elastic		In Elastic	
		New-tonian	Non Newtonian									
			Shear Dependant		Time Dependant		Bingham	Non Bingham	Hookean	Non Hookean	Visco-elastic	Visco plastic
			Pseudo-plastic	Dilatant	Thixo-tropic	Rheo-pectic						
3	Example	Water Milk (up to 30 per cent TS), Vegetable oils, Fruit juices, Sugar and salt solutions	Apple sauce, Banana, puree Juice concen-trate, Oyster sauce, CMC solution	Liquid chocolate 40 per cent, Corn starch solution	Gum solutions Set, stir red and drinking yogurt	No foods	Tomato ketchup, Mayon-naise, Whipped cream, Whipped egg, white Butter and Margarine	-	Firm green fruits and vege-tables Hard candy Eggs in the shell, *etc.* at very low strain	-	Practically most solid and semisolid foods	-

The magnitude of force is plotted against the time/distance, which is shown on the digital display. The measured forces are the function of various physical properties of food material as well as test parameters. Based on the types of products and properties to be measured, the texture measurement is of the following two types:

(i) **Texture analysis or single bite compression**.

(ii) **Texture profile analysis** - **TPA** or double bite compression. The whole process of texture analysis mimics the jaw movements that foods experience when we eat them.

Food Texture

As per ISO 8492 (Vocabulary for sensory evaluation), following is the official definition of Food Texture:

All the mechanical, geometrical and surface attributes of the product that are perceptible by mean of mechanical, tactile, visual and auditory receptor.

It is a mechanical analysis primarily concerned with measurement of the mechanical/physical properties of a food product/Dairy Product, as these products relate to their sensory properties detected by consumers. Because of its adaptability, texture analysis has become common place in many industries to measure a specific or range of characteristics or properties relating to the way a material behaves, breaks, flows, sticks, bends, *etc.*

Texture is essentially the effect of physical and chemical properties on our senses of sight, hearing and touch and is an important attribute in our perception of food. If texture is not right, we do not like the food - even if the flavour is OK. *e.g.* we do not like - i- Soggy crips and biscuits or ii- tough and over chewy meat. Because of the importance of the texture and texture analysis, major manufacturers of the food products routinely apply texture analysis techniques both in new product development and as part of quality control in all stages of manufacture.

More than fifty years of texture research has developed a set of definitions relating the sensory properties of a product to the instrumental properties which can be calculated from the results of a two cycle texture profile analysis test. Texture analyzers perform this test by applying controlled forces to the product and recording its response in the form of force, deformation and time.Consumer products succeed in the market place because their "textural characteristics" are pleasing to customers. This is certainly true with food products but it also applies to cosmetics, pharmaceuticals, packaging, industrial materials and even adhesive type materials.

Texture analysis is attempts to replicates our perception of texture. It involves material testing and rheology. Texture analyser can be employed for the measurement of rheological properties of almost all the solid and semisolid foods. It can measure and quantify **fundamental, empirical as well as imitative** tests in both compression and tension. The data generated using it can be easily correlated to those obtained during sensory evaluation of food products. Measurement can give Fundamental, Empirical and Initiative results.

Material Testing	Texture Analysis	Rheology
Elasticity	Chewiness	Visco- Elasticity
Plasticity	Hardness	Viscosity
Modulus	Crispiness	Powder Flow
	Gumminess	
	Stickiness	

Texture Measurement

Texture measurement may be done by the following ways:

1. Sensory Analysis

Sensory methods permit more complex attributes to be evaluated and cost more to administer. Sensory analysis of texture is based on the texture profiling which requires a trained panel who produce ratings based on three phases of chewing the food:

(i) **Initial effect**- perception on first bite,

(ii) **Masticatory** - perception of chewing and

(iii) **Residual**- changes perceived during chewing and swallowing.

Texture Profiling

The initial and masticatory effects are subdivided into mechanical, geometrical and other - essentially moisture and oiliness. The mechanical effects are further subdivided and are rated on a numeric scale.

Initial Effects

Hardness, Fracturability, Viscosity, Masticatory, Gumminess, Chewiness and Adhesiveness.

Residual Effects

Rate of breakdown, Type of breakdown, Moisture absorption and Mouth coating.

Sensory methods, properly conducted give reliable results but there are problems which include:

(i) Training and maintaining,

(ii) Sophisticated analytical tools are needed

(iii) Costs.

2. Instrumental Analysis

Instrumental methods, on the other hand, are reliable within their limitations and are more economical to use. These try and replicate some of the results of sensory methods. There are following three main approaches:

(i) **Fundamental methods**: based on measurement of physical properties such as stress/strain and rheology.

(ii) **Initiative methods**: aim to simulate chewing mechanically.

(iii) **Empirical methods**: tend to use specific devices to give specific measurements which have found to have practical value.

Fundamental Methods

Types of Force-Tensile forces, Compressive forces, Shear forces.

a: Tensile Forces Tend to Stretch the Material.

b: Compressive Forces Tend to Squash the Material.

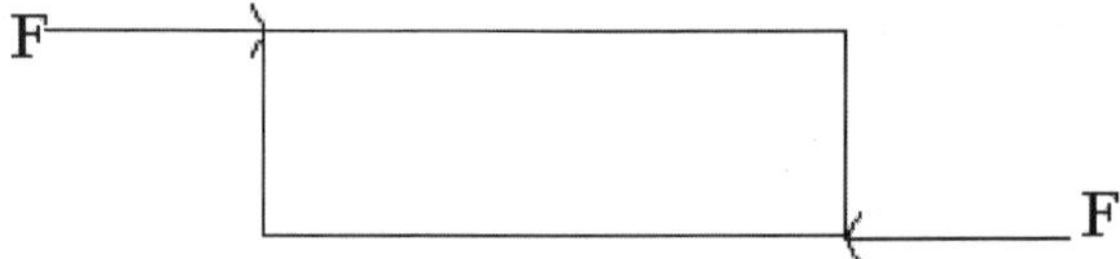

c: Shear Forces Tend to Distort or Displace the Material.

Elasticity

An elastic material is one which returns to its original shape and size when any deforming is removed, If A graph of load vs extension is a straight line, the material is said to obey **Hooke's Law.**

Stress and Strain

The extension resulting from the application of a force is a property of

(i) The dimensions of the material and

(ii) The physical properties of the material.

Stress and strain are quantities which are related to force and extension in such a way astom incorporate the material dimensions. Stress is defined as force per unit area whereas Strain is defined as extension per unit length. Tensile stress is defined as $\sigma = F/A$ (units = N m^{-2} or Pa). Tensile strain is defined as $\varepsilon = h/\Delta h$ (units: dimensionless).

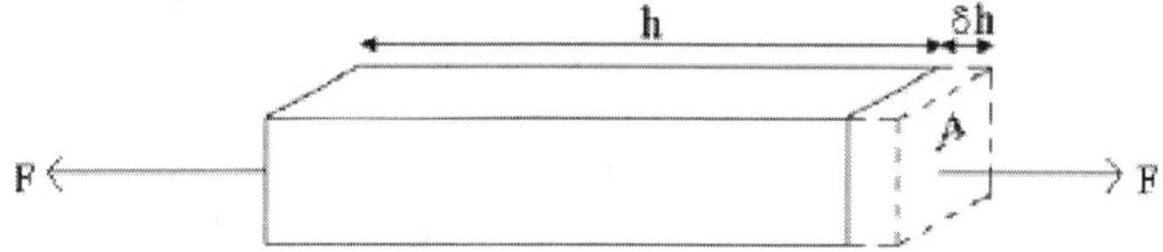

Stress-Strain Curves

If a graph of stress vs. strain is plotted a graph similar to that below may be produced.

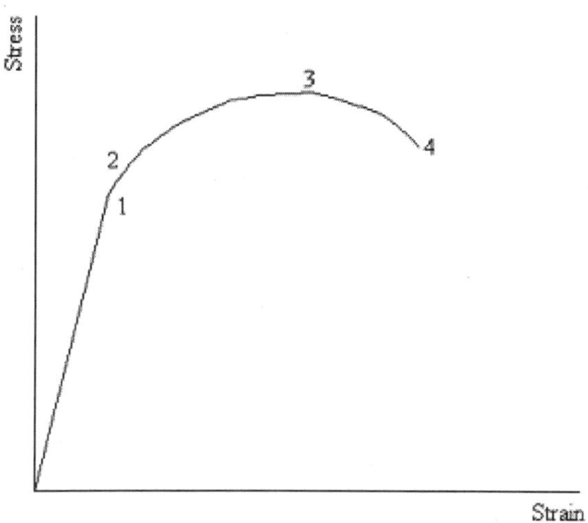

Typical Stress - Strain Curve.

The key points on the curve are:

(i) The elastic limit,

(ii) The yield point,

(iii) The Point of maximum stress (or failure), and

(iv) The point of fracture.

Parameters Required to be Measured during Texture Profile Analysis of Foods/Dairy Products [1]

The following parameters are required to be measured during Texture Profile analysis:

Sl.No.	*Parameter*	*Sensorial Definition*	*Instrumental Definition*
1	Hardness (H)	Force required to compress a food between molars. Defined as force necessary to attain given deformation.	Peak force of the first compression cycle.

Contd.

Sl.No.	Parameter	Sensorial Definition	Instrumental Definition
2	Brittleness (B)	Force at which a material fractures. Related to the primary parameters of hardness and cohesiveness, where brittle materials have low cohesiveness. Not all foods fracture and thus value may relate to hardness if only single peak is present. Brittle foods are never adhesive.	The first significant break in the first compression cycle.
3	Adhesiveness (A)	The work necessary to overcome the attractive forces between the surface of the food and the surface of other materials with which the food comes into contact (*e.g.* tongue, teeth, palate).Work required to pull food away from a surface.	The negative area for the first bite, representing the work necessary to pull compressing probe away from sample.
4	Cohesiveness (Co)	The strength of internal bonds making up the body of the product (greater the value the greater the cohesiveness).	The ratio of positive force during the second to that of the first compression cycle (downward strokes only).
5	Springiness (S)	Rate at which a deformed material goes back to its original condition after deforming force is removed.	Height that the food recovers during the time that elapses between the end of the first bite and the start of the second bite.
6	Chewiness (Ch)	Energy required to chew a Solid food product to a state where it is ready for swallowing. attribute is difficult to quantify precisely due to complexities of mastication (shear, compression, tearing and penetration)	Calculated Parameter: Product of Gumminess x Springiness (or Hardness x Cohesiveness x Springiness.
7	Gumminess (G)	Energy required to disintegrate a SEMI-SOLID food product to a state ready for swallowing. Related to foods with lowhardness levels.	Calculated parameter: Product of Hardness x Cohesiveness.

Source

1. *Sanket Borad and Ashish Kumar Singh, Instrumental Measurement of Rheological Properties of Dairy Foods, Compendium on Advanced Techniques and Novel Approaches for Quality and Safety Evaluation of Dairy Foods-2015, National Dairy Research Institute, Karnal.*

Advantages and Disadvantage of Instrumental Methods of Texture Analysis [1]

The Instrumental methods of Texture Analysis also has advantages and disadvantage. The details of advantages and disadvantages of instrumental analysis of texture, are summarized below:

Sl.No.	Types of Result	Advantages	Disadvantages
1	**Fundamental:** These tests do not correlate as well with sensory evaluation and are used mainly in research Labs.	Measurement of well defined mechanical and rheological properties.	Offers poor correlation with sensory methods. Depend upon uniform and Homogenous samples. Result interpretation can be difficult.

Sl.No.	*Types of Result*	*Advantages*	*Disadvantages*
2	**Empirical:** These tests measure parameters that may be poorly designed or depend upon the type of product containment. However they have been found to replace to textural quality and subsequently are the most widely used in food industry.	Simple to perform. Rapid. Suitable for routine quality control. Good correlation with sensory methods.	Often there is no fundamental understanding of the test. Arbitrary procedure- Frequently no recognized Standard is available.
3	**Initiative:** These tests initiates the correlation to which the sample is subjected to practice.	Close replication of mastication or other sensory methods. Good correlation with sensory methods.	Often there is no fundamental understanding of the test.

Source

1. *Technical literature of Body Texture Analyser of M/S Stable Micro Systems.*

Types of Texture Analysers

Followings are the few examples of the available Texture Analysers of different manufacturers:

1. **Stable Micro Systems:** TA.XT plus -Single Column, TA.HD plus -TwinColumn and TA.XT express -Single Column are the available models.
2. **Brookfield Engineering Lab, USA:** Different models of CT3 Texture Analysers are available.

The characteristics of different models of Texture Analysers of different manufacturers available in the market are given below in tabulated form:

Sl.No.	*TA.XT Plus*	*TA.HD Plus*	*TA.XT Express*
A	**M/S Stable Micro System**		
1	Most popular single column Texture Analyser for universal application.	High capacity Versatile twin column Texture Analyser.	Simple stand-alone Texture Analyser with limited capacity and testing stroke.
2	0.01-40mm/s Speed	0.01-13mm/s Speed.	0.01-10mm/s Speed.
3	Totally user programmable sequencing.	Totally user programmable sequencing.	Separate Keypad operation.
4	Multi channel Data acquisition.	Multi channel Data acquisition.	Operated using Expression software.
5	Integral Key boards.	Integral Key boards.	Menu selected Results.
6	Operated using Exponent 32-bit software	Operated using Exponent 32-bit software	-

Contd.

7	**Load cells-**1,5,30 and 50 kg. **Capacity-** 50 kg	**Load cells-**1,5,30,50, 100, 250, 500 and 750 kg. **Capacity-**750 kg	**Load cells-**1,5 and 10 kg. **Capacity-** 10 kg
8	**Max, Data Acquisition-** 500 pps	500pps	500pps
8	Features of Intelligent Load Cell – Automatic detection of load cell capacity. Load cell information such as capacity, calibration and serial numbers are stored within the load cell and are automatically detected upon installation.		

Sl.No.	*Characteristics*	*CT3 Models*						
		100	*1000*	*1500*	*4500*	*10 kg*	*25 kg*	*50 kg*
B	**M/S Brookfield**							
1	**Load cell range**	0.1 kg	1 kg	1.5 kg	4.5kg	10 kg	25 kg	50 kg
2	**Range (g)**	0-100	0-1000	0-1500	0-4500	1-10000	2-25000	5-50000
3	**Resolution (g)**	0.01	0-1	0.2	0.5	1.0	2.0	5.0
4	**Accuracy (FSR)***	±0.5 per cent	±0.5 per cent	±0.5 per cent	±0.5 per cent	±0.5 per cent	±0.5 per cent	±0.5 per cent
5	**Speed**	0.01 to 0.1 mm/s in increments of 0.01 mm/s 0.1 to 10 mm/s in increments of 0.1 mm/s Accuracy: ±0.1 per cent of set speed						
6	**Temperature**	-20°C to 120°C (Requires optional probe, DVP-94Y)						
7	**Output**	RS232 Compatible Serial Port, USB Port						
8	**Environmental Conditions**	0°C to 40°C temperature range (41°F to 104°F) 20 per cent - 80 per cent relative humidity, non-condensing atmosphere						
9	**Use**	Intended for indoor use only, Altitude: up to 2000m						

FSR: Full Scale Range.

This value applies to the full range of operating temperatures as defined later in this section under Environmental Conditions. Accuracy is 0.2 per cent FSR, when the instrument is operated in a stable ambient temperature (20°C - 25°C).

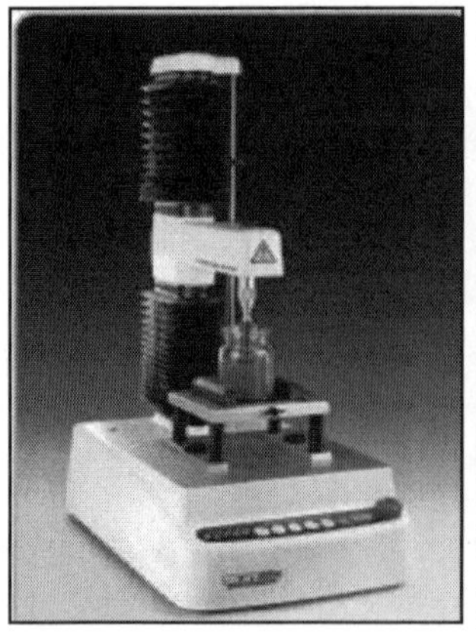

TA.XT Plus

TA.HD Plus

TA.XT Express

Methods of Measuring Texture by Analysers of M/S Stable Micro System

Stable Micro Systems' range of instruments can measure and analyse fundamental, empirical and imitative tests, covering those relating to texture analysis, materials science and rheology of solid, semi-solid, viscous liquid, powder and granulate materials. The said Texture analysers are instruments that measures the responses of a sample to:

(i) Tension

(ii) Compression

(iii) Penetration

(iv) Bend

Basic Principle of Texture Analysis

(i) Probe begin to move from start position towards sample at the pre-test speed.

(ii) When the probe measures the trigger force the speed changes to the test speed and force, distance and time is recorded.

(iii) Probe travels into the samples at the test speed until the specified force, distance or strain is reached.

(iv) The probe returns to the start at the post test speed.

(v) When the probe passes the original trigger point any surface stickiness will be detected.

(vi) The probe has returned to the start position.

Parameters to be Measured

The following parameters are to be measured:

(i) Compression

(ii) Texture Profile Analysis

(iii) Puncture and Penetration

(iv) Cutting and Shearing

(v) Fracture and Bending

(vi) Extrusion (Forward and Backward)

(vii) Tension

(viii) Adhesion.

Compression

This is used for testing of Breakfast cereals, Bread crumbs, Bruising potential of fruit, Tablet hardness, Elasticity/Springiness of Surimi and Freshness/Firmness of Cake/bread.

The sample being tested has surface area smaller than that of the probe in use. This is an empirical test. It is the AACC (74-09) International Standard to measure bread firmness.

Note

This test break the rule of product dimension. The probe has radiused circumference to minimize penetration.

Puncture and Penetration

- *(i)* The sample being tested must have a larger surface area than the contact area of the probe being used.
- *(ii)* Uses cylinder, conical and needle probe
- *(iii)* Causes irreversible changes in the sample.
- *(iv)* Measured force (F) has both compression and shear elements.

Typically used for testing of:

- *(i)* Biscuit dough consistency
- *(ii)* Cheese maturity/Hardness
- *(iii)* Compaction
- *(iv)* Ripeness. (Penetration of an Apple to measure ripeness).

i. Penetration of a Contained Sample

Determination of bloom strength of gelatin according to International Standard ISO 9665.

ii. Cone Penetration

Conical probe is used for situations where the stress may vary during the use of the products. *e.g.* spreadability (spreadability test for Margarine).

Cutting and Shearing

- *(a)* Warner-Bratzler blade- Uses a 60 deg. Notched Guillotine blade. Give a uniform cutting action.
- *(b)* Knife/Guillotine blade-Measure the force required to cut or slice through a sample *e.g.* Meat Products, cheese and Vegetables.
- *(c)* Bite Jaws- Simulate teeth when biting. Final Force and Force at yield indicate toughness. Toughness of Meat, Fibrousness of vegetables.
- *(d)* Wire cutter- Constant contact area with sample. Cutting is done by 0.3 mm dia wire. Firmness and spreadability of butter.
- *(e)* Kramer Shear cell- 5-10 Guillotine blade option. Shear and compression of mainly multi-particle products *e.g.* Cereals, Pickles in sauces, Fruit, Peas and Beans.

Craft Knife adapter and Perspex cutting blade (for cooked Pasta – complies with AACC Method 14-50) are also available.

Source

Operating Manual for Texture analysers, Stable Micro System.

Bookfield CT3 Body Texture Analyser

The CT3 Texture Analyzer is the most recent development of the original Boucher Jelly Tester. The original product was created as the Stevens Texture meter and evolved into the CNS Farnell LFRA TA (Leatherhead Food Research Association Texture Analyzer). Brookfield Engineering first enhanced the design of the LFRA, maintaining the characteristics of the original equipment, while expanding its measurement capabilities and ease of use by incorporating modern digital technology. This is the third generation of this venerable product, adding tension testing capability, increased load cell options, space for large samples, two options for base tables and a wider range of accessories for more flexibility.

The principle of operation of the CT3 Texture Analyzer is to subject a sample to controlled forces in compression using a probe, or in tension using grips. The resistance of the material to these forces is measured by a calibrated load cell and shown in either grams or Newtons. These forces are a function of the properties of the sample and the parameters of the test method.

Operating Menu

The CT3 Texture Analyzer offers following seven test modes, one calibration mode and one parameter selection table:

Test Modes

- *(i)* Normal-single compression cycle.
- *(ii)* Hold Time-compress and hold.
- *(iii)* Cycle Count-compress sample several times.
- *(iv)* Bloom-compression test measures gelatin bloom strength.
- *(v)* TPA - two cycle Texture Profile Analysis compression test.
- *(vi)* Tension - pulls apart a sample using tensile force.
- *(vii)* Surimi - single compression cycle, stops automatically after gel breaks.

Each test mode requires parameters to be set. All test mode selection and parameter setting is done using the Select/Scroll knob. This knob is both rotated and depressed during operation.

Calibration Mode

Static Load-monitors load while hanging weights for recalibration.

Parameter Selection Table

Global Parameter-allows setting pretest speed, post test speed and temperature units.

Note

When setting the test speed, pressing the RESET button will toggle the increments from 0.1mm/s to 1mm/s.

All parameters can be set within the ranges shown in following Table (A)

Parameter	*Load Cell Range*						
	100g	*1000g*	*1500g*	*4500g*	*10kg*	*25kg*	*50kg*
Trigger*	0.1 – 10g	0.1-100g	0.2-150g	0.5-500g	1 – 1000	2-2500g	5-5000g
Deformation (mm)	0.1-101.6	0.1-101.6	0.1-101.6	0.1-101.6	0.1-101.6	0.1-101.6	0.1-101.6
Speed (mm/s)	0.01-10	0.01-10	0.01-10	0.01-10	0.01-10	0.01-10	0.01-10
Hold Time	0 –9999 s	0 –9999 s	0–9999 s	0–9999 s	0–9999 s	0–9999 s	0–9999 s
Cycle Count	0- 99	0- 99	0- 99	0- 99	0- 99	0- 99	0- 99
Correction	5-99	5-99	5-99	5-99	N/A	N/A	N/A

* Minimum recommended values for trigger are shown in following Table (B):

Load Cell	*Recommended Minimum Trigger Value*	*Load Cell*	*Recommended MinimumTrigger Value*
100g	0.5g	10kg	10g
1000g	2g	25kg	25g
1500g	3g	50kg	50g
4500g	4.5g		

The specific parameters required will depend upon the type of test chosen:

Normal Test Requires- trigger,deformation, speed.

Cycle Count Test Requires-cycle count, trigger, deformation, speed.

TPA Test Requires- trigger, deformation, speed.

Surimi Test Requires- trigger, correction, speed.

Bloom Test Requires-All parameters are fixed according to industry standard

Hold Time Test Requires-hold time, trigger, deformation, speed.

Tension Test Requires- trigger, deformation, speed

Static Load Test Requires- none.

Global Parameters- pretest speed, post test speed, temperature units.

These parameters, once set, will be maintained by the CT3 even after power down to facilitate repetitive testing. The test mode used prior to powering down will be presented on the screen at the next power up. A test is started by depressing the start button.

The test parameters are defined as follows:

Trigger

The load, in grams, measured by the CT3 to indicate that the probe is in contact with the sample. Once the trigger value is reached, the test will begin at the defined speed. Brookfield recommends trigger values as specified in aforesaid Table (B).

Setting trigger value to zero disables trigger function so test will immediately start from current position.

Deformation

The total downward distance the probe will travel once the trigger value is reached.

Speed

The speed at which the probe will travel the specified distance.

Time

The number of seconds that the probe will be held at the defined distance during a Hold Time test.

Count

The number of cycles (Speed and Distance) that will be applied to the sample during a Cycle Count test.

Correction

Specifies the size of the load drop, in grams, that is required to recognize a gel rupture.

Source

Brookfield Engineering Labs., Inc., Manual No. M08-372-E0315.

Industry and International Standards

Many texture analysis tests, such as the Bloom Strength test for gelatin gels, are International Standards, whereas others are recognised as standard tests within an industry, such as Texture Profile Analysis for many food products.

Instrumental Profile Analysis

This technique was developed in the 1960's with the aim of reflecting sensory attributes of texture. It involves a "two bite" compression test which simulates first two chews on the food and the output is a curve of force vs time. From the texture profile curve, following seven parameters are determined:

- *(i)* **Fracturability**: The ease with which the material will break.
- *(ii)* **Hardness**: The force required to compress the material by a given amount.
- *(iii)* **Cohesiveness**: The strength of the internal bonds in the sample.
- *(iv)* **Adhesiveness**: The energy required to overcome attractive forces between the food and any surface it is in contact with.
- *(v)* **Springiness:** The elastic recovery that occurs when the compressive force is removed.
- *(vi)* **Gumminess**: The energy required to break down a semi-solid food ready for swallowing

(vii) **Chewiness**: The energy required to chew a solid food into a state ready for swallowing.

The properties of Gumminess and Chewiness are mutually exclusive. Not all the other properties will be found in a given food sample.One other parameter has subsequently been added; The Modulus of Deformability. defined as the initial slope of the force-deformation curve before the first break in the curve (*i.e.* before fracture of the sample).

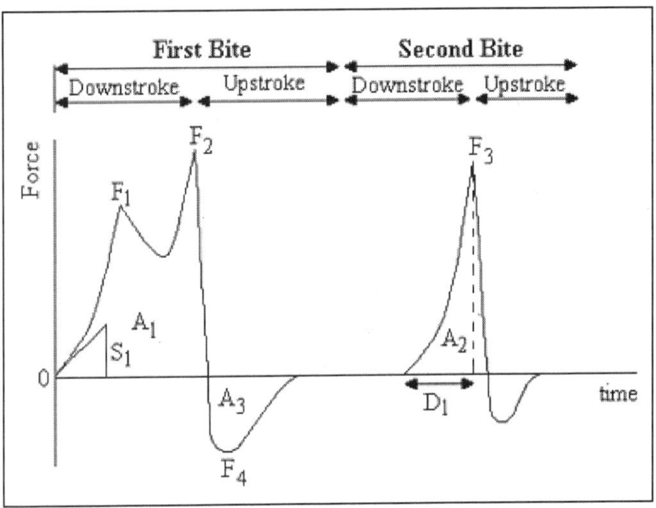

Texture Profile Curve.

Texture profile parameters are determined from:

i- Fracturability=F1, ii-Hardness =F2, iii-Cohesiveness = A2/A1, iv-Adhesiveness = (based on) A3, v- Springiness =D1, vi- Gumminess = hardness x cohesiveness = F2 x A2/A1, vii-Chewiness = hardness x cohesiveness x springiness = F2 x A2/A1 x D1 and viii- Modulus of deformability (based on) slope, S1.

Food Texture Properties

Strictly speaking the data from the curve should be either expressed as a force deformation curve or turned into a stress strain curve. For foods, however, Relatively large deformations mean large changes in area. For this reason, alternative definitions of stress and strain have been developed.

Alternative definitions of stress and strain:

$$\sigma_T = \frac{F}{A_0} \cdot \frac{h_0 - \Delta h}{h_0} \qquad \varepsilon_T = \ln\left(\frac{h_0}{h_0 - \Delta h}\right)$$

These are sometimes referred to as "True stress" (T) and "True strain" (T) or "Hencky' strain". TPA data based on this gives more reproducible data for comparing different foods since the stress and strain relationships are less dependant on experimental conditions.

Source

1. *sst-web.tees.ac.uk/external/U0000504/Notes/Sensory/Texture.html.*
2. *Geoff Walker*
3. *Teesside University.*

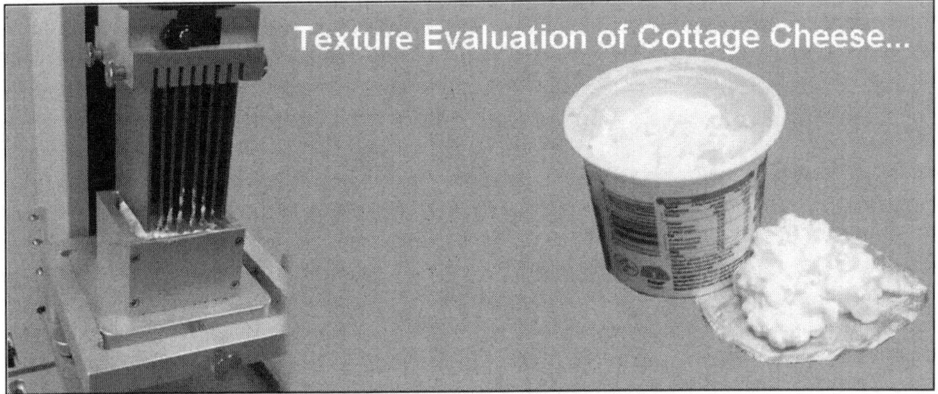

The following common properties are to be measured to judge the textural quality of Food/Dairy Products:

Adhesiveness, Apparent Modulus Breaking Point, Burst Strength, Chewiness Coefficient of Friction, Cohesiveness Consistency, Elasticity, Fracture Force, Gel Strength, Gumminess, Hardness, Pliability, Relaxation, Ripeness, Spreadability, Tackiness and Yield Point.

Texture analysers are used to determine the textural properties of the followings:

(i) Gels
(ii) Dairy Products
(iii) Biscuits, Breads and Dough
(iv) Confectionery
(vi) Meat, Fish and Poultry
(vii) Fruit and Vegetables
(viii) Cereals
(ix) Snack Foods.

B. Rheological Analysis of Liquid Foods [1,2]

There are following two parameters monitored during rheology of liquid foods:

(i) **Shear stress:** In the controlled shear stress (CSS) tests, the shear stress is set and the shear rate is measured.

(ii) **Shear rate:** Incase of controlled shear rate (CSR) tests, the shear rate is set and shear stress is measured.

Rotational Tests by Rotational Viscometers

In any of the method, the viscosity (η) is calculated from the values of shear stress at corresponding shear rate (ratio of shear stress to shear rate). It is possible to determine individual viscosity values or a complete viscosity curve, depending upon type of instrument. The latest instruments enable the measurement of flow curve and modeling of the data generated to fit the best suitable rheological model to characterize the flow behaviour of liquid foods. The best fit model (based on R2 value) is accepted to characterize the particular fluid food. Viscometers are also supplied with different geometry as in case of texture analysers. The different probes are used for measurement of viscosity of samples. Generally Cylinder measuring system and Cone/plate measuring system are used for low viscosity fluids while Plate/plate measuring system (PP) is used for high viscosity fluids.

Source

1. *User's Operating manual of Viscometer (M/S Brookfield).*
2. *Sanket Borad and Ashish Kumar Singh, Instrumental Measurement of Rheological Properties of Dairy Foods, Compendium on Advanced Techniques and Novel Approaches for Quality and Safety Evaluation of Dairy Foods-2015, National Dairy Research Institute, Karnal.*

C. Rheological Analysis of Solid and Liquid Foods by Oscillatory Rheometer [1,2]

In this rheometer the moving part of the measuring system oscillates, creating vibrations which are transferred to the sample. This provides details on the viscoelastic behaviour of the substance. The oscillation is usually carried out at very low deformations to stay in the elastic range. Oscillatory tests have the advantage that the samples are not destroyed during testing. An oscillatory test simultaneously determines both the viscous behavior (described by the loss modulus G") and the elastic behavior (described by the storage modulus G').

A variety of measuring systems allow both liquids and solids to be measured. This makes it possible to use one single instrument to investigate foods ranging from fluids such as water to milk and condensed milk and even semisolid and solid foods like cheese, Paneer, *etc.* Some examples are given below:

i. Amplitude Sweep Test (AST)

For AST, varying amplitude at constant frequency and temperature is set. This test provides information on the limit of the linear viscoelastic range and the structural character of the sample. The following characters are defined:

(a) $G' > G''$ defines Gel or paste character.

(b) $G'' > G'$ defines liquid character.

(c) The structural strength of a substance is expressed by the G' value.

The sample is analysed for the LVE range (linear viscoelasticity range) using AST. The LVE range obtained by ATS is then employed to perform the FST for the same sample.

ii. Frequency Sweep Test (FST)

It is carried out by varying the frequency (LVE range obtained from AST) at constant temperature and amplitude. Measurements at high frequencies represent the short-term behaviour of a sample. The long-term behaviour (storage stability) is described by measurement data at low frequencies.

iii. Temperature Cycle Test (TCT)

It is carried out by varying temperature at constant amplitude and frequency. Temperature tests can be used to determine the glass transition temperature and melting temperature.

Source

1. *User's Operating manual of Rheometer.*
2. *Sanket Borad and Ashish Kumar Singh, Instrumental Measurement of Rheological Properties of Dairy Foods, Compendium on Advanced Techniques and Novel Approaches for Quality and Safety Evaluation of Dairy Foods-2015, National Dairy Research Institute, Karnal.*

D. Tribo-rheometer

It is now being used as a new sensory tool for determining the tribological properties of foods. Tribological properties associated with friction, adhesion, and lubrication of interacting surfaces in relative motion in the mouth have been associated in the perception of food attributes like astringency, smoothness, roughness, and slipperiness. A tribo-rheometer is capable of differentiating dairy products with different fat contents both in liquid (milk) or semi-solid forms (yoghurt and soft-cheese) [1,2].

Tribology has a growing interest in oral texture and sensory research due to its ability to assess certain properties of the food during the complex oral processing that cannot be explained by bulk texture and rheology. During oral processing, food is squeezed and sheared between the tongue and other oral surfaces, such as teeth and palate and these forces applied in the mouth are mimicked by a tribometer [1,2].

There is a new triborheometry fixture that can be utilized with a commercial torsional rheometer in order to explore the coupled rheological and tribological properties of complex fluids and solid-liquid systems. The tribo-rheometer is a fixture that can be added to a commercial torsional rheometer to capture the tribological properties of the system (*i.e.* the frictional interactions of a fluid-surface pair) as well as the 'bulk' rheological properties of the fluid under shear. The fixture is self-leveling and both the normal load and the sample gap can be monitored or controlled. At large gaps, the fixture imposes an approximately constant shear rate on the sample and the bulk viscometric properties of the fluid can be measured. However, as the gap between the plates is reduced, the measured viscosity function becomes gap-dependent. Using new triborheometer fixture it is possible to obtain tribological information over a wider range of sliding velocities [3]. The data can be represented in the form of a classical Stribeck diagram or, by using a dimension less gap-dependent shift factor, it is possible to construct a more general 'friction map' of the gap and load dependent effective viscosity [4].

Application of Tribo-Rheometer (A New Triborheometry Fixture with a Torsional Rheometer)

There are a number of possible applications for this system in both industrial and academic research environments. Some examples are given below:

1. **Food Industry**
   (i) The food and consumer product industries also offer numerous future opportunities for triborheometry in quantifying concepts such as lubricity and astringency. High shear rates and narrow gaps are characteristic of many end-use applications including chewing of food stuffs [5].
   (ii) The triborheometrical studies of model chocolate systems at a standardized test temperature of 40°C are presently under way.
2. The device uses very small volumes of test fluid, and the construction material of the lower annulus or disk can be readily interchanged. It is thus possible to investigate the effects of surface patterning (lithographic or machined), the frictional properties of multilayer coatings and even the use of substrates comprised of living cells or tissues caffolds grown on perfusion membranes.

Source

1. *Operating manual of Tribo rheometer.*
2. *Sangeeta Prakash, School of Agriculture and Food Science, The University of Queensland, Australia.*
3. *H.P. Kavehpour and G.H.McKinley, Hatsopoulos Microfluids Laboratory, Massachusetts Institute of Technology, Cambridge, USA, Tribo-rheometry-from gap-dependent rheology to tribology.*
4. *G. Luengo, J.N. Israelachvili, and S. Granick. Generalized effects in confined fluids: new friction map for boundary lubrication. Wear, 200: 328–335, 1996.*
5. *S. Lee, P. Heuberger, M. and Rousset, and N.D. Spencer. Chocolate at a sliding interface. J. Food Science, 67 (7): 2712–2717, 2002.*

Application of Texture Analyser, Rheometer and Viscometers

The Texture Analyser, Rheometer and Viscometers are being used successfully in rheological analysis of dairy products and other food.

1. Applications in Dairy Industry

a. Applications of Texture Analyser

Body Texture Analyser is generally being used for analysis of following Dairy Products:

(i) Various Margarines and Butter (Cone Probe A, Cutting Wire B)

(ii) Cheddar Cheese -TPA and Relaxation, Puncture Test A, Cut Test B and Adhesiveness

(iii) Cottage and Cream Cheese (Spreadability)

(iv) Whipped and Sour Cream (Spreadability)

(v) Yoghurt (Spreadability and Cycle Test)

(vi) Ice cream (Spreadability)

(vii) Melted Cheese (Spreadability)-One of the most important characteristics of Mozzarella cheese used as pizza topping, but the most difficult to measure is the ability of the cheese to stretch when melted.

Stretch refers to the capacity of melted cheese to form fibrous strands that extend under tension. A test for measuring the stretchability of cheese was developed by using a texture-profile analyzer to pull strands of cheese upwards from a reservoir of melted cheese. The different cheeses were analyzed using the stretch test and also analyzed for apparent viscosity with a helical viscometer, for meltability using a tube melt test and for stretch using the pizza-fork test. Cheese was placed into a stainless steel cup and tempered in a water bath at 60, 70, 80, or 90°C for 30 min before analysis. The cup was then placed in a water-jacketed holder mounted on the base of the instrument. A three-pronged hook-shaped probe was lowered into the melted cheese and then pulled vertically until all cheese strands broke or 30 cm was reached. This produced a stretch profile as the probe was lifted through the reservoir of melted cheese and then pulled strands of cheese upwards.

The following three parameters were defined to characterize the stretchability of the cheese:

(i) **Melt strength (F_M)**: The maximum load, obtained as the probe was lifted through the cheese, was defined as melt strength.

(ii) **Stretch length (SL)**: The distance to which cheese strands were lifted was defined as Stretch length.

(iii) **Stretch quality (SQ)**: The load exerted on the probe as the strands of cheese were being stretched was defined as stretch quality.

There was a correlation between F_M and apparent viscosity. There was also some correlation between SL measured by the fork test and SL when the cheese was tested at 90°C, but no correlation occurred at lower temperatures. Differences in melted cheese performance were objectively measured by the Test, using a combination of descriptive parameters based on the load exerted on a three-pronged hook-shaped probe as the cheese was stretched. F_M (the maximum load exerted on the probe as the cheese was stretched) was correlated with apparent viscosity and provides an indication of how well the cheese would melt during baking on a pizza. How well the cheese stretches can be described in terms of distance (SL) and resistive load (SQ). Testing the cheese at different temperatures (*e.g.* 90°C and 70°C) may provide information on predicted performance of the cheese during baking, as well as at the time it is consumed. It is therefore proposed that this stretch test could be used as a means of obtaining more quantitative information on cheese performance than the fork test that is currently being used.

Source

1. *R.L.Fife, D.J.McMahonand, C.J.Oberg· WesternDairyCenter, Department of Nutrition and Food Sciences, Utah State University, Logan 84322 and Department of Micrppliobiology, Weber State University, Ogden UT 84408.*
2. *Ak, M.M., and S. Gunase-karan.995.Measuring elongational properties of Mozzarella cheese. J. Textural Studies.26: 147-160*
3. *Association of Official Analytical Chemists (AOAC).1990.Official Methods of Analysis. Vol.2.15th ed. AOAC, Arlington, VA.*
4 *Kindstedt, P.S.1995.Factors affecting the functional characteristics of unmelted and melted Mozzarella cheese. Pages 27–41 in Chemistry of Structure-Function Relationships in Cheese. E.L.Malin and M. H. Tunick, ed. Plenum Press, New York, NY.*
5. *Kindstedt, P.S., J. K. Rippe, and C. M. Duthie.1989a. Application of helical viscometry to study commercial Mozzarella cheese melting properties. J. Dairy Sci. 72: 3123–3128.*
6. *Texture Technologies, Scarsdale, NY10583.*

b. Application of Texture Analyser, Rheometer and Viscometers

The applications in dairy industry are tabulated below [1,2]:

Sl.No.	*Types of Dairy Foods*	*Parameters to be Analysed*	*Type of Tests*	*Instrument to be Applied*
1	Cheeses, Paneer, Desserts and Puddings, Khoa, Chhana and other TIDPs	H, S, Ad, Co, B, G, Ch.	Texture profile analysis	Texture analyser
		Viscoelastic behavior	AST, FST	OscillationViscometer
2	Ice cream and frozen desserts	Viscoelastic behavior	AST, FWT, TCT	OscillationViscometer
		Hardness	Penetration test	Texture analyser
3	Butter	Cutting strength	Wire cutting	Texture analyser
		Spreadability	Spreadability rig	Texture analyser
4	Set Yoghurt/Dahi and related products	Hardness, Adhesiveness, Work of shear, Work of adhesion	Single bite compression and Back extrusion	Texture analyser
		Viscoelastic behavior	AST, FST	OscillationViscometer
		Apparent viscosity	Viscosity curve	Rotational Viscometer
5	Cream, Table spreads,	Viscoelastic behavior	TCT	OscillationViscometer
		Apparent viscosity	Viscosity curve	Rotational Viscometer

2. Applications in Bakery Industry [1,2]

The application of Texture Analyser, Rheometer and Viscometers in Bakery industry are given below:

Sl.No.	*Types of Bakery Foods*	*Parameters to be Analysed*	*Type of Tests*	*Instrument to be Applied*
1	Bread dough, cake batter and other Gluten based dough	Elongation properties	Extension test	Texture analyser
		Viscoelastic behaviour	FST	OscillationViscometer
		Apparent Viscosity	Viscosity curve	Rotational Viscometer
2	Breads and Biscuits	Hardness	Penetration test	Texture analyser

3. Applications in Confectionery

The application of Texture Analyser and Viscometers in Confectionery are given below:

Sl.No.	*Types of Confectionery*	*Parameters to be Analysed*	*Type of Tests*	*Instrument to be Applied*
1	Chocolate liquor	Viscosity, Yield stress	Viscosity curve	Rotational Viscometer
2	Chocolate (brick)	Hardness, Stickiness	Single bite test	Texture analyser
		Cutting strength	Warner-Bratzler cutting test	Texture analyser

4. Applications in Edible Vegetable Oils and other Food Emulsions [1,2]

The application of Viscometers in aforesaid foods are given below:

Sl.No.	*Types of Foods*	*Parameters to be Analysed*	*Type of Tests*	*Instrument to be Applied*
1	Edible vegetable oils	Apparent Viscosity	Viscosity curve	Rotational Viscometer
2	Food emulsions	Viscoelastic behavior	TCT	OscillationViscometer
		Apparent viscosity	Viscosity curve	Rotational Viscometer

Source

1. *Operating Manual of Texture analysers.*
2. *Sanket Borad and Ashish Kumar Singh, Instrumental Measurement of Rheological Properties of Dairy Foods, Compendium on Advanced Techniques and Novel Approaches for Quality and Safety Evaluation of Dairy Foods-2015, National Dairy Research Institute, Karnal.*

Electronic Nose

In Food/Dairy industries, odour assessment is usually performed by human sensory analysis, by chemosensors, or by gas chromatography. The latter technique gives information about volatile organic compounds but the correlation between analytical results and actual odour perception is not direct due to potential interactions between several odourous components.

The sensor technology of artificial olfaction had its beginnings with the invention of the first gas multisensor array in 1982 [1]. Since 1982, research has been conducted to develop technologies, commonly referred to as electronic noses, that could detect and recognize odors and flavors. Advances in aroma-sensor technology, electronics, biochemistry and artificial intelligence made it possible to develop devices capable of measuring and characterizing volatile aroma as released from a multitude of sources for numerous applications. **These devices, known as electronic noses, were engineered to mimic the mammalian olfactory system within an instrument designed to obtain repeatable measurements, allowing identifications and classifications of aroma mixtures while eliminating operator fatigue** [2-7]. The electronic sensing refers to the capability of reproducing human senses using sensor arrays and pattern recognition systems. An electronic nose is a device intended to detect odours or flavours. It identifies the specific components of an odour and analyzes its chemical makeup to identify it.

Hundreds of different prototypes of artificial-nose devices have been developed. These prototypes collectively represent various electronic aroma detection (EAD) technologies that utilize different sensor types including metal-oxide [10-12], semi conductive polymers [13-14], conductive electro active polymers [9,15-17], optical [18], surface acoustic wave [18] and electrochemical gas sensors [19].

An electronic nose system typically consists of a multisensor array, an information-processing unit such as an artificial neural network (ANN), software with digital pattern-recognition algorithms, and reference-library databases [8,15,20-22].

The output from individual sensors are collectively assembled and integrated to produce a distinct digital response pattern. Identification and classification of an analyte mixture is accomplished through recognition of this unique aroma signature (electronic fingerprint) of collective sensor responses. The identity of a simple or complex mixture represented by a unique aroma signature pattern may be determined without having to separate the mixture into its individual components prior to or during analysis. A reference library of digital aroma signature patterns for known samples is constructed prior to analysis of unknowns. The ANN is configured through a learning process (neural net training) using pattern recognition algorithms that look for differences between the patterns of all analyte types included in the reference library. This process continues until a previously selected level of discrimination is met. The results are validated and assembled into the reference library to which unknown samples can be compared. Identification of unknowns is based on the distribution of aroma attributes or elements that the analyte pattern has in common with patterns present in databases of the reference library.

An odour is composed of molecules, each of which has a specific size and shape. Each of these molecules has a correspondingly sized and shaped receptor in the human nose. When a specific receptor receives a molecule, it sends a signal to the brain and the brain identifies the smell associated with that particular molecule. Electronic noses based on the biological model work in a similar manner, albeit substituting sensors for the receptors, and transmitting the signal to a program for processing, rather than to the brain. Electronic noses are one example of a growing research area called biomimetics, or biomimicry, which involves human-made applications patterned on natural phenomena.

Source

1. *Persaud, K.C.; Dodd, G. Analysis of discrimination mechanisms in the mammalian olfactory system using a model nose. Nature1982, 299, 352-355.*
2. *Davide, F. A.M.; Di Natale, C.; D'Amico, A. Self-organizing sensory maps in odour classification mimicking. Biosens. Bioelectron.1995, 10, 203-218.*
3. *Pelosi, P., Persaud, K.C. Gas sensors: towards an artificial nose. In Sensors and Sensory Systems for Advanced Robotics; Dario, P., Ed.; Springer-Verlag: Berlin, Germany, 1988, pp. 361-381.*
4. *Persaud, K.C. Electronic gas and odor detectors that mimic chemoreception in animals. Trend. Anal. Chem. 1992, 11, 61-67*
5. *Persaud, K.C., Bartlett, J., Pelosi, P.Design strategies for gas and odour sensors which mimic the olfactory system. In Robots and Biological Systems: Towards a New Bionics? Dario, P., Sandini, G., Aebisher, P., Eds., Springer-Verlag: Berlin, Germany, 1993, pp. 579-602.*
6. *Shirley, S.G., Persaud, K.C. The biochemistry of vertebrate olfaction and taste.Sernm. Neurosci.1990, 2,59-68.*
7. *Shurmer, H.V. An electronic nose: a sensitive and discrimination substitute for a mammalianolfactory system. Intern. Electric. Eng. Proc. 1990, 137, 197-204.*
8. *Gardner, J.W. Detection of vapours and odours from a multisensor array using pattern recognition: principal component and cluster analysis. Sens. Actuat. 1991, 4, 109-115.*
9. *Lonergan, M.C.; Severin, E.J.; Doleman, B.J.; Beaber, S. A.; Grubbs, R.H., Lewis, N.S. Arraybased vapor sensing using chemically sensitive, carbon black-polymer resistors. Chem. Mater.1996, 8, 2298-2312.*
10. *Egashira, M., Shimizu, Y. Odor sensing by semiconductor metal oxides. Sens. Actuat. 1993, 14,443-446.*
11. *Nanto, H., Sokooshi, H., Kawai, T. Aluminum-doped ZnO thin film gas sensor capable of detecting freshness of sea foods. Sens. Actuat. 1993, 14, 715-717.*
12. *Shurmer, H.V., Gardner, J.W.; Chan, H. T. The application of discrimination techniques to alcohols and tobacco using tin oxide sensors.Sens. Actuat. 1989, 18, 359-369.*
13. *Yim, H.S., Kibbey, C.E., Ma, S.C., Kliza, D.M.; Liu, D.; Park, S.B., Torre, C.E., Meyerhoff, M.E. Polymer membrane-based ion-, gas-, and bio-selective potentiometric sensors. Biosens.Bioelectron. 1993, 8, 1-38.*
14. *Pisanelli, A.M.; Qutob, A. A., Travers, P., Szyszko, S., Persaud, K.C. Applications of multi-array polymer sensors to food industries. Life Chem. Rep. 1994, 11, 303-308.*

15. *Freund, M.S., Lewis, N.S. A chemically diverse conducting polymer-based electronic nose.Proc.Natl. Acad. Sci. 1995, 92, 2652-2656.*

16. *Hatfield, J.V., Neaves, P., Hicks, P.J., Persaud, K.C., Tavers, P. Toward an integrated electronic nose using conducting polymer sensors. Sens. Actuat. 1994, 18, 221-228.*

17. *Persaud, K.C., Qutob, A. A.; Travers, P., Pisanelli, A.M.; Szyszko, S. Odor evaluation of foods using conducting polymer arrays and neural net pattern recognition. In Olfaction and Taste XI.; Kurihara, K., Suzuki, N., Ogawa, H., Eds., Springer-Verlag: Tokyo, Japan, 1994, pp. 708-710.*

18. *Staples, E.J. Electronic nose simulation of olfactory response containing 500 orthogonal sensors in 10 seconds. In Proceedings of the 1999 IEEE Ultrasonics Frequency Control and Ferroelectrics Symposium, Lake Tahoe, CA, USA, 2000, pp. 307-313,*

19. *Gardner, J.W., Bartlett, P.N. Electronic Noses. Principles and Applications; Oxford University Press: Oxford, UK, 1999, pp. 221-245.*

20. *Abe, H.; Kanaya, S.; Takahashi, Y., Sasaki, S.I. Extended studies of the automated odour sensing system based on plural semiconductor gas sensors with computerized pattern recognition techniques. Anal. Chim. Acta. 1988, 215, 155-168.*

21. *Gardner, J.W.; Shurmer, H.V. Odour discrimination with an electronic nose. Sens. Actuat.1992,8,1-11.*

22. *Kowaiski, B.R.; Bender, C.F.Pattern recognition: a powerful approach to interpreting chemical data. J. Amer. Chem. Soc. 1972, 94, 5632-5639.*

Aroma Types and Characteristics

Aroma is the most important factor influencing consumers' purchase intent, estimation of food quality and food acceptance (1,2). It is a specific sensory characteristic detected when volatile compounds enter the nasal portion of the olfactory system down through the back of the throat and get to be detected by olfactory receptors (3). Aromas are simple to complex mixtures of volatile compounds present in the air at concentration that may be detected by animals through the sense of olfaction. Aromas sometimes have been referred to as "smells" or "odors" when a particular connotation referring to the pleasantness or unpleasantness of an aroma is being expressed.

Aromas in general are characterized by the following four quantifiable qualitative dimensions:

i. Threshold

The detection threshold value is defined as the lowest concentration of aromatic compounds at which human subjects can detect the existence of the aroma [4]. The detection threshold is determined by diluting the aroma to the point where 50 per cent of the test population or human panel can no longer detect the aroma [5].

ii. Intensity

Intensity refers to the perceived strength of the aroma sensation, and increases as a function of concentration.

iii. Quality

Quality is the third dimension and is usually expressed by the use of descriptor types, or common-use words that describe the aroma by associating it to the aroma qualities of known substances; and usually referring to aromas released from plants or plant parts. McGinley and McGinley [6] proposed following eight aroma groups:

(a) **Earthy aromas** (musty, moldy, musk, stale, grassy, herbal, woody).

(b) **Floral aromas** (fragrant, flowery, perfume, eucalyptus, lavender).

(c) **Fruity aromas** (citrus, orange, lemon, apple, pear, pineapple, strawberry).

(d) **Spicy aromas** (cinnamon, mint, peppermint, onion, dill, garlic, pepper, cloves, vanilla, almond, pine).

(e) **Fishy aromas** (fishy, prawns, amine).

(f) **Sewage aromas** (septic, putrid, rancid, sulfurous, rotten, decayed, cadaverous, foul, sour, pungent, burnt, swampy).

(g) **Medicinal aromas** (disinfectant, phenol, camphor, soapy, ammonia, alcohol, ether, anesthetic, menthol).

(h) **Chemical aromas** (solvent, aromatic, varnish, turpentine, petroleum, creosote, tar, oily, plastic).

iv. Hedonic Assessment

The hedonic assessment dimension is associated with the relative pleasantness or unpleasantness of the aroma. Hedonic assessment may be quantified using scaled values that range from 1 (completely dislike) to 10 (very good, pleasant, and agreeable) or through objective judgments (excellent, terrible) using descriptor terms indicating relative satisfaction or agreeableness of the aroma (from very pleasant to completely unpleasant).

Source

1. *Bakker, J. (1997) Sensory evaluation of dairy fla-vours. In: Microbiology and biochemistry of cheese and fermented milk. izd. (Law, B. A., ed.), Black-ie Academic and Professional, London, 341-362. doi: dx.doi.org/10.1007/978-1-4613-1121-8_11*

2. *Santonico, M., Pittia, P., Pennazza, G., Martinelli, E., Bernabei, M., Paolesse, R., D'Amico, A., Compagnone, D., DiNatale, C. (2008): Study of the aroma of artificially flavoured custards by chemical sensor array fingerprinting. Sensors and Actuators B133, 345-351.*

3. *Gomez, A.H., Wang, J., Hu, G., Pereira, A.G. (2006): Electronic nose technique potential monitoring man-darin maturity. Sensors and Actuators B 113,347-353.*

4. *Yoshii, F.; Yamada, Y.; Hoshi, T.; Hagiwara, H. The creation of a database of odorous compounds focused on molecular rigidity and analysis of the molecular features of the compounds in the database. Chem.Senses 2002, 27,399-405.*

5. *Yuwono, A.S., Lammers, P.S. Performance test of a sensor array - based odor detection instrument. CIGR J. Sci. Res. Dev. 2004, 6, 1-16.*

6. *McGinley, C.M.; McGinley, M. A. Odor quantification methods and practices at MSW landfills. In Proceedings of the Air and Waste Management Association 91st Annual Meeting and Exhibition, Stillwater, MN, USA, 1998; pp.1-12.*

7. *Alphus D. Wilson 1, and Manuela Baietto, Review Applications and Advances in Electronic-Nose Technologies, Sensors 2009, 9, 5099-5148.*

Conceptual Development of the Electronic Nose

The term **"electronic nose"** comes from the similarities with the mammalian olfactory system. This instrument imitates the performance of the mammalian olfactory system and determines the aroma profile through the determination of the total profile of food volatile components. The first real tool for measuring aromas was developed by Hartman [1] in 1954. The sensing element was a microelectrode, a simple platinum wire of 0.8 mm in diameter, which measured the flow of current by a sensitive millivoltmeter.

The term **electronic nose** was coined in 1988 by **Gardner and Bartlett**, who later defined it as "**an instrument which comprises an array of electronic chemical sensors with partial specificity and appropriate pattern recognition system, capable of recognizing simple or complex odors**" [2].

In1991, scientific interest in the use and applications of electronic noses was sanctioned by the first advanced workshop on chemosensory information processing during a session of the **North Atlantic Treaty Organization (NATO).**

It is composed of an array of non-selective sensors which transform chemical information into an electrical (resistance, voltage, frequency) or optical one (colour); such an information then gets to be transformed into a digital form suitable for computer processing. Upon being sniffed or entered via the retronasal pathway if a product is tasted, volatile compounds reach the olfactory epithelium where, as a result of the interaction between volatile compounds and chemosensory receptors (olfactory neurons), electrical stimuli arise and are transmitted into the brain (3-4).

Each olfactory neuron has the ability to respond to several odorants; on the other hand, each odorant can be detected by multiple olfactory neurons (4-6). Electronic noses operate under the same principle. Products of similar aroma evoke similar sensory responses, meaning that their "fingerprints" are similar, while those of different aroma have mutually different "fingerprints".

Source

1. *Hartman, J.D. A possible method for the rapid estimation of flavours in vegetables. Proc. Amer.Soc.Hort.Sci. 1954, 64, 335-342.*

2. *Gardner, J.W.; Bartlett, P.N. A brief history of electronic noses. Sens. Actuat. B: Chem. 1994, 18,211-220.*

3. *Gostelow, P., Parsons, S. A., Stuetz, R.M. (2001): Odour measurements for sewage treatment works. Water Research 35 (3), 579-597.*

4. *Ampuero, S., Bosset, J.O. (2003): The electronic nose applied to dairy products: a review. Sensors and Actuators B 94, 1-12. doi: dx.doi.org/10.1016/S0925-4005 (03)00321-6.*

5. *Albert, K.J., Lewis, N.S., Schauer, C.L., Sotzing, G. A., Stitzel, S.E., Vaid, T.P., Walt, D.R. (2000): Cross-reactive chemical sensor arrays. Chemical Reviews 100, 2595-2626. doi: dx.doi.org/10.1021/cr980102w.*
6. *Milna Tudor Kalit, Ksenija Markovi, Samir Kalit, Nada Vahi and Jasmina Havranek, University of Zagreb, Zagreb, Croatia, Application of electronic nose and electronic tongue in the dairy industry, doi: 10.15567/mljekarstvo.2014.0402.*

Why Electronic Nose

(i) The **American Society for Testing and Materials (ASTM)** has classified 830 aroma descriptors [1]. Nevertheless, human panel tests have indicated that human subjects can correctly identify an average of less than 100 aromas after good training when discriminating familiar aromas [2].

(ii) The human thresholds for detection of some strong common odorants, such as citral (lemon) and butyric acid in air, can be quite low.

(iii) The threshold for recognition, the minimum concentration at which the subject can correctly identify the aroma qualities of the compound, is usually higher and varies from person to person according to level of training [3]. The perception threshold for aromas decreases upon extended contact with an aroma. After continued exposure, the perceived intensity falls exponentially, but by removal of the aroma, the intensity recovers [3].

Source

1. *Ohloff, G. Smelling Materials and Sense of Smell; Springer-Verlag: Berlin, Germany, 1990; pp. 1-37.*
2. *Desor, J. A.; Beauchamp, G.K. The human capacity to transmit olfactory information. Percept.Psychophys.1974, 16, 551-556.*
3. *Gardner, J.W.; Bartlett, P.N. Electronic Noses. Principles and Applications; Oxford University Press: Oxford, UK, 1999; pp. 221-245.*
4. *Alphus D. Wilson 1, and Manuela Baietto, Review Applications and Advances in Electronic-Nose Technologies, Sensors 2009, 9, 5099-5148.*

Working Principle and Electronic Nose Instrumentation

The electronic nose was developed in order to mimic human olfaction that functions as a non-separative mechanism: *i.e.* an odor/flavor is perceived as a global fingerprint. Essentially the instrument consists of head space sampling, sensor array, and pattern recognition modules, to generate signal pattern that are used for characterizing odors.

Gardner and Bartlett [1] provided a basic requisite definition of an electronic-nose device with a list of necessary components. The major components are as follows:

1. An aroma delivery system, which transfers the volatile aromatic molecules from the source material to the sensor array system.

2. A chamber where sensors are housed: this has usually fixed temperature and humidity, which otherwise would affect the aroma molecules adsorption.
3. An electronic transistor which converts the chemical signal into an electrical signal, amplifies and conditions it.
4. A digital converter that converts the signal from electrical (analog) to digital.
5. A computer microprocessor which reads the digital signal and displays the output after which the statistical analysis for sample classification or recognition is done.

As per the Gardner-Bartlett definition, a detection devise to be considered an electronic nose, it must contain an intelligent chemical-array sensor system that mimics the mammalian olfactory system and is used specifically to sense aromatic VOCs. It means under the aforesaid definition all sensing devices that have only one sensor or can detect only one compound or aroma (electronic aroma monitors) cannot be considered as electronic noses. Thus, electrochemical cells (ECs) that detect only one specific gas are not electronic noses.

Classical analytical instruments and detectors such as Electron capture detectors (ECD), Flame ionization detectors (FID), Flame photometry detector (FPD), Gas chromatographs (GC), Infrared spectrometers (IRS), Ion mobility spectrometers (IMS), Mass spectrometers (MS), Nuclear magnetic resonance spectrometers (NMRS), Photoionization detector (PID) and Quadrupole fingerprint mass spectrometers (QFMS) are not considered e-noses in the strictest sense because they do not provide a collective data output from a sensor array and are designed to detect and identify individual components of a gas mixture.

Source

1. *Gardner, J.W.; Bartlett, P.N. A brief history of electronic noses. Sens. Actuat. B: Chem. 1994, 18,211-220.*
2. *Alphus D. Wilson 1, and Manuela Baietto, Review Applications and Advances in Electronic-Nose Technologies, Sensors 2009, 9, 5099-5148.*

Apparatus

Generally electronic noses include three following major parts:

i. Sample Delivery System

The sample delivery system enables the generation of the headspace (volatile compounds) of a sample, which is the fraction analyzed. The system then injects this headspace into the detection system of the electronic nose. For constant operating conditions, the sample delivery system is important and essential.

ii. Detection System

It is a reactive part of the instrument and consists of a sensor set. The commonly used sensors are: metal-oxide-semiconductor (MOS) devices, conducting polymers, quartz crystal microbalance, surface acoustic wave (SAW-a class of micro electromechanical systems, MEMS) *etc.* When in contact with volatile compounds, the sensors react, which means they experience a change of electrical properties. Each sensor is sensitive to all volatile molecules but each in their specific way. Most electronic noses use sensor arrays that react to volatile compounds on contact. The adsorption of volatile compounds on the sensor surface causes a physical change of the sensor. A specific response is recorded by the electronic interface transforming the signal into a digital value. Recorded data are then computed based on statistical models. Other types of electronic noses have been developed that utilize mass spectrometry or ultra-fast gas chromatography as a detection system.

iii. A Computing System

The interaction between volatiles and sensors triggers a series of signals which are processed by the computer via an algorithm pattern recognition program. The computing system works to combine the responses of all of the sensors, which represents the input for the data treatment. This part of the instrument performs

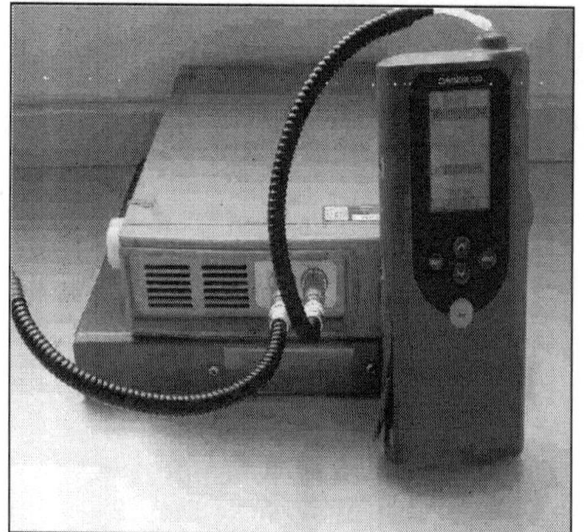

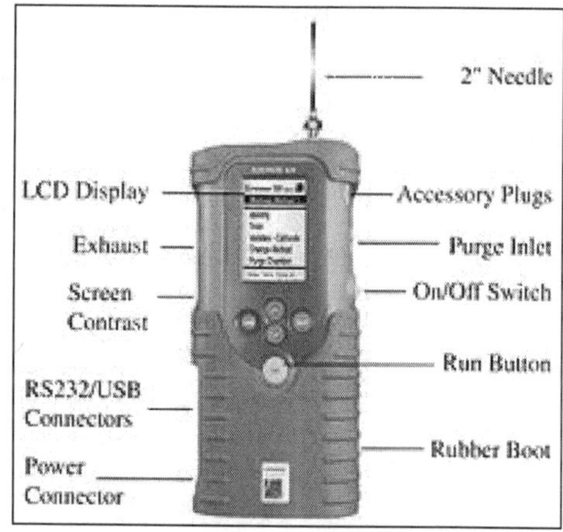

Cyranose 320, a handheld "electronic nose", developed by Cyrano Sciences of Pasadena, California.

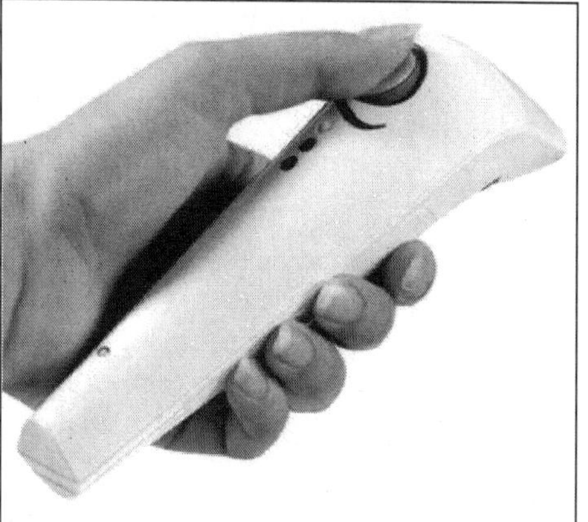

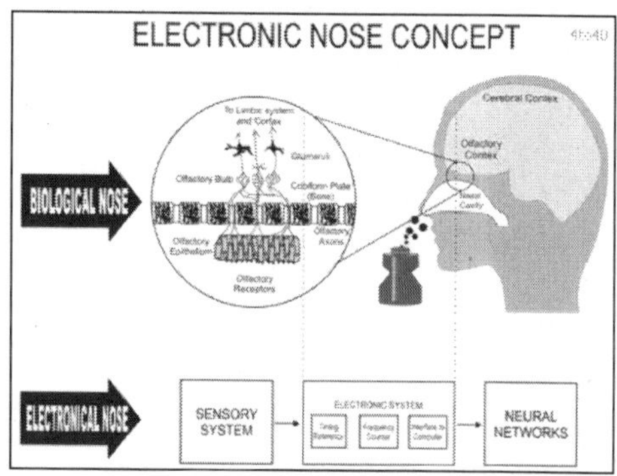

Different Models and Concept of Electronic Nose

global fingerprint analysis and provides results and representations that can be easily interpreted. Moreover, the electronic nose results can be correlated to those obtained from other techniques (sensory panel, GC, GC/MS).Many of the data interpretation systems are used for the analysis of results. The most widely used pattern recognition models are Principal Component Analysis (PCA), Linear Discriminant Analysis (LDA) and Artificial Neural Network (ANN) (1-3).

Source

1. *Giese, J. (2000): Electronic noses. Food Technology 54 (3),96-98.*
2. *Ampuero, S., Bosset, J.O. (2003): The electronic nose applied to dairy products: a review. Sensors and Actuators B 94, 1-12.*
3. *Deisingh, A.K., Stone, D.C., Thompson, M. (2004): Applications of electronic noses and tongues in food analysis. International Journal of Food Science and Technology 39, 587-604.*

Data Analysis for Electronic Noses

The digital outputs generated by e-nose sensors have to be analyzed and interpreted in order to provide useful information to the operator. The choice of method utilized depends on the type of available input data acquired from the sensors and the type of information that is sought. Commercially available analysis techniques fall into following three main categories as follows [1]:

i. Graphical Analyses (Bar chart, Profile, Polar and offset polar plots)

The simplest form of data reduction is graphical analysis useful for comparing samples or comparing aroma identification elements of unknown analytes relative to those of known sources in reference libraries.

ii. Multivariate Data Analyses (MDA)

Principal component analysis (PCA), Canonical discriminate analysis (CDA), Featured within (FW) and Cluster analysis (CA)-Multivariate data analysis comprises a set of techniques for the analysis of data sets with more than one variable by reducing high dimensionality in a multivariate problem when variables are partly correlated, so they can be displayed in two or three dimensions. Multivariate analysis can be divided into **untrained or trained** techniques.

a. Untrained Techniques

These are used when a database of known samples has not been previously built, therefore it is not necessary nor intended for recognizing the sample itself, but for making comparisons between different unknown samples to discriminate them. The simplest and most widely used untrained MDA technique is principal component analysis. PCA is most useful when no known sample is available, or when hidden relationships between samples or variables are suspected.

b. Trained or Supervised Learning Techniques

On the contrary these techniques classify unknown samples on the basis of characteristics of known samples or sets of samples with known properties that are usually maintained in a reference library that is accessed during analysis.

iii. Network Analyses

Artificial neural network (ANN) and Radial basis function (RBF)- The artificial neural network (ANN) in the best known and most evolved analysis techniques utilized in statistical software packages for commercially available electronic noses. Mimicking the cognitive processes of the human brain, it contains interconnected data processing algorithms that work in parallel. The result of ANN data analysis usually is in the form of a percentage match of identification elements in the sample with those of aroma patterns from known sources in the reference library.

Source

1. *Schaller, E.; Bosset, J.O.; Esher F. Electronic noses and their application to food. Lebensm.-Wiss. Ul.-Technol. 1998,31, 305-316.*
2. *Alphus D. Wilson 1, and Manuela Baietto, Review Applications and Advances in Electronic-Nose Technologies, Sensors 2009, 9, 5099-5148.*

Procedure

As a first step, an electronic nose need to be trained with qualified samples so as to build a database of reference. Then the instrument can recognize new samples by comparing volatile compounds fingerprint to those contained in its database. Thus they can perform qualitative or quantitative analysis.

In general, the analysis in the electronic nose is conducted by placing the product into an airtight glass jar (concentration chamber), where the product should be kept for approximately one hour to release volatile components. By a syringe through the jar cover, an aliquot of the sample headspace is pumped over the electronic nose sensors. The measurement phase lasts for 60 seconds; after that, a stand-by phase is activated so as to clean the system and re-set the sensors back to their baseline level. This is accomplished by virtue of pumping fresh air into the system; the air first crosses the sensors and then reaches the empty concentration chamber [1]. The scheme of working of an electronic nose is shown in following Figures:

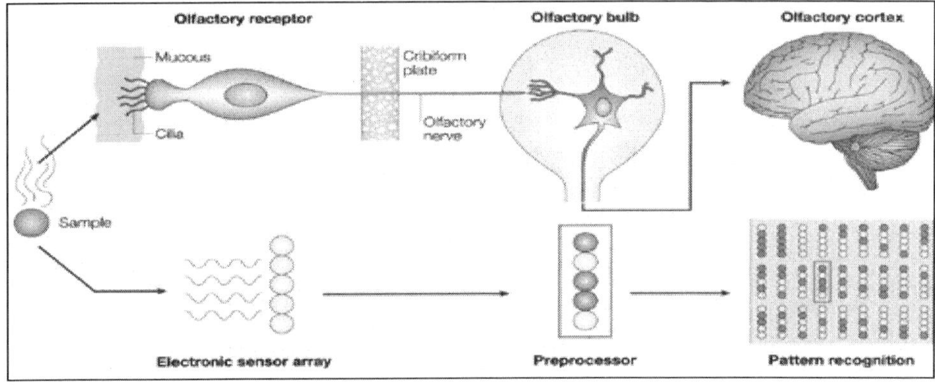

Comparison of an Electronic Nose and Human Brain Performance [2,4].

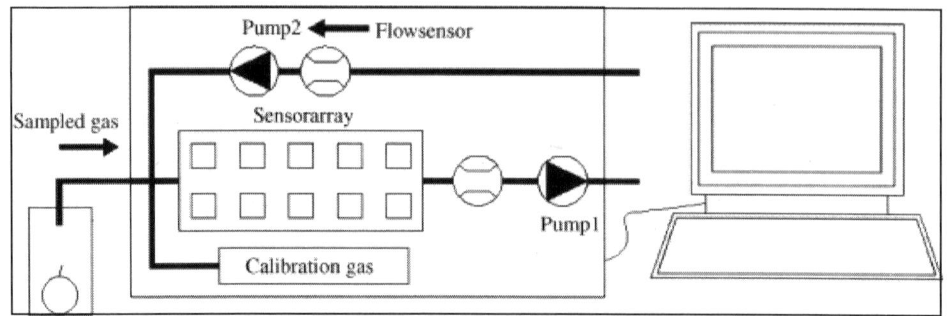

Electronic Nose Scheme [3,4].

Source

1. *Gomez, A.H., Wang, J., Hu, G., Pereira, A.G. (2007): Discrimination of storage shelf-life for mandarin by electronic nose technique, LWT 40, 681-689.*
2. *Turner, A.P.F., Magan, N. (2004): Electronic noses and disease diagnostics. Nature Reviews Microbiology 2, 161-166.*
3. *Gomez, A.H., Wang, J., Hu, G., Pereira, A.G. (2008): Monitoring storage shelf life of tomato using electronic nose technique. Journal of Food Engineering 85, 625-631.*
4. *M. Tudor Kalit et al.: Electronic nose and electronic tongue in the dairy industry, Mljekarstvo 64 (4), 228-244 (2014).*

Quality of Sensors

A good sensor should fulfill a number of criteria. The details of some of these criteria are given below:

(i) First, the sensor should have highest sensitivity to the target group of chemical compound (s) intended for detection and with a threshold of detection similar to that of the human nose, down to about 10-12 g mL-1 [1].

(ii) To be most useful with diverse detection capabilities, e-nose sensors should have relatively low selectivity in order to be sensitive to a wide number of different chemical compounds.

(iii) Sensors should be capable of operating at relatively low temperatures when necessary, have short calibration and training requirements, fast recovery time between runs and maintenance procedures to maintain low operating costs.

(iv) They must also have short recording and analysis times, particularly when used as on-line systems, and high sensor array stability.

Source

1. *Schaller, E.; Bosset, J.O.; Esher F. Electronic noses and their application to food. Lebensm.-Wiss.Ul.-Technol. 1998, 31, 305-316.*

2. *Alphus D. Wilson 1, and Manuela Baietto, Review Applications and Advances in Electronic-Nose Technologies, Sensors 2009, 9, 5099-5148.*

Electronic Aroma-Detection Sensor Types [24]

The basis of electrochemical gas sensor operation involves interactions between gaseous molecules and sensor-coating materials which modulate electrical current passing through the sensor, detectable by a transducer that converts the modulation into a recordable electronic signal [1]. There are many different types of electrochemical sensors (*e.g.* metal-oxide gas sensors, metal-oxide semiconductor field effect transistors, conducting polymer gas sensors, acoustic wave gas sensors, quartz crystal microbalance sensors, surface acoustic wave devices, field-effect gas sensors, electrochemical gas sensors, pellistors, fiber-optic gas sensors) and many different types of sensor-coating materials which are classified according to additive doping materials, the type and nature of the chemical interactions, the reversibility of the chemical reactions and running temperature. A summary of the types and mechanisms involved with some common gas sensor technologies are given in following Table [24]:

Sensor Type	*Sensitive Material*	*Detection Principle*
Acoustic sensors: Quartz crystal microbalance (QMB); surface and bulk acoustic wave (SAW, BAW).	Organic or inorganic film layers.	Mass change (frequency shift).
Calorimetric; catalytic bead (CB)	Pellistor	Temp. or heat change (from chemical reactions)
Catalytic field-effect sensors (MOSFET)	Catalytic metals	Electric field change.
Colorimetric sensors	Organic dyes	Color changes, absorbance.
Conducting polymer sensors	Modified conducting polymers.	Resistance change.
Electrochemical sensors	Solid or liquid electrolytes.	Current or voltage change.
Fluorescence sensors	Fluorescence-sensitive detector.	Fluorescent-light emissions.
Infrared sensors	IR-sensitive detector.	I R-radiation absorption
Metal oxides semi-conducting (MOS, Taguchi)	Doped semi-conducting metal oxides (SnO2, GaO).	Resistance change.
Optical sensors	Photodiode, light-sensitive.	Light modulation, optical changes.

Metal Oxide Sensors [24]

Metal-oxide sensors have very high sensitivity (sub-ppm levels for some gases) and respond to oxidizing compounds (zinc-oxide, tin-dioxide, titanium-dioxide, iron oxide) and some reducing compounds, mainly nickel-oxide or cobalt-oxide [2]. Metal oxide sensors consist of a ceramic substrate coated with metal-oxide semiconducting film (*e.g.* SnO_2, TiO_2, ZnO, ZrO_2) which reacts with oxygen from the air. From a chemical point of view, the sensing reaction is based on an oxygen exchange between the volatile gas molecules and the metal coating material.

Electrons are attracted to the loaded oxygen and result in decreases in sensor conductivity [3]. During this reaction, metal-oxide electrical resistance is reset to the background level. During the measurement, when sensors are exposed to volatile compounds, the resistance of metal-oxide is changed due to the adsorption of volatile compound onto the metal-oxide surface, where it reacts with oxygen through oxidation/reduction. This change in electrical resistance is considered to be a response of the sensor to a particular sample (4-5). MOS sensors respond to molecules like alcohols, ammonia, butane, carbon monoxide, chlorine, ethylene, heptane, hexane, hydrogen, hydrogen sulphide, methane, nitrogen dioxide, ozone, propane, sulphur dioxide and toluene (6).

The metal-oxide semiconductor field effect transistors **(MOSFET)** were firstly reported by Lundstrom *et al.* [7] in 1975 based on the tendency of a number of metals to adsorb and dissolve hydrogen [8]. The first hydrogen-sensitive MOSFET used palladium film as gate electrodes [7]. Since then, other combinations of metal oxides and catalysts such as palladium-titanium dioxide [6], palladium-gallium arsenide [10], and palladium-zinc oxide [11] have been investigated. MOFSET sensors rely on the change of electrostatic potential. They consist of three parts: silicon semiconductor, silicon oxide insulator and catalytic metal "gate" (usually palladium, platinum, iridium or rhodium)through which the applied voltage creates an electric field. When polar compounds interact with the metal, the electric field is modulated and recorded by the transistor [12]. The doping metal (or gate) can be a thick (100-200 nm) or thin (6-20 nm) film. In the first case, the sensor can only respond to dissociated hydrogen. Thus, sensor sensitivity to hydrogen non-releasing molecules such as ammonia or carbon monoxide is very low. A thinner layer of metal on the sensor changes the catalytic activity towards these kinds of molecules [13-15].

Conducting or Conductive Polymer Gas Sensors [24]

These sensors operate based on changes in electrical resistance caused by adsorption of gases onto the sensor surface. Conductive electro active polymers have attracted much interest for use as electronic noses since the early 1980s [16], particularly because they have high sensitivities, short response times, are easily synthesized, have good mechanical properties and are particularly useful because they operate at room temperature [17]. Conductive polymer gas sensors consist of a substrate, usually silicon, a pair of gold-plated electrodes and a conducting organic polymer coating as the sensing element [12]. The sensitivity of conductive polymers to VOCs is measured as changes in electrical resistance. Conducting polymers are usually synthesized by chemical or electrochemical oxidizing of the corresponding monomers. The most widely used sensor coating monomers are polypyrrole, polyaniline and polythiophene [1], but polyacetylene, poly (phenylvinylene), poly (3,4-ethylene dioxythiophene), poly (*N*-vinylcarbazone), poly (thienylenevinylene) and many others have been investigated [18,17]. The common feature of conductive polymer materials is the presence of a conjugated pi-electron system which extends over the whole polymer.

Acoustic Wave Gas Sensors [24]

These sensors use a mechanical (acoustic) wave as the sensing mechanism. As the acoustic wave propagates through or on the surface of the sensor coating material, any changes to the characteristics of the propagation path, due to the sorption of VOCs, affect the velocity and/or amplitude of the wave [19,20]. They consist of a piezoelectric substrate, usually quartz (SiO_2), lithiumniobate ($LiNbO_3$), lithium tantalite ($LiTaO_3$) or zinc oxide, doped with a suitable sorptive material. material.

Electrochemical Gas (EC) Sensors [24]

These sensors operate at room temperature, have low power consumption and are very robust, but still quite bulky [1]. Their sensing methodology is based on the electrochemical oxidation or reduction of volatile molecules at a catalytic electrode surface [21]. This technology has a good relevance when applied to the detection and measurement of electrochemically active gases, but they are not very sensitive to a wide diversity of compounds, especially aromatic hydrocarbons [18]. The sensitive electrode typically is composed of a layer of precious metal, such as gold combined with carbon, coated onto a hydrophobic membrane. Commercial electrochemical gas sensors are currently available for a wide range of toxic gases. In each case, the concentration of gas is determined by measuring the current flowing in the sensors. They are commonly used in mining operations for personnel-protection monitoring, tunneling and other industrial applications. The disadvantages of these sensors, from the standpoint of their application within electronic nose, are their size and relatively high selectivity for a limited number of simple gases [22].

Calorimetric or Catalytic Bead (CB) Sensor [24]

It consists of two coils of fine platinum wire embedded in a bead of alumina connected in a Wheatstone bridge circuit. One of the pellistors is impregnated with a special catalyst that promotes oxidation. The other pellistor is treated to inhibit oxidation. Current is passed through the coils to heat the bead until oxidation of the sample gas occurs at 500-550°C. Combustion of the gas raises the temperature further and increases the resistance of the platinum coil in the catalyzed bead, leading to an imbalance of the bridge. This output change in resistance is linear for most gases and response time is a few seconds. The sample gas must contain at least 12 per cent oxygen by volume for oxidation. The Area REA monitor produced by Rae Systems is an example of combined-technology e-nose that contains a CB sensor.

Optical Sensor [24]

These systems are some what more complex than typical sensor-array systems having transduction mechanisms based on changes in electrical resistance. Optical sensors work by means of light modulation measurements and consist of an assortment of technologies ranging from diverse light sources with optical fibers to various photodiode and light-sensitive photo detectors. Various operational modes have been developed that measure changes in absorbance, fluorescence, light polarization, optical layer thickness, or colorimetric dye response. The simplest optic sensors use color- changing indicators, such as metalloporphyrins, to measure absorbance with a LED and photodetector system upon exposure to gas analytes. Two specialized types of optical sensors are the colorimetric and fluorescence sensors. Colorimetric sensors use thin films of chemically-responsive dyes as a colorimetric sensor array. Fluorescence sensors detect fluorescent light emissions from the gas analyte at a lower wavelength and are more sensitive than colorimetric sensor arrays.

Piezoelectric Sensors

These types of sensors consist of a membrane-coated piezoelectric quartz crystal and two gold electrodes that form an electric field. Adsorption of volatile molecules onto the surface of the membrane increases the mass of the device, resulting in the change of its resonance frequency (5).

Mass Spectrometry-Based Sensors

When it comes to food quality control, in the last few years mass spectrometry-based (MS) electronic noses have become an increasingly used alternative (or complement) to gas sensor-based electronic noses (23). Mass detection-based e-noses use mass spectrometer as a sensor array. Upon injection, an MS pattern of the unresolved volatile com-pounds mixture is created. This means that each mass-to-charge ratio acts as a sensor. In this way, sensors detect any compound having a particular mass-to-charge ratio, so that MS-based electronic noses potentially have hundreds of sensors. A huge advantage of this kind of sensors is that they operate under well-known technology, with reproducibility, stability and sensitivity of mass spectrometers being long well-established (5).

Source

1. *Gardner, J.W., Bartlett, P.N.Electronic Noses. Principles and Applications, Oxford University Press: Oxford, UK, 1999, pp. 221-245.*
2. *Mielle, P. Electronic noses: towards the objective instrumental characterization of food aroma. Trends Food Sci. Technol.1996,7,432-438.*
3. *Demarne, V., Grisel, A., Sanjinks, R., Rosenfeld, D., Levy, F.Electrical transport properties of thin polycrystalline SnO2 film sensors. Sens. Actuat. B: Chem. 1992, 7, 704-708.*
4. *Albert, K.J., Lewis, N.S., Schauer, C.L., Sotzing, G. A., Stitzel, S.E., Vaid, T.P., Walt, D.R. (2000): Cross-reactive chemical sensor arrays. Chemical Reviews 100, 2595-2626.*
5. *Ampuero, S., Bosset, J.O. (2003): The electronic nose applied to dairy products: a review. Sensors and Actuators B 94,1-12.*
6. *Cosio, M.S., Ballabio, D., Benedetti, S., Gi-gliotti, C. (2006): Geographical origin and authentication of extra virgin olive oils by an electronic nose in combination with artificial neural networks. Analytica Chimica Acta 567, 202-210.*
7. *Lundstrom, K.I.; Shivaraman, M.S.; Svenson, C.M. A hydrogen-sensitive Pd gate MOS transistor. J. App. Phys. 1975, 46,3876-3881.*
8. *Lundstrom, I.; Shivaraman, M.S.; Svenson, C.S.; Lundkvist, L. Hydrogen sensitive MOSFET. Appl. Phys. Lett. 1975, 26, 55-57.*

9. *Kobayashia, H.; Kishimotoa, K.; Nakatoa, Y. Reactions of hydrogen at the interface of palladium-titanium dioxide Schottky diodes as hydrogen sensors, studied by work function and electrical characteristic measurements. Surf. Sci. 1994, 306, 393-405.*

10. *Lechuga, L.M.; Calle, A.; Golmayo, D.; Briones, F. Hydrogen sensor based on a Pt/ GaAs Schottky diode. Sens. Actuat. B: Chem. 1991, 4, 515-518.*

11. *Ito, K. Hydrogen-sensitive Schottky barrier diodes. Surf. Sci. 1979, 86, 345-352.*

12. *Schaller, E., Bosset, J.O.; Esher F. Electronic noses and their application to food. Lebensm.-Wiss.Ul.-Technol. 1998, 31, 305-316.*

13. *Winquist, F.; Spetz, A.; Armgarth, M.; Lundstrom, I.; Danielsson, B. Biosensors based on ammonia sensitive metal-oxide-semiconductor structures. Sens. Actuat. 1985, 8, 91-100.*

14. *Åbom, A.E.; Comini, E.; Sberveglieri, G., Hultman, L.; Eriksson, M. Thin oxide films as surface modifiers of MIS field effect gas sensors. Sens. Actuat. B: Chem. 2002, 85, 109-119.*

15. *Åbom, A.E.; Persson, P.; Hultman, L.; Eriksson, M. Influence of gate metal film growth parameters on the properties of gas sensitive field-effect devices. Thin Solid Films 2002, 409,233-242.*

16. *Nylander, C.; Armgarth, M.; Lundstrom, I. An ammonia detector based on a conducting polymer. Anal. Chem. Symp. Ser. 1983, 17, 203-207.*

17. *Bai, H.; Li, C.; Chen, F.; Shi, G. Aligned three-dimensional micro-structures of conducting polymer composites. Polymer 2007, 48, 5259-5267.*

18. *Strike, D.J.; Meijerink, M.G.H.; Koudelka-Hep, M. Electronic noses – A mini-review. Fres. J. Anal. Chem. 1999, 364, 499-505.*

19. *Draft, B. Acoustic wave technology sensors. IEEE Trans. 2001, 49, 795-802.*

20. *Abdollahi, A.; Jiang, Z., Arabshahi, S. A. Evaluation on mass sensitivity of SAW sensors for different piezoelectric materials using finite-element analysis. IEEE Trans. 2007, 54, 2446-2455.*

21. *Hobbs, B.S.; Tantrum, A.D.S., Chan-Henry, R.Liquid electrolyte fuel cells. In Techniques and Mechanisms in Gas Sensing; Moseley, P. T., Norris, I.O.W., Williams, D.E., Eds.; Adam Hilger: Bristol, UK, 1991; pp.161-188.*

22. *Kiesele, H.; Wittich, M.H. Electro-chemical gas sensors for use under extreme climatic conditions.Dräger Rev. 2000, 85, 10-13.*

23. *Vinaixaa, M., Vergaraa, A., Durana, C., Llobeta, E., Badiab, C., Brezmesa, J., Vilanovaa, X., Correiga, X. (2005) Fast detection of rancidity in potato crisps using e-noses based on mass spectrometry or gas sensors. Sensors and Actuators B: Chemical 106 (1),67-75.*

24. *Alphus D. Wilson 1, and Manuela Baietto, Review Applications and Advances in Electronic-Nose Technologies, Sensors 2009, 9, 5099-5148.*

Advantages and Disadvantages of Different Sensors [1]

There are a variety of advantages and disadvantages of using various e-nose sensors based on their response and recovery times, sensitivities, detection range, operating limitations, physical size, inactivation by certain poisoning agents, and other limitations that are specific to individual sensor types. The types and categories of advantages and limitations associated with individual e-nose sensor types are closely linked with the nature of the technology that determines the principle for detection and the types of gas analytes that may be detected with each sensor type. The major advantages and disadvantages associated with each e-nose sensor type are given in following table:

Sensor Type	*Advantages*	*Disadvantages*
Calorimetric or catalytic bead (CB)	Fast response and recovery time, high specificity for oxidized compounds	High temperature operation, only sensitive to oxygen-containing compounds
Catalytic field-effect Sensors (MOSFET)	Small sensor size, inexpensive operating costs	Requires environmental control, Base line drift, low sensitivity to ammonia and carbon dioxide
Conducting polymer sensors	Ambient temperature operation, sensitive to many VOCs, short response time, diverse sensor coatings, inexpensive, resistance to sensor poisoning	Sensitive to humidity and temperature, sensors can be overloaded by certain analytes, sensor life is limited
Electrochemical sensors (EC)	Ambient temperature operation, low power consumption, very sensitive to diverse VOCs	Bulky size, limited sensitivity to simple or low mol. wt. gases
Metal oxides semi conducting (MOS)	Very high sensitivity, limited sensing range, rapid response and recovery times for low mol. wt. compounds (not high)	High temperature operation, high power consumption, sulfur and weak acid poisoning, limited sensor coatings, sensitive to humidity, poor precision.
Optical sensors	Very high sensitivity, capable of Identification of individual compounds in mixtures, multi parameter detection capabilities	Complex sensor-array systems, more expensive to operate, low portability due to delicate optics and electrical components
Quartz crystal microbalance (QMB)	Good precision, diverse range of sensor coatings, high sensitivity.	Complex circuitry, poor signal-to noise ratio, sensitive to humidity and temperature

Source

1. *Alphus D. Wilson 1, and Manuela Baietto, Review Applications and Advances in Electronic-Nose Technologies, Sensors 2009, 9, 5099-5148.*

Types of Electronic Noses [1]

Many electronic noses are commercially available. A summary of some of the most widely used electronic noses with manufacturers, models available and technological basis are given in following table:

Instrument Type	*Manufacturer*	*Models Produced*	*Technology Basis*
Single-technology (e-nose sensors only)	Airsense Analytics	i-Pen, PEN2, PEN3	MOS sensors
	Alpha MOS	FOX 2000, 3000, 4000	MOS sensors
	Applied Sensor	Air quality module	MOS sensors
	Chemsensing	ChemSensing Sensor array	Colorimetric optical
	CogniScent Inc.	ScenTrak	Dye polymer sensors
	CSIRO	Cybernose	Receptor-based array
	Dr. Födisch AG	OMD 98, 1.10	MOS sensors
	Forschungszentrum Karlsruhe	SAGAS	SAW sensors
	Gerstel GmbH Co.	QSC	MOS sensors
	GSG Mess- und Analysengeräte	MOSES II	Modular gas sensors
	Illumina Inc.	oNose	Fluorescence optical
	Microsensor Systems Inc	Hazmatcad,Fuel Sniffer, SAW Mini CAD mk II	SAW sensors
	Osmetech Plc	Aromascan A32S	Conducting polymers
	Sacmi	EOS 835, Ambiente	Gas sensor array
	Scensive Technol	Bloodhound ST214	Conducting polymers
	Smiths Group plc	Cyranose 320	Carbon black-polymers
	Sysca AG	Artinose	MOS sensors
	Technobiochip	LibraNose 2.1	QMB sensors
Combined-technology (e-nose + other types)	Airsense Analytics	GDA 2	MOS, EC, IMS, PID
	Alpha MOS	RQ Box, Prometheus	MOS, EC, PID, MS
	Electronic Sensor Technology	ZNose 4200, 4300, 7100	SAW, GC
	Microsensor Syst	Hazmatcad Plus	SAW, EC
		CW Sentry 3G	SAW, EC
	Rae Systems	Area RAE monitor	CB, O2, EC, PID Thermistor, EC, PID,CO2, humidity
		IAQRAE	
	RST Rostock	FF2, GFD1	MOS, QMB, SAW

The **Alpha-MOS** (Toulouse, France) Fox electronic nose was designed in collaboration with the Universities of Warwick and Southampton. It employs either six (Fox 2000), 12 (Fox 3000) or 18 (Fox 4000) metal oxide gas sensors. **The Aromascan A32S** (Osmetech Plc, UK)is an organic matrix-coated polymer-type 32-detector e-nose of Electronic noses:

Source

1. *Alphus D. Wilson 1, and Manuela Baietto, Review Applications and Advances in Electronic-Nose Technologies, Sensors 2009, 9, 5099-5148.*

Advantages and Disadvantages of Electronic Noses [1,2]

The followings are advantages and disadvantages of electronic noses:

Advantages

Artificial or electronic noses with diverse sensor arrays have a number of advantages over traditional analytical instruments.

(i) Electronic nose sensors do not require chemical reagents.
(ii) Have good sensitivity and specificity.
(iii) Provide rapid results
(iv) Allow non-destructive sampling of odorants or analytes [1].

Further more, e-noses generally are:

(i) Far less expensive than analytical systems.
(ii) Easier and cheaper to operate
(iii) Have greater potential for portability and field use compared with complex analytical laboratory instruments.

Thus, electronic noses have far greater potential to be used eventually by unskilled consumers for innumerable practical applications in residential and public settings.

Disadvantages

Some disadvantages of e-nose sensing include:

(i) Problems with reproducibility and recovery.
(ii) Negative effects of humidity and temperature on the sensor responses.
(iii) Inability to identify individual chemical species with sample gases.

Thus, electronic noses will never completely replace complex analytical equipment or odor panels for all applications, but offer quick real-time detection and discrimination solutions for applications requiring accurate, rapid and repeated determinations.

Source

1. *García-González, D.L.; Aparicio, R. Sensors: From biosensors to the electronic nose. Grasasy Aceites 2002, 53, 96-114.*

2. *Alphus D. Wilson 1, and Manuela Baietto, Review Applications and Advances in Electronic-Nose Technologies, Sensors 2009, 9, 5099-5148; doi: 10.3390/s90705099.*

Applications of Electronic Nose

Electronic-nose systems have been designed specifically to be used for numerous applications in many different industrial production processes. These have a wide range of applications in various markets and industries ranging from food processing, industrial manufacturing, quality control, environmental protection, security, safety and military applications to various pharmaceutical, medical, microbiological and diagnostic applications [7].

Sl.No.	Industry Sector	Application Area	Specific Use Types and Examples
1	Agriculture	Crop protection	Homeland security, safe food supply
		Harvest timing and storage	Crop ripeness, preservation treatments
		Meat, seafood, and fish products	Freshness, contamination, spoilage
		Plant production	Cultivar selection, variety characteristics
		Pre- and post-harvest diseases	Plant disease diagnoses, pest identification
2	Food and beverage	Consumer fraud prevention	Ingredient confirmation, content standards
		Quality control assessments	Brand recognition, product consistency
		Ripeness, food contamination	Marketable condition, spoilage, shelf life
		Taste, smell characteristics	Off-flavors, product variety assessments
3	Manufacturing	Processing controls	Product characteristics and consistency
		Product uniformity	Aroma and flavor characteristics
		Safety, security, work conditions	Fire alarms, toxic gas leak detection
4	Environmental	Air and water quality monitoring	Pollution detection, effluents, toxic spills
		Indoor air quality control	Malodor emissions, toxic/hazardous gases
		Pollution abatement regulations	Control of point-source pollution releases
5	Regulatory	Consumer protection	Product safety, hazardous characteristics
		Environmental protection	Air, water, and soil contamination tests
6	Scientific research	Botany, ecological studies	Chemotaxonomy,ecosystem functions
		Engineering, material properties	Machine design, chemical processes
		Microbiology, pathology	Microbe and metabolite identifications

Electronic noses are not designed to be universally appropriate sensor systems for every conceivable gas-sensing application nor are they capable of serving every possible analytical need. Thus, the suitability of an electronic nose for a specific application is highly dependent on the required operating conditions of the sensors in the array and the composition of the analyte gases being detected. A proper selection of an appropriate e-nose system for a particular application must involve an evaluation of systems on a case by case basis [1,2,3,4].

1-Electronic nose instruments are used by Research and Development laboratories, Quality Control laboratories and process and production departments for various purposes [5]:

(i) **In R&D laboratories**-Formulation or reformulation of products, Benchmarking with competitive products, Shelf life and stability studies, Selection of raw materials, Packaging interaction effects and Simplification of consumer preference test [5].

(ii) **In Quality Control laboratories for at line quality control such as-** Conformity of raw materials, intermediate and final products, Batch to batch consistency, Detection of contamination, spoilage, adulteration, Origin or vendor selection and Monitoring of storage conditions [5].

(iii) **In process and production departments for-**Managing raw material variability, Comparison with a reference product, Measurement and comparison of the effects of manufacturing process on products, Following-up cleaning in place process efficiency, Scale-up monitoring and Cleaning in place monitoring [5].

Source

1. *Persaud, K.; Dodd, G.Nature1982; 299; 352 - 355.*
2. *"Wasp Hound". Science Central.http://www. scien central. com/articles/iew.php3? article_id= 218392717.Retrieved 23 February 2011.*
3. *Alpha M.O.S, a full range of solutions for odor and taste profiling in R&D and QC.*
4. *Alpha M.O.S, Applications Notes.*
5. *Wikipedia.*
6. *Pima County Marks 1 Year of Odor Management Innovation*
7. *Alphus D. Wilson 1, and Manuela Baietto, Review Applications and Advances in Electronic-Nose Technologies, Sensors 2009, 9, 5099-5148.*

2. Industry-wise Application

The details related to some individual examples of electronic nose applications in each of these individual industries and product areas are given below:

1. Application in Dairy Industry

The aroma profiles of Milk and Milk products, are defined by a complex mixture of volatiles at different levels. It can be affected by milk composition that depends, among other, on dairy cattle management (genetics, herd health and feeding regimens) and milk manufacturing process [1]. The type of milk used for its production and the ripening process are the main factors that determine the characteristic cheese aroma [2]. Industrial quality and commercial success of dairy products can be severely affected by microbial milk spoilage, which can lead to the production of undesirable off-flavours, physical defects and secondary metabolite toxicity [3]. Dairy products contain off-flavor compounds created by a variety of mechanisms such as through the action of natural and microbial enzymes and chemical changes catalyzed by light or heavy metals.

Source

1. *Biolatto, A., Grigioni, G., Irurueta, M., Sancho, A.M., Taverna, M., Pensel, N. (2007): Seasonal variation in the odour characteristics of whole milk powder. Food Chemistry 103, 960-967.*

2. *Cabezas, L., Sánchez, I., Poveda, J.M., Sesena, S., Palop, M.L.L. (2007): Comparison of microflora, chemical and sensory characteristics of artisanal Manchego cheeses from two dairies. Food Control 18, 11-17.*

3. *Magan, N., Pavlou, A., Chrysan-thakis, I. (2001): Milksense: avolatile sensing system recognise spoilage bacteria and yeasts in milk. Sensors and Actuators B, 28-34.*

4. *Milna Tudor Kalit, Ksenija Markovi, Samir Kalit1, Nada Vahcic and Jasmina Havranek, University of Zagreb, Zagreb, Croatia, Application of electronic nose and electronic tongue in the dairy industry.*

The details of recently reported applications of electronic noses to dairy products is given below:

i. Ageing of Milk and Shelf-life Prediction

(a) The Headspace of milk typically presents a complex mixture of organic volatiles at varying concentrations and with a high percentage of relative humidity and the milk matrix is highly heterogeneous containing different levels of lipids, proteins and carbohydrates (1).

(b) Groups of UHT milk aged 1–8 days, and 1–3 days of pasteurized milk were classified by PCA performed on normalised data. A home-made MOS sensors array was used. 10 ml of milk were placed in 20 ml vials, four to five measurements per sample. Samples were incubated 15 min at 30°C during analysis, the headspace was carried into the injector with a flow of N_2, sensor temperature was setat 250°C. Sensors showed a response and recovery time of 2–3 min.

(c) The possibility of practical application of an electronic nose in dairy industry for the raw (starting) milk quality control and rancidity monitoring in different production stages was demonstrated [2]. The results shown in following figures proved the ability of a semiconductor thin films-based electronic nose to well distinguish UHT and pasteurised milk during days in which the milk has different rancidity values:

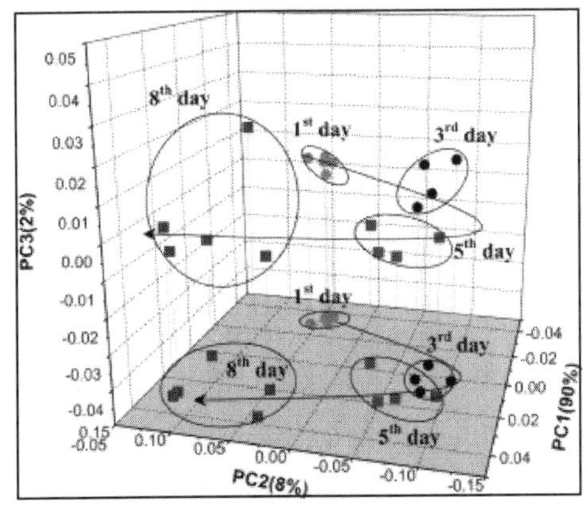

PCA Plot of UHT Milk on different Days.

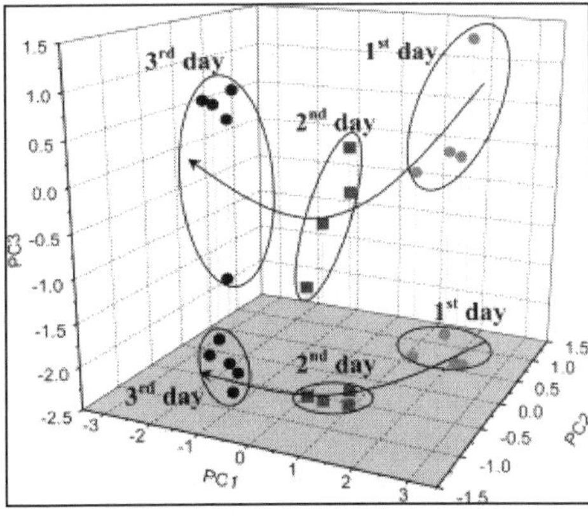

PCA Plot of UHT Milk on different Days (Capone).

(d) Shelf-life prediction [3] of 2 per cent -fat pasteurised milk and whole-fat chocolate milk was obtained with a home assembled SPME–MS–MVA system (solid phase micro-extraction, mass spectrometry, multivariate analysis). In this,3 ml of milk sample and 5 ml internal standard (10_g/ ml chlorobenzene) were placed in 6ml vials. Samples were incubated 20 min at 50 °C during fibre exposure, otherwise they were stored at 7.2 °C. The injector temperature was set at 275 °C and the transfer line followed a temperature program between 150 and 180°C. Twenty samples of each set were used for model validation solely. Mass intensities were normalised by the intensity of chlorobenzene. Partial least-squares modeling predicted sample shelf-life with respect to sensory results with an accuracy of ±0.62 and ±0.88 days and a correlation coefficient of 0.9801 and 0.9832 for milk and chocolate milk, respectively. A significant increase in concentration with ageing was detected by GC–MS for certain volatiles such as: dimethyl sulphide, 2-heptanone, ethyl acetate, pentanal, pyrrolidine, hexanoic acid,2-methyl-butanal, furfuraldehyde, *etc.* [4].

(e) As mastitis is a disease that causes significant changes in milk composition and its volatile substance content, Eriksson *et al.* [5] investigated into the possibility of using an electronic nose (equipped with MOS and MOFSET sensors) for the discrimination between the milk excreted from the mastitis-affected breast and the healthy one. The results were supported by the GC-MS analysis which identifies 103 volatile compounds. Mastitic milk contained higher levels of sulphides, ketones, amines and acids.

(f) It was found that the electronic nose has a lower detection limit (0.5 mg/kg) and better precision compared to dynamic headspace gas chromatography (GC) for determining the presence of trimethylamine in Swedish milk samples [6].

(g) **The shelf-life of milk** also has been studied [7]. A Fox 4000 electronic nose equipped with 18 sensors and an auto sampler was used to evaluate the growth of total bacteria in milk stored at ambient temperature and 5 °C. The results showed that measurements generated by the electronic nose could be used to detect both bacterial growth in milk and shelf-life.

Source

1. *S.Capone, P.Siciliano, F. Quaranta, R. Rella, M. Epifani, L.Vasanelli, Analysis of vapours and foods by means of an electronicnose based on a sol-gel metal oxide sensors array, Sens. ActuatorsB 69 (2000) 230–235.*

2. *S.Capone, M. Epifani, F. Quaranta, P. Siciliano, A. Taurino, L.Vasanelli, Monitoring of rancidity of milk by means of an electronic nose and a dynamic PCA analysis, Sens. Actuators B 78 (1-3) (2001)174-179.*

3. *R. T.Marsili, Shelf-life prediction of processed milk by solid-phase microextraction, mass spectrometry, and multivariate analysis, J. Agric. Food Chem. 48 (2000) 3470–3475.*

4. *R. T. Marsili, N. Miller, Determination of the cause of off-flavors in milk by dynamic headspace GC–MS and multivariate data analysis, in: A.M. Spanier, F. Shahidi, T.H.*

Parliament, C. Mussinan, C.-T.Ho (Eds.), Food Flavours and Chemistry: Advances of the New Millennium, Conti E. T., RSC, 2001, pp. 118–129.

5. *Eriksson, A., Persson Waller, K., Svennersten-Sjaunja, K., Haugen, J-E., Lundby, F., Lind, O. (2005): Detection of mastitic milk using a gas-sensor array system (electronic nose). International Dairy Journal 15, 1193-1201.*
6. *Ampuero, S.; Zesiger, T.; Gustafsson, V.; Lunden, A.; Bosset, J.O. Determination of trimethylamine in milk using an MS based electronic nose. Eur. Food Res. Techn. 2002, 214, 163-167.*
7. *Labreche, S; Bazzo, S.; Cade, S.; Chanie, E. Shelf life determination by electronic nose: application to milk. Sens. Actuat. B: Chem. 2005, 106, 199-206.*

ii. Classification of Off-flavours in Milk

The off-flavours (due to oxidation) in milk originate mostly from enzymatic activity, bacterial metabolism, heat, photo oxidation and oxidation catalysed by chemicals (sanitizers for production lines or pro-oxidant metals -copper, iron and nickel). Exposing milk to light induces two major effects:

(a) The first 2–3 days a burnt oxidised flavour develops probably due to the degradation of sulphur-containing amino acids from the whey into methional, mercaptans, sulphides and disulphides. After the second day a persistent metallic, cardboard-like off-flavour occurs (attributed to the autoxidation of unsaturated fatty acids).

(b) On the other hand, heat provokes a typical boiled off-flavour (probably due to the formation of sulphur compound).In general, low molecular weight aldehydes, ketones and fatty acids are responsible for most off-flavours observed in foods/Dairy Products and beverages upon exposure to light. Marsili [1-2] has worked on the discrimination of off-flavours in milk. Prior to analysis, samples were conditioned at 19°C for 16 h except for fresh milk control samples, and then heated at 45–50°C during the 12–15 min of fibre exposure (75 _m Carboxen/PDMS SPME fibre). Masses corresponding to volatiles normally present in non-defective milk were disregarded (*e.g.* acetone, 2-butanone) and intensity of masses from target compounds were amplified. The intensities of all masses were normalised by the intensity of the internal standard. A fairly steady increase in pentanal and hexanal was observed with increasing exposure time to 200 ft^3 of fluorescent light. **Typically, hexanal and dimethyl disulphide were found to be good indicators of light damaged milk, whereas pentanal, isopentanal, hexanal, heptanal, octanal, nonanal, and 1-octen-3-one usually indicated copper induced oxidation in milk.** The application of an electronic nose to the study of boiled off-flavour in UHT milk [3] was performed with an NST 3220 which typically uses MOSFET, MOS, and QMB sensors (CSP, UK). PLS analysis (partial least square) was able to discriminate down to10 per cent of boiled milk into reference milk, compared to 30 per cent for a sensory panel. Di Natale [4] has reported a correct discrimination between UHT and pasteurised milk with a home made electronic tongue

based on metaloporphyrin whereas a QMB sensors array, based also on metaloporphyrin, failed. An Electronic Nose 4048 with 12 CP sensors (Neotronics, UK), was used to distinguish between different intermediate products during process of block-milk (milk and sugar mixture evaporated to powder), used in the preparation of milk-chocolate. Parametric analysis of trimethylamine concentration in milk (fishy off-flavour) performed with an electronic nose based on mass spectrometry. No control of the relative humidity in the electronic nose was undertaken. The results correlate well with GC–MS and sensory analysis. PCA analysis showed some association to descriptors such as caramel, nutty, burnt and chocolate, typical of volatiles generated by Maillard reactions [5] during heating. An unusual type of off-flavour was detected in milk from cows bearing a genetic defect. The loss of FMO_3 enzyme activity apparently hinders the oxidation of trimethylamine (TMA) into a non-odoriferous compound. Thus, TMA concentrates in milk to significant levels resulting in a disagreeable fishy off-flavour. Parametric classification of trimethylamine concentration in pasteurised milk was obtained with an MS-based electronic nose, SMart NoseTM (LDZ, CH).

Source

1. *R. T.Marsili, SPME-MS-MVA as a rapid technique for assessing oxidation off-flavors in food, in: R.L. Rouseff, K.R. Cadwallader (Eds.), Headspace Analysis of Foods and Flavours-Theory and Practice, Plenum Press, New York, 2001.*

2. *R. T. Marsili, Comparison of solid-phase microextraction and dynamic headspace methods for the gas chromatographic-mass spectrometricanalysis of light-induced lipid oxidation products in milk, J.Chromatogr. Sci.37 (1999) 17-23.*

3. *T. Mulville, UHT the Nose Knows, Food Manufacture, March 2000, pp. 27-28.*

4. *C. Di Natale, R. Paolesse, A. Macagnano, A. Mantini, A. D'Amico, A.Legin, L. Lvova, A.Rudnitskaya, Y.Vlasov, Electronic nose and electronic tongue integration for improved classification of clinical and food samples, Sens. Actuators B 64 (1-3) (2000) 15-21.*

5. *Ch. Zondervan, S. Muresan, H.G. de Jonge, E.U. T. van Velzen, C.Wilkinson, H.H. Nijhuis, T. Leguijt, Controlling Maillard reactions in the heating process of blockmilk using an electronic nose, J. Agric. Food Chem. 47 (1999) 4746–4749.*

iii. Microbiological Quality of Milk

In the recent years electronic noses have been successfully used in monitoring of the milk microbial count, its quality and shelf-life.

The details related to the classification of bacterial cultures in milk and Screening of aroma-producing lactic acid bacteria done by using Electronic Noses, are given below:

a. Classification of Bacteria Cultures in Milk

(i) Microbial milk spoilage causes not only undesirable off-odours but can even develop some toxicity. Some bacterial species such as Bacillus cereus, Pseudomonas fragi, Pseudomonas perolens and Bacillus pumilus produce volatile compounds such as 3-methyl-1-butanol, ethyl butyrate, ethyl 3-methylbutanoate, ethyl hexanoate, acetyldehyde, acetic acid, ethanol and other alcohols, *etc.* which can be used for an early detection of microbial milk spoilage and the differentiation of spoilage-causing microbial species [1].

(ii) **Magan *et al.*** [2] used an electronic nose composed of 14 CP sensors, BH-114, to investigate the early detection of spoilage bacteria and yeast in milk-based media, such as Bacillus cereus or *Staphylococcus aureus,* or yeasts like *Klyveromyces lactis* or *Candida pseudotropicalis* in different concentrations. A correct classification of individual strains after 5 h incubation was performed by DFA on *C. pseudotropicalis, S. aureus, K. lactis, B. cereus, Pseudomonas* spp., unspoiled skimmilk and butanol control-milk.

(iii) **L abreche *et al*.** [3] showed that measurements generated by an electronic nose can be used to detect bacteria growth in milk and to determine its shelf-life, and can therefore be used for milk freshness control. An electronic nose equipped with 18 metal oxide sensors was used in an analysis of volatile compounds present in milk during 52 days of storage at two temperatures: the ambient temperature and the temperature of 5 °C. In parallel, total bacteria growth in milk was measured as well.

(iv) **Haugen [4]** used an NST 3220 instrument consisting of10 MOSFET, 5 MOS and 1 IR-based CO_2 sensor (NordicSensor Technologies, S) for the discrimination of three different disinfection–resistant bacteria as well as a mixture of all three strains cultured in milk. Only the Pseudomonas culture showed clear differences from the others.

(v) **Rapid detection of Food Borne Pathogens [5]** -Electronic nose is also helpful in detection of **food borne pathogens.** In microbiology the smell of the cultural bacteria often provides a clue to the identification of the organisms present which requires skill. Large amounts of different gaseous components are released from substrates contaminated with spoilage organisms. The traditional approach has been sample extraction followed by gas chromatography, which is tedious and requires some knowledge of the molecules involved. Electronic nose can he applied either in monitoring factors influencing spoilage or factors indicating spoilage. The samples from headspace are passed into the sensor which contains several odor sensors. A computer collects the sensory signals from the sensors where first pretreatment of data is done. The data are further analyzed by software and the results displayed.

Source

1. *Alphus D.Wilson and Manuela Baietto, Review Applications and Advances in Electronic-Nose Technologies, Sensors 2009, 9, 5099-5148.*
2. *N.Magan A.Pavlou, I.Chrysanthakis, Milk-sense: a volatile sensing system recognises spoilage bacteria and yeasts in milk, Sens. Actuators B 72 (2001) 28–34.*
3. *Labreche, S; Bazzo, S.; Cade, S.; Chanie, E. Shelf life determination by electronic nose: application to milk. Sens. Actuat. B: Chem. 2005, 106, 199-206.*
4. *J.E. Haugen, Electronic noses in food analysis, in: R.L. Rouseff, K.R. Cadwallader (Eds.), Headspace Analysis of Foods and Flavours-Theory and Practice, Plenum Press, New York, 2001.*
5. *P.K. Mandal, A.K. Biswas, K. Choiand U.K. Pal, Methods for Rapid Detection of Foodborne Pathogens: An Overview, Journal of Food Technology, 6: 87-102.*

b. Screening of Aroma-Producing Lactic Acid Bacteria

Consumer demands are constantly increasing, which requires the development of new types of cheeses. The formation of the typical cheese aroma comes as a result of complex biochemical reactions mediated mainly through cheese micro-flora. Therefore, the usage of different new cultures of lactic acid bacteria contributes to the innovativeness and attractiveness of cheese products. Classical molecular methods of bacterial strain classification have limited usage in selection of new lactic acid bacteria that will facilitate the development of better and more attractive cheese aroma, since they give no indication of the capacity to produce flavour compounds [1-10]. Maril-ley *et al.* [11] showed that MS-based electronic nose system can be very efficient in discriminating closely related strains and different bacterial species based on their production of volatile compounds. Strains of the same genotype were grouped together in conformity with the classification obtained by the Polymerase Chain Reaction (PCR) molecular method. In contrast to other electronic noses, which only detect changes in physical properties of gas sensors, MS-based electronic sensors generate a mass spectrum and therefore can be used to screen for new aroma-producing strains.

As Lactococcus lactis are widely spread in nature and represent the lactic acid bacteria most commonly used in cheese production, Gutiérrez- Méndez *et al.* [12] screened for aroma generated by Lactococcus lactis strains isolated from different sources, so as to explore the possibility of their use in starter cheese cultures. To that goal, an electronic nose equipped with 12 MOS sensors was employed. Lactococcus lactis strains showed clear phenotypic differences (lactose fermentation, proteolytic activity and citrate utilization) related to their isolation source, but no clear relationship between the source of isolation (dairy source, non-dairy source and industrial starter culture) and aroma-producing ability. Additionally, an aroma-potentiating effect was observed when mixtures of two Lactococci strains isolated from different sources acted together, which could be explained by the cooperation between their metabolic pathways [13].

Source

1. *Ampuero, S., Zesiger, T., Gustafsson, V.; Lunden, A., Bosset, J.O. Determination of trimethylamine in milk using an MS based electronic nose. Eur. Food Res. Techn. 2002, 214, 163-167.*
2. *Labreche, S; Bazzo, S.; Cade, S.; Chanie, E. Shelf life determination by electronic nose: application to milk. Sens. Actuat. B: Chem. 2005, 106, 199-206.*
3. *Russell, P. Sensory analysis. Milk Ind. Int. 1995, 97, 11-12.*
4. *Riva, M., Benedetti, S., Sinelli, N. Combined techniques of NIRS and Electronic Nose for the study of the shelf life of lattiero-caseari products. Ingred. Aliment. 2004, 3, 19-23.*
5. *Zondevan, C., Muresan, S.; de Jonge, H.G.; Thoden van Velzen, E.U., Wilkinson, C.; Nijhus, H.H., Leguijt, T. Controlling Maillard reactions in the heating process of blockmilk using an electronic nose. J. Agric. Food Chem.1999,47,4746-4749.*
6. *Navràtil, M.; Cimander, C., Mandenius, C.F.On-line multisensor monitoring of yogurt and Filjölk fermentations of production scale. J. Agric. Food Chem. 2004, 52, 415-420.*
7. *Benedetti, S., Toppino, P.M.; Riva, M. Study of the shelf life of manufactured Taleggio cheese: 2. Applications of the electronic nose. Sci. Tech. Lattiero-Casearia 2002, 53, 259-282.*
8. *Trihaas, J.; Nielsen, V. Electronic nose technology in quality assessment: monitoring the ripening process of Danish Blue cheese. J. Food Sci. 2005, 70, E44-E49.*
9. *Biolatto, A.; Grigioni, G.; Irurueta, M.; Rancho, A.M.; Taverna, M.; Pensel, N.Seasonal variation in the odour characteristics of whole milk powder. Food Chem.2007,103,960-967.*
10. *Irmler, S.; Heusler, M.L.; Raboud, S.; Schlichtherle-Cerny, H.; Casey, M.G.; Eugster-Meier, E. Rapid volatile metabolite profiling of Lactobacillus casei strains: selection of flavour producing cultures. Austral. J. Dairy Techn. 2006, 61.*
11. *Marilley, L., Ampuero, S., Zesiger, T., Casey, M.G. (2004): Screening of aroma-producing lactic acid bacteria with an electronic nose. International Dairy Journal 14, 849-856.*
12. *Gutiérrez-Méndez, N., Vallejo-Cordoba, B., GonzálezCórdova, A.F., Nevárez-Moorillón, G. V., Rivera Chavira, B. (2008): Evaluation of Aroma Generation of Lactococcus lactis with an Electronic Nose and Sensory Analysis. Journal of Dairy Science 91, 49-57.*
13. *Alphus D. Wilson 1, and Manuela Baietto, Review Applications and Advances in Electronic-Nose Technologies, Sensors 2009, 9, 5099-5148.*

iv. Differentiation of Cheeses

Classification based on the cheese variety, its geographical origin, ripening stage and manufacturing procedure –In cheeses, quality, flavor and taste are closely connected to the ripening process which depends on the growth of bacteria, lipid degradation and oxidation, and proteolysis. Traditionally, sensory analysis was used to determine the product identity of cheese. However, detection of aroma compounds using electronic noses has become more and more important.

A. Classification by the Cheese Variety

(a) A Fox 2000 with six MOS sensors (Alpha MOS, F) was used for the discrimination of four different Swiss cheese samples (0 per cent -fat, 33 per cent -fat, sharp and bland) together with a Jarlsburg cheese [1-2]. Discriminatory analysis produced a correct classification in agreement with sensory and SPME-GC–FID analyses. Detection of "rind-taste" in Swiss Emmental cheese was reported. Schaller *et al.*, unsuccessfully tried to detect defective (rind taste) Swiss Emmental cheese samples [3] with an NST 3320 instrument (Nordic Sensor Technologies AB, Linköping, S). This electronic nose was equipped with 10 MOSFET and 12 MOS sensors. The rind taste in defective cheese samples was evident only by tasting probably due to a negligible volatility of the compounds responsible for the taint.

Source

1. *W.J. Harper, The strengths and weaknesses of the electronic nose, in: R.L. Rouseff, K.R. Cadwallader (Eds.), Headspace Analysis of Foods and Flavours: Theory and Practice, Plenum Press, New York,2001.*
2. *K.D. Jou, W.J. Harper, Pattern recognition of Swiss cheese aroma compounds by SPME/ GC and an electronic nose, Milchwissenschaft 53 (5) (1998) 259–263.*
3. *E. Schaller, J.O. Bosset, F. Escher, Feasibility study: detection of rind taste off-flavour in Swiss Emmental cheese using an electronic nose and a GC–MS, Mitt. Lebensm. Hyg. 91 (2000) 610–615.*

(b) **O'Riordan and Delahunty [1]** showed that an electronic nose based on eight MOS sensors has the ability to discriminate cheeses of the same variety manufactured under standardised cheese-making conditions, but using different starter cultures (bulk starter culture *vs* commercial DVS starter culture). PC1, which accounted for 95 per cent of differences between cheeses, grouped the majority of cheeses based on the type of starter culture used during man-ufacturing. Data obtained by an electronic nose were compared to the outcomes of cheese-graders assessment. The results showed an electronic nose to be capable of discriminating between various classes of cheese in a manner similar to that of a cheese-grader.

Source

1. *O'Riordan, J.P., Delahunty, C.M. (2003): Characterisation of commercial Cheddar cheese flavour. 1: traditional and electronic nose approach to quality assesment and market classification. International Dairy Journal 13, 355-370.*

B. Classification by the Geographical Origin of Cheeses and other Dairy Product

This is to classify samples of a said dairy product by their place of production. The differences in the composition of volatiles are expected to originate from minor variations in the production technology, raw materials or produce conditioning. Some examples are:

(*a*) **Samples of caseinate** from three different suppliers were classified with a FOX 4000 based on MOS sensors (Alpha MOS, F).PCA and DFA analyses correctly classified samples from the different suppliers. Results were validated with unknown samples. A sensory analysis found differences in quality between the different suppliers [1-2].

(*b*) **Classification of Emmental cheese** [3] by the geographic origin performed with an electronic nose based on mass spectrometry. Gursoy *et al.* [4] investigated into the possible use of an electronic nose (based on ion mobility spectrometry [IMS] and metal oxide sensors) for the discrimination of hard cheeses based either on their origin or their variety. Nine different Emmental cheese samples stemming from different countries (Germany, France, the Netherlands, Switzerland and Finland) were used to evaluate the potential of an electronic nose to discriminate between cheese samples based on their geographical origin. The results showed that such an e-nose is capable of clearly distinguishing between Emmental cheeses produced in different countries. It is known that Emmental cheese can be produced in more than one central European region, but Swiss Emmental cheese is definitely regarded as the most expensive. Such an electronic nose could be a rapid, simple and low-cost tool capable of checking for possible adulteration and therefore of verifying the authenticity of Swiss Emmental cheese. Within the frame of the same research, seven different cheese samples obtained from various markets in Finland, France or Belgium were used to attempt variety-based discrimination. These samples were of an extra-hard (Parmigiano Reggiano, Pecorino Romano, Grana and Grana Padano) and hard category (Old Master, Emmental and Gruyere). The results of PCA analysis of data provided by an e-nose indicated that the device is capable of clearly separating hard from extra-hard cheeses. As well as, a clear discrimination was obtained for cheese samples of Parmigiano Reggiano and Pecorino Romano, the first thereby being made of ewe and the second being made of cow milk. Grana (Finland) and Grana Padano (Italy) cheeses are both cow milk cheeses produced in a similar manner. Despite differences in their origin, their positions in the score plot were very close, which shows that these cheese varieties are of similar characteristics.

c. A good classification was also performed by another Fox system on two different brands of whole UHT, skimmed UHT and half-creamed UHT milk. In addition, a whole and a half-cream pasteurised milk were identified as a separate group from the set of UHT samples. The classification of Emmental cheese based on geographical origin was explored with an MS-based electronic nose, SMart NoseTM (LDZ, CH).

Source

1. *Application note 55, Dairy products: comparison of caseinate suppliers, http://www.alpha-mos.com.*
2. *Application note 34, Assessing quality in the dairy industry, http://www.alpha-mos.com.*
3. *L.Pillonel, S. Ampuero, R. Tabacchi, J.O. Bosset, Analytical methods for the determination of the geographic origin of Emmental cheese, Volatile compounds by GC–MS–FID and electronic nose, Eur.J.Food Res. Technol.216 (2003) 179–183.*
4. *Gursoy, O., Somervuo, P., Alatossava, T. (2009): Preliminary study of ion mobility based electronic nose MGD-1 for discrimination of hard cheeses. Journal of Food Engineering 92 (2), 202-207.*

C. Classification of Cheese by the Ripening Stage

(a) An AromaScan electronic nose (Electra House, UK) with 32 PC sensors was used to discriminate two different Pecorino Toscano cheeses (ewe's cheese), which differed mainly in their maturation time. These results are comparable to those obtained with a purge and trap GC–MS–PCA whereas Curie-point pyrolysis mass spectrometry failed. The GC–MS–PC analysis was based on hexanal, acetic acid, 2-butanone and tetra-hydrofuran peaks. Several instruments were tested for the classification of Swiss Emmental cheese by ripening stage. A SMart NoseTM based on massspectrometry (LDZ, CH) showed only partial classification, whereas the eight MOS sensors of the eNose 5000 and the 10MOSFET +5 MOS sensors of a NST 3220 (Nordic SensorTechnologies, S) showed correct classifications of samples ripened for 1, 21 and 98 days [1-3].

Source

1. *E.Schaller, J.O. Bosset, F. Escher, Practical experience with electronic nose systems for monitoring the quality of dairy products, Chimia 53 (1999) 98–102.*
2. *G. Contarini, M. Povolo, P.M. Toppino, B. Radovic, M. Lipp, E. Anklam, Comparison of three different techniques for the discrimination of cheese: application to the ewe's cheese, Milchwissenschaft 56 (3) (2001) 136–140*
3. *S. Ampuero, J.O.Bosset, Swiss Federal Dairy Research Station, FAM, Schwarzen burgstrasse 161, Liebefeld, CH-3003 Bern, Switzerland.*

(b) An electronic nose based on an array of six MOS sensors in combination with artificial neural network (ANN) can be very effective in classification of Pecorino cheeses (typical Italian Pecorino and "fossa" Pecorino cheeses) based on their ripening time and manufacturing techniques. The results of electronic nose analysis were compared to the results obtained by SPME-GC-MS analysis; the authors claimed the best results are to be obtained by e-nose signals' analysis. This could be explained by the fact that SPME-GC-MS can identify only individual chemical compounds, while an electronic

nose can recognise even the combination of such compounds responsible for odour generation [1].

Source

1. *Cevoli, C., Cerretani, L., Gori, A., Caboni, M.F., Gallina Toschi, T., Fabbri, A. (2011): Classification of Pecorino cheeses using electronic nose combined with artificial neural network and comparison with GC-MS analysis of volatile compounds. Food Chemistry 129 (3), 1315-1319.*

(c) Upon monitoring the ripening process of 96 Danish blue cheeses during 12 weeks, Trihaas *et al.* [1] showed that an e-nose can be successful in determining the maturity of cheese. At the beginning (Day 0) and at the end of the ripening process (Week 12) cheeses can be very well differentiated by an electronic nose, but, contrast to that, after 2 and 4 weeks of the ripening the e-nose could not clearly separate between them due to their similar aroma profile, confirmed also by chemical analyses. In these ripening stages, volatile compounds and 4.6-pH soluble profiles are very similar.

Source

1. *Trihaas, J., Vognsen, L., Nielsen, P.V. (2005): Electronic nose: New tool in modeling the ripening of Danish blue cheese. International Dairy Journal 15, 679-691.*

(d) It was investigated into the potential use of ion mobility spectrometry (IMS)-based and metal oxide sensor-based e-nose to discriminate Emmental cheeses of different ripening age [1]. Twenty-four Emmental cheese samples of ripening ages of 3, 6 and 9 months were purchased from local markets in Finland. PCA score plot showed that cheeses ripened for 9 months can be clearly discriminated from those ripened for 3 or 6 months, leading also to the conclusion that the latter two cheese clusters are very similar when it comes to their volatile compound content. The results of the research showed that an electronic nose (based on conducting polymer sensors) is capable of not only distinguishing between the Pecorino Toscano cheeses of various ripening age (20 days *vs* 4 months), but also of detecting samples having a higher concentration of contaminating bacteria [2]. The sample (2F) which belongs to neither of the two groups showed the same anomalous behaviour after a dynamic headspace analysis. The microbiological analysis showed this sample to have a content of contaminating bacteria higher than all other cheeses.

Source

1. *Gursoy, O., Somervuo, P., Alatossava, T. (2009): Preliminary study of ion mobility based electronic nose MGD-1 for discrimination of hard cheeses. Journal of Food Engineering 92 (2), 202-207.*

2. *G. Contarini, M. Povolo, P.M. Toppino, B. Radovic, M. Lipp, E. Anklam, Comparison of three different techniques for the discrimination of cheese: application to the ewe's cheese, Milchwissenschaft 56 (3) (2001) 136–140.*

D. Manufacturing Procedures and Others [7]

Some examples of the application of e-nose in said field are tabulated below:

Sl.No.	Scientists	Studies/Observations
1	Russell	First focused on the classification of Parmesan cheeses with differing rates of maturity. **The Aromascan e-nose** successfully distinguished the two types. They also were able to characterized Gorgonzola and Cottage cheese using polypyrrole semi-conductor sensors [1].
2	Zondevan *et al.*	Used the electronic nose to classify block milk products, subjected to various heating processes, in order to predict the most favorable heating method. In this case, the results were moderate to good [2].
3	Riva *et al.*	Compared with near-infrared spectroscopy (NIRS), the electronic nose has shown better results. Demonstrated that NIR spectra (NIRS) collected from Crescenza cheese were influenced both by raw material and by its high moisture content. NIRS, like the classical measures such as acidity and texture chances, was able to detect only chemical modifications that occurred during the first stages of storage, whereas the results obtained by spectroscopy and electronic nose analyses were more useful because they could be used to monitor the shelf-life of these products as online sensors [3].
4	Gonzalez de Llano *et al.*	**Modelling of cheese ripening-** An electronic nose found its application in predicting the quality of cheese during ripening. The changes witnessed during a 12 week-ripening of the Danish blue cheese using commercial electronic nose which employs 12 MOS sensors was analysed [6].
5	Gonzalez de Llano *et al.* Gallois, A. *et al.* Trihaas, J. *et al.*	**Modelling of cheese ripening-** In parallel, chemical analyses (of volatile compounds, pH 4.6 soluble peptides and free fatty acids) of the cheese samples were performed, those results later being used to indicate changes during maturation. E-nose measurements gave data suitable to model and predict chemical changes that are strongly related to the observed quality parameters. The results showed that an e-nose is very successful in predicting free fatty acid and pH 4.6 soluble peptide contents, but unable to give a satisfactory prediction of volatile compound contents. The reasons behind such an outcome could be the fact that volatile compounds do not change linearly over time, since they are both produced and decomposed during the ripening period, as well as the fact that different cheese homogenization methods were used with e-nose employment and GC-MS analysis [4-6].

Source

1. *Russell, P. Sensory analysis. Milk Ind. Int. 1995, 97, 11-12.*
2. *Zondevan, C.; Muresan, S.; de Jonge, H.G.; Thoden van Velzen, E.U.; Wilkinson, C.; Nijhus, H.H.; Leguijt, T. Controlling Maillard reactions in the heating process of block milk using anelectronic nose. J. Agric. Food Chem. 1999, 47, 4746-4749*
3. *Riva, M., Benedetti, S., Sinelli, N. Combined techniques of NIRS and Electronic Nose for the study of the shelf life of lattiero-caseari products. Ingred. Aliment. 2004, 3, 19-23.*
4. *Gonzalez de Llano, D., Ramos, M., Polo, C., Sanz, J., Martinez-Castro, I. (1990): Evolution of the volatile components of an artisanal Blue cheese during ripening, Journal of Dairy Science 73, 1676-1683.*

5. *Gallois, A., Langlois, D. (1990): New results in the volatile odorous compounds of French cheeses. Le Lait 70, 89-106.*

6. *Trihaas, J., Vognsen, L., Nielsen, P.V. (2005): Electronic nose: New tool in modelling the ripening of Danish blue cheese. International Dairy Journal 15, 679-691.*

7. *Alphus D. Wilson 1, and Manuela Baietto, Review Applications and Advances in Electronic-Nose Technologies, Sensors 2009, 9, 5099-5148.*

2. Meat Products

Much work has been done in the electronic detection of quality characteristics of meat products within the food industry. These are tabulated below [4]:

Sl.No.	*Scientists*	*Studies/Observations*
1	Berdague and Talou	Studied instruments based on MOS sensors, starting with the Alabaster-UV(Have a stainless steel measurement chamber containing one semiconductor gas sensor, a UV-lamp, air inputs, and outputs connected to a fan. The signal obtained was displayed as dead sorption curves in "Alabaster" units.) as one of the first commercially available e-nose instruments. It was shown that this simple instrument could differentiate maxima in aroma perception resulting from the maturation of dry non-spiced meats, and rapidly detected sex-linked differences in meat product composition [1].
2	Rajamaki *et al.*	i-Studied the applicability of an electronic nose for the quality control of modified-atmosphere packaged broiler chicken cuts with different temperature regimes. The results were compared with those obtained by microbiological, sensory, and headspace GC analyses. The e-nose could clearly distinguish broiler chicken packages with deterioration from fresh packages [2].
		ii-The electronic nose was capable of detecting even early signs of spoilage in modified atmosphere packed poultry meat [2].
3	Vestergaard *et al.*	Found the storage time of a pork-meat pizza topping product was predictable using an electronic nose [3].

Source

1. *Berdague, J.L., Talou, T. Examples of applications for meat products of semiconductor gas sensors. Sci. Alim. 1993, 13, 141-148.*

2. *Rajamäki, T., Alatomi, H.; Titvanen, T.; Skytta, E.; Smolander, M.; Ahvenainen, R. Application of an electronic nose for quality assessment of modified atmosphere packaged poultry meat.Food Contr. 2004, 17,5-13.*

3. *Vestergaard, J.S.; Martens, M.; Turkky, P. Application of an electronic nose system for prediction of sensory quality changes of a meat product (pizza topping) during storage. LWT -Food Sci. Technol. 2007, 40, 1095-1101.*

4. *Alphus D. Wilson 1, and Manuela Baietto, Review Applications and Advances in Electronic-Nose Technologies, Sensors 2009, 9, 5099-5148.*

3. Fish and Seafood Products [7]

The following work has been done in the electronic detection of quality characteristics of Fish and Seafood Products within the food industry:

Sl.No.	*Scientists*	*Studies/Observations*
1	Olafsson *et al.*	First evaluated spoilage in three different fish species (haddock, cod and redfish) using two to six MOS sensors. The samples were kept at room temperature or on ice, and the results compared with sensory analyses [1].
2	Winquist *et al.*	Trained a system based on MOSFET sensors to predict the age of cod filet. This study showed some very convincing results on the feasibility of the system [2].
3	Jonsdottir *et al.*	Attempted to standardize cod-roe products by using an electronic nose. Two compounds, 3-methyl-1-butanol and 3-methylbutanal, detected with the electronic nose were found to be quality indicators for objectively assessing the ripening of roe [3].
4	Olafsdottir *et al.*	Focused on the feasibility of a prototype gas-sensor array system, the Fish Nose, built to study the VOCs of fish [4].
5	Haugen *et al.*	Confirmed the feasibility of the FishNose on direct quality measurements of smoked salmon [5].
6	Chantarachoti *et al.*	Evaluated the capability of a portable electronic nose in detecting spoilage of whole Alaskan pink salmon stored at 14 °C and in slush ice. In 92 per cent of the samples, the instrument could correctly classify the fish as either fresh or spoiled [6].

Source

1. *Olafsson, R.; Martinsdottir, E.; Olafsdüttir, G.; Sigfússon, T.I.; Gardner, J.W. Monitoring of fish freshness using tin oxide sensors. In Sensors and Sensory Systems for an Electronic Nose; Gardner, J.W., Bartlett, P.N., Eds.; Kluwer: Dordrecht, The Netherlands, 1992; pp. 257-272.*

2. *Winquist, F.; Sundgren, H.; Lundstrom, I. A practical use of electronic nose: quality estimation of cod fillet bought over the counter. In Digest of Technical Papers, 8th International Conference on Solid-State Sensors and Actuators and Eurosensors IX; Royal Swedish Academy of Engineering Sciences: Stockholm, Sweden, 1995; Volume 1, pp. 695-698.*

3. *Jonsdottir, R.; Olafsdottir, G.; Martinsdottir, E.; Stefansson, G. Flavor characterization of ripened cod roe by gas chromatography; sensory analysis; and electronic nose. J. Agric. Food Chem.2004, 52, 6250-6256.*

4. *Olafsdottir, G.; Chanie, E.; Westad, F.; Jonsdottir, R.; Thalmann, C.R.; Bazzo, S.; Labreche, S.; Marcq, P.; Lunby, F.; Haugen, J.E. Prediction of microbial and sensory quality of cold smoked Atlantic salmon (Salmo salar) by electronic nose. J. Food Sci. 2005, 70, S563-S574.*

5. *Haugen, J.E.; Chanie, E.; Westad, F.; Jonsdottir, R.; Bazzo, S.; Labreche, S.; Marcq, P.; Lundby, F.; Olafsdottir, G. Rapid control of smoked Atlantic salmon (Salmo salar) quality by electronic nose: correlation with classical evaluation methods. Sens. Actuat. B: Chem. 2005, 116, 72-77.*

6. *Chantarachoti, J.; Oliveira, A.C.M.; Himelbloom, B.H.; Crapo, C. A.; McLachlan, D.G. Portable electronic nose for detection of spoiling Alaska pink salmon (Oncorhynchus gorbuscha). J. Food Sci. 2006, 71, S414-S421.*

7. *Alphus D. Wilson 1, and Manuela Baietto, Review Applications and Advances in Electronic-Nose Technologies, Sensors 2009, 9, 5099-5148.*

4. Freshness, Quality, Ripeness and Shelf-Life of Food/Fruits [5]

The age of fruits (ripeness or maturity level) determines the shelf life and future rate of quality loss due to changes in flavor, firmness and color. Traditional measuring techniques such as the starch conversion index and flesh firmness or pressure test are used to determine fruit quality. These testing methods are destructive and involve random sampling to assess fruit quality.

Utilizing e-noses as a means of monitoring fruit freshness and shelf-life prior to marketing can have a number of benefits that maximize corporate profits and optimize customer satisfaction. Test result Information from e-noses on fruit physiological states, based on changes in released volatiles, can be applied to retard the ripening process through exposure of the fruit to ripening inhibitors at the appropriate time, adjustments in fruit storage conditions to preclude ethylene accumulation and removal of bruised or damaged fruits that enhance ripening of surrounding fruits and contribute to storage losses due to rots, decays, and various fruit diseases.Some examples of the application of e-nose to assess fruit quality, are given below:

Sl.No.	*Scientists*	*Studies/Observations*
1	Shaller *et al.*	Provided a very good review of electronic-nose systems that have been used for various applications with different types of foods [1].
2	Pathange *et al.*	Used maturity indices such as starch index and puncture strength to categorize fruit of the "Gala" apple variety into three maturity groups referred to as immature, mature and over-mature fruits. Multivariate analysis of variance of the e-nose sensor data indicated that the instrument could classify the fruit into the correct group in 87 per cent of the samples [2].
3	Oshita *et al.*	Examined the shelf-life of "La France" pears that were grouped into three storage treatments and judged using e-nose data as either immature, mature or over-mature. The aromas of pears from the three treatments were classified into three classes based on their physiological states determined from distinct aroma patterns derived from a 32-sensor array output [3].
4	Gòmez *et al.*	Studied volatile production of unripe, half-ripe, full-ripe and over-ripe tomatoes using the PEN 2 E-nose (10 different metal oxide sensors) with principal component analysis (PCA) and linear discriminant analysis (LDA). The results demonstrated that the electronic nose could differentiate among the ripeness states of tomatoes and classify them with 100 per cent reliability in each ripeness group [4].

Source

1. *Schaller, E., Bosset, J.O., Esher F. Electronic noses and their application to food. Lebensm.-Wiss.Ul.-Technol. 1998,31,305-316.*

2. *Pathange, L.P., Mallikar-junan, P., Marini, R.P., O'Keefe, S.; Vaughan, D. Non-destructive evaluation of apple maturity using an electronic nose system. J. Food Engin. 2006, 77,1018-1023.*

3. *Oshita, S.; Shima, K.; Haruta, T.; Seo, Y.; Kawagoe, Y., Nakayama, S.; Takahara, H.Discrimination of odors emanating from "La France" pear by semi-conducting polymer sensors.Computers Electr. Agric. 2000, 26, 209-216.*

4. *Gòmez, A.H., Hu, G., Wang, J., Pereira, A.G.Evaluation of tomato maturity by electronic nose.Computers Electr. Agric. 2006, 54, 44-52.*

5. *Alphus D. Wilson 1, and Manuela Baietto, Review Applications and Advances in Electronic-Nose Technologies, Sensors 2009, 9, 5099-5148.*

5. Agricultural Plant Production [4]

The aroma of grains is the primary criterion of fitness for consumption.How ever, the sniffing of grain lots for quality grading is potentially hazardous to humans and should be avoided because of inhalation of toxic or pathogenic mold spores such as from Aspergillus species. Some examples of the application of e-nose to assess grain quality, are given below:

Sl.No.	*Scientists*	*Studies/Observations*
1	Jonsson *et al.*	Used an electronic nose with three different complementary sensors (ten gas sensitive metal-oxide-semiconductor field effect transistors, four tin dioxide-based sensors and a CO2 sensor) to test samples of oats, rye and barley with different aromas, and wheat with different levels of ergosterol, fungal and bacterial contamination [1].
2	Di Natale *et al.*	Used a sensor array formed by five different metal-oxide semiconductor thin film-based sensors in order to recognize five different vintage years of the same wine. In those years, the quality and percentage composition of the grapes were almost constant according to the wine-making procedure [2].
3	Campagnoli *et al.*	tested the feasibility of using the electronic nose for detecting processed animal proteins in feedstuffs [3].

Source

1. *Jonsson, A.; Winquist, F.; Schnürer, J.; Sundgren, H.; Lundtröm, I. Electronic nose for microbial quality classification of grains. Internatl. J. Food Microbiol. 1997, 35, 187-193.*

2. *Di Natale, C.; Davide, F. A.M.; D'Amico, A.; Sberveglieri, G.; Nelli, P.; Faglia, G.; Perego, C.Complex chemical pattern recognition with sensor array: the discrimination of vintage years of wine. Sens. Actuat. 1995, 25, 801-804.*

3. *Campagnoli, A.; Pinotti, L.; Tognon, G.; Cheli, F.; Baldi, A.; Dell'Orto, V. Potential application of electonic nose in processed animal proteins (PAP) detection in feedstuffs. Biotechnol. Agron.Soc. Environm. 2004, 8, 253-255.*

4. *Alphus D. Wilson 1, and Manuela Baietto, Review Applications and Advances in Electronic-Nose Technologies, Sensors 2009, 9, 5099-5148.*

Section B

Instrumentation in Chemical Analysis of Food/Milk Products

Chapter 2

Types of Analytical Methods and Development of Analytical Methods

Instrumental Methods of Analysis [1]

The interesting and important branch of chemistry is analytical chemistry, which is the study of the separation, identification, and quantification of the chemical components of natural and artificial materials. Instrumental methods use an apparatus to measure physical quantities of the analyte such as light absorption, fluorescence, or conductivity. The separation of materials is accomplished using chromatography, electrophoresis or Field Flow Fractionation methods.

Classification of Analytical Methods [1]

On the basis of the operation, analytical methods can be classified into the following categories:

A. Qualitative and Quantitative Analytical Methods

(i) **Qualitative analysis** gives an indication of the identity of the chemical species in the sample. Qualitative instrumental analysis is that measured property indicates presence of analyte in matrix.

(ii) **Quantitative analysis** determines the amount of one or more of these components. The separation of components is often performed prior to analysis. Quantitative instrumental analysis is that magnitude of measured property is proportional to concentration of analyte in matrix.

B. Classical and Instrumental Analytical Methods

Analytical methods can be separated into classical and instrumental methods [1]. The dividing line between instrumental and non-instrumental methods is very fine and often fades out. Even the classical methods make use of devices such as analytical balance or volumetric apparatus. It can be said that instrumental techniques involve use of electro-mechanical gadgets comprising of electronic components and circuits, optical elements and mechanical parts [2]. We should not be confused with the term "instrumental". It does not mean that classical methods do not use instruments. Classical methods also use instruments but these equipments are not as modern or highly specialized than the instrumental methods [4].

a. Classical Methods

Classical methods (also known as wet chemistry methods) use separations such as precipitation, extraction and distillation [1]. Classical methods were time-consuming and were limited by **errors both determinate and indeterminate [2].** These methods are the techniques which are the fundamentals of laboratory practices. These are the traditional method of chemical analysis which is still being used by scientists even up to this date [4]. Classical methods can be of qualitative or quantitative types.

1. **Qualitative**-Identification by color, indicators, boiling points, melting points, odors [1]. Qualitative method is the determination of the identity of the unknown based on the criteria provided and the presence or absence of a material in a sample [4].
2. **Quantitative**-Quantitative analysis is achieved by measurement of Mass (weight) or volume (*e.g.* gravimetric, volumetric) [1,2]. Quantitative method focuses on the determination of the measurable properties of the unknown [4].

b. Instrumental Methods

In simple words, instrumental analysis is the analysis done by instruments. The instruments and techniques developed by the physicist for the determination of physical constants have furnished the chemical analyst with new devices which can be used for the quantitative and qualitative determination. The Instrumental/physical methods have enabled the analyst to broaden the scope of analysis, since in many cases accurate measurements can be made without destruction of the sample [1]. **Instrumental methods** are sometimes coined as the modern method of analysis. The reason is that these analytical techniques use modern equipments such as computers and the like. Examples of instrumental method are the spectroscopy, chromatography and microscopy [4]. Modern methods of instrumental analysis have evolved from the earlier classical methods which were dependent on measurement of mass or volumes [2]. Physical and physical-chemical methods of the analysis **(PCMA)** are based on dependence application between measured physical properties of substances and qualitative (quantitative) composition. PCMA are divided on the followings [1]:

(i) **Optical methods** are based on measurement of optical properties of substances.

(ii) **Chromatographic methods** are based on usage of ability of different substances to selective sorption.

(iii) **Electrochemical methods** are based on measurement of electrochemical properties of substances.

(iv) **Radiometric methods** are based on measurement of radioactive properties of substances.

(v) **Thermal methods** are based on measurement of heat effects of substances.

(vi) **Mass spectrometric methods** are based on studying of the ionized fragments ("splinters") of substances.

Advantage of PCMA/Modern Analytical Methods [1,2,3]

(i) High sensitivity (a low limit of detection (10-9 mg) and definition}- Improvement in sensitivity leading to trace analysis at ppt or even femto- mole levels. Such sensitivities could not be imagined using the conventional classical approach.

(ii) High selectivity.

(iii) Rapid analysis methods (Time saving)- Analysis time has been reduced from several hours or days to even minutes

(iv) Automation and computerization is possibility- Fast decisions in case of online analysis during manufacturing operations or deciding on viability before taking commercial decisions.

(v) Possibility of analysis on distance and Possibility of the local analysis.

(vi) Possibility of the analysis without destruction of the sample.

(vii) Improvement in accuracy of results by elimination of errors introduced due to personal bias. Sufficient reliability and accuracy of results are obtained by instrumental method.

(viii) A small amount of a sample is needed for analysis.

(ix) Complex mixture can be analysed either with or without their separation.

(x) When non-instrumental method is not possible, instrumental method is the only answer to the problem.

Lacks/Limitation of PCMA/Instrumental Methods [1,3]

(i) Definition error is near ± **5 per cent** (on occasion to 20 per cent), whereas - 0,01-0,005 per cent for gravimetry and 0,1-0,05 per cent for titrimetry.

(ii) Reproducibility of results in separate methods is worse, than in classical methods of the analysis.

(iii) It is necessary of usage of standards and standard solutions, graduation of equipment and plotting of calibration charts.

(iv) The sensitivity and accuracy depends upon the type of instrument.

(v) There is frequent need of checking results with other methods.

(vi) Complexity of used equipment, its high cost, high cost of standard substances and trained personnel requirement (Specialized training for handling instrument).

(vii) In some cases, instrumental method may not be specific.

Instrumental Methods can also be of qualitative or quantitative types.

1. **Qualitative-**Chromatography, Electrophoresis and Identification by measuring physical property (*e.g.* Spectroscopy, Electrode potential).
2. **Quantitative -**Measuring property and determining relationship to concentration (*e.g.* Spectrophotometry, Mass spectrometry) Often, same instrumental method used for qualitative and quantitative analysis.

Source

1. *Vogel's Text Book on Quantitative chemical analysis revised by G.H.Jeffery, J. Bassett, J.Mendham and R.C.Denney.*
2. *Dr. Deepak bhanot on lab.training.com*
3. *Pharmastuff 4U.*
4. *Jerichorayel, College Teacher/Senior Educator, Posted on January 9, 2016 on answer.*

Types of Instrumental Methods

The technology of using instruments to measure and control the physical and chemical properties of material is called instrumentation. The following Instrumental methods are generally being used for Qualitative and Quantitative analysis:

Property	*Example Method*
Radiation emission	Emission spectroscopy - fluorescence, phosphorescence, luminescence
Radiation absorption	Absorption spectroscopy - spectrophotometry, photometry, nuclear magnetic resonance, electron spin resonance
Radiation scattering	Turbidity, Raman
Radiation refraction	Refractometry, interferometry
Radiation diffraction	X-ray, electron
Radiation rotation	Polarimetry, circular dichroism
Electrical potential	Potentiometry
Electrical charge	Coulometry
Electrical current	Voltammetry - amperometry, polarography
Electrical resistance	Conductometry
Mass	Gravimetry

Selection of Analytical Methods [1,2,3]

Different analytical techniques are available to determine a particular property of a Dairy products/food material. Hence selection of most appropriate technique for the specific application, is highly important. The selection of analytical technique depends on the property to be measured, the type of food to be analyzed and the reason for carrying out the analysis. The details of some of the important criteria helpful inselecting an instrumental analytical technique are given below:

i. Precision [2,3]

How reproducible?, A measure of the ability to reproduce a result by a specific analyst is using the same equipment and experimental approach keeping other conditions unchanged. When a sample is analyzed several times, the individual results are rarely the same. Instead, the results are randomly scattered. **Precision** is a measure of this variability. The closer the agreement between individual analyses, the more precise the results. A method's precision depends on several factors, including the uncertainty in measuring the signal and the ease of handling samples reproducibly [3]. Indeterminate or random errors.

$$\text{Absolute standard deviation: } s = \sqrt{\frac{\sum_{i-0}^{i=N} (x_i - \bar{x})^2}{N-1}}$$

$$\text{Variance: } s^2$$

$$\text{Relative standard deviation: } RSD = \frac{s}{\bar{x}}$$

$$\text{Standard deviation of mean: } s_m = \frac{s}{\sqrt{N}}$$

ii. Accuracy/Bias [1,2,3]

A measure of how close one can actually measure the value to the true value or expected result of the parameter being measured. Determinate errors (operator, method, instrumental).

$$\text{bias} = \bar{x} - x_{true}$$

We can express accuracy as an absolute error, *e* [3]

e = obtained result - expected result

or as a percentage relative error, per cent e_r

per cent e_r = ((obtained result – expected result)/expected result) x 100

iii. Sensitivity [2,3]

How small a difference can be measured?. A measure of the lowest concentration of the component that can be detected by a given procedure.

$$\text{Calibration sensitivity: } S = \frac{dSignal}{dc} c + Signal_{blank} = mc + Signal_{blank}$$

(larger slope of calibration curve m, more sensitive measurement).

The ability to demonstrate that two samples have different amounts of analyte is an essential part of many analyses. A method's **sensitivity** is a measure of its ability to establish that such differences are significant. Sensitivity is often confused with a method's **detection limit**, which is the smallest amount of analyte that we can determine with confidence [3].

iv. Reproducibility [2]

A measure of the ability to reproduce result using the same experimental approach in same as well as different laboratories using same/different equipment.

v. Simplicity of Operation [2]

A measure of the ease with which relatively unskilled workers may carryout the analysis.

vi. Detection Limit

What range of amounts?.Signal must be bigger than random noise of blank.

$$\text{Minimum signal: } \text{Signal}_{min} = \text{Av. Signal}_{blank} + ks_{blank}$$

From statistics k=3 or more (at 95 per cent confidence level).

vii. Speed [2]

Analysis of single sample or the number of samples in a given time.

viii. Specificity [2,3]

A measure of the ability to detect and quantify specific components within a food material, even in the presence of other similar components *e.g.*, Fructose in the presence of sucrose or glucose. An analytical method is specific if its signal depends only on the analyte. Although **specificity** is the ideal, few analytical methods are completely free from the influence of interfering species [3].

ix. Dynamic Range [1,2]

At detection limit we can say confidently analyte is present but cannot perform reliable quantitation

Level of quantitation (LOQ): k=10

Limit of linearity (LOL): when signal is no longer proportional to concentration

$$\text{Dynamic range: } \frac{\text{LOL}}{\text{LOQ}} \qquad 10^2 \text{ to} > 10^6$$

x. Selectivity [1,2,3]

How much interference?.No analytical method is completely free from interference by concomitants. Best method is more sensitive to analyte than interfering species (interferent).

$$\text{Matrix with species A\&B:}\quad \text{Signal} = m_A c_A + m_B c_B + \text{Signal}_{blank}$$

$$\text{Selectivity coefficient:}\quad k_{B,A} = \frac{m_B}{m_A}$$

k's vary between 0 (no selectivity) and large number (very selective).

Selectivity is a measure of a method's freedom from interferences. The selectivity of a method for the interferent relative to the analyte is defined by a **selectivity coefficient [3].**

xi. Calibration Methods [1]

Basis of quantitative analysis is magnitude of measured property is proportional to concentration of analyte

$$\text{Signal} \propto [x] \quad \text{or} \quad \text{Signal} = m[x] + \text{Signal}_{blank}$$

$$[x] = \frac{\text{Signal} - \text{Signal}_{blank}}{m}$$

Calibration curves (working or analytical curves)

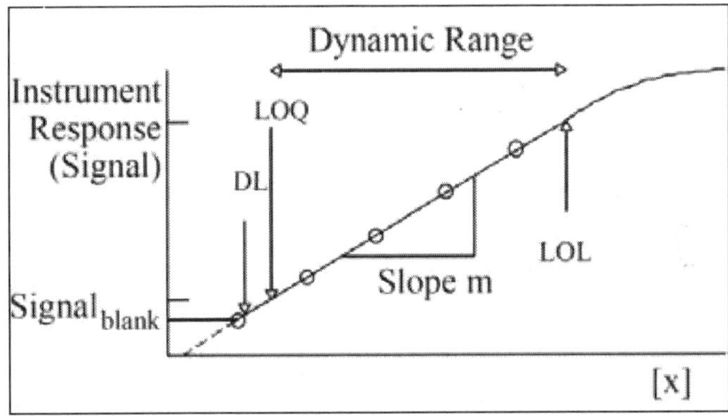

Calibration expression is

Absorbance = slope [Analyte (ppm)] + intercept

xii. Nature of Food Matrix [2]

The composition, physical properties (solid or liquid, transparent or opaque, polar or nonpolar) *etc.* of the food matrix material surrounding the analyte influences the type of method that can be used to perform an analysis.

Source

1. *Pomeranz, Y., and Meloan, C.E. (2004). Food Analysis- Theory and Practice, 3rd ed. Chapman and Hall Inc., New York.*

2. *Bimlesh Mann, Rajan Sharma and Richa Singh, Instrumentation for Quality Analysis of Dairy Products: An Overview, Dairy Chemistry Division ICAR-National Dairy Research Institute, Karnal.*

3. *http://chemwiki.ucdavis.edu/analytical_chemistry/analytical_ chemistry_2.0/03_ the_ vacubulary_of_analytical_chemistry/3D_Selecting_ an_analytical_method.*

Development/Advancement in Analytical Techniques for Analysis of Milk and Milk Products

The Physical, Chemical and Microbiological analysis of milk products is an important way for effectively controlling the manufacturing procedures of milk products, for assessing and controlling the quality of raw, intermediate and final products and for determining the product composition. The analytical chemical methods used for determining the quality of milk products ranges from traditional procedures based on wet chemistry to modern instrumental procedures. To overcome the problems raised by using the traditional wet chemistry methods (*e.g.* Time consuming and suitability for determining the major component only *etc.*), several instrumental techniques have been developed. Dairy Industry has seen several significant advances in testing of milk and milk products.

1. Milk Testing (Component Analysis) [1]

Milk component testing (*i.e.*, fat, protein, lactose, total solids, solids-not-fat, and other solids) impacts every one related to dairy *e.g.* Milk producers, Lab staff, Dairy Technologists *etc.* Milk component analysis is relatively unusual in the field of quantitative analytical chemistry because an analytical test result determines the allocation of very large amounts of money between buyers and sellers of milk [2]. Therefore, there is high incentive to develop and refine these methods to achieve a level of analytical performance rarely demanded of most methods or laboratory staff working in analytical chemistry. Well-defined statistical methods to characterize and validate analytical method performance combined with significant improvements in both the chemical and instrumental methods have allowed achievement of improved analytical performance for testing. The development of instruments based on IR, MIR and Ultrasony for milk analysis and improvement in the performance of the chemical methods used to calibrate aforesaid techniques have enabled the dairy industry to shift from paying for milk based on weight and fat content, to valuing milk based on the weight of each milk component (*e.g.*, fat, true protein, other solids)delivered.

The value added dairy foods for specific market segments has created a need for instrumental and sensory approaches and quantitative data to support product development and marketing. Bringing together sensory data from quantitative descriptive analysis (QDA) and analytical data from gas chromatography olfactometry (GCO) for identification of odor-active compounds in complex natural dairy foods has enabled the sensory scientist and analytical chemist to work together to improve the consistency and quality of dairy food flavors [3]. The details of advancements in milk testing are given below:

i. Harmonized Method-Validation Procedures [1]

The development of internationally harmonized guidelines for inter laboratory studies (in the 1980s and1990s) for validation of analytical method performance that has impacted milk component testing was an important development in the field of routine methods of quantitative analysis. The methods validation organizations, including the **AOACI,** the International Organization for Standardization (**ISO**), and the International Dairy Federation **(IDF)** have originally adopted the guidelines in 1987 which was published by Association of Official Analytical Chemists International (**AOACI**) in 1989 and revised in 1995. The guidelines, which have been particularly useful in validation of analytical method performance in the dairy industry, provide a framework for inter laboratory study design, minimum requirements for the number of sample materials and laboratories, and standardized statistical procedures for removal of outliers and calculation of statistical indices of method performance. The statistical indices (*e.g.*, r-value and R-value) that are published as part of each method provide clear validated benchmarks for the analyst for expected within laboratory agreement on replicates (Repeatability) and expected between-laboratory agreement (Reproducibility)if the method is being carried out correctly.

ii. Traceability and Uncertainty in Analytical Results [1]

The analytical communities worldwide have been actively engaged in developing harmonized standards and procedures for ensuring the quality of analytical results as described by ISO in 1999 and in the Eurachem Guide in 2000, followed by anupdate in 2003. The reliability of analytical results is inextricably linked to the concepts of traceability and uncertainty. Eurachem indicated in 2003 that traceability in analytical chemistry refers to the link between an analytical result and some agreed-upon reference standard (an accuracy benchmark) by means of a series of unbroken comparisons all having a stated uncertainty (an estimate of variability based on the dispersion of the results, including method bias).

iii. Chemical Methods of Analysis (Primary Methods) [1]

The results from validated chemical methods (so called old fashioned methods) performed on a small number of samples form the basis for calibration of the high-speed testing. Thus the accuracy of these primary, or reference, tests is of paramount importance because all MIR milk testing is dependent on their accuracy. Current versions of these reference methods are available in the 2000 edition of **Official Methods of Analysis** published by AOAC. Some examples of reference methods are:

(a) For fat -**AOACI method 995.19**, ether extraction.

(b) For Protein- Kjeldahl nitrogen analysis, true protein - **AOACI method 991.22**.

(c) crude protein -**AOACI method 991.20.** The improvement in accuracy of MIR measurement of all milk components that could be expected when true protein rather than crude protein is used as a reference was described because MIR only detects true protein at the classical protein measurement wavelength [2].

(*d*) Lactose measurement by polarimetry is the classical approach, but **the AOAC method 896.01** is very difficult and generally does not perform. Spectrophotometric-based enzymatic methods such as **AOAC method 984.15** have become more common for measurement of lactose.

(*e*) The **AOAC, air oven drying method 990.20** is used as a reference method for total solids measurement.

The following were major improvements made by scientists:

a. *Ether Extraction of Fat*

The ether extraction method was reported [4a] in 1988. The use of a color indicator was reported to help the analyst do a better job removing the ether layer without contamination; addition of water to adjust the level of the interface to facilitate the removal of the ether layer; inclusion of a third extraction; and inter laboratory studies documenting the performance of the method that an analyst should be able to achieve.

b. *Protein Estimation*

One of the major improvements in the Kjeldahl method made in 1990 [4] was the substitution for mercury catalyst by an optimized level of copper sulfate which is friendly to workers and the environment, and achieves equal or superior recovery of nitrogen. Another advance (1991) was the development of a sample preparation procedure to determine true protein directly by Kjeldahl, instead of calculating the difference between separate total nitrogen and nonprotein measurements [5]. This made it practical to use true protein measurement by Kjeldahl as a reference method for calibration of MIR milk analyzers.

c. *Total Solids Estimation*

The major improvements in the total solids method made were reported [6,7,8]. The standardization of sample size, routine use of disposable oil-free aluminum weighing dishes (allows the milk to spread more evenly over the bottom of the pan), drying temperature and time, the initiation of drying starting from a hot oven, the generic definition of the temperature recovery rate that the oven must have at the beginning of the test; and interlaboratory studies documenting the performance of the method were reported.

iv. Electronic Methods of Analysis (Secondary Methods) [1]

Secondary methods require calibration using a set of samples with established reference values. The following electronic instruments are now being used:

a. *Milk-o-Testers (Based on the scattering of light)*

It is a commercially successful electronic milk testing instrument which emerged in the1960s and worked on light scattering. Milk-o-testers of M/S Foss Electric, Denmark, M/S REIL Jaipur, M/S NETCO, New Delhi *etc.* are being used.

b. Lactostar Automatic Milk Analyzer

It applies a combined thermo-optical method to analyze the constituents of milk. The optical measuring method (Tubidimetric) determines the sum of fat and protein contents. Fat and SNF contents are determined by Thermal Analysis, whereas the protein and lactose contents by computational analysis. The freezing point of milk is predicted on the basis of protein content and fat-free dry matter [9].

c. MIR Based Instruments

Mid infrared milk analysis is a rapid, cost-effective secondary testing method. The rapid development in quality and design of the optical bench, electronics, and computer handling of data for MIR analysis has almost completely replaced the MilkoTester. The MIR method measures infrared light absorbance at classical vibration frequencies in the MIR spectra for the analysis of milk components. The MIR filter-based instruments use pairs (sample and reference) of optical filters to select a band of wavelengths for measurement of fat, protein, and lactose. The MIR estimation of total solids is done by calculation based on the values determined for fat, protein, and lactose and a regression correlation to reference values for total solids determined by the oven-drying method. The first generation of MIR instruments used the carbonyl stretch to measure fat (sometimes called Fat A), whereas the second generation used the carbon-hydrogen (sometimes called Fat B) as well as the carbonyl stretches. The more advanced **Fourier transform infrared instruments** that use interferograms to produce full spectral data are now available. Fourier transform infrared is capable of collecting more spectral information and theoretically should enable more accurate testing. **Milk analyzer (**based on the ultrasonic technique), **Milkoscan (**134 A and B,33B.133AA and 132 A, Based on IR) **and Milkoscan Minor** (Based on IR), **Milkoscan FT** (Based on FTIR) are now available for compositional analysis of milk and adulterants in milk. Details of different milk analysers were given in author's another book [10].

Recently following two equipments are made available for compositional analysis and detection of adulterants in milk:

(i) **MilkoscreenTM (Based on FTIR) [11,12]** -In MilkoScreenTM (Indian Dealer- M/S Indifoss Analytical Ltd. Ahmadabad and Manufacturer- FOSS, Denmark), the detection is based on the concept of FT-IR (Fourier Transform infrared spectroscopy). The instrument has been claimed to detect urea (≥ 0.25 per cent), melamine (≥ 0.1 per cent), ammonium sulphate (≥ 0.1 per cent), sucrose (≥ 0.7 per cent) and water (≥ 20 per cent).

(ii) **EMAT (Electronic Milk Adulteration Tester) [11,12] -** In EMAT (Electronic Milk Adulteration Tester- M/S REIL, Jaipur), the principal of detection is based on identification of specific ions of adulterants in milk sample. The instrument has been claimed to detect urea (≥ 0.1 per cent), salt (≥ 0.2 per cent), liquid soap (≥ 0.2 per cent), detergent (≥ 0.3 per cent), caustic soda (≥0.3 per cent) and hydrogen peroxide (≥ 0.3 per cent).

d. Protein Analysis Instruments

Pro- milk MK II, Automated Kjeldahl for Protein Estimation *etc.* are now available to estimate the protein accurately.

e. Pro Milk Tester (MK II) or Pro Milk Automatic or Spectrophotometer

The Pro-Milk instrument is used for the determination of protein in milk by **dye binding method**. The basis of this method is that certain dyes (Amido Black-10 B, Acid Orange-12and Orange –G) react the amino groups of arginine, lysine and histidine of milk proteins. The protein and the dye that it binds precipitate and are separated from the solution of unbound dye by centrifuging or filtration. The optical density of the filtrate is measured in a colorimeter fitted with galvanometer type readout [9,10].

f. Automated Kjeldahl for Protein Estimation

The different Automated Kjeldahl apparatus for protein estimation are now available *e.g.* M/S N- Foss (Kjeltec), M/S J.P.Selecta, M/S Pelican *etc.*

These instruments include systems for sample digestion and distillation. These are designed to attain the best possible accuracy and precision. It also sets safety standards in kjeldahl analysis. Operational safety and convenience are enhanced by the unattended operation and featured such as automatic reagent addition, tube emptying and fully integrated safety systems.

g. Polarimeter [9,10]

It is mainly used to estimate the sugars in food industry. It measures Optical Rotation, Specific Rotation, Concentration, and °Z Sugar Degrees. Polarimeter can be used to measure various optical properties of a material, including linear birefriengence, circular birefriengence (also known as optical rotation or optical rotary dispersion), linear dichroism, circular dichroism and scattering.

h. Butyro Refractometer and Digital Butyro Refractometers [9,10]

An instrument that measures the critical angle of the sodium D line (589 nm)at 20°C. Butyro-refractometer (BR) readings, characteristic for the particular liquid or solid, are inter-convertible and are concerned with the degree of bending of light waves passing through a liquid or transparent solid. Generally, it is expressed as the ratio between the sine of the angle of incidence to the sine of the angle of refraction when a ray of light of a definite known wavelength (usually 589.3 nm, the mean of the D-lines of sodium) passes from air into oils and fats. Now a days **Digital Refractometers** of M/S Ataco, Rudolph *etc.* are available to determine the BR reading of milk fat.

i. Rancimat (M/S Metrohm)

The rancimat (modern PC controlled instrument) determines the oxidative stability. The Rancimat method developed as an automated version of extremely demanding AOM method (active oxygen method) for determination of induction time of fats and oils. A rough estimation of the shelf life of a product is possible with unique temperature exploration [9]. It is being used for estimation of rancidity of ghee, butter oil *etc.* in dairy/oil industry.

Source

1. *D. M. Barbano and J. M. Lynch, Major Advances in Testing of Dairy Products: Milk Component and Dairy Product Attribute Testing, J. Dairy Sci. 89: 1189–1194.*
2. *Lynch, J. M., D. M. Barbano, J. R. Fleming, and D. Nicholson. 2004.Component testing, the dairy industry, and AOAC International. Inside Laboratory Management: A Publication of AOAC International 8 (4): 24–28.*
3. *Lawless, H. T., and H. Heymann. 1999. Sensory Evaluation of Foods: Principles and Practices. Aspen Publishers, Inc. Gaithersburg, MD.*
4. *Barbano, D. M., J. L. Clark, C. E. Dunham, and J. R. Fleming. 1990.Kjeldahl method for determination of total nitrogen content of milk: Collaborative study. J. AOAC 73: 849–859.*

4a. *Barbano, D. M., J. L. Clark, and C. E. Dunham. 1988. Comparisonof the Babcock and ether extraction methods for determination of fat content of milk: Collaborative study. J. AOAC 71: 898–914.*

5. *Barbano, D. M., J. M. Lynch, and J. R. Fleming. 1991. Direct and indirect determination of true protein content of milk by Kjeldahl analysis: Collaborative study. J. AOAC 74: 281–288.*
6. *Clark, J. L., D. M. Barbano, and C. E. Dunham. 1989a. Comparison of two methods for determining total solids content of raw milk: Collaborative study. J. AOAC 72: 712–718.*
7. *Clark, J. L., D. M. Barbano, and C. E. Dunham. 1989b. Combination of total solids determined by oven drying and fat determined by Mojonnier extraction for measurement of solids-not-fat content of raw milk: Collaborative study. J. AOAC 72: 719–72.*
8. *Lynch, J. M., D. M. Barbano, and J. R. Fleming. 1997. Performance evaluation of direct forced-air total solids and Kjeldahl total nitrogen methods: 1990 through 1995. J. AOAC 80: 1038–1043.*
9. *Bimlesh Mann, Rajan Sharma and Richa Singh, Instrumentation for Quality Analysis of Dairy Products: An Overview, Dairy Chemistry Division ICAR-National Dairy Research Institute, Karnal.*
10. *M.K.Srivastava, Hand Book on analysis of milk, CBS publishers, New Delhi, ISBN 978-81-239-2668-1.*
11. *Company Technical Brochures.*
12. *Rajan Sharma, Y.S Rajput and Bimlesh Mann, Detection of Adulteration in Milk and Milk Products– Present Status, ICAR- National Dairy Research Institute, Karnal.*

2. Milk Product Composition and Attribute Testing [1]

Increasing trend in dairy product manufacturing has also driven the development of new, quick methods of analysis for composition and quality control.

Sensory and instrumental analyses of the attributes of dairy foods are used to judge the success of quality control before the product is sent to the consumer, with the ultimate sensory evaluation done by the consumer.

i. Compositional Analysis

The electronics and equipment design has allowed the emergence of inline MIR process control hardware for real-time standardization of milk composition for cheese manufacture [2]. For composition control of liquid dairy products (*e.g.*, fluid milk, milk for cheese making), inline MIR and near infrared (**NIR**) milk analyzers have been developed and are commonly used. Near infrared reflectance methods have also used for solid products like cheese.

ii. Sensory Analysis

Earlier dairy product judging and the scorecard system have been used as the basis for sensory evaluation of dairy products but these approaches did not serve researchers and product developers very well. Sensory analysis of dairy foods has evolved from a defect identification-oriented system to an attribute intensity scaling approach that quantitatively describes many dimensions of a product's characteristics. It was indicated in 1981that the score-card is relatively subjective [3]. It was suggested that the scorecard approach (ADSA-an American system) may not be very useful in quantitatively characterizing the relative intensity of positive attributes of a product or detecting interactions [3,4,]. Quantitative descriptive analysis (QDA) was developed in the 1970s and it was discussed in1999 as a quantitative method yielding data suitable for statistical analysis and therefore of value for research, development, and basic sensory science [5].

Instrumental Sensory Analysis

The rapid development of instrumental methods for olfactory sensory detection has brought a new dimension to the coupling of sensory science and chemical analysis. It was reported in 1984 that the development of **GCO approaches** to characterization and quantification of the aroma profile of food products has brought a powerful new tool to sensory science [6].

Gas Chromatography

A food may contain hundreds of different volatile compounds that can be separated and detected by gas chromatography but only a small number of those compounds may elicit a sensory response in a human. Flow splitters and column effluent conditioners have allowed persons to evaluate each peak by sniffing as it is eluted from the chromatograph.

Mass Spectrophotometry

In addition, the individual peaks that have a detectable aroma can be isolated and analyzed by mass spectrophotometry to determine their structure and identification [7]. In 2004, this approach was used to identify specific compounds and relate them to the results of QDA analysis [8,9].

GCO Combined with GC/MS

The development of GCO, combined with gas chromatography/mass spectrometry is an example of the development of analytical tools that can be correlated with aroma and odor perception of foods.

GCO Combined with LC/MS

The development of rapid analytical methods, such as liquid chromatography coupled with mass spectrometry for protein and peptide sequence analysis, will help in understanding the complex processes of production of peptides and flavor development in natural cheeses. Degradation of proteins and amino acids ultimately produce aroma compounds detected by GCO.

Besides common equipments now-a-days sophisticated equipments and techniques (**Chromatography-**Paper, TLC, HPTLC, HPLC, GLC, **Spectroscopy-**UV/VIS, IR, NMR, Mass, Atomic, Ultrasonic, Mictowave, **Biosensors, Immunoassays, Polarimetry, Polarography and Voltammetry, Electrophoresis, PCR, Flowcytometry**) are also available that helps in determining the components present in milk and milk products/Food Products in small quantities. These sophisticated equipments helped in purification and quantification of bio-actives and also in determination of adulteration.

Source

1. *D. M. Barbano and J. M. Lynch, Northeast Dairy Foods Research Center, Department of Food Science, Cornell University, Ithaca, NY 14853.*
2. *McKenna, D. 1999. Standardization in cheese plants. Chapter 4 in Practical Guide for Control of Cheese Yield. Int. Dairy Fed., Brussels, Belgium.*
3. *McBride, R. L., and C. Hall. 1979. Cheese grading versus consumer acceptability: An inevitable discrepancy. Aust. J. Dairy Technol.34: 66–68*
4. *Lawless, H. T., and M. R. Claasen. 1993. Validity of descriptive and defect-oriented terminology systems for sensory analysis of fluid milk. J. Food Sci. 58: 108–112, 119.*
5. *Lawless, H. T., and H. Heymann. 1999. Sensory Evaluation of Foods: Principles and Practices. Aspen Publishers, Inc. Gaithersburg, MD.*
6. *Acree, T.E., J. Barnard, and D. G. Cunningham. 1984. A procedure for sensory analysis of gas chromatograph effluents. Food Chem.14: 273–286.*
7. *Karagu¨ l-Yu¨ ceer, Y., M. A. Drake, and K. R. Cadwallader. 2003. Aroma-active components of liquid Cheddar whey. J. Food Sci.68: 1215–1219.*
8. *Carpino, S., J. Horne, C. Melilli, G. Licitra, and D. M. Barbano.2004a. Contribution of native pasture to the sensory properties of Ragusano cheese. J. Dairy Sci. 87: 308–315.*
9. *Carpino, S., S. Mallia, S. LaTerra, C. Melilli, G. Licitra, T. E. Acree, D. M. Barbano, and P. J. Van Soest. 2004b. Composition and aroma compounds of Ragusano cheese: Native pasture and TMR.J. Dairy Sci. 87: 816–830.*

Chapter 3

Spectroscopic Techniques

Spectroscopy deals with the production, measurements and interpretation of spectra arising from the interactions of electromagnetic radiation with matter. Spectroscopy is the measurement of such Electromagnetic radiation/Ultrasonic waves as a means of obtaining information about the system and their components. Spectroscopic methods based on the adsorption or emission of radiation in the ultraviolet, visible, infrared and radio (nuclear magnetic resonance) frequency ranges are most commonly encountered in traditional food analysis. The following Spectroscopic techniques are being used in analysis of dairy products/Foods:

1. UV-VIS (Ultraviolet and Visible) Spectroscopy.
2. IR (Infrared) Spectroscopy -NIR (Near Infrared), MIR (Mid Infrared) FTIR spectroscopy are also used.
3. NMR (Nuclear Magnetic Resonance) Spectroscopy.
4. Microwave Spectroscopy.
5. Atomic Spectroscopy.
6. Ultrasound Spectroscopy.
7. Mass Spectroscopy.

1. UV-VIS (Ultraviolet and Visible) Spectroscopy [1]

Ultraviolet-visible spectroscopy (ultraviolet-visible spectrophotometry) refers to absorption spectroscopy or reflectance spectroscopy in the ultraviolet-visible spectral region. It uses light in the visible and adjacent ranges. The absorption or reflectance in the visible range directly affects the perceived color of chemicals involved. In this region of the electromagnetic spectrum, molecules undergo electronic transitions. This technique is complementary to fluorescence spectroscopy. This spectroscopy is routinely used in analytical chemistry for the quantitative determination of different analytes. Determination is usually carried out in solutions.

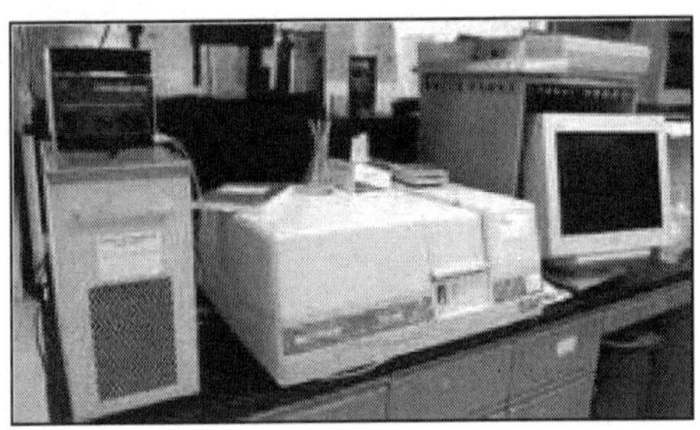

Beckman DU640 UV/Vis Spectrophotometer.

Source

1. *Wikipedia.*

Electromagnetic Radiations

The visible part of the electromagnetic spectrum is, by definition, radiation visible to the human eye. Other detection systems reveal radiation beyond the visible regions of the spectrum and these are classified as **radiowave, microwave, infrared, ultraviolet, X-ray and γ-ray [1].** These regions together with the processes involved in the interaction of the radiation of these regions with matter, are illustrated in following Table [1a]:

Region of Radiation	*Radiowave*	*Microwave*	*Infrared*	*Visible and Ultraviolet*	*X-ray*	*γ- ray*
Process invoved	Change of spin	Change of orientation	Change of configu-ration	Change of electron distribution	Change of electron distribution	Change of Nuclear configu-ration
Energy (J mol^{-1})	-	10	10^3	10^5	10^7	10^9

The nature of the various radiations shown above have been interpreted by Maxwell's classical theory of electro- and magneto-dynamics -hence, the term **electromagnetic radiation**. According to this theory, radiation is considered as two mutually perpendicular electric and magnetic fields, oscillating in single planes at right angles to each other. These fields are in phase and are being propagated as a sine wave [1].

Ultraviolet/Visible Region and Colour of Visible Light [2]

Light consists of radiation to which the human eye is sensitive, wave of different wavelengths giving rise to light of different colours, while a mixture of light of these wavelengths constitutes white light. White light covers the entire visible spectrum 400-780. For Colorimetry and spectrophotometry, the Visible region and the adjacent Ultraviolet region are of major importance. The following table helps in understanding the composition of Ultraviolet and visible light:

	A-UV Region	
Less than 400 nm	Ultra Violet Region	
	B- Visible Region	
Wavelength of Max. Absorption (nm)	*Colour Absorbed*	*Colour Observed*
400-420	Violet	Green- Yellow
420-440	Violet –Blue	Yellow
440-470	Blue	Orange
470-500	Blue –Green	Red
500-520	Green	Purple
520-550	Yellow- Green	Violet
550-580	Yellow	Violet -Blue
580-620	Orange	Blue
620-680	Red	Blue -Green
680-780	Red	Green

The visual perception of colour arises from the selective absorption of certain wavelength of incident light by the coloured object. The other wavelengths are either reflected or transmitted, according to the nature of the object, and are perceived by the eye as the colour of the object. If a solid opaque object appears white, all wavelengths are reflected equally, if the object appears black, very little light of any wavelength is reflected, if it appears blue, the wavelengths that give blue stimulus are reflected *etc.*

Electromagnetic waves are usually described in terms of - a- wavelength λ (Distance between the peaks of waves in cm unless otherwise specified), b- wave number à (number of waves per cm) and c-the frequency v (number of waves per second). The relation between aforesaid three quantities are as follows:

$$\frac{1}{\text{Wavelength }(\lambda)} = \text{Wave Number (à)} = \frac{\text{Frequency (v)}}{\text{Velocity of Light (c)}}$$

The units commonly being used are as follows:

1 Angstrom unit = 1Å = 10^{-10} meter = 10^{-8}cm

1 nanometer = 1nm= 10Å =10^{-7}cm

1 micrometer = 1μm= 10^{4}Å=10^{-4}cm

Wave Number (à) = $1/\lambda$ waves per cm

Frequency v = $c/\lambda \approx 3 \times 10^{10}/\lambda$ waves per second

Source

1. *Barbara Stuart, Infrared Spectroscopy- Fundamentals and Applications, 2004 John Wiley and Sons, Ltd ISBNs: 0-470-85427-8 (HB); 0-470-85428-6 (PB).*

1a. *Barbara Stuart, Biological Applications of Infrared Spectroscopy, ACOL Series, Wiley, Chichester, UK, 1997.*

2. *Vogel's Text Book on Quantitaive chemical analysis revised by G.H.Jeffery, J. Bassett, J.Mendham and R.C.Denney.*

Principles of Spectrophotometry [1,2,3]

UV radiations have enough energy to **induce electronic transitions.** Molecules containing π-electrons or non-bonding electrons (n-electrons) can absorb the energy in the form of **ultraviolet or visible light to excite these electrons to higher anti-bonding molecular orbitals [4].** The more easily excited the electrons (*i.e.* lower energy gap between the HOMO and the LUMO), the longer the wavelength of light it can absorb. There are four possible types of transitions (π-π*, n-π*, σ-σ*, and n-σ*), and their can be ordered as follows: σ-σ* > n-σ* > π-π* > n-π.

A spectrophotometer consists of two instruments, namely a spectrometer for producing light of any selected color (wavelength), and a photometer for measuring the intensity of transmitted light/radiation. The instruments are arranged so that liquid in a cuvette can be placed between the spectrometer beam and the photometer. The amount of light passing through the tube is measured by the photometer. The photometer delivers a voltage signal to a display device, normally a galvanometer. The signal changes as the amount of light absorbed by the liquid changes.

If development of color is linked to the concentration of a substance in solution then that concentration can be measured by determining the extent of absorption of light at the appropriate wavelength. For example hemoglobin appears red because the hemoglobin absorbs blue and green light rays much more effectively than red. The degree of absorbance of blue or green light is proportional to the concentration of hemoglobin.

When light (monochromatic or heterogeneous) falls upon the homogeneous medium, a portion of the incident light is reflected, a portion is absorbed within the medium and the remainder is transmitted.

$$I_o = I_a + I_t + I_r$$

Where- I_o - Intensity of incident light, I_a - Intensity of absorbed light, I_t - Intensity of transmitted light and I_r – Intensity of reflected light.

For a air-glass interfaces arising from the use of glass cells, it may be stated that 4 per cent of the incident light is reflected. I_t is usually eliminated by the use of control (comparison cell) and hence

$$Io = Ia + It$$

The change of absorption of light with the thickness of the medium can be easily understandable by the following Laws:

A. Lambert's Law

It states that when monochromatic light passes through a transparent medium, the rate of decrease in intensity with the thickness of the medium is proportional to the intensity of light.

Light decreases exponentially as the thickness of the absorbing medium increases arithmetically or that any layer of given thickness of the medium absorbs the same fraction of the light incident upon it. This law can be expressed by the differential equation:

$$-\frac{dI}{d} = kI$$

Where - I - Intensity of incident light of wave length λ, l-is the thickness of the medium and k-is a proportionality factor.

By putting $I = I_0$ when l=0, we get

$$\text{In}\ \frac{I_0}{It} = kl$$

The ratio It/I_0 is the fraction of the incident light transmitted by a thickness l of the medium and is termed as **transmittance** T. Its reciprocal I_0/It is the opacity, and the **absorbance** A of the medium (formerly called the **optical density** D or **Extinction** E) is given by

$$A = \log I_0/It$$

Thus a medium with absorbance 1 for given wavelength transmits 10 per cent of the incident light at the same wavelength.

B. Beer's Law

Beer studied the effect of concentration of coloured constituent in solution upon the light transmission or absorption. He found the same relation between transmission and concentration as Lambert had discovered between transmission and thickness of the layer. When monochromatic light (light of a specific wavelength) passes through a solution there is usually a quantitative relationship (Beer's law) between the solute concentration and the intensity of the transmitted light, that is, the intensity of a beam of monochromatic light decreases exponentially as the concentration of the absorbing substance increases arithmetically which may be written as:

$$I = I_0 X 10^{-kcl}$$

where I_0 is the intensity of transmitted light using the pure solvent, I is the intensity of the transmitted light when the colored compound is added, c is concentration of the colored compound, l is the distance the light passes through the solution, and k is a constant. If the light path l is a constant, as is the case with a spectrophotometer, **Beer's law** may be written,

$$I \div I_0 = 10^{-kc} = T$$

where k is a new constant and T is the transmittance of the solution. There is a logarithmic relationship between transmittance and the concentration of the colored compound. Thus,

$$-\log T = \log 1/T = kc = \text{Optical Density (O.D.)}$$

Most spectrophotometers have a scale that reads both in O.D. (absorbance) units, which is a logarithmic scale, and in per cent transmittance, which is an arithmetic scale. As suggested by the above relationships, the absorbance scale is the most useful for colorimetric assays.

Source

1. *w.ruf.rice.edu/~bioslabs/methods/protein/spectrophotometer.htm.*
2. *David R. Caprette (caprette@rice.edu), Rice University 16 Sep 96.*
3. *Vogel's Text Book on Quantitaive chemical analysis revised by G.H.Jeffery, J. Bassett, J.Mendham and R.C.Denney.*
4. *Metha, Akul (13 Dec 2011). "Principle". PharmaXChange.info.*

C. Beer-Lambert's Law [3,4]

This states that the absorbance of a solution is directly proportional to the concentration of the absorbing species (coloured compounds) in the solution and the path length [4] and absorbance versus concentration plot is a straight line. The intensity of light decreases exponentially as it passes through an absorbing medium. Beer-Lambert Law quantifies the process of absorption. At a given wavelength,

$$A = \varepsilon c l$$

where -ε - Molar extinction co-efficient. It is a constant for the absorbing species and defines the absorption of species at particular wavelength, **c**- concentration of sample in cuvette and **l**-length of light path through the sample.

This the fundamental equation of colorimetry and spectrophotometry and often spoken of as **Beer-Lambert Law**. This equation is being used for determination of concentration from Absorbance measurement. Absorbance is directly proportional to the concentration, c, and the path length, *b*, of the absorbing species:

$A \propto c$ and $A \propto b$

so $A \propto b \propto c$

$A = a \propto b \propto c$

Proportionality constant, a = absorptivity- units (L/g·cm)

If units of concentration are M (mol/L) then molar absorptivity ε is used.

$$A = \varepsilon bc = -\log T = -\log \frac{I}{I_0} = \log \frac{I_0}{I}$$

Absorbance is an **additive property**.

$$A_{total} = A_1 + A_2 \ldots$$
$$= \varepsilon_1 bc_1 + \varepsilon_2 bc_2 \ldots$$

in a mixture of 2 components

$$A_{\lambda 1} = \varepsilon_{1,\lambda 1} \cdot b \cdot c_1 + \varepsilon_{2,\lambda 1} \cdot b \cdot c_2$$
$$A_{\lambda 2} = \varepsilon_{1,\lambda 2} \cdot b \cdot c_1 + \varepsilon_{2,\lambda 2} \cdot b \cdot c_2$$

The Absorbance of a mixture of n species is given by –

$$\mathbf{A}_{\text{mixture}} = \Sigma^n_{i=0} \mathbf{A}_i$$

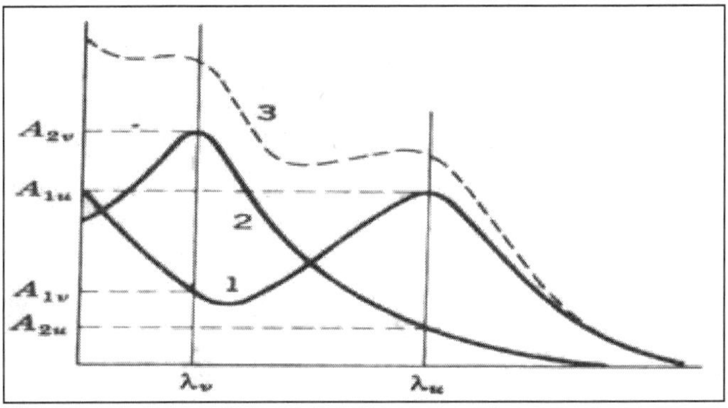

So **Beer–Lambert law** is used to calculate the molar absorption coefficient, ε and λ **max** which is the wavelength at which maximum absorption occurs.

Transmittance $T = \dfrac{I}{I_0}$

Percent Transmittance $\%T = \dfrac{I}{I_0} \times 100\%$

Absorbance $A = -\log T$

$= -\log \dfrac{I}{I_0}$

$= \log \dfrac{I_0}{I}$

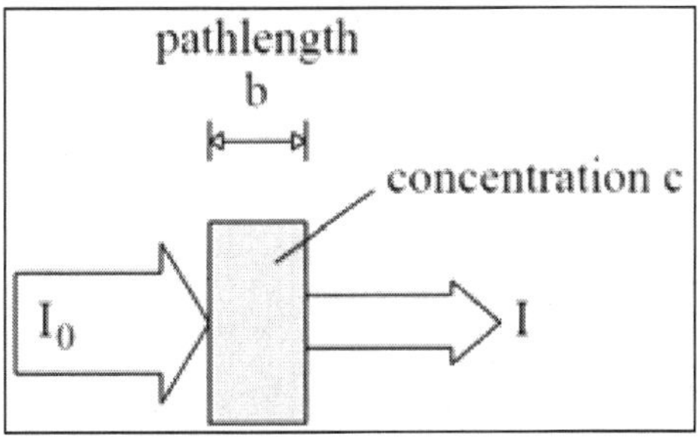

The Beer-Lambert Law is useful for characterizing many compounds but does not hold as a universal relationship for the concentration and absorption of all substances. A 2nd order polynomial relationship between absorption and concentration is sometimes encountered for very large, complex molecules such as organic dyes (Xylenol Orange or Neutral Red) [3].

Photometric Error

(i) At low absorbance values, I (incident) ≈ I (transmitted)Thus, error is relatively large.

(ii) At high absorbance values, I (transmitted) too small not measured accurately.

Relative error in A vs per cent T of $KMnO_4$solutions at 523 nm where ε = 2400.

Lowest error needs per cent T = 38,6 per cent = A value ≈ 0.43 A.

Working is possible within ± 15 per cent which is absorbance 0.2 – 0.8 as best measurement range.

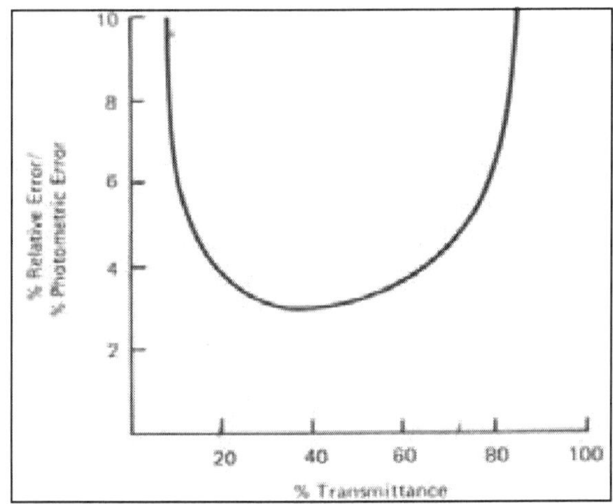

Source

1. *David R. Caprette (caprette@rice.edu), Rice University 16 Sep 96.*
2. *Vogel's Text Book on Quantitaive chemical analysis revised by G.H.Jeffery, J. Bassett, J.Mendham and R.C.Denney.*
3. *Wikipedia.*
4. *Metha, Akul (22 Apr 2012). "Derivation of Beer-Lambert Law". PharmaXChange. info.*

Instrumentations

The instrument used in ultraviolet-visible spectroscopy is called a UV/Vis **spectrophotometer**. It measures the intensity of light passing through a sample and compares it to the intensity of light before it passes through the sample. The ratio I/I o is called the transmittance, and is usually expressed as a percentage (per cent T). The absorbance, A, is based on the transmittance. The basic parts of a spectrophotometer are a light source (radiation source), a holder for the sample (sample container), a diffraction grating in a monochromator or a prism to separate the different wavelengths of light, a detector (radiation detector)and Signal processor/Readout. The details are given below:

i. Radiation Sources [1,2]

Tungsten Lamp (Visible), Deuterium lamp (UV) and Globar Silicon Carbide (IR) are generally used as light/radiation sources. The radiation source is often a Tungsten filament (300-2500 nm), a deuterium arc lamp, which is continuous over the ultraviolet region (190-400 nm), Xenon arc lamp, which is continuous from 160-2,000 nm; or more recently, light emitting diodes (LED) [3] for the visible wavelengths [1].

ii. Sample Containers and Optics [1,2]

Samples for UV/Vis spectrophotometry are most often liquids, although the absorbance of gases and even of solids can also be measured [1]. To determine the absorption of radiation by given solution, the solution must be placed in a suitable transparent cell/container called Cuvettes (Cells) which can be accurately located in the beam of radiation. Cuvettes are of glass or Silica/quartz *etc.* depending upon the wave length to be covered *e.g.* Glass cell for 400-3000 nm (vis-near IR), Fused Silica/quartz cell for 200-3000 nm (UV-near IR) and NaCl cell for 200-15,000 nm (UV-far IR).Glass and plastic cuvettes are also common, although glass and most plastics absorb in the UV, which limits their usefulness to visible wavelengths [3]. Standard cells are of rectangular form commonly with an internal width of 1 cm. (This width becomes the path length, L, in the Beer-Lambert law). Test tubes can also be used as cuvettes in some instruments. The type of sample container used must allow radiation to pass over the spectral region of interest [1]. Larger cells are required when solution of low absorbance are to be examined likewise cells with short path length are required when solutions of high absorbance are to be examined. In addition to the usual rectangular pattern cells, continuous flow cells and sampling cells are also available.

iii. Wavelength Selector

Filters-Absorption Filter

Colored glass or dye between two glass plates, and

Interference Filter

Two thin sheets of metal sandwiched between glass plates, separated by transparent material. Interference for transmitted wave through 1st layer and reflected from 2nd layer.

Monochromators

Consists of Entrance slit, Collimating lens or mirror, Dispersion element (prism or grating), Focusing lens or mirror and Exit slit. Typical Prism Monochromator is shown below:

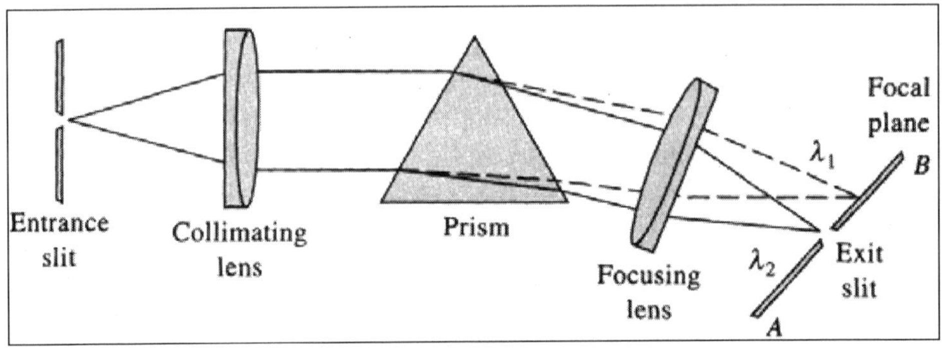

The scanning monochromator moves the diffraction grating to "step-through" each wavelength so that its intensity may be measured as a function of wavelength. Fixed monochromators are used with CCDs and photodiode arrays. As both of these devices consist of many detectors grouped into one or two dimensional arrays, they are able to collect light of different wavelengths on different pixels or groups of pixels simultaneously [1].

iv. Radiation Detector

The detector is typically a photo multiplier tube (PMT), a photodiode, a photodiode array or a charge-coupled device (CCD). Single photodiode detectors and photomultiplier tubes are used with scanning monochromators, which filter the light so that only light of a single wavelength reaches the detector at one time [1]. The following other detectors can also be used:

(i) Photovoltaic Cells

(ii) Phototube-radiation falls on a sensitive Plate which produces electron,.

(iii) Thermal detectors - sensitive to IR (λ>750 nm).

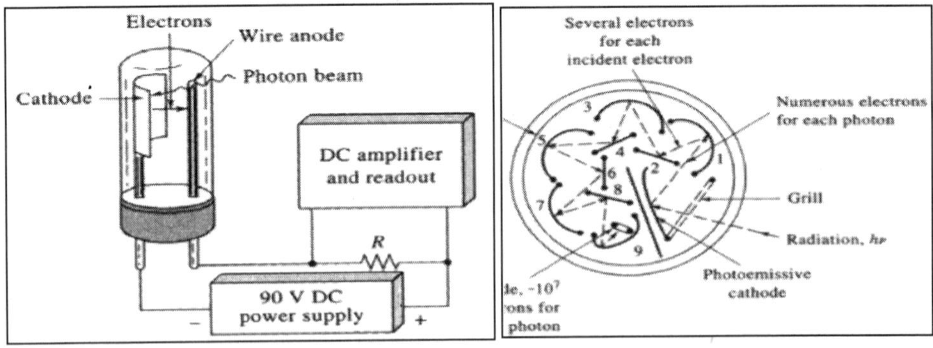

v. Signal Processor and Readout

Source

1. *Wikipedia.*

2. *Vogel's Text Book on Quantitaive chemical analysis revised by G.H.Jeffery, J. Bassett, J.Mendham and R.C.Denney.*

3. *Skoog, Douglas A.; Holler, F. James; Crouch, Stanley R. (2007). Principles of Instrumental Analysis (6th ed.). Belmont, CA: Thomson Brooks/Cole. pp. 169–173. ISBN 9780495012016.*

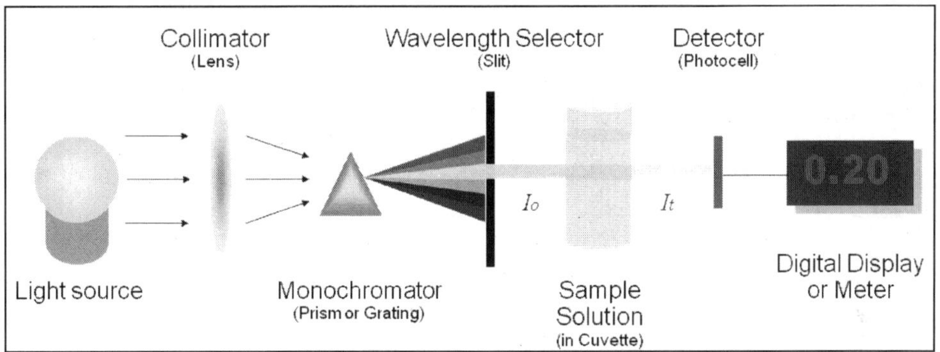

Outline of Working of UV/Vis Spectrophotometer (*Source*: Online2PDF.com).

Types of UV/Vis Spectrophotometers [1,2]

UV/Vis Spectrophotometers are Single beam or Double beam.

i. Single Beam UV/Vis Spectrophotometer

In a single beam instrument (Spectronic 20), all of the light passes through the sample cell. Io must be measured by removing the sample [2]. This was the earliest design and is still in common use in both teaching and industrial labs. In this instrument, firstly a blank reference solution is placed in the unit to set 100 per cent Absorbance. Thereafter, measurements are taken for standard reference/Unknown sample.Subsequent data processing gives per cent T/ABS/Con. of the unknown.

ii. Double Beam UV/Vis Spectrophotometer

The blank reference solution and the standard reference/unknown sample are measured simultaneously. This nullifies the effect of undesired Changes in the operating conditions in the system if occur.

Most modern general purpose UV/Vis Spectrophotometers are double beam instruments. In this instruments the monochromated beam of radiation from tungsten and deuterium lamps, is divided (Splitting) into two identical beams, one of which passes through the reference cell and other through the sample cell. The splitting and recombination of beams are accomplished by the following way:

(a) Light from the source is segmented by a rotating segmented mirror which allows radiation through to the sample and reflects it through the reference cell alternately.

(b) The two beams are recombined by a rotating mirror in phase with the first rotating mirror.

(c) The combined beam is allowed to fall on the detector in which the reference signal is electronically subtracted from the sample signal before it is sent to the recording system.

The reference beam intensity is taken as 100 per cent Transmission (or 0 Absorbance), and the measurement displayed is the ratio of the two beam intensities. Some double-beam instruments have two detectors (photodiodes), and the sample and reference beam are measured at the same time. In other instruments, the two beams pass through a beam chopper, which blocks one beam at a time. The detector alternates between measuring the sample beam and the reference beam in synchronism with the chopper. There may also be one or more dark intervals in the chopper cycle. In this case, the measured beam intensities may be corrected by subtracting the intensity measured in the dark interval before the ratio is taken.

iii. Specialized Instruments [2]

These have also been made. These include attaching spectrophotometers to telescopes to measure the spectra of astronomical features. UV-visible micro spectrophotometers consist of a UV-visible microscope integrated with a UV-visible spectrophotometer.

A complete spectrum of the absorption at all wavelengths of interest can often be produced directly by a more sophisticated spectrophotometer. In simpler instruments the absorption is determined one wavelength at a time and then compiled into a spectrum by the operator. By removing the concentration dependence, the extinction coefficient (ε) can be determined as a function of wavelength.

Different Models and Characteristics

The details related to Different Models and Characteristics are given below:

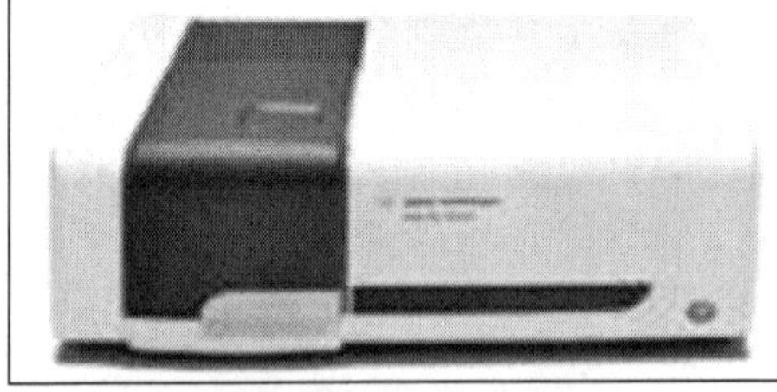

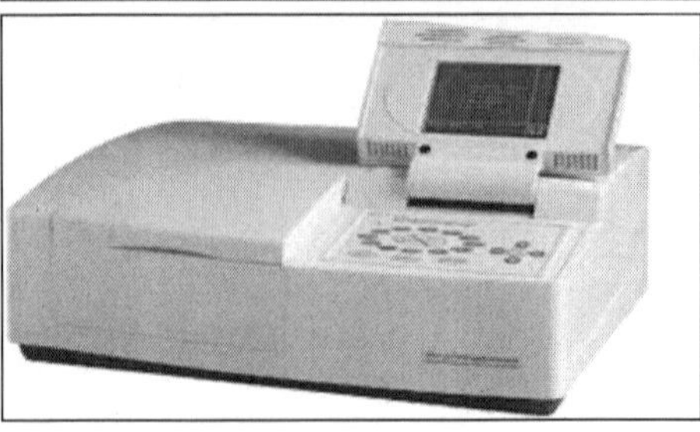

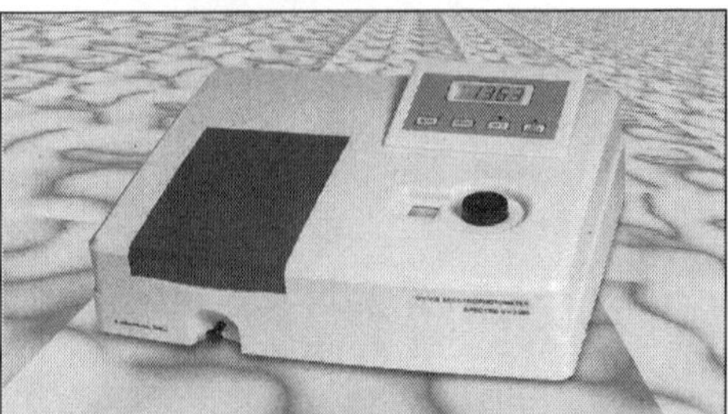

Model UVD 2950 **Spectro 2905**

Different Models of UV/Vis Spectrophotometers.

Characteristics of Some UV/Vis Spectrophotometers

Sl. No.	*Characteristics*	*Different models of UV/Vis Spectrophotometers*			
		Double Beam PC Scanning Labomed, USA UVD2950	*Shimadzu UV-1800*	*Double Beam Spectro-photometer - (UV 3200)*	*7315 UV/ Visible Spectro-photometer. Jenway*
1	Wave length	190-1100 nm	190-1100nm	190nm to 1100nm	198 to 1000nm
2	Spectral band width	1nm	1nm(190-1100nm)	Wavelength Display - 0.1nm	5nm
3	Resolution	1nm	-	Wavelength Setting - 0.1nm	1nm
4	Stray light	>2.1Abs(200nm)	Less than 0.02 per cent NaI at 220 nm, NaNO2 at 340 nm, Less than 1 per cent KCl at 198 nm	0.02 per cent @ 340nm for NaNO2, 0.25 per cent @ 198nm for KCl	<0.5 per cent at 340nm and 220nm
5	Wave length accuracy	±0.3nm (with automatic wavelength)	±0.1nm at 656.1nmD2 ±0.3 nm (190-1100nm)	± 0.1 nm @656.1 nm D2, ± 0.3nm (190 to 1100nm)	± 2nm
6	Wave length clone	±0.2 nm	Repeatability- ±0.1nm	Repeatability- 0.1nm	Repeatability-± 0.5nm
7	Photovoltaic system	Double beam optical system	Double Beam	Double Beam	
8	Optical methods	Transmittance, Absorbance, Energy and Focus	Absorbance and Transmittance	Absorbance and Transmittance	
9	Range	-0.3 ~ 3.0 Abs	Absorbance: -4 to 4Abs Transmittance- 0-400 per cent	0 - 400 per cent T, Absorbance -4 to 4	T-0 to 199.9 per cent A--0.300 to 2.500A
10	Optical Resolution	±0.002 Abs (0~0.5). 0.004 Abs ± (0.5~1.0). ±0.3T (0~100 T).	Accuracy-(±) 0.002 Abs (0.5 Abs) 0.004 Abs (1.0 Abs) 0.006 Abs (2.0 Abs) Repeatability- Less Than ± 0.001 Abs (0.5 Abs). Less Than ± 0.001 Abs (1.0 Abs). Less Than± 0.003 Abs (1.0 Abs)	± 0.002 Abs (0.5), ± 0.004 Abs (1.0), ± 0.006 Abs (2.0) Reproducibility- 0.001 Abs (0.5 Abs), 0.001 Abs (1.0 Abs), 0.003 Abs (2.0 Abs)	Accuracy ±1 per cent T, ±0.01Abs at 1.000 Absorbance Ressolution-0.1 per cent T, 0.001A

Contd.

Sl. No.	*Characteristics*	*Different models of UV/Vis Spectrophotometers*			
		Double Beam PC Scanning Labomed, USA UVD2950	*Shimadzu UV-1800*	*Double Beam Spectro-photometer - (UV 3200)*	*7315 UV/ Visible Spectro-photometer. Jenway*
11	Cloning Optical	±0.001 Abs (0~0.5). 0.002 Abs ± (0.5~1.0). ±0.15T(0~100 T).	-	-	-
12	Optical Display	-9.999 9.999---	-	-	-
13	Optical Noise	±0.001 Abs (500nm to 2 nm spectral band width Abs 0)	Within 0.00005 Abs RMS value (at 700nm)	0.000016 Abs RMS @ 500 nm	-
14	Scanning Speed	1400 nm/min	-	2 - 3000 nm per Minute	-
15	Light source	Socket Deuterium Lamp and Socket Tungsten Halogen Lamp	-	Tungsten and Deuterium Lamp, pre-aligned	Xenon lamp
16	Detector	Double Beam	-	Double beam hollow graphic gratings 1200 lines/mm	-
17	Show room	2Cell holder	-	Standard 10mm Single Cell Holder 2pcs	-
18	Screen	Liquid Crystal Display(310-240 LCD Dot matrix	-	-	-
19	PC Interface	PC Interface: RS-232, USB	PC, Memory, Printer- *(i)* DPU, ESC/P, PCL printers, USB I/F Windows-compliant printers are avail-able with USB memory and PC software. *(ii)* USB memory (option) Saved as text and UVPC file.	USB Port and Parallel Port (Printer) *(i)* Online software upgrade via internet helps to keep it updated. *(ii)* Data Download-to-PC software expands	USB, Analogue, RS232, Internal printer. Jenway 73 series PC software (735 100) and interface cable (013 203)

Contd.

Sl. No.	Characteristics	Different models of UV/Vis Spectrophotometers			
		Double Beam PC Scanning Labomed, USA UVD2950	*Shimadzu UV-1800*	*Double Beam Spectro-photometer - (UV 3200)*	*7315 UV/ Visible Spectro-photometer. Jenway*
			(*iii*) USB memory + UV Probe(Standard) Win XP.	the data storage to unlimited.	
20	Other	Can be used alone or linked to PC	Can be used alone or linked to PC	PC Models come standard with windows based application software UV/Vis Analyst	Limited by attached mass storage device

Source

1. *Technical Brochures of Equipment manufacturers.*
2. *Wikipedia.*

Operational Terminologies of UV/Vis Spectrophotometer

Operating Modes

Some UV/VIS Spectrophotometers have the following operating modes:

(*i*) Single wavelength.
(*ii*) Multi wavelength
(*iii*) Spectrum Scan
(*iv*) Time Scan.

i. Single Wavelength

In this mode, spectrophotometer takes measurement at one wavelength selected by user and the following analysis can be done:

(*a*) Absorbance
(*b*) Per cent Transmittance
(*c*) Concentration.

Monitor displays the selected wave length, sample no., cell no. and measurements (Abs, T and Conc. as selected)

ii. Multi Wavelength

In this mode, readings are taken at minimum one and maximum 15 -20 (depending upon the model of instrument) user selected wavelengths. Monitor displays the selected wave lengths, sample no., cell no. and the measurements (Abs and per cent T as selected).

Multi Component Analysis

This mode is used to determine the concentration of each component of a mixture solution (containing multiple components) from the measured mixture absorption curve.

Concentration

This mode is used to determine the concentration of sample (s) either by multiplying by a factor or by establishing a curve in memory using standard or using the saved data.

iii. Spectrum Scan

In this mode, the sample is scanned between the selected wavelength range. The monitor displays the scan in either per cent T or ABS as selected. The scan can be easily switched between per cent T and ABS mode of display without rescanning the sample. The readings taken in this scan can be further processed, if required for peak pick, valley pick, expansion, derivatives, smoothening, MWL anal. averaging subtraction and concentration by compression. It is possible to see reading at any point on graph by moving curser to that point.

iv. Time Scan

This allow the user to record Absorbance or per cent Transmittance of the solution with time. The time scan is performed in single wavelength operating mode only. Time interval between two readings and total number of the readings to be taken are pre-determined. The monitor displays the scan in either per cent T or ABS as selected. The Time scan can be easily switched between per cent T and ABS mode of display as well as in tabular or graphic form.

Note–The storage and displays of data depends upon the model of instrument.

Measuring Modes

UV/Vis Spectrophotometer can be set to make following measurements.:

(i) Absorbance.

(ii) Per cent Transmittance

(iii) Concentration

(iv) Multi Component Analysis.

Absorbance and per cent Transmittance

These parameters of the solution can be measured in all the four operating modes (Time scan - only single wavelength).

Concentration

The concentration can be measured in single wavelength, Multi wave length and scan range operating mode.

Multi Component Analysis

It can be done only in multi wave length mode.

Source

1. *Systronic Technical Brochure (Systronic).*

Factors affecting the Measurement [1]

The test results of UV/Vis Spectro-photometer can be effected by the following important factors:

i. Monochromaticity and Spectral Bandwidth

The monochromatic source of radiation for the light incident on the sample cell, is very important [2]. Monochromaticity is measured as the width of the "triangle" formed by the intensity spike, at one half of the peak intensity. A spectrometer has a spectral **bandwidth** (that characterizes how **monochromatic** the incident light is) and if this bandwidth is comparable to (or more than) the **width** of the absorption line, then the measured extinction coefficient will be mistaken. In reference measurements, the instrument bandwidth is kept below the width of the spectral lines. When a test material is being measured, the bandwidth of the incident light should also be sufficiently narrow. Reducing the spectral bandwidth reduces the energy passed to the detector and will, therefore, require a longer measurement time to achieve the same signal to noise ratio.

ii. Wavelength Error

In liquid samples, the extinction coefficient usually changes slowly with wavelength. A peak of the absorbance curve (a wavelength where the absorbance reaches a maximum) is where the rate of change in absorbance with wavelength is smallest [2]. Measurements are usually made at a peak to minimize errors produced by errors in wavelength in the instrument, that is errors due to having a different extinction coefficient than assumed.

iii. Stray Light

The purity of the light used, is another important factor. The most important factor affecting this is the **stray light level of the monochromator** [2]. The detector used is broadband which responds to all the light that reaches it. If a significant amount of the light passed through the sample contains wavelengths that have much lower extinction coefficients than the nominal one, the instrument will report an incorrectly low absorbance. A instrument will reach a point where an increase in sample concentration will not result in an increase in the reported absorbance, because the detector is simply responding to the stray light. The stray light level approximately depends on the following:

(a) An instrument with a single monochromator would typically have a stray light level corresponding to about 3 Absorbance Units (AU), which would make measurements above about 2 AU problematic.

(*b*) A instrument with a **double monochromator** would have a stray light level corresponding to about 6 AU, which would therefore allow measuring a much wider absorbance range.

iv. Absorption Flattening and Deviations from the Beer–Lambert Law

At sufficiently high concentrations, the absorption bands will saturate and show **absorption flattening**. The absorption peak appears to flatten because close to 100 per cent of the light is already being absorbed. The concentration at which this occurs depends on the particular compound being measured. Solutions that are not homogeneous can show deviations from the Beer-Lambert law because of the phenomenon of absorption flattening. This can happen, for instance, where the absorbing substance is located within suspended particles [5,6]. The deviations will be most noticeable under conditions of low concentration and high absorbance.

v. Measurement Ucertainty Sources

The above factors contribute to the **measurement uncertainty** of the results obtained with UV/Vis spectrophotometry. If UV/Vis spectrophotometry is used in quantitative chemical analysis then the results are additionally affected by uncertainty sources arising from the nature of the compounds and/or solutions that are measured. These include spectral interferences caused by absorption band overlap, fading of the color of the absorbing species and possible composition mismatch between the sample and the **calibration solution [4].**

Source

1. *Wikipedia.*
2. *Metha, Akul (14 May 2012)."Limitations and Deviations of Beer-Lambert Law". PharmaXChange.info.*
3. *Ansell, S.; Tromp, R. H.; Neilson, G. W. (1995). "The solute and aquaion structure in a concentrated aqueous solution of copper (II) chloride". J. Phys.: Condens. Matter. 7 (8): 1513-1524.*
4. *Soovali, L.; Room, E.-I.; Kutt, A.; et al. (2006)."Uncertainty sources in UV-Vis spectrophotometric measurement". Accreditation and Quality Assurance. 11: 246–255.*
5. *Berberan-Santos,"Beer's law revisited", J. Chem. Educ. 67 (1990) 757.*
6. *Wittung, Kajanus, Kubista, Malmström,"Absorption flattening in the optical spectra of liposome-entrapped substances", FEBS Lett 352 (1994) 37).*

UV-VIS-Spectroscopy Based Equipment for Dairy

Chemspec 150 (Bentley Instrument), an automatic, fast (upto 150 samples/hour) and compact instrument being used by dairy for determination of Urea (or MUN) in milk.

Source.

1. *Technews, No.74, May-June-2008, NDDB.*
2. *Bentley Instruments (Website: www.bentleyinstruments.com).*

Application of UV-VIS Spectroscopy

UV/Vis spectroscopy is routinely used in analytical chemistry for the quantitative determination of different analytes, such as transition metal ions, highly conjugated organic compounds, and biological macromolecules. Spectroscopic analysis is commonly carried out in solutions but solids and gases may also be studied. UV/Vis can also be applied to determine the kinetics or rate constant of a chemical reaction and an equilibrium constant.In addition to the above applications, UV-VIS Spectroscopy is now frequently being used in Food/Dairy Industries. Some of the examples are given below:

i. Determination of Protein

A number of methods have been devised to measure protein concentration, which are based on UV-visible spectroscopy. These methods use either the natural ability of proteins to absorb (or scatter) light in the UV-visible region of the electromagnetic spectrum, or they chemically or physically modify proteins to make them absorb (or scatter) light in this region. The basic principle behind each of these tests is similar. First of all a calibration curve of absorbance (or turbidity) versus protein concentration is prepared using a series of protein solutions of known concentration. The absorbance (or turbidity) of the solution being analyzed is then measured at the same wavelength, and its protein concentration determined from the calibration curve. The main difference between the tests are the chemical groups which are responsible for the absorption or scattering of radiation, *e.g.*, peptide bonds, aromatic side-groups, basic groups and aggregated proteins.

ii. Visual Detection of Melamine in Milk Products

A simple, rapid, field-portable colorimetric method for the detection of melamine based on melamine-induced color change of label-free gold nanoparticles (Au NPs) was developed in this study. Melamine can induced the aggregation of Au NPs and results in the color change from wine-red to purple, which provided a platform for rapid and field-portable colorimetric detection of melamine. The proposed method can be used to detect melamine in liquid milk and infant formula with a detection limit of 1.0 and 4.2 ppm, respectively, within 30 min by naked eyes observation without the aid of any advanced instrument and the need of any complex pretreatment, and detect as low as 0.15 ppm of melamine in liquid milk and 2.5 ppm of melamine in infant formula with UV–vis-spectroscopy. The proposed method is promising for on-site screening of melamine adulterant in milk products.

Source

1. *Liangqia Guo, Jianhai Zhong, Jinmei Wu, FengFu Fu, Guonan Chen, Xiaoyan Zheng and Song, Ministry of Education Key Laboratory of Analysis and Detection for Food Safety (Fuzhou University), Fuzhou, Fujian 350108, China.*

iii. Determination of Uric Acid in Cereal and Cereal Products

(AOAC Method No.970.24 can be used with applicable levels of more than or equal to 4 mg/100g.):

The method is based on the precipitation of proteins and treatment of protein free filtrate with uric acid and sodium cyanide and measuring the resultant blue colour colorimetrically.

Apparatus

(i) Photo electric colorimeter/spectrophotometer

(ii) Volumetric flask– 50 ml capacity.

Reagents

(i) Sodium Tungstate solution- 10 per cent (w/v).

(ii) Standard Sulphuric Acid solution–0.667N.

(iii) Benedicts Uric acid reagent (Dissolve 100 gm of pure SodiumTungstate in 600 ml water. Add5 gm of Arsenic acid (As_2O_3) followed by 25 ml of 85 per cent phosphoric acid and 20 ml of conc HCl. Boil the mixture for 20 minutes, cool and make volume upto 1 liter).

(iv) Sodium Cyanide solution -5 per cent containing 2 ml of ammonia per liter. This solution requires to be prepared fresh after about six weeks.

(v) Standard Uric acid solution (Benedicts) stock solution (Dissolve 9 gm of Sodium dihydrogen phosphate in about 200-300 ml water. If the solution is not clear, filter and make upto 500 ml with hot water. Weigh 200 mg of pure uric acid in1 litre volumetric flask and add a few mls of water to suspend the uric acid. Now add the solution made earlier and shake till the uric acid dissolves completely. Cool, add 1.4 ml of glacial acetic acid, dilute to mark and mix. Add 5 ml chloroform to prevent bacterial growth.**5 ml of stock solution contains 1 mg uric acid**).

(vi) Working Standard uric acid solution (Dilute 50 ml of stock solution containing 10 mg of uric acid with 400 ml distilled water in a 500 ml volumetric flask. Add 25 ml dilute HCl (1+ 9). Make the solution upto mark and mix.)The working solution should be prepared from stock solution which is more than 10 days old.

Procedure

Weigh 50 gm sample and grind it finely. Take between 4-20 gm powder expected to contain 1to5 mg uric acid and suspend in 200 ml water. Allow the mixture to stand for 2 hours and then mix in a Waring blender for 10 minutes and centrifuge at about 2000 r.p.m for 10 minutes. To 100 ml of clear centrifugate add 10 ml Sodium tungstate solution and mix. Then add 10 ml standard sulphuric acid solution to precipitate the proteins present in the extract. Mix and allow to stand for 5 minutes and filter. Take an aliquot of the filtrate containing between 0.15-0.3 mg uric acid per 10 ml filtrate in the 50 ml volumetric flask and add 5 ml of sodium cyanide solution followed by

1 ml of Benedicts uric acid reagent. Shake gently and make upto mark with distilled water. Determine the intensity of the colour in a potentiometric colorimeter using 520 nm filter. The optical density obtained can be recorded as OD1. Take 10 ml of standard uric acid solution containing 0.2 mg of uric acid in a 50ml flask, add5 ml of sodium cyanide followed by1 ml of Benedicts uric acid reagent. Dilute to mark after 5 minutes and determine the intensity of colour in a photo electric colorimeter using a 520 nm filter. A parallel test using the same quantity of good un infested sample as the sample under test should be run as a control.

Calculation

$$\frac{(\text{OD1} - \text{OD2}) \times 10 \times 2}{\text{Weight of sample in g}}$$

Source

1. *I.S 4333 (Part 5) 1970–Methods of Analysis for Food grains Part 5-Determination of Uric acid/ISI Hand book of Food Analysis, Part 4, page 84.*
2. *Manual of Methods of analysis of Foods (Cereals and Cereal products), Lab Manual-3, FSSAI, New Delhi.*

iv. Determination of Synthetic Colour in Biscuits, Cake *etc.*

Apparatus

(*i*) Glass pestle and mortar.
(*ii*) Beakers 100 and 250 ml capacity.
(*iii*) Chromatographic Chamber 30 cm X 20cmX10 cm.
(*iv*) Test tubes
(*v*) Spectrophotometer.
(*vi*) Water bath
(*vii*) Porcelain dish

Reagents

(*i*) 2 per cent ammonia in 70 per cent alcohol.
(*ii*) 100 per cent pure white knitting wool (Boil in1 per cent sodium hydroxide solution and then in water to remove alkali. Wash repeatedly with distilled water and dry).
(*iii*) Chromatographic paper Whatman No 1.
(*iv*) Hydrochloric acid– 0.1N–5 ml of conc. HCl diluted to1 litre with water.

Procedure

Thoroughly grind 10 gm of powdered material with 50 ml of 2 per cent ammonia in 70 per cent alcohol. Allow to stand for few hours and centrifuge. Pour the clear supernatant liquid in the dish and evaporate on the water bath. Dissolve the residue in 30 ml water acidified with acetic acid. Add a 20 cm strip of pure white wool to the solution and boil. When the wool takes up the colour fairly completely, take out and wash with water. Transfer the washed wool to a small beaker and boil gently with dilute ammonia (1+4). If the colour is stripped, the presence of an acid dye is indicated. Remove the wool. Make the liquid slightly acidic and add a fresh piece of wool and boil until the colour is removed. Extract the dye from the wool again with a small volume of dilute ammonia. This double stripping technique usually gives a pure colour. Natural colour may also dye the wool during the first treatment but the colour is not removed by ammonia. Transfer the solution to a volumetric flask and make the volume to 50 ml. with water.

Note

1. Basic dyes can be separated by making the food alkaline with ammonia, boiling with wool and then stripping with dilute acetic acid. All the present Permitted water soluble colours are acidic and the presence of a basic dye would indicate presence of non permitted dye.
2. The method given is sensitive only beyond 20-25 ppm. Pearsons chemical analysis of foods is recommended to be followed.

Separation of Colours by Paper Chromatography

Take a Whatman No 1 filter paper sheet (15 cm x 30 cm) and draw a line parallel to the bottom edge of the sheet about 2 cm away from it. Pipette 0.5 ml of extracted dye with the help of a graduated pipette and apply it on the filter paper in the form of a band on the line. Prepare 0.1 per cent solutions of permitted dyes and with the help of a capillary tube, apply spots of all these dyes on the line leaving about 1.5 cm distance between two spots. Allow the coloured spots to dry and subsequently suspend the paper sheet in the chromatographic chamber such that the lower edge of the sheet remains dipped in the solvent placed in the chamber. The following solvent systems may be used for separation of colours. Solvent 5 has been found to give good resolution.

Apparatus

(i) 1 per cent ammonia-1 ml ammonia (sp gr 0. 88)+99 ml water.

(ii) 2.5 per cent Sodium chloride.

(iii) 2 per cent Sodium Chloride in 50 per cent alcohol.

(iv) Isobutanol: Etahnol: Water (1: 2: 1 (v/v)).

(v) n-Butanol: Water: Acetic acid (20: 12: 5).

(vi) Isobutanol: Ethanol: Acetic acid (3: 12: 5).

Close the chromatographic chamber tightly and let the solvent rise. When the solvent front has reached about 20 cm from the base line, remove the filter paper sheet and allow it to dry. Mark coloured bands and carefully cut the cloured strips from the paper. Cut the coloured strips into small pieces and transfer to a test tube and add about 1 ml 0.1 N HCl. Allow the colour to extract and decant the coloured extract into a volumetric flask. Repeat the process of extraction and decanting till all the colour is removed from the paper. Make upto volume. Determine absorbance maxima and read the optical density at absorbance maximum against a blank prepared by cutting an equivalent strip plain portion of chromatogram and extracting it with 0.1 N HCl. From the absorbance values compute the concentration of the dye by reference to the plot of concentration versus optical density.

Plotting of Standard Curve

Prepare 0.1 per cent solution of the dye in 0.1 N HCl. Take 0.25, 0.50, 0.75,1.0, 1.25 and 1.5 ml aliquot of this and dilute to 100 ml with 0.1 N HCl. Read their absorbance at respective absorbance maxima. Plot absorbance values against concentration of the dye. The Rf values vary slightly owing to variation of temperature, solvent purity and solvent saturation of the Chromatographic chamber. It is thus essential that known dyes are applied with the sample as control.

Source

1. *I.S 12711: 1989 Bakery Products – Methods of Analysis.*
2. *Manual of Methods of analysis of Foods (Cereals and Cereal products), Lab Manual-3, FSSAI, New Delhi.*

v. Determination of Non-Fat Milk Solids in Milk Bread (Colorimetric/ Spectrophotometric)

The method is a colorimetric one for estimation of non fat milk solids in bread based on the orotic acid (2, 6 dihydroxy pyrimidine- 4 carboxylic acid) content. The mean orotic acid content of non fat milk solids is 62.5 mg/100 gm (range 48.0 – 74.5 mg/100 gm).

Apparatus

(i) Air oven.
(ii) Homogeniser.
(iii) Pipettes 5, 10 and 25 ml
(iv) Glass stoppered test tubes.
(v) Volumetric flasks 10, 50, 100, 500 ml capacity.
(vi) Water bath
(vii) Colorimeter.

Reagents

(i) Zinc sulphate 23 per cent (w/v).
(ii) Potassium Hexacyanoferrate 15.0 per cent (w/v).
(iii) p-Dimethyl amino benzaldehyde in propanol 3 per cent (w/v).
(iv) Standard Orotic acid -Dissolve 50 mg orotic acid in a mixture of 1ml of 0.88 ammonia and 10 ml water. Dilute to 500 ml with water. Take 10 ml aliquot and dilute to100 ml with water. Further dilute 2.5, 5, 10, and 15 ml of this solution to 50 ml to produce solutions containing 2.5, 5, 10,15µg of orotic acid per 5 ml.

(v) Saturated Bromine water.

(vi) Ascorbic acid solution 10 per cent (g)

(vii) n- Butyl acetate.

(viii) Anhydrous Sodium Sulphate.

Procedure

Weigh 5 gm of dried sample obtained after determination of moisture, transfer to the homogeniser, add 100 ml water and mix at the maximum speed for 1 minute. Filter the supernatant liquid through a 15 cm Whatman filter paper No 541, rejecting the first 10 ml. Only 5 ml is required for the determination. Into a series of glass stoppered tubes, add by pipette 5 ml of test solution (containing 2-15 µg orotic acid), 5 ml of each of the standard orotic acid solutions and 5 ml of water to act as a blank. Add to each tube 1.5 ml of saturated bromine water and allow the mixture to stand at room temperature for not more than 5 minutes.

As the addition of bromine water is made to the series of tubes, the times will vary slightly between each, the time of reaction is not critical provided it is between 1 and 5 minute. Add 2 ml of 10 per cent Ascorbic acid solution to each tube and place the tubes in a water bath at 40°C for 5 minutes. Cool to room temperature, add to each tube 4 ml n-butyl acetate and shake vigorously for 15 seconds. Transfer the upper separated layers to dry test tubes containing 1 gm anhydrous Sodium sulphate. Mix gently.

Add another gram of anhydrous Sodium sulphate. Mix gently and allow to separate. Transfer the clear butyl acetate layer to 1 cm cell and measure the optical density at 461 - 462 nm against the blank.

Calculation

Draw a calibration graph of the standard orotic acid solution plotting the optical density on the X-axis against the concentration of orotic acid on the Y–axis. Determine the orotic acid content in 5 ml of sample extract by interpolation of the colorimetric reading on the calibration graph and hence the amount in the dry sample. For converting to milk assume that skim milk powder contains 62.5mg orotic acid per 100 gm.

Source

1. *IS: 12711: 1989 Bakery products–Methods of Analysis/Pearson Composition and Analysis of Foods 9th edn, page 316.*
2. *Manual of Methods of analysis of Foods (Cereals and Cereal products), Lab Manual-3, FSSAI, New Delhi.*

vi. Determination of Caffeine Content-Chromatographic-Spectrophotometric

Method (Applicable to roasted, soluble (instant) and decaffeinated coffee):

Reagents

(i) Ammonium hydroxide solution-1: 2 (V/V).

(ii) Sulphuric acid -4N.

(iii) Diethyl ether-Water Saturated.

(iv) Chloroform -Water Saturated.

(v) Celite 545.

(vi) Caffeine standard solution -10, 20, 30 ug/ml in Chloroform-Accurately weigh 100 mg of caffeine (USP, anhydrous) into 100 ml volumetric flask, dissolve in CHCl3 and make upto volume. Dilute 10 ml aliquot to 100 ml with chloroform. Further dilute 10, 20, and 15ml aliquots to 100, 100 and 50 ml respectively with chloroform to obtain standard solutions of 10, 20, and 30 µg/ml.

Apparatus

(i) Glass column-25x250mm size.

(ii) Recording UV-VIS Spectro-photometer-To record 250-350 mm range with matched1cm cells.

Preparation of Sample

(a) **For Roasted Coffee**: Accurately weigh about1gm ground sample and transfer to100 ml beaker. Add 5ml NH_4OH (1+2) and warm on boiling water-bath for 2 minutes. Cool, transfer to100ml volumetric flask and make upto volume with water. To 5ml aliquot of the turbid solution add 6gm celite545 and mix carefully.

(b) **For soluble Coffee:** Proceed as in (a) taking 0.5g sample and an aliquot of 3ml.

(c) **For decaffeinated roasted coffee:** Accurately weigh1gm of ground sample. Transfer to100 ml beaker, add 5ml NH_4OH (1+2) and warm on boiling waterbath for 2 minutes. Add 6gm celite545 and mix carefully.

(d) **For decaffeinated soluble coffee**: Proceed as in (c) taking 0.5gm sample.

Preparation of Columns

(a) **Acid column:** Place fine glass wool and plugin to the base of 25x 250mm column. Add 3ml 4N H_2SO_4 to 3gm celite and mix well by kneading with spatula. Transfer into the tube and tampusing gentle pressure and place small glass wool above the surface.

(b) **Basic Column: Layer I-**Mix 3 gm celite545 and 2ml 2N NaOH and place in 25x250 mm tube. Transfer over glass wool plug as in (a) into the column. LayerII-Transfer sample plus celite mixture in about 2g portions to tube directly overlayer I, tamping before adding mixture portion of sample until homogenous and compact layer is obtained. Dry wash beaker with about 1gm celite 545, transfer to tube and tamp to uniform mass. Dry wash beaker with wad of glasswool and transfer to top of basic column.

Determination

Mount basic column above acid column. Pass 150 ml water saturated ether sequentially through basic column to acid column and discard ether. Then pass 50ml water saturated ether through acid column and discard ether. Place 50 ml volumetric flask under acid column. Pass 48 ml, water saturated $CHCl_3$ through acid column washing tip of basic column with first portions. Dilute contents of volumetric flask (100ml) to volume with water saturated chloroform, mix, and read O.D. at 275 against water saturated chloroform $CHCl_3$ blank, by scanning from 350 to 250 nm. Determine O.D of standards and use this value to calculate the caffeine percentage.

Source

1. *A.O. A.C 17thedn,2000 Official Method 979.11 Caffeine in Roasted Coffee, Chromatographic-Spectrophotometer method.*
2. *Manual of Methods of analysis of Foods (Beverages - Coffee, Tea, Sugar, Sugar Products etc.), Lab Manual-4, FSSAI, New Delhi.*

vii. Determination of Titanium Di-oxide in Chewing Gum and Bubble Gum

Reagents

(i) Titanium di-oxide standard solution: (0.1 mg/ml) -Weigh accurately 50 mg titanium di-oxide in a 250 ml beaker. Add 15 g anhydrous sodium sulphate and 50 ml conc. Sulphuric acid. Add 1 or 2 glass beads, cover with watch glass and heat on a hot plate to boil and dissolve. Cool and add 100 ml distilled water accurately with stirring. (If the solution is cloudy warm on a steam bath to clarify) Cool and transfer to 500-ml volumetric flask containing200-ml water. Make up to volume.

(ii) Sulphuric Acid -1: 9 volume.

(iii) Hydrogen peroxide- 30 per cent grade

(iv) Sodium Sulphate -Analar grade.

Procedure

(i) Accurately weigh 2-3 grams of prepared sample in a 100-ml silica dish.

(ii) Char the material on a burner and ash in muffle furnace at 800°C for 3-4 hours.

(iii) Cool, add 2-g anhydrous sodium sulphate and 10 ml conc. H_2SO_4, cover with watch glass and bring to boiling on a hot plate and dissolve.

(iv) Cool thoroughly and rinse the watch glass carefully with 30-ml water. Transfer to 100-ml volumetric flask. (If solution is cloudy heat on steam bath to clarify) cool and dilute to volume with water.

(v) Transfer 3-ml aliquot of the sample solution to 5 ml volumetric flask or graduated cylinder. Dilute to volume with 10 per cent H_2SO_4. Add 0.2 ml 30 per cent H_2SO_4. Mix well.

(vi) Measure the absorbance at 408 nm against a prepared blank. Determine the concentration of TiO_2 in sample using a standard curve.

Preparation of Standard Curve

Transfer 0, 1, 2,3 and 4 ml of TiO2 standard solution (0.1 mg/ml) to 5-ml volumetric flask or graduated cylinder. Dilute to volume with 1+9 H_2SO_4. Mix well. Measure the absorbance of the colour at 408 nm in a spectrophotometer and prepare a standard curve.

Calculation:

$$\text{TiO2 (in g per cent)} = \frac{\text{Dilution x mg TiO2 x Dilution x Absorbance of sample x 100}}{\text{Wt. of sample x Dilution x Absorbance of standard x 1000}}$$

Source

1. *A.O. A.C 17 th edn, 2000 Official Method 973.38 Titanium Dioxide in Cheese.*
2. *Manual of Methods of analysis of Foods (Beverages - Coffee, Tea, Sugar, Sugar Products etc.), Lab Manual-4, FSSAI, New Delhi.*

viii. Spectrophotometric Determination of Higher Alcohols in Alcoholic Beverages (Rum, Gin, Whisky, Brandy, Beer, Vodka, Wine, Toddy and Fenny-Cashew and Coconut)

Apparatus

(i) Spectrophotometer, double beam

(ii) Steam bath.

(iii) Test tube, stoppered, 15 ml capacity.

Reagents

(i) p-dimethylaminobenzaldehyde solution – Dissolve 1 g in a mixture of 5ml sulphuric acid and 90 ml distilled water and transfer to a 100ml volumetric flask and make up to the mark.

(ii) Iso-butyl alcohol, GR grade.

(iii) Iso-amyl alcohol, GR grade.

(iv) Ethyl alcohol, redistilled, middle 50 per cent fraction.

Preparation of Synthetic Standard of Higher Alcohols

Weigh 2 g isobutyl alcohol and 8 g iso-amyl alcohol into 1 L volumetric flask and dilute to mark with water. Pipette two 10ml portions into 100ml volumetric flasks and dilute to mark, one with water and other with ethyl alcohol. Prepare working standards for products in the range of 1.0 to 6.0 g synthetic higher alcohol per 100L by diluting 1.0 to 6.0 ml aliquots of alcohol standards solution to 100 ml with alcohol solution. (Solution containing 6 ml synthetic standard would give an absorbance of 0.83±0.03 at 530 nm).

Procedure

It involves preparation of sample and determination.

(*a*) **Preparation of sample**: i-Transfer 200 ml of alcoholic drink into a 500 ml distillation flask containing about 25 ml of distilled water and a few pieces of pumice stone. Distil the contents in about 35 min and collect the distillate in a 200 ml volumetric flask till the volume almost reaches the mark. Bring the distillate to room temperature and make up to volume with distilled water and mix thoroughly and ii-For samples containing 6 g fusel oil per 100 L, dilute the distilled sample with distilled water to concentrations of 2.0 to 5.0 g/100L.

(*b*) **Determination**: Pipette 2 ml of aliquot of sample (or diluted sample), 2 ml of distilled water (for reagent blank) and 2 ml of synthetic standard to each of the test tubes (15mm x 150mm-with stoppers). Stopper and place it in ice-bath in a rack. Pipette 1 ml p-dimethylaminobenzaldehyde solution into each tube; shake and replace in ice-bath for 3 min. With tubes retained in ice-bath, add 10 ml sulphuric acid and shake the tubes and replace in ice-bath for 3 min. Transfer the rack containing tubes into steam bath for 3 to 5 min. and bring it to room temperature. Read the per cent T or Absorbance (OD) of developed colour of samples and series of standards in spectrophotometer at 530/535 nm against reagent blank as reference. Plot higher alcohol g/100 L Concentrations of Standards Vs. per cent T or OD. From the OD of the sample find out the concentration of Higher alcohol g/100L using the standard curve.

Source

1. *Manual of Methods of analysis of Foods (Alcoholic Beverages), Lab Manual-16, FSSAI, New Delhi.*

ix. Spectrophotometric Determination of Methyl Alcohol in Alcoholic Beverages

Apparatus

(*i*) Separating funnel.

(*ii*) Spectrophometer.

Reagents

(*i*) Potassium permanganate solution-3.0 g $KMnO_4$ and 15.0 ml H_3PO_4 shall be dissolved in 100ml water. The solution shall be prepared monthly.

(*ii*) Sodium salt of chromotropic acid (sodium 1,8- dihydroxynaphthalene - 3,6 disulfonate) 5 per cent aqueous solution (w/v). If not clear, the sodium salt of chromotropic acid shall be filtered. It shall be prepared weekly.

Purification of Chromotropic Acid

If absorbance of blank is greater than 0.05, the reagent shall be purified as follows: 10 g chromotropic acid or its Na salt shall be dissolved in 25 ml water (2 ml H_2SO_4 shall be added to the aqueous solution of the salt to convert it to free acid).

Add 50 ml of methanol and heat to just boiling and filter. Add 100 ml isopropyl alcohol to precipitate free chromotropic acid. More isopropyl alcohol may be added to increase yield of purified acid.

Methanol Stock Solution

Dilute 1.0 g methanol (99.99 per cent pure) to 100 ml with 40 per cent (v/v) ethanol, methanol free. Dilute to 10 ml of this solution to 100 ml with 40 per cent ethanol.

Methanol Standard Solution

Dilute appropriate volume of methanol (12.1.4) to 100 ml vol. flasks with40 per cent ethanol to get final concentration of 20, 40, 60, 80 and 100 ppm of methanol.

Procedure

Take 50 ml of sample in a simple still and distil, collecting about 40ml of distillate.Dilute 1 ml of distillate to 5ml with distilled water and shake well. Take 1 ml of this solution, 1 ml of distilled water (for blank) and 1 ml of each of the methanol standards in to 50 ml stoppered test tubes and keep them in an ice-cold water bath. Add to each test tube, 2 ml of $KMnO_4$reagent and keep aside for 30 min. Decolourize the solution by adding a little sodium bisulphite and add 1 ml of chromotropic acid solution. Mix well and add 15ml of sulphuric acid slowly with swirling and place in hot water bath maintaining 80°C for 20 min. Observe the colour development from violet to red. Cool the mixture and measure the absorbance at 575 nm using 1cm cuvette cell.

Calculations

Calculate methanol content in g/100 Liters of absolute alcohol as follows:

$$\text{Methanol} = \frac{A2 \times C \times D \times 1000 \times 100 \times 100}{A1 \times S}$$

where- A2 = absorbance of sample solution, C = concentration of methanol std. solution, D = dilution factor for sample solution, A1 = absorbance of methanol std. solution and S = ethanol content of liquor sample in per cent (v/v).

Source

1. *Manual of Methods of analysis of Foods (Alcoholic Beverages), Lab Manual-16, FSSAI, New Delhi.*

X. Colorimetric/Spectrophotometric Determination of Nitrite in Meat

Extraction of a test portion in hot water, precipitation of the proteins and filtration. In the presence of nitrite development of a red colour by the addition of sulphanilamide and N - napthylethylenediamine dihydrochloride to the filtrate and photometric measurement at 538 nm

Reagents

I. Solutions for Precipitation of Proteins

(i) Dissolve 106 gm of Potassium ferrocyanide trihydrate in water and dilute to 1000 ml.

(ii) Dissolve 220 gm of Zinc acetate dihydrate and 30 ml glacial acetic acid in water and dilute to 1000 ml

(iii) Dissolve 50 gm of disodium tetraborate decahydrate in 1000 ml of tepid water and cool to room temperature.

II. Standard Sodium Nitrite Solution

Dissolve 1.000 gm pure sodium nitrite in water and dilute to 100 ml in a volumetric flask. Pipette 5 ml of the solution into a 1000 ml volumetric flask and make upto volume. Prepare a series of standard solutions by pipetting 5 ml, 10 and 20 ml of the solution into 100 ml volumetric flasks and diluting to mark with water. These standard solutions contain 2.5 µg, 5.0 µg, and 10 µg sodium nitrite respectively. The standard solutions and the 0.05 gm/1 solution from which they are prepared shall be made on the day of the use.

III. Solution for Colour Development

(i) Dissolve by heating on a water bath, 2 gm of sulphanilamide in 800 ml water. Cool, filter if necessary and add 100 ml of cone HC1 while stirring. Dilute to 1000 ml with water.

(ii) Dissolve 0.25gm of N – napthyl ethylenediamine dihydrochloride in water. Dilute to 250 ml with water Store in a stoppered brown bottle in a refrigerator for not more than one week.

(iii) Dilute 445 ml of Concentrated HCl (sp.gr 1.19) to 1000 ml with water.

Apparatus

(i) Meat mincer - fitted with a perforated plate with holes not greater than 4 mm in diameter.

(ii) Analytical Balance.

(iii) Volumetic flasks - 100 ml, 250 and 1000 ml.

(iv) Pipette 10 ml.

(v) Conical flask.

(vi) Boiling water bath

(vii) Fluted filter paper

(viii) Photoelectric colorimeter or spectrophotometer.

Procedure

Weigh to the nearest 0.00 l gm, about 10 gm of the test sample, transfer quantitatively to a 300 ml conical flask and add successively 5 ml of saturated borax solution and 100 ml water at a temperature not below 70°C Heat the flask for 15 minutes on the boiling water bath and shake repeatedly. Allow the flask and its contents to cool to room temperature and add successively 2 ml of Pot ferrocyanide followed by 2 ml of zinc acetate. Mix thoroughly after each addition.

Transfer the contents to a 200 ml volumetric flask. Dilute to mark with water and mix. Allow the flask to stand for 30 minutes at room temperature. Carefully decant the supernatant liquid and filter it through fluted filter paper to obtain clear solution.

Colour Development

Pipette an aliquot of the filtrate (v ml) not more than 25 ml into a 100 ml volumetric flask and add water to make upto 60 ml. Add 10 ml of sulphanilamide solution followed by 6 ml of cone HC1 and leave the solution in the dark for 5 minutes. Add 2 ml of N -Napthylethylenediamine solution and leave for 5- 10 minutes in the dark. Dilute to mark with water. Measure the absorbance of the solution in a 1 cm cell using a photoelectric colorimeter or spectrophotometer at a wave length of about 538 nm.

Preparation of a Calibration Curve

Prepare a calibration curve by taking 10ml water in 4 separate volumetric flasks, adding 10 ml each of the standard sodium nitrite solution containing 2.5, 5.0 and 10 µg of nitrite/ml, developing the colour and measuring as above.

Calculation

It is calculated by using the following formula:

$$\text{Nitrite content expressed as } NaNO_2 = \frac{c \times 2000}{M \times V}$$

where, V - volume in ml of aliquot portion of filtrate taken for test, M - mass in gm of sample taken and c - Concentration of sodium nitrite in µg/ml read from the calibration curve that corresponds with the absorbance of the solution prepared from the sample.

Source

1. *I.S 5960 (Part VII): 1996/I.S.O 2918: 1975 Meat and Meat Products Methods of Test - Determination of Nitrite content.*
2. *Manual of Methods of Analysis of Foods- Meat Products and Fish Products (Lab Manual-6), FSSAI, New Delhi.*

2. IR (Infrared) Spectroscopy

Infrared spectroscopy (IR spectroscopy or Vibrational spectroscopy) deals with the infrared **region of the** electromagnetic spectrum, that is light with a longer wavelength and lower frequency than visible light [1]. Infrared spectroscopy involves the interaction of infrared radiation with matter and it is a technique based on the vibrations of the atoms of a molecule. An infrared spectrum is commonly obtained by passing infrared radiation through a sample and determining what fraction of the incident radiation is absorbed at a particular energy. The energy at which any peak in an absorption spectrum appears corresponds to the frequency of a vibration of a part of a sample molecule [1a]. It covers a range of techniques, mostly based on absorption spectroscopy. The method or technique of infrared spectroscopy is conducted with an instrument called an **infrared spectrometer** (or spectrophotometer) to produce an **infrared spectrum**. An IR spectrum is essentially a graph of infrared light absorbance (or transmittance) on the vertical axis vs. frequency or wavelength on the horizontal axis. Typical units of frequency used in IR spectra are reciprocal centimeters (sometimes called wave numbers), with the symbol cm^{-1}. Units of IR wavelength are commonly given in micrometers (formerly called "microns"), symbol μm, which are related to wave numbers in a reciprocal way [2].

Infrared Spectroscopy is the most important and widely used rapid quantitative analytical tool and it is certainly one of the most important analytical techniques available to scientists and Industries (dairy/food industries *etc.*). The technique is based on the absorption of radiation by the sample and is characterized by the wavelength of the light used, commonly mid infrared (MIR) and near infrared (NIR) [3]. Infrared energy is the electromagnetic energy of molecular vibration and it is characterised by the wave length of the light used. It covers a range of techniques, mostly based on absorption spectroscopy. As with all spectroscopic techniques, it can be used to identify and study chemicals. Samples may be solid, liquid, or gas [2]. IR radiation **does not have enough energy to induce electronic transitions as seen with UV. Absorption of IR is restricted to compounds with small energy differences in the possible vibrational and rotational states [4].** The advantages of the technique include the following:

- *(i)* The method is inexpensive, fast, reliable, accurate and environment friendly.
- *(ii)* Provides for simultaneous determination of various major milk constituents [3].
- *(iii)* One of the great advantages of infrared spectroscopy is that virtually any sample in virtually any state may be studied [1a].

Source

1. *Vogel's Text Book on Quantitative chemical analysis revised by G.H.Jeffery, J. Bassett, J.Mendham and R.C.Denney.*

1a. *Barbara Stuart, Infrared Spectroscopy- Fundamentals and Applications, 2004 John Wiley and Sons, Ltd ISBNs: 0-470-85427-8 (HB); 0-470-85428-6 (PB).*

2. *Wikipedia.*

3. *Technews, No.74, May-June-2008, NDDB.*

4. *https://teaching.shu.ac.uk/hwb/chemistry/tutorials/irspec1.htm.*

History of IR Spectroscopy [1,2]

(i) The discovery of the infrared region in 1800 is credited to William F. Hershel's famous work, "Experiments on the Refrangibility of the Invisible Rays of the Sun", read April 24, 1800 at the Royal Society [1].

(ii) In 1880, Langley introduced bolometer.

(iii) In 1905, first IR spectra were published by Coblentz.

(iv) In 1937, E. Lehrer first fully-automated spectrometer.

(v) In1940, developments in USA were made.

(vi) In 1950, well-resolved spectra were developed.

(vii) In 1946, first spectra catalogue was made.

(viii) In 1960s, introduction of Fourier Transform technique, first use of Michelson interferometer, first commercial spectrometer.

(ix) After 1965, technique found much useful. Infrared spectrometers have been commercially available since the 1940s [3].

Source

1. *Phil. Transact. Roy. Soc. 90, 284-292 [2].*

2. *Sylvia-Monique Thomas, Infrared and Raman Spectroscopy,* *University of Nevada Las Vegas.*

3. *Barbara Stuart, Infrared Spectroscopy- Fundamentals and Applications, 2004 John Wiley and Sons, Ltd ISBNs: 0-470-85427-8 (HB); 0-470-85428-6 (PB).*

Division of the Electromagnetic Spectrum [1,2,3,4]

The term "infra red" covers the range of the electromagnetic spectrum between 0.78 and 1000 μm. The most useful I.R. region lies between 4000- 670cm^{-1}.In the context of infra red spectroscopy, wavelength is measured in "wave numbers", which have the units cm^{-1}.

Wave number = 1/wavelength in centimeters

The infrared portion of the electromagnetic spectrum is usually divided into three regions:

1-the near-IR (NIR),2- mid-IR (MIR) and 3-far- infrared.

The higher energy **near-IR,** can excite overtone **or** harmonic **vibrations**. The **mid-infrared,** may be used to study the fundamental vibrations and associated rotational-vibrational **structure**. The **far-infrared** lying adjacent to the microwave region, has low energy and may be used for rotational spectroscopy. **The approximate wave number and Wave Length of aforesaid infrared portion are given in following table.** This classifications are only loosely based on the relative molecular or electromagnetic properties. **NIR and MIR Spectroscopy** are commonly used in dairy Industry.

Region	Energy(kJ/mol)	Wave Number (cm^{-1})	Wave Length (μm)
Near IR (Overtone region)	150-50	12800-4000/ 14000-400	0.78-2.5/0.8-2.5
Mid IR (Vibration-rotation region	50-2.5	4000-200/4000-400	2.5-50/2.5-25
Far IR (Rotation region)	2.5-0.1	200-10/400-10	50-1000/2.5-1000

Source

1. *Wikipedia.*
2. *https://teaching.shu.ac.uk/hwb/chemistry/tutorials/irspec1.ht.*
3. *Vogel's Text Book on Quantitative chemical analysis revised by G.H.Jeffery, J.Bassett, J.Mendham and R.C.Denney.*
4. *Sylvia-Monique Thomas, Infrared and Raman Spectroscopy, University of Nevada Las Vegas.*

Principle

The main region of interest for analytical purposes is from 2.5-25 μm (4000-400 wave numbers-waves per centimetre, cm-1 and $1cm^{-1}$= 10000/wavelength). Normal optical materials such as glass or quartz absorb strongly in the infrared, so instruments for carrying out measurement in this region differ from those used form electronic (UV/VIS) region.

A spectrophotometric method for the determination of composition and structure of compounds depending on the absorption of radiation in the **INFRARED** region of wavelengths,2-50μm (μm=10^{-6} m). It is preferable to express the region by frequency units.**IR mostly used for rapid qualitative but not quantitative analysis.**

Infrared spectra originate from the different modes of vibration and rotation of a molecule. At wavelengths below 25μm the radiation has sufficient energy to cause changes in the vibrational energy levels of molecule and these are accompanied by changes in rotational energy levels. For a molecule to absorb IR, the vibrations or rotations within a molecule must cause a net change in the dipole moment of the molecule. The alternating electrical field of the radiation (remember that electromagnetic radiation consists of an oscillating electrical field and an oscillating magnetic field, perpendicular to each other) interacts with fluctuations in the dipole moment of the molecule. If the frequency of the radiation matches the vibrational frequency of the molecule then radiation will be absorbed, causing a change in the amplitude of molecular vibration [1].

For a molecule to show infrared absorptions it must possess a specific feature, *i.e.* an electric dipole moment of the molecule must change during the vibration. This is the **selection rule** for infrared spectroscopy. **An 'infrared-inactive'** molecule is a **homonuclear diatomic molecule** because its dipole moment remains zero no matter how long the bond. Vibration of two similar atoms against each other, for example Oxygen or Nitrogen molecules, will not result in a change of electrical symmetry or dipole moment of the molecule and such molecules will not absorb in the IR Region. Molecules having no permanent dipole moments (O_2, H_2, Cl_2etc) are **IR inactive. An 'infrared-active'** molecule is a **heteronuclear diatomic molecule**. The dipole moment of such a molecule changes as the bond expands and contracts. Molecules with permanent dipole moments –μ (HCl, H_2O, NO) are **IR active**.

IR Active (Having Dipole moment)	*IR Inactive (No Dipole moment)*
δ+ δ-	O_2 - O-O
H – Cl	H_2 - H-H
→	Cl_2 - Cl-Cl

Molecular Rotations [1]

Rotational transitions are of little use. Rotational levels are quantized, and absorption of IR by gases yields line spectra. However, in liquids or solids, these lines broaden into a continuum due to molecular collisions and other interactions.

Molecular Vibrations [1,2]

The positions of atoms in a molecules are not fixed; they are subject to a number of different vibrations. The number and types of vibrational modes are given below:

A. Normal Modes of Vibration

The interactions of infrared radiation with matter may be understood in terms of changes in molecular dipoles associated with vibrations and rotations.

a. Number of Vibrational Modes

In order for a vibrational mode in a molecule to be "IR active," it must be associated with changes in the dipole. A permanent dipole is not necessary, as the rule requires only a change in dipole moment [3]. A molecule can vibrate in many ways, and each way is called a vibrational mode. A molecule of n atoms has 3 N degrees of freedom. In a non-linear molecule 3 of these are rotational and 3 of these degrees are translational, and the remaining belong to fundamental vibrations (normal modes). In a linear molecule 2 degrees are rotational and 3 degrees are translational [5]. The degree of freedom in linear and Nonlinear diatomic and Polyatomic molecules are summarised below [6,7]:

Sl.No.	*No. of Atoms in Molecules*	*Types of Degree of Freedom*	*Types of molecule*		*Vibration due to Stretching or Bending*
			Linear	*Nonlinear*	
1	Diatomic Molecule	Translational	-	3	-
		Rotational	-	2	-
		Vibrational	-	1	Corresponds to a stretching motion

Contd.

Sl.No.	No. of Atoms in Molecules	Types of Degree of Freedom	Types of molecule		Vibration due to Stretching or Bending
			Linear	Nonlinear	
2	Polyatomic Molecule	Translational	3	3	-
		Rotational	2	3	-
		Vibrational	3*N*–5	3N – 6	Triatomic Molecule – Nonlinear- two correspond to stretching motions, with the remainder corresponding to a bending motion. Linear - four modes, two of which have the same frequency, and are said to be degenerate.
		Total	3N	3N	-

The degrees of vibrational modes depends upon the linearity of the molecule for example:

Linear Molecules (CO_2 etc.)

For molecules with N atoms in them, linear molecules have 3N- 5 degrees of vibrational modes.

Nonlinear Molecules (H_2O etc.)

For molecules with N atoms in them, nonlinear molecules have 3N - 6 degrees of vibrational modes [2,5].

(i) **H_2O**, a non-linear molecule, will have 3 × 3 - 6 = 3 degrees of vibrational freedom, or modes [2,5]. Water has three degrees of translational freedom and three degrees of rotational freedom.

(ii) Simple diatomic molecules have only one bond and only one vibrational band. If the molecule is symmetrical, *e.g.* N_2, the band is not observed in the IR spectrum, but only in the Raman spectrum [2].

(iii) The **CO_2** molecule is an example for a linear molecule and has 4 fundamental vibrations (3 x 3 - 5) [5]. CO_2 has three degrees of translational freedom and two degrees of rotational freedom since no detectable energy is involved in rotation around the O=C=O axis [6].

(iv) Asymmetrical diatomic molecules, *e.g.* CO, absorb in the IR spectrum. More complex molecules have many bonds, and their vibrational spectra are correspondingly more complex, *i.e.* big molecules have many peaks in their IR spectra [2].

b. Types of Vibrations

Vibrations can involve either a change in bond length (stretching) or bond angle (bending). Vibrations fall into the following two main categories:

i. Stretching

Change in inter-atomic distance along bond axis or Change in bond length. Some bonds can stretch in-phase (symmetrical stretching) or out-of-phase (asymmetric stretching) as shown below:

Stretching Vibrations [6]

```
  ↔
  O        R- - - - -O- - -H
 /  \
H    H
Symmetric   Asymmetric
```

Stretching Vibrations [1]

Symmetric   Asymmetric

In covalent bonds, atoms are not joined by rigid links - the two atoms are held together because both nuclei are attracted to the same pair of electrons. The two nuclei can vibrate backwards and forwards - towards and away from each other - around an average position. The Following diagram shows the stretching that happens in a carbon-oxygen single bond [4]:

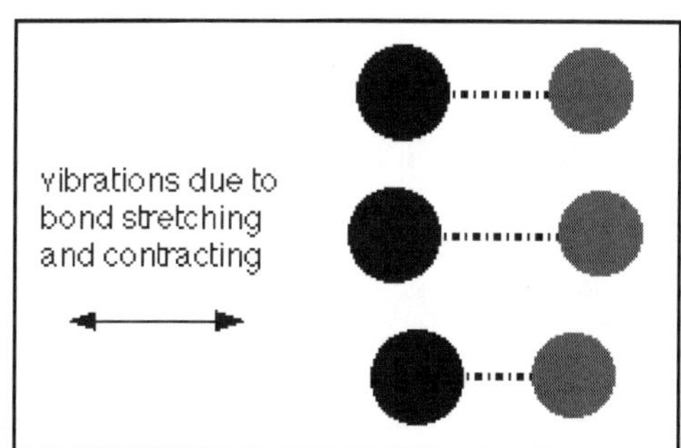

ii. Bending

The bonds can also bend. There is change in angle between two bonds, which is called bend. There are following four types of bend:

(a) Rocking

(b) Scissoring

(c) Wagging

(d) Twisting/Torsion.

Examples of Bending

(a) **In ν (C-O)**- i-methanol 1034 cm^{-1}, ii- ethanol 1053 cm^{-1} and iii- butanol 1105 cm^{-1}.

(b) The atoms in a CH_2X_2 group, commonly found in organic compounds and where X can represent any other atom, can vibrate in nine different ways.

Six of these involve only the CH_2 portion: symmetric and anti-symmetric stretching, scissoring, rocking, wagging and twisting, as shown below [2].

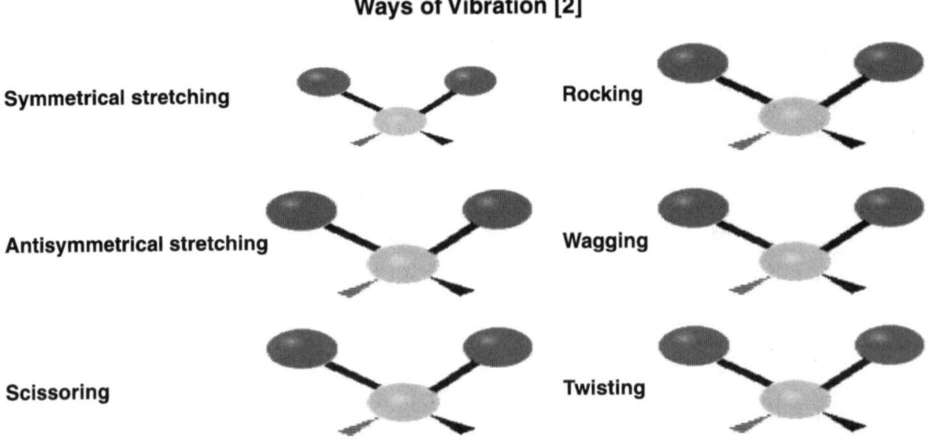

These figures do not represent the "recoil" of the C atoms, which, though necessarily present to balance the overall movements of the molecule, are much smaller than the movements of the lighter H atoms.

(c) Water Molecule. Because CH_2 is attached to X_2 hence CH_2X_2 has 6 modes, unlike H_2O, which only has 3 modes. The rocking, wagging, and twisting modes do not exist for H_2O, since they are rigid body translations and no relative displacements exist [2]. The following diagram shows the bending of the bonds in a water molecule. The effect of this, of course, is that the bond angle between the two hydrogen-oxygen bonds fluctuates slightly around its average value. Imagine a lab model of a water molecule where the atoms are joined together with springs [4].

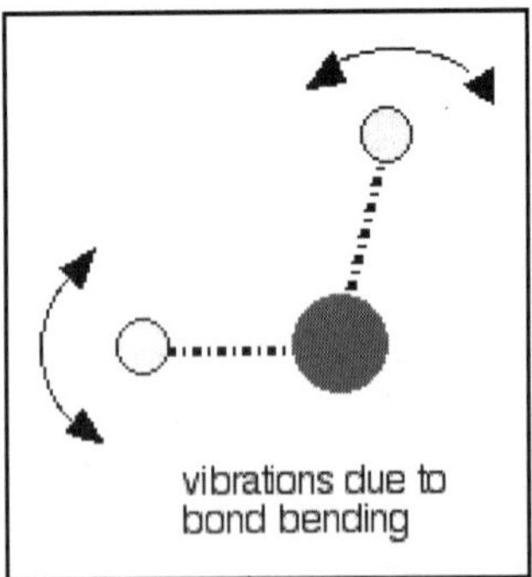

Infrared spectroscopy exploits the fact that molecules absorb specific frequencies that are characteristic of their structure. These absorptions are resonant frequencies, *i.e.* the frequency of the absorbed radiation matches the transition energy of the bond or group that vibrates. The energies are determined by the shape of the molecular potential energy surfaces, the masses of the atoms, and the associated vibronic coupling [2].

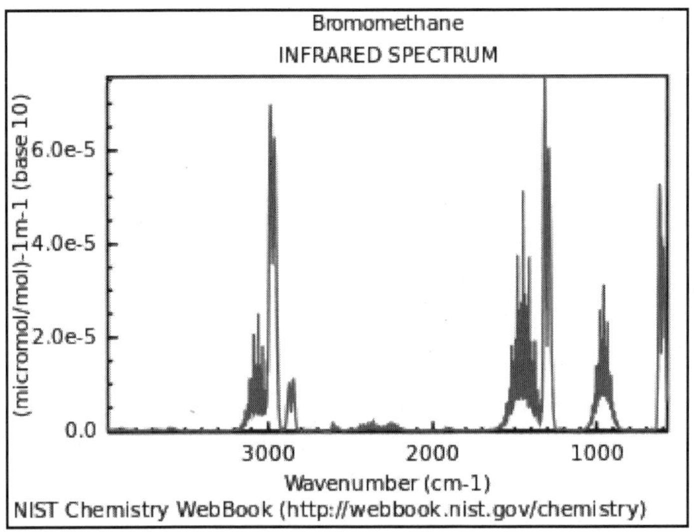

Sample of an IR Spec. Reading; this One is from Bromomethane, Showing Peaks around 3000,1300 and 1000 (On the X basis) [2].

B. Special Effects [1,2,5]

The most important and simplest Infra red bands arise from the simplest distortions of the molecule (**Normal modes**). There are a number of factors that may complicate the interpretation of infrared spectra. Some special effects are also observed and details of these special effects are as follows:

a. Vibrational Coupling [1,6]

In addition to the aforesaid vibrations, interaction between vibrations can occur **(coupling)** if the vibrating bonds are joined to a single, central atom and this ia called as **Vibrational coupling**. bands with slightly different positions can occur due to the network of molecules in a structure. The energy levels mix, hence resulting in the same number of vibrational modes, but at different frequencies, and bands can no longer be assigned to one bond. This is very common and occurs when adjacent bonds have similar frequencies. Coupling commonly occurs between C–C stretching, C–O stretching, C–N stretching, C–H rocking and C–H wagging motions. The Vibrational coupling is influenced by the following factors:

(i) Coupling is greatest when the coupled groups have approximately equal energies.

(ii) No coupling is seen between groups separated by two or more bonds.

(iii) Strong coupling of stretching vibrations occurs when there is a common atom between the two vibrating bonds.

(iv) Coupling of bending vibrations occurs when there is a common bond between vibrating groups.

(v) Coupling between a stretching vibration and a bending vibration occurs if the stretching bond is one side of an angle varied by bending vibration.

b. Overtone Bands [2,5,6]

In some cases, **"overtone bands"** arised from the absorption of a photon that leads to a doubly excited vibrational state, also observed. The Overtone bands in an IR spectrum are analogous and are multiples of the fundamental absorption frequency. Such bands appear at approximately twice the energy of the normal mode.Since the energy is proportional to the frequency absorbed and this is proportional to the wave number, the first overtone will appear in the spectrum at twice the wave number of the fundamental.

c. Combination Modes [2,5,6]

Some vibrations, so-called combination modes, involve more than one normal mode. The phenomenon of Fermi resonance can arise when two modes are similar in energy. Fermi resonance results in an unexpected shift in energy and intensity of the bands *etc.* When an overtone or a combination band has the same frequency as, or a similar frequency to, a fundamental, two bands appear, split either side of the expected value and are of about equal intensity. The effect is greatest when the frequencies match, but it is also present when there is a mismatch of a few tens of wave numbers. The two bands are referred to as a Fermi doublet. Sum or difference of two vibrations, they can be helpful to determine the speciation of a molecule in sample.

d. Degenerated Vibrations

Symmetrically equivalent vibrations having the same wavenumber, *e.g.*, CO_2 bending vibrations [5].

e. Vibration–Rotation Bands [6,8]

While analyzing the infrared spectra of gaseous hetero-nuclear molecules at high resolution, a series of closely spaced components are observed. This type of structure is due to the excitation of rotational motion during a vibrational transition and is referred to as an vibration-rotation spectrum [8]. The absorptions fall into groups (called branches) and are labeled (P, Q and R) according to the change in the rotational quantum number associated with the transition.

Source

1. *https://teaching.shu.ac.uk/hwb/chemistry/tutorials/irspec1.htm.*
2. *Wikipedia.*
3. *Paula, Peter Atkins, Julio de -2009.*
4. *https://www2. chemistry. msu.edu/faculty/reusch/virttxtjml/./infrared/infrared.htm.*

5. *Sylvia-Monique Thomas, Infrared and Raman Spectroscopy, University of Nevada Las Vegas.*

6. *Barbara Stuart, Infrared Spectroscopy- Fundamentals and Applications, 2004 John Wiley and Sons, Ltd ISBNs: 0-470-85427-8 (HB); 0-470-85428-6 (PB).*

7. *B.Stuart, Modern Infrared Spectroscopy, ACOL Series, Wiley, Chichester, UK, 1996.*

8. *Atkins, P. and de Paula, J., Physical Chemistry, 7th Edn, Oxford University Press, Oxford, UK,2002.*

Sample Preparation [1,2]

Infrared spectroscopy is a versatile experimental technique and it is relatively easy to obtain spectra from samples in solution or in the liquid, solid or gaseous states. The sampling procedure depends very much on the type of sample to be examined, for instance, whether it is a solid, liquid or gas. Liquids, solutions, pastes, powders, films, fibres, gases and surfaces can all be examined with a judicious choice of sampling technique.

i. Gaseous Samples

Gases have densities which are several orders of magnitude less than liquids, and hence pathlengths must be correspondingly greater. These samples require a sample cell with a long pathlength to compensate for the diluteness. The pathlength of the sample cell depends on the concentration of the compound to be measured.

(*a*) **Concentrations down to several hundred ppm-** A simple glass tube with length of 5 to 10 cm equipped with infrared windows at the both ends of the tube can be used for this range.

(*b*) **Concentrations well below ppm**-Sample gas having this range of concentrations can be measured with a **White's cell** in which the infrared light is guided with mirrors to travel through the gas. White's cells are available with optical pathlength starting from 0.5 m up to hundred meters.

ii. Liquid Samples

These can be sandwiched between two plates of a salt (sodium chloride, or common salt, although a other salts such as potassium bromide or calcium fluoride are also used) [2]. The plates are transparent to the infrared light and do not introduce any lines onto the spectra.

iii. Solid Samples

These can be prepared in following ways:

a. Common Method

Sample is to crush with an oily mulling agent (usually Nujol) in a marble or agate mortar, with a pestle. A thin film of the mull is smeared onto salt plates and measured.

b. Second Method

In this method, grind a quantity of the sample with a specially purified salt (usually potassium bromide) finely (to remove scattering effects from large crystals). This powder mixture is then pressed in a mechanical press to form a translucent pellet through which the beam of the spectrometer can pass [2].

c. Cast Film Method/Technique

It is used mainly for polymeric materials. The sample is first dissolved in a suitable, non hygroscopic solvent. A drop of this solution is deposited on surface of KBr or NaCl cell. The solution is then evaporated to dryness and the film formed on the cell is analysed directly. This technique is suitable for qualitative analysis.

d. The Final Method

It is to use **microtomy** to cut a thin (20-100 μm) film from a solid sample. This is one of the most important ways of analysing failed plastic products for example because the integrity of the solid is preserved.

e. Other Method

In photoacoustic spectroscopy, the minimal sample treatment is needed. The sample, liquid or solid, is placed into the sample cup which is inserted into the photoacoustic cell which is then sealed for the measurement. The sample may be one solid piece, powder or basically in any form for the measurement. For example, a piece of rock can be inserted into the sample cup and the spectrum measured from it.

The spectra obtained from different sample preparation methods will look slightly different from each other due to differences in the samples' physical states.

f. Attenuated Total Reflectance Spectroscopy [1]

A useful way of analysing solid samples without the need for cutting samples uses ATR or **attenuated total reflectance** spectroscopy. Using this approach, samples are pressed against the face of a single crystal. The infrared radiation passes through the crystal and only interacts with the sample at the interface between the two materials.

With increasing technology in computer filtering and manipulation of the results, samples in solution can now be measured accurately (water produces a broad absorbance across the range of interest, and thus renders the spectra unreadable without this computer treatment).Some instruments will also automatically tell you what substance is being measured from a store of thousands of reference spectra held in storage.

Source

1. *Wikipedia.*
2. *Laurence M. Harwood; Christopher J. Moody (1989). Experimental organic chemistry: Principles and Practice (Illustrated ed.). Wiley-Blackwell. p. 292. ISBN 0-632-02017-2.*

Method for Analysis by IR Spectroscopy (1,2)

The dispersive element in dispersive instruments is contained within a monochromator. Dispersion occurs when energy falling on the entrance slit is collimated onto the dispersive element and the dispersed radiation is then reflected back to the exit slit, beyond which lies the detector. The dispersed spectrum is scanned across the exit slit by rotating a suitable component within the monochromator [2].

The infrared spectrum of a sample is recorded by passing a beam of infrared light through the sample. When the frequency of the IR is the same as the vibrational frequency of a bond, absorption occurs. Examination of the transmitted light reveals how much energy was absorbed at each frequency (or wavelength). This can be achieved by scanning the wavelength range using a **monochromator**. Alternatively, the whole wavelength range is measured at once using a **Fourier transform** instrument and then a transmittance or **absorbance** spectrum is generated using a dedicated procedure. Analysis of the position, shape and intensity of peaks in this spectrum reveals details about the molecular structure of the sample [1].

This technique works almost exclusively on samples with covalent bonds. Simple spectra are obtained from samples with few IR active bonds and high levels of purity. More complex molecular structures lead to more absorption bands and more complex spectra. The technique has been used for the characterization of very complex mixtures [1].

Source

1. *Wikipedia.*
2. *Barbara Stuart, Infrared Spectroscopy- Fundamentals and Applications, 2004 John Wiley and Sons, Ltd ISBNs: 0-470-85427-8 (HB); 0-470-85428-6 (PB).*

Instrumentation [1-5]

There are differences between the instruments used for UV/Vis Spectrophotometry and those designed form Infrared spectrophotometry because of the facts that glass and quartz absorb strongly in IR region and photomultipliers are insensitive to the radiation. Recent developments include a miniature IR-spectrometer that's linked to a cloud based database and suitable for personal everyday use [1a], and NIR-spectroscopic chips [1b] that can be embedded in smartphones and various gadgets. There are following types of spectrometers:

i. Non-dispersive Spectrometer

(No variable wavelength selection possible).

ii. Dispersive Instruments

(Variable wavelength selection using gratings or series of filters)-Traditionally, dispersive instruments, available since the 1940s, were used to obtain infrared spectra. The first dispersive infrared instruments employed prisms made of materials such as sodium chloride [5]. Dispersive spectrometers consist of a monochromator splitting radiation into its parts, which then separately exit to the detector by turning the grating or mirrors. With the Echelle grating earlier problems of low light intensity were overcome by using specially formed grooves. Traditional IR Spectrophotometer were constructed with monochromation being carried out using sodium chloride or potassium bromide prisms but in modern IR Spectrophotometer diffraction grattings have replaced prisms. As means of monochromation, prisms have disadvantage because they are hygroscopic and MIR range normally necessitated the use of two different prisms in order to get adequate dispersion over the whole range. Grattings provide higher resolving power than do prisms and can be designed to operate effectively over a wider spectral range. Now two grattings are also being used with a automatic change of grattings. **Dispersive Grating IR Instruments** is similar to UV-Vis spectro-photometer, but sample after source and before monochromator in IR (sample after monochromator in UV-Vis - less incident light).Double beam is used and useful to eliminate atmospheric gas interference [2,3].

iii. Fourier-transform Infrared Spectrometers

(Spectral splitting via interfero-meter allows wavelength-dependent radiation modulation)- In recent decades, a very different method of obtaining an infrared spectrum has superceded the dispersive instrument. Fourier-transform infrared spectrometers are now predominantly used and have improved the acquisition of infrared spectra dramatically [4]. Virtually all modern infrared spectrometers are FTIR instruments.

The term **Fourier transform infrared spectroscopy originates** from the fact that a Fourier transform (a mathematical algorithm) is required to convert the raw data into the actual spectrum. This technique is used to obtain an infrared spectrum of absorption, **emission**, photoconductivity or Raman scattering of a solid, liquid or gas. An FTIR spectrometer simultaneously collects spectral data in a wide spectral range (Significant advantage) where as Dispersive **spectrometer** measures intensity over a narrow range of wavelengths at a time and because of this FTIR technique has made dispersive infrared spectrometers obsolete (except sometimes in the near infrared) and opened up new applications of infrared spectroscopy [1].

Fourier transform infrared (FTIR) spectroscopy is a measurement technique that allows one to record infrared spectra [1]. It is based on the idea of the interference of radiation between two beams to yield an **interferogram [1,5].** The latter is a signal produced as a function of the change of pathlength between the two beams. The two domains of distance and frequency are interconvertible by the mathematical method of **Fourier-transformation [5].** FTIR technique has been used to distinguish successfully a number of biological systems including proteins, fats and carbohydrates, food and pharmaceutical products as well as cells and tissues from plants and animals. However, recent developments in FTIR spectroscopy have made FTIR technique a powerful tool for authentication of food products [7].

Principle [1]

Infrared light (The radiation emerging from the source) is guided through an interferometer and then through the sample (or vice versa) before reaching a detector. A moving mirror inside the apparatus alters the distribution of infrared light that passes through the interferometer. The signal directly recorded, called an "interferogram", represents light output as a function of mirror position. A data-processing technique called Fourier transform turns this raw data into the desired result (the sample's spectrum)-Light output as a function of infrared wavelength (or equivalently, wavenumber). Upon amplification of the signal, in which high-frequency contributions have been eliminated by a filter, the data are converted to digital form by an analog-to-digital converter and transferred to the computer for Fourier-transformation. As described above, the sample's spectrum is always compared to a reference [1,5].

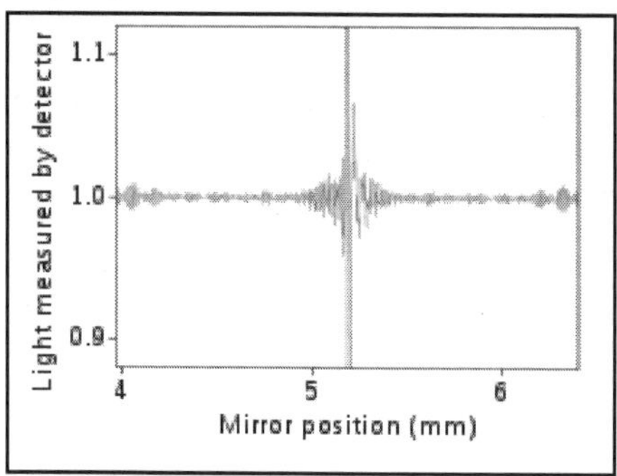

An Interferogram from an FTIR Measurement
(Horizontal axis - position of the mirror, and Vertical axis - amount of light detected).

Sample Preparation

Liquids are placed in liquid cell of short pathlength (0.015-1 mm). Solids are dissolved in a solvent. Solids (2 - 4 mg) is grinded and mixed with KBr to make semi-transparent pellet; or: grind and mix with Nujol (hydrocarbon oil) to form mull. One drop is placed between NaCl plates.

Instrumentation-Fourier Transform (FTIR) Instruments [6]

The instrumentation involves the source of IR, Inferometer, sample compartment, detector and computer. The main component in the FTIR spectrometer is an interferometer. This device splits and recombines a beam of light such that the recombined beam produces a wavelength-dependent interference pattern or an **interferogram**. The Michelson interferometer is most commonly used.

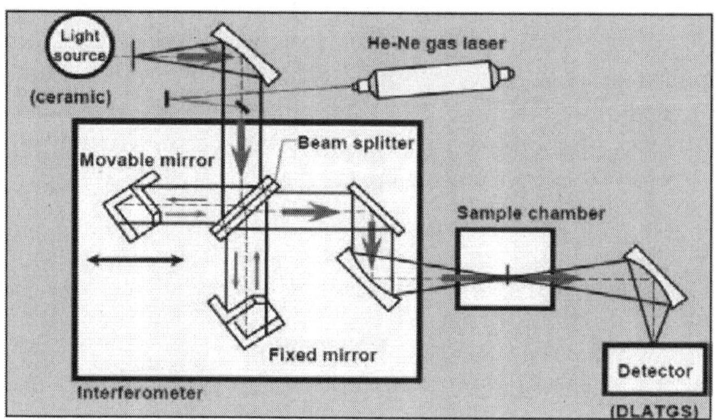

a. A Michelson Interferometer

It is the most common interferometer used in FTIR spectrometry. This consists of three components, a beam splitter, a moving mirror and a fixed mirror. It consists of two perpendicularly plane mirrors, one of which can travel in a direction perpendicular to the plane. The beam splitter is a semi reflecting transparent film that splits the beam into two halves with 50 per cent of radiation each. The beam splitter material has to be selected according to the region to be examined [5] *e.g.*:

(i) **For the mid- or near-infrared regions**: Materials such as germanium or iron oxide are coated onto an 'infrared-transparent' substrate such as potassium bromide or caesium iodide to produce beam splitters for the mid (MIR)- or near-infrared (NIR) regions.

(ii) **Far-infrared region**: Thin organic films, such as poly (ethyleneterephthalate), are used in the far-infrared region.

b. Computer

It is needed to turn complex signal into spectrum. The measured signal is digitized and sent to the computer where the Fourier transformation takes place. The final infrared spectrum is then presented to the user for interpretation and any further manipulation. Computers have two advantages:

1. Few optics, no slits = high intensity
2. Simultaneously measure all spectrum at once saves time.

Advantages of FTIR Spectrometers

There is an alternate method for taking spectra (the"dispersive" or "scanning mono-chromator" method), where one wavelength at a time passes through the sample. The dispersive method is more common in UV-Vis spectroscopy, but is less practical in the infrared than the FTIR method. FTIR instruments have several significant advantages over older dispersive instruments. Some of these are the Fellgett (or multiplex) advantage and the Jacquinot (or throughput) advantage [5].

a. Fellgett's Advantage/Multiplex Advantage

One reason that FTIR is favored is called "Fellgett's advantage" or the "multiplex advantage". The information at all frequencies is collected simultaneously, improving both speed and signal-to-noise ratio [1]. The Fellgett advantageis due to an improvement in the SNR per unit time, proportional to the square root of the number of resolution elements being monitored. This results from the large number of resolution elements being monitored simultaneously [5].

b. Jacquinot Advantage

Another is called Jacquinot's Throughput Advantage. A dispersive measurement requires detecting much lower light levels than an FTIR measurement [4a]. There are other advantages, as well as some disadvantages, [4a] but virtually all modern infrared spectrometers are FTIR instruments. In addition, because FTIR spectrometry does not require the use of a slit or other restricting device, the total source output can be passed through the sample continuously. This results in a substantial gain in energy at the detector, hence translating to higher signals and improved SNRs [5].

c. Speed Advantage

Another strength of FTIR spectrometry is its speed advantage. The mirror has the ability to move short distances quite rapidly, and this, make it possible to obtain spectra on a milli second timescale [5].

In short, the advantages of FT spectrometers over dispersive ones, are concluded as follows [4]:

(i) Wavelengths are measured simultaneously.

(ii) Shorter measurement time at same signal-to-noise ratio.

(iii) Higher light throughput at same spectral resolution.

(iv) Wave number stability is higher.

(v) Scattering can be neglected, because sample is place behind the interferometer.

Source

1. *Wikipedia.*

1a. *"What Happened When We Took the SCiO Food Analyzer Grocery Shopping". IEEE Spectrum: Technology, Engineering, and Science News. Retrieved 2017-03-23.*

1b. *"A Review of New Small-Scale Technologies for Near Infrared Measurements". www. american pharmaceutical review.com. Retrieved 2017-03-23.*

2. *Rudrik, K. and Wust (2002), Spectrophotometric techniques -IR Spectroscopy, Encyclopedia of Dairy Sciences (Volume-1), pp-45-50, Elsevier Science Limited.*

3. *Vogel's Text Book on Quantitaive chemical analysis revised by G.H.Jeffery, J. Bassett, J.Mendham and R.C.Denney.*

4. *Sylvia-Monique Thomas, Infrared and Raman Spectroscopy, University of Nevada Las Vegas.*

4a. *Chromatography/Fourier transform infrared spectroscopy and its applications, by Robert White, p7.*

5. *Barbara Stuart, Infrared Spectroscopy- Fundamentals and Applications, 2004 John Wiley and Sons, Ltd ISBNs: 0-470-85427-8 (HB); 0-470-85428-6 (PB).*

6. *Technews, No.74, May-June-2008, NDDB.*

7. *Rajan Sharma, Bimlesh Mann and Vikrant, Fourier Transform Infrared Spectroscopy – Concept and its Application in Quality Assessment of Dairy Foods, NDRI, Karnal.*

Components of IR Spectro-photometers [1-4]

Generally the typical setup of an IR Spectrophotometer consists of Source of IR radiation/light, Cells (Reference and sample cells), Prism/Diffraction gratting (the spectral apparatus), Detector and a computer [3,4].

A. Radiation Sources

The main source of radiation are Nernst Glower heated rare earth oxide rod (~1500 K) 1-10 μm, A nichrome wire wound on a ceramic support and Globar, a bonded silicon carbide rod. These are heated electrically to temperatures with in the range 1200-2000° C when they will glow and produce the IR radiation [3]. FTIR spectrometers use:

- *(i)* aAGlobar or Nernst source for the mid-infrared region.
- *(ii)* A high-pressure mercury lamp for the far-infrared region.
- *(iii)* Tungsten–halogen lamps for the near-infrared [5].

The details of different types of radiation sources are given below [4]:

i. Nernst Rod

It works at a temperature of 1900 K and is used for the MIR region. It consists of zirconium oxide rods with additives of yttrium oxides or oxides of other REE. The Nernst rod is a non-conductor at room temperature, initial heating is required for ignition. It has following **disadvantages:**

- *(a)* Sensitive.
- *(b)* Deforms easily.
- *(c)* Oxide mix has negative temperature coefficient of the electrical resistance, *i.e.* electrical conductivity increases with temperature.

ii. Globar

(Most frequently used Light source). It is silicon carbide in form of rods or helixes. A globar can be directly ignited, and has a burning temperature of 1500 K, which needs to be water-cooled. The globar has the **advantage of a relatively high emissivity** to about 100 cm^{-1}, so it can be used into the far IR region.

iii. Metallic Helices

Chromium nickel alloys or tungsten with operating temperatures of about 1300K, are usually air-cooled and used in NIR. For NIR, **tungsten-halogen lamps can also be used**.

iv. Mercury High-Pressure Lamps

For FIR, below 100 cm^{-1}, this can be used.It's plasma emission surpasses the spectral radiation power of a black radiator of the same temperature. They must be started by a high-voltage pulse.

v. Lasers

These can be used as radiation sources

- *(a)* **Gas lasers** scan rather narrow spectral regions (CO_2 laser 1100 - 900 cm^{-1}, CO laser 2000-1800 cm^{-1}, NO_2 laser 900-910 cm^{-1}, He-Ne laser (632, 1.152 and 3.391 μm).

(*b*) **Semi-conductor diode lasers** are larger tunable, and can cover almost the whole MIR region, *e.g.*, PbSnSe-type lasers (disadvantage - low operation temperature of 15 - 90 K). The gallium arsenide -, or gallium aluminium arsenide lasers work in the IR region, and can be operated at room temperature.

B. Absorption Cells [2,3]

The cells and window should be fabricated from IR transparent materials e.g Lithium fluoride (2.5-5.9μm/4000-1695cm^{-1}), Calcium fluoride (2.4-7.7μm/4167-1299cm^{-1}), Sodium Chloride (2.0-15.4μm/5000-649cm^{-1}), Potassium bromide (9.0-26.0μm/1111- 385cm^{-1}) and Calcium bromide (9.0-26.0μm/1111- 385cm-1).

Cells used for accurate quantitative analysis must have smooth, polished, parallel window surfaces and be of fixed path length, Cells are commercially available with fixed path length and with a variable path length. As many solutions for IR spectrophotometry involve organic solvents, it is necessary to use sample cells which can not only be stoppered to prevent solvent evaporation, but also be readily dismantled for cleaning and polishing [3].

C. Prism/Diffraction Gratting - The Spectral Apparatus [4]

Infrared microscopes use reflecting optics such as Cassegrain objectives with 15 fold or 36 fold magnification instead of glass or quartz. Using a microscope makes it possible to select a distinct measurement area on the sample and exclude contaminations from the measurement.

The optical system in a spectrometer is used to transmit radiation to the detector. Lens systems of glass or quartz as used in UV and VIS range are useless in the IR range, because they absorb most of the radiation below 5000 cm^{-1}. Thus, IR spectrometers are equipped with mirror optics.

D. Detectors [3,4]

Detectors convert the optical signal into electrical signals [4]. The detection of IR signal is of prime importance and hence a range of detectors is available for the same. The type of detector used in particular IR Instrument depends upon the type and quality of that instrument. When IR radiation is incident on the detector there is a change in polarization which can be employed to produce an electric signal. The detector will only produce a signal when the intensity of incident radiation changes. There is a variety of different types, such as thermal, pneumatic, pyroelectric or photoelectric detectors. The example of few detectors are as follows [3]:

(*i*) Golay pneumatic detector.

(*ii*) Pyro electric detectors.

(*iii*) Photoconducting PbS, HgCdTe light sensitive, resistance thermometer at 77 K fast and sensitive.

Now-a-days most detectors are photoelectric detectors because of their higher sensitivity. Incident light alters the electrical conductivity in an irradiated semi conductor material. The photosignal is measured as a change in voltage via the resistance or current [4]. The details of some common types of detectors are given below:

(*i*) In Sb-photoconductive or photodiode, 10000 – 1500 cm^{-1}, liquid-nitrogen-cooled.

(*ii*) MCT- photoconductive, spectral range 12500 – 400 cm^{-1}, liquid-nitrogen-cooled.

(*iii*) DTGS- pyroelectric, MIR region, works at room temperature.

The use of detectors depends upon the regions in which measurements are to be made [5] *e.g.*:

i. For Mid Infrared Region

There are two commonly used detectors employed for the mid-infrared region. The normal detector for routine use is a pyroelectric device incorporating deuterium tryglycine sulfate (DTGS) in a temperature-resistant alkali halide window. For more sensitive work, mercury cadmium telluride (MCT) can be used, but this has to be cooled to liquid nitrogen temperatures.

ii. In the Far-Infrared Region

Germanium or Indium–antimony detectors are employed, operating at liquid helium temperatures.

iii. For the Near-Infrared Region

The detectors used are generally lead sulfide photoconductors.

Chart Recorder/Computer

Earlier, all IR Spectrophotometers were provided with chart recorders which will present the complete IR Spectrum on a single continuous sheet [3] but now more advanced instruments also possess visual display unit/Computer on which the spectra can be displayed and analysed. The computer forms a crucial component of modern infrared instruments and it controls the instrument on the following aspects:

(*i*) Sets scan speeds and scanning limits.

(*ii*) Starts and stops scanning.

(*iii*) Reads spectra into the computer memory from the instrument as the spectrum is scanned (spectrum is digitized).

(*iv*) The computer is also used to plot the spectra,

(*v*) Spectra may be manipulated using the computer (by adding and subtracting spectra or expanding areas of the spectrum of interest) and

(*vi*) The computer is also used to scan the spectra continuously and average or add the result in the computer memory. Complex analyses may be automatically carried out by following a set of pre-programmed commands [5].

Source

1. *Wikipedia.*
2. *Rudrik, K. and Wust (2002), Spectrophotometric techniques -IR Spectroscopy, Encyclopedia of Dairy Sciences (Volume-1), pp-45-50, Elsevier Science Limited.*
3. *Vogel's Text Book on Quantitaive chemical analysis revised by G.H.Jeffery, J. Bassett, J.Mendham and R.C.Denney.*
4. *Sylvia-Monique Thomas, Infrared and Raman Spectroscopy, University of Nevada Las Vegas.*
5. *Barbara Stuart, Infrared Spectroscopy- Fundamentals and Applications, 2004 John Wiley and Sons, Ltd ISBNs: 0-470-85427-8 (HB); 0-470-85428-6 (PB).*

IR Spectrophotometer.

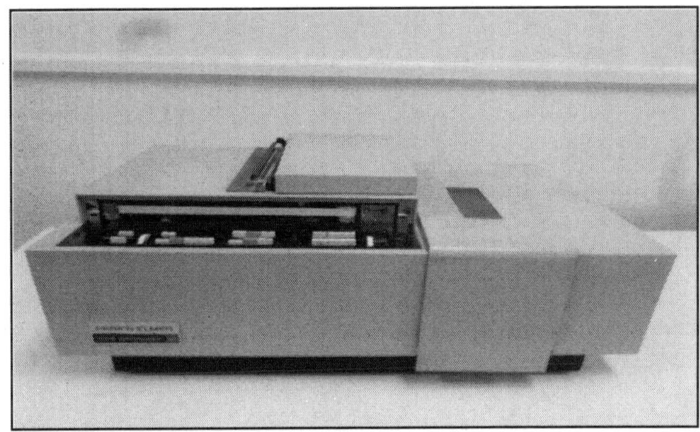

Perkin Elmer 597 Infrared Spectrophoto meter 5102-0014.

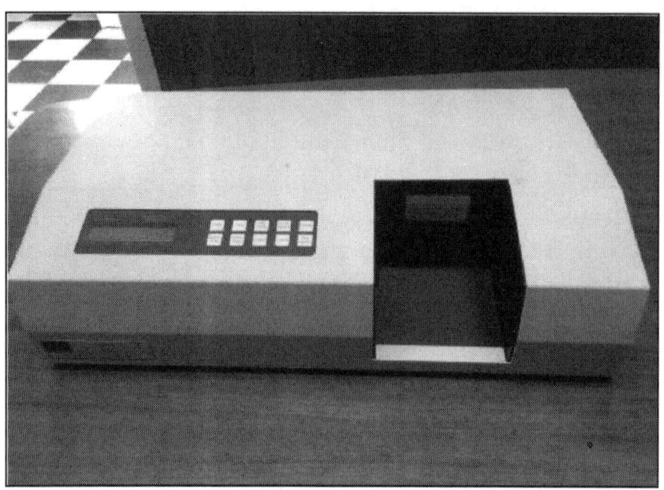

M500 Infrared Spectrophotometer, Buck Scientific.

Good Laboratory Practice (GLP) and International Standards for IR Spectroscopy [1,2]

1. Good Laboratory Practice (GLP) for Calibration

As IR Spectroscopy is an indirect method and hence it requires calibration with a reference chemical method first and then subsequent regular check on the performance. There is need to establish following GLP guidelines for IR instrument:

(i) Application of infrared spectra of a certified material of known concentration (an inert) -Take a standard spectrum periodically and compare it with the spectrum got with instrument. The IR instrument must be recalibrated, if the difference between standard spectrum and the spectrum obtained with the instrument is unacceptable.

(ii) Application of reference method if reference material is not available- In this case, the performance of instrument is monitored by analysing corresponding samples by a reference. In case, difference between the results obtained by reference method and infrared spectrometer beyond acceptable limits, is observed then the infrared equipment needs new calibration.

2. International Standards

Some international standards have been specified for the use of Infrared spectroscopic techniques for the analysis of dairy products. The details of aforesaid international standards are given below:

a. International Standards for IR Spectroscopy

(i) AOAC972.16-Fat, Lactose, Protein and solids in milk (Mid-Infrared spectroscopic method).

(ii) AOAC 975.18-Protein in milk.

(iii) AOAC 975.19- Lactose in milk.

(iv) IDF 141C: 2000- Whole milk-Determination of milk fat, protein and lactose content- Guide for the operation of MIR instruments.

(v) ISO 21543/IDF 201: 2006-Milk products-Guidelines for the application of NIR Spectroscopy.

b. International Standard for NIR Spectroscopy

Milk products- Guidelines for the application of Near Inftared spectrometry-IDF 201/ISO 21543.

Source

1. *ISO/IDF/AOAC.*
2. *Technews, No.74, May-June-2008, NDDB.*

Two Beam Absorption Spectrometer and Comparison with Reference [1]

a. Two Beam Absorption Spectrometer

A beam of infrared light is produced, passed through an interferometer and then split into two separate beams. One is passed through the sample, the other passed through a reference. The beams are both reflected back towards a detector, however first they pass through a splitter, which quickly alternates which of the two beams enters the detector. The two signals are then compared and a printout is obtained. This "two-beam" setup gives accurate spectra even if the intensity of the light source drifts over time.

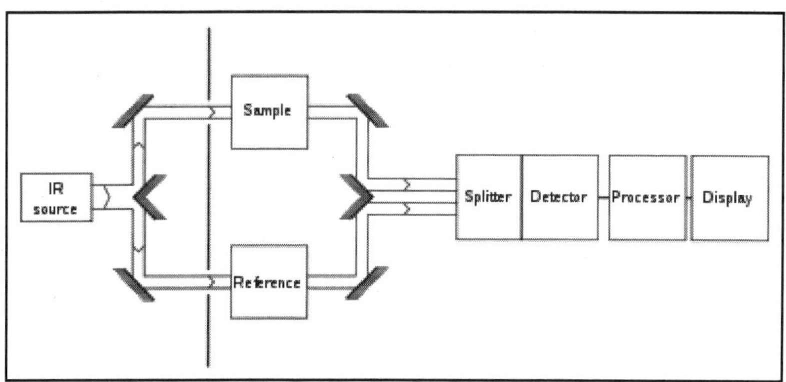

Schematics of a Two-Beam Absorption Spectrometer.

b. Comparison with Reference

It is necessary to measure both the sample and a "reference" (or "control") while taking the infrared spectrum of a sample. This is because each measurement is affected by not only the light-absorption properties of the sample, but also the properties of the instrument. The reference measurement helps in making it possible to eliminate the instrument influence. Mathematically, the sample transmission spectrum is divided by the reference transmission spectrum.

The appropriate "reference" depends on the measurement and its goal.

(i) The simplest reference measurement is to simply remove the sample by replacing it by air.

(ii) Sometimes a different reference is more useful. if the sample is a dilute solute dissolved in water in a beaker, then a good reference measurement might be to measure pure water in the same beaker. Then the reference measurement would cancel out not only all the instrumental properties, but also the light-absorbing and light-reflecting properties of the water and beaker, and the final result would just show the properties of the solute.

A common way to compare to a reference is sequentially-first measure the reference, then replace the reference by the sample and measure the sample. This technique is not perfectly reliable; if the infrared lamp is a bit brighter during the reference measurement, then a bit dimmer during the sample measurement, the measurement will be distorted. A "two-beam" setup, can correct for these types of effects to give very accurate results. The Standard addition method can be used to statistically cancel these errors.

Source

1. *Wikipedia.*

Two-Dimensional IR [1,2,3,4]

Two-dimensional infrared correlation spectroscopy analysis is the application of 2D correlation analysis on infrared spectra. The 2D synchronous and 2D asynchronous spectra represent a graphical overview of the spectral changes due to a perturbation (changing concentration or changing temperature) as well as the relationship between the spectral changes at two different wave numbers.

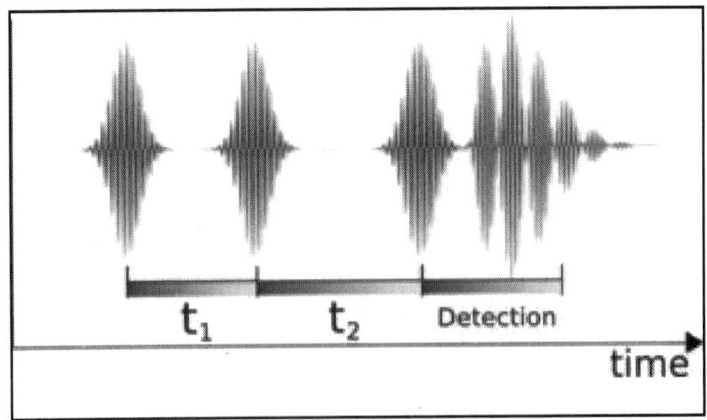

Pulse Sequence used to obtain a two-dimensional Fourier transform infrared spectrum. The time period is usually referred to as the coherence time and the second time period is known as the waiting time. The excitation frequency is obtained by Fourier transforming along the axis.

Nonlinear 2D IR Spectroscopy [2,3]

It is the infrared version of **correlation spectroscopy** and it is a technique that has become available with the development of **femtosecond** infrared laser pulses. In this experiment, first a set of pump pulses is applied to the sample. This is followed by a waiting time during which the system is allowed to relax. The typical waiting time lasts from zero to several picoseconds, and the duration can be controlled with a resolution of tens of femtoseconds. A probe pulse is then applied, resulting in the emission of a signal from the sample. The nonlinear two-dimensional infrared spectrum is a two-dimensional correlation plot of the frequency ω_1 that was excited by the initial pump pulses and the frequency ω_3 excited by the probe pulse after the waiting time. This allows the observation of coupling between different vibrational modes; because of its extremely fine time resolution, it can be used to monitor molecular dynamics on a picosecond timescale. It is still a largely unexplored technique and is becoming increasingly popular for fundamental research.

Nonlinear Two-dimensional Infrared Spectroscopy and Two-dimensional Nuclear Magnetic Resonance (2DNMR) Spectroscopy

(i) As with 2DNMR spectroscopy, this technique (Non Linear 2D IR) spreads the spectrum in two dimensions and allows for the observation of cross peaks that contain information on the coupling between different modes.

(ii) In contrast to 2DNMR, Non Linear 2D IR spectroscopy also involves the excitation to overtones. These excitations result in excited state absorption peaks located below the diagonal and cross peaks.

(iii) In **2DNMR, two distinct techniques (COSY and NOESY) are frequently used. The cross peaks in the first are related to the scalar coupling, while in the latter they are related to the spin transfer between** different nuclei. In Non Linear 2D spectroscopy, analogs have been drawn to these 2DNMR techniques. Non Linear 2D IR spectroscopy with zero waiting time corresponds to COSY, and Non Linear 2D IRspectroscopy with finite waiting time allowing vibrational population transfer corresponds to NOESY. The COSY variant of Non Linear 2D IR spectroscopy has been used **for determination of the secondary structure content of proteins [4].**

Source

1. *Wikipedia.*

2. *P.Hamm, M.H. Lim, R. M.Hoch-strasser (1998). "Structure of the amide I band of peptides measured by femtosecond nonlinear-infrared spectroscopy".J.Phys. Chem. B 102 (31): 6123.*

3. *S.Mukamel (2000). "Multidimensional Fentosecond Correlation Spectroscopies of Electronic and Vibrational Excitations". Annual Review of Physics and Chemistry 51 (1): 691.Bibcode: 2000 ARPC.51.691M.*

4. *N. Demirdoven, C. M. Cheatum, H.S.Chung, M. Khalil, J. Knoester, A. Tokmakoff (2004)."Two-dimensional infrared spectroscopy of antiparallel beta-sheet secondary structure". Journal of the American Chemical Society 126 (25): 7981.*

Isotope Effects [1,2]

The different isotopes in a particular species may give fine detail in infrared spectroscopy. For example, the O–O stretching frequency (in reciprocal centimeters) of **oxyhemocyanin** is experimentally determined to be 832 and 788 cm^{-1} for í (^{16}O–^{16}O) and í (^{18}O–^{18}O), respectively.

By considering the O–O bond as a spring, the wavenumber of absorbance, ν can be calculated:

$$\nu = \frac{1}{2\pi c}\sqrt{\frac{k}{\mu}}$$

where: k -the spring constant for the bond, c - the speed of light, and μ -the reduced mass of the A–B system.

$$\mu = \frac{m_A m_B}{m_A + m_B}$$

(m_i is the mass of atom i).

The reduced masses for ^{16}O–^{16}O and ^{18}O–^{18}O can be approximated as 8 and 9 respectively. Thus

$$\frac{\nu(^{16}O)}{\nu(^{18}O)} = \sqrt{\frac{9}{8}} \approx \frac{832}{788}.$$

where: v - the wavenumber; [wavenumber = frequency/ (speed of light)]

The effect of isotopes, both on the vibration and the decay dynamics, has been found to be stronger than previously thought. In some systems, such as silicon and germanium, the decay of the anti-symmetric stretch mode of interstitial oxygen involves the symmetric stretch mode with a strong isotope dependence. This can be easily understood by the following example:

(i) For a natural silicon sample, the lifetime of the anti-symmetric vibration - 11.4 ps.

(ii) When the isotope of one of the silicon atoms is increased to ^{29}Si, the lifetime increases to 19 ps.

(iii) When the silicon atom is changed to ^{30}Si, the lifetime becomes 27ps [2].

Source

1. *Wikipedia.*
2. *Kohli, K.; Davies, Gordon; Vinh, N.; West, D.; Estreicher, S.; Gregorkiewicz, T.; Izeddin, I.; Itoh, K. (2006). "Isotope Dependence of the Lifetime of the 1136-cm-1 Vibration of Oxygen in Silicon". Physical Review Letters. 96 (22): 225503.*

Near Infrared (NIR) Spectroscopy

The absorptions observed in the near-infrared region (13000–4000 cm^{-1}) are overtones or combinations of the fundamental stretching bands which occur in the 3000–1700 cm^{-1} region [4]. The bands involved are usually due to C–H, N–H or O–H stretching [3]. The NIR spectral region is information rich. For example, positions C-H stretch combination bands occur four times, the C-H overtones occur four times, the O-H combination bands (three times), the O-H overtones (three times), and the N-H overtones (three times) [1]. The resulting bands in the near infrared are usually weak in intensity and the intensity generally decreases by a factor of 10 from one overtone to the next. The bands in the near infrared are often overlapped, making them less useful than the mid-infrared region for qualitative analysis. However, there are important differences between the near-infrared position of different functional groups and these differences can often be exploited for quantitative analysis [3]. The important molecules for NIR measurements have most often been water (as O-H stretch) and associated solutes, proteins, carbohydrates, fats and oils, and hydrocarbon classes. NIR continues to provide a valuable measurement technique, applicable to both natural and synthetic materials, for use in continuous or real-time process monitoring.

a. History [1]

O.W.Wheeler described the near infrared region as extending "from about 2 m into the visible at about 0.7 m." R.F. Goddu and D. A. Delker demonstrated the spectra-structure correlations and average molar absorptivity for a number of functional groups for the NIR region (1.0 to 3.1 m), and the maximum recommended path lengths for twelve solvents (useful for NIR spectroscopy) over the wavelength region 1.0 to 3.1 m. Several Scientists have reviewed the work done on NIR spectroscopy and summary of aforesaid review work is given below:

Name of Scientists	*Nature of Work*	*Period*
Professor J.W. Ellis	Reviewed work below 3 microns for absorption of organic liquids.	
W. Kaye	Review work in NIR	1920 to 1954.
R.F. Goddu	Compiled an extensive review of NIR	Prior to 1960
K.B. Whetsel	Review work in NIR	Prior to 1968
Stark *et al.*	Review work in NIR	Prior to 1986
W.W. Coblentz	Research investigations into the molecular structures of organic compounds	Since 1912

b. NIR Spectral Features

a. Predominant Near-Infrared Spectral Features [1]

(i) The methyl C-H stretching vibrations.

(ii) Methylene C-H stretching vibrations.

(iii) Aromatic C-H stretching vibrations.

(iv) O-H stretching vibrations.

b. Minor but Still Important Spectral Features

Methoxy C-H stretching, carbonyl associated C-H stretching; N-H from primary amides, secondary amides (both alkyl, and aryl group associations), N-H from primary, secondary, and tertiary amines, and N-H from amine salts.

c. Advantages [1,2]

The following advantages of NIR measurements over other vibration techniques are found true:

1. General Advantages

(i) C-H associated vibrational information is repeated 8 times from 690 nm to 3000 nm.

(ii) Simple harmonics may be selected or more information rich combination regions.

(iii) Low cost instruments with high signal-to-noise (SNR) are simple to make and typically exhibit signal-to-noise ratios (SNR) of 25000-100000: 1.

(iv) High NIR throughput is possible, even when employing low cost fiber optics.

(v) Variable path lengths for industrial use are possible, typically 1 mm to 10 cm or more using different NIR spectral regions.

(vi) NIR Light penetrates plant and animal tissue easily for biomedical applications (when using 900 nm and longer).

2. Advantages in Food and Dairy Industry [2]

NIR spectroscopy is a proven and accepted method of food analysis and is used to monitor product quality parameters within the food, beverage, and dairy industries and offers significant advantages:

(i) Real-time analysis.

(ii) Measure multiple constituents simultaneously.

(iii) Non-destructive testing.

(iv) No sample prep.

(v) No chemical waste stream.

(vi) Environmentally friendly.

(vii) Less sample waste.

(viii) Decrease operation costs, increase yield.

Source

1. *Wikipedia.*
2. *Technews, No.74, May-June-2008, NDDB.*
3. *Barbara Stuart, Infrared Spectroscopy- Fundamentals and Applications, 2004 John Wiley and Sons, Ltd ISBNs: 0-470-85427-8 (HB); 0-470-85428-6 (PB).*
4. *Weyer, L. G. and Lo, S. C., 'Spectra–Structure Correlations in the Near-infrared', in Handbook of Vibrational Spectroscopy, Vol. 3, Chalmers, J. M. and Griffiths, P. R. (Eds), Wiley, Chichester, UK, 2002, pp. 1817–1837.*

Far-Infrared (FIR) Spectroscopy [1,2]

This deals in far-infrared region which is in between 400 and 100 cm–1 [2]. This region is more limited than the mid infrared for spectra–structure correlations but does provide information regarding the followings [1]:

i. Vibrations of Molecules Containing Heavy Atoms

Intramolecular stretching modes involving heavy atoms can be helpful for characterizing compounds containing halogen atoms, organometallic compounds and inorganic compounds.

ii. Molecular Skeleton Vibrations

Skeletal bending modes involving an entire molecule occur in the far infrared for molecules containing heavier atoms because bending modes are usually no more than one-half of the wave number of the corresponding stretching mode.

iii. Molecular Torsions

Torsional modes arise because the rotation about single bonds is not 'free'. As a result, when certain small groups are bonded to a large group, they undergo a motion with respect to the heavier'anchor' group.

iv. Crystallattice Vibrations

Crystal lattice vibrations are associated with the movement of whole molecular chains with respect to each other in crystalline solids.

Source

1. *Barbara Stuart, Infrared Spectroscopy- Fundamentals and Applications, 2004 John Wiley and Sons, Ltd ISBNs: 0-470-85427-8 (HB); 0-470-85428-6 (PB).*
2. *Griffiths, P. R., 'Far-Infrared Spectroscopy', in Handbook of Vibrational Spectroscopy, Vol. 1, Chalmers, J. M. and Griffiths, P. R. (Eds), Wiley, Chichester, UK, 2002, pp. 229–239.*

Mid-Infrared (MIR) Spectroscopy [1]

The MIR spectrum (4000–400 cm^{-1}) can be approximately divided into four regions and the nature of a group frequency may generally be determined by the region in which it is located. The regions are generalized as follows:

(i) The X–H stretching region (4000–2500 cm^{-1})

(ii) The triple-bondregion (2500-2000 cm^{-1}).

(iii) The double-bond region (2000–1500 cm^{-1})

(iv) The fingerprint region (1500–600 cm^{-1}).

It has been assumed so far that each band in an infrared spectrum can be assigned to a particular deformation of the molecule, the movement of a group of atoms, or the bending or stretching of a particular bond. This is possible for many bands, particularly stretching vibrations of multiple bonds that are'well behaved'. However, many vibrations are not so well behaved and may vary by hundreds of wavenumbers, even for similar molecules. This applies to most bending and skeletal vibrations, which absorb in the 1500–650 cm^{-1} region, for which small steric or electronic effects in the molecule lead to large shifts. A spectrum of a molecule may have a hundred or more absorption bands present, but there is no need to assign the vast majority. The spectrum can be regarded as a 'fingerprint' of the molecule and so this region is referred to as the **fingerprint region.**

Source

1. *Barbara Stuart, Infrared Spectroscopy- Fundamentals and Applications, 2004 John Wiley and Sons, Ltd ISBNs: 0-470-85427-8 (HB); 0-470-85428-6 (PB).*

IR, FTIR and NIR Technology based Equipments available for Quality Analysis in Dairy

The following IR Based equipments are being used for rapid analysis of milk and milk products:

1. IR based Equipments [1,2,3]

(i) **Milkoscan™ Minor**: It is IR based equipment and being used for analysis of Fat, Protein, Lactose, Total Solids, Solids-not-fat and Freezing point depression (Six parameters) on milk.

(ii) **Lactoscope™**: Different models including C3, C4 and FTIR (Both manual and automatic) are available and this instrument is capable of analysing the Fat, Protein, Lactose and Total solids (Composition) of milk and milk products.

(iii) **Food Scan™**: It is Dairy Analyser for rapid analysis of moisture, fat, protein *etc.* in solid and semi solid dairy products (Butter and Cheese).

2. FTIR Based Equipments [1,2,3]

The following FTIR Based equipments are being used for rapid analysis and process control of milk and milk products:

a. FTIR Spectrometers for Quality Analysis

1. MilkoscanTM

It is a important and valuable lab equipment for dairy Industry. The details of different models of Milkoscan (M/S Foss) are given below:

(i) **Milkoscan™ FT-** It is FTIR based equipment and being used for high speed automatic multiple component analysis including Urea, Casein, Free fatty acids, Saturated and unsaturated fatty acids, Freezing point depression of milk.

(ii) **MilkoScan™ FT120–**It works on the FTIR measuring principle, in compliance with IDF and AOAC standards. MilkoScan FT 120 is suitable for both process control and sophisticated analysis in the lab. It is versatile and user-friendly equipment for analysis of raw milk, intermediates and final products and has wide range of ready to use calibration and low testing cost per sample. A complete analysis of product composition with up to 24 parameters can be presented in just 30 seconds. It offers determination of fat, protein, lactose, total solids and solids-non-fat in milk, cream and simple dairy products. All routine operation takes place from one single screen picture, samples can be analyzed without pre-heating and data are stored automatically. The details related to the operation of Milkoscan are given in chapter -04 of author's another book entitled Hand Book on Analysis of milk.

MilkoScan™ FT120.
(*Source*: Wikipedia)

b. Process Scan™ FT

It is FTIR Based equipment suitable for real time on line standardisation of milk and other liquid dairy products.

Source

1. *Foss Instruments.*
2. *Delta Instruments (www.aitests.com/Delta CD/diproduct.htm).*
3. *Technews, No.74, May-June-2008, NDDB.*

3. NIR Technology based Equipments [1,2]

The following NIR Based equipments are being used for rapid analysis of milk and milk products:

(*i*) **InfraXact™:** It can be used in Lab or in processing section. It is NIR Based equipment for compositional analysis of Food and Feed.

(*ii*) **Process Analytics™Dairy:** It is NIR Based equipment developed and designed for real time, in line control of quality of Milk powder, Processed cheese, Cream cheese, Butter and other dairy products.

Source

1. *Foss Instruments.*
2. *Technews, No.74, May-June-2008, NDDB.*

Application of IR, MIR, NIR Spectroscopy

The details related to applications of IR, MIR, NIR spectroscopy are given below:

Application of IR Spectroscopy

A. Dairy Industry [1,2,3]

This procedure is reliable, accurate inexpensive, rapid and environment friendly and provides simultaneous estimation of various major constituents of milk. It can be used for compositional analysis and on line peocess controls.

a. Application in Compositional Analysis

IR Spectroscopy is used to determine nearly all major constituents in all dairy products. Some examples are given in following table:

Products	*Constituents*	*Products*	*Constituents*
Cream	Fat, Dry matter	Milk powder	Moisture, Fat, Protein, Lactose
Creamer	Moisture (water), Fat	Ice cream mix	Fat, Dry matter
Coffee creamer	Moisture, Fat, Protein	Raw/Market milk	Fat, Protein, Lactose, Drymatter
Evaporated milk	Fat, Dry matter	Skim Milk	Drymatter, Protein, Casein
Skim milk powder	Moisture, Fat	Feeding stuffs	Various constituents

b. Application in Process Control

It is also used in the process line for better process control. The two important examples showing the application of IR Spectroscopy in process control, are as follows:

(i) MIR spectroscopy could be used to standardize fat and Protein in milk on line.

(ii) NIR spectroscopy could be used to determine water in milk powder directly after the drying chamber.

Source

1. *Wikipedia.*
2. *Rudrik, K. and Wust (2002), Spectrophotometric techniques–IR Spectroscopy, Encyclopedia of Dairy Sciences (Volume-1), pp-45-50, Elsevier Science Limited.*
3. *Technews, No.74, May-June-2008, NDDB.*

B. Biological and Food Industry Application [1]

Infrared spectroscopy has proved to be a powerful tool for the study of biological molecules.Biological systems, including lipids, proteins, peptides, biomembranes, nucleic acids, animal tissues, microbial cells, plants and clinical samples, have all been successfully studied by using infrared spectroscopy.

a. Lipids

Infrared spectroscopy can provide valuable structural information about lipids. Many lipids contain phosphorus and are classified as phospholipids.Lipids are organized in bilayers of about 40–80°A in thickness, where the polar head group points towards the aqueous phase and the hydrophobic tails point towards the tails of a second layer. The chains can be in an all-*trans* conformation which is referred to as the **gel phase, or a liquid crystalline phase** is obtained when the chain also contains gauche C–C groups (8). The infrared spectra of phospholipids can be divided into the spectral regions that originate from the molecular vibrations of the hydrocarbon tail, the interface region and the headgroup [9,10].

b. Proteins and Peptides

The infrared spectra of proteins exhibit absorption bands associated with their characteristic amide group [11,12,13]. In-plane modes are due to C=O stretching, C–N stretching, N–H stretching and O–C–N bending, while an out-of-plane mode is due to C–N torsion. The characteristic bands of the amide groups of protein chains are similar to the absorption bands exhibited by secondary amides in general, and are labelled as amide bands. There are nine such bands, called amide A, amide B and amides I–VII, Fourier-transform infrared (FTIR) spectroscopy is particularly useful for probing the structures of membrane proteins [14,15]. This technique can be used to study the secondary structures of proteins, both in their native environment as well as after reconstitution into model membranes. The details of some important IR Bands of Lipids and Proteins are given in following table [16]:

Sl.No.	*Biological Molecules*	*Type of Vibration*	*Major Infrared Bands of Lipids*	
			Wave Number (cm^{-1})	*Types of Stretching/Bending*
1	Lipids	Stretching	3010, 2956, 2920, 2870 and 2850 respectively	=C–H stretching, CH_3 asymmetric stretching, CH_2 asymmetric stretching, CH_3 symmetric stretching and CH_2 symmetric stretching respectively.
			1730	C=O stretching
			1228 and 085 respectively	PO_2^- asymmetric stretching and PO_2^- symmetric stretching respectively.
			1170	CO–O–C asymmetric stretching.
			1047	C–O–P stretching.
			1047	C–O–P stretching.
		Bending	1485 and 1405 respectively	$(CH_3)3N^+$ asymmetric bending and $(CH_3)3N^+$ symmetric bending respectively.
			1378	CH_3 symmetric bending.
		Wagging band progression	1400–1200	CH_2 wagging band progression.
		Scissoring	1473, 1472, 1468, 1463	CH_2 scissoring.
		Rocking	730, 720, 718	CH_2 rocking.
2	Proteins and Peptides	A	3300	N–H stretching in resonance with overtone.
		B	3110	(2 × amide II).
		I	1653	80 per cent C=O stretching; 10 per cent C–N stretching; 10 per cent N–H bending.
		II	1567	60 per cent N–H bending; 40 per cent C–N stretching.
		III	1299	30 per cent C–N stretching; 30 per cent N–H bending; 10 per cent C=O stretching; 10 per cent O=C–N bending; 20 per cent other.
		IV	627	40 per cent O=C–N bending; 60 per cent other.
		V, VI	725 and 600 respectively.	N–H bending and C=O bending respectively.
		VII	200	C–N torsion.

C. Food Science/Industry

Another important application of Infrared Spectroscopy is in the food industry to measure the concentration of various compounds in different food products [2,3]. The instruments are now small, and can be transported, even for use in field trials. Both mid-infrared and near-infrared techniques may be used to obtain qualitative and quantitative information about food samples [8, 17,18]. Foods are complex mixtures, with the main components being water, proteins, fats and carbohydrates.

(*a*) One important infrared method is the determination of **trans-**isomers in fats and oils. It is commercially important for labelling to determine the **trans-content** in oils and Fats [8].

(*b*) Near-infrared (NIR) spectroscopy is widely used in the food industry as a fast routine analytical method for the quantitative measurement of water, proteins, fats and carbohydrates [18]. Although the near-infrared bands are less useful for qualitative analysis of foods because of their broad overlapped appearance, these bands are suitable for quantitative analysis when using chemometric techniques [8].

(*c*) **Liquor Industry:** Wine analysis is also feasible with near-infrared spectroscopy [19]. This approach provides a relatively straight forward method for analysing ethanol in wine. The O–H bond in ethanol produces a near-infrared band that is easily distinguished from the O–H band of water [8]. Other alcoholic beverages, including beers and spirits, can be analysed by using NIR spectroscopy.

D. Chemical Compound Analysis [1]

Infrared absorption spectra can be used for the identification of pure compounds or for the detection and identification of impurities. Most of the applications are concerned with organic compounds, primarily because water, the chief solvent for inorganic compounds, absorbs strongly beyond 1.5 μm. Inorganic compounds give broad absorption bands, whereas organic compounds may give rise to numerous narrower bands. The Infra red spectrum of a compound may be regarded as a sort of **finger- print** of that compound. Thus for the identification of a pure compound the spectrum of unknown substance is compared with the spectra of a limited number of possible substances suggested by other properties and when a match between spectra is obtained, identification is complete. This procedure is valuable for structural isomers and not for optical isomers. Aliphatic Hydrocarbons, Aromatic Compounds, Oxygen-Containing Compounds (Alcohols and phenols, Ethers, Aldehydes, Ketones, carboxylic acid, Esters *etc.*), Nitrogen-Containing Compounds (Amines, Amides), Halogen-Containing Compounds, Heterocyclic Compounds, Boron Compounds, Silicon Compounds, Phosphorus Compounds, and Sulfur Compounds can be indentified by this technique [8].

E. Research and Development/Quality Control in Industry [1]

Infrared spectroscopy is a simple and reliable technique widely used in both organic and inorganic chemistry, in research and industry. It is used in quality control, dynamic measurement, and monitoring applications such as the long-term unattended measurement of CO_2 concentrations in green houses and growth chambers by infrared gas analyzers.

F. Forensic Science [1]

It is also used in **forensic analysis** in both criminal and civil cases, for example in identifying **polymer degradation**. It can be used in detecting how much alcohol is in the blood of a suspected drunk driver measured as 1/10,000 g/mL = 100 μg/mL.

G. Polymer Industry [1]

Polymers play an enormously important role in modern society; these materials are fundamental to most aspects of modern life, such as building, communication, transportation, clothing and packaging.Infrared spectroscopy is also useful in measuring the degree of polymerization in polymer manufacture. As most polymers are organically based, the spectral assignments made for organic molecules are helpful when interpreting the infrared spectra of polymers. The different types of polymers are determined by the following ways:

(*i*) **If the polymer is a thermoplastic**-It can be softened by warming and formed in a hydraulic press into a thin film.

(*ii*) Alternatively, the polymer may be dissolved in a volatile solvent and the solution allowed to evaporate to give a thin film on an alkali halide plate.

(*iii*) Some polymers, such as cross-linked synthetic rubbers, can be microtomed [8].

Infrared spectroscopy is also readily employed for the quantitative analysis of additives or contaminants in polymers, study of structure, orientation, surface, polymerisation. The structural properties of polymers, such as tacticity, branching, crystallinity, hydrogen bonding and orientation, may be readily investigated by using infrared techniques.Changes in the character or quantity of a particular bond are assessed by measuring at a specific frequency over time. Modern research instruments can take infrared measurements across the range of interest as frequently as 32 times a second. This can be done whilst simultaneous measurements are made using other techniques. This makes the observations of chemical reactions and processes quicker and more accurate.

H. Semi Conduct or Microelectronics [1,4]

Infrared spectroscopy has also been successfully utilized in the field of semiconduct or microelectronics- for example, infrared spectroscopy can be applied to semiconductors like **silicon, gallium arsenide, gallium nitride, zinc selenide,** amorphous silicon, **silicon nitride,** ***etc.*** The instruments are now small, and can be transported, even for use in field trials.

I. Painting and Other [1,5,6,7]

IR has been successfully used in analysis and identification of pigments in paintings [5] and other art objects [6] such as illuminated manuscripts [7]. Infrared spectroscopy is used in the paint industry for quality control, product improvement and failure analysis, and for forensic identification purposes [8,20]. Paints are mainly comprised of polymeric binders and pigments in a dispersive medium. The binders are commonly alkyds (oil-modified polyesters), acrylics and vinyl polymers. Titanium oxide is the most commonly used pigment, while water or organic solvents are used as the medium. Paints may also contain other additives such as fillers (*e.g.* calcium carbonate) and stabilizers (*e.g.*lead oxide).

Source

1. *Wikipedia.*
2. *Osborne, Brian G. (2006). Near-Infrared Spectroscopy in Food Analysis. John Wiley and Sons.*
3. *Villar, A.; Gorritxategi, E.; Aranzabe, E.; Fernandez, S.; Otaduy, D.; Fernandez, L. A. (2012). "Low-cost visible–near infrared sensor for on-line monitoring of fat and fatty acids content during the manufacturing process of the milk". Food Chemistry. 135 (4): 2756–2760.*
4. *Lau, W.S. (1999). Infrared characterization for microelectronics. World Scientific. ISBN 981-02-2352-8.*
5. *Infrared spectroscopy at ColourLex. Retrieved December 11,2015.*
6. *Derrick, M.R., Stulik, D. and Landry J.M., Infrared Spectroscopy in Conservation Science, Scientific Tools for Conservation, Getty Publications, 2000. Retrieved December 11, 2015.*
7. *Paola Ricciardi, Unlocking the secrets of illuminated manuscripts. Retrieved December 11, 2015.*
8. *Barbara Stuart, Infrared Spectroscopy- Fundamentals and Applications, 2004 John Wiley and Sons, Ltd ISBNs: 0-470-85427-8 (HB); 0-470-85428-6 (PB).*
9. *Watts A. and De Pont J. J. H. H. M. (Eds), Progress in Protein–Lipid Interactions, 2, Elsevier, Amsterdam, The Netherlands,1986.*
10. *Lewis, R. N. A. H. and McElhaney, R. N., 'Vibrational Spectroscopy of Lipids', in Handbook of Vibrational Spectroscopy, Vol. 5, Chalmers, J. M. and Griffiths, P. R. (Eds), Wiley, Chichester, UK, 2002, pp. 3447-3464.*
11. *Bandekar, J., Biochim. Biophys. Acta, 1120, 123–143 (1992).*
12. *Fabian, H. and Mantele, W.,'Infrared Spectroscopy of Proteins', in Handbook of Vibrational Spectroscopy, Vol. 5, Chalmers, J. M. and Griffiths, P. R. (Eds), Wiley, Chichester, UK, 2002, pp. 3399–3425.*
13. *Fabian, H., 'Fourier Transform Infrared Spectroscopy in Peptide and Protein Analysis', in Encyclopedia of Analytical Chemistry, Vol. 7, Meyers, R. A. (Ed.), Wiley, Chichester, UK, 2000, pp.5779-5803.*
14. *Jackson, M., Moore, D. J., Mantsch, H. H. and Mendelsohn, R., Vibrational Spectroscopy of Membranes', in Handbook of Vibrational Spectroscopy, Vol. 5, Chalmers, J. M. and Griffiths, P. R. (Eds), Wiley, Chichester, UK, 2002, pp. 3508–35180*
15. *Arrondo, J. L. R. and Goni, F. M., Prog. Biophys. Mol. Biol., 72, 367–405 (1999).*
16. *Stuart, B., Biological, Applications of Infrared Spectroscopy, ACOL Series, Wiley, Chichester, UK, 1997.*
17. *Li-Chen, E. C. Y., Ismail, A. A., Sedman, J. and van de Voort, F. R., 'Vibrational Spectroscopy of Food and Food Products', in Handbook of Vibrational Spectroscopy, Vol. 5, Chalmers, J. M. and Griffiths, P. R. (Eds), Wiley, Chichester, UK, 2002, pp. 3629–3662.*

18. *Osborne, B. G. and Fearn, T., Near-Infrared Spectroscopy in Food Analysis, Wiley, New York,1988.*

19. *McClure, W. F. and Stanfield, D. L., 'Near-infrared Spectroscopy of Biomaterials', in Handbook of Vibrational Spectroscopy, Vol. 1, Chalmers, J. M. and Griffiths, P. R. (Eds), Wiley, Chichester, UK, 2002, pp. 212–225.*

20. *Carr, C., 'Vibrational Spectroscopy in the Paint Industry', in Handbook of Vibrational Spectroscopy, Vol. 4, Chalmers, J. M. and Griffiths, P. R. (Eds), Wiley, Chichester, UK, 2002, pp.2935–2951.*

Application of FTIR Spectrophotometer in Dairy Industry

The advancement (Statistical software-chemometric technique) in FTIR spectroscopy makes this technique useful in the identity of dairy products (milk, ice cream, yogurt, butter, cheese, *etc.*). Some of the examples of application of FTIR in dairy industry are given below:

i. Quality Analysis and Process Control

Milkoscan based on FTIR is being used for determination of Fat, SNF, Protein, Lactose, Total solids *etc.* Some models are equipped for detection of adulterants in milk. FTIR Based equipments are suitable for real time on line standardisation of milk and other liquid dairy products

ii. Rapid Detection of Food Borne Pathogens [1]

Fourier transform infrared spectroscopy (FT-IR) is used to generate bacterial spectral scans based on the molecular composition of a sample and mainly consists of the infrared source, the sample and the detector. It is a non destructive rapid method and sample identification depends on the available spectral library. When IR is absorbed or transmitted through the sample to the detector, it generates a scan or fingerprint profile. A library of spectral scans can be generated for different bacterial species and strains, which can be used for future comparison. This method requires transfer of cells (biomass) from the growth media to an IR reflecting substrate for spectral collection. FT-IR has been used for classification or identification of several foodborne pathogens: Yersinia, Staphylococcus, Salmonella, Listeria, Klebsiella, Escherichia, Enterobacter, Citrobacter, *etc.* [2,3,4]. FT-IR photoacoustic spectroscopy was used for the identification of spores of several Bacillus species with 100 per cent accuracy [5].

Source

1. *P.K. Mandal, A.K. Biswas, K. Choi and U.K. Pal, 2011. Methods for Rapid Detection of Foodborne Pathogens: An Overview. American Journal of Food Technology, 6: 87-102.*

2. *Gupta, M.J., J.M. Irudayaraj, C. Debroy, Z. Schmilovitch and A. Mizrach, 2005. Differentiation of food pathogens using FTIR and artificial neural networks. Trans. ASAE., 48: 1889-1892.*

3. *Mossoba, M.M., S.F. Al-Khaldi, J. Kirkwood, F.S. Fry, J. Sedman and A. A. Ismail, 2005. Printing microarrays of bacteria for identification by infrared micro spectroscopy. Vibrational Spectroscopy,38: 229-235.*

4. *Sivakesava, S., J. Irudayaraj and C. Deb Roy, 2004. Differentiation of microorganisms by FTIR-ATR and NIR spectroscopy. Trans. ASAE., 47: 951-957.*

5. *Thompson, S.E., N.S. Foster, T.J. Johnson, N.B. Valentine and J.E. Amonette, 2003. Identification of bacterial spores using statistical analysis of Fourier transform infrared photoacoustic spectroscopy data. Applied Spectroscopy, 57: 893-899.*

iii. Other Application

FTIR is successfully being used for detection of adulterants in milk, determination of SNF, determination of cholesterol *etc.* and these applications are tabulated below [8]:

Sl.No.		*Application*	*Instruments/ Wavelengths Used*	*Chemometric Technique Used*	*Significant Results*
1		Detection of Adulterants in milk.			
	a	Estimation of milk urea nitrogen [1]	CombiScope FTIR	-	Good precision and accuracy in repeatability and reproducibility for the measurement of raw milk urea nitrogen. However failed to detect addition of 40 mg urea in milk per 100ml milk.
	b	Detection of soymilk in cow–buffalo milk [2]	Bruker with ATR/1472–1241 cm^{-1}	Principal Component Analysis	The presence of soya milk can be predicted in the range of 1472-1241 cm^{-1} using multiple linear regression with coefficient of determination (R2) of 0.99 and 0.92 for calibration and validation respectively.
	c	To identify Adulterated rawmilk [5]	Foss Analytical/ 5000 – 1000 cm^{-1}	Principal Component Analysis	FT-IR can be used for the identification of adulterated milk with 0,05 per cent and 0,075 per cent of sodium bicarbonate and citrate respectively.
	d	Presence of beef fat in butter [6]	Nicolet 6700 FTIR with ATR/1500-1000 cm^{-1}	Partial least square (PLS) regression	2.42 per cent beef could be detected in butter fat.
	e	Detection of Melamine in milk [7]	Thermo Nicolet Avatar 330 FTIR Spectrometer/ 4000–650 cm^{-1}	partial least squares models	FTIR in combination with partial least-squares models could detect melamine at 2.5 ppm level.
2		Determination of SNF in milk [3].	Bruker FTIR Spectrophoto-meter/3000–920 cm^{-1} and 1800–920 cm^{-1}	Partial least Squares regression (PLS) and support vector machine (SVM) regression	FTIR coupled to chemometrics Techniques (like PLS and SVM regressions) is an adequate method for the Quantitative analysis of SNF.

Contd.

Sl.No.	Application	Instruments/ Wavelengths Used	Chemometric Technique Used	Significant Results
3	Determination of Cholesterol in Dairy products [4].	Bio-Rad FTS 6000/2800 and 3200 cm^{-1}	Partial least square (PLS) and principal component regression (PCR)	Results indicate that FTIR spectroscopy can determine the cholesterol content in dairy products in approximately 5 min.

Source

1. *Oliveira, M. C. P. P., Silva, N. M. A., Bastos, L. P. F., Fonseca, L. M., Cerqueira, M. M. O. P., Leite, M. O., and Conrrado, R. S. (2012). Fourier Transform Infrared Spectroscopy (FTIR) for MUN analysis in normal and adulterated Milk. Arquivo Brasileiro de Medicina Veterinaria e Zootecnia, 64 (5), 1360-1366.*
2. *Jaiswal, P., Jha, S. N., Borah, A., Gautam, A., Grewal, M. K., and Jindal, G. (2015). Detection and quantification of soymilk in cow–buffalo milk using Attenuated Total Reflectance Fourier Transform Infrared spectroscopy (ATR-FTIR). Food Chemistry, 168, 41-47.*
3. *Bassbasi, M., Platikanov, S., Tauler, R., and Oussama, A. (2014). FTIR-ATR determination of solid non fat (SNF) in rawmilk using PLS and SVM chemometric methods. Food Chemistry, 146, 250-254.*
4. *Paradkar, M. M., and Irudayaraj, J. (2002). Determination of cholesterol in dairy products using infrared techniques: 1.FTIR spectroscopy. International Journal of Dairy Technology, 55 (3), 127-132.*
5. *Cassoli, L. D., Sartori, B., Zampar, A., and Machado, P. F. (2011). An assessment of Fourier Transform Infrared spectroscopy to identify adulterated raw milk in Brazil. International Journal of Dairy Technology, 64 (4), 480-485.*
6. *Nurrulhidayah, A. F., Che Man, Y. B., Rohman, A., Amin, I., Shuhaimi, M., and Khatib, A. (2013). Authentication analysis of butter from beef fat using Fourier Transform Infrared (FTIR) spectroscopy coupled with chemometrics.International Food Research Journal, 20 (3), 1383-1388.*
7. *Jawaid et al., 2013.*
8. *Rajan Sharma, Bimlesh Mann and Vikrant, Fourier Transform Infrared Spectroscopy-Concept and its Application in Quality Assessment of Dairy Foods, NDRI, Karnal.*

iv. Detection of Milk Fat Adulteration [1]

The adulteration of butter fat with foreign fat could be detected by observing the FTIR spectra at the specific wavelength due to the ratios of *cis*-unsaturations of fatty acid moieties as reported [2]. FTIR coupled with chemomatrics and PLS analysis was employed [3] and successfully detected adulteration of milk fat with beef fat and chicken fat at the frequency region of 1500-1000 cm^{-1} (beef fat adulteration) and 1200-1000 cm^{-1} (chicken fat adulteration), respectively.

Source

1. *Vivek Sharma and Tanmay Hazara, Detection of Milk Fat Adulteration using Modern Approaches, Dairy Chemistry Division, NDRI, Karnal.*
2. *Sato, T., Kawano, S. and Iwamoto, M. 1990. Detection of foreign fat adulteration of milk fat by near infra-red, spectroscopic method. J. Dairy Sci. 73: 3408-3413.*
3. *A. Nurrulhidayah. F.; Rohman A.; Amin I.; Shuhaimi M. and Khatib A. (2013) Analysis of chicken fat as adulterant in butter using fourier transform infrared spectroscopy and chemometrics. Grasasy Aceites, 64 (4), 349-355.*

Application of NIR Spectroscopy [1]

The methods based on Qualitative and quantitative near infrared (NIR) spectroscopy require the application of multivariate calibration algorithms and statistical methods to model NIR spectral response to chemical or physical properties of the samples used for calibration. The NIR method relies on the spectra-structure correlations existing between a measured spectral response caused by harmonics of the fundamental vibrations occurring at infrared frequencies. The harmonic vibrations that occur at unique frequencies depend upon the quantity of absorber (analyte), type of absorbing molecules present within the sample, and the sample thickness. The important points in NIR spectroscopy are summarised below:

(a) Quantitative methods are possible in that cases where changes in the response of the near infrared spectrometer are proportional to changes in the concentration of chemical components, or in the physical characteristics of samples.

(b) Recent advancement in NIR measurement technique include the emergence of chemometrics and the diminishing distinction between **near infrared**, and **infrared** as measurement techniques.

(c) The near infrared spectroscopy is used as a technique for identification, quantitative analysis, and structure-correlation assessment of NIR functional group frequencies.

(d) Traditional near infrared spectroscopy has been most often used for analysis of proteins (2180 nm), carbohydrates (2100 nm), and moisture (1440 and 1940 nm).

(e) The basic uses of near infrared spectroscopy have been for process control, for quality assessment, for identification of raw materials and process byproducts, and for chemical quantitative analysis of complex mixtures.

(f) When analyzing synthetic and natural materials NIR spectroscopy has shown unprecedented industrial success in multiple applications utilizing tens of thousands of instruments in **grains**, forages, baking products, flour, beverages, feeds, pharmaceuticals, **dairy products**, *etc.*

Source

1. *Wikipedia.*

Some examples of application of this spectroscopy are given below:

A. Grain and Cereal Analysis

Commercial grains are commonly analysed by using NIR spectroscopy [6,7]. The major constituents of grains are water, protein, oil, fibre, minerals and carbohydrates and it is commercially important to quantitate the composition. The NIR spectra of such materials are, thus, dominated by the overtones and combination bands of C–H, N–H, O–H and C=O bonds.Some of the examples are as follows:

- *(i)* P. Williams (In 1973) reported the use of a commercial NIR grain analyzer for analyses of cereal products following the pioneer work of Norris and others. B. Osborne and T. Fearn described the uses of NIR in the food and beverage industries. P. Williams and co-workers described the work in flour milling using near-infrared spectroscopy for the determination of moisture in cereals and cereal grains.
- *(ii)* Forage analysis using NIR measurement has been a major application of the technique largely due to the work of J.S. Shenk, M. Westerhaus, W. Barton, G. Marten, N. Martin [1,3].

Source

1. *H. Martens and T. Naes, Multivariate Calibration, John Wiley and Sons, Chichester, 1992.*
2. *Wikipedia.*
3. *The handbook edited by G.C. Marten, University of Minnesota; J.S. Shenk, Pennsylvania State University and F.E. Barton, Richard Russell Research Center, USDA.*
4. *A book edited by C. Roberts, J. Workman, and J. Reeves.*
5. *ASD Technial Literature.*
6. *McClure, W. F. and Stanfield, D. L., 'Near-infrared Spectroscopy of Biomaterials', in Handbook of Vibrational Spectroscopy, Vol. 1, Chalmers, J. M. and Griffiths, P. R. (Eds), Wiley, Chichester, UK, 2002, pp. 212–225.*
7. *Williams, P., 'Near-infrared Spectroscopy of Cereals', in Handbook of Vibrational Spectroscopy, Vol. 5, Chalmers, J. M. and Griffiths, P. R. (Eds), Wiley, Chichester, UK, 2002, pp. 3693–3719.*

B. Dairy Industry

NIR spectroscopy has been used for many years in the dairy industry to examine milk. While milk consists of more than 80 per cent water, the concentrations of the fat and protein components have a significant effect on the near-infrared spectrum. The spectra of milk samples shift to lower wavelengths as the fat and protein concentrations decrease. This is due to changes in the amount of light scattering by the fat globules and protein micelles. The following are some examples of applications in dairy industry:

i. In Analysis of Cheese and Cheese Slices [1,2]

In the NIR region there are three major chemical entities measured, C-H = fat, N-H = protein, O-H = water and C-O-H = sugar. These happen to be the major components of dairy, food and agricultural products [2]. As such analysis of fat, protein, water and sugars in milk and cheese are well established.

Near Infrared Transmission spectroscopy (NIT), measures fat, protein and water in cheese by passing light through a 10mm thick sample of the cheese. NIT offers the advantage over NIR reflectance measurements, in that the spectra are collected through the sample and not just the surface. The ability to develop NIT calibrations for pH and Moisture in Cheese Slices and the bulk cheese used in the preparation of the slices, is used. pH is not measured directly in the NIR region, however the pH of the sample affects the water peak and this can be used to calibrate for pH.

The **NIT-38 Dairy Analyser** can measure pH and Moisture in cheese. Approximately 30-50 samples for both Bulk Cheese and Cheese Slices would be required to develop a more accurate and robust calibration, however the calibrations developed with these samples can be used as a starter calibration to which more spectra can be added.

Source

1. *A Study in Australia.*
2. *Barbara Stuart, Infrared Spectroscopy- Fundamentals and Applications, 2004 John Wiley and Sons, Ltd ISBNs: 0-470-85427-8 (HB); 0-470-85428-6 (PB).*

ii. Determination of Cheese Quality and Authenticity (NIR and MIR) [1]

The near-infrared (NIR) and mid-infrared (MIR) spectroscopic techniques for process monitoring, quality control, and authenticity determination in cheese processing, are developed. Infrared spectroscopy has been identified as an ideal process analytical technology tool, and recent publications have demonstrated the potential of both NIR and MIR spectroscopy, coupled with chemometric techniques, for monitoring coagulation, syneresis, and ripening as well as determination of authenticity, composition, sensory, and rheological parameters. Emerging research areas for these technologies, such as cheese authenticity and food chain traceability, are also established.

Source

1. *Tony Woodcock, Colette C.Fagan, Colm P.O'DonnellandGerard Downey, Food and Bioprocess Technology, Springer New York, Volume 1, Number 2/June, 2008.*

iii. VIS/NIR Spectroscopy

Vis/NIR spectroscopy has been successfully used in the production of cheese, meat and meat products, fish, egg products, milk, butter, beer, wine, fruit juices, baked goods, cereal, produce, jams, jellies, nutritional supplements and more. Some of the more common applications include measuring of the followings:

(i) Fat levels.

(ii) Protein content.

(iii) Moisture content.

(iv) Oil content.

(v) Sugar content

(vi) Starch content.

(vii) Blend uniformity analysis.

Food scientists/Technologists are also utilizing Vis/NIR spectroscopy to develop specific ingredients to enhance the nutritional quality of foods and beverages. With an ASD NIR food analyzer, anyone can generate increased information for tighter production controls, decreased off-spec product, better overall product quality, and a boost to bottom line.

ASD's Portable, Benchtop and Online Vis/NIR Food Analyser.

ASD's portable, benchtop and online Vis/NIR food analysis instrumentation provides the flexibility to measure where and when it is most advantageous. ASD's Vis/NIR measurement technology offers the cost effective solution on the following:

(i) Easily analyze product on the production floor or manufacturing facilities, on the receiving dock, in the warehouse, in many cases even through glass and clear plastic bags and bottles.

(ii) Monitor in-process and finished product to ensure quality and consistency.

(iii) Minimize risk of product and/or material waste.

Source

1. *ASD Technical Literature.*

Application of Mid Infrared Spectroscopy

IDF has recently published an bulletin IDF 406/2006 which consists of New Application of Mid–Infra-red Spectroscopy for the analysis of Milk and Milk Porducts. Some examples are given below:

i. Detection of Milk Fat Adulteration by Temperature Controlled Attenuated Total Reflectance-Mid-Infrared (ATR-MIR) Spectroscopy [1]

This is a spectroscopic technique used for the rapid estimation of butter adulteration. The methodology is typically based on the infra-red spectroscopic technique [2]. Here the temperature was controlled, which allowed the stabilization of analysis temperature at 65± 2°C. The data was analyzed by using statistical tool namely Multivariate data analysis and calibration models were developed covering all possible adulteration ratios. In this case adulteration of butter with margarine @ 2.5 per cent could be predicted.

Source

1. *Vivek Sharma and Tanmay Hazara, Detection of Milk Fat Adulteration using Modern Approaches, Compendium on on Advanced Techniques and Novel Approaches for Quality and Safety Evaluation of Dairy Foods-2015, NDRI, Karnal.*
2. *Koca.N; Kocaogulu-Vurma.N. A; Harper.W.J; Rodriguez-Saona. L.E (2010) Application of temperature controlled attenuated total reflectance – mid- infrared (ATR-MIR) spectroscopy for rapid estimation of butter adulteration.Food Chemistry.121: 778- 782.*

3. Nuclear Magnetic Resonance (NMR) Spectroscopy [1,2,3]

Nuclear Magnetic Resonance (NMR) spectroscopy is an analytical chemistry technique used in quality control and reserach for determining the content and purity of a sample as well as its molecular structure [2].

Nuclear Magnetic Resonance (NMR) Spectroscopy is a non-destructive analytical technique that is used to probe the nature and characteristics of molecular structure. A simple NMR experiment produces information in the form of a spectrum, which is able to provide following details [3]:

(i) The types of atoms present in the sample.
(ii) The relative amounts of atoms present in a sample.
(iii) The specific environments of atoms within a molecule.
(iv) The purity and composition of a sample.
(v) Structural information about a molecule, including constitutional and conformational isomerisation.

Definition and Key Features of Nuclear Magnetic Resonance

Definition

Nuclear magnetic resonance is defined as a condition when the frequency of the rotating magnetic field becomes equal to the frequency of the processing nucleus [4]. NMR is an effect where by magnetic nuclei in a magnetic field absorb and re-emit electromagnetic (EM) energy. This energy is at a specific resonance frequency which depends on the strength of the magnetic field and other factors. This allows the observation of specific quantum mechanical magnetic properties of an atomic nucleus [1].

Many scientific techniques exploit NMR phenomena to study molecular physics, crystals and non-crystalline materials through NMR spectroscopy.NMR is also routinely used in advanced medical imaging techniques, such as in magnetic resonance imaging (MRI) [1].

Key Feature

A key feature of NMR is that the resonance frequency of a particular substance is directly proportional to the strength of the applied magnetic field. It is this feature that is exploited in imaging techniques; if a sample is placed in a non-uniform magnetic field then the resonance frequencies of the sample's nuclei depend on where in the field they are located. Since the resolution of the imaging technique depends on the magnitude of magnetic field gradient, many efforts are made to develop increased field strength, often using superconductors. The effectiveness of NMR can also be improved using hyperpolarization, and/or using two-dimensional, three-dimensional and higher-dimensional multi-frequency techniques [1].

Units of NMR [4]

The nuclear magnetic resonance values are expressed in any of following three ways:

(i) δ-the reference compound be quoted (δ denotes that chemical shift is independent of oscillator frequency).

(ii) **Cps**- the reference compound must be quoted and the oscillator frequency given.

(iii) **τ-TMS** (tetra methylsilane) or DSS (2, 2 dimethyl-2 silapentane-5 sulphonate) is assumed independent of both oscillator frequency and reference compound.

Characteristics of NMR Spectroscopy [3]

There are a number of important characteristics of NMR spectroscopy which makes it favourable to a wide array of industrial, commercial and research applications:

i. Commonly Used Nuclei in NMR [1,3]

All stable isotopes that contain an odd number of protons and/or of neutrons have an intrinsic magnetic moment and angular momentum (a non zero spin), while all nuclides with even numbers of protons and/or of neutrons have zero spin. There are many nuclei which are NMR active. (i) Some of the more common NMR active nuclei are: **The most commonly studied nuclei are ^{1}H and ^{13} C.** The nuclei from isotopes of many other elements *e.g.*^{2}H,^{10}B,^{11}B, ^{14}N, ^{15}N, ^{17}O, **^{19}F, ^{23}Na, ^{29}Si,^{31}P, ^{35}Cl, ^{113}Cd, ^{129}Xe,^{195}Pt**, are studied by high-field NMR spectroscopy [1]

(ii) NMR spectroscopy is generally a non-destructive technique, meaning that samples can be recovered

(iii) Only a small quantity of material is required for analysis; sample sizes of 5-20mg are generally sufficient for most NMR experiments.

(iv) Sample preparation is simple and minimal. Typically, samples are simply dissolved in an appropriate solvent [3].

Source

1. *Wikipedia.*
2. *chem.ch.huji.ac.il/nmr/whatisnmr/whatisnmr.html.*
3. *NMR Spectroscopy, Application of NMR, School of chemistry, University of Sydeny html.*
4. *www.environmentalpollution.in.*

Principle [1,2]

The principle behind NMR is that many nuclei have spin and all nuclei are electrically charged. If an external magnetic field is applied, an energy transfer is possible between the base energy to a higher energy. The energy transfer takes place at a wavelength that corresponds to radio frequencies and when the spin returns to its base level, energy is emitted at the same frequency. The signal that matches this transfer is measured in many ways and processed in order to yield an NMR spectrum for the nucleus concerned [1].

The principle of nuclear magnetic resonance is based on the spins of atomic nuclei. The magnetic measurements depend upon the spin of unpaired electron whereas nuclear magnetic resonance measures magnetic effect caused by the spin of protons and neutrons. Both these nucleons have intrinsic angular momenta or spins and hence act as elementary magnet [5].

The principle of NMR involves following two sequential steps:

(i) The alignment (polarization) of the magnetic nuclear spins in an applied, constant magnetic field H_0.

(ii) The perturbation of this alignment of the nuclear spins by employing an electro-magnetic (usually radio frequency –RF pulse). The required perturbing frequency is dependent upon the static magnetic field (H_0) and the nuclei of observation [2].

The two fields are usually chosen to be perpendicular to each other as this maximizes the NMR signal strength. The resulting response by the total magnetization (M) of the nuclear spins is the phenomenon that is exploited in NMR spectroscopy and magnetic resonance imaging. Both use intense applied magnetic fields (H_0) in order to achieve dispersion and very high stability to deliver spectral resolution.NMR phenomena are also utilized in low-field NMR (In analysis of dairy products) [2].

Theory of NMR and Nuclear Spin/Angular Momentum [3,4,5]

Theory [5] -The hydrogen nucleus or protons can be regarded as a spinning positively charged unit and so it will generate a tiny magnetic field H_0 along its spinning. Now if this nucleus is placed in an external magnetic field H_0, it will naturally line up either parallel A or antiparallel B to the direction of external field. The A will be more stable, being of lower energy.

The energy difference AE between two states will be absorbed or emitted as the nucleus flips from one orientation to the other. Then:

$$AE = hv$$

where - v - a radiation frequency and h -Planck's constant.

If correct frequency is applied to the sample containing hydrogen nuclie and sample is placed in the external field H_Q, then low energy nuclie A will absorb AE = hv, and flips to B. Thus on flipping back down, they remit hv as a radiation signal which is picked up by the instrument. In other words, if both radio frequency and magnetic field are simultaneously applied to the nucleus, transition from lower to higher level will occur when equation -1 will be equal to equation-2

$$\delta \mathbf{E} = \partial \mathbf{H}/2\pi \text{ Equ. -1}$$

$$\delta \mathbf{E} = \mathbf{hv} \text{ Equ. -2}$$

$$\mathbf{or\ v} = \delta \mathbf{H}/2\pi \text{ Equ. -3}$$

where - δ - Gyromagnctic ratio, a constant characteristic of a particular nucleus, δE - energy difference between two spin states, h - Planck's constt, v - frequency of resonance absorption, H - strength of applied magnetic field at nucleus.

The system at this condition is said to be in resonance and hence the name nuclear magnetic resonance. The observed value of H is therefore a function of molecular environment of proton affording the signal.

Nuclear Spin/Angular Momentum

Some atomic nuclei have an intrinsic property called "spin". This was first demonstrated in 1922 by the Stern-Gerlach experiment, in which a beam of silver atoms were passed through a magnet field and split in to two beams. These two beams represent the two states, α and β, of the silver (spin 1/2) nuclei.

The nuclear spin has an intrinsic angular momentum, a vector that is represented by the symbol **I** (vectors will be in bold). Vectors have 3 orientations (x, y, and z) and a length. However, the HeisenbergUncertainty Principle tells that we can only know one orientation and the length simultaneously. By convention in NMR, we say that we know the z-orientation of the angular momentum of a nucleus in a magnetic field. The square of the magnitude of **I** is given by

$$\mathbf{I}^2 = \hbar^2 [\mathbf{I}\ (\mathbf{I} + 1)$$

where– I- the spin quantum number (*e.g.* 1/2) and **ħ**- Planck's constant divided by 2π. The z-component of **I** is given by

$$\mathbf{I_Z} = \hbar \mathbf{m}$$

where- *m* - the quantum number with values $m = (-I, -I+1,, I-1, I)$. In the case of spin 1/2 nuclei, $I = 1/2$ and *m* can be -1/2 and 1/2.

Nuclei that have spin angular momentum also have a magnetic moment. NMR transition energies are very small. These small energies translate into low sensitivity. In the absence of a magnetic field, the α and β states are equally populated, leading to no net magnetization (that is why we need large static magnets to do NMR). When samples are placed into a magnetic field, *a* small excess of nuclei fall into the a state. This excess of spins in the α over the β states accounts for the entire net magnetization which is used in the NMR experiment. The ratio of the number of spins in the α state to those in the β state is given by a **Boltzman distribution [3].**

Source

1. *chem.ch.huji.ac.il/nmr/whatisnmr/whatisnmr.html.*
2. *Wikipedia.*
3. *Dr. Arthur S. Edison, Theory and Applications of NMR Spectroscopy, WWW: http://ascaris.health.ufl.edu.*
4. *"Principles of Nuclear Magnetic Resonance in One and Two Dimensions" Ernst, Bodenhausen and Wokaun, Oxford (1987).*
5. *www.environmentalpollution.in.*

Chemical Shift [1,2]

The most important molecular parameter determined by NMR is **the chemical shift**. The chemical shift is defined as a measure of the resonance frequency of the nuclei in a given chemical environment. The magnitude of the chemical shift is proportional to the strength of applied field and is caused by the circulations of surrounding electrons about the protons [2].

The precise resonant frequency of the energy transition is dependent on the effective magnetic field at the nucleus. This field is affected by electron shielding which is in turn dependent on the chemical environment. As a result, information about the nucleus' chemical environment can be derived from its resonant frequency. In general, the more electronegative the nucleus is, the higher the resonant frequency. Other factors such as ring currents (anisotropy) and bond strain affect the frequency shift. It is customary to adopt tetramethylsilane (TMS) as the **proton reference frequency**. This is because the precise resonant frequency shift of each nucleus depends on the magnetic field used. The frequency is not easy to remember so it was decided to define chemical shift as follows to yield a more convenient number such as 7.17 ppm [1].

$$\delta = (\nu - \nu_0)/\nu_0$$

The chemical shift, using this equation, is not dependent on the magnetic field and it is convenient to express it in ppm where (for proton) TMS is set to ν_0 thereby giving it a chemical shift of zero. For other nuclei, ν_0 is defined as $\Xi\ \nu_{TMS}$ where Ξ (Greek letter Xsi) is the frequency ratio of the nucleus (*e.g.*, 25.145020 per cent for ^{13}C).

The chemical shift is easily understandable by the following example [1]:

i. Ethyl Benzene

In ^1H NMR spectrum of ethyl benzene

(*a*) The methyl (CH_3) group is the most electron withdrawing (electronegative) and therefore resonates at the lowest chemical shift.

(*b*) The aromatic phenyl group is the most electron donating (electropositive) so has the highest chemical shift.

(*c*) The methylene (CH_2) falls somewhere in the middle. However, if the chemical shift of the aromatics were due to electropositivity alone, then they would resonate between four and five ppm. The increased chemical shift is due to the delocalized ring current of the phenyl group.

Chemical shifts are affected slightly by isotopic substitution, an effect that is known as an isotope shift.

Spin-spin Coupling [1]

The effective magnetic field is also affected by the orientation of neighboring nuclei. This effect is known as spin-spin coupling which can cause splitting of the signal for each type of nucleus into two or more lines. The size of the splitting (coupling constant or *J*) is independent of the magnetic field and is therefore measured as an absolute frequency (usually Hertz). The number of splittings indicates the number of chemically bonded nuclei in the vicinity of the observed nucleus. There are two types of couplings:

i. First Order Coupling

The first order approximation are correct provided that all the coupled spins have widely separated chemical shifts. The different nuclei are labeled with the letters A and X (in a system of this type the letters come from widely separated parts of the alphabet). If the chemical shifts are similar then distortions in peak height occur for example-In ethylbenzene, the methyl (CH_3) group has a coupling pattern in the form of A_3X_2 which to a first order approximation looks like an AX_2 multiplet. Likewise, the methylene (CH_2) group has the form A_2X_3 that is equivalent to AX_3. The first order approximation works because the groups are widely separated in the spectrum.

ii. Second Order Coupling

For more than two spins, extra signals may appear. These effects are called second order coupling. For example- In ethylbenzene, the aromatic signals are close together and display second order effects. The **ortho** signal is a doublet AX while the **meta** and **para** signals are triplets.

Source

1. *chem.ch.huji.ac.il/nmr/whatisnmr/whatisnmr.html.*

2. *www. environmental pollution.in.*

Instrumention

The basic elements of a typical NMR spectrometer consist of the following main parts [1]:

(i) **Magnet** - It is with strong, stable homogeneous field. The field must be constant over the area of the sample.

(ii) **Radio frequency oscillator (transmitter)-**It is connected to coil which transmits energy to the sample in a direction perpendicular to the magnetic field.

(iii) **Sample container-**It is usually a glass tube spun by an air driven turbine to average the magnetic field over the sample dimensions.

(iv) **Radio frequency receiver-** It is connected to a coil encircling the sample. The two coils are perpendicular to each other and to the magnetic field.

(v) **A read out system-** The other supporting parts are - consisting of an amplifier, recorder and additional components for increasing sensitivity, accuracy or convenience.

(vi) **Sweep generator-** It supplies a variable d-c current to a secondary magnet so that the total applied magnetic field can be varied (swept) over a limited range.

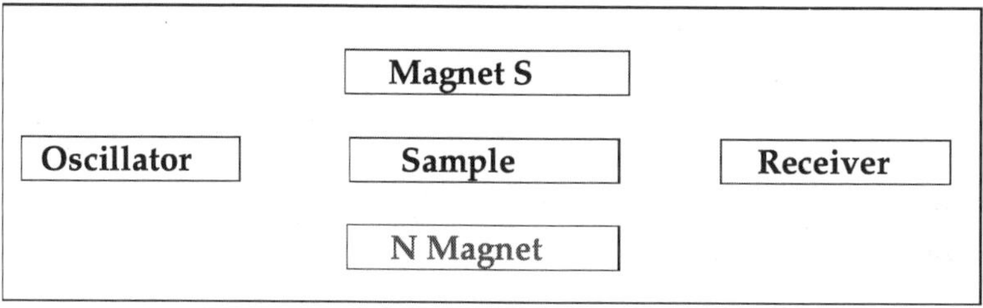

Working Line of NMR Spectrometer.

Source

1. *www.environmentalpollution.in*

Operation of NMR Spectrometer [1]

Always a dilute solution is analysed. The compound to be studied is generally mixed with a solvent like CC_{14} or etramclhyl silane and the dilute solution is filled in a tube.Now when a sample under investigation is placed in the magnetic field and subjected to rf field of oscillator then at particular combinations of the oscillator frequency and field strength, the rf. energy is absorbed by certain nuclei and an rf. signal is picked up by the detector.

The following two ways have been employed in NMR experiments for getting the desired particular combinations:

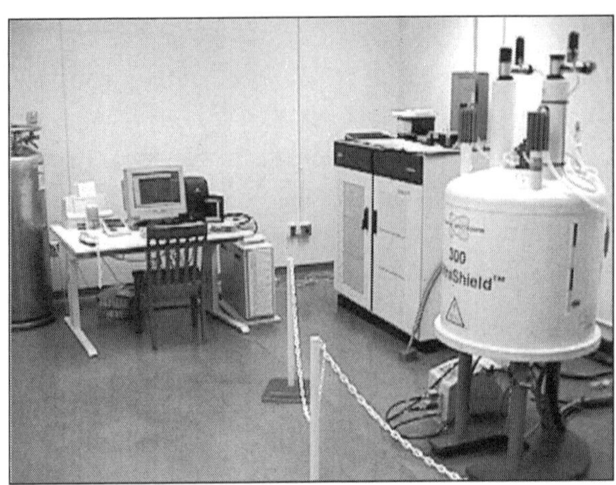

GHz NMR Spectrometer (1000 MHz, 23.5 T)
(*Source*: Centre de RMN à Très Hauts Champs' in Lyon, France).

A 900MHz NMR Instrument with 21.2T Magnet
(*Source*: HWB-NMR, Birmingham, UK).

(i) In one way, the magnetic field remains constant and radio frequency is varied.

(ii) In second, the radio frequency remains unchanged and magnetic field is varied till resonance conditions are obtained and there is detectable absorption by the nucleus.

Source

1. *www.environmentalpollution.in*

Interpretation of NMR Spectrum [1]

The number spectrum gives several kinds of information. Some of these information are discussed below:

(i) The number of signals (peaks) tells us how many kinds of protons are present in a molecule.

(ii) The area under each signal tells us how many protons of each kind are in the molecule.

(iii) The position (chemical shift) of the signal informs about the bonding environment of each proton.

(iv) All hydrogens with identical environments in a molecule have same chemical shift (*e.g.*, a- all the three protons of a methyl CH_3, b-the protons of a methylene-CH_2, c- one identical).

(v) Protons on heteroatoms (H-S, H -N, H-O *etc.*) show highly variable chemical shifts and sometimes broad peaks.

(vi) Hydrogen on different carbons yields the same absorptional signal if they are structurally indistinguishable.

(vii) Sometimes a proton exhibits an absorption signal which is split into several peaks because of coupling with its neighbouring protons. In such cases a coupling constant J is calculated.

(viii) The number of peaks (N) into which a proton signal is split equals one more than the number of vicinal protons (n) (number of equivalent neighbours causing splitting):

$$N = n + 1$$

N = 2 (one vicinal H) = doublet (d), = 3 (two vicinal H's) = triplet (t), = 4 (three vicinal H's) = quartet (q).

The following example of spectrum of-isomers-dimethyl ether and ethanol is helpful in understanding the above aspect:

In the low resolution spectra of isomers dimethyl ether, only one signal is shown because all 6H atoms are equivalent. The 3H atoms in a given CH_3 group are indistinguishable, as are the two CH_3 groups. The spectrum of ethanol has three signals-one each for CH_3, CH_2 and OH protons. The relative areas under the peak for the ethanol are 3: 2: 1 for CH_3, CH_2, and OH groups respectively. CH_2 signal is far away than CH_3 signal because of electron withdrawing effect of two adjacent atoms.

Source

1. *www.environmentalpollution.in.*

Types of NMR Techniques [1,2]

On the basis of the different experimental equipment employed, NMR techniques can be classified in following varieties:

i. Low-Resolution NMR

It is concerned with the quantitative determination of the different components in a sample by the analysis of the amplitudes and the decay rates of the NMR signal and is the technique that has most extensively been used for food analysis [2]. This is mainly due to the possibility of studying the properties of water - a very abundant probe, whose amount, dynamic state and interaction with solid matrixes affect many of the properties of food products.

ii. High-Resolution NMR

It is the area of NMR, which are most familiar and is concerned with the elucidation of the chemical structure of molecules in solution, so that its use in the study of food products is somewhat limited by the need to have the component to be studied insolution.

iii. Solid-State High-Resolution NMR

The use of this **NMR** to study dairy products is in its infancy and is concerned with the elucidation of chemical structure and conformation of solid components (proteins and carbohydrates) in the sample.

iv. NMR Imaging

It enables one to describe the internal morphology of the materials to be investigated with an on-plane resolution of up to 10μm and slice thickness of about 1 mm. Its potential lies in the possibility of obtaining resolved spectra from different regions within the image and of monitoring dynamic events, such as the migration of water, during the processing, ripening and storage of cheese.

Source

1. *Elvino Brosio and Renato Barbieri, Nuclear Magnetic Resonance in the Analysis of Dairy Products, Reviews in Analytical Chemistry, Vol. 15, No. 4, 1996.*
2. *E. Brosio and A. Di Nola, Trends in Analytical Chemistry, 1,284 (1982).*

Application of NMR Spectroscopy

Common applications of NMR Spectroscopy are guven below [1]:

(i) Structure elucidation.

(ii) Chemical composition determination.

(iii) Formulations investigation.

(iv) Raw materials fingerprinting.

(v) Mixture analysis.

(vi) Sample purity determination.

(vii) Quality assurance and control.

(viii) Quantitative analysis.

(ix) Compound identification and confirmation.

(x) Analysis of inter- and intramolecular exchange processes.

(xi) Molecular characterization.

(xii) Reaction kinetics examination.

(xiii) Reaction mechanism investigation.

Methods in NMR spectroscopy have particular relevance to the following fieldds [1]:

(i) Chemical research and development-organic, inorganic and physical chemistry.

(ii) Chemical manufacturing industry.

(iii) Biological and biochemical research.

(iv) Food industry.

(v) Pharmaceutical development and production.

(vi) Agrochemical development and production

(vii) Polymer industry.

Source

1. *NMR Spectroscopy, Application of NMR, School of chemistry, University of Sydeny html.*

Application in Food/Dairy Industry [1]

NMR is a powerful technique for determining the chemical structure, chemical composition and physical properties of different components in a sample. The possibility of using a non-invasive technique to obtain information on the chemical and physical structure of complex, heterogeneous materials makes NMR particularly suitable for the study of food products.

The intrinsic low sensitivity is only problem with NMR and because of this reason:

(i) NMR is used to obtain information on the most abundant components of foods - water, lipids, proteins and carbohydrates.

(ii) NMR is not well suited for the study of components present at only trace level, such as colorants and flavours.

However, other characteristics of NMR (it is completely non-invasive, it is not time consuming, it does not need any pretreatment of the sample) and the wealth of detailed information that can be obtained, largely compensate for the low sensitivity of the technique.

The inherent insensitivity of NMR is the reason why 'H NMR is used for the analysis of food systems, even if in some cases less sensitive nuclei like^{13}C, ^{31}P, ^{17}0 and ^{23}Na are used.

Source

1. *Elvino Brosio and Renato Barbieri, Nuclear Magnetic Resonance in the Analysis of Dairy Products, Reviews in Analytical Chemistry, Vol. 15, No. 4, 1996.*

Application in Dairy Industry [1,2,3,4]

NMR was first used to investigate the properties of milk in the 1950s and has since been employed in a wide range of studies; including properties analysis of specific milk proteins to metabolomics techniques used to monitor the health of dairy cows [2]. The applications of NMR to dairy products are numerous and can be divided into following three groups [1]:

i. The First Group (High-Resolution NMR Spectroscopy)

It comprises the use of high-resolution NMR spectroscopy for the molecular study of dairy compounds such as casein and whey protein.

a. Ricotta Cheese HR SOLID-STATE NMR Analysis [3]

Proton-decoupled ^{13}C NMR spectra of Ricotta cheese (preliminarily freeze-dried, in order to avoid phase separation while spinning in the probe) were obtained (using a **Bruker AM400 spectrometer** operating at 100 MHz for 13C, equipped with a broad band MAS probe) and analysed. The carbonyl signal is characterized by two intense side bands, arising from an incomplete averaging of the large chemical shift anisotropy of the carbonyl group. Signals between 15 and 40 ppm were assigned to side chain methylene and methyl groups, while resonances around 55 ppm corresponded to α-CH carbon atoms. The peaks (signal) at 70-80 ppm and 90-105 ppm were observed and assigned to -CHOH and anomeric saccharide carbon atoms, respectively.^{13}C CPMAS can only be used as a fingerprint of the protein component in cheese.

b. ^{31}Ñ NMR Analysis of Milk [3]

High-resolution NMR generally involves^{1}H and ^{13}C nuclei. However, with the advent of high-field Fourier Transform NMR instruments many other nuclei can be measured (*e.g.*^{31}Ñ *etc.*). In the ^{31}P NMR spectrum of liquid whole or skim milk [4] only three resonance signals are observed. Of these the most intense signal corresponds to inorganic phosphate, Pi, in solution; the other two peaks are assigned to seryl phosphate residues of casein and to a phosphate diester. The presence of the phosphocreatine signal at 2.32 ppm represents an unexpected result of this investigation because, while creatine and creatinine are known to be present in milk, the presence of phosphocreatine had previously not been reported in milk. Further more, some precursors of phospholipids have also been detected by analysis of the 31P NMR spectrum, probably as a result of the enzymatic action of phospholipase on phospholipids or originating in leakage from the secretory cells.

ii. The Second Group (Magnetic Resonance Imaging)

It involves the use of Magnetic resonance imaging applied to dairy processing.

iii. The Third Group (NMR Relaxation Measurements)

It concerns studies based on NMR relaxation measurements. The NMR relaxation applications, including pulsed field gradient NMR for the measurement of molecular diffusion in dairying plays significant role. Relaxation measurements on dairy products have usually been performed at low field and the main aim was to propose a non-destructive technique to determine water or fat content and solid fat content/Solid not fat (SFC/SNF).

a. Low Resolution NMR in the Study of Milk and Milk Derivatives [3]

Water is the most abundant component in milk (~85 per cent) and it is extensively present in all types of cheese (from 25 to 75 per cent). It plays an essential role in determining the structure of proteins, polysaccharides and lipid aggregates, influencing the texture and stability of the final product. **NMR relaxation and diffusion measurements** have been shown to give detailed information on the state of water, to be a sensitive probe in the study of structure changes associated with protein and/or lipid components. The quantitative variations in water **transverse relaxation** rate occurring after heating of whey protein solutions can be used to measure the degree of protein denaturation.Due to the importance of acidification of milk in different manufacturing processes like yogurt and cheese, transverse relaxation time has also been measured to investigate the effect of pH on skim milk [5,6]. NMR transverse relaxation times of water have been measured in skim milk after renneting, which causes a gel network to be formed, and during the syneresis of the curd [7]. The water mobility can be obtained by measuring the **self-diffusion coefficient** of water molecules.Unlike relaxation times, experimental values of self diffusion coefficients are easy to interpret, as they are directly related to the molecular displacement of water molecules. The **pulsed field gradient NMR** technique is a reliable method for determining molecular self-diffusion coefficients with good precision and relatively short measuring times [8]. The pulsed field gradient technique was used to measure water and fat self-diffusion coefficients in Cheddar and Swiss cheese [9].

It is a technique in increasing use for dairy research and provides us with unique information that can be applied to research or to quality control of dairy samples. In addition, it is a non-destructive and very versatile technique, providing data on the same sample under different parameters. NMR spectroscopy is being used for qualitative and quantitative analysis, monitoring reactions in vivo, isotopic analysis, study of the physical state of milk fat and water, structural characterization and studies on the conformational and aggregation state of proteins [1].

Source

1. *Josefina Belloque and Mercedes Ramos, Instituto de Fermentaciones Industriales, CSIC, Juan de la Cierva 3, 28006 Madrid, Spain.*
2. *Anthony D. Maher and Simone J. Rochfort, Applications of NMR in Dairy Research, Metabolites. 2014 Mar; 4 (1): 131–141.*
3. *Elvino Brosio and Renato Barbieri, Nuclear Magnetic Resonance in theAnalysis of Dairy Products, Reviews in Analytical Chemistry, Vol.15, No.4,1996.*
4. *P.S.Belton, R.L.J. Lyster and C.P. Richards, J. Dairy Research, 52, 47 (1985).*
5. *S.P.F.M. Roefs, H. Van As and T. Van Vliet, J. Food Science,54, 704 (1989).*
6. *F. Mariette, C. Tellier, G. Bηiŕε and P. Marchai, J. Dairy Research,60, 175 (1993).*
7. *C. Tellier, F. Mariette, J.P. Guillement and P. Marchai, J. Agric. FoodChem., 41, 2259 (1993).*
8. *O.S. Stejskal and J.E. Tanner, J. Chem. Phys., 42, 288 (1965).*
9. *P. T. Callaghan, K.W. Jolley and R.S. Humphrey, J. Colloid InterfaceSei., 93, 521 (1983).*

Other Specific Application in Dairy Industry

Some of the applications of NMR spectroscopy in dairy are as follows:

a. Characterization of Buffalo Milk by ^{31}P-nuclear Magnetic Resonance Spectroscopy [1]

The nucleous^{31}P nuclear magnetic resonance (^{31}P-NMR) was used in investigation of buffalo milk, milk ultrafiltrate, and phospholipids from milk fat and for comparison corresponding data were also acquired for cow milk samples. During aforesaid investigation, following results were found:

i. In Buffalo Milk Samples

Glycerophos- phorylcoline, glycerophosphoryl ethanolamine and inorganic orthophosphate resonances were identified. A broad peak assigned to serylphosphate residues of casein was also identified.

ii. Buffalo Milk Ultrafiltrate

It showed the presence of several phosporous signals, and ten of them (inorganic phosphate, phosphocreatine, glycerophos- phorylcoline, glycero-phosphoryl-ethanolamine, *N*-acetylglucos-amine-1-phosphate, lucose-6-phosphate, galactose-1-phosphate, phosphorylethanol-amine, phosphorylcoline, and glycerol-1-phosphate) were identified by addition of pure standards.

iii. In Phospholipids Fractions from Buffalo Milk

Phosphatidylcholine, phosphatidylethanolamine, phosphatidylserine, sphingomyelin and phosphatidyl-inositol were identified. A preliminary estimation (mol per cent) of each phosphorylated compound was obtained. By comparing ^{31}P-NMR data from milk samples from buffaloes and cows, it was revealed that these milks are rather similar as far as the phosporous distribution in small molecules is concerned.

Source

1. *Giuseppina Andreotti, Enrico Trivellone and Andrea Motta, Istituto di Chimica Biomolecolare del CNR, Comprensorio Olivetti, Edificio 70, Via Campi Flegrei 34, I-80078 Pozzuoli, Napoli, Italy.*

b. Characterisation of Pressure-Treated Skimmed Milk Powder Dispersions [1]

Skimmed milk powder (bovine) and skimmed milk powder/sucrose, before, during and after high-pressure treatment have been studied by nuclear magnetic resonance (NMR) spectroscopy using P-31nucleous. The P-31 spectra of samples at 293K before and after pressure (400–600MPa) treatment were not significantly different from one another, whether or not sucrose was present. This indicates that physicochemical changes relating to phosphorus-containing components occurring during a pressure cycle are either reversible or result in re-immobilisation, while protein components may be rearranged. The P-31 NMR spectra of these dispersions at aforesaid conditions showed that pressure caused a considerable increase in the free inorganic phosphate concentration and that the increase was proportional to the magnitude of pressure. Decompression to ambient caused an exact reversal of this trend.

Source

1. *Colin D Hubbard, David Caswell, Hans-Dietrich Lüdemann, Martin Arnold, Journal of the Science of Food and Agriculture, Volume 82, Issue 10, pages 1107–1114, August 2002.*

c. Determination of the Casein Content in Bovine Milk by ^{31}P-NMR [1]

The relative proportion of caseins to total protein is a parameter that can be used to control the protein quality in standardised milk. ^{31}P-NMR was used to analyse the casein content of milk, by the quantitation of the area under the resonances belonging to SerP, and using methylenediphosphonic acid as internal standard. The results obtained by this procedure are similar to values of caseins obtained from Kjeldahl method and NMR analysis for slightly heated milk samples. Heating at 95°C for 15 min did not alter the casein content results. Casein content of raw, pasteurised and UHT milks (25.6±1.4, 26.4±1.8, 25.5±1.6 g casein/l milk, respectively with an average of 25.8±1.6 g casein/l), obtained by NMR, were not significantly different. Hence, it is concluded that ^{31}P-NMR could be used as an alternative method to determine casein in raw, pasteurised, dry and UHT milks.

Source

1. *Josefina Belloque and Mercedes Ramos, Instituto de Fermentaciones Industriales (C.S.I.C.), Juan De La Cierva 3, 28006 Madrid, Spain.*

d. Milk Identification of Different Species -^{13}C-NMR Spectroscopy of Triacylglycerols from Cow and Buffalo Milk [1]

Triacylglycerols from cow and buffalo milk fat were investigated by 13C nuclear magnetic resonance (NMR) spectroscopy. By the addition of pure triacylglycerols standards, the resonances of both milk fats were identified, and the peaks were used for qualitative and quantitative analysis of acyl groups.Multivariate analysis treatment of triacyl-glycerols distribution and composition parameters helped in identifying the milk. Hence NMR can safely be used to quantitate milk fatty acid content and to identify the milk of different animal species [2,3,4].

Source

1. *G. Andreotti, Istituto Nazionale di Fisica della Materia, Unita' di Ricerca di Salerno, Universita' degli, Studi di Salerno, I-84081 Baronissi (SA), Italy.*
2. *E. Trivellone and A. Motta, Istituto di Chimica delle Molecole di Interesse Biologico and Istituto Nazionale diChimica dei Sistemi Biologici del CNR, I-80072 Arco Felice (NA), Italy.*
3. *R.Lamanna, Area della Ricerca del CNR di Napoli, I-80072 Arco Felice (NA), Italy.*
4. *A. Di Luccia, Istituto di Scienze dell'Alimentazione del CNR, I-83100 Avellino, Italy.*

e. Water Mobility in Acidified Milk Drinks Studied by Low-field ^{1}H NMR [1,2]

Low-field nuclear magnetic resonance (LF-NMR) transverse relaxation (T_2) was used to characterise acidified milk drinks (AMDs) with varying composition.

Pectin was added to the AMDs at concentrations of 0.0 per cent, 0.1 per cent, 0.3 per cent or 0.5 per cent and the protein used was either skim milk powder (SMP) or SMP with added whey protein concentrate. Distributed exponential analysis of the T_2 relaxation revealed that the AMDs contained a single water component and that the T_2 relaxation times and distributions differed significantly with respect to pectin concentration and protein type. Temperature profiles (5–25–5°C) showed that pectin concentrations of 0.3 per cent or 0.5 per cent prevented phase separations in the AMDs as a consequence of heating and cooling, whereas a new free water component emerged in the samples with pectin concentrations of 0.0 per cent or 0.1 per cent pectin. Overall, LF-NMR provides a powerful tool for elucidating industrially relevant texture problems such as whey separation.

Source

1. *Tina Salomonsen, Marie Tholstrup Sejersen, Nanna Viereck, and Søren Balling Engelsen, Quality and Technology, Department of Food Science, The Royal Veterinary and Agricultural University, Rolighedsvej 30, DK- 1958 Frederiksberg C, Denmark.*

2. *Richard Ipsen, Dairy Technology, Department of Food Science, The Royal Veterinary and Agricultural University, Rolighedsvej 30, DK- 1958 Frederiksberg C, Denmark, Received 13 September 2005; accepted 4 April 2006.*

f. ^{1}*H NMR Properties of Peptides and Amino Acids [1,2]*

The initial strategies of NMR analysis and structure determination of peptides and proteins was developed [1]. Each amino acid has characteristic patterns of NMR peaks. The term randomcoil is meant to mean a completely random (not ordered) state.Recent studies on unfolded proteins or partially folded proteins have shown that often times there is no true random coil state. However, for the purposes of NMR analysis and assignments, the random coil properties are useful and a good starting point for identification of amino acids.

(a) The following Two types of experiments (both of which correlate protons through scalar (J) coupling), form the basis of analysis of peptides and proteins by 1H NMR:

  (i) **Correlation Spectroscopy (COSY)-** It only correlates peaks with non-zero coupling constants (usually 2 or 3 covalent bonds apart).

  (ii) **Total Correlation Spectroscopy (TOCSY)**-It on the other hand, is capable of correlating all peaks in a coupled spin network. That is, if "I" is coupled to "S" and "S" is coupled to "T", then there will be a TOCSY interaction between"I" and "T", even if they are not directly coupled.

(b) **NOESY (Nuclear Overhauser effect spectroscopy) or ROESY (Rotating frame Overhauser effect spectroscopy)-** These experiments correlate protons that are close together (within about 5 Å inspace). There will be more about the NOE next week. The set of NOESY or ROESY peaks will often contain peaks that are covalently connected but will also include peaks that are not. The TOCSY spectrum correlates peaks within an amino acid but does not cross outside of that amino acid. NOESY or ROESY, on the other hand, correlate both within and between amino acids.

Source

1. *Kurt Wüthrich,"NMR of Proteins and Nucleic Acids", Wiley Interscience (1986).*
2. *Dr. Arthur S. Edison, Theory and Applications of NMR Spectroscopy, WWW: http:// ascaris.health.ufl.edu.*

g. Measuring Fat Content of Dairy Products by NMR using an MQC-23 [1]

Nuclear Magnetic Resonance (NMR) provides a fast, direct and user friendly method for determination of the fat and oil content in foodstuffs. This technique is based on measurement of the NMR response obtained from fat (oil) in the product, and quantification of the fat (oil) content by calibration. A variety of dairy products with a range of fat contents between 1 per cent and 82 per cent were analysed by NMR using an MQC-23 fitted with a 26 mm hydrogen probe. Analyses were conducted using the EasyCal 'Fat in Foodstuffs' application software package. The functionality of the instrument may be increased to include the determination of Solid Fat Content (SFC), by the addition of a hydrogen probe (10 mm) and conditioning/tempering hardware as detailed in the AOCS/ISO SFC protocols.

Method

Samples were prepared in duplicate by spreading 3–5 g of each sample on glass fibre filter paper and placing either of the followings:

(i) On a metal dish or

(ii) Directly inside a glass vial.

All samples were dried in an oven at 103°C for ~5 hours (drying time dependent on amount of sample). After drying, the open filter samples were placed inside glass vials. All samples were conditioned at 40°C for one hour, prior to analysis. NMR measurement time was ~32 seconds.

The advantage of this method over extraction techniques is that a large number of samples can be prepared in batches.

Calibration

A calibration can be generated using either one 100 per cent fat (oil) sample or a set of 3-6 samples of real products with predefined fat (oil) contents spanning the range of concentrations of interest.Samples of dried macaroni cheese, baked cheese, chicken powders, chicken sandwich fillers, milk powders and meat pastes were used to generate the calibration. NMR gives an excellent linear correlation between the NMR response and the concentration of fats in the products.Since the hydrogen density of the fat components will vary with food type (*e.g.* butter, milk, vegetable oils and modified/blended oils) it is recommended that a calibration is generated from fat derived from the same source material. Such a calibration may be generated by a 100 per cent calibration method or a multi-point calibration covering the range of interest.

Advantages of Benchtop NMR

(i) It can be calibrated to cover a concentration range from 0 to 100 per cent fat (oil).

(ii) A primary calibration can be produced using a single sample of fat (oil).

(iii) NMR is very stable over the long-term, therefore requires little recalibration.

(iv) The sample measurement time is short (typically 20 seconds).

(v) Minimal sample preparation is required.

(vi) It is virtually insensitive to sample matrix granularity and additives such as spices, flavours, colours and salt.

Source

1. *Application note-010, Oxford instruments, Industrial Analysis, UK.*

h. Rapid Determination of Moisture/Solids and Fat in Dairy Products by Microwave and Nuclear Magnetic Resonance Analysis [1]

A peer-verified method is presented for the determination of per cent moisture/ solids and fat in dairy products by microwave drying and nuclear magnetic resonance (NMR) analysis. The method involves determining the moisture/solids content of dairy samples by microwave drying and using the dried sample to determine the fat content by NMR analysis. The laboratories analyzed various dairy products by using a CEM SMART system (moisture) and the SMART Trac (fat). The samples included milks, creams, ice cream mix, sour cream, yogurt, cream cheese, and mozzarella, Swiss, and cheddar cheeses. These samples represented a range of products that processors deal with in daily plant operations. The results were compared with moisture/solids and fat values derived from AOAC-approved.

Source

1. *Cartwright G, McManus BH, Leffler TP, Moser CR, J AOAC Int. 2005 Jan-Feb; 88 (1): 107-20.*

Application of NMR in Microbiology [1]

The application of NMR in microbiology depends upon the following characteristics of NMR instruments:

Sensitivity

The sensitivity of the NMR experiment should probably be the foremost concern of prospective users. It depends on the Nature of the nucleus, Operating magnetic field: , Active volume, Concentration of resonant nuclei, Line-width, Accumulation time, Relaxation delay and Decoupling.

Resolution

In the early applications of NMR to microbial physiology and metabolism, spectral resolution was rarely an issue. The ranges of ^{13}C and ^{15}N chemical shifts is such that few spectral overlaps were observed. Phosphorus spectra are somewhat crowded, for instance in the phosphomonoester region. This favourable state of affairs is changing because more complex systems are being investigated and also because proton NMR, which has a smaller frequency dispersion, is being used more often. In some cases higher operating fields will alleviate the problem, but the almost definitive solution is two- or multi-dimensional NMR, using multiple labelling and one- or multi-dimensional acquisition and processing.

NMR is used for for the investigation of microbial physiology and metabolism. The first report on the application of NMR to micro-organisms is that of Eakin *et al.*, on the yeast, Candida utilis [2]. The current applications of NMR are as follows:

i. Qualitative Applications

These include compound identification and localisation.

a. Identification of Compounds

NMR is called upon to help in identifying these compounds. This will be particularly true for recently discovered organisms, or for some branches of secondary metabolism. NMR, contrary to other characterisation methods, is not specific of any one class of molecules are clearly advantageous.

b. Identification of Metabolic Pathways or Branches in Pathways

Molecules involved in metabolism have a carbon backbone. It should therefore not be surprising to learn that the most powerful method to investigate a pathway consists in labelling carbon atoms in order to unravel their origin and fate. This can be done with high sensitivity with radioactive ^{14}C labelling, NMR spectroscopy can be used to identify new pathways or unknown branches. The details related to some examples are tabulated below:

Sl.No.	Scientists	ScientificWorks
1	Lelait and Grivet [3]	Used the labelling patterns of amino-acids (obtained after isolation and hydrolysis of proteins) to confirm the main metabolic pathways of acetogenic bacteria growing on ^{13}C labelled glucose or carbonate.
2	Hajjaj *et al.* [4].	Grew the filamentous fungus *Monascus ruber* on $^{13}C1$, $^{13}C2$ and $^{13}C1$-$^{13}C2$ labelled acetate to delineate the complex biosynthesis of citrinin.
3	Lajoie *et al.* [5]	NMR is also useful for investigations of ecosystems or complex bacterial populations. The presence of homoacetogens in the microflora of human feces was proven. Detected doubly labelled acetate formed from H2 and $^{13}CO2$.

c. Metabolic Expression in Genetically Modified Organisms

Ansanay *et al.* [6] inserted the gene coding for the malolactic enzyme (from Lactococcus lactis) into the genomes of Saccharomyces cerevisiae and Schizosaccharomyces pombe. Biochemical analyses, ^{13}C labelling (from [1-^{13}C] -glucose) and comparison of the two species showed that **S. cerevisiae** was indeed able to perform the malolactic fermentation, but with low yield due to a very inefficient malate transport system.

d. Distinction of Inside and Outside of Cell (Number of Cell Compartments)

Several NMR parameters have been used to locate solutes either inside or outside of cells.

(a) **pH measurement:** The first method relies on the pH dependence of the chemical shift, specially that of phosphorylated compounds. The appearance of several "inside" inorganic phosphate signals is a proof of the existence of several cellular compartments, each with a distinctive pH.

(b) Another approach makes use of a non-toxic, non-permeating shift and/ or relaxation reagent. It is also possible to recognise atoms bound to the outside of the cell membrane.

(c) **Diffusion weighted spectra:** A more recent technique is based on diffusion weighted spectra. The "inside" molecules have a rather low effective

diffusion coefficient, while the "outside" ones diffuse faster. It is possible to completely suppress extracellular contributions to the proton spectrum.

ii. Quantitative Application

The quantitative aspects, such as pH, concentration and flux measurements are examined, as well as the corresponding experimental constraints.

a. Cell Volume

A non-permeating inert solute is added to the culture medium and a spectroscopic signal is recorded in the absence and in the presence of the cells. The difference is proportional to the excluded volume, occupied by the cells. On the other hand, the total water volume, both in the presence and absence of cells, is determined from the signal intensities of a freely diffusable compound. A careful determination of the intracellular volume for a suspension of **Enterococcus faecalis** by proton NMR is reported.

b. Diffusion Coefficients

Spin echo NMR, in the presence of pulsed field gradients, was sensitive to the diffusion of solute molecules. A quantitative determination of diffusion coefficients, both inside and outside the cell is possible.

c. Concentrations

Non-invasive, real time, continuous determinations of concentrations represent the main uses of metabolic NMR. The solute concentrations in the culture medium (or supernatant) and cell extracts are measured.In contrast to the direct concentration measurements referred to in the previous section, it is also possible to assay NMR invisible compounds through the effect they have on visible solutes.

e. Metabolic Flux

Metabolic fluxes (or ratio of fluxes at branch points) are the main quantitative data sought by workers interested in metabolism. These parameters are the equivalent of the Vmax and Km of enzymology and they cannot be determined from NMR measurements alone when using a single label. It is sometimes masked in NMR, because of its low sensitivity, this analytical technique operates not with tracer amounts but with labelled compounds which form a notable proportion of the substrate. The isotopic equilibrium (or flux) is different from substrate equilibrium (or flux). Some examples of the application of NMR in this field are given in following table:

Sl.No.	*Scientists*	*Scientific works*
1	Dominguez *et al.* [7]	Combined acidic and basic extraction procedures, HPLC and fluorescence assays with ^{13}C NMR analysis of labelling patterns to determine carbon fluxes in the central metabolism of Corynebacterium glutamicum.
2	Lutz *et al.* [8]	Have taken advantage of ^{31}P-^{13}C couplings to determine the specific enrichment of phosphorylated metabolites.

Contd.

Sl.No.	Scientists	Scientific works
3	Roberts *et al.* [9]	Used ^{15}N NMR to measure the free amino-acid turnover rate in methanogens.
4	Wolin *et al.* [10]	Examined the colonisation of the digestive tracts of infants by Bifidobacteria. These micro-organisms use a unique pathway of hexose catabolism to produce acetate and lactate.

Source

1. *Jean-Philippe Grivet, NMR in Microbiology: Theory and Applications, Publisher: Horizon Scientific Press, Editors- Jean-Noël Barbotin and Jean-Charles Portais Amiens, France, ISBN: 978-1-898486-21-3 (hbk).*
2. *Eakin et al., NMR in Microbiology: Theory Applications, ISBN: 978-1-898486-21-3 (hbk).*
3. *Lelait and Grivet, NMR in Microbiology: Theory and Applications, ISBN: 978-1-898486-21-3 (hbk).*
4. *Hajjaj et al., NMR in Microbiology: Theory and Applications, ISBN: 978-1-898486-21-3 (hbk).*
5. *Lajoie et al., NMR in Microbiology: Theory and Applications, ISBN: 978-1-898486-21-3 (hbk).*
6. *Ansanay et al., NMR in Microbiology: Theory and Applications, ISBN: 978-1-898486-21-3 (hbk).*
7. *Dominguez et al., NMR in Microbiology: Theory and Applications, ISBN: 978-1-898486-21-3 (hbk).*
8. *Lutz et al., NMR in Microbiology: Theory and Applications, ISBN: 978-1-898486-21-3 (hbk).*
9. *Roberts et al., NMR in Microbiology: Theory and Applications, ISBN: 978-1-898486-21-3 (hbk).*
10. *Wolin et al., NMR in Microbiology: Theory and Applications, ISBN: 978-1-898486-21-3 (hbk),*

4. Microwave Spectroscopy or Rotational Spectroscopy

The interaction of microwaves (with matter) can be detected by observing the attenuation or phase shift of a microwave field as it passes through matter. These are determined by the imaginary or real parts of the microwave susceptibility (the index of refraction). The absorption of microwaves may also trigger a much more easily observed event like the emission of an optical photon in an optical double-resonance experiment or the deflection of a radioactive atom in an atomic beam [1].

Microwave spectroscopy is an effective method of physical and chemical research. Measurement of the frequencies of the band spectra of molecules makes possible highly accurate determination of the structure of molecules, and also the study of the nature of the chemical bond.

It studies the absorption and emission of electromagnetic radiation in the microwave region of the electromagnetic spectrum by molecules associated with a corresponding change in the rotational quantum number of the molecule [1]. The development of microwave technology for RADAR during World War II, enables us to use microwaves in spectroscopy.

Theory [3]

It is concerned with transitions between rotational energy levels in the molecules, the molecule gives a rotational spectrum only If it has a permanent dipole moment.Rotational spectroscopy is only really practical in the gas phase where the rotational motion is quantized. In solids or liquids the rotational motion is usually quenched due to collisions between their molecules.Rotational spectrum from a molecule requires that the molecule have a dipole moment and that there be a difference between its center of charge and its center of mass, or equivalently a separation between two unlike charges. It is this dipole moment that enables the electric field of the light (microwave) to exert a torque on the molecule, causing it to rotate more quickly (in excitation) or slowly (in de-excitation) [2]. Details related to dipole moment and homonuclear/heteronuclear diatomic molecules are given below:

i. Homonuclear Diatomic Molecules

These molecules (such as H_2, O_2, N_2, Cl_2) have zero dipole (non polar) and have zero change of dipole during the rotation, hence no interaction with radiation. Therfore homonuclear diatomic molecules are **microwave inactive**.H-H and Cl-Cl don't give rotational spectrum

ii. Heteronuclear Diatomic Molecules

These molecules (such as HCl, HF, CO) have permanent dipole moment (polar compound) and change of dipole occurs during the rotation, hence interaction with radiation takes place. Therefore, heteronuclear diatomic molecules are **microwave active**.H-Cl, and C=O give rotational spectrum.

Classes of Rotating Molecules [3]

Molecules can be classified into following main groups depending on their moments of inertia.

Sl.No.	*Moments of Inertia*	*Groups of Molecules*
1	IC = IB, IA = 0	Linear Molecules
2	IC = IB = IA	Spherical top
3	IC = IB > IA	Symmetric top
4	IC > IB > IA	Asymmetric top

The band absorption spectrum of a molecule depends on its configuration (that is, on whether the molecule is a linear, spherical, symmetric, or asymmetric rotating body). The band spectrum of any molecule may be calculated if its moments of inertia, which depend on the configuration and dimensions of the molecule, are known. Comparison of theoretically computed band spectra of molecules with experimentally observed spectra makes possible determination of the configuration of a molecule, the length of its bonds, and the angles between them [1].

The Microwave spectra of some molecules are given below [1]:

(i) Diatomic molecule -Carbon monoxide (CO), has one of the simplest rotational spectra.

(ii) Tri-atomic molecules, hydrogen cyanide (HCa″N) has a simple rotational spectrum for a linear molecule and hydrogen isocyanide (HN=C:) for a non-linear molecule.

(iii) Homonuclear diatomic molecules such as dioxygen (O_2), dihydrogen (H_2), *etc.* do not have a dipole moment and, hence, no purely rotational spectrum. However, electronic excitations can lead to asymmetric charge distributions and thus provide a net dipole moment to the molecule. Under such circumstances, these molecules will exhibit a rotational spectrum.

Population of Energy Levels (Intensity of Rotational Spectral Lines)

It is affected by the following factors [3]:

i. Boltzmann Distribution

The Boltzmann distribution alone does not fit the shape of the spectra so, degeneracy of the states is required.Suppose the number of molecules in first state is N_0 and number of molecules in any higher state is N_J then

$$\frac{NJ}{N_0} = e^{-BJ\,(J+1)hc/kT}$$

The population of the state decreases as the J –value increases. For example- For HCl, 2B = 21.2 →B = 10.6, T=300 K, h=6.626x10^{-34} J.s, k=1.38x10^{-23} J/K, c = 3x 10^{10} cm/s, then the population of state are as follows:

(i) J= 0 → $N_0/N_0 = e^0 = 1$

(ii) J= 1 → $N_1/N_0 = 0.903$

(iii) J= 2 → $N_2/N_0 = 0.7367$

ii. Number of Degenerated Energy Levels (levels which have the same energies)

The existence of more than two energy states having the same energy is called **Degeneracy.**

Factors Affecting the Energy of Levels [3]

The following factors are affecting the energy of levels:

(i) **Effect of isotopic substitution-**On changing from $^{12}C^{16}O$ to $^{13}C^{16}O$, atomic mass increases, B decreases, so energy of levels becomes lower.

(ii) **Effect of Bond Length:** Bond length and mass effect the rotational constant of diatomic molecules.

Source

1. *Microwave Spectroacopy - Free All dictionary.*
2. *Wikipedia.*
3. *Tarek A. Fayed, Microwave (Rotational) Spectroscopy.*

Apparatus

There are following two types of microwave spectrometer in use-:

i. Conventional Stark-Modulated Spectrometer

In this, the sample is contained in a long (1- to 3-metre, or 3.3- to 9.8-foot) section of a rectangular wave guide, sealed at each end with a microwave transmitting window (*e.g.*, mica or Mylar), and connected to a vacuum line for evacuation and sample introduction. The radiation from the source passes through a gaseous sample and is detected by a crystal diode detector that is followed by an amplifier and display system (chart recorder). In order to increase the sensitivity of the instrument, signal modulation by application of a high-voltage square wave across the sample is used.

ii. Fourier-Transform Spectrometer

In this, radiation is confined in an evacuated cavity between a pair of spherical mirrors and the sample is introduced by a pulsed nozzle that lowers the temperature of the sample to less than 10 K. The sample is subjected to rotational energy excitation by application of a pulsed microwave signal, and the resulting emission signal is detected and Fourier-transformed to an absorption versus frequency spectrum. In both instruments the energy absorbed or emitted as the molecules undergo transitions from one quantized rotational state to another is observed.

The Fourier-transform instrument has the **advantage** of providing higher resolution and of exhibiting a much simpler spectrum due to the low sample temperature that insures that the majority of the molecules are in the few lowest energy states.

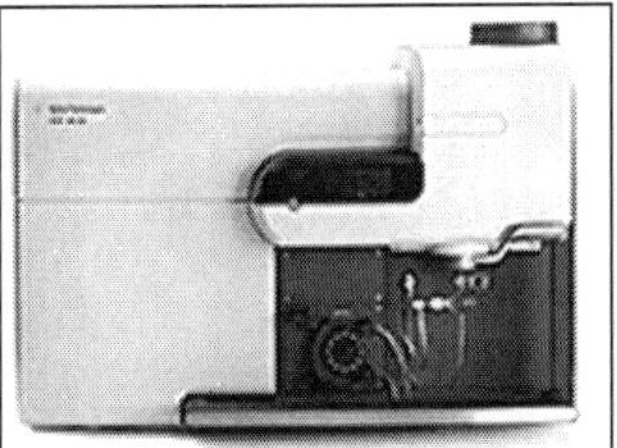

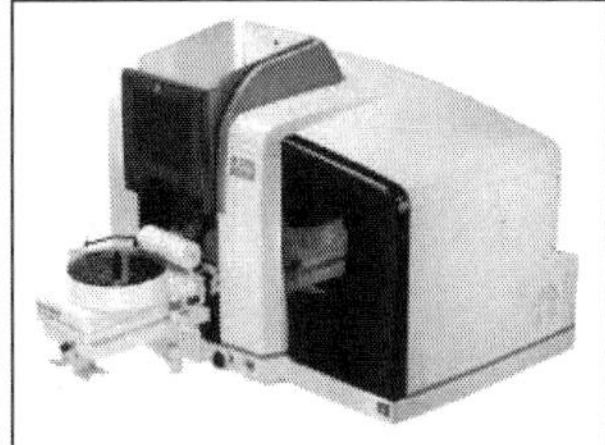

Different Models of Microwave Spectrophotometer (*Source*: Wikipedia).

Applications of Microwave Spectroscopy

It has been used in monitoring and control of industrial processes. It is an ideal process analyzer as it has following characteristics [1]:

- *(i)* Non-invasive: The measurement can be made outside of the reaction chamber, eliminates the need for sampling or physical removal of sample.
- *(ii)* Non-destructive.
- *(iii)* It can be used for solids, liquids, gases and suspensions.
- *(iv)* It can be used for dark coloured samples.
- *(v)* Analyses large sample volumes, as microwaves diffuse out from the transmitter though the entire sample becomes lower.

Source

1. *Tarek A. Fayed, Microwave (Rotational) Spectroscopy.*

Microwave spectroscopy has been used in monitoring and control of industrial processes as well as in quality analysis of food/dairy products. The details are given below:

A. Monitoring and Control of Industrial Processes

Microwave spectroscopy is being successfully used in monitoring the processing of the following:

- *(i)* Materials with low dielectric constants, such as plastics, glass, ceramics and composite materials.
- *(ii)* Determination of moisture in various tobacco types.
- *(iii)* Monitoring of a batch esterification reaction as in the esterification of butanol by acetic acid.
- *(iv)* Monitoring of the drying process in industry as it is one that is hard to monitor. For example, huge cakes of wet material when dried in big vessels.
- *(v)* Astrophysical applications, radioastronomy: probe of the molecular universe (molecular clouds) using Telescope.

Source

1. *Tarek A. Fayed, Microwave (Rotational) Spectroscopy.*

B. Quality Analysis [1,2,3]

Microwave spectroscopy is also being successfully used in quality analysis. Some examples are given below:

i. Determination of Structure of Small Molecules

Microwave spectroscopy is commonly used in physical chemistry to determine the structure of small molecules *e.g.* ozone, methanol, or water with high precision.

Other common techniques (X-ray crystallography) for determining molecular structure, do not work very well for some of these molecules (especially the gases) and are not as precise. However, microwave spectroscopy is not useful for determining the structures of large molecules such as proteins. Modern microwave spectrometers have very high resolution. When hyperfine structure can be observed, the technique can also provide information on the electronic structures of molecules [1,2].

ii. Rapid Determination of the Moisture Content of Milk Powder by Microwave Sensor

A rapid method of determining the moisture content of milk powder by microwave sensor was investigated.Calibration models of milk powder moisture content were established using the linear least squares method. The results predicted by the microwave technique were compared with the results determined by standard methodology and it was found that results showed the maximal measurement deviation of 0.2 per cent. Therefore, the **microwave technique may be a powerful tool for determining the moisture content of milk powder** in the dairy industry.

Source

1. *Jingjing Yang, Ming Huang, and Jihong Shi, School of Information Science and Engineering, Yunnan University, Kunming, unnan 650091, PR China.*
2. *Jinhui Peng, Faculty of Materials and Metallurgical Engineering, Kunming University of Science and Technology, Kunming 650093, PR China.*
3. *Wikipedia.*

iii. Analysis on Determining Quality of Vegetable Oil [1]

Microwave spectroscopy is established as a technique for distinguishing various oil types, with the long-term aim of authenticity determination. The studies [2,] have demonstrated that nonconventional methods, such as microwave spectrometry, are able to offer real-time measurement of oils adequate for determining product authenticity. Various oil samples of oils (Sun flower Oil -100 per cent, Grape-seed Oil -100 per cent, Groundnut Oil-100 per cent, Olive Oil-mixed, Olive Oil-medium flavour, Olive Oil-fruity flavor, Olive Oil -Extra Virgin) were analysed and the results show that each produces a unique signal spectra. This paves way for the application of this method in the detection of adulteration in edible oils. This analysis has included differing additives and combinations of oils to determine the capability of the method in distinguishing common fraud techniques. The results prove that any dilution of any of the sample could alter the characteristics of the spectrum, proving satisfactory.In addition, associated hardware and software can be helpful to make the entire system portable and affordable for industry use.

Laboratory testing was carried outusing a **Vector Network Analyser (VNA)**. The microwave sensor applied for the test, is made of polyimide laminate substrate of 2mm thick with printed pattern conductive area of 14mm x 5mm. The MUT cavity is14mm x 14mm x 2mm. the 2 terminal plates of the printed pattern are terminated on a male F connector.

Source

1 *S. B. Osman, O. Korostynka, A. Mason, J. D. Cullen, A. I. Al-Shamma, Application of Microwave Spectroscopy Analysis on Determining Quality of Vegetable Oil, Proceedings of the 8th international conference on sensing technology, sep.2-4,2014, Liverpol, U.K.*

2. *E. Bader, A. Attar, A. Mason, L. Wendling, A. Al-Shamma'a"Investigation of an Embedded Microwave Spectrometer for Alcohol Detection and Measurement", School of Built Environment, LiverpoolJohn Moores University.*

iv. Microwave Spectroscopy to Chemical Analysis [1]

The most analytical techniques generally can only be applied to "well-behaved" chemical systems. Since gas phase reaction products do not always follow the "well-behaved" guideline, an alternate analytical tool, such as microwave spectroscopy, could prove beneficial for analyzing such systems and may well provide some new routes to synthetic chemistry. The present state-of-the-art for applications of microwave techniques to analytical studies were presented. An attempt was made to describe the limitations and advantages of microwave spectroscopy for probing chemical systems for product identification and for optimization of the efficiency of gas phase chemical reactions.

Source

1. *F.J.Lovas, Application of microwave spectroscopy to chemical analysis, PubMed, ISA Trans. 1975; 14 (2): 145-51.*

5. Atomic Spectroscopy [1,2]

This spectroscopy involves the placement of the sample in an environment that is hot enough to break molecular bonds resulting production of atoms and these atoms then identified and their concentration measured by emission or absorption of characteristic radiation. As this involves emission or absorption of radiation and hence this spectroscopy can further be categorised as per the followings:

i. Atomic Absorption Spectroscopy (AAS)

It is a technique which studies absorption of electromagnetic radiations in relationship to molecular structure and is used for measuring the concentration of various elements in the sample through their absorption of light. It is a relatively simple and reliable technique which uses absorption of optical radiation by free atoms for determining the contents of different elements [1]. The technique makes use of absorption spectrometry to assess the concentration of an analyte in a sample. It requires standards with known analyte content to establish the relation between the measured absorbance and the analyte concentration and relies therefore on the Beer-Lambert Law **[5]**.

It is a spectro analytical procedure for the quantitative determination of chemical elements using the absorption of optical radiation (light) by free atoms in the gaseous state [5]. Atomic absorption spectroscopy was first used as an analytical technique, and the underlying principles were established in the second half of the

19th century by Robert Wilhelm Bunsen and Gustav Robert Kirchhoff, University of Heidelberg, Germany [6]. The modern form of AAS was largely developed during the 1950s by chemists led by Sir Alan Walsh at the Commonwealth Scientific and Industrial Research Organisation (CSIRO), Division of Chemical Physics, in Melbourne, Australia [7,8].

In this, a light source emitting radiation characteristic of specific element is passed through the atomised sample and the transmitted radiation is measured. The method is based on the absorption of radiation by free atoms. In AAS, a light source emitting radiation characteristic of specific element is passed through the atomised sample and the transmitted radiation is measured. Thus, the most important step is to produce free atoms of metals (**M**) from sample solution (or Solid).Double beam spectrometers are successfully used in AAS [2,3].

Principle and Theory [1]

Atomic absorption spectroscopy is based on the principle that when a beam of electromagnetic radiation is passed through a substance, the radiation may either be absorbed or transmitted depending upon the wavelength of the radiation. The absorption of radiation would bring about an increase in the energy of the molecule. The energy gained by the molecule is directly proportional to the wavelength of radiation. The increase in the energy of the molecule leads to the electronic excitations where electrons jump to higher energy levels. A particular wavelength that a given molecule can absorb depends upon the changes in vibrational, or rotational or electronic states.

When a monochromatic radiation of frequency v is incident on a sample/ molecule, the molecule in the gaseous state (E_1) absorbs a photon of energy hν and it undergoes a transition from lower energy level to higher energy level. A detector (placed to collect the radiation after interaction with the molecule) shows that intensity has reduced. With wide range of frequencies, the detector shows the energy has been absorbed only from the frequency.

$$\nu = (\Delta E)/h$$

Hence we get an absorption spectrum which is defined as a record of the radiation absorbed by the given sample as a function of wavelength of radiation. **The energy difference between the levels is given as:**

$$\Delta E = E_2 - E_1 = h\nu = hc/\lambda.$$

ii. Atomic Emission Spectroscopy (AES)

Max Planck (In the 20thcentury) concluded that energy could be absorbed or emitted discontinuously in the form of packets of energy called quantas. When the atom absorbs energy and jumps from lower stable state the higher state and while coming back to their ground state energy level, they emit the energy equal to the difference between the energies of two levels. This emission radiation is recorded in the atomic emission spectra and is called **emission spectroscopy [4].**

In AES, the sample is heated to sufficiently high temperature that the electrons of the atoms excited. When the electrons return to their normal position, they emit radiation characteristic of element present. The radiation transmitted through the atomizer is detected by either photon detectors or mass spectrometers (provides Hugh sensitivity). AES is a method of chemical analysis that uses the intensity of light emitted from a flame, plasma, arc, or spark at a particular wavelength to determine the quantity of an element in a sample. The wavelength of the atomic spectral line gives the identity of the element while the intensity of the emitted light is proportional to the number of atoms of the element [5].

Definition and Principle [4]

When the molecule jumps from the higher energy level to lower energy level, a photon of energy is emitted. The spectrum obtained is called emission spectrum and the spectroscopy is called atomic emission spectroscopy. Atomic emission spectroscopy is a technique to determine the quantities of elements in the sample by using the intensity of light emitted by plasma, spark, flame or arc.

In atomic emission spectrum, the sample is applied with energy in the form of heat or light. The source of energy can be flame, plasma or electrical arc. The excited molecule jumps from lower to higher energy level. At this state the molecules are unstable and comes back from higher energy level to lower energy level emitting radiations of certain wavelength with following characteristics:

i-The radiations are emitted in the form of photons. The wavelengths of photons emitted are recorded, **ii**-These radiations are recorded in the emission spectrometer. The level of emissions of a substance is the difference between the substance in excited and lower stateand **iii**-Each element has different and unique level of emission which helps the scientist to detect the elements. The frequencies are recorded as light spectra in the emission spectrometer.

Suppose the molecule comes back from the energy level E_2 to lower energy level E_1 emitting radiations. Then **the frequency is given by,**

$$v = \Delta E/h.$$

The energy difference between the levels is given as:

$$\delta e = e2 - e1 = hv = hc/\gamma.$$

Types of AES [5]

Flame emission spectroscopy, inductively coupled plasma atomic emission spectroscopy and Spark and arc atomic emission spectroscopy are being used and the details of these spectroscopies are given below:

i. Flame Emission Spectroscopy

A sample of a material/analyte is brought into the flame as either a gas, sprayed solution, or directly inserted into the flame by use of a small loop of wire, usually platinum. The heat from the flame evaporates the solvent and breaks chemical bonds to create free atoms. The thermal energy also excites the atoms into excited electronic states that subsequently emit light when they return to the ground electronic state. Each element emits light at a characteristic wavelength, which is dispersed by a grating or prism and detected in the spectrometer.

ii. Inductively Coupled Plasma Atomic Emission Spectroscopy (ICP-AES)

It uses an inductively coupled plasma to produce excited atoms and ions that emit electromagnetic radiation at wavelengths characteristic of a particular element [9,10].

ICP (plasma) is an ionized gas at extremely high temperature. The energy in the plasma is transferred by collision of Ar+ with the atoms of interest (sample) enough to ionize many metals with *I.E.* 7-8 eV.

$$Ar^{+} \rightarrow Ar^{+} + e^{-}$$

Most metals are ionizable emitting in the UV range, nonmetals do not form ions in the ICP (require a vacuum). Plasma torch inner tube contains the sample aerosol and Ar support gas. Radio frequency generator produces a magnetic field which sets up an oscillating current in the ions and electrons of the support gas (Ar) produces high temperatures (up to10,000 K). At this temperature, sample gets ionized and ionic transitions which are observable at UV and visible wavelengths happens. Excited atoms and ions emit their characteristic radiation, which are collected by a device that sorts the radiation by wavelength.Intensity of the emission is detected and turned into a signal that is output as concentration [5a].

Advantages of ICP-AES

(i) Excellent limit of detection and linear dynamic range.

(ii) Multi-element capability.

(iii) Low chemical interference and a stable and reproducible signal.

Disadvantages of ICP-AES

(i) Spectral interferences

(ii) Cost and operating expense.

(iii) The fact that samples typically must be in a liquid solution.

ICP Atomic Emission can theoretically analyze almost all elements and it can also measure multiple elements simultaneously.Flame Atomic Absorption Spectrometry and Flame Emission Spectrometry can only measure around 70 elements and flame techniques canonly measure one at a time. Same process occur as with flame atomic emission [5a].

iii. Spark and Arc Atomic Emission Spectroscopy

It is used for the analysis of metallic elements in solid samples. For non-conductive materials, the sample is ground with graphite powder to make it conductive. In traditional arc spectroscopy methods, a sample of the solid was commonly ground up and destroyed during analysis. An electric arc or spark is passed through the sample, heating it to a high temperature to excite the atoms within it. The excited analyte atoms emit light at characteristic wavelengths that can be dispersed with a monochromator and detected. Earlier, because of uncontrolled the spark or arc conditions, the analysis for the elements were qualitative but now with the modern spark sources with controlled discharges can be considered quantitative. **Both qualitative and quantitative spark analysis are widely used for production quality control in foundry and metal casting facilities.**

Source

1. *Atomic Absorption Spectroscopy, Principle, Instrumentation and application, Chemistry@TutorVis.*
2. *CMM Smith, Atomic spectrometric Techniques, Encyclopedia of Dairy Science, Vol.1, pp-51-55, Elsevier Science Limited, 2002.*
3. *Technews, No.74, May-June-2008, NDDB.*
4. *Atomic Emission Spectroscopy, Principle, Instrumentation and application, Chemistry @ TutorVis.*
5. *Wikipedia.*

5a. *K.P.Indumathi, Rajan Sharma and Manju Singh, Various Techniques for Mineral Analysis in Food., Compendium of Lectures on Advanced Techniques and Novel Approaches for Quality and Safety Evaluation of Dairy Foods -2015, NDRI, Lucknow.*

6. *Robert Bunsen and Gustav Kirchhoff. Chemical Heritage, Foundation. Retrieved 2014-07-29.*
7. *McCarthy, G.J. "Walsh, Alan - Biographical entry". Encyclopedia of Australian Science.Retrieved 22 May 2012.*
8. *Koirtyohann, S. R. (1991). "A history of atomic absorption spectrometry". Analytical Chemistry. 63 (21): 1024A–1031A. ISSN 0003-2700.*
9. *Stefánsson A, Gunnarsson I, Giroud N (2007)."New methods for the direct determination of dissolved inorganic, organic and total carbon in natural waters by Reagent-Free Ion Chromatography and inductively coupled plasma atomic emission spectrometry". Anal. Chim. Acta. 582 (1): 69–74.*
10. *Mermet, J. M. (2005). "Is it still possible, necessary and beneficial to perform research in ICP-atomic emission spectrometry". J. Anal. At. Spectrom. 20: 11–16.*

Difference between Atomic Emission Spectroscopy (AES) and Atomic Absorption Spectroscopy (AAS) [1]

The difference between AAS and are tabulated below:

Sl.No.	AAS	AES
1	In AAS,when the atom or molecule absorbs a photon of energy, it undergoes a transition from lower energy state to higher energy state.	In AES, when the molecule jumps from higher level to lower energy levels, a photon of energy is emitted.
2	The spectrum obtained is called absorption spectrum.	The spectrum obtained is called emission spectrum.
3	In AAS, there are only few dark bands.	In AES, there are generally only few coloured bands.
4	The atomic absorption spectrum is easy to interpret.	The Atomic emission spectra is not as easy to interpret.

Contd.

Sl.No.	AAS	AES
5	In absorption spectroscopy, in the spectrometer there is analyzer to sense the source.	In emission spectroscopy, in the spectrometer, there is no need for the analyzer as the source being in a sense, its own analyzer.
6	It is less sensitive to temperature fluctuations.	It is very sensitive to temperature fluctuations.

Source

1. *Atomic Absorption Spectroscopy, Principle, Instrumentation and application, Chemistry @ TutorVis.*

Instrumentation

A. Atomic Emission Spectroscopy [1]

The instrument used for recording a spectrum is known as spectrometer or spectrophotometer.

(i) The first step is **atomization or excitation**. In this solid liquid or solution analyte is converted into gaseous atom. The sample is then made excited and the excitation in emission spectrometer can be thermal or electrical, but often takes the form of electromagnetic radiation. The most common methods of excitations are flames and plasmas. If the sample is solid, it can be dissolved in solvent for analyzing.

(ii) In Atomic emission spectroscopy, in the spectrometer, the sample, after undergoing excitation, is its own source. It is necessary that only the emitted radiation is collected and analyzed and then recorded in the usual way. In case of emission spectrometer, no analyser is necessary, the source being its own analyser. A monochromatic is employed which separates radiations into individual wavelengths. The dispersing element is used which is generally a glass prism.

(iii) **Detector**-The detector is the device which converts spectral radiation into electrical signal that is transmitted to a recording device called a recorder. The recorder produces a print on the chart. It is essential that the detector does not receive radiation directly from the exciting beam and the two are placed at right angles.

(iv) **A modulator**–It is placed between the source of excitation and the sample, ensures that only the emission is recorded which directly arises from the excitation. Any other spontaneous emission in this way is ignored.

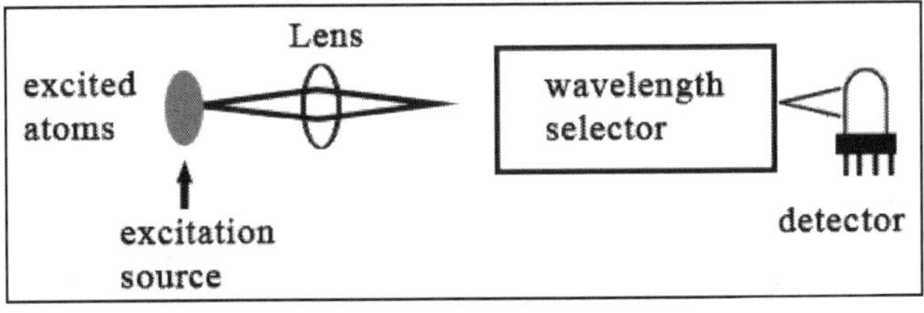

Schematic Layout of AE Spectrophotometer.

Source

1. *Atomic Emission Spectroscopy, Principle, Instrumentation and application, Chemistry @ TutorVis.*

B. Atomic Absorption Spectroscopy

In the atomic absorption spectrometer, the source of radiation in the spectrometer is the tungsten filament emitting white light or hydrogen discharge lamp. The radiation from the source is directed by some device (for example in this case a mirror) on the sample. The radiation then passes through an analyser (the grating in this case), which selects the frequency reaching the detector at any given time. The signal from the detector passes then to a recorder which is attached to the analyser so as to produce a trace of the absorbance of varying frequencies [1]. The details of different components are given below:

i. Atomization

In order to analyze a sample for its atomic constituents, it has to be atomized. The atoms should then be irradiated by optical radiation, and the radiation source could be an element-specific line radiation source or a continuum radiation source. The radiation then passes through a monochromator in order to separate the element-specific radiation from any other radiation emitted by the radiation source, which is finally measured by a detector.

A. Commonly Used Atomization

The atomizers most commonly used now-a-days are **flames and electrothermal** (graphite tube) atomizers. Other atomizers, such as glow-discharge atomization, hydride atomization, or cold-vapor atomization might be used for special purposes.

i. Flame Atomizers

The oldest and most commonly used atomizers in AAS are flames, principally the air-acetylene flame with a temperature of about 2300 °C and the nitrous oxide [2] system (N_2O)-acetylene flame with a temperature of about 2700 °C. Flame, premix burner, sample (1–2 mL) + oxidant + fuel: air + acetylene; 2400–2700 K acetylene + nitrous oxide, 2900-3100 K, refractory elements.Liquid or dissolved samples are typically used with flame atomizers

Nebulization-The sample solution is aspirated by a pneumatic analytical nebulizer, transformed into an aerosol (narrow size distribution and smallest sizes are desirable). A flexible capillary tube connects the solution to the nebuliser. At the tip of the capillary, the solution is 'nebulised' – *ie* broken into small drops. The larger drops fall out and drain off while smaller ones vaporise in the flame. An aerosol is introduced into a spray chamber, where it is mixed with the flame gases and conditioned in a way that only the finest aerosol droplets (< 10 μm) enter the flame. This conditioning process is responsible that only about 5 per cent of the aspirated sample solution reaches the flame, but it also guarantees a relatively high freedom from interference [3]. On top of the spray chamber is a burner head that produces a flame that is laterally long and only a few mm deep. The radiation beam passes through this flame at its longest axis, and the flame gas flow-rates may be adjusted to produce the highest concentration of free atoms. The burner height may also be adjusted, so that the radiation beam passes through the zone of highest atom cloud density in the flame, resulting in the highest sensitivity. This technique is typically used for determinations in the mg L^{-1} range, and may be extended down to a few µg L^{-1} for some elements.

The processes in a flame include the followings [3]:

(i) **Stages of desolvation (drying)** in which the solvent is evaporated and the dry sample nano-particles remain,

(ii) Vaporization (transfer to the gaseous phase) in which the solid particles are converted into gaseous molecule,

(iii) **Atomization** in which the molecules are dissociated into free atoms,

(iv) Ionization where (depending on the ionization potential of the analyte atoms and the energy available in a particular flame) atoms may be in part converted to gaseous ions.

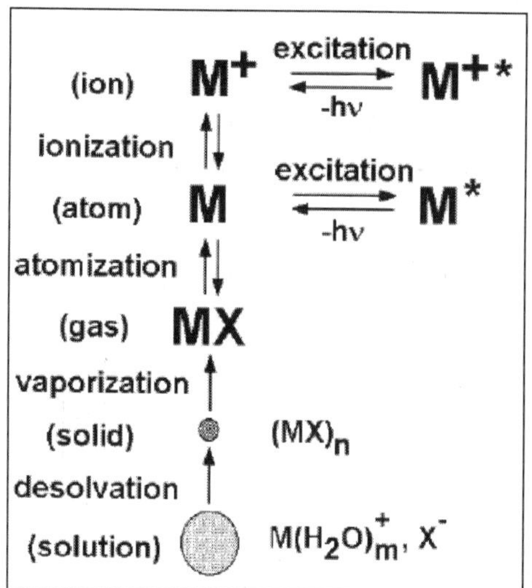

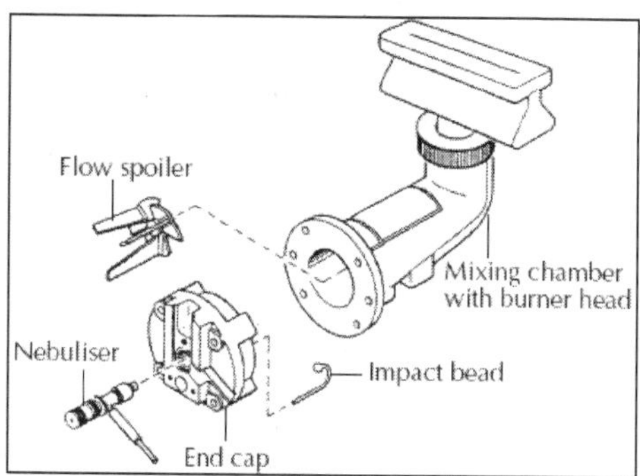

ii. Electrothermal Atomizers

Electrothermal AAS (ET AAS) using graphite tube atomizers was pioneered by Russian Scientist [4] since the late 1950s, and investigated in parallel by German Scientist [5]. Now-a-days typically 20–25 mm in length and 5–6 mm inner diameter graphite tube are being used.With this technique liquid/dissolved, solid and gaseous samples may be analyzed directly. A measured volume (typically 10–50 µL) or a weighed mass (typically around 1 mg) of a solid sample are introduced into the graphite tube and The tube is heated electrically by passing a current through it in a pre-programmed series of steps:

(i) **Drying-** the solvent is evaporated- 30 – 40 sec at 150 °C

(ii) Pyrolysis-30 sec at 600 °C (Ashing and charring of organic material and convert sample into oxides at fast heating rate). The majority of the matrix constituents are removed.

(iii) **Atomization-**5-10 sec at 2000 - 2500 °C, To produce free atoms. The analyte element is released to the gaseous phase.

(iv) **Finally heating the tube to a still higher temperature** -2700 °C- cleans it ready for the next sample. During this heating cycle the graphite tube is flushed with argon gas to prevent the tube burning away. In electrothermal atomisation almost 100 per cent of the sample is atomised.

Advantages [3]

(i) This technique has the advantage that any kind of sample, solid, liquid or gaseous, can be analyzed directly.

(ii) In ET AAS a transient signal is generated, the area of which is directly proportional to the mass of analyte (not its concentration) introduced into the graphite tube.

(iii) Its sensitivity is 2–3 orders of magnitude higher than that of flame AAS, so that determinations in the low μg L^{-1} range (for a typical sample volume of 20 μL) and ng g^{-1} range (for a typical sample mass of 1 mg) can be carried out. For improved sensitivity atomization is performed by a graphite furnace. **This makes the technique much more sensitive than flame AAS.**

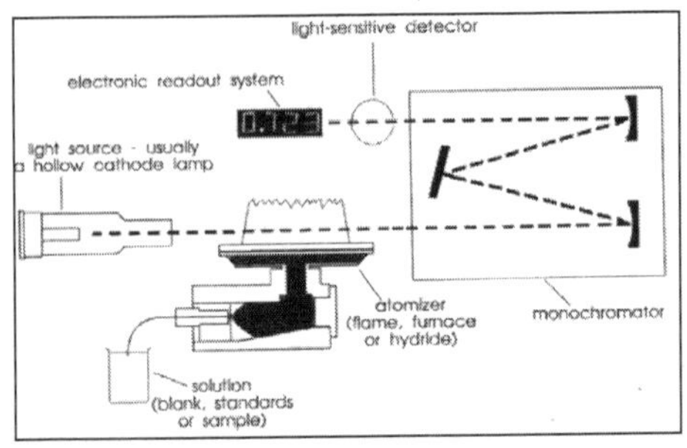

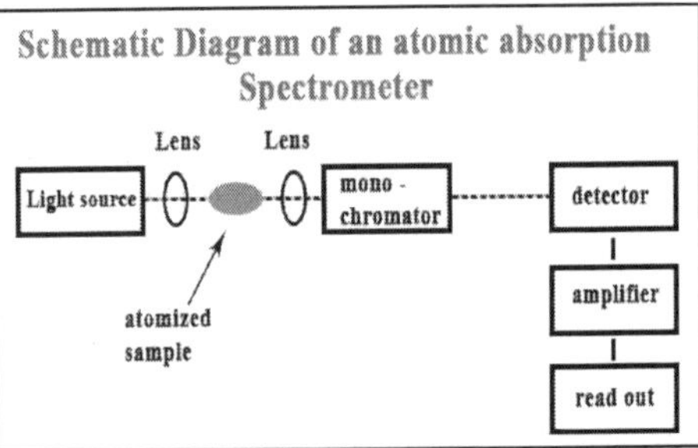

Source: [1]

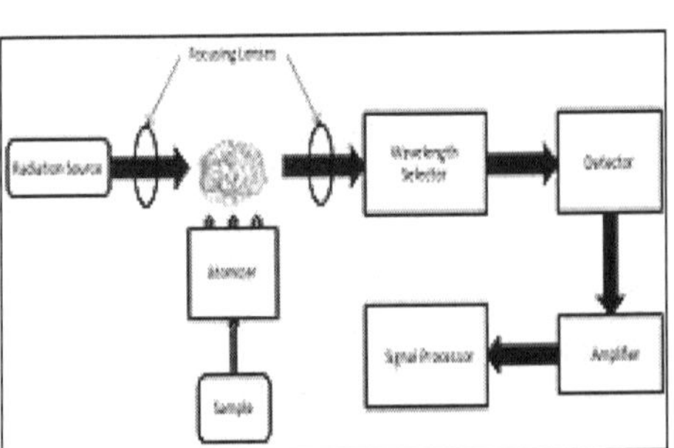

Atomic Absorption Spectrometer Layout. *Source*: [3]

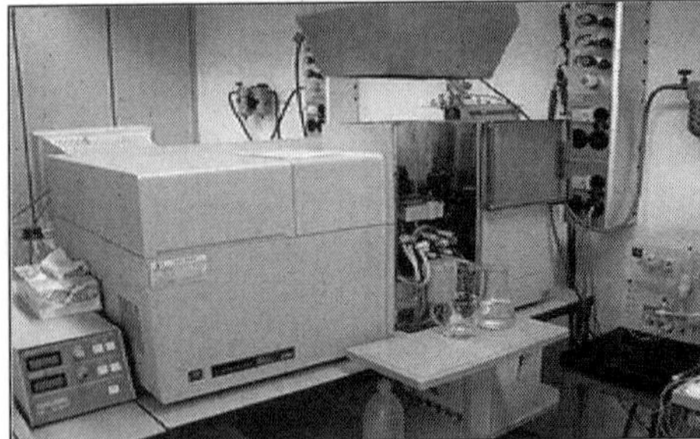

Flame Atomic Absorption Spectroscopy Instrument. *Source*: [3]

Instrumental Set up.

B. Specialized Atomization

Glow-discharge atomization, Hydride atomization, Cold-vapor atomization are the atomizations used as specialised atomization:

i. Hydride Generator/Hydride Atomization [3]

Hydride generation techniques are specialized in solutions of specific elements. The technique provides a means of introducing samples containing arsenic, antimony (As), tin (Sn), selenium (Se), bismuth (Bs), and lead (Pb) into an atomizer in the gas phase. They are useful to improve sensitivity of AAS determination of these elements. With these elements, hydride atomization enhances detection limits by a factor of 10 to 100 compared to alternative methods.Hydride generation occurs by adding an acidified aqueous solution of the sample to a 1 per cent aqueous solution of sodium borohydride, all of which is contained in a glass vessel.

$$As_3+ + NaBH_4 \rightarrow AsH_3 \text{ (Arsenic trihydride, Arsine)}$$

The liberated hydride is transported to an atomizer by an inert carrier gas or by other means, where it undergoes decomposition. This process forms an atomized form of the analyte, which can then be measured by absorption or emission spectrometry.

The advantage of this technique over flame AAS is the separation and enrichment of the element to be determined and the significantly lower detection limit resulting from greater efficiency of sample introduction. This method may have more interference effects than the flame AAS system.

ii. Cold-Vapor Atomization [3]

The cold-vapor technique an atomization method limited to only the determination of mercury, due to it being the only metallic element to have a large enough vapor pressure at ambient temperature. The method initiates by converting mercury into Hg^{2+} by oxidation from nitric and sulfuric acids, followed by a reduction of Hg^{2+} with tin (II) chloride. The mercury, is then swept into a long-pass absorption tube by bubbling a stream of inert gas through the reaction mixture. The concentration is determined by measuring the absorbance of this gas at 253.7 nm.

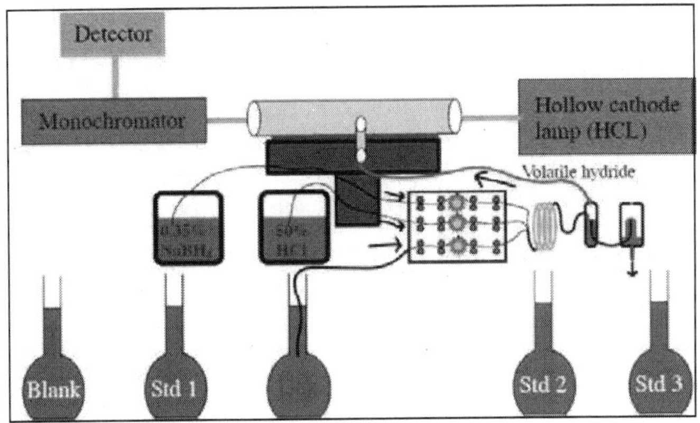

Mercury is converted first into Hg (II) and then allowed to react with a reducing agent such as $SnCl2$ to produce elemental Hg. Air or nitrogen is allowed to pass through the solution to carry the hg produced into the absorption cell (glass tube with quartz windows placed in the radiation path and AA is measured.

iii. Glow-Discharge Atomization [3]

A glow-discharge device (GD) serves as a versatile source, as it can simultaneously introduce and atomize the sample. The glow discharge occurs in a low-pressure argon gas atmosphere between 1 and 10 torr. In this atmosphere lies a pair of electrodes applying a DC voltage of 250 to 1000 V to break down the argon gas into positively charged ions and electrons. The requirement for samples of glow discharge atomizers is that they are electrical conductors.

2. Light Source [3]

There are line source AAS (LS AAS) and continuum source AAS (CS AAS).

A. LS AAS

In classical LS AAS, as it has been proposed by Alan Walsh, the high spectral resolution required for AAS measurements is provided by the radiation source itself that emits the spectrum of the analyte in the form of lines that are narrower than the absorption lines.

B. CS AAS

Continuum sources, such as deuterium lamps, are only used for background correction purposes. The advantage of this technique is that only a medium-resolution monochromator is necessary for measuring AAS; however, it has the disadvantage that usually a separate lamp is required for each element that has to be determined. In CS AAS, in contrast, a single lamp, emitting a continuum spectrum over the entire spectral range of interest is used for all elements. Obviously, a high-resolution monochromator is required for this technique.

A. Light Source for Line Source AAS [3]

i. Hollow Cathode Lamps (HCL) [3]

These are the most common radiation source in LS AAS. Inside the sealed lamp, filled with argon or neon gas at low pressure, is a cylindrical metal cathode containing the element of interest and an anode as shown on next page. A high voltage (300V) is applied across the anode and metal (Na, Fe, Mn) cathode, resulting in an ionization of the fill gas. The gas ions (Ar ions) are accelerated towards the cathode and, upon impact on the cathode, sputter cathode material that is excited in the glow discharge to emit the radiation of the sputtered material.

ii. Electrodeless Discharge Lamps (EDL) [3]

It contain a small quantity of the analyte as a metal or a salt in a quartz bulb together with an inert gas, typically argon gas, at low pressure. The bulb is inserted into a coil that is generating an electromagnetic radio frequency field, resulting in a low-pressure inductively coupled discharge in the lamp. An EDL emission is higher than that from an HCL but EDLs need a separate power supply and might need a longer time to stabilize.

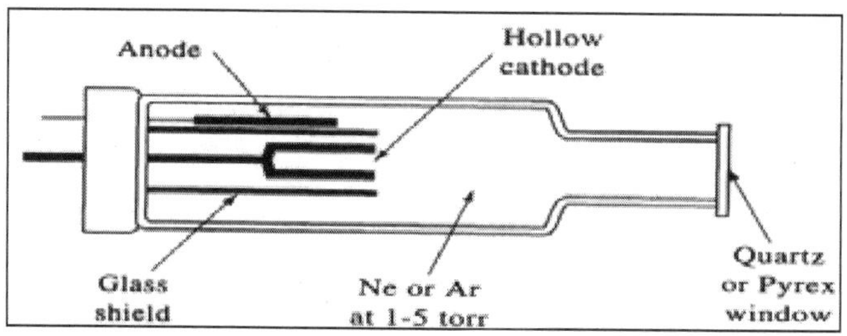

Cathode Lamp

iii. Deuterium Lamps

Deuterium HCL or even hydrogen HCL and deuterium discharge lamps are used in LS AAS for background correction purposes [6]. The radiation intensity emitted by these lamps decreases significantly with increasing wavelength, so that they can be only used in the wavelength range between 190 and about 320 nm.

B. Continuum Light Sources [3]

When a continuum radiation source is used for AAS, it is necessary to use a high-resolution monochromator. In addition, it is necessary that the lamp emits radiation of intensity at least an order of magnitude above that of a typical HCL over the entire wavelength range from 190 nm to 900 nm. A special **high-pressure xenon short arc lamp**, operating in a hot-spot mode has been developed to fulfill these requirements.

3. Spectrometer [3]

The medium-resolution spectrometers are used for LS AAS and high-resolution spectrometers are designed for CS AAS and there is a difference between these two types of spectrometers. The spectrometer includes the spectral sorting device (monochromator) and the detector.

i. Spectrometers for LS AAS- [3]

In LS AAS the high resolution that is required for the measurement of atomic absorption is provided by the narrow line emission of the radiation source, and the monochromator simply has to resolve the analytical line from other radiation emitted by the lamp. Another feature to make LS AAS element-specific is modulation of the primary radiation and the use of a selective amplifier that is tuned to the same modulation frequency, as already postulated by Alan Walsh. Simple monochromators of the Littrow or (better) the Czerny-Turner design are typically used for LS AAS.Photomultiplier tubes are the most frequently used detectors in LS AAS, although solid state detectors might be preferred because of their better **signal-to-noise ratio.**

ii. Spectrometers for CS AAS [3]

When a continuum radiation source is used for AAS measurement it is indispensable to work with a high-resolution mono-chromator. The resolution has to be equal to or better than the half width of an atomic absorption line. The first commercial equipment for HR-CS AAS was introduced by **Analytik Jena** (Germany), based on the design proposed by Becker-Ross and Florek.

These spectrometers use the followings:

(*a*) **Monochromators**: **i**-a compact double monochromator with a prism pre-monochromator and **ii**-an echelle grating monochromator for high resolution.

(*b*) **Detector**: A linear charge coupled device (CCD) array with 200 pixels is used as the detector.

The second monochromator does not have an exit slit; hence the spectral environment at both sides of the analytical line becomes visible at high resolution.

Source

1. *Atomic Absorption Spectroscopy, Principle, Instrumentation and application, Chemistry @ TutorVis.*
2. *Koirtyohann, S. R. (1991). "A history of atomic absorption spectrometry". Analytical Chemistry. 63 (21): 1024A–1031A. ISSN 0003-2700*
3 *Wikipedia.*
4. *L'vov, Boris (1990). Recent advances in absolute analysis by graphite furnace atomic absorption spectrometry. Spectrochimica Acta Part B: Atomic Spectroscopy. 45. pp. 633–655.*
5. *"Analytical Methods for Graphite Tube Atomizers" (PDF). http://www.agilent.com. Agilent Technologies.*
6. *Rakshit, Amitava. "Basics of Laboratory Safety: Common laboratory rules and regulations". The International Association for Ecology. Intecol. Retrieved 26 September 2016.*

Types of Interference

Spectral interference and chemical interference are the two important interferences:

i. Spectral Interference

(*a*) Overlapping signal from absorption or emission of unwanted species (molecules or radicals). It can be corrected by choosing another wavelength or use D2 lamp for flame signals.

(*b*) **Ionization-**Higher atomization temperature cause ionization of analytes like Na or K;

$$Na \rightarrow Na^{+} + e$$

It can also be corrected by adding ionization suppresser, easily ionizable atoms like CsCl (to increase electrons in the flame to reverse the direction of ionization reaction of the analyte ionization, Le Chatelier's principle).

ii. Chemical Interference

Some metals cannot be determined in the presence of other species due to the formation of thermally stable compounds and can not be atomized.

Example

Ca can not be determined in the presence of phosphate, because of the formation of refractory calcium phosphate and decreases the atomization of Ca and the amount of atoms produced are very low. It can be corrected by addition of a releasing agents is added like La**3+**, salt solution $LaCl_3$ to allow chemical exchange

$$Ca_3\ (PO4)_2 + 2LaCl_3 \rightarrow 3CaCl_2 + 2LaPO_4$$

$CaCl_2$ is easily atomized.

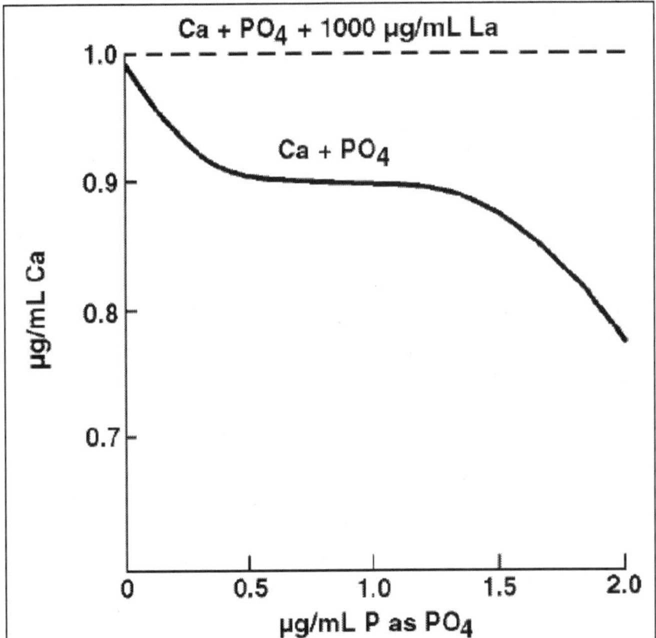

EDTA may also be used for other metals. Rich flame may also be used to give high temperature able to atomize refractory materials.

Matrix Modification

With electrothermal atomization, chemical modifiers can be added which react with an interfering substance in the sample to make it more volatile than the analyte compound. This volatile component vaporizes at a relatively low temperature and is removed during the low and medium temperature stages of electrothermal atomization.

Example

For the determination of heavy metals in tobacco leaves, 0.5 -1.0 g of the tobacco leaves are digested with a mixture of 1: 10 $HClO_4$: HNO_3 until complete dissolution. The resulted solutions are diluted into 100 mL with distilled water, in volumetric flasks. The instrument is calibrated by analyzing standard solutions of the metals: Na, K, Ca, Mg, Cu, Mn, Fe, Ni, Co and Zn. The standard solutions are prepared by serial dilution from a 1000 ppm (mg/mL). A calibration graph is prepared by aspiration of the standard solutions (1,2,4,10 and 20 ppm) and plotting the absorbance versus concentration. When the concentration is very low, preconcentration procedures are applied like solvent extraction, evaporation. When the matrix is complex the technique of standard addition is used.

Sample Preparation

Sample preparation is often simple, and the chemical form of the element is usually unimportant. This is because atomisation converts the sample into free atoms irrespective of its initial state. The sample is weighed and made into a solution by suitable dilution. Elements in biological fluids such as urine and blood are often measured simply after a dilution of the original sample.

Calibration

Standard solutions are measured first aspirated and their AA signals are measured The unknown sample solution is then analysed. The value of the unknown is used to determine the concentration of the metal in the unknown solution.

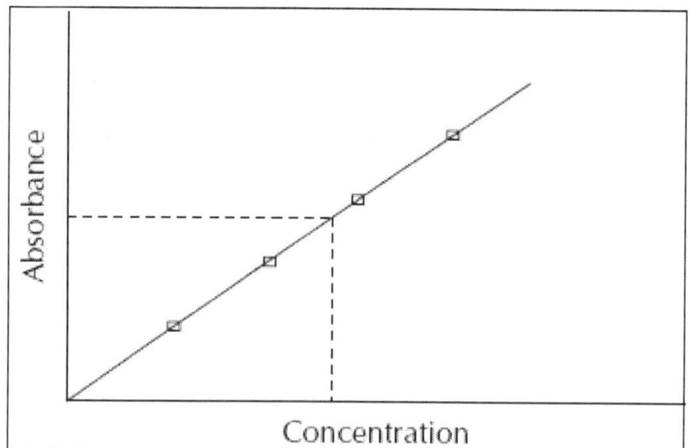

AOAC Standards [1]

AOAC has specified a standard *e.g.* **AOAC.991.25** for estimation of Ca, Mg, and P in cheese by AAS where as **IDF** has prescribed a method for Determination of Ca, Na, K- Atomic absorption Spectrophotometric method for milk and milk products under **IDF 119/ISO 18070.**

Source

1. *CMM Smith, Atomic spectrometric Techniques, Encyclopedia of Dairy Science, Vol.1, pp-51-55, Elsevier Science Limited, 2002.*
2. *Technews, No.74, May-June-2008, NDDB.*

Method Validation [1]

The atomic spectroscopy based analytical procedures are validated through the following approaches:

(i) Analysis of reference materials.

(ii) Comparison of results obtained using a second analytical procedure (Reference Method)

(iii) Comparison of results obtained from another analytical Laboratory (Reference Method).

Source

1. *CMM Smith, Atomic spectrometric Techniques, Encyclopedia of Dairy Science, Vol.1, pp-51-55, Elsevier Science Limited, 2002.*
2. *Technews, No.74, May-June-2008, NDDB.*

Atomic Spectroscopy Based Equipments for Analysis of Dairy Products [1]

The analytical characteristics and other details of different commonly used instruments based on atomic spectroscopy (Emission/Absorption *e.g.*FAAS, FAES, ETAAS, ICPAES, ICPMS *etc.*) for analysis of trace elements in milk and milk products, are given below:

a. Analytical Characteristics

The analytical characteristics of FAAS, FAES, ETAAS, ICPAES, ICPMS are as follows:

Sl.No.	*Absorption/ Emission Techniques*	*Sample Volume*	*Limit of Detection*	*Multi-element*
1	FAAS	5-10	1-1000	Possible
2	FAES	5-10	1-100	Yes
3	ETAAS,	0.01-0.1	0.01-0.1	Possible
4	ICPAES	1-10	0.05-10	Yes
5	ICPMS	1-10	0.001-0.01	Yes

b. Other Details

The other details of FAAS, FAES, ETAAS, ICPAES, ICPMS are as follows:

(i) **Flame Atomic Absorption Spectrometry (FAAS):** It is useful technique for detection/determination of Cu, Zn, Ca, Fe, Mg and Mn in dairy products. It is good for liquid samples *e.g.* milk *etc.* and does not need

sample preparation.

(ii) **Flame Atomic Emission Spectrometry (FAES):** A low cost and useful technique for determination of relatively high concentrations of Sodium (Na) and Potassium (K) in milk products.

(iii) **Electro thermal Atomic Absorption Spectrometry (ETAAS):** It is capable of **in situ** sample preparation and hence it can determine several elements in solid sample directly. The use of slurries or suspension of powdered samples with conventional liquid sample handling system facilitates solid sample analysis.

(iv) **Inductively Coupled Plasma Flame Atomic Emission Spectrometry (ICPAES):** It is capable of doing fast- sequential and simultaneous multi element analysis with lower limit of detection. It also provides wide linear dynamic range of calibration for analysis of milk and milk products.

(v) **Inductively Coupled Plasma Mass Spectrometry (ICPMS):** It is capable of multi-element and Isotopic analysis with enhanced sensitivity. Detection of Non metal *e.g.* P and I (Iodine)in milk and milk products, can be easily done by using this technique.

Source

1. *Technews, No.74, May-June-2008, NDDB.*

International Standards for Determination of Lead in Milk and Milk Products [1,2,3]

The following international standards are available for estimation of Lead:

Sl.No.	*Parameters*	*Products*	*Description/Principle*	*International Standards*
1	Lead	Milk and Milk Products	Digestion, Atomic absorption spectro-photometry	AOAC 972.25
			Ashing, Anodic stripping voltametry; estimated quantification limit based on 10 g sample is 0.01 ppm (ppm=mg/kg)	AOAC 982.23
			Digestion, Anodic stripping voltametry	AOAC 986.15
			Microwave digestion, Atomic absorption spectrophotometry; approximate limit of detection is 0.1 ppm	AOAC 999.10
			Ashing, Atomic absorption spectro-photometry; formula provided for calculation of detection limit	AOAC 999.11
			Atomic absorption spectro-photometry	NMKLb 139 (1991)

Source

1. *Technews, No.47, Nov-Dec-2003, NDDB.*
2. *Codex (1999). Methods of Analysis and Sampling. Standard no. XOT 16. Codex*

Alimentarius Commission, Rome.

3. *19. IDF (2000). Inventory of IDF/ISO/AOAC International Adopted Methods ofAnalysis and Sampling for Milk and Milk Products. Bulletin no. 350. International Dairy Federation, Brussels, pp. 4-21.*

Application of Atomic Spectroscopy

The applications of AAS and AES are given below:

A. Application of Atomic Emission Spectroscopy [1]

Atomic emission spectroscopy is used for the followings:

(i) It is used for regulation of alkali metals in the pharmaceutical processes,

(ii) It is also used for analysing motor oils.

(iii) AES helps in the detection of elements in the given sample and hence for structure elucidation.

(iv) It is also used for detection of trace metals in different samples

(v) It is applicable in the smelting process of ores, in the process of extraction of metals.

B. Application of Atomic Absorption Spectroscopy

It has various applications in following various fields:

(i) **Determination of structural elucidation of various substances [4]:** This technique is more beneficial as a very small quantity of substance is required for analyzing.

(ii) **Analysis of water [4]** for its metal quantity if present.

(iii) **Pharmaceutical manufacturing processes [2,4]:** AAS is used in many pharmaceutical manufacturing processes where small quantities of catalyst remain in the final product. By using AAS the amount of catalyst present can be determined.

(iv) **Biological tissues/Clinical analysis [2,4]:** AAS can also be used in biological tissues such as blood, liver, brain tissue, muscle tissue and fluids for analyzing metals.

(v) **Environmental analysis [2]:** Monitoring our environment *e.g.* finding out the levels of various elements in rivers, seawater, drinking water, air, petrol and drinks such as wine, beer and fruit drinks.

(vi) **Industry:** Many raw materials are examined and AAS is widely used to check that the major elements are present and that toxic impurities are lower than specified-*eg* in concrete, where calcium is a major constituent, the lead level should be low because it is toxic. It is being successfully used in Dairy/Food Industries.

a. In Dairy Industry

It has been used successfully for the analysis of trace and major elements (Nutritionally important- Fe, Ca, P, Mg, Cu, Zn, I *etc.* or toxic - Pb, Hg, As *etc.*) in milk and milk products. The details related to the Principle of Atomic Spectroscopy and determination of aforesaid elements by AAS are given in chapter- 08 of author's another book- Hand book on analysis of milk [3].

Source

1. *Atomic Emission Spectroscopy, Principle, Instrumentation and application, Chemistry@TutorVis.*
2. *CMM Smith, Atomic spectrometric Techniques, Encyclopedia of Dairy Science, Vol.1, pp-51-55, Elsevier Science Limited, 2002.*
3. *M.K.Srivastava, Hand Book on analysis of milk, IBDC, 2010.*
4. *Atomic Absorption Spectroscopy, Principle, Instrumentation and application, Chemistry@TutorVis.*

b. In Food Industry

Lead, Copper, Zinc *etc.* are being determined by AAS in different foods. The details of the some test methods are given below:

i. AAS Method for Determination of Lead, Copper and Zinc in Confectionary Products

Equipment

1 Atomic absorption spectrophotometer (Flame), 2. Crucibles, 3. Hot plate, 4. Calibrated weighing balance and 5. Muffle Furnace.

Reagents

1. Nitric acid
2. Water
3. Hydrochloric acid- (1: 1)
4. Standard stock solution (Certified Reference Materials)–1000 mg/L of Pb, Cu and Zn (traceable to NIST).

Procedure

(i) **Sample preparation**: Homogenize product, if necessary using non-contaminating equipment.

(ii) Weigh about 10-20 g of the sample to nearest 0.01g in a crucible, and dry on a hot plate at 100°C.

(iii) Place the dish in muffle furnace at initial temperature not higher than 100°C. Increase temperature at a maximum rate of 50°C per hour to 450°C. Let dish stand for at least 8 hours or overnight.

(iv) Take crucible out of furnace and let cool.

(v) Wet ash with 1-3 ml of water and evaporate on hot plate.

(vi) Put crucible back in furnace at no more than 200°C and raise temperature (50-100°C per hour) to 450°C. Proceed for 1-2h or longer.

(vii) Repeat procedure until product is completely ash *i.e.* ash should be white/ grey or slightly covered.

(viii) Add 5 ml (1: 1) HCl to crucible ensuring that all ash comes into contact with acid. Evaporate acid on hot plate.

(ix) Dissolve residue in 10-30 ml 0.1M HNO_3. Cover with watch glass and let stand for 1-2 hours. Make up to a particular volume.

(x) Treat blank in the same way as prepared sample.

(xi) Standard preparation: Add 10 ml of standard stock solution into100 ml volumetric flask and make up the volume using deionized water. (Pb/ Cu/Zn 100 mg/L solution).

(xii) Preparation of standard working solution: To a series of 100 ml one-mark volumetric flasks, pipette 0.2, 0.5, 1.0, 2.0 and 5.0 ml of the standard working solution and 1ml of concentrated nitric acid and then dilute to mark with deionized water. These solutions then contain lead concentration 0.2, 0.5, 1.0, 2.0 and 5.0 µg of Pb/Cu/Zn per ml of the solution.

(xiii) Aspirate both standard and sample solutions into the AAS.

Calculation

Determine the Pb/Cu/Zn concentration in the sample solution by extrapolating the value from the standard curve plotted using conc. of different standard solution vs. absorbance reading.

$$\text{Pb/Cu/Zn, mg/kg} = \frac{\text{(S-B) X Dilution}}{\text{Weight of Sample}}$$

where- S- Concentration of Pb/Cu/Zn in sample and B-Concentration of Pb/ Cu/Zn in Blank.

Source

1. *A.O. A.C 18th edn, 2005 Official Method 999.11.*
2. *Manual of methods of Analysis of Food (Confectionary Products), Lab Manual -4, FSSAI, Govt. of India.*

ii. AAS Method for Determination of Copper in Alcoholic Bevearges (Rum, Gin, Whisky, Brandy, Beer, Vodka, Wine, Toddy, Fenny-Cashew and Coconut)

Apparatus

1. Atomic absorption Spectrophotometer -Double beam.
2. Hollow Cathode Lamp–Copper.
3. Follow operating instructions of manufacturer for the selection of optimum gas flow and wavelength settings, beam alignment.
4. Microwave Digestor with Quartz tubes for digestion.

5. Muffle furnace.
6. Fume Hood.
7. Steam bath.
8. Silica crucible.
9. Acetylene Ultra pure grade.

Reagents

1. Nitrogen-Ultra pure grade.
2. Water-triple distilled or Milli-Q/18Ω.
3. Copper SRM (100 µg/ml) traceable to NIST.
4. Alcohol-distilled.

Preparation of Cu Working Standard Solutions

Take suitable aliquots from copper SRM to prepare 0.25, 0.50 and 1.00µg/ml (ppm) Cu solutions and make up to known volume with 1N HNO_3.

Procedure

It involves preparation of ash and determination.

Preparation of Ash Solution

It can be done either dry or wet ashing.

(i) **Dry Ashing:** Take 50 to100 ml of liquor sample in a clean pre-acid washed silica crucible and evaporate to dryness in a steam bath in a fume hood. Ignite at 500°C for 3 to 4 hrs. Cool and add 5ml of ultra pure nitric acid and warm on steam bath covering a watch glass in a fume hood. Filter through whatman no.42 filter circle (pre-acid washed) and wash the residue with hot water.Collect the filtrate and make up to 25 ml volume in a volumetric flask.

(ii) **Wet Ashing**: Take 50 to 100 ml of liquor sample in a glass bowl and evaporate to dryness. Add 5 ml of ultra pure nitric acid and transfer to the quartz tube of microwave digestor using little distilled water. Pressure Digest the solution in microwave digestion apparatus for 30 min. Cool and make up to 25 ml volume.

(iii) **Blank Solution**: Prepare a blank by taking 5 ml of ultrapure nitric acid and make up to 25 ml volume.

Determination

Aspirate the blank into the AAS flame and set the instrument for zero absorbance. Aspirate the Cu Std. solutions sequentially for absorbance data acquisition.Now aspirate the sample-acid digest into the flame to record the absorbance and in turn note down the displayed concentration of Cu in µg. Calculate the concentration in the test sample involving the dilutions made.

Calculation

Calculate the concentration by using the following formula:

$$\text{Copper content} = \frac{\text{Reading (in µg) displayed x Dilution}}{\text{(in µg/ml or ppm) Weight of sample}}$$

Source

1. *Manual of methods of Analysis of Food (Alcoholic Beverages), Lab Manual -16, FSSAI, Govt. of India.*

6. Ultrasound Spectroscopy [2,3,4]

Ultrasonic spectroscopy (**Low intensity Ultra-sonic**) is a non-destructive analytical technique based on the measurements of parameters of low energy ultrasonic waves propagating through analysed samples.Ultrasonic spectroscopy is based on the measurements of following two independent parameters:

i. Ultrasonic Attenuation

Ultrasonic attenuation is determined by the energy losses in ultrasonic waves and can be expressed in terms of viscosity of the medium or its longitudinal loss modulus. It provide information about physico chemical properties by allowing analysis of kinetics of fast chemical reactions, microstructure of materials including particle sizing, aggregation, gelation, crystallization and other processes and characteristics.

ii. Ultrasonic Velocity

Ultrasonic velocity is determined by the density and the elastic response of the sample to the oscillating pressure in the ultrasonic wave and can be expressed in terms of compressibility or storage modulus (longitudinal). This parameter is extremely sensitive to the molecular organisation, composition and intermolecular interactions in the analysed medium and is responsible for the major portion of applications of **high-resolution ultrasonic spectroscopy** for analysis of chemical properties of materials.

Ultrasonic techniques are now finding increasing applicaton in the food/ Dairy Industries for analysis of Food/Dairy Products. The technique can provide information about physicochemical properties of food products, such as composition, physical state and flow rate [2,3].

This technique is well known through its successful application in medicine and a number of fields of material analysis. However, in the past the limited resolution of the measurements and complicated cleaning and sample handling procedures prevented a broad spread of this technique in research and analytical laboratories [3,4].

Source

1. *Wikipedia.*

2. *Smith, C., OD riscoll, B., Karyashov, E. and Buckin, v., High resolution ultrasonic spectroscopy for analysis of dairy system.*
3. *Evgeny Kuryashov and Vitaly Buckin, Department of Chemistry at University College Dublin.*
4. *Cormac Smyth and Breda O'Driscoll, Ultrasonic Scientific Limited, GEC, Dublin. Ireland, www.ultrasonic-scientific.com.*

Advantages of Ultrasonic Analysis [2,3]

The some of the advantages of Ultrasonic analysis are given below:

(i) Non-destructive analysis of intrinsic properties of materials includes measurements of signals (waves) travelling through the analysed sample. Until now only one wave has dominated in the field of material analysis the electromagnetic wave. Ultrasound provides an alternative wave, which is the same as an acoustical one, only at higher frequency (or tone in acoustical terms). In this wave oscillating pressure causes oscillation of compressions and therefore by its nature it is a rheological wave.

Compression in the ultrasonic wave changes the distances between the molecules of the sample, which responds by intermolecular repulsions. Therefore ultrasonic spectroscopy allows us to probe intermolecular forces in the sample, thus providing new information about its interior. Amplitudes of deformations in the ultrasonic wave employed in analytical ultrasound are extremely small, making the ultrasonic analysis non-destructive. As most materials are ultrasonically transparent, ultrasound allows the analysis of a broad variety of samples including opaque materials.

(ii) Another advantage of the technology employed in the HR-US101 ultrasonic spectrometer is the absence of large actuators as in dynamic rheology or bulky light sources and other optical parts or expensive consumables, thus representing a robust and multipurpose instrument, which performs a broad range of analytical functions of fast, non-destructive and non-expensive analysis.

Source

1. *Technical literatures of Instrument.*
2. *Evgeny Kuryashov and Vitaly Buckin, Department of Chemistry at University College Dublin.*
3. *Cormac Smyth and Breda O'Driscoll, Ultrasonic Scientific Limited, GEC, Dublin. Ireland, www.ultrasonic-scientific.com.*

Ultrasonic Spectroscopy Based Equipments for Dairy [1,2,3]

The following Ultrasound Spectroscopy based equipments are being used for rapid analysis of milk and milk products:

A. High-Resolution Spectrometers

i. HR-US101 High-Resolution Spectrometer [1,2]

The uptake of ultrasonic spectroscopy has been hindered by its technical complexity and limited resolution. A new device has overcome these problems is finding applications in various dairy processes. HR-US101 high-resolution spectrometer overcomes these limitations. This instrument can perform analysis in a broad range of sample volume (1ml typically, 0.03ml to 200ml and higher when custom made) and performs analysis in various regimes with a record resolution. It is used in the analysis of gelation processes in milks, melting of milk fat in milk and butter, composition analysis, particle sizing, creaming effects and coagulation in milks and milk based drinks. The details related to these applications are given below:

a. Ultrasonic Monitoring of Gelation Process in Milk

At time 0 the ultrasonic cell was filled with 1ml milk and bacterial culture was added. The gelation starts at one hour as indicated by a sharp increase in ultrasonic attenuation. Ultrasonic velocity increases at a pregelation stage showing a change in milk composition as a result of bacterial activity. This is followed by an additional increase in ultrasonic velocity at gelation stage. The levelling off of ultrasonic velocity after 1.7 hours shows a decline in bacterial activity. The structure of the gel however changes until 2.5 hours as can be seen from ultrasonic attenuation curves. In the time interval between 1.7 and 2.5 hours the ultrasonic velocity curves are frequency dependent, which is a good indication of synerisis.

b. Optimisation of Composition of Calcium-Fortified Milks

Calcium fortified milk drinks is one of the industry's responses to today's consumers needs for a healthy lifestyle. The addition of calcium changes the balance of colloid interactions in milk thus leading to coagulation during heat treatment or when it is added to hot beverages.

c. Coagulation

It is indicated by a sharp decrease in the ultrasonic velocity and increase in the ultrasonic attenuation. This allows us to determine the coagulation temperature (0.1°) with a resolution, which is even higher than required for developing the optimal milk composition (no coagulation at temperature below 100°C). The results of ultrasonic assessments enable us to study the effects of the addition of different ingredients/stabilisers on the coagulation temperature of milk.

Source

1. *Evgeny Kuryashov and Vitaly Buckin, Department of Chemistry at University College Dublin.*
2. *Cormac Smyth and Breda O'Driscoll, Ultrasonic Scientific Limited, GEC, Dublin. Ireland, www.ultrasonic-scientific.com.*
3. *Technews, No.74, May-June-2008, NDDB.*

ii. HR-US 102 High-Resolution Ultrasonic Spectrometer

The spectrometer is equipped with precision temperature controller and digital stirring system and can perform measurements in dilute and concentrated solutions, suspensions and emulsions in various regimes: kinetics, temperature ramp, titrations and others. The following different models of HR-US 102 are used for analysis of liquid milk:

HR-US 102

PHR-US 102 FT

HR-US 102 SS

a. HR-US 102 P

It has the same configuration as the HR-US 102 with additional interface, which offers extended capabilities by allowing user to attach external modules such as cells for semi-solid materials, measurements in flow, *etc.*

Computerised High-Resolution Ultrasonic Spectrometer.
(*Source*: Sonas Technologies Ltd, 42 Ailesbury Grove, Dundrum, Dublin 16, Ireland).

b. HR-US 102 SS

It contains two 1.5 ml cells optimized for measurements in semi-solid materials such as biological tissue, pastes, waxes, milkfats *etc.* The spectrometer is equipped with precision temperature controller and allows kinetics, temperature ramp and other measurements.

c. *HR-US 102 FT*

It is fitted with two 1 mL flow through cells. The spectrometer allows analysis of liquids in flow. Two cells can be used for simultaneous analysis of two liquids or in differential regime, *i.e.* when one cell is permanently filled with a reference sample.

B. LactiCheck (Quadra chem Lab Ltd)

It provides milk composition results for fat, solids, protein, added water and density in just 85 seconds. Compact, light weight and portable, the LactiCheck provides fast, precise and affordable analysis of raw or processed milk and fluid milk products. LactiCheck is available in single (LC01) and dual (LC02) channel models and along with Cow milk options allow measurement of Goat, Sheep and Buffalo milk. **For Farm Testing-**Portable, precise, simple measurement of raw co-mingled milk facilitates feed and breeding decisions. User-friendly, robust and light-weight, the LactiCheck gives farmers feedback critical to practical herd management.

C. Milk-Lab COMPACT

It is small and automatic equipment suitable for analysing fat, SNF, protein, total solids and temperature in raw milk. It is meant for use in milk collection centres, dairy farms, milk collection trucks and small dairies.

D. Milk-Lab PRO

This equipment is suitable for analysing fat, SNF, protein, total solids, lactose, freezing point, added water conductivity, pH and temperature in milk. It is also available different product types which is suitable for Milk shake, Flavoured milk, Evaporated milk, Reconstituted milk, whey, Cream and Ice cream.

Source

1. *Product showcase, Dairy Industries International, May-2008, pp-40-41.*
2. *Technews, No.74, May-June-2008, NDDB*
3. *Smith, C., O'Driscoll, B., Kuryashov, E., and Buckin, V., High resolution ultrasonic spectroscopy for analysis of dairy systems.*

7. Mass Spectrophotometry [1,2]

It is an analytical technique that measures the mass- to-charge ratio of charged particles. In mass spectrometry, a compound is ionized and the resulting ions are separated according to their mass-to-charge ratio, and the ion abundances are measured. It is used for determining masses of particles, for determining the elemental composition of a sample or molecule, and for elucidating the chemical structures of molecules, such as peptides and other chemical compounds. Although there are many different kinds of mass spectrometers, all of them make use of electric or magnetic fields to manipulate the motion of ions produced from an analyte of interest and determine their m/z [3].

The mass spectrum can be used to determine the mass of the analytes, their elemental and isotopic composition, or to elucidate the chemical structure of the sample [4]. MS is an experiment that must take place in gas phase and under vacuum (1.33 * 10^{-2} to 1.33 * 10^{-6} pascal). Therefore, the development of devices facilitating the transition from samples at higher pressure and in condensed phase (solid or liquid) into a vacuum system has been essential to develop MS as a potent tool for identification and quantification of organic compounds and peptides [5].

Source

1. *A.Sparkman, O. David (2000). Mass spectrometry desk reference. Pitts- burge, Global View Pub. ISBN 0-9660813-2-3.*
2. *Wikipedia.*
3. *Encyclopedia of Biophysics-Springer.*
4. *Dass, Chhabil (2007-01-01). Fundamentals of Contemporary Mass Spectrometry.John Wiley and Sons, Inc. pp. 151–194.ISBN 9780470 118498.*
5. *Sharp, Thomas R. (2009-01-01). Nassar, Ala F.; Collegiateessor, Paul F. Hollenberg; VP, JoAnn Scatina, eds. Drug Metabolism Handbook. John Wiley and Sons, Inc. pp. 167–227. ISBN 9780470439265.*

Apparatus

Initially, the analyzed sample has to be introduced into the ion source of the instrument, where the sample molecules undergo the ionization process. These ions are then extracted into the analyzer where they are separated according to their m/z. The separated ions are detected and the signal containing information about the m/z values is stored together with their relative abundance (intensity) as a mass spectrum. To allow the ions traveling from one end of the instrument to the other part without any hindrance originated from air molecules, both the analyzer and detector (and often the ion sourceas well) are kept under high vacuum [1]. The separated ions MS instruments consist of following three important and key parts:

1. Ion Source [1-4]

The ion source is the part of the mass spectrometer that ionizes the material under analysis (the analyte). The ions are then transported by magnetic or electric fields to the mass analyzer. It can convert gas phase sample molecules into ions (or, in the case of electrospray ionization, move ions that exist in solution into the gas phase). Electron ionization and chemical ionization are used for gases and vapors. In chemical ionization sources, the analyte is ionized by chemical ion-molecule reactions during collisions in the source. For liquid and solid biological samples, Electrospray ionization (invented by John Fenn) and Matrix-assisted laser desorption/ionization (MALDI, developed by K. Tanaka), are used. Inductively (ICP) sources are used primarily for cation analysis of a wide array of sample types. Others include glow Discharge, field desorption (FD), Fast atom bombardment (FAB), Thermospray, Desorption/ionization on silicon (DIOS), Direct Analysis in Real Time (DART), Atmospheric pressure chemical ionization (APCI), Secondary ion mass spectrometry (SIMS), Spark ionization and Thermal ionization (TIMS).Ion Attachment Ionization (newer technique) allows for fragmentation free analysis.

Source

Mass Spectrometry and Hyphenated Instruments in Food Analysis by Tomas Cajka, Jana Hajslova, and Katerina Mastovska.

Types of Ionizations/Ion Sources for Gas Chromatography and Liquid Chromatography Coupled with Mass Spectrosopy [1]

The sample can be introduced to the ion source directly, or after its previous separation involving either GC or high-performance LC (HPLC); thus, the sample is separated into a series of components, which then enter the mass spectrometer for their subsequent analysis. There are several ionization methods available, each having its own advantages and limitations. The fitness of the ion source depends on factors such as the polarity, molecular weight, thermal lability of the analytes, and complexity of the examined sample. The Ion sources for GC -MS, LC–MS and other techniques are explained below:

A. Ion Sources for GC-MS

Electron ionization and Chemical ionization represent the fundamental ionization techniques. On the basis of the scientific literature abstracted in SciFinder Scholar the electron ionization (EI) was used in approx. 95 per cent of all food GC-MS applications, while the rest of applications (5 per cent) employed chemical ionization (CI) [1].

i. Electron Ionization (EI)

This process starts by the acceleration of electrons through an electric field. These energetic electrons interact with neutral molecules (M). Upon collision, the molecule loses an electron and becomes a particle with an odd number of electrons and a positive charge. This radical cation (M^+) is typically called molecular ion. If the molecular ion contains sufficient internal energy, it undergoes fragmentation.

The energy required for the ionization process is called the ionization energy and can vary between 0 and over 100 eV. Under standardized conditions, the electrons are produced in the EI source from a heated filament. The moving electrons directed across the source are forced into a beam by a magnetic field. This electron beam then interacts with sample molecules that have been vaporized. The ions are extracted from the source by an electric field and passed into the analyzer as an ion beam or as an ion packet.EI is preferred for the identification of unknowns, determination of molecular structure, and confirmation of target analyte identity through consistent ion abundance ratios and library spectra matching.

ii. Chemical Ionization (CI)

It represents a low energy or soft ionization technique. It is very suitable for those less volatile or thermally labile molecules that do not yield molecular ions by EI. For CI, a suitable reagent gas is introduced into the ion source at a concentration that largely excesses the amount of the analytes. The reagent gas is usually ionized as in EI. The formation of primary ions is followed by reactions between those primary ions and the noncharged molecules of gas producing the CI reagent ions as well as the thermal electrons.

i. In Positive Chemical Ionization (PCI)

The ion source is filled with a reagent gas (*e.g.*, methane), at a relatively high pressure (0.1–100 Pa), which undergoes EI, producing an excess of reagent ions. PCI can offer both increased sensitivity and improved detectability due to reduced chemical noise from background or co-eluting analytes, resulting in increased signal to noise ratio (S=N).

ii. Negative Chemical Ionization (NCI)

The production of a large population of low-energy electrons during the CI operation utilizes another ionization technique: negative chemical ionization (NCI). The main advantages of NCI compared to EI and PCI include a possibility of up to 100-fold improvement in sensitivity, and higher degree of selectivity.

B. Ion Sources for LC-MS

Electrospray ionization (ESI), Atmospheric pressure chemical ionization (APCI), and Atmospheric pressure photoionization (APPI) represent ionization techniques widely employed. On the basis of SciFinder Scholar, ESI was the most often reported ionization technique in food LC-MS applications (approx. 80 per cent), while APCI and APPI were more rarely employed (18 per cent and 2 per cent, respectively). In addition, the direct ionization techniques such as matrix-assisted laser desorption ionization (MALDI), desorption electrospray ionization (DESI), desorption atmospheric pressure chemical ionization (DAPCI), and direct analysis in real time (DART) can be employed for sample characterization [1].

i. Electrospray Ionization (ESI)

It represents a very soft ionization technique, resulting typically in a single peak of (de) protonated molecular ion or molecular ion adducts with, *e.g.*, sodium, potassium, or ammonia. It is suitable for ionic analytes and other relatively polar compounds with MW ranging from less than 100 Da to more than 1,000,000 Da. In the analysis of large molecules, such as proteins, ESI generates multiply charged ions, which reduces the m=z ratio, effectively extending the mass range of mass spectrometers. During ESI, the sample is dissolved in a polar solvent and introduced though a narrow stainless steel capillary (nebulizing needle, ESI needle) either from a syringe pump or as the effluent flow from a LC.

ii. Atmospheric Pressure Chemical Ionization (APCI)

It is a somewhat similar ionization technique to CI in GC–MS. The chemical ionization requires a vacuum, the APCI occurs at atmospheric pressure. In APCI, the analyte solution is introduced from a direct inlet probe or an LC eluate at a flow rate between 2 mL/min and 2 mL/min into a pneumatic nebulizer, where it is converted into a vapor by a nitrogen beam. Similarly to ESI, the APCI spectra are characterized by predominant molecular species and adduct ions with very little fragmentation. The ions then enter the mass spectrometer through a tiny inlet and are focused towards the analyzer.

iii. Atmospheric Pressure Photoionization (APPI)

It represents a novel, alternative ionization method for LC–MS. The ionization process in APPI is initiated by 10 eV photons, which are emitted by a krypton discharge lamp.

C. Ion Source for ICP-MS [1]

The sample for inductively coupled plasma (ICP)-MS has to be introduced into the plasma as an aerosol, usually as a liquid sprayed through a nebulizer. Under the conditions of high temperature (6,000–10,000 K) in the plasma, the sample is atomized and ionized, creating positively charged atomic ions. These atoms are converted to positively charged ions before their extraction into the vacuum system for the detection creating the ICP-MS spectrum, which represents the elemental composition of the sample. The most important features of ICP-MS, compared to other elemental analysis techniques, are extreme sensitivity, selectivity, and simultaneous multi-element capability.

D. Ion Sources for Direct MS Analysis

The following types of ionizations are ion sources for Direct MS analysis:

i. Matrix-Assisted Laser Desorption Ionization (MALDI)

It is mainly suitable for peptides, proteins, glycoproteins, oligoproteins, and oligonucleotides which are thermolabile and/or nonvolatile organic compounds with high molecular mass. It involves the following steps:

- *(a)* Before the analysis, the sample is pre-mixed with a solvent containing highly UV-absorbing matrix compound.
- *(b)* This mixture is dried before the analysis removing the liquid solvent used in preparation.
- *(c)* The obtained deposit of matrix–analyte solid solution then undergoes the ablation of bulk portions of this solid solution by intense pulses of laser for a short duration. The irradiation by the laser induces rapid heating of the crystals as a result of the accumulation of a large amount of energy in the condensed phase through excitation of the matrix molecules. The rapid heating leads to sublimation of the deposits and their expansion into the gas phase followed by evaporation of the matrix molecules away from the clusters to leave the free analyte ions in the gas phase.
- *(d)* The analyte ions, created by proton transfer from ionized matrix molecules, are then extracted into the mass spectrometer for the analysis [5].
- *(e)* In most MALDI instruments, the desorption/ionization process takes place in vacuum, thus the sample is not accessible during the analysis. Atmospheric pressure MALDI (APMALDI) still requires sample dilution/

coating with a UV-absorbing matrix compound and operation under enclosed conditions to protect the operator from potential exposure to laser radiation [6].

ii. Desorption Electrospray Ionization (DESI) and Desorption Atmospheric Pressure Chemical Ionization (DAPCI)

The DESI and DAPCI Ionisations are explained below:

a. DESI

It is carried out by directing electrosprayed charged droplets and ions of solvent on to the analyzed sample surface. The spray is directed at an insulating sample or an analyte deposited on an insulating surface such as polytetra- fluoroethylene. The desorbed ions are sampled with a commercial mass spectrometer equipped with an atmospheric interface connected via a transfer line made either of metal or an insulator.

b. DAPCI

Those compounds that do not provide sufficient ion intensity by DESI then DAPCI offers an alternative/better option because it provide increased sensitivity for compounds of moderate polarity. The DAPCI technique uses nitrogen sheath gas and a solvent from which ions are produced by a corona discharge. Reagent ions formed in the corona discharge region react with desorbed analyte molecules, creating protonated or deprotonated molecular ions depending on the polarity of the ionization mode [7].

iii. Direct Analysis in Real Time (DART)

It is based on the atmospheric pressure interactions of long-lived electronic excited-state atoms or vibrionic excited-state molecules with the sample and atmospheric gases. A gas (Helium or Nitrogen) flows through a chamber where an electrical discharge produces ions, electrons, and excited-state (metastable) atoms and molecules. Most of the charged particles are removed as the gas passes through perforated lenses or grids and only the neutral gas molecules (including metastable species) remain. A perforated lens or grid at the exit of DART acts as an electrode to promote ion drift toward the orifice of the atmospheric pressure interface of the mass spectrometer.DART does not use any solvent where as DESI uses solvent.

Source

1. *Mass Spectrometry and Hyphenated Instruments in Food Analysis by Tomas Cajka, Jana Hajslova, and Katerina Mastovska.*
2. *Fenn, J.B., Mann, M., Meng, C.K., Wong, S.F., Whitehouse, C.M. (1989)."Electrospray ionization for mass spectrometry of large bio-molecules".Science 246 (4926),64-71.*
3. *Tanaka, K., Waki, H., Ido, Y., Akita, S., Yoshida, Y., Yoshida, T. (1988)."Protein and Polymer Analyses up to m/z 100 000 by Laser Ionization Time-of flight Mass Spectrometry". Rapid Common Mass Spectrom,2 (20),151-3.*

4. *A.P.Bruins (1991)."Mass spectrometry with ion sources operating at atmospheric pressure". Mass Spectrometry Reviews 10 (1): 53-7.*

5. *Karas, M. and Hillenkamp, F., Laser desorption ionization of proteins with molecular masses exceeding 10,000 daltons, Anal. Chem., 60, 2299, 1988.*

6. *Laiko, V.V., Baldwin, M. A., and Burlingame, A.L., Atmospheric pressure matrix-assisted laser desorption/ionization mass spectrometry, Anal. Chem., 72, 652, 2000.*

7. *Cotte-Rodriguez, I. et al., Desorption electrospray ionization of explosives on surfaces: Sensitivity and selectivity enhancement by reactive desorption electrospray ionization, Anal. Chem., 77, 6755, 2005.*

2. Mass Analyzer

Once the molecules are converted into charged particles (ions), they undergo mass analysis in a mass analyzer. It sorts the ions by their masses by applying electromagnetic fields. It measures the value of an indicator quantity and thus provides data for calculating the abundances of each ion present. The following two laws govern the dynamics of charged particles in electric and magnetic fields in vacuum:

(a) Lorentz force law.

(b) Newton's second law of motion in non-relativistic case, *i.e.* valid only at ion velocity much lower than the speed of light.

There are many types of mass analyzers, using either static or dynamic fields, and magnetic or electric fields, but all operate according to the above differential equation. Each analyzer type has its strengths and weaknesses. Many mass spectrometers use two or more mass analyzers for tandem mass spectrometry (MS/MS) [1]. Mass analyzers are nothing but sieve. A range of DC/Rf voltages are applied across the mass analyzer quadrupole. These quadrupole at a specific DC/Rf voltage have selectivity for one particular mass to charge ratio (m/z) particles. At one point in time one particular DC/Rf voltage limits the entry of other mass/charge ratio particles and selectively allows one mass/charge ratio particles to pass through the quadrupole which is then forced to the detector. The signal corresponding to the particular mass to charge ratio gets registered as the ion abundance of that ion [1a].

A wide range of mass analyzers are now-a-days available. i-While quadrupole, quadrupole ion trap, linear quadrupole ion trap, and magnetic sector represent scanning instruments, in which the mass spectrum is recorded by gradual changing of electrical and magnetic field, ii- the other group of mass analyzers such as time-of-flight (TOF), A sector field mass analyzer, Fourier transform ion cyclotron resonance (FT-ICR) analyzer, and orbitrap are nonscanning instruments, in which entire mass spectra are obtained simultaneously [2].

a. Key Operating Parameters

The choice of an optimal mass analyzer is determined by following several key operating parameters [2,3,4,5]:

Sl.No.	*Key Operating Parameters Responsible for Choice of Mass Analyser*	*Details of Key Operating Parameters*
1	**Mass range** (Range of m/z over which a mass analyzer can separate and then Detect/record ions)	*(i)* The majority of mass spectrometers combined with GC operate typically up to m/z 1000 because in GC, analyte volatility/thermolability effectively dictates the upper mass limit. *(ii)* In LC, the mass range is often increased up to several orders higher m/z.
2	**Mass resolution/mass resolving power** (The ability of a mass analyzer to separate two ions of similar mass)	*(i)* As per recent IUPAC recommendation on standard definitions of MS terms (drafted in 2006) [6], mass resolution is defined as the smallest mass difference Δm between two equal magnitude peaks, so that the valley between them is a specified fraction of the peak height. Mass resolving power is then defined as the observed mass divided by the difference between two masses that can be separated, $m/\Delta m$. *(ii)* The most mass analyzers operate at so-called unit mass resolution (or actually at mass resolving power <1000). High-resolution MS instruments provide improved selectivity, thus reduced chemical noise and, consequently, increased S/N in complex samples
3	**Mass accuracy** (the deviation between measured mass (accurate mass) and calculated mass (exact mass) of an ion expressed as an error value (mDa, ppm).	*(I)* The unit-resolution mass spectrometers provide mass accuracy of approx. 0.1–0.2 Da (*e.g.*, 200.0 and 200.1 Da),while the high-resolution instruments operate at mass accuracy <5 ppm (*e.g.*, 200.000 and 200.001 Da).
4	**Spectral acquisition speed** [time required for recording of a mass spectrum or selected ion(s)].	*(i)* Maximal spectral acquisition speed is a critical parameter in detection of very narrow peaks generated during fast chromatographic separation. *(ii)* For scanning instruments terms scans/s, s/scan, or s/decade are commonly used where as for nonscanning instruments a term spectra/s is preferred.
5	**Selection of acquisition mode** (allows the use of different modes of mass spectra collectionin scanning instruments).	*(i)* Both magnetic sector and quadrupole instruments can be operated in full scan and selected ion monitoring (SIM); in the latter case structural information is sacrificed in favor of detection sensitivity.
6	**Detectability or often expressed as sensitivity of the instrument**(which is characterized as minimal S/N ratio at a given concentration of a reference compound).	*(i)* Compounds such as octafluoronaphthalene or hexachlorobenzene are typically used in GC - MS, and reserpine in LC -MS.
7	**Linear dynamic range** (the range of analyte concentrations, over which the detector proportionally responds to concentration changes).	*(i)* For common scanning instruments, this range varies between 5–6 orders of magnitude. *(ii)* The nonscanning instruments (TOFMS), offer linear dynamic range of typically 4 orders of magnitude in maximum.

Contd.

Sl.No.	Key Operating Parameters Responsible for Choice of Mass Analyser	Details of Key Operating Parameters
8	**Availability of tandem MS function** (which is a method involving at least two stages of mass analysis, either in conjunction with a dissociation process or a chemical reaction that causes a change in the mass or charge of particular ion).	i. The main tandem MS/MS scan modes are- 1. Product ion scan, which involves selection of an ion of interest, its activation, and mass analysis of the productions in full scan mode. 2. Precursor ion scan represents opposite process compared to the product ion scan, providing information about all precursor ions that react to produce a selected product ion. 3. Neutral loss scan is a scan that determines all precursor ions that react to the loss of a selected neutral mass. 4. Selected react ion monitoring (SRM) monitors a single transition from a precursor to a product ion. 5. Multiple reaction monitoring (MRM) is used if several different reactions are monitored in one time window. 6. MS ^{n} scans, which is commonly applied on ion trap analyzers.
9	**Versatility**	*(i)* Versatility, in terms of the use of alternative ionization techniques, extends the scope of applications.
10	**Cost**	*(i)* In routine practice, less expensive scanning instruments (quadr upole and ion trap analyzers) are widely used. *(ii)* The cost TOF is approx. 2–3 times and cost of Sector is approx. 6 times compared to low resolution quadrupole or ion trap instruments.

b. Characteristics of some Common Mass Analysers

Mass resolving power, Mass accuracy, Mass range, Speed and Linear dynamic range are the important characteristics of analysers. The characteristics of some common mass analysers are tabulated below:

Sl.No.	Types of Mass Analysers		Characteristics
1	**One group of mass analyzers**		
	a	**Quadrupole**	*(i)* Most popular mass analyzer mainly due to its relatively low cost,ruggedness, reliability, and the simplicity of operation. *(ii)* This mass analyzer (mass filter) consists of four (hyperbolic or cylindrical) rods placed in a radial array. Opposite rods are charged by positive direct-current (DC) voltage,while adjacent rods have the opposite (negative) charge applied. Ions are introduced into the quadrupole field by means of a low accelerating potential. *(iii)* It can be operated in two modes-1- full scan of a selected mass range and 2- SIM.

Contd.

Sl.No.	*Types of Mass Analysers*		*Characteristics*
	b	**Quadrupole ion trap(3D Trap)**	*(i)* It consists of three cylindrically symmetric electrodes (two end-caps and a ring). The ions of selected m/z range are trappe d by an RF potential, which is formed between end-cap electrodes and a ring electrode.
			(ii) Although the quadrupole ion traps allow both full scan and SIM acquisition, there is no increase in detection sensitivity in SIM mode that is typically observed with quadrupole or sector instruments.
	c	**Linear quadrupole ion trap(2DTrap)**	*(i)* It differs from the 3D ion trap as it traps ions in the radial dimension by an RF quadrupole field and by static DC potentials applied to the ends of a four- rod structure.ii-It has several advantages over the 3D trap:
			(a) Since there is no quadrupole electric field in the z direction the ion injection and extraction efficiencies can be nearly 100 per cent.
			(b) 3D trap has a limited capacity for ion storage due to its small strapping volume, the 2D trap has almost one order of magnitude increase in ion capacity.
	d	**Magnetic sector**	*(i)* The single focusing magnetic sector mass analyzer consists of a magnetic sector, in which the magnetic field is used for separation of the ions with different m/z ratios. The ions entering the mass analyzer are initially accelerated to a high velocity using an electric field. The ions then pass through a magnetic sector where the magnetic field is applied in a direction perpendicular to the direction of ion motion.
			(ii) The simplest way to operate a magnetic sector mass analyzer is to keep the accelerating potential and the electric sector at a constant potential and vary the magnetic field.
			(iii) A double focussing magnetic sector mass analyzer has been employed in a number of applications.
2	**Other group of mass analyzers**		
	a	**Time-of-flight (TOF) [orthogonal acceleration, Non scanning mass analyzer]**	*(i)* This mass analyzer consists of a pulsing electrode, a flight tube, and a reflectron. The generated ions are accelerated to get constant kinetic energy. followed by their ejection into a mass analyzer's flight (drift) tube using pulsed electric-field gradient, which is oriented orthogonally to the ion beam.
			(ii) The mass analyzer efficiency of TOFMS instruments is as high as 25 per cent in full spectra acquisition. This value is significantly higher as compared to that obtained by scanning instruments such as quadrupole (~0.1 per cent) [7].
			(iii) The TOF is non scanning mass analyzer hence all ions are recorded simultaneously, thus there are no changes in the ratios of analyte ions across the peak during the acquisition of the mass spectrum and, consequently, no spectral skewing is encountered.

Contd.

Sl.No.	Types of Mass Analysers	Characteristics
		(iv) Two types of TOFMS instruments differing in their basic characteristics are available.One type uses high- resoluti on analyzers (7000 FWHM) providing only moderate acquisition speed (up to 20 spectra/s), and the second type are unit -resolution instruments that feature high acquisition speeds (up to 500 spectra/s) [8].
b	**Fourier transform ion cyclotron resonance (FT-ICR) analyzer**	*(i)* It performs the mass analysis in a cubic cell that consists of pair of trapping plates, excitation plates, and detector plates.
		(ii) After trapping the ions in the cell, they are excited to a larger cyclotron radius by an oscillating electric field perpendicular to the fixed magnetic field. A packet of the ions is also formed during the excitation. The signal is then detected as an image current on a pair of receivers. The resulting signal consists of a superposition of sine waves, from which a regular mass spectrum is obtained by applying Fourier transformation.
		(iii) Have extremely high mass resolving power and sensitivity.
C	**Orbitrap**	*(i)* It represents a special case of an ion trap. Contrary to a conventional ion trap, which uses RF to hold the ions inside the trap, the moving ions in orbitrap are trapped in an orbit around a spindle-shaped electrode by an electrostatic field.
		(ii) The features include high mass resolving power, large space charge capacity, high mass accuracy and high mass range [9,10].
3	**Hybrid Instruments**	
a	**Triple quadrupole (QqQ)**	*(i)* Different types of mass analyzers can be coupled together forming hybrid instruments. It represents the most common known example.
		(ii) It consists of two single quadrupoles with a collision cell in between. The ions are directed from the ion source into the first quadrupole, where the precursor ion is selected for MS/MS reaction in the collision cell (usually a hexapole). The product ions are then separated in the second quadrupole and recorded by the detector.
b	**Quadrupole/time-of-flight (Q-TOF)**	*(i)* Last quadrupole in the series is replaced by TOF mass analyzer, the quadrupole/time-of-flight mass spectrometer is obtained, which combines the simplicity of quadrupole with high efficiency of a TOF mass analyzer.
		(ii) The Q-TOF systems offer high-resolution capability of up to 10,000 FWHM mass resolving power, allowing high-resolution and accurate mass measurement of the product ions generated in the collision cell [11].
c	**Magnetic sector/ time-of-flight**	These types of hybrids instruments are also being used successfully.
d	**Time-of-flight/time-of-flight(TOF-TOF)**	
e	**Quadrupole/linear ion trap (Q-LIT)**	

c. Specifications and Features of Mass Analysers

The general specifications and features of selected mass analyzers hyphenated to both GC and LC are shown in following tables [2]:

Sl. No.	*Criteria*	*Ion Trap*	*Quadrupole*	*Triple Quadrupole*	*High-Speed TOF*	*High-Resolution TOF*
1	Mass range	Up to 2,000-6,000 Da	Up to 1,000-3,000 Da	Up to 1,000-3,000 Da	Up to 6,000 Da	Up to 30,000 Da
2	**Mass accuracy**	0.1–0.2 Da	0.1–0.2 Da	0.1–0.2 Da	<15 ppm	<3 ppm
3	**Acquisition mode**	Full scan, SRM, MRM.	Full scan, SIM.	SIM, Full scan, SRM, MRM.	Full spectra	Full spectra
4	**Maximal spectral acquisition speed**	≈2 scans/s (m/z50–1,000 Da)	≈6 scans/s (m/z50–1,000 Da)	-	100 spectra/s (m/z50 -1,000 Da)	20 spectra/s (m/z50-1,000 Da)
5	**Sensitivity**	>pg in full scan	>pg in full scan <pg in SIM	>pg in full scan <pg in MS/MS	>pg	pg
6	**Linear dynamic range**	4–5	>5	>5	>3	4
7	**Versatility**	APPI,ESI	ESI, APCI	ESI, APCI, APPI	ESI	ESI, APCI, APPI.
8	**Mass resolution/ mass resolving power**	Unit mass	Unit mass	Unit mass	>2,000 FWHM	>12,000 FWHM
9	**MS/MS**	MS^n n -2–10	None[a]	MS^2	None[a]	None[a]

Note: APCI: Atmospheric pressure chemical ionization, APPI: Atmospheric pressure photoionization, ESI: Electrospray ionization, FWHM: Full width at half maximum, MS: Mass spectrometry, SIM: Selected ion monitoring, TOF: time-of-flight, SRM: Selected reaction monitoring, MRM: Multiple reaction monitoring, pg: Picogram and fg: Femtogram.

SIM-Selected Ion Monitoring and MRM [1a]

Multiple reaction monitoring: The details of SIM and MRM are tabulated below:

SIM	*MRM*
When there are lot of interferences from the sample and analyte of interest is getting lost in the crowd of ions which reflects in lower Signal to Noise ratio to quantify then there is a provision of Single Ion Monitoring. Where i- in first, standard solution of analyte runs in order to know the Target ion specific for the analyte and its fragment ions which is specific for the anlayte.	Multiple Reaction Mode. When confirmation of the presence of an analyte which is not available in the library,is required then its necessary to know the fragmentation pattern which is specific to every analyte. Then to select the collision energy at which maximum abundance of the fragment ions can be obtained needs to be selected, which can only be done by Multiple Reaction Mode.

SIM	*MRM*
ii-Then sample containing analyte runs and select the option of Single Ion Monitoring, The information of target ion to be acquired by the MS, is given and for confirmation of the analyte, the information of fragment ions based on the abundance is given. The mass analyzer of MS will selectively monitor the presence of the target analyte molecule only and will not monitor other ions. There by preventing noise and increasing the signal to noise ratio and hence the resolution.	Where in a selected target ion of the analyte of interest is allowed inside the Quadrupole 1 of mass analyzer which then at different collision energies is fragmented in the Reaction Cell (Quadrupole 2) and further analyzed for its mass based on abundance.

Source

1. *Wikipedia.*

1a. *K.P. Indumathi, Rajan Sharma and Jyotika, GC-MS Application in Food Analysis, Dairy Chemistry Division, Compendium of Lectures on Advanced Techniques and Novel Approaches for Quality and Safety Evaluation of Dairy Foods-2015, NDRI, Karnal.*

2. *Mass Spectrometry andHyphenated Instruments in Food Analysis by Tomas Cajka, Jana Hajslova, and Katerina Mastovska.*

3. *Mastovska, K. and Lehotay, S.J., Practical approaches to fast gas chromatography- mass spectrometry, J.Chromatogr. A,1000,153,2003.*

4. *Mastovska, K., Instrumental aspects and application of (ultra)fast gas chromate-graphy–mass spectrometry, in Niessen, W. (Ed.), Encyclopedia of Mass Spectrometry, Elsevier, Oxford, 2006, p. 73.*

5. *Cajka, T. and Hajslova, J., Gas chromatography–time-of-flight mass spectro-metry in food analysis, LC GCEur., 20, 25, 2007.*

6. *Murray, K.K. et al., Standard Definition of Terms relating to Mass Spectrometry (UIPAC Recommendations2006). Available: http: =www.iupac.org= reports=provisional= abstract06= murray_prs. pdfvia the Internet. Accessed on Feb 1 2007.*

7. *Guilhaus, M., Selby, D., and Mlynski, V., Orthogonal. acceleration time-of- flight mass spectrometry, MassSpec. Rev., 19, 65, 2000.*

8. *Cajka, T. and Hajslova, J., Gas chromatography–time-of-flight mass spectrometry in food analysis, LC GC Eur., 20, 25, 2007.*

9. *Scigelova, M. and Makarov, A., Orbitrap mass analyzer-overview and applications in proteomics, Proteomics, 6, 16, 2006.*

10. *Hu, Q. et al., The Orbitrap: A new mass spectrometer, J. Mass Spec., 40, 430, 2005.*

11. *Ferrer, I. and Thurman, E.M., Liquid chromatography/time-of-flight/mass spectrometry (LC/TOF/MS)for the analysis of emerging contaminants, TrAC-Trend Anal. Chem., 22, 750, 2003.*

3. Detector [1,2]

It measures the value of an indicator quantity and thus provides data for calculating the abundances of each ion present [1]. The detector allows to monitor the ion current and, after its amplification, to record the data in the form of mass spectra. The ions separated by mass analysers for their individual m/z values, are deteted by Ion Detectors. The ion detectors can be classified into following three classes [2]:

(a) Point ion collectors that detect the arrival of all ions sequentially at one point *e.g.*, **electron multiplier, photon multiplier**.

(b) Array collectors that detect the arrival of all ions simultaneously along a plane *e.g.*, **Microchannel plate (MCP) detector**.

(c) Detectors that consists of a pair of metal surfaces within the mass analyzer region, which the ions only pass near as they oscillate and only a weak image current is produced in a circuit between the electrodes (this principle is employed only for FT-ICR MS and orbitrap).

The most common types of MS detectors include the **electron multiplier**, the **photon multiplier**, and the **microchannel plate (MCP) detector [2]** and their haracteristics are given in following table:

Sl.No.	*Types of Detectors*	*Charateristics*
1	Electron multiplier	*(i)* In this, the ions reach the first plate (dynode) of an electron multiplier and then the ejected electrons are accelerated through an electric potential to a second dynode. This process is typically repeated 10–12 times, which gives the amplifications of 10^6. The final flow of electrons provides an electric current that can be further increased by electronic amplification.
		(ii) The final flow of electrons provides an electric current that can be further increased by electronic amplification.
		(iii) It is typically used for quadrupole, ion trap, and sector instruments.
2	Photon multiplier	*(i)* It is made up of two conversion dynodes, a phosphorescent screen, and a photomultiplier.
		(ii) Considering the positive mode, secondary ions are accelerated towards the dynode that holds the negative potential. Secondary electrons that are generated are accelerated towards the phosphorescent screen, where conversion into photons occurs, followed by their detection by the photomultiplier.
		(iii) It is typically used for quadrupole, ion trap, and sector instruments.
3	MCP detector	*(i)* It consists of a large number of miniature electron multiplier elements placed side by side over a plane. The surface of the MCP is metal coated, serving as electrodes.
		(ii) A voltage applied between the electrodes produces an electric-field gradient. When an ion hits the front part of the inner walls of the channel at this point, multiple secondary electrons are emitted. These electrons are then accelerated by the electric-field gradient inside the channel, and repeatedly hit the walls on the opposite side, emitting secondary electrons.
		(iii) At the end of this process, the electrons are captured by the anode, which produces an electrical signal.
		(iv) It is commonly used for detection of ions in TOFMS.

Source

1. *Wikipedia.*
2. *Mass Spectrometry and Hyphenated Instruments in Food Analysis by Tomas Cajka, Jana Hajslova, and Katerina Mastovska.*

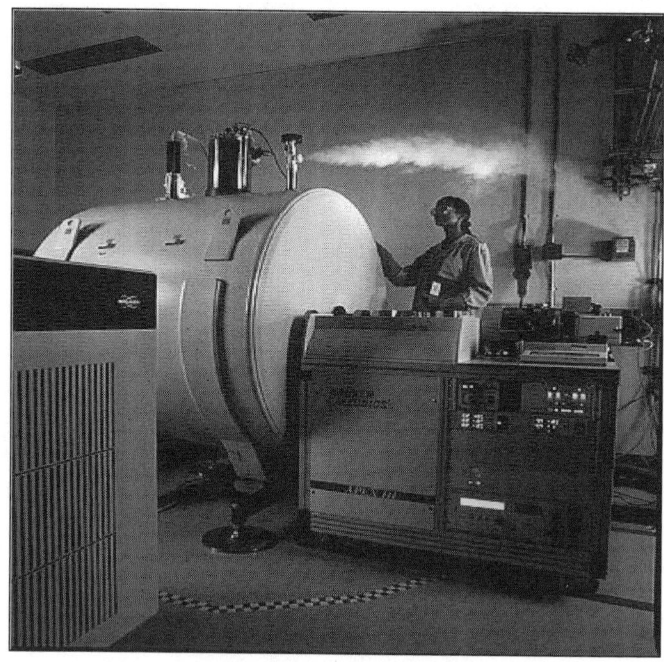

Gas Chromatograph (Right) Directly Coupled to a Mass Spectrometer (Left) FT-ICR Mass Spectrometer (*Source*: Wikipedia).

Procedure

The MS principle consists of ionizing chemical compounds to generate charged molecules or molecule fragments and measuring their mass- to -charge ratio.MS is an instrumental technique based on the separation of ions in vacuum, in the gas phase, according to their mass-to- charge ratios (m/z). The analyzed molecules have to be first ionized, forming a molecular or pseudo molecular ions, depending on the ionization technique. If the (pseudo) molecular ion contains sufficient internal energy, it undergoes further fragmentation, which leads to formation of fragmention (s).Generally, each fragment ion derived from the molecular ion can again undergo fragmentation. All these formed ions are separated in the mass spectrometer according to their m/z and are detected. A mass spectrum of the molecule is then produced, which represents a distribution chart of the abundances (y -axis) of ions versus their m/z values (x-axis). The peak with the highest intensity in the spectrum is called the base peak and the spectrum is generally normalized to the abundance (intensity) of this peak (*i.e.*, its intensity is assigned 100 per cent) [1].

A typical MS procedure involves the different steps which include- i- A sample is loaded onto the MS instrument, and undergoes vaporization. ii-The components of the sample are ionized (formation of charged particles- Ions) by one of a various methods *e.g.*, by impacting them with an electron beam, iii-The ions are separated according to their mass-to-charge ratio in an analyzer by electromagnetic fields, iv-The ions are detected, usually by a quantitative method and v-The ion signal is processed into mass spectra [2].

Source

1. *Mass Spectrometry and Hyphenated Instruments in Food Analysis by Tomas Cajka, Jana Hajslova, and Katerina Mastovska.*
2. *Wikipedia.*

Applications of Mass Spectroscopy in Food/Dairy Industries

A food matrix is very complex, in addition to major components such as lipids, proteins, and saccharides, a wide range of other natural minor compounds are contained (*e.g.*, vitamins, aroma and flavor compounds, pigments). Under certain conditions, contaminants and other hazardous compounds may be present in food matrices as a consequence of human activities or due processing practices. MS techniques have proved to be an excellent tool for qualitative characterization and quantitative determination of various food components because of their high sensitivity and specificity [1].

The technique has both qualitative and quantitative uses. These include identifying unknown compounds, determining the isotopic composition of elements in a molecule, and determining the structure of a compound by observing its fragmentation. Other uses include quantifying the amount of a compound in a sample or studying the fundamentals of gas phase ion chemistry (the chemistry of ions and neutrals in a vacuum).MS is now in very common use in analytical laboratories that study physical, chemical, or biological properties of a great variety of compounds/Dairy products/Food [2].

MS-based strategies for food and nutrition proteomics are now capable of addressing a wide range of analytical questions *e.g.* food safety, certification and traceability of products. MS and proteomic techniques can be used in the field of food safety. **In addition to characterisation of proteins, mass spectroscopy is also successfully being used for three areas related to food safety- i-the detection of microorganisms, ii-the determination of modifications induced by food processing and iii- the identification of protein components responsible for food allergy and intolerance [2].** Some important application of Mass spectroscopy in dairy/ food industries for determination of Natural Substances in Food, Migrants from Packaging, Food Contaminants, Processing Contaminants are given below [1,2].

Source

1. *Mass Spectrometry and Hyphenated Instruments in Food Analysis by Tomas Cajka, Jana Hajslova, and Katerina Mastovska.*
2. *Wikipedia.*

1. Natural Substances in Food [1,2]

The applicability of MS, LC–APPI-MS and others has been successfully evaluated in the analysis of natural substances *e.g.*lipids, Saccharides, Peptide and Proteins, Vitamins, Aroma and Flavour compounds in food:

i. Characterisation/Determination of Proteins

Proteins and peptides are major constituents of Dairy products/food and play a very important and decisive role in determining its nutritional and functional properties.Mass spectrometry (MS) has gained increasing popularity as part of the toolkit of techniques developed to investigate proteins at the proteome-wide scale. **a-The structural characterisation of proteins is the one of the important areawhere Mass spectroscopy is particularly useful, b-**The application of mass spectrometry to protein analysis has been revolutionised in recent years by the development of soft ionisation techniques such as Electrospray Ionisation (ESI-MS) and Matrix Assisted Laser Desorption/Ionisation (**MALDI-TOF-MS**).Proteins separated by two-dimensional gel electrophoresis (2DE), and capillary HPLC-fractionated peptides, can be characterised by MS techniques. This is crucial in the study of food proteins from sources such as meat, cereals, or milk [1]. The MALDI-MS is useful for obtaining fingerprint of the protein composition of various food samples, thus, rapid and accurate evaluation of the authenticity is possible, **c-**The LC–ESI-MS employing TOF, represents a tool for accurate mass determination of proteins. In addition, LC–ESI-MS=MS (or MALDI-TOF/TOF) allows analysis of proteins for elucidation purposes as well as the characterization of low-molecular mass peptides [2,3].

Proteomics [4]

It is the large-scale study of protein expression, protein-protein interactions, or post-translational modifications. Unlike other methodologies that analyze a few proteins at a time, proteomics can analyzethousands of proteins in a single experiment. Proteomics is a powerful tool for the identification of proteins and study their localisation, functions, modifications and possible interactions or complexes they can form. Proteomics aims to the analysis of a number of proteins at a time unlike traditional identification of one protein at a time. Unlike "the genome" there is no single, static proteome in any organism instead there are dynamic collections of proteins in different cells and tissues that display variations in response to various conditions such as stress or infectious processes.

The classical method of quantitative and qualitative expression proteomics combines protein separation by high-resolution 2D gel electrophoresis with MS (Mass Spectrometer) or MS/MS identification of selected protein spots. Because, even the best 2D gels can routinely separate no more than 1500 proteins, this technique is limited to the most abundant proteins if a crude protein mixture (whole-cell lysate) is used. Two-Dimesional SDS-PAGE separation method has become synonymous with proteomics and remains the single best method for resolving highly complex protein mixtures.

There are following two basic mass spectrometry methods for the identification of proteins:

i. Peptide Finger Printing

It is often associated with 2D-PAGE protein separation scheme. Individual spots from the 2D-PAGE gel are isolated and the proteins digested with trypsin. These proteins are typically analyzed with a MALDI time of flight mass spectrometer. The mass spectrometer will record all the peptide m/z detected in the gel spot. Identification of a protein is based on the measurement of multiple peptides that can come from that protein. For example, after a mass spectrometer has determined the m/z for the peptides from a gel spot, this information will be matched to a protein database. A successful protein match will be based on the number of peptides matched to the protein and the accuracy of the matches.

ii. Tandem Mass Spectrometers

The second method to identify proteins involves use of tandem mass spectrometers that allow for sequencing of individual peptides

Proteomics Approches for Milk [4]

Several proteomic approaches have been explored for the analysis of milk and whey proteins. The advantage of the method of 2-DE in combination with massspectrometry over other methods such as ELISA and immunoassays is that it can simultaneously detect more differentially expressed proteins. The method of 2–DE combined with LC/MS/MS is capable of identifying differences of low abundant proteins in each lactation stage. Comparative studies involving human and bovine milk proteome revealed 44 proteins in human milk and 51 proteins in bovine milk related to host defence functions.

Source

1. *Wikipedia.*
2. *Mass Spectrometry and Hyphenated Instruments in Food Analysis by Tomas Cajka, Jana Hajslova, and Katerina Mastovska.*
3. *Careri, M. and Mangia, A., Analysis of food proteins and peptides by chromatography and massspectrometry, J. Chromatogr. A, 1000, 609, 2003.*
4. *Ashok K. Mohanty, Introduction to Proteomics and Mass Spectrometry for application in Dairy Processing and Food Safety, Animal Biotechnology Centre, Compendium of Lectures on Advanced Techniques and Novel Approaches for Quality and Safety Evaluation of Dairy Foods- 2015, NDRI, Karnal.*

ii. Determination of Lipids

Both LC–APCI-MS and LC–APPI-MS has been successfully evaluated in the analysis of mixtures of neutral lipids, such as triacylglycerols and sterols. For the analysis of polar lipids (*e.g.*, phospholipids), both LC–ESI-MS and LC–APCI-MS are the methods of choice. Applications of GC–EI-MS to lipid analysis include mainly the characterization of acylglycerol and sterol fractions [1,2,3,4].

Source

1. *Mass Spectrometry and Hyphenated Instruments in Food Analysis by Tomas Cajka, Jana Hajslova, and Katerina Mastovska.*
2. *Careri, M., Bianchi, F., and Corradini, C., Recent advances in the application of mass spectrometry in food-related analysis, J.Chromatogr. A,970,3,2002.*
3. *Cai, S. and Syage, J. A., Atmospheric pressure photoionization mass spectrometry for analysis of fatty acid and acylglycerol lipids, J. Chromatogr. A, 1110, 15, 2006.*
4. *Canabate-Diaz, B. et al., Separation and determination of sterols in olive oil by HPLC-MS, Food Chem., 102, 593, 2007.*

iii. Determination/Characterization of Carbohydrates/Saccharides

Mass Spectroscopic techniques with different ionization modes and the hyphenated techniques play an significant role in the structural elucidation of Carbohydrtaes/Sugars. **a**-MALDI-TOFMS is being used as powerful tool for the characterization of polysaccharides in various food samples, **b**-LC-MS with fast atom bombardment (FAB) interface has been proved valuable for precise isotopic measurements for saccharides and for their structural characterization, **c**-For the characterization of mono- and oligo-saccharides by GC-MS, a derivatization procedure is required to remove hydrogen bonding [1,2].

Source

1. *Mass Spectrometry and Hyphenated Instruments in Food Analysis by Tomas Cajka, Jana Hajslova, and Katerina Mastovska.*
2. *Careri, M., Bianchi, F., and Corradini, C., Recent advances in the application of mass spectrometry in food-related analysis, J. Chromatogr. A, 970, 3, 2002.*

iv. Determination/Identification of Vitamins

Many HPLC-based methods with conventional detectors sensitivity and/or selectivity whereas the LC–MS technique with ESI and APCI followed by their MS detection enables reliable identification/quantification of these nonvolatile, thermally labile biologically active compounds [1,2].

Source

1. *Mass Spectrometry and Hyphenated Instruments in Food Analysis by Tomas Cajka, Jana Hajslova, and Katerina Mastovska.*
2. *Careri, M., Bianchi, F., and Corradini, C., Recent advances in the application of mass spectrometry in food-related analysis, J. Chromatogr. A, 970, 3, 2002.*

v. Separation and Characterization of Aroma and Flavor Compounds

GC-MS has a considerable potential in the separation and characterization of food aroma and flavor compounds and it is being used. The coupling with solid phase microextraction (SPME) further increases the potential of this technique.**a**-GC–MS equipped with various analyzers, such as quadrupole and ion trap, allowed obtaining information on aroma and flavor compounds [1,2], **b**- Recently, the application of high-speed TOFmass analyzer combined with comprehensive two-dimensional (GC_GC) has been used as a tool for rapid and comprehensive analysis of these compounds [3,4]. Electron Ionization (EI) is preferred for the characterization of the volatile fraction, since it allows library searching based on the EI mass spectra. Chemical Ionization (CI) is used mostly to confirm MW of the compounds of interest.

vi. Detection of Milk Fat Adulteration using Matrix-Assisted Laser Desorption/ Ionization Time of Flight Mass Spectrometry (MALDI-TOF MS) [5]

Mass spectrometry is being increasingly used to identify adulterations in food and milk due to its simplicity and the need for minimum sample preparation [6]. MALDI-QTOF MS provided rapid and unambiguous profiles of the TAG composition, and that the TAG are detected mostly as sodiated molecules [TAG+Na] + [7]. It was reported that in case of pure milk fat the ions of m/z 881.8, 907.8, 909.8 and 911.8 corresponds to [TAG+Na] + with the POO, OOO, OSO and SSO compositions (P, palmitic acid; O, oleic acid and S, stearic acid), respectively, which are changed during adulteration with any foreign fat. It was reported [8] that during adulteration of milk fat with vegetable oil MALDI-QTOF MS spectra data changed due to change in the triacylglycerol characteristic, this phenomena due to the illegal addition of non-milk fat to milk is shown to mostly increase the relative abundance of the ion clusters centred on charge by mass ratio of triacylglycerol.

Source

1. *Mass Spectrometry and Hyphenated Instruments in Food Analysis by Tomas Cajka, Jana Hajslova, and Katerina Mastovska.*
2. *Kataoka, H., Lord, H.L., and Pawliszyn, J., Applications of solid-phase microextraction in food analysis, J. Chromatogr. A, 880, 35, 2000.*
3. *Adahchour, M. et al., Comprehensive two-dimensional gas chromatography with time-of-flight massspectrometric detection for the trace analysis of flavour compounds in food, J. Chromatogr. A, 1019,157, 2003.*
4. *Cajka, T. et al., Solid phase microextraction–comprehensive two-dimensional gas chromatography–timeof-flight mass spectrometry for the analysis of honey volatiles, J. Sep. Sci., 30, 534, 2007.*
5. *Vivek Sharma and Tanmay Hazara, Detection of Milk Fat Adulteration using Modern Approaches, Dairy Chemistry Division, NDRI, Karnal.*
6. *Sanvido, G. B., Garcia, J. S., Corilo, Y. E., Vaz, B. G., Zacca, J. J., Cosso, R. G., et al. (2010). Fast screening and secure confirmation of milk powder adulteration with maltodextrin via ESI (+)-MS and selective enzymatic hydrolysis.Journal of Agricultural and Food Chemistry, 58, 9407–9412.*
7. *Saraiva, S. A., Catharino, R. R., Cabral, E. C., and Eberlin, M. N. (2009). Amazonian vegetable oils and fats: Fast typification and quality control via triacylglycerol (TAG) profiles from dry matrix-assisted laser desorption/ionization time-of flight (MALDI-TOF) mass spectrometry fingerprinting. Journal of Agricultural and Food Chemistry, 57, 4030–4034.*

8. *Garcia. Jerusa Simone; Gustavo Braga Sanvido; Sergio Adriano Saraiva; Jorge Jardim Zacca; Ricardo Guanaes Cosso; Marcos Nogueira Eberlin (2012) Bovine milk powder adulteration with vegetable oils or fats revealed by MALDIQTOF MS. Food Chemistry 131 (2012) 722–726.*

2. Detection of Microorganisms/Microbial Contamination in Food

a. By Bacterial Profiling

Traditional means of controlling microbial spoilage and safety hazards in foods include freezing, blanching, sterilisation, curing and use of preservatives where as MS profiling methods (Bacterial profiling through MALDI-TOF-MS and ESI-MS/MS fingerprinting of bacterial proteins) are ideal for fast and sensitive detection of pathogens or spoilage microorganisms that affect food quality and safety during processing and storage. MS is applied in order to control and detect foodborne microorganisms. The major application of MS is the detection of microorganisms, which may be the cause of food spoilage or be hazardous to human health, and the safety evaluation of food components. These techniques allow different species and, in some cases strains, to be identified..

b. By Quantitative Identification of Specific Metabolites Produced by the various Spoilage Microorganisms

(i) A more accurate description of contaminating microorganisms has been achieved by integrating proteomics with methodologies that are able to provide either structural or quantitative identification of specific metabolites produced by the various spoilage microorganisms.

(ii) This information can also be used to identify areas of the food chain that are the most vulnerable to microbial contamination. The risks associated with food contamination are not limited to the presence and virulence of a microorganism. Protein toxins can be released into food, which are able to survive for long periods of time even after the source of contamination has been removed.

(iii) A procedure combining a proteomic approach with immunochemical, chromatographic, electrophoretic and MS techniques was developed. This procedure monitored production and levels of Enterotoxin A (SEA) and B (SEB) of *Listeria monocytogenes, Staphylococcus aureus, Escherichia coli, Clostridium botulinum* A and Salmonella species. These pathogenic bacteria excrete a variety of virulence factors into the extra cellular medium and at the cell surface. This mode of action plays an essential role in the colonisation of the host cells, and thus reflects the degree of bacterial pathogenicity. The toxins, being heat-stable and resistant to proteases, can be a danger for the consumer.

Source

1. *Wikipedia*

3. Modifications Induced by Food Processing

(i) Heat treatment of milk can lead to the denaturation of whey proteins. The complex series of covalent reactions that take place (known as Maillard reactions) produce a decrease in nutritional quality and the formation of possible toxic compounds. A well-controlled Maillard reaction can improve the physico-chemical properties of whey proteins. The novel compounds formed have been proposed as useful markers to demonstrate either uncorrected heat treatment or the presence of heated milk/milk powders added to fresh milk.

(ii) Some markers of heated milk have been identified, including lysinoalanine (LAL), formed by the reaction of abundant lysine residues and dehydroalanine (produced by heat-induced dephosphorylation of the phospho-serine residues of casein).

(iii) An analytical method involving detection by LC/ESI-MS without any sample pre-fractionation step has been developed recently, making LAL a useful marker for the detection of heat treated products in a variety of foodstuffs, ranging from milk for drinking to milk powders and dairy products.

Source

1. *Wikipedia.*

4. Detection of Food Contaminants [1]

Several MS methods associated other techniques are being used for detection of food contaminants *e.g.*- Pesticide residues, Veterinary drug residues, Mycotoxins, Food allergens.

i. Pesticide Residues

Now-a-days, both GC–MS and LC–MS/MS are of a growing popularity in this field. The types of MS instruments used for pesticide analysis include single mass analyzers (quadrupole, ion trap, and TOF) as well as hybrid instruments (triple quadrupole, quadrupole/time-of-flight). When employing SIM and MS/MS modes, settings of time segments are typically needed, which limits the number of targeted analytes that can be detected in a particular time.

GCMS [1a]

Gas chromatography (GC) is a common type of chromatography used in analytical chemistry for separating and analyzing compounds that can be vaporized without decomposition.GC-FID is being used for sepeartion, detection and further analysis. The identification using GC-FID may also be hampered by contaminants or coeluting compounds. Another limitation of using GC-FID is obtaining adequate standards.In GC-MS, the separated compounds from GC are detected and further analyzed by mass spectrometric technique. Various ionization techniques (EI, CI *etc.*) are also possible in GC–MS. EI are commonly preferred. CI (both PCI and NCI) as a softer ionization technique tends to give lower LODs depending on the pesticide.

Typical applications of GC/MS are Pesticide residues and pollutants in water agricultural products foodstuff, organic solvents in packaging materials, ink, *etc.*, Drugs of abuse in urine, blood, tablets fatty acid contents in edible oils fat, Essential oil characterization, Allergens in fragrance, polymer characterization, *etc.*

Method

The analysis by GC-MS involves the following [1a]:

i. Sample Preparation and Derivatization

Samples for GC are organic compounds with sufficient volatility, and sufficient thermal stability. In practice, some compounds which do not fulfill the criteria above can be made to fulfill the criteria by means of chemical reaction ('derivatization'), thus making it possible for these compounds to be analyzed by GC.

ii. Analysis in GC-MS

GC-MS has two components- GC for separation of the fatty acids based on volatility, polarity and molecular weight of the analytes which is then detected by Mass Spectrometry. GC contains-**Injection port** (Sample as soon as it is introduced into the injection port can be diluted using carrier gas by Split mode), **Column** (Based on the nature of the compound to be separated the column can be of -Polar, -Mid Polar, -Non polar and of varying length)and **Detector** (The response obtained by the detector is converted into signals corresponding to the concentration. MS has an ionization port where the analyte enters right after eluting out of the column). Mass analyzers are nothing but sieve. A range of DC/Rf voltages are applied across the mass analyzer quadrupole. These quadrupole at a specific DC/Rf voltage have selectivity for one particular mass to charge ratio (m/z) particles.

LC-MS/MS

Regarding the LC–MS/MS, among the various interfacing systems developed, ESI and APCI have improved the feasibility of identification of pesticides of different chemical structures in foods at concentrations lower or comparable to those obtainable by GC–MS [2,3]. LC–MS/MS provides a complimentary tool to GC–MS for the analysis of more polar or thermolabile pesticides, which were previously analyzed by GC–MS with a various degree of difficulty or only after derivatization.

The details of different other food contaminants, MS Techniques, Detection limits *etc.* are tabulated below:

Sl.No.	*Name of Pesticides*	*Techniques*	*Remarks*
1	Polychlorinated biphenyls (PCBs)	GC–MS	*(i)* Reliable technique for PCBs quantification, particularly given by the availability of 13C-labeled PCB standards.
			(ii) EI mode, Both single quadrupole (SIM) and ion trap (MS/MS) are frequently employed in the routine analysis of PCBs [4].
		GC-GC	*(i)* the application of both high-resolution and high-speed TOFMS (the latter in combination with GC-GC) represents a tool successfully applied in the analysis of PCBs [5,6].

Contd.

Sl.No.	Name of Pesticides	Techniques	Remarks
2	Polychlorinated dibenzo-p-dioxins and dibenzofurans (PCDDs/PCDFs)	GC coupled to high-resolution MS (sector analyzer)	*(i)* Quantitative determination of PCDDs/PCDFs occurring in biotical matrices at ultra-trace levels. *(ii)* Although PCI/NCI techniques can be used for their determination, a majority of laboratories employ EI. High-resolution systems (mass resolving power of 10,000) provide higher selectivity. *(iii)* Very expensive, bulky, and requires operation by a highly trained specialist.
		GC–MS/MS employing ion trap analyser	*(i)* Valuable technique for improved selectivity in dioxin analysis and less expensive [7,8]. *(ii)* The high selectivity is obtained due to formation of characteristic dioxin fragment ions produced by the secondary ionization,
		GC-GC–TOFMS	*(i)* Valuable technique for improved selectivity in dioxin analysis and less expensive [9,10]. *(ii)* The improvement of selectivity is achieved employing the secondary column with different polarity.
3	Brominated flame retardants (BFRs)-		
	a Polybrominated diphenyl ethers (PBDEs)	GC-MS operated either in EI or NCI mode.	*(i)* The low-resolution MS(NCI, in addition of EI), The high-resolution MS [(sector) (EI mode)].
		High-resolut -ion TOFMS under NCI conditions	*(i)* Demonstrated in the analysis of PBDEs [11].
	b Hexabromocy-clododecane (HBCD)	GC–MS	*(i)* Traditionally, analysis of HBCD has been carried out using GC–MS, typically in NCI mode by monitoring the [Br] ions for a higher sensitivity. However, this technique does not allow quantification of individual diastereomers of HBCD (a-, b-, and g-HBCD).
		LC–ESI-MS/MS or LC–APCIMS/MS	*(i)* These are employing nonpolar (C18, C30) or chiral HPLC columns for their separation represent a versatile tool for the isomer-specific determination of HBCD isomers [12].
	c Tetrabromo-bisphenol A (TBBPA)	GC–MS	*(i)* Acidification and derivatization are required before the GC–MS analysis [12].
		LC–ESI-MS	*(ii)* It allows its direct determination [12].
4	Polycyclic aromatic hydro-carbons (PAHs)	GC–EI-MS	*(i)* Operated in SIM mode represents probably the most common GC technique for determination of PAHs in food matrices [13]. *(ii)* The problem encountered in the analysis of PAHs is separation of isomers and limited EI fragmentation, which does not allow reliable confirmation at ultra-trace levels.
		LC–APCI-MS	*(i)* Alternatively, it allows determination of PAHs without derivatization (post column), which is typically required in LC–ESI-MS [14].

Source

1. *Mass Spectrometry and Hyphenated Instruments in Food Analysis by Tomas Cajka, Jana Hajslova, and Katerina Mastovska.*

1a. *K.P. Indumathi, Rajan Sharma and Jyotika, GC-MS Application in Food Analysis, Dairy Chemistry Division, Compendium of Lectures on Advanced Techniques and Novel Approaches for Quality and Safety Evaluation of Dairy Foods- 2015, NDRI, Karnal.*

2. *Pico, Y. et al., Control of pesticide residues by liquid chromatography-mass spectrometry to ensure foodsafety, Mass Spec. Rev., 25, 917, 2006.*

3. *Alder, L. et al., Residue analysis of 500 high priority pesticides: Better by GC–MS or LC–MS/MS? MassSpec. Rev., 25, 838, 2006.*

4. *Ahmed, F.E., Analysis of polychlorinated biphenyls in food products, TrAC-Trend Anal. Chem., 22,170, 2003.*

5. *Dallüge, J., Roose, P., and Brinkman, U. A. Th., Evaluation of a high-resolution time-of-flight massspectrometer for the gas chromatographic determination of selected environmental contaminants, J. Chromatogr. A,970,213, 2002.*

6. *Focant, J. et al., Comprehensive two-dimensional gas chromatography with isotope dilution time-of-flightmass spectrometry for the measurement of dioxins and polychlorinated biphenyls in foodstuffs: Comparisonwith other methods, J. Chromatogr. A, 1086, 45, 2005.*

7. *Eppe, G. et al., PTV-LV-GC/MS/MS as screening and complementary method to HRMS for the monitoring of dioxin levels in food and feed, Talanta, 63, 1135, 2004.*

8. *Loran, S. et al., Evaluation of GC-ion trap-MS/MS methodology for monitoring PCDD/ Fs in infantformulas, Chemosphere, 67, 513, 2007.*

9. *Focant, J. et al., Comprehensive two-dimensional gas chromatography with isotope dilution time-of-flight mass spectrometry for the measurement of dioxins and polychlorinated biphenyls in foodstuffs: Comparison with other methods, J. Chromatogr. A, 1086, 45, 2005.*

10. *Hoh, E., Mastovska, K., and Lehotay, S.J., Optimization of separation and detection conditions for comprehensive two-dimensional gas chromatography–time-of-flight mass spectrometry analysis of polychlorinated dibenzo-p-dioxins and dibenzo-furans, J.Chromatogr. A,1145,210,2007.*

11. *Cajka, T. et al., Challenges of gas chromatography–high-resolution time-of-flight mass spectrometry for simultaneous analysis of polybrominated diphenyl ethers and other halogenated persistent organic pollutants in environmental samples, J. Sep. Sci., 28, 601, 2005.*

12. *Covaci, A. et al., Recent developments in the analysis of brominated flame retardants and brominated natural compounds, J. Chromatogr. A, 1153, 145, 2007.*

13. *Stolyhwo, A. and Sikorski, Z.E., Polycyclic aromatic hydro-carbons in smoked fish-a critical review, Food Chem., 91,303, 2005.*

14. *Takino, M. et al., Determination of polycyclic aromatic hydrocarbons by liquid chromatography-electrospray ionization mass spectrometry using silver nitrate as a post-column reagent, J. Chromatogr. A,928, 53, 2001.*

ii. Detection/Determination of Veterinary Drug Residues [1]

LC–MS-based strategies have been widely applied for determination and confirmation of veterinary drug residues in food samples. LC–MS revolutionizes this field, enabling quantitative and confirmatory multiresidue, multiclass analysis of veterinary drug residues, independent of their chemical structure or biological activity, thus replacing traditional single residue/class immunochemical or microbial methods [2]. The use of GC–MS-based methods is also possible, although their main limitation is the need of extensive cleanup and time-consuming derivatization procedures prior to their GC analysis since most of these compounds are relatively polar [3].

Source

1. *Mass Spectrometry and Hyphenated Instruments in Food Analysis by Tomas Cajka, Jana Hajslova, and Katerina Mastovska.*
2. *Gentili, A., Perret, D., and Marchese, S., Liquid chromatography-tandem mass spectrometry for performing confirmatory analysis of veterinary drugs in animal-food products, TrAC-Trend Anal. Chem., 24,705, 2005.*
3. *Careri, M., Bianchi, F., and Corradini, C., Recent advances in the application of mass spectrometry in food-related analysis, J. Chromatogr. A, 970, 3, 2002.*

iii. Detection/Determination of Mycotoxins [1]

The GC–MS determination of mycotoxins is performed either by direct analyser or (mostly) employing various chemical derivatizations. The derivates can be detected selectively to very low levels using NCI, but also EI or PCI. Although quadrupole MS (SIM) is mostly used, the high-molecular weight of derivatized analytes predetermines their molecules for MS^n-measurements, especially in complex matrices like foodstuff and beverages. Recently, LC–ESI-MS/MS and LC–APCI-MS/MS methods, allowing the screening of mycotoxins of various classes within a single run, have been developed. These LC–MS-based methods represent a popular tool in mycotoxin analysis due to the improvement of sensitivity and avoiding extensive cleanup procedures and derivatization steps compared to GC–MS methods [2,3,4].

Source

1. *Mass Spectrometry and Hyphenated Instruments in Food Analysis by Tomas Cajka, Jana Hajslova, and Katerina Mastovska.*
2. *Nielsen, K.F. and Thrane, U., Fast methods for screening of trichothecenes in fungal cultures using gas chromatography–tandem mass spectrometry, J.Chromatogr. A, 929, 75, 2001.*
3. *Sforza, S., Dall'Asta, C., and Marchelli, R., Recent advances in mycotoxin determination in food and feed by hyphenated chromatographic techniques/mass spectrometry, Mass Spec. Rev., 25, 54, 2006.*

4. *Zöllner, P. and Mayer-Helm, B., Trace mycotoxin analysis in complex biological and food matrices by liquid chromatography–atmospheric pressure ionisation mass spectrometry, J. Chromatogr. A, 1136,123, 2006.*

iv. Food Allergy and Intolerance [1-5]

Detection of Food Allergens and Intolerance are the two important fields in which MS plays significant roles.

(a) Food allergy is an increasingly important issue for the food industry. The detection of the presence of toxic or allergenic compounds in food, is the another important application of MS.

(i) Toxic compounds may originally be present in the raw material (and therefore need to be eliminated by the manufacturing process), or they may be generated during the production process. In the case of food-related allergy and intolerance, MS analysis is focused on compounds which are harmful for specific groups of consumers. Mass spectrometry-driven identification of either genes for allergic diseases or allergenic proteins is currently underway.

(ii) The systematic analysis of cereal and legume species, such as wheat, rice, pea, soy and peanut, using high resolution separation techniques (in combination with MS) is leading to the detection and identification of several previously uncharacterised allergenic proteins in seed samples. MS-based methods have been developed to provide good detection limits (1ppm) for almonds, pecan nuts, hazelnuts and walnut antigens in food ingredients.

(iii) **Allergy to cow's milk is one of the more prevailing food allergies in infants, and in several cases, is due to intolerance of the main milk proteins- casein and beta-lactoglobulin.** Despite the enzymatic hydrolysis of major milk proteins being carried out in order to increase food tolerance, residual antigens in hydrolysed milk formulae have been reported. Research has lead to a high-tolerability of dairy products via limited proteolysis of milk proteins. This intervention removes protein sequences involved in an allergic reaction.

(b) Food allergy and intolerance is another field in which MS has proven to be a key player. This includes the removal of epitopes from a certain food. The correct MS technique can provide the means to identify an immunodominant epitope in the target food. This is a starting point for designing antibodies against specific epitopes, and for standardising the quantitative detection of trace amounts of allergen by the development of protein chips.Future developments in protein array technology (Driven by MS-derived information on proteins) will also lead to an improvement in food safety.

Source

1. *Han X, Aslanian A, Yates JR.Curr Opin Chem Biol 2008; 12: 483.*

2. *Ferranti P, Proceedings of the workshop on Quality, Salubrity and Safety of some Typical Italian Cheeses, National Research Council/Italian Ministry of University and Research. Rome,5th May 2005.*

3. *Tareke F, Rydberg P, Karlsson P, Eriksson Tornqvist SM.ChemResToxicol2007,13: 517.*

4. *Ferranti P, Mamone G, Picariello G, Addeo F. J. Mass Spectrom 2007, 42: 1531.*

5. *Antonella Nasi and Pasquale Ferranti, Department of Food Science, University of Naples "Federico II", Italy.*

5. Detection of Processing Contaminats [1]

Some processing contaminats are responsible not only for deteriorating the quality of food but also cause bad effect on human health.MS techniques can be used for detection of some of the processing contaminants and the details of which are tabulated below:

Sl.No.	*Name of Pesticides*	*Techniques*	*Remarks*
1	Acrylamide	GC–MS methods	*(i)* Employing either quadrupole or ion trap/triple quadrupole (tandem MS) can be used in EI or CI mode for the analysis of acrylamide in foods.
			(ii) A direct GC–MS determination without a derivatization step is also possible when a suitable sample preparation method is employed removing acrylamide precursors (asparagine and reducing sugars) from the analyzed sample.
			(iii) Alternatively, HRMS (sector or TOF) instruments are applied in the acrylamide analysis.
		LC–ESI-MS/MS	*(i)* These techniques with triple quadrupole for the routine analysis because this instrumental technique applied for the quantitative analysis of acrylamide has high sensitivity and avoids potential problems with acrylamide formation that may occur in the direct GC–MS analysis [2].
2	Chloropropanols.(1,3-dichloro propan-2-ol (1,3-DCP), 3-mono chloropropane-1,2-diol (3-MCPD), and 2-monochloro propane-1,2-diol (2-MCPD)	GC–MS	*(i)* It can be used after their previous derivatization [3,4].
		MS/MS	*(ii)* Quadrupole mass analyser operated in SIM mode or ion trap analyser in MS/MS are typically used for their detection [3,4].
3	Heterocyclic amines.(mutagenic and carcinogenic substances classified as the IQ type (amino imidazoazaarenes) and non-IQ type (pyrolytic hetero cyclic amines)	GC–MS methods	*(i)* Since most of these compounds are polar and nonvolatile, a derivatization step is usually performed before their analysis.

Contd.

Sl.No.	Name of Pesticides	Techniques	Remarks
		LC–MS	(i) Derivatization step can be effectively avoided using LC–MS with a single quadrupole or HRTOFMS, which allows selective and sensitive detection.
		LC–MS/MS	(i) The selectivity can be further enhanced by using LC–MS/MS with either ion trap ortriple quadrupole MS instruments. Both ESI and APCI can be used for ionization [5,6].

Source

1. *Mass Spectrometry and Hyphenated Instruments in Food Analysis by Tomas Cajka, Jana Hajslova, and Katerina Mastovska.*
2. *Dunovska, L. et al., Direct determination of acrylamide in food by gas chromatography–high-resolution time-of-flight mass spectrometry, Anal. Chim. Acta, 578, 234, 2006.*
3. *Chung, W., Hui, K., and Cheng, S., Sensitive method for the determination of 1,3-dichloropropan-2-ol and 3-chloropropane-1,2-diol in soy sauce by capillary gas chromatography with mass spectrometric detection, J. Chromatogr. A, 952 185, 2002.*
4. *Hamlet, C.G. and Sutton, P.G., Determination of the chloropropanols, 3-chloro-1,2-propandiol and2-chloro-1,3-propandiol, in hydrolysed vegetable proteins and seasonings by gas chromatography/iontrap tandem mass spectrometry, Rapid Com. Mass Spec.,11,1417, 1997.*
5. *Barcelo-Barrachina, E. et al., Evaluation of reversed-phase columns for the analysis of heterocyclic aromatic amines by liquid chromatography-electrospray mass spectrometry, J. Chromatogr. B, 802,45,2004.*
6. *Toribio, F.et al., Ion-trap tandem mass spectrometry for the determination of heterocyclic amines in food, J. Chromatogr. A, 948, 267, 2002.*

6. Assessment of Quality of Milk-Based Products by MS

The contribution and potentialities of mass spectrometry-based techniques to assess quality of milk-based products, are well established. The Knowledge and understanding of the chemistry of milk and its components is essential for designing processing equipment and conditions needed for the manufacture and distribution of high-quality dairy products. In spite of having complex milk matrix, the mass spectrometry, coupled or not to separation techniques, constitutes a key tool for analysis of milk. **Mass spectrometry is a key analytical tool used to address a wide range of questions/problems in food and nutrition analysis and research.**

Source

1. *Guy PA, Fenaille F., Department of Quality and Safety Assurance, Nestlé Research Center, Lausanne, Switzerland. philippe.guy@rdls.nestle.com.*

7. Migrants from Packaging [1]

Generally food grade paking materials are being used for packing of Dairy/ Food products.If packing materials are not food grade then there are chances of migration of chemicals (Additives, chemicals originated due to recyclisation, ingredients of ink used for printing *etc.*) from packing materials to food which causes health problems.Hence detection/determination of these migrants are very essential and some of the MS methods for determination of migrants are given in following table:

Sl.No.	*Name of Pesticides*	*Techniques*	*Remarks*
1	Phthalate and adipate esters	GC–MS methods	*(i)* Represents the key method for the determination of these packaging contaminants [2].
2	Epoxy compounds (bisphenol A diglycidyl ether (BADGE), bisphenol F diglycidyl ether (BFDGE), and their hydrolyzed and chlorinated derivates)	GC–MS methods	*(i)* These methods were developed to confirm the presence of these compounds. However, this technique requires a derivation step. *(ii)* The use of HPLC methods with either APCI or ESI for the identification of these substances has been recently reported [3].

Source

1. *Mass Spectrometry and Hyphenated Instruments in Food Analysis by Tomas Cajka, Jana Hajslova, and Katerina Mastovska.*
2. *Balafas, D., Shaw, K.J., and Whitfield, F.B., Phthalate and adipate esters in Australian packaging materials, Food Chem., 65, 279, 1999.*
3. *Pardo, O. et al., Determination of bisphenol diglycidyl ether residues in canned foods by pressurized liquid extraction and LC-tandem MS, J. Chromatogr. A, 1107, 70, 2006.*

Chapter 4

Chromatography Techniques

The chromatography is a physical method of separation in which the components to be separated are distributed between two phases. It is a technique in which solutes are resolved by differential rates of elution as they pass through a chromatographic column. Their separation is governed by their distribution between the mobile and stationary phases. Chromatography, a group of methods for separating very small quantities of complex mixtures, with very high resolution, is one of the most important techniques. The chromatography involves passing a mixture dissolved in a mobile phase through a stationary phase, which separates the analyte to be measured from other molecules in the mixture based on differential partitioning between the mobile and stationary phases. In chromatography the analytes partition between different phases that move with different velocities relative to each other. Chromatography may be **preparative or analytical.** The purpose of preparative chromatography is to separate the components of a mixture for further use (a form of purification) whereas Analytical chromatography is done normally with smaller amounts of material and is for measuring the relative proportions of analytes in a mixture [1,2,3]. The chromatographic methods are very useful for analysis of complex foods such as Milk and milk products [1,2].

History of Chromatography

The science of chromatography began early in the **twentieth century [3], with the Russian botanist Mikhail Tswett,** who used a column packed with calcium carbonate to separate plant pigments. New types of chromatography developed during the 1930s and 1940s made the technique useful for many types of separation process [4]. The substantial development of Chromatography was due to the work of Archer John Porter Martin **and** Richard Laurence Millington Synge during the 1940s and 1950s, for which they won the 1952 Nobel Prize in Chemistry [5] and their work encouraged the rapid development of several types of chromatography methods.

The method was developed rapidly in the years after World War II, and began to be applied with the invention of the electron capture detector (**ECD) in 1960 by James Lovelock**. This detector (with its specificity and very high sensitivity toward halogenated organic compounds) was needed to determine traces of pesticides in soils, food and water and halocarbon gases in the atmosphere. The pernicious effects of long lived, bioaccumulating pesticides, such as DDT, would have been very difficult to detect without the use of the ECD [6,7,8].

The successful use of liquid Chromatography (LC) for a given problem requires the right combination of a variety of operating conditions such as the type of column, packing and mobile phase, column length and diameter, mobile phase flow rate, column temperature and sample size. In order to select the best combination of chromatographic conditions, a basic understanding of the various factors that affect LC separations is necessary [9,10].

Source

1. *Rajesh Kumar Bajaj and Priyanka Singh Rao, HPLC and its Application in Food Analysis, Dairy Chemistry Division, Compendium of Lectures on Advanced Techniques and Novel Approaches for Quality and Safety Evaluation of Dairy Foods- 2015, NDRI, Karnal.*
2. *Wikipedia.*
3. *Majors, Ronald E. (2010-09-07) Fast and Ultrafast HPLC on sub-2 µm Porous Particles - Where Do We Go From Here? – LC-GC Europe. Lcgceurope.com.*
4. *Xiang, Y.; Liu Y. and Lee M.L. (2006). "Ultrahigh pressure liquid chromatography using elevated temperature". Journal of Chromatography A 1104 (1–2): 198–202.*
5. *Horváth, Cs.; Preiss B. A. and Lipsky S.R. (1967). "Fast liquid chromatography. Investigation of operating parameters and the separation of nucleotides on pellicular ion exchangers". Analytical Chemistry 39 (12): 1422–1428.*
6. *F: \ Instrumental Analysis\ Chromatography.htm.*
7. *Chromatography Today, C.F. Poole and S.K.Poole, Elsevier, 1991.*
8. *Ardo, Y. and Kristiansen, K.R. (2002), Chromatographic methods, Encyclopedia of Dairy Sciences (Volume-1), pp-55-62, Elsevier science Limited.*
9. *AH.El-hamdy and NK.El-fizga, Journal of Chromatography 1995; 708: 351.*
10. *M. A. Rounds, and S.S. Nielson, (1994). Basic Principles of Chromatography. In Introduction to Chemical Analysis of Foods. Nielson, S.S. (Ed.) Jones and Barlett Publishers, Boston, pp. 403-424.*

Principle of Chromatography/Chromatographic Separation [1,2,3]

The basis of all types of chromatography is the partition of the sample compounds between a stationary phase and a mobile phase which flows over or through the stationary phase. Different combinations of gaseous or liquid phases give rise to the types of chromatography used in analysis, namely gas chromatography (GC), liquid chromatography (LC), thin layer chromatography (TLC), and supercritical fluid chromatography (SFC).

Chromatography is a separations method based on the concept of partition coefficient. Any solute will partition between two immissible solvents. Chromatography relies on differences in partitioning behavior between a flowing mobile phase and a stationary phase to separate the the components in a mixture. All chromatographic systems rely on the aforesaid fact [1]. If matrix support is polar (*e.g.* paper, silica *etc.*) it is forward phase chromatography, and if it is non polar (C-18) it is reverse phase [2]. A reproducible fraction will partition into each phase, depending on the relative affinity of the substance for each phase. A substance which has affinity for the moving or **mobile phase** will be moved rapidly through the system. A material which has a stronger affinity for the **stationary phase,** on the other hand, will spend more time immobilized in that phase, and will take a longer time to pass through the system. Therefore, it will be separated from the first substance. By definition, chromatography is a separation technique in which a sample is equilibrated between a mobile and a stationary phase.

(i) A column (or other support for TLC) holds the stationary phase and the mobile phase carries the sample through it. Sample components that partition strongly into the stationary phase spend a greater amount of time in the column and are separated from components that stay predominantly in the mobile phase and pass through the column faster. The sample is separated into zones or bands.

(ii) Gas chromatography employs an inert gas as the mobile phase, and either a solid adsorbent or a nonvolatile liquid coated on a solid support as the stationary phase. The solid or coated support is packed into a tube, with the gas flowing through it. Separation depends on the relative partial pressures of the sample components above the stationary phase

(iii) Liquid chromatography uses similar packed tubular columns and usually a pump to force a liquid mobile phase through the column.

(iv) Supercritical fluid chromatography occupies a middle ground between gas and liquid chromatography. The mobile phase is a supercritical fluid, *i.e.*, a fluid above its critical temperature and pressure. This allows the use of GC type detectors, since the mobile phase has gas-like properties, but also allows continuous variation in such mobile phase properties as viscosity and density, by changing temperature and pressure.

Finally, chromatography may be done on a planar surface. The sample is transported over a solid surface such as cellulose or silica gel, coated on a plate. The sample components are moved over the surface by the mobile phase which is usually allowed to travel through the adsorbent layer by capillary action.

The movement of analytes in the column can be described mathematically. The basis of chromatography is the equilibrium of each analyte between the mobile and stationary phase. This can be expressed by a simple equilibrium equation, where K_x is partition coefficient,

$K_x = [C]_s/[C]_m$ *Equation 1*

It means that the concentration in the stationary phase ([Cs]) is directly related to the concentration in the mobile phase ([Cm]), at least when the concentrations are low. Chromatographic separations are best done with a small amount of analyte, which keeps either phase from becoming saturated with analyte, so that the concentrations in the two phases are directly proportional.

The quantities in the equilibrium expression for Kx, [C] s and [C] m are not easy to measure and hence a new constant, the capacity factor, k' can be defined as:

$$\mathbf{k'x} = \frac{\text{(moles of X in stationary phase)}}{\text{(moles of X in mobile phase)}}$$ *Equation 2*

Since the number of moles can be expressed as the concentration multiplied by the volume, Equations 1 and 2 can be combined and reduced to:

$k'x = Kx\ (Vs/Vm)$ *Equation 3*

All sample molecules spend the same amount of time in the mobile phase. If they were completely unretained by the stationary phase they would exit the column in the time it takes for one volume of mobile phase to pass through the column. This is equal to the void volume of the column. Molecules pass through the column in the time equal to the passage of one void volume, Vm, plus the time spent in the stationary phase, expressed by k'x. Therefore, the volume of eluent which will pass through the column before the sample elutes (the retention volume, Vr) can be expressed as:

$$Vr = Vm\ (1 + k'x)$$

The retention volume, Vr, is related to the column flow Fc, and the retention time, tr. Likewise, the volume of the mobile phase, Vm, is related to the flow and the time the void volume takes to pass through the column, to.

$$V_r = t_r F_c$$

$$V_m = t_o F_c$$

Substituting these into equation 3.0 and rearranging gives the following:

$$k'x = (t_r - t_o)/t_o$$

Source

1. *F: \Instrumental Analysis\Chromatography.htm.*
2. *Wikipedia.*
3. *M. A. Rounds, and S.S. Nielson, (1994). Basic Principles of Chromatography. In Introduction to Chemical Analysis of Foods. Nielson, S.S. (Ed.) Jones and Barlett Publishers, Boston, pp. 403-424.*

Quantitation in Chromatography [1]

The amount of each component in the sample is measured as it issues from the column, by passing the effluent through a detector and integrating the detector signal over time. For quantitation in both gas and liquid chromatography, the detector signal is usually fed into following a digital or analog recorder:

i. The Strip Chart Recorder

It produces the typical chromatogram, a plot of signal versus time, as a series of peaks. The peak area is related to the quantity of each material, and the location on the time axis or the retention time is used to identify the compound.

(a) Peak areas obtained on strip chart recorders may be determined graphically by drawing tangent lines to the peak sides and measuring the widths and heights. The area (A) is calculated from:

$$A = h_{max} \times w_{½} \text{ or } (h_{max} \times w_b)/2$$

(b) In addition, a peak area may be measured by cutting the peak out of the chart paper with sharp scissors, and weighing it on an analytical balance.

ii. Digital Recorder

Modern chromatographs have high precision in retention time as well as peak area. Consequently, chromatograms provide excellent qualitative and quantitative information. Digital processing of the signal allows the peak areas to be integrated automatically, with the beginning and ending points of peaks indicated. Baselines are usually set automatically and may sometimes be adjusted by the operator.

Calculation of quantity of sample-In order to calculate the quantity of sample from the peak area or peak height, the detector must be first calibrated by running standards. Quantitation may be done by either peak height or peak area, with most accurate results being achieved with use of areas, although, when chromatograms are being manually interpreted, heights may be more readily measured. The following external standards, internal standards or a combination of both can be used:

a. External Standard Method

A series of standards containing known quantities of the analytes are prepared and run. The peak heights or areas are plotted versus the quantities or concentrations, and the samples are run in the same way. A major source of error in this method is the reproducibility of the injection, especially if manual injections are being made using a syringe. Since most chromatographic samples are only a microliter or less, accurate measurement is difficult. Automatic injectors and sampling valves can reduce this error to a few per cent.

The concentration of the sample is calculated from the calibration graph or from:

$$C_{unk}/C_{std} = A_{unk}/A_{std}$$

where, C - the concentration and A- the peak area for the standard (std) and the unknown (unk).

This simple calculation is usually good enough if the sample and standard are of the same order of magnitude and the detector response is linear over the range of concentrations covered by the sample and standard. If sample concentrations vary over a wider range, a calibration curve should be constructed over the entire range.

Since it is not always possible to obtain standards over a suitable range or for every compound one needs, detector response factors may be used. These allow calibration with one compound as a surrogate for others.

$$C_{unk}/C_{std} = f_{unk}\ A_{unk}/A_{std}$$

where, f_{unk} - the detector response factor which relates the detector response to a quantity of the sample compound to its response to the same amount of the surrogate standard compound. These factors can be determined experimentally. Then the chromatograph can be calibrated daily against the surrogate compound, and the detector response factors used to calibrate for other compounds.

b. The Internal Standard Method

An internal standard is added to the sample before analysis. This is composed of one or more compounds with sufficient similarity to the target analytes, so that they behave similarly, but must surely not appear in the samples. Internal standard compounds must also be readily separated from any of the compounds in the sample. If an internal standard is added before such steps as concentration, extraction, or dilution, it is then not as necessary to make accurate volume measurements, and the injection volume is also not as critical. By examination of the peak area of the internal standard in sequential injections, the reproducibility of the injection volumes can be seen. Small changes in the volume show up as changes in the internal standard peak. Again, it is important to keep the concentrations of the internal standard and the target compounds in a similar range, to keep linearity problems from arising. The area of the internal standard (A_{is}) is used to normalize the areas of all other sample peaks, thus eliminating the effect of differences in injection volumes or dilutions.

$$C_{unk} = (A_{unk}/A_{is})\ C_{is}\ f_{unk}$$

Source

1. *F: \Instrumental Analysis\Chromatography.htm.*
2. *Wikipedia.*

Different Chromatographic Terms Used [1]

The following terms are important in chromatography and these are to be well understood before analysis:

(i) **Analyte:** It is the substance to be separated during chromatography.

(ii) **Analytical Chromatography:** It is used to determine the existence and possibly also the concentration of analyte (s) in a **sample.**

(iii) **A bonded phase:** It is a stationary phase that is covalently bonded to the support particles or to the inside wall of the column tubing.

(iv) **A chromatogram:** It is the visual output of the chromatograph. In the case of an optimal separation, different peaks or patterns on the chromatogram correspond to different components of the separated mixture.

(v) **A chromatograph:** It is equipment that enables a sophisticated separation *e.g.* gas chromatographic or liquid chromatographic separation.

(vi) **The eluate:** It is the mobile phase leaving the column.

(vii) **The eluent:** It is the solvent that will carry the analyte.

(viii) **An eluotropic series:** It is a list of solvents ranked according to their eluting power.

(ix) **An immobilized phase:** It is a stationary phase which is immobilized on the support particles, or on the inner wall of the column tubing.

(x) **The mobile phase:** It is the phase which moves in a definite direction. It may be a liquid (LC and CEC), a gas (GC), or a supercritical fluid (supercritical-fluid chromatography, SFC). In the case of **HPLC** the mobile phase consists of a non-polar solvent (*e.g.* hexane) in normal phase or polar solvents in reverse phase chromatography and the sample being separated. The mobile phase consists of the sample being separated/analyzed and the solvent that moves the sample through the column The mobile phase moves through the chromatography column (the stationary phase) where the sample interacts with the stationary phase and isseparated.

(xi) **The retention time:** It is the characteristic time it takes for a particular analyte to pass through the system (from the column inlet to the detector) under set conditions.

(xii) **The sample:** It is the matter analyzed in chromatography. It may consist of a single component or it may be a mixture of components. When the sample is treated in the course of an analysis, the phase or the phases containing the analytes of interest is/are referred to as the sample whereas everything out of interest separated from the sample before or in the course of the analysis is referred to as waste.

(xiii) **The solute:** It refers to the sample components in partition chromatography.

(xiv) **The solvent:** It refers to any substance capable of solubilizing other substance, and especially the liquid mobile phase in LC.

(xv) **The stationary phase:** It is the substance which is fixed in place for the chromatography procedure (*e.g.* silica layer in TLC).

Source

1. *Wikipedia.*

Types of Chromatography [1,2]

On the basis of the conditions of Stationary and Mobile phase, the chromatographic techniques are broadly classified in to the following categories:

a. Broader Classification

1. **Elution Chromatography:** Flushing of sample through column by continual mobile phase (eluent) addition, migration rate µ fraction time spent in mobile phase.
2. **Planar chromatography:** Flat stationary phase, mobile phase, moves through capillary action or gravity.

3. **Column chromatography:** Tube of stationary phase, mobile phase, moves by pressure or gravity.

b. Sub Classification

Paper Chromatography, Thin Layer Chromatograhy, Gas Chromatography.

Source

1. *Wikipedia.*
2. *IUPAC Nomenclature for Chromatography IUPAC Recommendations 1993, Pure and Appl. Chem., Vol. 65, No. 4, pp.819–872, 1993,*

Planar Chromatography [3]

It is a separation technique in which the stationary phase is present as or on a plane. The plane can be a paper, serving as such or impregnated by a substance as the stationary bed or a layer of solid particles spread on a support such as a glass plate.When plane is paper, it is then called as **Paper chromatography** and when plane is layer of solid particles spread on a support such as a glass plate, then it is called as Thin layer chromatography. Different compounds in the sample mixture travel different distances according to how strongly they interact with the stationary phase as compared to the mobile phase. The specific Retention factor (R_f) of each chemical can be used to aid in the identification of an unknown substance.

1. Paper Chromatography (PC) [1,2,3]

Paper chromatography is a technique that involves placing a small dot or line of sample solution onto a strip of chromatography paper (made of cellulose). The paper is placed in a jar containing a shallow layer of solvent and sealed. The solvent during rising through the paper, meets the sample mixture which starts to travel up the paper with the solvent. This paper is made of cellulose, a polar substance, and the compounds within the mixture travel farther if they are non-polar [3].

It is an analytical method that is used to separate coloured chemicals or substances, especially pigments. This can also be used in secondary or primary colours in ink experiments. **Double-way paper chromatography**, also called two-dimensional chromatography, involves using two solvents and rotating the paper 90° in between. This is useful for separating complex mixtures of compounds (Amino acids) having similar polarity. This method has been largely replaced by thin layer chromatography, but is still a powerful teaching tool. If a filter paper is used, it should be of a high quality paper. The mobile phase is developing solutions that can travel up to the stationary phase carrying the sample along with it.Paper chromatography is one method for testing the purity of compounds and identifying substances. Paper chromatography is a useful technique because it is relatively quick and requires small quantities of material. It is a **planar chromatography systems** where in a cellulose filter paper acts as a stationary phase on which separation of compounds occurs.

Principle [1,2]

In paper chromatography, like thin layer chromatography, substances are distributed between a stationary phase and a mobile phase. The stationary phase is usually a piece of high quality filter paper. The mobile phase is a developing solution that travels up the stationary phase, carrying the samples with it. Components of the sample will separate readily according to how strongly they adsorb onto the stationary phase versus how readily they dissolve in the mobile phase.

The principle involved is **partition chromatography** where in the substances are distributed or partitioned between to liquid phases. One phase is the water which is held in pores of filter paper used and other phase is that of mobile phase which moves over the paper. The compounds in the mixture get separated due to differences in their affinity towards water (in stationary phase) and mobile phase solvents during the movement of mobile phase under the capillary action of pores in the paper.

The principle can also be **adsorption chromatography** between solid and liquid phases, where in the stationary phase is the solid surface of paper and the liquid phase is of mobile phase. But most of the applications of paper chromatography work on the principle of partition chromatography *i.e.* partitioned between to liquid phases.

When a colored chemical sample is placed on a filter paper, the colors separate from the sample by placing one end of the paper in a solvent. The solvent diffuses up the paper, dissolving the various molecules in the sample according to the polarities of the molecules and the solvent. If the sample contains more than one colour, that means it must have more than one kind of molecule. Because of the different chemical structures of each kind of molecule, the chances are very high that each molecule will have at least a slightly different polarity, giving each molecule a different solubility in the solvent. The unequal solubilities cause the various color molecules to leave solution at different places as the solvent continues to move up the paper. The more soluble a molecule is, the higher it will migrate up the paper. It is important to note that when using water (a very polar substance) as a solvent, the more polar the colour, the higher it will rise on the paper.**Note-**If a chemical is very nonpolar it will not dissolve at all in a very polar solvent. This is the same for a very polar chemical and a very nonpolar solvent.

R_f Value

The retention factor (R_f) may be defined as the ratio of the distance traveled by the substance to the distance traveled by the solvent. R_f values are usually expressed as a fraction of two decimal places. If R_f value of a solution is zero, the solute remains in the stationary phase and thus it is immobile. If R_f value = 1 then the solute has no affinity for the stationary phase and travels with the solvent front. To calculate the R_f value, take the distance traveled by the substance divided by the distance traveled by the solvent (as mentioned earlier in terms of ratios). For example, if a compound travels 2.1 cm and the solvent front travels 2.8 cm, (2.1/2.8) the R_f value = 0.75. R_f value depends on temperature and the solvent used in experiment, so several solvents offer several R_f values for the same mixture of compound.

Types of Paper Chromatography [1,2]

Based on the way the development of chromatogram on the paper is done in the procedures, broadly following 5 types of Paper Chromatographic techniques are being used:

(i) **Descending Paper Chromatography:** In this type, development of the chromatogram is done by allowing the solvent to travel down the paper. The mobile phase is present in the upper portion.

(ii) **Ascending Paper Chromatography:** In this type, the solvent travel upward direction of the Chromatographic paper.

Note: Both the Descending and Ascending Paper Chromatography are used for separation of Organic and Inorganic substances.

(iii) **Ascending-Descending Paper Chromatography:** It is the hybrid of both the above techniques. The upper part of the Ascending chromatography can be folded over a rod and allowing the paper to become descending after crossing the rod.

(iv) **Radial Paper (Circular) Chromatography:** In this, a circular filter paper is taken and the sample is given at the center of the paper. After drying the spot the filter paper is tied horizontally on a Petridish containing solvent. So that Wick of the paper is dipped inside the solvent. The solvent rises through the wick and the component get separated in form of concentrate circular zone.

(v) **Two-Dimensional Paper Chromatography:** In this technique a square or rectangular paper is used. Here the sample is applied (Spotted) to one of the corners and allowed for first development. Then the paper is again immersed in mobile phase at right angle to previous development for second chromatogram.

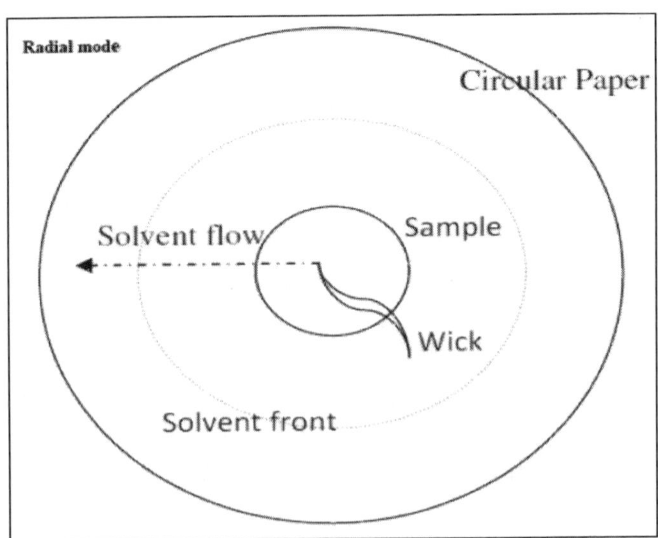

Ascending Paper Chromatography Radial Paper Chromatography.

Source

1. *Haslam, Edwin (2007)."Vegetable tannins– Lessons of a phyto-chemical lifetime". Phytochemistry 68 (22–24): 2713–21.*
2. *bheem.hubpages.com.*
3. *Wikipedia.*

Applications of Paper Chromatography

It is used for the following analysis:

A. Clinical and Pharmaceutical Industries [1]

It is specially used for separation of mixture having polar and non polar compounds *e.g.* [1].

(*i*) Separation of amino acids.
(*ii*) To determine organic compounds, biochemicals in urine *etc.*
(*iii*) In pharma sector for determination of hormones, drugs.
(*iv*) Sometimes used for evaluation of inorganic compounds like salts and complexes.

Source

1. *Wikipedia*

B. Food Industry

Few examples for application in Food Industry are given below:

I. Detection of Kesari Dal Powder (Lathyrus sativus) in Besan

Principle

The presence of kesari dal powder is detected on the basis of the presence of an unusual amino acid namely beta-N-oxalyl L amino alanine which is not present in the seeds of other legumes.

Apparatus

(*i*) Steam bath/water bath.
(*ii*) Air Oven.
(*iii*) Chromatographic paper Whatman No. 1.

Reagents

(*i*) Ethyl alcohol 70 per cent.
(*ii*) Isoproponol solution–10 per cent.
(*iii*) Liquified Distilled Phenol-water solution - (4: 1).
(*iv*) Ninhydrin solution - 0.1 per cent in acetone or ethanol.
(*v*) Buffer-Pyridine: Acetic acid: water (0.5: 5: 95) pH 3.60.

Procedure

Weigh approximately 5 gm powdered sample and extract it with 100 ml ethyl alcohol (70 per cent) by keeping overnight, with occasional shaking. Filter the extract and evaporate to dryness on a steam/water bath. Extract the residue with 10 ml of iso-proponol solution, filter and use this solution for chromatography. Spot 20 µl of the extract using a haemoglobin pipette or capillary tube at a distance of 1 cm from the bottom of the chromatographic filter paper. Keep in a solvent chamber saturated with phenol-water solution overnight. Remove from the chamber, dry filter paper in a current of air at room temperature for 4-5 hours or in an oven at 80°C for 1 hour and spray with ninhydrin solution. Dry the chromatogram in the oven for 15 minutes. The appearance of bluish–purple spot at about Rf value 0.1 shows presence of BOAA which is present only in Lathrus sativus. Other proteins extracted simultaneously also give similar colour but at different Rf values. Always run known sample of kesari dal powder simultaneously and compare the spot of control with sample under test.

Source

1. *I.S.I Handbook of Food Analysis (Part IV) - 1984 Page121).*
2. *Manual of Methods of analysis of Foods (Cereals and Cereal Products), Lab Manual.*
3. *FSSAI, GOI, New Delhi.*

II. Test for the Presence of Added Synthetic Colour in Confectionery Products (Sugar, Boiled Confectionery, Lozenges) by Paper Chromatographic Separation

The general scheme for identifying synthetic food colours present in foods normally involve preliminary treatment of the food, extraction of the colour from the prepared solution of the food, separation of colours in case of mixtures and identification of the separated colours.

Apparatus

(i) Pipette.

(ii) Beaker

(iii) Flask.

Reagents

(i) White knitting wool- Extract pure white wool in a soxhlet extractor with petroleum ether for 2-3 hrs to remove fat. Boil in very dilute solution of sodium hydroxide and then in water to free it from alkali.

(ii) Paper: Whatman No. 1 chromatographic paper.

(iii) Solvents- 1 ml (0.88 sp. gr) ammonia + 99 ml water,2.5 per cent aqueous sodium chloride, 2 per cent sodium chloride in 50 per cent ethanol, Acetic acid solution in water (1: 3), Iso-butanol-ethanol-water (1: 2: 1, v/v), n-butanol-water-glacial acetic acid (20: 12: 5, v/v), Iso-butanol-ethanol-water (3: 2: 2, v/v). to 99 ml of this add1ml of (0.88 sp gr.) ammonia and 80 gm phenol in 20 gm water.

Procedure

a. Preliminary Treatment of Food

Assuming that an acidic colour is present, the preliminary treatment involves removing interfering substances and obtaining the dye in acid solution prior to boiling with wool.

(i) **Non-alcoholic beverages** *e.g.* soft drinks-As most foods in this group are acidic they can be usually treated directly with wool, otherwise, slightly acidify the food with acetic acid.

(ii) **Alcoholic liquids** (*e.g.*Wine)-Boil to remove alcohol and acidify if necessary as in (a).

(iii) **Starch based foods** (*e.g.*cakes, custard powder *etc.*)-Grind10 g of sample thoroughly with 50ml of 2 per cent ammonia in 70 per cent alcohol, and allow it to stand for an hour and centrifuge.Pour the separated liquid into a dish and evaporate on water bath. Take up the residue in 30ml dilute acetic acid.

(iv) **Candied fruits-**Treat as in (c)

(v) **Products with high fat content** (*e.g.*Sausages, meat, fish paste)- De-fat the sample with light petroleum and extract the hotwater (acidify *etc.* as usual).Note that oil soluble colour tend to organics olvents.If the extraction is difficult treat with warm 50-90 per cent acetone or alcohol (which precipitates starch) containing 2 per cent ammonia. The organic solvent should be removed before acidifying as in (c).

b. Extraction of the Colour from the Food

Introduce about 20 cm length **of** woollen thread into a beaker containing about 35 ml of the prepared acidified solution of the sample and boil for a few minutes till the woollen thread is dyed. Take out the woollen thread and wash it with tap water. Transfer the washed woollen thread to a small beaker containing dilute ammonia and heat again. If the colour is stripped by the alkali, the presence of an acid coal-tar dye is indicated. Remove the woollen thread. Make the liquid slightly acidic and boil with a fresh piece of woollen thread. Continue boiling until the colour is taken by the woollen thread. Extract the dye from the woollen thread again with a small volume of dilute ammonia, filter through a small plug of cotton and concentrate the filtrate over a hot water bath. This double stripping technique usually gives a pure colour extract. Natural colours may also dye the wool during the first treatment, but the colour is not usually removed by ammonia. Basic dyes can be extracted by making the food alkaline. with ammonia, boiling with wool and then stripping with dilute acetic-acid. At present, all the permitted water soluble coal-tar dyes are acidic, hence an indication of the presence of a basic dye suggests that an unpermitted colour is present.

c. Identification of the Separated Food Colours by Paper Chromatography

Draw a pencil-line parallel to the bottom edge of the paper (Whatman No.1) at about 2 cm distance. Spot the concentrated solution of the unknown dye on the line together with a series of spots (about 2 cm apart) of aqueous solutions of standard permitted dyes of similar colour and dry.Run the chromatogram, by ascending technique, using a selected solvent. Solvent No.5 is often helpful for general purposes. Identify the colour in the sample by matching its spot with the spot of the standard colour and confirm by co spotting.

d. Determination of Synthetic Food Colours in Food Products

1. For Samples Containing Single Colour

a. Preparation of Standard Curve

Stock solution-Weigh 0.1 g m of each reference colour and dissolve in 0.1N HCl in separate 100 ml volumetric flasks and make up the volume with 0.1NHCl in each case.Working standard: Pipette 0.25, 0.5, 0.75, 1.0, 1.25 and 1.5 ml of stock solution of each of the reference colours into series of clean and dry 100 ml volumetric flasks and dilute to volume with 0.1N HCl. Determine the optical densities of each of the reference colours at the respective wave length of maximum absorption (refer table) Obtain the standard curve for each colour by plotting optical density against concentration.

b. Determination in Sample by Column Chromatography

Transfer a known weight of the sample (approximately 5-10 gm) into a glass stoppered separatory funnel. Extract the colour with 70 per cent acetone. Shake acetone extract with petroleum ether (40-60°C) in order to remove carotenoids and other natural pigments, if any. Continue extraction with petroleum ether until petroleum ether extract is colourless. Pass the acetone extract containing only coal-tar food colours through a column (2.1·45 cm) containing aluminium oxide acidified with 1 per cent HCl. Elute the adsorbed colour with 1 per cent ammonia. Evaporate the eluate to dryness on a hot water bath, dissolve the residue with 0.1 N HCl, transfer quantitatively to a 100 ml volumetric flask and make up the volume with 0/1N HCl. Determine the optical density of the dye solution at the wavelength of maximum absorption. Calculate the dye concentration from the standard curve.

Showing Absorption Maxima of Permitted Food Colours

Name of the Colour	*Absorption Maxima (nm)*
Carmosine	516
Ponceau 4R	507
Erythrosine	527
Green FCF	624
Indigo Carmine	609
Brilliant Blue FCF	630
Tartrazine	427
Sunset yellow FCF	482

2. For Samples Containing Mixture of Colours

i. Paper Chromatography

Extract the colours present in the samples and isolate as described under column chromatography.Make up the purified dye solution to a known volume with water (5 ml).Spot an aliquot (approximately 0.5 to 1.0 ml) of the purified dye on Whatman No.1 filter paper as a band and develop the chromatogram using butanol- acetic acid: water (20: 5: 12) solvent system. After drying, cut out the coloured spots on the chromatogram and elute with 0.1 N HCl. Prepare a blank by cutting an equivalent strip from plain portions of the chromatogram and elute with 0.1 N HCI. Make up the eluate to a known volume (100 ml) with 0.1 N HCI and determine the dye content as described under column chromatography.

ii. Thin Layer Chromatography

Extract the colours present in the sample as described under column chromatography. Concentrate the eluate and make up to known volume with water (5 ml).

Preparation of TLC Plate

To 50 g of silica gel without binder add 50 ml of starch solution (0.6 gm of soluble starch dispersed in 100 ml or glass distilled water heated to boiling to gelatinise starch) and 50 ml of 1.25 per cent solution of disodium salt of ethylene diamine tetra acetic acid, Mix the slurry well. Spread slurry using applicator on glass plates (20 ·20 cm) to thickness of 0.5mm. Allow the plate to air dry and then dry at 120°C for two hours. Spot an aliquot of the purified dye on TLC plate and develop the chromatogram using isoamyl: glacial acetic acid: water (40: 20: 20) solvent system. Remove the plate and dry. Scrape out the colour spots on the plate and transfer to test tubes. Elute the colour using 0.1N HCl. Prepare a blank by scraping from the plain portions of the plate. Make up the eluates to a known volume (100 ml) with 0.1N HCl. Determine the dye content as described under column chromatography.

Source

1. *Manual Methods of Analysis for Adulterants and Contaminants in Food, I.C.M.R 1990 Page 56).*
2. *Manual of Methods of analysis of Foods (Beverage, Sugar, Confectionery Products.), Lab Manual -4, FSSAI, New Delhi.*

2. Thin Layer Chromatography (TLC)

It is a chromatography technique used to separate non-volatile mixtures [1]. It is a widely used laboratory technique and is similar to paper chromatography with the advantage of faster runs, better separations, and the choice between different stationary phases. It involves a stationary phase of a thin layer of adsorbent material (silica gel, alumina, or cellulose) coated on a flat, inert substrate (sheet of glass, plastic, or aluminium foil). The layer of adsorbent is known as the stationary phase. After the sample has been applied on the plate, a solvent or solvent mixture (known as the mobile phase) is drawn up the plate via capillary action. Because different analytes ascend the TLC plate at different rates, separation is achieved [2].

Advantages

It has the advantage of faster runs, better separations, better quantitative analysis, and the choice between different adsorbents. For even better resolution and faster separation that utilizes less solvent. TLC is very versatile; multiple samples can be separated simultaneously on the same layer, making it very useful for screening applications such as testing drug levels and water purity [12].

For even better resolution and to allow for quantification, high-performance TLC can be used [13]. A number of enhancements can be made to the original method to automate the different steps, to increase the resolution achieved with TLC and to allow more accurate quantitative analysis. This method is referred to as **HPTLC, or "high-performance TLC"**.

Analytical Procedure

It involves the following steps:

i. Plate Preparation

TLC plates are usually commercially available, with standard particle size ranges to improve reproducibility. They are prepared by mixing the adsorbent, such as silica gel, with a small amount of inert binder like calcium sulfate (gypsum) and water. This mixture is spread as a thick slurry on an unreactive carrier sheet, usually glass, thick aluminum foil, or plastic. The resultant plate is dried and activated by heating in an oven for thirty minutes at 110°C. The thickness of the absorbent layer is typically around 0.1- 0.25 mm for analytical purposes and around 0.5 -2.0 mm for preparative TLC [3].

ii. Running of Plates [4]

(a) A small spot of solution containing the sample is applied to a plate, about 1.5 centimeters from the bottom edge. The solvent is allowed to completely evaporate off, otherwise a very poor or no separation will be achieved. If a non-volatile solvent was used to apply the sample, the plate needs to be dried in a vacuum chamber.

(b) A small amount of an appropriate solvent (eluent) is poured into a glass beaker or any other suitable transparent container (separation chamber) to a depth of less than 1 centimeter. A strip of filter paper is put into the chamber so that its bottom touches the solvent and the paper lies on the chamber wall and reaches almost to the top of the container. The container is closed with a cover glass or any other lid and is left for a few minutes to let the solvent vapors ascend the filter paper and saturate the air in the chamber.

(c) The TLC plate is then placed in the chamber so that the spot (s) of the sample do not touch the surface of the eluent in the chamber, and the lid is closed. The solvent moves up the plate by capillary action, meets the

sample mixture and carries it up the plate (elutes the sample). The plate should be removed from the chamber before the solvent front reaches the top of the stationary phase and dried.

Separation Process and Principle

Different compounds in the sample mixture travel at different rates due to the differences in their attraction to the stationary phase and because of differences in solubility in the solvent [5]. By changing the solvent, or perhaps using a mixture, the separation of components (measured by the R_f value) can be adjusted. Also, the separation achieved with a TLC plate can be used to estimate the separation of a flash chromatography column [6].

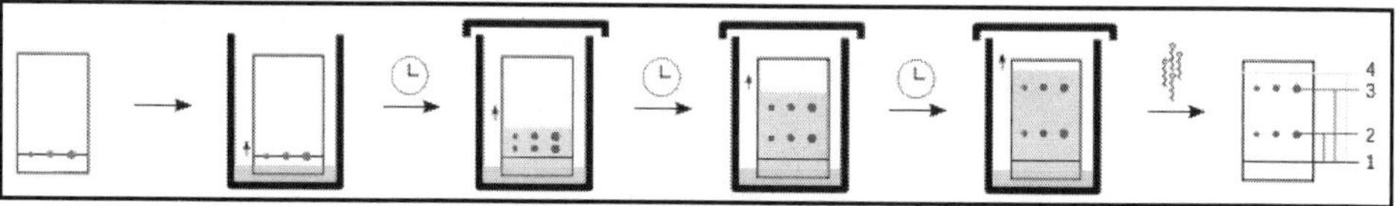

Development of a TLC Plate, a Purple Spot Separates into a Red and Blue Spot.

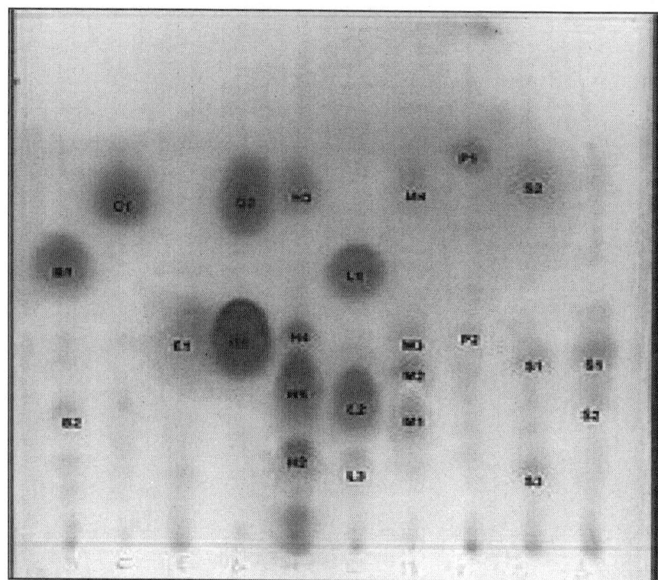

Chromatogram of 10 Essential Oils Coloured with Vanillin Reagent.

Separation of compounds is based on the competition of the solute and the mobile phase for binding places on the stationary phase [7]. For instance, if normal-phase silica gel is used as the stationary phase, it can be considered polar. Given two compounds that differ in polarity, the more polar compound has a stronger interaction with the silica and is, therefore, more capable to dispel the mobile phase from the binding places. As a consequence, the less polar compound moves higher

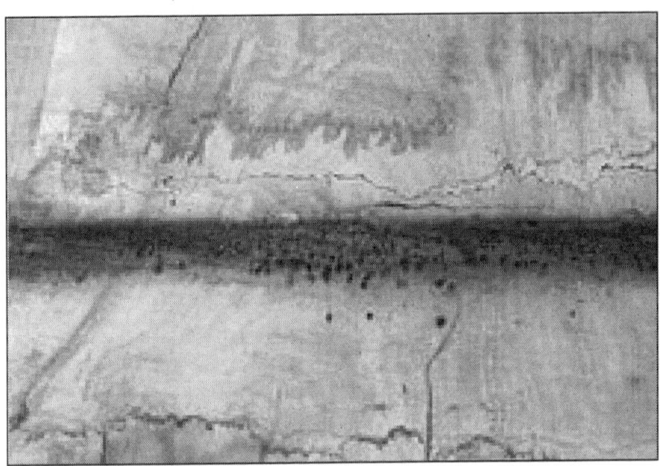

Surface of a Freshly Cut Plank of *Eucalyptus camaldulensis* Displaying Thin-Layer Chromatography. The horizontal blue strip is from a reaction between the iron bandsaw supports and the acidic timber.

up the plate (resulting in a higher R_f value) [6]. If the mobile phase is changed to a more polar solvent or mixture of solvents, it is more capable of dispelling solutes from the silica binding places, and all compounds on the TLC plate will move higher up the plate. It is commonly said that "strong" solvents (eluents) push the analyzed compounds up the plate, whereas "weak" eluents barely move them. The order of strength/weakness depends on the coating (stationary phase) of the TLC plate. For silica gel-coated TLC plates, the eluent strength increases in the following order: perfluoroalkane (weakest), hexane, pentane, carbon tetrachloride, benzene/toluene, dichloromethane, diethyl ether, ethylacetate, acetonitrile, acetone, 2-propanol/*n*-butanol, water, methanol, triethylamine, acetic acid, formic acid (strongest). For C18-coated plates the order is reverse. This means that if a mixture of ethyl acetate and hexane as the mobile phase is used, adding more ethyl acetate results in higher R_f values for all compounds on the TLC plate. Changing the polarity of the mobile phase will normally not result in reversed order of running of the compounds on the TLC plate. An eluotropic series can be used as a guide in selecting a mobile phase. If a reversed order of running of the compounds is desired, an apolar stationary phase should be used, such as C18-functionalized silica.

Analysis

As the chemicals being separated may be colorless then the following methods are being used to visualize the spots:

(i) fluorescent analytes like quinine may be detected under blacklight (366 nm).

(ii) Often a small amount of a fluorescent compound, usually manganese-activated zinc silicate, is added to the adsorbent that allows the visualization of spots under UV-C light (254 nm). The adsorbent layer

will thus fluoresce light-green by itself, but spots of analyte quench this fluorescence.

(iii) Iodine vapors are a general unspecific color reagent.

(iv) Specific color reagents into which the TLC plate is dipped or which are sprayed onto the plate exist [8,9,10] - Potassium permanganate-oxidation and Bromine.

(v) In the case of lipids, the chromatogram may be transferred to a PVDF membrane and then subjected to further analysis, for example mass spectrometry, a technique known as Far-Eastern blotting.

Once visible, the R_f value, or retardation factor, of each spot can be determined by dividing the distance the product traveled by the distance the solvent front traveled using the initial spotting site as reference. These values depend on the solvent used and the type of TLC plate and are not physical constants.

Applications of TLC

Because of its simplicity and speed, TLC is often used for monitoring chemical reactions and for the qualitative analysis of reaction products. It can be used to monitor the progress of a reaction, identify compounds present in a given mixture, and determine the purity of a substance. Specific examples of these applications include- analyzing ceramides and fatty acids, detection of pesticides or insecticides in food and water, analyzing the dye composition of fibers in forensics, assaying the radiochemical purity of radio pharmaceuticals, or identification of medicinal plants and their constituents [7].

1. In Characterization and Isolation of Chemical Compounds

i. Characterization

In organic chemistry, reactions are qualitatively monitored with TLC. Spots sampled with a capillary tube are placed on the plate: a spot of starting material, a spot from the reaction mixture, and a cross-spot with both. A small (3 by 7 cm) TLC plate takes a couple of minutes to run. The analysis is qualitative, and it will show if the starting material has disappeared, *i.e.* the reaction is complete, if any product has appeared, and how many products are generated (although this might be underestimated due to co-elution). Unfortunately, TLCs from low-temperature reactions may give misleading results, because the sample is warmed to room temperature in the capillary, which can alter the reaction-the warmed sample analyzed by TLC is not the same as what is in the low-temperature flask. One such reaction is the DIBALH reduction of ester to aldehyde.

In one study TLC has been applied in the screening of organic reactions [11], for example in the fine-tuning of BINAP synthesis from 2-naphthol. In this method, the alcohol and catalyst solution (for instance iron (III) chloride) are placed separately on the base line, then reacted, and then instantly analyzed.

ii. Isolation

Since different compounds will travel a different distance in the stationary phase, chromatography can in effect be used as an isolation technique. The separated compounds each occupying a specific area on the plate, they can be scraped away, put in another solvent to separate them from the stationary phase and used for further analysis.

Example

In the chromatography of an extract of green leaves (spinach), there are 7 stages of development, Carotene elutes quickly and is only visible until step 2. Chlorophyll A and B are half way in the final step and elute in the first compound staining yellow. Once the chromatography is over, the carotene can be removed from the plate, put back into a solvent and ran into a spectrophotometer to characterize its wavelength absorption.

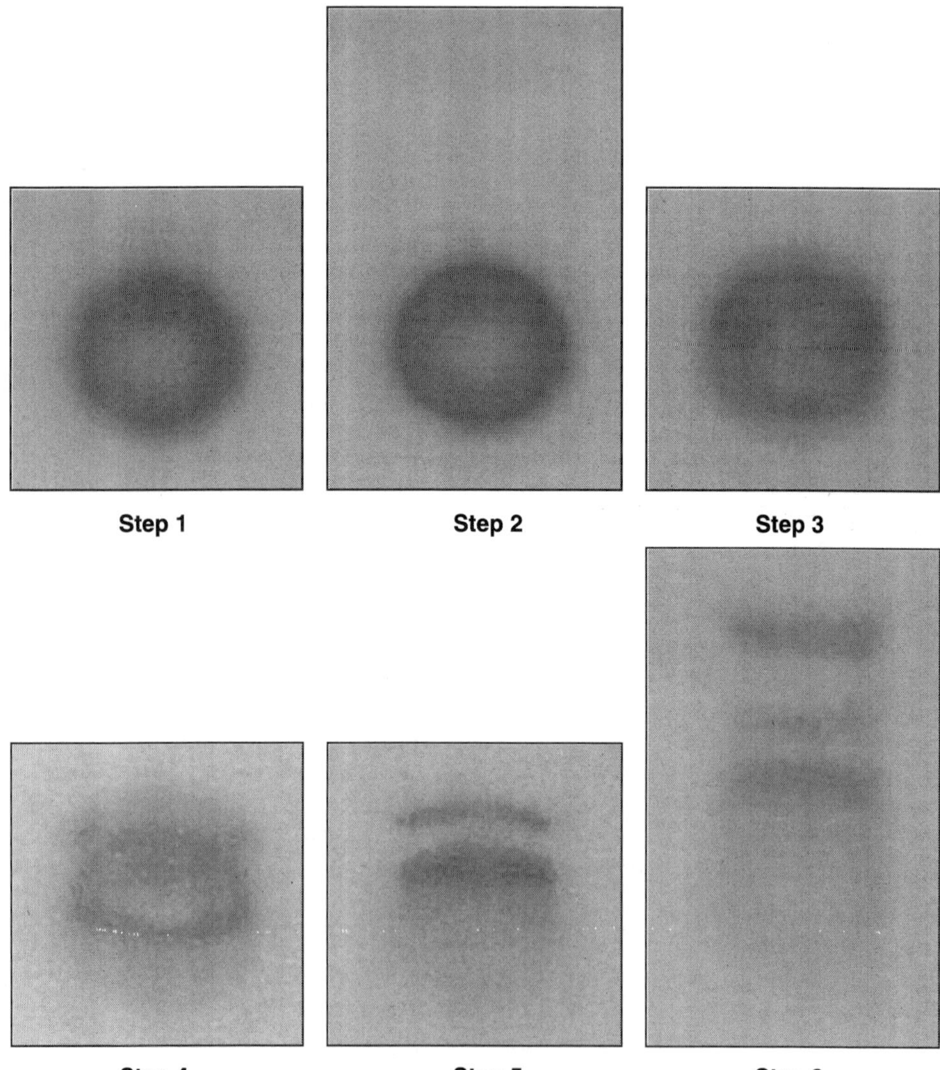

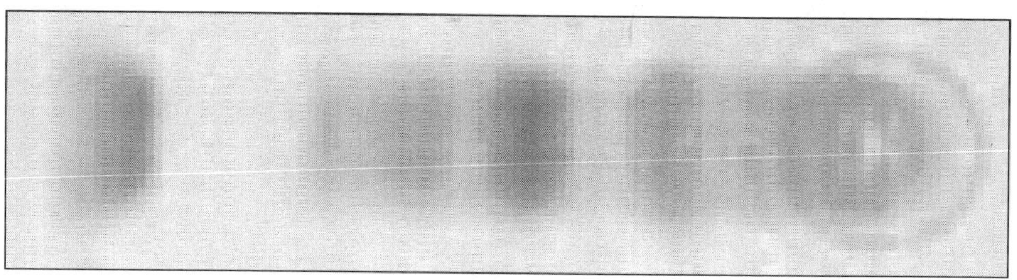

Step 7

Source

1. *Harry W. Lewis and Christopher J. Moody (13 Jun 1989). Experimental Organic Chemistry: Principles and Practice (Illustrated ed.). WileyBlackwell. pp. 159-173.*
2. *A.I. Vogel, A.R. Tatchell, B.S. Furnis, A.J. Hannaford, and P.W.G. Smith.Vogel's Textbook of Practical Organic Chemistry (5th ed.).*
3. *Tables showing the thickness value of commercial regular and preparative Thin Layer Chromatography plates.*
4. *Thin Layer Chromatography: How To http://www. reach devices.com/TLC.html.*
5. *Thin Layer Chromatography (TLC): Principle with animation.*
6. *Fair, J. D.; Kormos, C. M. J. Chromatogr. A2008, 1211 (1-2),49-54.*
7. *Reich, E.; Schibli A. (2007). High-performance thin-layer chromatography for the analysis of medicinal plants (Illustrated ed.). New York: Thieme.*
8. *Thin Layer Chromatography stains http://www. reachdevices.com/TLC_stains.html.*
9. *Jork, H., Funk, W., Fischer, W., Wimmer, H. (1990).*
10. *Jork, H., Funk, W., Fischer, W., Wimmer, H. (1994).*
11. *TLC plates as a convenient platform for solvent-free reactions Jonathan M. Stoddard, Lien Nguyen.*
12. *Bernard., Fried, (2003). Handbook of Thin-Layer Chromatography. Marcel Dekker Inc. ISBN 08247 48662.*
13. *Wikipedia.*

2. Application in Food Analysis

I. Test for Presence of Rice-Bran Oil

Oryzanolin rice-bran oil is isolated using 30 per cent aqueous potassium hydroxide solution and detected on thin-layer Chromatographic plate.

Apparatus

(i) Conical flasks, 250 ml capacity - 4 Nos., 100 ml capacity - 2 Nos.

(ii) Thin layer Chromatographic plates (0.25 mm) prepared by coating slurry of silica gelG. on glass plate of 20 x10cm dimension.

(iii) Iodine chamber for visualization of spots

(iv) Spotting capillaries.
(v) Separating funnel (100ml capacity)
(vi) Hot water bath.

Reagents

(i) Aqueous potassium hydroxide solution 30 per cent.
(ii) Hydrochloric acid dilute.
(iii) Blue litmus paper.
(iv) Diethylether AR grade.
(v) Sodium sulphate anhydrous.
(vi) Benzene - acetic acid (100: 1v/v)
(vii) Chloroform AR grade.

Procedure

Take 20ml of the oil in a100ml capacity separating funnel and add to it equal volume of aqueous potassium hydroxide solution.Shake the contents gently but constantly for 10 min.Keep the separating funnel on a stand for about 45min to allow the separation of alkali layer. Draw the alkali layer and neutralise with dilute hydrochloric acid solution. Confirm the neutralization with blue litmus paper.

Extract this salt solution with diethyl ether (20 ml x 3 times). Wash the diethyl ether extract with distilled water and dry on anhydrous sodium sulphate. Evaporate the solvent on hotwater bath and spot the residue in chloroform on TLC plate. Develop theTLC plate in pure benzene: acetic acid mixture. Allow the solvent front to move a distance of 15cm.Visualise the spots in iodine chamber. Appearance of a spot between Rf 0.7 to 0.75 indicates the presence of rice bran oil.Run a control by taking a sample of rice bran oil and compare the spot given by test sample under identical conditions.

Sensitivity

The above method can detect rice-bran oil in the redible vegetable Oils upto the minimum of 5 per cent level.

Source

1. *Manual methods of Analysis for Adulterants and Contaminates in foods I.C.M.R (1990) Page5.*

2. *Manual of Methods of analysis of Foods (Oils and Fats.), Lab Manual -2, FSSAI, New Delhi.*

II. Separation of Cholesterol by Reversed Phase Thin Layer Chromatography

A preliminary separation of total sterols from the unsaponifiable matter is achieved on silica gel-G thin layer chromatography. Subsequently the sterols are separated by reversed phase chromatography on Kieselghur-G using liquid paraffin as stationary phase and aqueous acetone saturated with liquid paraffin as the mobile phase.

Procedure

Separation of total sterols from unsaponifiable matter-Extract the unsaponifiable matter from the fat sample as per the method described else where. Evaporate the ether and dissolve the residue in 5ml of chloroform.

Prepare 20 x 20 cm glass plates coated with 0.5 mm thick silica Gel-G (BDH). Air dry the plates and activate at 110°Cfor 2 hr. Cool the plates to room temperature, spot the unsaponifiable matter along with the standard cholesterol on the plate. Develop the plates in diethyl ether: petroleum ether (1: 1) solvent system.Remove the plates when the solvent front reaches 14 cm height (it takes about 30 min time).

Air-dry the plates and expose to iodine vapours for a while. Total Sterols spot corresponding to standard spot of cholesterol appear as brown Colour spots.Mark spots and scrap off with stainless steel blade into a test tube. Extract the sterols using chloroform from silica gel. Separate the sterols by reversed phase thin-layer chromatography.

Preparation of Equilibrated Aqueous Acetone with Liquid Paraffin

Take 300 ml of 4: 1 acetone: water in a separating funnel and add 30 ml of liquid paraffin (heavy grade).Shake well and keep for18hr at room temperature for equilibration.Separate the lower layer of liquid paraffin and dilute to 5 proportion with petroleum ether.Use this for treating Kieselghur-G (Merck) coated thin-layer chromatographic plates. The upper acetone-water mixture serves solvent system to develop the paraffin treated plates.

Preparation of the Plates for Reversed Phase TLC

Coat 20x20 cm glass plates of 0.5 mm thick layers with Kieseghur-G (Merck) and water (1: 2) slurry. Air-dry the plates and then activate at 110°C for an hour. Cool the plates to room temp.in a desiccator. For treatment of TLC plates with liquid paraffin, carefully dip the plate holding horizontally for a few seconds in a tray containing 5 per cent liquid paraffin

Solution in petroleum ether as described above. Air-dry the plates. Spot the sterols in chloroform isolated from unsaponifiable matter by a preliminary separation on silica gel-G thin layer chromatography on paraffin treated plates along with standard cholesterol. Develop the plate using the solvent system of acetone: water (4: 1) which was earlier equilibrated with paraffin. After the solvent front has ascended to a height of 15cm remove the plate and air-dry.Spray with p-Anisaldehyde reagent (1.5 g. p-anisaldehyde and 1.5 ml conc. sulphuric acid in 27 ml ethyl alcohol) followed by heating at110°C for 5 min. The sterol spots appear as blue spots on pale pink background.Cholesterol appears at Rf 0.48 distinctly separated from other closely related sterols.

Source

1. *Manual Methods of Analysis of Adulterants and Contaminants in Foods. I.C.M.R (1990) Page 7).*
2. *Manual of Methods of analysis of Foods (Oils and Fats.), Lab Manual -2, FSSAI, New Delhi.*

III. Test for Presence of Animal Body Fat in Vegetable Fat Based on the Presence of Unusual Fatty Acids in Animal Fats by Gas Liquid Chromatography

Preparation of Fatty Acid Methylesters

Take 30 to 50mg of melted fat (1 drop)in a glass-stoppered test tube and add1ml of dichloromethane/benzene followed by 2ml of 1 per cent sodium methoxide solution (1g sodium dissolved in 100ml of anhydrous methanol).Hold the testtube at 60°C for 10 min. Cool and add 0.1m lof glacial acetic acid followed by 5ml of distilled water and 5ml petroleum ether (40°-60°C).Mix the contents. Allow the layers to separate. Take out about 2 ml of the upper layer containing the methylesters in a small tube and concentrate it by passing nitrogen gas before injecting to gas chromatograph.

Gas Chromatograph

The instrument fitted with flame ionisation detector and stainless steel column of 10 ft packed with15 per cent diethylene glycol succinate on chromosorb W (80-100 mesh), or any other intermediate polar stationary phase column. Maintain the column temperature at 185°C, flow rate of carrier gas nitrogen at 2.8kg/cm^2 (25ml/min) and chart speed at 1cm/min.

The fatty acid composition of animal fat (beef tallow) and Vanaspati containing animal fat show the presence of odd chain fatty acids and branched chain fatty acids namely C15: 0, C15: 1, C17: 0 and C17: 1. These fatty acids are absent in vegetable fats.On the basis of this fact it is possible to detect the presence of animal fat (beef tallow) in vegetable fats.

Source

1. *Manual Methods of Analysis for Adulterants and Contaminants in Foods. I.C.M.R (1990) Page 8.*
2. *Manual of Methods of analysis of Foods (Oils and Fats.), Lab Manual -2, FSSAI, New Delhi.*

3. Column Chromatography [1,2,3]

It is a separation technique in which the stationary bed is within a tube. The particles of the solid stationary phase or the support coated with a liquid stationary phase may fill the whole inside volume of the tube (packed column) or be concentrated on or along the inside tube wall leaving an open, unrestricted path for the mobile phase in the middle part of the tube (open tubular column). Differences in rates of movement through the medium are calculated to different retention times of the sample [2,3].

The modified version of column chromatography called **flash column chromatography** (flash) [9,10] was introduced by W. C. Still in1978 [4]. This process is very similar to the traditional column chromatography, except that here the solvent is driven through the column by applying positive pressure and because of this it allows most separations to be performed in less than 20 minutes, with improved separations compared to the old method.

Modern flash chromatography systems are available in the market as pre-packed plastic cartridges and the solvent is pumped through the cartridge. Systems may also be linked with detectors and fraction collectors providing automation. The introduction of gradient pumps resulted in quicker separations and less solvent usage [5,6].

Source

1. *Wikipedia.*
2. *Ettre, L. S. (1993). "Nomenclature for chromatography (IUPAC Recommendations 1993)". Pure and Applied Chemistry. 65 (4).*
3. *T, Manish. "How does column chromatography work?". BrightMags. Retrieved 7 April 2017.*
4. *Still, W. C., Kahn, M., Mitra, A., J. Org. Chem.1978, 43 (14), 2923-2925.*
5. *Still, W. C.; Kahn, M.; Mitra, A. (1978). "Rapid chromatographic technique for preparative separations with moderate resolution" (PDF). J. Org. Chem. 43 (14): 2923–2925.*
6. *Harwood, Laurence M.; Moody, Christopher J. (1989). Experimental organic chemistry: Principles and Practice (Illustrated ed.). WileyBlackwell. pp. 180–185. ISBN 978-0-632-02017-1.*

4. Gas Chromatography (GC)

It is widely used in analytical chemistry and is also known as **Gas-Liquid chromatography, (GLC).** Gas chromatography (GC) is a common type of chromatography used in analytical chemistry for separating and analyzing compounds that can be vaporized without decomposition. It is a separation technique in which the mobile phase is a gas. It is always carried out in a column, which is typically "packed" or "capillary" and it is based on a partition equilibrium of analyte between a solid stationary phase (often a liquid silicone-based material) and a mobile gas (most often Helium). The stationary phase is adhered to the inside of a small-diameter glass tube (a capillary column) or a solid matrix inside a larger metal tube (a packed column). The components that are volatile or could be made volatile by heating or derivatization, can be analysed by GC [1].

In gas chromatography, the eluent is an inert gas, often helium, hydrogen or nitrogen. The eluent actually has little effect on the separation process, which is governed more by the volatility of each sample component and its interaction with the stationary phase. Stationary phases are either solids or liquids and are contained in columns which range in internal diameter from 100 micrometers to 4 mm. The chromatographic system consists of three essential elements: an injection system, a temperature controlled column, and a detector [2].

Types of Gas Chromatography [2]

Chromatographic stationary phases may be either solid or liquid.

1. Gas-Solid Chromatography (Stationary Phase-Solid)

When a solid stationary phase is used, adsorption is the retention mechanism, and the technique is called gas solid chromatography. It is used mainly for low molecular weight volatile organic compounds and gases such as CO, CO_2 or SO_2 or H_2S.

2. Gas-Liquid Chromatography (Stationary phase liquid that is fixed or immobilized on a certain support material)

The stationary phase may also be a viscous liquid coated or chemically bonded to the surface of a packing material, or on the inside wall of a open tubular capillary column. This is called gas liquid chromatography (GLC). The mechanism of retention in a liquid stationary phase involves partitioning of the analyte between the gas and the stationary phase. Liquid phases must have good solvent characteristics, should be inert and stable at relatively high temperatures.

Some common Stationary phases for gas liquid chromatography, are given in following table:

Sl.No.	*Stationary Phase*	*Common Trade Name*	*Max. Temp. ºC*	*Common Application*
1	Poly dimethyl siloxane	OV-1,SE-30	350	General purpose, non polar phase, hydrocarbons, poly nuclear aromatics, drugs, steroids, PCBs.
2	Poly(phenyl methyl dimethyl) siloxane(10 per cent phenyl)	OV-3,SE-52	350	Fatty acid methyl esters, alkaloids, drugs, halogenated compounds.
3	Poly(phenyl methyl) siloxane (50 per cent phenyl)	OV-17	250	Drugs, steroids, pesticides, glycols
4	Poly (trifluoro propyl di-methyl) siloxane	OV-210	200	Chlorinated aromatics, nitro aromatics, alkyl substituted benzenes.
5	Poly ethylene glycol	Carbowax20M	250	Free acids, alcohols, ethers, essential oils, glycols.
6	Poly (dicynoallyl dimethyl) siloxane	OV-275	240	Poly unsaturated fatty acids, rosin acids, free acids, alcohols

Source

1. *Wikipedia.*
2. *F: \Instrumental Analysis\Chromatography.htm.*

Working of Gas Chromatography

It involves the following steps:

(i) Injection of sample mixture into a column packed with certain materials (stationary phase).

(ii) Passing a fluid/gas (the mobile phase) through the column.

(iii) Components of the mixture held strongly or weakly to the stationary phase.

(iv) Mobile phase pass to other end of the column taking away the weakly bound component first to the outlet of the column. Thus, the components will be separated according to the boiling point, molecular weight, polarity, ability to form hydrogen bonding with the stationary phase. The components will appear at the end of the column at different times, tR, the retention time.

Retention Volume, **VR**-The volume of the carrier gas which succeed to separate the component

$$\underbrace{V_R = t_R \cdot F}_{\text{retained}}$$

F = average volumetric flow rate (mL/min).

F can be estimated by measuring flow rate exiting the column using soap bubble meter (some gases dissolving in soap solution).

- often scales with vapor pressure (constant polarity analytes).

Sample Preparation [1]

The method of sample preparation involves following two important steps:

i. Extraction

Samples for GC are organic compounds with sufficient volatility and sufficient thermal stability. In practice, some compounds which do not fulfill the criteria above can be made to fulfill the criteria by mean of chemical reaction thus making it possible for these compounds to be analysed by GC.

ii. Derivatization

Derivatization is the process of chemically modifying a compound to produce a new compound which has properties that are suitable for analysis using a GC. The main reason for derivatizing is to impart volatility to otherwise nonvolatile compounds. To permit analysis of compounds not directly amenable to analysis due to, for example, inadequate volatility or stability and to improve chromatographic behavior or detectability. The low volatility may result from the size of the molecule and the resultant large dispersion forces holding the molecule together. Smaller molecules may have a low volatility due to the strong intermolecular attractions between polar groups. In the latter case, masking the polar groups by derivatization can yield dramatic increases in volatility. The following types of dervatisation processes are being used:

(i) Silylation

(ii) Alkylation

(iii) Acylation.

Source

1. *K. P. Indumathi, Rajan Sharma and Jyotika, GC-MS Application in Food Analysis, Dairy Chemistry Division, Compendium of Lectures on Advanced Techniques and Novel Approaches for Quality and Safety Evaluation of Dairy Foods- 2015, NDRI, Karnal.*

Instrumentation for Gas Chromatography

It includes the followings:

i. Carrier Gas

He (common), N_2, H_2, F=25-150 mL/min packed column, F=1-25 mL/min open tubular column,

ii. Injection Devices [1]

For good chromatography, it is essential to have an injector which can place a small, narrow, and reproducible band of sample on the head of the column. The following injecting devices are being used:

a. A Syringe

It is the traditional injection device in GC. Sample in liquid form is injected using microsyringe. A heated injection port, equipped with a soft polymeric septum, is located at the head of the column. The injector must be kept at some high temperature to change into vapour phase for easy analysis. The sample is injected through the septum into the heated carrier gas stream, and vaporizes. The carrier gas flows through the injection port and sweeps the sample onto the column. A glass liner is often incorporated into the injection port, so that it may be removed and cleaned or replaced if samples leave a non-volatile residue. The microsyringes of following sizes are used:

(i) 1-20 µL packed column

(ii) 10-3 µL capillary column.

b. A Splitting Injector Port

Since a high efficiency column may require a sample of only a few nanograms, a syringe may not produce adequately reproducible samples. A splitting injector port may be used. In this case the sample is injected into a chamber where it is mixed with a volume of carrier gas. Then a small part of this mixture is allowed to flow into the column, and the rest is vented.

c. A Gas Sampling Valve

For injection of gas samples, a gas sampling valve is often used to inject a measured portion into the column. This valve contains a loop of known volume which is filled in the 'load position' by allowing the sample gas to flow through it. When the valve is turned to the 'inject position', the carrier gas is diverted through the loop, and the sample is carried onto the column.

d. Others

A thermal desorber is employed for transferring samples from an adsorbent trap into the column. A purge-and-trap apparatus is used for stripping volatile compounds from water or sludge samples, and injecting them into the column.

iii. Column [1,2]

Column is the heart of the chromatography. Gas chromatography column is maintained inside an oven and the temperature of the column can be varied as per the need of the analyte to be analyzed by software enabled programme. Based on the nature of the compound to be separated the column can be of i-Polar, ii-Mid Polar, iii -Non polar and of varying length [3]. There is a vast selection of GC columns available. Many of them are so slightly different from others that they are readily interchangeable. The columns differ in physical dimensions. Large diameter packed columns, having a larger volume of stationary phase, are able to accommodate large samples. The smallest columns, open tubular capillary columns, have the highest efficiency, but are limited to tiny partcles.2-50 m coiled stainless steel/glass/Teflon are used.

The hundreds of stationary phases were made available for gas chromatography. To characterize these in a quantitative fashion, Rohrschneider developed a series of constants to describe the selectivity of a column toward different types of sample. A suitable phase for a particular sample can be selected on a rational basis by using these constants. McReynolds has extended and refined this work and **McReynolds constants** for stationary phases are listed.

The McReynolds constants are based on the assumption that different molecular interactions (dispersion, orientation, induced dipole, and donor-acceptor complexation) between functional groups on the column material and those of the sample are additive. A series of probe compounds was chosen to represent each type of interaction, and the retention time for each probe is determined on the column in question. Since retention times change with phase loading, temperature, and flow, a retention index for each probe is determined by injecting the probe compound along with a set of normal hydrocarbons which will bracket the probe compound's retention time. The retention index is calculated from the following equation:

$$I = 100z + 100\frac{\log(t'_{R(x)} - t'_{R(z)})}{\log(t'_{R(z+1)} + t'_{R(z)})}$$

where,

t'R is an adjusted retention time. The unknown is indicated as x, and z is the number of carbons in the normal hydrocarbon eluting just before the sample. The hydrocarbon, z+1, is that which elutes just after the sample. This index is not entirely independent of column conditions, but is still useful. **The McReynolds constant is expressed as the retention index of each probe compound on the tested column minus that on squalene** and these constants are useful in comparing columns, and in selecting a new column to improve a particular separation.

Types of Columns [1]

The details of some columns are given below:

a. Packed Columns

These columns vary in internal diameter, with most being constructed from 1/4 or 1/8 inch OD tubing. They also vary in the particle size of the packing, and the amount of stationary phase coated on the support. The loading is determined by the mass of stationary phase coated on a given mass of support. Loading ranges from a few per cent to 20 or 30 per cent. Stationary phases are usually coated onto an inert support material, which should be of uniform size and good mechanical strength. The efficiency of the finished column depends on both the particle size and uniformity of the size. Smaller particles give more efficient columns. Usually column packings in the 60-80 mesh or 80-100 mesh ranges are used. These are often made from treated diatomaceous earth, a silica based material.

Stainless steel or glass tubing is used to form the columns. Stainless steel is considerably easier to handle. Glass columns are usually deactivated by being silanized, and have a lower surface activity, so may be needed for especially sensitive samples.

For gas-solid chromatography, several types of solid column packings are available. These may be porous polymeric materials, silica, or carbon based solids. Solid phases may be more durable, and able to stand up to some fairly aggressive samples, since they have no surface coating.

b. Open Tubular Columns

Open tubular columns are often used for complex mixtures, where large numbers of components must be separated. These are made of fused silica tubing, coated with a thin film of stationary phase on the interior, and covered with a polyimide coating on the outside surface. Recently, open tubular columns are also being made with silica lined stainless steel tubing.

Source

1. *F: \Instrumental Analysis\Chromatography.htm.*
2. *Wikipedia.*
3. *K. P. Indumathi, Rajan Sharma and Jyotika, GC-MS Application in Food Analysis, Dairy Chemistry Division, Compendium of Lectures on Advanced Techniques and Novel Approaches for Quality and Safety Evaluation of Dairy Foods- 2015, NDRI, Karnal.*

iii. Oven

0-400°C ~ average boiling point of sample accurate to <1°C. A column has a minimum and maximum use temperature, which depends mostly upon the stationary phase. The minimum temperature is usually the freezing point of the stationary phase, since the solid has quite different properties from the liquid, and does not always partition the sample adequately. The maximum temperature is usually that at which the column bleed becomes intolerable. This is usually due to thermal breakdown of the stationary phase, or to the boiling off of the lower molecular weight fraction of the phase.

iv. Detectors

The eluent from the column is directed to one or more detectors. These produce signals which are proportional to either the amount of sample present in the detector at any moment, or the concentration of sample in the detector. The detector signal is usually displayed as a plot of signal magnitude versus time, giving the classic chromatogram FID, TCD, ECD, (MS).

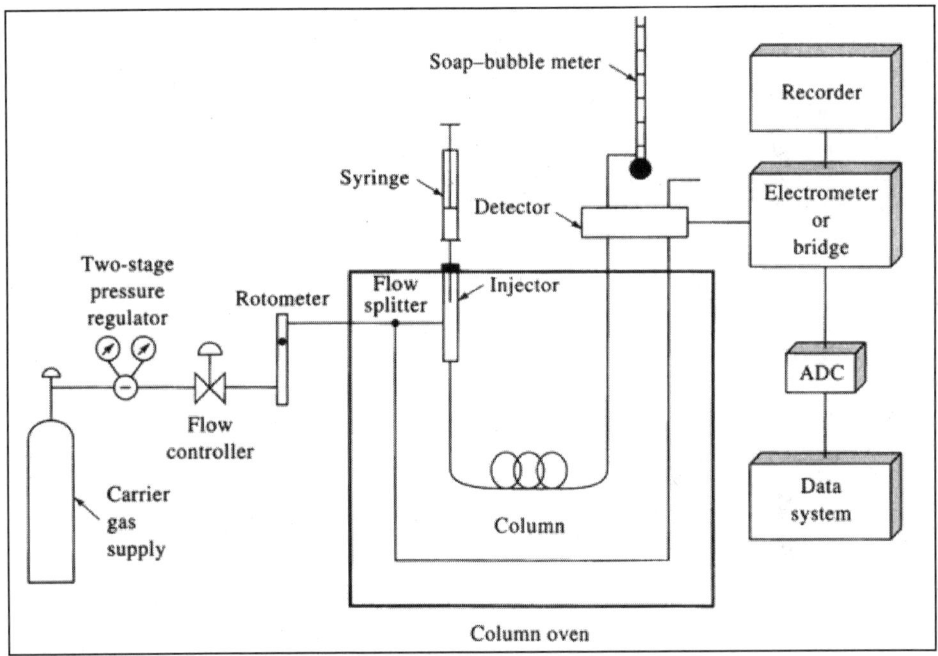

Outline of Gas Chromatograph.

GC Detectors [1,2,3]

For reasonable GC analysis, the detector need to be:

(i) Sensitive (10-8-10-15 g solute/s).

(ii) Operate at high T (0-400°C)

(iii) Stable and reproducible

(iv) Linear response.

For better perfortmance, detectors are desired to have the followings:

(i) Wide dynamic range.

(ii) Fast response.

(iii) Simple (reliable).

(iv) Nondestructive.

(iv) Uniform response to all analytes.

There are different types of detectors *e.g.* Thermal Conductivity Detector -TCD, Flame Ionization Detector -FID, Electron Capture Detector-ECD, Photoionization Detector-PID, Electrolytic Conductivity Detector-ElCD, Flame Photometric Detector-FPD, Thermionic or Nitrogen-Phosphorous Detector-NPD, Mass Selective Detector -MSD are being used in Gas Chromatiography. Details of some common detectors are given below:

Flame Ionization Detector (FID)

The separated components are allowed to enter a flame (Hydrogen/air) to be burned and changed into radicals which will be ionized by the flame. The ions will be detected by electrical conduction to give signal that a component is separated [3]. The response is roughly proportional to the number of carbon atoms in the flame at any time, although certain substituent atoms, such as chlorine, reduce the response. It is nearly universally sensitive to organic compounds, and shows good sensitivity and excellent linearity. The FID has following chracteristics/advantages [1,3]:

(i) Sensitive (10–13 g/s).

(ii) The detector has a very wide linear (Dymanic) range, over about 7 orders of magnitude (10^7).

(iii) Weakly sensitive to carbonyl, amine, alcohol, amine groups.

(iv) Not sensitive to non-combustibles gases, H_2O, CO_2, SO_2, NO_x, inorganic gases.

(v) The detector a low dead volume of about 1 ml, and is relatively noise free and easy to operate [1].

Disadvantage [1]

Destructive - Its major disadvantage is that it destroys the sample, so it cannot be passed on to another detector. Also, the fact that it requires both compressed air and hydrogen, as well as carrier gas, can be an inconvenience, especially when instrumental portability is an issue.

Thermal Conductivity Detector (TCD)

The TCD is a truly universal detector. It consists of a heated sensor in a thermostated chamber, through which the effluent flows. Helium is usually used as a carrier gas, as it has the highest thermal conductivity of any gas, except for hydrogen [1]. The conduction of heat from certain filament by carrier gas (mobile phase) He N_2 or H_2, will be lowered when the organics reached the detector because the thermal conductivity of them is lower than that of the carrier gas. Thus, the organics will cause temperature rise in filament. The TCD is characterized by following:

(i) Wide dynamic range (105).

(ii) Nondestructive.

(iii) Insensitive (10-8 g/s) - non-uniform [3].

Electron Capture Detector (ECD) [1]

The response of ECD depends on the electron-capturing properties of the sample. The detector is highly sensitive to the presence of electron capturing substituents, such as halogens, peroxides, and nitro- groups, on the sample molecules. ECD consists of a chamber containing a ß-emitting foil, usually nickel-65. This radioactive source emits electrons which ionize the carrier gas, and form a small current between the electrodes in the chamber. When electron- capturing species are present, they reduce this current. The reduction is detected by the electronics and measured. The detector response varies widely, depending on the electronegativity of the species being detected. Since ECD's contain a radioactive isotope, they are subject to governmental regulation. The sensitivity of the ECD toward halogenated hydrocarbons and many pesticides is extremely high. The response to a particular compound is usually quite temperature sensitive, and the detector's linearity varies with conditions and analyte.

Source

1. *F: \Instrumental Analysis\Chromatography.htm.*
2. *Wikipedia.*
3. *Technews Issue No.74 (May-June 2008), NDDB.*

International Standards for Analysis of Milk Products by GC/GLC Methods [1,3]

(a) Some important international standards that have been specified for GC/GLC analysis of specific components of milk and milk products are given in the following table:

Sl.No.	*Component*	*Product*	*International Standard*
1	Vegetable fat	Milk fat	IDF 054:1970
2	Sterol composition	Anhydrous milk fat	ISO 12078/IDF 159:2006 and ISO 18252/ IDF 200:2006
3	Fatty acid composition	Milk fat	ISO 15885/IDF 184:2002
4	Poly chlorinated biphenyls	Milk and milk products	IDF 130A:1991
5	Pesticides (organochlorine compounds)	Milk and milk products	IDF 144:1990 and 75C:199

(b) In addition to above GC is also being used for the determination of pesticide residues in milk and milk products. Some examples are given below [2,3]:

Sl.No.	Pesticide residues	Methods
1	Halogenated Hydrocarbon [Organochlorine] pesticides (Lindane, Chlordane, Aldrin, Heptachlor, DDT, Dieldrin).	GC with Electron Capture detector or Electrolyte Conductivity Detector.
2	Organophosphates (Chlorpyrifos, Fenthion, Parathion, Malathion).	GC with Electron Capture Detector.
3	Pyrethrins (Cypermethrin, Permethrin, Allethrin, Fenvalerate).	GC with Electron Capture Detector. HPLC also used.

Source

1. *Technews Issue No.74 (May-June 2008), NDDB.*
2. *Technews Issue No.47 (Nov- Dec 2003), NDDB.*
3. *Jeon, I.J. and Ikins, W.G. (1995). Analyzing Food for Nutrition Labeling and Hazardous Contaminants. Marcel Dekker, Inc. USA, pp. 333-362.*

Application in Food Industry

Typical uses of GC include testing the purity of a particular substance, or separating the different components of a mixture (the relative amounts of such components can also be determined). In some situations, GC may help in identifying a compound. In preparative chromatography, GC can be used to prepare pure compounds from a mixture. Some examples of GC applications in food and dairy products analysis, are given below:

I. Determination of Total Residual Hexane in Solvent Extracted Oil Seed Flours

(Soya Flour, Groundnut Flour, Sesame Flour, Coconut Flour, Cottonseed Flour)- This is a method for the determination of total amount of volatile hydrocarbons, referred to as Hexane remaining in oilseed residues after extraction with hydrocarbon based solvents.

Principle

Desorption of Hexane by heating at 110°C with water in a closed vessel, and determination of the Hexane in the headspace by gas chromatography using capillary or packed columns and expressing the results as Hexane.

Reagents and Materials

(i) Technical n-Hexane or light petroleum, with a composition similar to that used in the industrial extraction of oilseeds, failing that n–Hexane.

(ii) Carrier Gas: - Hydrogen or Nitrogen, Helium *etc.*, dry and containing less than 10 mg/Kg of Oxygen.

(iii) Auxillary gases-a- Hydrogen 99.9 per cent pure, containing no organic impurities, b-Air containing no organic impurities.

Apparatus

(i) Gas Chromatograph with flame ionization detector equipped with glass capillary column approx 30 metres long and 0.3 mm in diameter coated with methyl polysiloxane film of 0.2 µm thickness or failing this a packed column of at least 1.7 metre length with 2-4 mm internal diameter packed with acid washed diatomaceous earth of particle size of 150-180 µm and coated with methyl polysiloxane. If a capillary column is used the apparatus shall have a 1/100 input divider.

(ii) Electric oven, capable of being maintained at 110°C.

(iii) Gas Syringe, graduated capacity 1 ml, preferably with a valve.

(iv) Penicillin type flasks of capacity 50- 60 ml all with the same volume to within 2 per cent.

(v) Septa, inert to solvents, of approximately 3 mm thickness capable of producing a hermetic seal after crimping.

(vi) Metallic foil caps of Aluminum.

(vii) Crimping pliers.

(viii) Liquid syringe 10 µm capacity.

Sampling and Sample Storage

It is essential that the loss of hexane from sample be prevented. The sample shall fill a completely sealed container (preferably a crimped metal box and shall be stored at – 20°C below in a deep freezer. Plastic containers shall not be used. The determination of residual Hexane shall be carried out as soon as the container has been brought to room temperature and opened.

Procedure

Weigh to the nearest 0.1 gm, 5 gm of the laboratory sample into a flask (iv). Add 2.5 ml water, seal the flask with a septum, cover with a foil cap (vi) and crimp with the pliers. All these operations should be performed rapidly. Place the flask in the oven maintained at 110°C for 90 minutes, remove the flask from the oven and let it cool for 2 minutes. Agitate by inverting. It is important to leave the flasks in the oven for the same length of time for each sample. Set the oven temperature at 40°C, injector and detector temperature at 120°C, Carrier gas pressure at 0.3 bar (30 kPa).Using the gas syringe previously heated to 50-60°C take exactly 0.5 ml of gaseous phase and inject quickly into the chromatograph.

Calibration

Three points with 2.5, 5.0, 10.0 µl of solvent are usually sufficient to construct the calibration curve, they correspond to 264, 660, 1320 mg/kg of Hexane if the test portion is 5 gm.

Prepare a calibration series using flasks of the same capacity as used for the determination. Add to the flasks 6 ml of water followed immediately by various quantities of n–Hexane measured accurately with the help of the syringe. Seal each flask with the septum, cover with the foil cap and crimp with the plier place the various flasks for the establishment of one calibration graph in the oven for 15 minutes at 110°C. At the end of this time remove the flasks from the oven and leave to cool for 2 minutes. With the gas syringe heated between 50 –60°C take exactly 0.5 ml of headspace and inject quickly into the chromatograph. Carry out two determinations on the sample.

Expression of Results

Construct the calibration graph by plotting the area of the solvent peak as a function of the mass of the solvent introduced into the flask (1 µl corresponding to 660 µg). Determine the sum of the peak areas of Hexane and various Hydrocarbons which usually make up the technical solvent. Do not include peaks due to oxidation products if present in significant amounts but report calibration these separately. Read off from the graph the mass m_1 in microgram of Hexane present in the flask.

The total residual Hexane in the residue expressed in microgram of hexane per kilogram is equal to m_1/m_0.

where,

m_0 = the mass in gms of the test portion and m_1 = the mass in microgram of solvent present in the flask. Take as the result the arithmetic mean of two determinations.

Source

1. *I.S 12983: 1990/I.S.O 8892: 1987, Oilseed Residues -Determination of Total Residual Hexane.*
2. *Manual of Methods of analysis of Foods (Cereals and Cereal products), Lab Manual-3, FSSAI, New Delhi.*

II. Determination of Hexane Residues in Oils and Fats

The residual hexane content is the quantity of volatile hydrocarbons remaining in the fats and oils following processing involving the use of solvents. The volatile hydrocarbons are desorbed by heating the sample at 80°C in a closed vessel after addition of an internal standard. After determination of a calibration factor, hydrocarbons in the headspace are determined by gas chromatography using packed or capillary columns. Results are expressed as hexane in mg/kilogram (mg/kg, or ppm). The method is applicable to the determination of, free volatile hydrocarbons" expressed in terms of hexane remaining in animal and vegetable fats and oils after extraction with hydrocarbon based solvents. It is suitable for determination of quantities of hexane between 10 and 1500 mg/kg in fats and oils.

Apparatus

1. **Gas Chromatograph** having **a-**Thermostatic column capable of maintaining the desired column temperature within ±1°C, **b-**Sample inlet system, separately thermostated which can be maintained at a minimum temperature of 100°C. If a capillary column is used, the inlet system must be capable of a 1/100 split injection. For serial analysis a headspace gas

chromatograph with automatic sample injection and tempering bath is satisfactory, **c**-Flame ionization detector-can be separately thermostated and maintained at a min.of 100°C.

2. **Recorder**-An electronic recorder (high precision) is required, if a recorder trace is to be used for calculating the composition of the samples analysed,
3. **Electronic Integrator** (preferred) which permits rapid and accurate calculations.
4. **Chromatographic Column**-Either packed or capillary column with the following minimum requirements- a-Packed Column-stainless steel or glass, approx. 2 metres long and 1/8 inch internal diameter with acid washed and silanised diatomaceous earth, 150-180 mu particle size (80-100 meshChromosorbWAW is suitable), stationery phase–squalene consisting of 10 per cent of packing, b-Capillary column-glass or fused silica approx.30 metres long and 0.3 mm internal diameter. Stationery phase-Methylpolysiloxane (film thickness 0.2mu).
5. **Syringe**-1µl,10µl,1000µl capacity, gas tight.
6. **Septum vial**-20ml capacity.
7. **Septa and Aluminium caps** suitable for septum vials together with crimping pliers. The septa must be resistant to oils and solvents (butyl rubber or red rubber is recommended).
8. **Tongs** suitable for holding septum vials.
9. **Heating bath** with clamps for holding septum vials, thermostatically regulated and capable of maintaining a temperature of 80°C. For continuous operation glycerol is recommended as heating liquid.
10. **Shaking machine.**

Reagents

1. **Gases**-a-Carrier-Helium (preferred for better resolution) or Nitrogen 99.99 per cent pure, dried and containing a maximum of 10 mg O_2/kg, b-Flame Ionization Detector-Hydrogen, minimum purity 99.95 per cent, Air or Oxygen, dry, hydrocarbon free (less than 2ppm hydrocarbon equivalent to CH_4).
2. **Technical Hexane or light Petroleum** with a composition similar to that used in industrial extraction or failing these n–hexane. For calibration, technical extraction hexane is preferred.
3. n- Heptane (internal standard) analytical reagent grade.
4. Vegetable Oil-, solvent free, freshly refined and deodorized. The oil is to be used for calibration and should be of a similar nature as the sample. It should be free from extraction solvent (less than 0. 01 per cent).

Sampling and Sample Storage

It is essential that loss of solvent from the sample be prevented. The Laboratory sample should be in a completely sealed condition and stored at 4°C. Plastic containers should not be used. Sample analysis should be carried out immediately when the sample container is opened.

GC Operating Conditions

Carrier gas flow depends on the carrier gas and the type of column being used for analysis and should be optimized accordingly.

The flow of hydrogen and air or oxygen to the FID should be optimized according to the manufacturer's recommendation.Injector and detector temperatures should be set at about120 degree C. The column should be maintained at 40°C.

Procedure

1-Determination of the calibration factor-Weigh to the nearest 0.01gm,5 gm of solvent free vegetable oil (reagent4) into each of the 7 septum vials. Seal each vial with a septum and cap.By means of a syringe add technical Hexane to 6 of the seven vials (the vial with no added solvent is the blank) according to the following table:

µl/5gm	0.5	1	2	4	7	10
mg/100gm	67	134	268	536	938	1340

One vial remains without the addition of solvent.

If n –hexane is used for calibration the following table applies

µl/5gm	0.5	1	2	4	7	10
mg/100gm	66	132	264	528	924	1320

Shake the 6 vials containing the solvent in the shaking machine vigorously for 1 hour. Using the syringe add 5 µl of internal standard (reagent 3) to each of the 7 vials. Successively immerse the vials upto the neck in the heating bath at 80°C at intervals of approx 15 minutes. This time interval depends on the duration of the GC analysis which is complete on the elution of the internal standard (n-heptane). The samples must be placed in the heating at intervals such that each sample is tempered for exactly 60 minutes.

Warm the gas tight syringe to 60°C. After tempering at 80°C for exactly 60 minutes and without removing the vial from the heating bath, use the gas tight syringe and withdraw through the septum 1000 µl (1 ml) of the head space above the oil. Inject immediately into the gas chromatograph. For each of the vial containing added solvent a calibration factor F may be determined by the formula.

$$F = \frac{C_S \times A_1}{(A_H - A_B - A_1) \times C_1}$$

where,

A_H: Total peak area of solvent hydrocarbons including the area of internal standard present in the spiked oil. For identification purposes a typical chromatogram of solvent composition should be obtained.Hydrocarbons which usually make up the technical hexane are 2 Methylpentane, 3 Methyl pentane. Methylcyclopentane, cyclohexane *etc.* Donot include peaks due to oxidation products which may be present in significant amounts.

A_B: Peak area of the solvent hydrocarbons present in the oil to which solvent has not been added (blank) less the peak area of the internal standard.

A_1: Peak area corresponding to the internal standard in the spiked samples.

C_1: Quantityof the internals tandard added expressed in mg/kg of the oil.

C_S: Quantity of technicall hexane added to the oil present in the vial expressed in mg/kg of the oil.

Express the results to the third decimal place. Calibration factors of the six standards should be approx the same. The mean calibration factor should be 0.45 if n–heptanes is used and 0.57 if cyclohexane is used.

The factor (F) so evaluated can be used for determining vial quantities of hexane less than 60mg/kg.If the value of F found for the vial containing 0.5µL of hexane is significantly below the mean value, this deviation is probably due to difficulty in introducing exactly 0.5µL and this determination must be either eliminated or repeated. For quantities of hexane between 10 and 20 mg/kg it is better to prepare calibration standards by adding 2µl of internal standard in stead of 0.5 µl.

Sample Analysis

Weigh to the nearest 0.01gm, 5gm of the test sample into a septum vial as quickly as possible and close immediately with a septum and cap. Using a syringe add through the septum exactly 5µl of the internal standard. Shake vigorously by hand for about 1 minute and then immerse the vial upto the neck in the heating bath at 80 degree C for exactly 60 minutes.Warm the gas tight syringe to 60°C. After tempering at 80°C for exactly 60 minutes use the gas tight syringe and take from the vial without removing it from the bath 1000µl (1ml) of the headspace above the sample.Immediately inject into the gas chromatograph. Carry out two determinations in rapid succession on each sample.

Calculation-The residual solvent expressed in mg/kg (ppm) is given by the formula:

$$W = \frac{(A_H - A_1) \text{ X F X } C_1}{A_1}$$

where,

A_H: Total peak area of solvent hydrocarbons including the area of internal standard.Hydrocarbons which usually make up the technical solvents are 2 methylpentane,3methylpentane, methylcyclopentane, cyclohexaneetc. Donot include peaks due to the oxidation products.Some of these products may be present in significant amount.

A_1: Peak area corresponding to internal standard in the sample.

C_1: Quantity of the internal standard added in mg/kg.

(Note: -For an addition of 5µL of heptanes/5gm of sample C_1 = 680mg/kg and C_1 = 750mg/kg if cyclohexane isused) and F = Calibration factor obtained in procedure.Report as the final result the mean of the results of two determinations.

Source

1. *A.O.C.S (1989) Official Method Ca3b.-87.*
2. *Manual of Methods of Analysis of Foods (Oils and Fat), Lab Manual-2, FSSAI, NewDelhi.*

III. Determination of Fatty Acid Composition of Oils and Fats by Gas Liquid Chromatography

The methylesters are formed using boron trifloride or methanol and alkali and separated by gas–liquid chromatography using a flame ionization detector. The pattern of methylesters can be compared with authentic oils for identification.

Apparatus

1. **Gas liquid chromatograph** with the following characteristics:
   - *(a)* Injection system heated to a temperature of 20-50°C higher than the column.
   - *(b)* Oven–capable of heating the column to atleast 220°C and maintaining the temperature to within ±1°C. If temperature programming is to be employed, twin columns are recommended.
   - *(c)* Packed column- may be glass or stainless steel. However glass is preferred as steel may decompose polyunsaturated fatty acids having more than 3 double bonds. Some successful column packings with column length, internal diameter and operating temperature are as follows:

     i-12-15 per cent ethylene glycolsuccinate on 100/120 mesh gas chromP (2mx4mm, at180 °C), ii-2-10 per cent Apizon–Lon 80/100mesh ChromosorbW or Celite (2m x4 mmat 220 degree C), iii-10 per cent Butan-1-4 diolsuccinate on 80/100mesh Chromo-sorbW or celite (2mx 4mm at 175 °C), iv-3 per cent SE–30 on 100/120 mesh Chromosorb–G silanised (2mx 3mm at190 degree C), Condition the newly prepared column by disconnecting the detector and heating the column in the oven to the normal operating temperature for 16 hours while running the carrier gas at a rate of 20–60ml/min amd v-Detector-Flame ionization detector-capable of being heated to a temperature above that of the column.
2. **Syringe**–10µl graduated in1/10th of a microlitre.
3. **Recorder**–electronic with high precision with rate of response below1.5 second, width of paper 25cm, paper Speed 25-150cm/hr,4- Integrator or calculator for rapid and accurate calculations, 5-50 and100ml boiling flasks, 6-Reflux condenser,7-Graduated pipette-10ml,8-Test tubes with ground stoppers and 9-250ml Separating funnels.

Reagents

1. Carrier Gas-Inert gas (nitrogen, helium, argon) thoroughly dried and containing less than10mg/kg of oxygen.
2. Auxillary gas Hydrogen 99.9 per cent minimum purity. Free from organic impurities, air or oxygen.
3. Reference standards- a mixture of methylesters or methylesters of oils of known purity preferably similar to the fatty matter being analysed.
4. Methanolic Sodium hydroxide solution-approx 0.5N. Dissolve 2gm of Sod.Hydroxide in100ml methanol containing not more than 0.5 per cent m/m water.When the solution has to be stored for considerable time, a small amount of white ppt of Sod.Carbonate may be formed. This has no effecton the preparation of the methylesters.
5. Methanolic Solution of Borontrifloride-12-15 per cent m/m, 14 and 50 per cent solutions are commercially available. The metahnolic solution of boron trifloride should be stored in a refrigerator.
6. Heptane- Chromatographic quality.
7. Redistilled pet. Ether 40-60°C.
8. Anhydrous Sodium sulphate.
9. Saturated solution of Sodium chloride.
10. Methyl red-1gm/litre in 60 per cent alcohol.

Procedure

Prepare the methylesters of the fatty acids. The method using boron trifloride gives good results and is preferable to alternative methods which may be used when boron trifloride is not available. Because of the toxic character of borontrifloride various operations must be performed under a ventilated hood. All glassware must be washed with water immediately after use. If the oil or fatty acids include fatty acids containing more than 2 double bonds it is advisable to purge the air from the methanol and the flask by passing a stream of nitrogen in to the methanol for a few minutes.

Transfer about 350mg of clear oil to a 50 ml conical flask, add 6ml of 0.5 N methanolic sod. hydroxide solution, 7 ml of boron trifloride solution and a boiling chip. Fit the condenser to the flask. Boil under reflux until the droplets of oil disappear (5-10 minutes). Add the appropriate amount of boron trifloride solution with a bulb or automatic pipette through the top of the condenser.Continue boiling for 2 minutes. Add 2-5ml of heptane to the boiling mixture through the top of the condenser. Continue boiling for 1 minute.Withdraw the source of heat and then remove the condenser. Add a small amount of saturated Sod. Chloride solution to the flask in order to bring the level of liquid in to the neck of the flask. Transfer about 1 ml of the upper layer (Heptane solution) into a test tube with a ground glass neck and add a little anhydrous Sod. Sulphate to remove any trace of water. This solution will contain about 5-10 per cent of methylesters and may be injected directly into the column of gas liquid chromatograph.

Alternate Method for Preparation of Methylesters

Methylesters can also be prepared without the use of borontrifloride. This involves methylesterification of the fatty acids in an alkaline medium and is suitable for neutral oils and fats with an acid value less than 2.

Reagents

(i) Methanol containing not more than 0.5 per cent water.

(ii) Methanolic potassium hydroxide solution-approx1N. Dissolve 5.6gm Potassium hydroxide in100ml of methanol containing not more than 0.5 per cent m/m water (anhydrous methanol).

(iii) Heptane chromatographic quality.

(iv) Anhydrous Sod.Sulphate

(v) Nitrogen, containing not more than 0.5mg/Kg of oxygen.

Procedure

If the oil includes fatty acids containing more than 2 double bonds, it is advisable to purge the air from the methanol and the flask by passing stream of nitrogen into the methanol for a few minutes.

Transfer about 4 gm of clear sample oil into a 100ml round bottomed or conical flask. Add about 40ml of methanol, 0.5ml of methanolic Pot hydroxide solution and a boiling chip. Fit under a reflux condenser, stir and bring to boil. The solution should be come clear (5-10 minutes). Cool under running water and transfer the contents to a 125 ml separating funnel, rinsing the flask with 20ml of heptane. Add about 40 ml water, shake and allow to separate. The esters pass into the upper heptanes layer. Separate. Extract the aqueous layer again with 20ml heptane.

Combine the two extracts and wash them with several 20ml portions of water. Separate and dry the ester solution over anhydrous Sod. Sulphate.Filter through cotton wool in to a 50ml conical flask and evaporate solution to approx.20 ml on a water bath while passing a stream of nitrogen.

Determination of Fatty Acids

Programme GC to maintain column temperature of 185°C and detector temperature at 200°C. Inject 0.1-2 µl of 5-10 per cent of heptane solution of methylesters by piercing the septum of the inlet port. Withdraw needle and note formation of a small peak on the chart paper due to solvent making start reference point. It is possible to work with lower column temperature where the determination of acids below C_{12} is required or higher temperature when determining fatty acids above C_{20}.

It is also possible to employ temperature programming to take care of both situations. Analyse reference standard mixture of known composition in the same operating conditions as those employed for the sample and measure the retention distances or retention times for the common fattyesters. Identify the peaks for the sample from the graph. If an integerator is used obtain the figures from it.

Fattyacids appear on the chart in increasing number of carbon atoms and increasing unsaturation. Thus C_{16} appears before C_{18}, $C_{18:1}$ before $C_{18:2}$ and so on.

Source

1. *IUPAC2.301,2.302 (1979)/F. A.O Manuals of Food quality Control 14/8, Pages 274-281.*
2. *A.O. A.C17th edn, 2000 Official method 969.33 and 969.22 Fatty acids in oils and fats.Preparation of methylesters/Gas chromatographic method.*
3. *Manual of Methods of Analysis of Foods (Oils and Fat), Lab Manual-2, FSSAI, NewDelhi.*

IV. Detection of Foreign Fats in Ghee

The presence of foreign fat in ghee can be detected by Triglyceride Profiling using Low-Resolution Gas -Liquid- Chromatography (Fit a capillary column of 2.5 mts in the GC oven).Milk fat is an important component in the dairy industry, which plays a significant role in economic, nutrition, physical and chemical properties of milk and milk products. It involves the following steps:

a. Sample Preparation

Take 25mg of molten and filtered (Whatman 1or 4 filter) ghee in a 5 ml volumetric flask and make the volume to 5.0 ml with hexane to get a solution of 5mg/ml concentration and dissolve the contents. Use this solution for GC- injection.

b. Conditions of GLC analysis

(i) Injector condition- Temp of injector-350°C; Split ratio-1: 500

(ii) Carrier gas- Nitrogen; Total Flow 79 mL/min; Column Flow 1.49 mL/min; Purge flow 3mL/min.

(iii) Injection volume-1 µl.

(iv) Column-Capillary type, 5 per cent Phenyl/95 per cent Dimethyl Polysiloxane (BP5/RTx5 *etc.*); Length- 2.3 m, Inner Dia-0.25 mm, Film Thickness-0.25 µm; Max temp-335°C.

(v) Temperature programme for TAG profiling of ghee- Injection at 80°C and holding for 50 sec then increase the temperature to 190°C @ 50°C and hold for 1 min and finally increase the temperature to 335°C @ 6°C and hold for 6 min.

(vi) FID condition-Temperature 370°C.

(vii) **Other parameters-**Make up flow 30 mL/min, H2 flow 40 mL/min, Air flow-400 mL/min.

c. Triglyceride Standard

Prepare a solution of the standard in hexane to get the concentration of 5 mg/ml as in case of ghee. Use Supelco TAG mix (C24, C30, C36, C42, C48) from Sigma Aldrich as standard.

d. Results

The following TAG (per cent w/w) profile of fat sample is helpful in detection of foreign fat in ghee:

Triacyl Glycerol	*Cow Ghee*	*Buffalo Ghee*	*Lard*	*Triacyl Glycerol*	*Cow Ghee*	*Buffalo Ghee*	*Lard*
C24	0.17	0.019	-	C28	0.324	0.434	-
C32	1.582	1.879	-	C36	9.613	9.2133	-
C40	11.80	7.353	-	C44	6.315	7.927	0.0626
C48	8.366	7.60	1.900	C50	11.312	11.92	15.542
C52	11.044	11.84	59.420	C54	4.960	4.3	22.918

Source

1. *Vivek Sharma and Tanmay Hazra, Detection of Foreign Fats in Ghee -by Triglyceride Profiling using Low-Resolution Gas -Liquid- Chromatography, Dairy Chemistry Division, Compendium of Lectures on Advanced Techniques and Novel Approaches for Quality and Safety Evaluation of Dairy Foods-2015, NDRI, Karnal.*

Applications of different Chromatographic Techniques in Food Analysis

The following parameters in different Foods and other allied products are being tested by Chromatographic methods:

1. Determination of Aflatoxins (For ground nuts and Ground nut Products, oil seeds and Food grains)–CB Method

Apparatus

Stoppered Conical Flask, Measuring Cylinders-25, 50, 250 ml, Chromatography column -25 mm (i.d) X 300 mm length, High speed Blender, Funnel – 7.5 cm dia or Buchner Funnel with Whatman No1 filter paper, Wrist action shaker, Rotary evaporator, UV light Chamber, Micropipette -5 µl -100 µl, adjustable and Vials, Borosilicate-screw cap lined with foil or teflon.

Extraction

50 gm of powdered representative sample is taken in a 500 ml conical flask and 25ml water, 25 gm diatomaceous earth (Celite) and 250 ml chloroform are added to it. The flask is securely stoppered with masking tape and shaken on a wrist action shaker for 30 minutes to extract the toxin. Filter through fluted filter paper. If filteration is slow, transfer to a Buchner funnel pre-coated with about 5 mm layer of diatomaceous earth and filter using light vacuum. Collect first 50 ml of filtrate.

Column Chromatography

Place ball of glass wool loosely in bottom of 22 x 300 mm chromatographic column and add 5 gm of anhydrous Sod Sulphate to give base for silica gel. Add Chloroform until tube is about ½ full. Pour 10 gm of silica gel made into a slurry with chloroform. Wash sides of the tube with about 20 ml Chloroform and stir to disperse silica gel. When rate of settling slows, drain some $CHCl_3$ to aid settling, leaving 5-7 cm above silica gel. Slowly add 15 gm anhydrous Sod sulphate. Drain $CHCl_3$to top of Sod. Sulphate. Add 50 ml of sample extract to column, elute at maximum flow rate with 150 ml hexane followed by 150 ml anhydrous ether and discard. Elute Aflatoxin with 150 ml methanol-Chloroform (3+97) collecting this fraction from time of addition till flow stops. The elute is evaporated on water bath preferably under a gentle stream of nitrogen. The residue is quantitatively transferred to a vial, solvent evaporated and redissolved in a known volume of chloroform (0.2- 1.0 ml) and kept in a vial for quantification.

Preliminary TLC

Uncap vial containing extract residue, add 200 µl benzene -acetonitrile (98+ 2) and reseal with polythene stopper. Shake vigorously to dissolve. Puncture polythene stopper to accommodate needle of 10 µl syringe. In subdued incandescent light and as rapidly as possible spot 2, 5 and 10 µl spots on imaginary line 4 cm from bottom of TLC plate. Keep vial for quantitative analysis. On same plate spot 2.5 and 10 µl of aflatoxin standards Place 50 ml acetone-$CHCl_3$ (1+9) in trough of unlined developing tank. Use only 1 plate per tank, placing trough to one side to permit maximum exposure of coated surface to tank volume. Immediately insert plate into tank and seal tank. The chamber must be saturated with solvent before use.

Develop plate 40 minutes or until aflatoxins reach Rf 0.4-0.7. remove from tank, evaporate solvent at room temperature and view under long wave UV lamp in a viewing chamber. Observe pattern of 4 florescent spots.

Quantitative TLC

If preliminary TLC shows need for further dilution of test solution, evaporate to dryness on a steam bath and redissolve in calculated amount of benzene-acetonitrile. Spot successively 3.5, 5.0, and 6.5 µl of test solution. All spots should be approx same size and equal to 0.5 cm diameter. On the same plate spot 3.5, 5.0,6.5 µl aflatoxin standards. After development of the plate, dry the plate in subdued light. Compare florescent intensities of the sample spot with those of the standard spots and determine which of the sample spot matches one of the standards. If the spots of the smallest quantity of sample are too intense to match standards, the sample should be diluted and re chromatographed.

Calculation

$$\text{Calculate concentration of Aflatoxin B1 in } \mu g/kg = \frac{SxYxV}{XxW}$$

where,

S - µl aflatoxin standard which matches the unknown, Y - Concentration of Aflatoxin B1 standard µg/ml, V - µl of final dilution of sample extract, X - µl of sample extract spotted giving florescent intensity equivalent to S (B1 standard), W-wt in gm of the sample contained in final extract (10 gm if 50 ml $CHCl_3$ extract is used).

Note-Calculate Aflatoxin B2, G1, and G2 similarly.

Source

1. *A.O.A.C. 17th edn, 2000 Official Method 968.22 Aflatoxins in Peanuts and Peanut Products CB Method).*

2. Determination of Aflatoxins (For Peanut and Peanut Products, Cereals and Pulses-BF Method)

Extraction

Weigh 100 gm of peanut meal or powder or 50 gm peanut butter into a blender jar. Add 250 ml methanol- water (55+45) and 100 ml hexane to peanut butter and 500 ml methanol -water (55+45) 200 ml hexane and 4 gm Sod. Chloride to peanut powder. Blend 1 minute at high speed. Transfer to 250 ml centrifuge bottle and centrifuge 5 minutes at 2000 rpm, otherwise let mixture stand undisturbed in blender jar when separation will occur within 30 minutes. Pipette 25 ml of lower aqueous methanol phase in a separating funnel, add 25 ml chloroform stopper and shake 30-60 seconds. Let layers separate and drain bottom $CHCl_3$ layer through anhydrous sod. sulphate into a 250 ml beaker. Repeat extraction with two 25 ml portions of chloroform. Collect in a beaker and evaporate combined chloroform extract to between 2 ml and just dryness. Do not leave beaker on hot plate after solvent has evaporated. Transfer extract with careful washing to screw capped borosilicate vial and evaporate to dryness under gentle stream of nitrogen, dissolve extract in 200 µl benzene-acetonitrile (98 + 2), spot on TLC or HPTLC along with known aflatoxin standards. Develop the plate in chloroform- acetone (9+1) and compare florescence intensity under long wave UV lamp as in clause 2.5 above.Calculate as shown in 2.5 above.

Source

1. *A.O.A.C. 17th edn,2000 Official Method 970.45 Aflatoxins in Peanuts and Peanut Products BF Method, AOCS - AOAC method).*

Determination of Aflatoxins-Romer Mini Column Method

Apparatus

High Speed Blender, Ultraviolet light - Long wave UV with intensity of 430 u watt/cm2. at 15 cm at 365 nm, Minicolumn - Borosilicate standard wall tubing 6 (i.d) x 150 mm, tapered at1 end to 2 cm, Minicolumn Support rack- Test tube rack may be used, Rubber bulb – with 7 mm bulb at one end.

Reagents

Solvents–Chloroform and Acetone, Pot. Hydroxide wash solution -0.02 M KOH with 1 per cent KCL- Dissolve 1.12 gm KOH pellets and 19 gm KCL in 1 litre water, Sod. Hydroxide Solution - 0.02 M - 8.0 gm NaOH/litre, Sulphuric acid Solution 0.03 per cent - Dissolve 0.3 ml H_2SO_4 in 1 litre, Precipitating reagents- (1) Copper carbonate -Basic (2) Ferric Chloride Slurry-Mix 20 gm of $FeCL_3$ with 300 ml water. Diatomaceous Earth, Column packing (a) Florisil 100-200 mesh (b) Silica gel 60 for column chromatography (c) Alumina Neutral 80 -200 mesh - activate 2 hrs at 110°C (d) Calcium Sulphate anhydrous - 20 - 40 mesh. Dry packing material 1- 2 hrs at 110°C. Store all packing materials and packed columns in vapour tight containers and Aflatoxin solution for spiking - Dilute solutions of B1 and G1 to final concentration of 2 µg/ml.

Preparation of Mini Column

Trap small plug of glass wool into tapered end of column. To column add to height indicated in the following order – 5-7 mm $CaSO_4$, 5-7 mm Florisil, 18-20 mm silica gel, 8-10 neutral alumina and 5-7 mm $CaSO_4$. Tamp small plug of glass wool on top of column. Tamp column after each addition to settle packing and maintain interfaces as level as possible. After packing apply pressure to top glass wool plug with 5 mm dia glass rod.

Extraction

Weigh 50 gm test sample into the blender jar, add 250 ml acetone -water (85+ 15) and blend 3 minutes. Alternatively use 500 ml glass stoppered Erlenmeyer flask and shake 45 minutes on mechanical shaker. Filter through Whatman filter paper No 4 into 250 ml graduated cylinder. Collect 150 ml filtrate and transfer to 400 ml beaker.

Purification

To a 600 ml beaker quantitatively add 170 ml of 0.02N Sod Hydro-xide and 30 ml $FeCl_3$ slurry and mix well. To the filtrate in 400 ml beaker add about 3 gm basic Copper carbonate, mix well and add to the mixture in 600 ml beaker. To this add 150 ml diatomaceous earth and mix well. Filter using 160 mm funnel or Buchner funnel using Whatman filter paper No 4. Quantitatively transfer 150 ml filtrate to 500 ml separator, add 150 ml 0.03 per cent Sulphuric acid and 10 ml Chloroform. Shake vigorously for about 2 minutes and let separate. Transfer lower $CHCl_3$ layer (13-14 ml) to 125 ml separator. Add 100 ml KOH wash solution, swirl gently 30 seconds and let separate. If emulsion occurs drain emulsion into 10 ml glass stoppered flask, add about 1 gm anhydrous Sod Sulphate, stopper shake 30 seconds and let separate ($CHCl_3$ phase need not be completely clear). If emulsion is not broken, transfer emulsion to 125 ml separator and wash with50 ml 0.03 per cent H_2SO_4. Collect 3 ml Chloroform layer in 10 ml glass stoppered cylinder for chromatography.

Chromatography

Transfer 2 ml of $CHCl_3$ solution to minicolumn using 5 ml syringe with 5 in, 15 gauge needle. Let drain by gravity (15 – 30 mins). When solvent reaches top of adsorbent, add 3 ml elution solvent ($CHCl_3$–acetone (9 + 1). Let drain by gravity until solvent again reaches top of adsorbent. Do not let columns run dry during determination. Examine columns in darkened room under UV lamp. Look for blue fluorescent band at the top of florisil layer (ca 2.5 cm from bottom of column) indicative of aflatoxin. Perform analysis with -clean test portion and with test portion spiked with known amounts of aflatoxin to obtain comparison standards. Some uncontaminated products show white, yellow or brown fluorescence at top of florisil in sample column. If band has no definite bluish tint test portion is negative.

Source

1. *A.O. A.C 17th edn, 2000 Official Method 975. 36. Aflatoxins in Food and Feed - AACC- AOAC Method)*

3. Determination of Aflatoxins in Corn and Peanut Powder/Butter–(Liquid Chromatographic Method)

Principle

Aflatoxins are extracted and purified, derivatised with trifluoroacetic acid (aflatoxin B1 and G1 to B2 and G2 respectively) separated by reverse phase liquid chromatography and detected by fluorescence. Method can measure 0.1 ng of aflatoxin B1, B2, G1, and G2. Detection limit is about 0.3 ng/gm.

Apparatus

1. Liquid chromatograph-with Rheodyne septumless injector, Fluorichrom-fluorescence detector, 7.54 and 7.60 excitation filters (360nm) 3-73 and 4-76 glass emission filters (440nm) fitted with flow cell integrator or recorder, 0.5 cm/min chart speed. Flow rate 1.0 ml/min. Set up detector preferably with tungsten source, using low lamp, high gain attenuation 20 or adjust range to give minimum half scale deflection with 1.25 ng aflatoxin B1 or G1. For optimum performance detector should be left on continuously.
2. LC column -5 cm x 4.6 mm i.d. Supercosil LC-18 No5–8230 (Note -New LC columns or those that have been stored in methanol for extended periods require conditioning with concentrated standards in order to achieve optimum resolution and sensitivity to aflatoxin B1 and G1).
3. Clean Up Column - 20 cm x 1cm i.d with Teflon stopcock and coarse frit bed support, detachable glass solvent reservoir with 24/40 fitting.
4. Adjustable pipettes-10-100 and100-200 µl with disposable tips.
5. Filter tube -glass 15 x 2.5 cm i.d with coarse frit bed support (glass wool not recommended)

Reagents

1. Solvents-distilled-in-glass grade methanol, hexane, methylene chloride, benzene, acetone, acetonitrile. Anhydrous ethyl ether stored in metallic container (Glass bottled ether forms peroxides soon after opening which degrades aflatoxins).
2. LC elution solvents–H_2O-CH_3CN-CH_3OH (700 + 170 + 170). Adjust ratio of water to obtain baseline resolution of aflatoxin B2 and G2.
3. Silica gel for Column chromatography–Silica gel 60, (0.063-0.2mm). activate by drying at 100°C. Cool to room temperature. Weigh desired quantity (100 gm) into glass stoppered container. Add 1ml water in small increments, agitate silica gel between additions. Shake or tumble mechanically 4 -6 hrs. Let stand 16 hrs.

4. Trifluoroacetic acid (TFA)-Assay by titration– equal to or more than 98.5 per cent. Transfer 1-2 ml TFA to a 1 dram vial with Teflon lined cap. Keep in freezer when not in use. Discard if discoloration appears.
5. Sodium sulphate, anhydrous. Sift out fines to obtain 20 - 40 mesh. Heat 2-3 hrs at 600°C to remove organic impurities.
6. Aflatoxin standard solutions–a-Aflatoxin stock solution-10µg/ml. Prepare individual stock solution in benzene-CH_3CN (98+2) and determine concentration of each by measuring UV absoption if desired and b- Working standard solutions - Use Eppendorf pipette to transfer appropriate quantity stock solution to each 4 dram vial to obtain final concentrations of aflatoxins in each vial.

Vial	*B1 and G1 ng*	*B2 and G2 ng*
1	250	125
2	500	250
3	1000	500
4	2000	1000

Evaporate solutions to dryness under gentle stream of nitrogen (drying may be facilitated by warming to 40° C). Using Eppendorf pipette add 200 µl hexane and 50 µl of TFA to each vial cap and vortex mix 30 seconds. Let solutions stand 5 minutes, then add 10 ml H_2O-CH_3CN (9+1) and vortex mix 30 seconds. Let layers separate 5 -10 min or centrifuge at 1000 rpm for 30 sec. Final concentration of aflatoxins shall be:

Vial	*B1 and G1 ng/10.05ml*	*B2and G2 ng/10.05ml*
1	0.25	0.125
2	0.5	0.25
3	1.0	0.50
4	2.0	1.0

Extraction and Partition

Transfer 50 gm prepared corn, or peanut powder or peanut butter to 1 litre blender jar, add 200 ml of methanol followed by 50 ml of 0.1M HCl and blend 3 minutes at high speed. Filter through 24 cm whatman No 1 filter paper. Filtrate may not be completely clear. Collect 50 ml filtrate. Transfer to 250 ml separatory funnel. Add 50 ml 10 per cent NaCl solution, swirl, add 50 ml hexane and shake gently about 30 seconds. Let phases separate and drain lower aqueous layer into another 250 ml separatory funnel. Discard hexane layer. Add 25 ml CH_2Cl_2 and shake moderately 30 seconds. If emulsion occurs break up with clean pipette. Let phases separate and drain lower CH_2Cl_2 layer through coarse granular anhydrous sod. sulphate in glass filter tube. Collect elute in 250 ml beaker. Evaporate elute on steam bath under gentle stream of nitrogen to 2 -3 ml.

Column Chromatography

Slurry 2 gm silica gel with about 10 ml ether-hexane (3+1) in 30 ml beaker, pour slurry into clean up column and wash beaker with additional 5 ml ether- hexane solvent to effect transfer. Keep stop cock closed and let silica gel settle without tamping. Wash sides of column with 2-3 ml ether-hexane using wash bottle. After gel settles, open stop cock and while column drains add about 1 cm anhydrous sod. sulphate. Transfer eluate collected after extraction to column. Wash beaker with about 2 ml of CH_2Cl_2 and add wash to column. Do not use more than 5- 6 ml CH_2Cl_2 to transfer eluate to column. With stop cock fully open, add 25 ml benzene – acetic acid (9+1) and the 30 ml ether-hexane (3+1) to column draining each wash to top of sod. sulphate. Discard washes. Elute aflatoxin with 100 ml CH_2Cl_{2-} acetone (90+10) and collect elute in 250 ml beaker. Evaporate elute on steam bath under gentle stream of nitrogen to about 6 ml. Quantitatively transfer to 3 dram vial. Evaporate elute to dryness to steam bath in an aluminium block under gentle stream of nitrogen. Evaporate remaining 200 µl just to dryness under gentle stream of nitrogen by holding vial in palm of hand and slowly rotating vial.

Derivatisation

Add 200 µl hexane to the residue obtained above. Then add 50 µl of TFA using Eppendorf pipette, cap vial and vortex mix vigorously for 30 seconds. This procedure must be followed closely to ensure consistent reaction yields. Let mixture stand 5 minutes. Using Eppendorf pipette add 1.950 ml H_2O -CH_3CN (9+1). Vortex mix vigorously for exactly 30 seconds and let layers separate 10 mm. Concentration is 10 gm/2 ml aqueous CH_3CN.

Determination

Using instrument parameters mentioned under apparatus successively inject 25 µl of derivatised standard solution. Prepare standard curve to check linearity of responses. Inject 25 µl of TFA treated test solution (lower aqueous phase). If test peaks are outside the linear range, dilute aliquot of TFA treated test solution to suitable volume with water -CH_3CN, remix on vortex mixer and inject another 25 µl portion.

Calculation

Calculate individual aflatoxin concentration as follows:

Use responses of standard containing 500 ng B1 and G1 and 250 ng B2 and G2 for calculation.

$$\text{Aflatoxin ng/gm} = (P/P') \times C \times (2/10) \times 1000 \times D$$

where,

P and P′ - peak areas or heights for test solution and standard per 25 ul injection, C - concentration of individual aflatoxin in standard solution (0.5 or 0.25 ug/10.05 ml, D - dilution factor if 2 ml test solution for injection is iluted.

Source

1. *A.O. A.C 17th edn, 2000 Official Method 990.33 Aflatoxins in Corn and Peanut Butter, Liquid Chromatographic Method).*

4. Determination of Aflatoxin M1 in Milk, Cheese and Khoya

Aflatoxin M1 is hydroxylated metabolite of Aflatoxin B1 secreted in the milk of animals receiving aflatoxin B1. It is a potential hepatocarcinogen. Handle with same care as other aflatoxins.

Apparatus

Separating funnel 250 ml, Centrifuge, Chromatographic column-22mm i.d X 300 mm length with a tap., TLC plates, Filter paper No1, Micropipettes 5-100 µl, adjustable and Borosilicate vials, screw capped with Al or Teflon lining.

Reagents

1. Solvents–Glass distilled acetic acid, acetone, acetonitrile, chloroform, ether– peroxide free, ethyl alcohol, hexane, isopropanol, toluene.
2. Sod Chloride solution-Saturated solution, about 40 gm/100 ml.
3. Silica gel for column chromatography- E Merck silica gel No 60 (70 -230Mesh) or equivalent. Stir 1hr in methanol, filter, treat similarly with $CHCl_3$. Activate by drying 1 hour at 105 °C. Add water 1 ml/100 gm, shake until thoroughly mixed and store 15 hrs in air tight container.
4. Sod Sulphate, anhydrous, granular.
5. Diatomaceous earth-Celite.
6. Aflatoxin M1 reference standard- Make standards by dilution of stock solution with benzene – acetonitrile (9+ 1).
7. Densitometer.

Extraction

Shake 50 ml milk, 10 ml of saturated salt solution (40 gm Na Cl/100 ml water), and 120 ml chloroform at 30 °C in a 250 ml separating funnel and allow to separate for 2 minutes. For milk powder reconstitute 5 gm with 50 ml water, for cheese and Khoya blend 15 gm sample with 1 ml saturated salt solution, 5 gm diatomaceous earth or celite and 100 ml chloroform for 60 seconds in a blender jar. Drain lower $CHCl_3$ layer into125 ml erlenmeyer flask. Centrifuge if layers do not separate (15 minutes at 2000 rpm). Add 10 gm anhydrous. Sod sulphate to $CHCl_3$ with stirring. Filter into 100 ml graduated cylinder. Save filtrate for column chromatography.

Column Chromatography

Fill column half full with $CHCl_3$. Add 2 gm silica gel slurry made with silica off column sides with $CHCl_3$. Add sample extract and and drain entire solution through column by gravity if flow rate slows stir sod. sulphate gently.Rinse graduated cylinder with $CHCl_3$ and add rinsings to column. Wash column with 25 ml toluene - acetic acid (9 + 1) to remove coloured compounds.Wash with 5 ml of hexane-ether-acetonitrile (5+3+2)to remove fat. Elute Aflatoxin M1 with $CHCl_3$. Add 2 gm Sod sulphate above silica gel. Drain off excess $CHCl_3$ and rinse 40 ml $CHCl_3$-acetone (4 + 1).Evaporate to dryness and use for TLC or HPTLC as desired.

Thin Layer Chromatography

Dissolve sample residue in 100 µl of benzene -actonitrile (9 + 1), mix well and spot on TLC or HPTLC. Spot 20 µl of test solution and 2, 4, 6, 8, and 10 µl M1 standard (0.25 ug/ml). Develop plate in chloroform- actone-isopropanol (87+10 + 3) and calculate Aflatoxin M1 in µg/kg or ppb as follows:

$$\text{AflatoxinM1 } (\mu g/Kg) = \frac{SxYxV}{XxW}$$

where,

S-volume m1 standard which matches with sample, Y-Concentration of M1 standard in ug/ml, V -Volume final dilution of sample extract in ul, X - volume of sample extract spotted which matches with florescence intensity equivalent to M1 standard (S), W -Volume in ml of sample contained in final extract and W - Original volume or wt test portion X filtrate volume/120).

Source

1. *A, O, A, C 17th edn, 2000 Official Method 980.21 Aflatoxin M1 in Milk and Cheese Thin Layer Chromatography Method.*

5. Liquid Chromatography (LC)

Liquid chromatography is a method of physical separation in which the components of a liquid mixture are distributed between two immiscible phases, *i.e.,* stationary and mobile [1]. It is a separation technique in which the mobile phase is a liquid. It can be carried out either in a column or a plane. Present day liquid chromatography that generally utilizes very small packing particles and a relatively high pressure is referred to as High performance liquid chromatography **(HPLC) [1].**

The practice of LC can be divided into the following five categories:

(i) Adsorption chromatography.

(ii) Partition chromatography.

(iii) Ion-exchange chromatography.

(iv) Size-exclusion chromatography.

(v) Affinity chromatography.

Among these, the most widely used variant is the reverse-phase (RP) mode of the partition chromatography technique, which makes use of a non-polar (hydrophobic) stationary phase and a polar mobile phase. In common applications, the mobile phase is a mixture of water and other polar solvents (*e.g.,* methanol, isopropanol, and acetonitrile), and the stationary matrix is prepared by attaching long-chain alkyl groups (*e.g.,* n-octadecyl or C_{18}) to the surface of irregularly or spherically shaped 5 µm diameter silica particles [4].

High Performance Liquid Chromatography- (HPLC) or High Pressure Liquid Chromatography [1-4]

HPLC is an abbreviation for High Performance Liquid Chromatography. It has also been referred to as High Pressure LC. It is chromatographic technique of great versatility and analytical power. Separation can be achieved by utilizing the principles of ion exchange, size exclusion, and affinity chromatography. HPLC has been around for about 35 years and is the largest separations technique used. HPLC is optimum for the separation of chemical and biological compounds that are non-volatile. If a compound is volatile (*i.e.* a gas, fragrance, hydrocarbon in gasoline, *etc.*), gas chromatography is a better separation technique [1]. High-performance liquid chromatography (HPLC) developed as a direct off shoot of classical column liquid chromatography through improvements in the technology of columns and instrumental components (pumps, injection valves, and detectors). HPLC, alone or in combination with other methods such as MS, has proven to be a key method for identifying compounds that can be used for the authentication of dairy products [2].

When a material is non-volatile or when it is so thermally fragile that it cannot be analyzed by gas chromatography, then liquid chromatography may be appropriate for the analysis. High performance columns are constructed of packing materials with very small particle diameters, on the order of 3-10 micrometers. Therefore, the eluent cannot flow through under gravity flow as was done in early liquid chromatographic analyses. A high pressure pump is necessary to force the eluent through the column at the rate which delivers the maximum number of theoretical plates. High pressure, high performance liquid chromatography is one of the essential tools for foods, phaemceuticals, environmental *etc.* analysis. It can readily handle high molecular weight compounds, such as polynuclear aromatic hydrocarbons, highly polar compounds such as phenols or organic acids, and even inorganic ions [3].

A large number of compounds can be separated rapidly and at high sensitivity by HPLC. The performance characteristics and applicability of the technique are continually improving, principally due to the large and ever increasing choice of column packing, and the growing number of solvent combinations. The combination of HPLC and mass spectrometry (MS) has increased the range of applications. The apparatus used for such analyses has become progressively smaller and smaller [2].

Source

1. *Wikipedia.*
2. *Rajesh Kumar Bajaj and Priyanka Singh Rao, HPLC and its Application in Food Analysis, Dairy Chemistry Division, Compendium of Lectures on Advanced Techniques and Novel Approaches for Quality and Safety Evaluation of Dairy Foods -2015, NDRI, Karnal.*
3. */F:/Instrumental Analysis/Chromatography.htm.*
4. *Dass, Chhabil (2007-01-01). Fundamentals of Contemporary Mass Spectrometry. John Wiley and Sons, Inc. pp. 151-194. ISBN 9780470118498.*

History of HPLC [2]

(*i*) High-performance liquid chromatography (HPLC) developed during the 1960s.Beginning of the 60's: start of HPLC as High Pressure Liquid Chromatography.

(ii) End of the 70's improvements of column material and instrumentation –High Performance Liquid Chromatography.

(iii) Since beginning of the 80's: "boom" in HPLC started

(iv) Since 2006 new terms popped up like UPLC, RRLC, UFLC, RSLC.

The Principle of HPLC [2,3]

In the HPLC technique, the sample is forced through a column that is packed with a stationary phase composed of irregularly or spherically shaped particles, a porous monolithic layer, or a porous membrane by a liquid (mobile phase) at high pressure [2]. It is a separation technique that involves the followings:

(i) The injection of a small volume of liquid sample.

(ii) Into a tube packed with tiny particles (3 to 5 micron (μm) in diameter called the **stationary phase**).

(iii) Where individual components of the sample are moved down the packed tube (**column**) with a liquid (**mobile phase**) forced through the column by high pressure delivered by a pump.

(iv) These **components are separated** from one another by the column packing that involves various chemical and/or physical interactions between their molecules and the packing particles.

(v) These separated components are detected at the exit of this tube (**column**) by a flow-through device (**detector**) that measures their amount. An output from this detector is called a "**liquid chromatogram**".

Note

In principle, LC and HPLC work the same way except the speed, efficiency, sensitivity and ease of operation of HPLC is vastly superior.

The basic principle of HPLC is anything that can be dissolved can be analysed by HPLC. In this technique, the sample is forced through a column that is packed with a stationary phase composed of irregularly or spherically shaped particles, a porous monolithic layer, or a porous membrane by a liquid (mobile phase) at high pressure.

Mode of Separation [1,2]

HPLC is based on the separation of components of a mixture in relatively short columns (15 to 30cm) packed with very uniform micro-particulate silica (3 to 10 μm), chemically modified to obtain a surface coating with different functional groups. The use of high pressure and automatic equipment enables an efficient and fast separation of components [1]. The correct selection of the column packing and the mobile phase are the most important factors in successful HPLC.

HPLC can be divided into several related techniques, depending on the separation mechanism and the column type. HPLC is historically divided into two different sub-classes (two chromatographic techniques-**Normal phase liquid chromatography and Reversed phase liquid chromatography**) based on the polarity of the mobile and stationary phases and now there are following four major separation modes that are used to separate most compounds. The details of aforesaid techniques are given below [2,3]:

a. Normal Phase (Polar Stationary Phase and Nonpolar Mobile Phase) Liquid Chromatography (NPLC)

Normal phase chromatography relies on such column packings as silica and alumina. Modern silica packings with polar bonded coatings are also available and are more reproducible and easy to use. The difficulty with silica is its high affinity for water. Any trace of water in the solvent will be adsorbed onto the column, thereby changing its characteristics. This makes reproducible chromatography harder to achieve [3]. This has following characteristics:

- *(i)* Methods in which the stationary phase is more polar than the mobile phase (*e.g.* toluene as the mobile phase, silica as the stationary phase) and has fewer application. In this mode, the column packing is polar (*e.g.* silica gel, cyanopropyl-bonded, amino-bonded, *etc.*) and the mobile phase is non-polar (*e.g.* hexane, iso-octane, methylene chloride, ethyl acetate).
- *(ii)* Normal phase separations are performed less than 10 per cent of the time,
- *(iii)* The technique is useful for-a-water-sensitive compounds, b- geometric isomers, c- cis-trans isomers, d- class separations and e- chiral compounds [2].

b. Reversed Phase (Nonpolar Stationary Phase and Polar Mobile Phase) Liquid Chromatography (RPLC)

The most useful types in food and environmental analysis are reverse phase, normal phase and ion chromatography. Reverse phase liquid chromatography is probably the most frequently used, and the most versatile. It is called 'reverse' because of the comparison with 'normal phase', which is only called normal because it was invented first.

- *(i)* It has the opposite polarity (*e.g.* water-methanol mixture as the mobile phase and C18 = octadecylsilyl as the stationary phase). The column packing is non-polar (*e.g.* C18, C8, C3, phenyl, *etc.*) and the mobile phase is water (buffer)+water-miscible organic solvent (*e.g.* methanol, acetonitrile). It is a chromatographic technique that can separate a mixture of compounds and is used in biochemistry and analytical chemistry to identify, quantify and purify the individual components of the mixture [1,2]. It is the most popular mode and over 90 per cent of chromatographers use this mode.
- *(ii)* RPLC is mostly used in analysis of foods. The stationary phases for this type of HPLC are prepared by covalently binding hydrophobic ligands such as C4-, C8-, C18-alkyl chains or aromatic functions to the surface of a rigid siliceous or polymeric support. The stationary phase has strong hydrophobic character and because of this, proteins and peptides are bound very strongly from an undiluted aqueous mobile phase so that

their dilution requires the use of hydro-organic eluents. The separation of peptides and protein in RFC is typically carried out by gradient elution with increasing concentration of an organic modifier such as acetonitril, methanol, tetrahydrofuran, and isopropanol. The mobile phase usually contains low levels of trifluoroacetic or phosphoric acids, the role of which is to protonate the residual silanol groups at the surface of the siliceous support and the carboxyl groups of the elutes as well as to form ion pairs with the charged amino groups of the substances to be separated.

(iii) The technique can be used for non-polar, polar, ionizable and ionic molecules.

(iv) Making RPC very versatile.

(v) For samples containing a wide range of compounds, gradient elution is often used.

(vi) One begins with a predominantly water-based mobile phase and then adds organic solvent as a function of time.

(vii) The organic solvent increases the solvent strength and elutes compounds that are very strongly retained on the RPC packing [2].

c. *Ion Exchange Chromatography (IEC) [2,4]*

It is usually referred to as ion chromatography. It uses an ion exchange mechanism to separate analytes based on their respective charges. It is performed in columns but can also be useful in planar mode. IEC uses a charged stationary phase to separate charged compounds including **anions, cations, amino acids, peptides,** and **proteins.** In conventional methods the stationary phase is an ion exchange resin that carries charged **functional groups** which interact with oppositely charged groups of the compound to be retained. There are following two types of ion exchange chromatography:

(i) **Cation-Exchange** -In the Cation-Exchange Chromatography the stationary phase has negative charge and the exchangeable ion is a cation,

ii0 **Anion-Exchange**-In the Anion-Exchange Chromatography the stationary phase has positive charge and the exchangeable ion is an anion [4].

Ion exchange chromatography **is commonly used to purify proteins using FPLC.**

Ion Exchange Chromatography has following characteritics:

(i) In ion exchange, the column packing contains ionic groups (*e.g.* sulfonic, tetra alkyl ammonium) and the mobile phase is an aqueous buffer (*e.g.*phosphate, formate, *etc.*).

(ii) Ion exchange is used by about 20 per cent of the liquid chromatographers.

(iii) The technique is well suited for:

(a) The separation of inorganic and organic anions and cations in aqueous solution.

(*b*) Ionic dyes, amino acids, and proteins can be separated by ion exchange because such compounds are salt in brine water [2].

Application Example of Ion Exchange Chromatography

Basic proteins on strong cation exchanger (-SO3-)-1-RNA polymerase, 2-Chymotrypsinogen and 3- Lysozyme HPLC.

d. Size Exclusion Chromatography (SEC)

It is also known as **gel permeation chromatography (GPC- Non-aqueous SEC) or gel filtration chromatography** (sometimes referred Aqueous SEC) and separates molecules according to their size (or more accurately according to their hydrodynamic diameter or hydrodynamic volume).Smaller molecules are able to enter the pores of the media and, therefore, molecules are trapped and removed from the flow of the mobile phase. The average residence time in the pores depends upon the effective size of the analyte molecules. In SEC, there is no interaction between the sample compounds and the column packing material. Instead, molecules diffuse into pores of a porous medium. Depending on their size relative to the pore size, molecules are separated. Molecules larger than the pore opening do not diffuse into the particles, while molecules smaller than the pore opening enter the particle and are separated. Large molecules elute first. The SEC technique is used by 10-15 per cent of chromatographers, mainly for polymer characterization and for proteins. Smaller molecules elute later. It is generally used for the following:

(*i*) It is a low-resolution chromatography technique and thus it is often reserved for the final, "polishing" step of a purification.

(*ii*) It is also useful for determining the **tertiary structure** and **quaternary structure** of purified proteins, especially since it can be carried out under native **solution** conditions.

Source

1. *Rajesh Kumar Bajaj and Priyanka Singh Rao, HPLC and its Application in Food Analysis, Dairy Chemistry Division, Compendium of Lectures on Advanced Techniques and Novel Approaches for Quality and Safety Evaluation of Dairy Foods- 2015, NDRI, Karnal.*
2. *Wikipedia.*
3. *F: \Instrumental Analysis\ Chromatography.htm.*
4. *Ninfa, Alexander J (2009). Fundamental Laboratory Approaches for Biochemistry and Biotechnology. ISBN 978-0-470-47131-9.*

Apparatus [1,2]

A complete apparatus consists of a pumping system, either for a single eluent or for a gradient, an injector, a column, one or more detectors and a data handling system. It typically utilizes different types of stationary phases, a pump that moves the mobile phase (s) and analyte through the column, and a detector to provide a characteristic retention time for the analyte. The detector may also provide additional information related to the analyte. Analyte retention time varies depending on the strength of its interactions with the stationary phase, the ratio/composition of solvent (s) used, and the flow rate of the mobile phase. It is a form of liquid chromatography that utilizes smaller column size, smaller media inside the column, and higher mobile phase pressures. With HPLC, a pump (rather than gravity) provides the higher pressure required to move the mobile phase and analyte through the densely packed column. The increased density arises from smaller particle sizes. This allows for a better separation on columns of shorter length when compared to ordinary column chromatography [1]. HPLC instrumentation is made up of eight basic components: mobile phase reservoir, solvent delivery system, sample introduction device, column, detector, waste reservoir, connective tubing and a computer, integrator or recorder. The details related to the 5 major HPLC components and their functions, are given below:

1. Solvent Delivery Systems [2]

Before being fed to the pumping system, the solvents should be filtered and degassed.

(i) **Filtreration**-Any particulate material in the solvent must be removed, because particulates may damage the pumps, and will, in time, collect at the top of the column and cause plugging.

(ii) **Degassing** – It is important, because the dissolved gases may form bubbles when the pressure drops as the solvent enters the detector. Many detectors are severely disrupted by bubbles. Dissolved gases can be removed by purging the solvents with helium, which is quite insoluble in most solvents, or by passing the solvent through a microporous filter under vacuum.

Vacuum filtration is the most common technique, since it accomplishes both the filtration and degassing processes at the same time.

2. Pumping System [1,2]

The characteristics of a HPLC pump which are of highest importance are its ability to deliver a constant, pulse-free flow, over a wide range of different flows. The materials of the pump system must be resistant to attack by the wide variety of mobile phases to be used. The pumping system should have the following characteristics:

(i) The role of the pump is to force a liquid (called the mobile phase) through the liquid chromatograph at a specific flow rate, expressed in milliliters per min (mL/min) in a controlled, accurate and precise manner.

(ii) Normal flow rates in HPLC are in the 1- to 2-mL/min range.

(iii) Typical pressures are 700 – 1500 PSI (4.816 – 10.32 MPa; 1 Mpa = 150 PSI; 1 bar= 0.1 Mpa). Typical pumps can reach pressures in the range of 6000-9000 psi (400 to 600-bar). Pressures of up to 10000 psi (Appreox. 60 Mpa or 600 atmospheres) are generated by the pump. The Ultra High

Performance Liquid Chromatography systems or UHPLCs can work at up to 120 MPa (17,405 lbf/in^2), or about 1200 atmospheres.

(iv) Another consideration is the ease of changing solvents, which is related to the hold-up volume of the system.

(v) During the chromatographic experiment, a pump can deliver a constant mobile phase composition (isocratic) or an increasing mobile phase composition (gradient- ability to generate a gradient of two or even three solvents, in a reproducible fashion).

Types of Pump Modules and Isocratic Flow/Gradient Elution [1,2,3]

i. Isocratic Pump

(i) Delivers constant mobile phase composition.

(ii) Solvent must be pre-mixed.

(iii) Lowest cost pump.

Isocratic Flow [1,2]

A separation in which the mobile phase composition remains constant throughout the procedure is termed **isocratic** (meaning constant composition). In isocratic elution, peak width increases with retention time linearly according to the equation for N, the number of theoretical plates. This leads to the disadvantage that late-eluting peaks get very flat and broad. Their shape and width may keep them from being recognized as peaks. In isocratic elution, the selectivity does not change if the column dimensions (length and inner diameter) change -that is, the peaks elute in the same order.

ii. Gradient Pump/Solvent Gradient System

Solvent gradient systems require a method of mixing a constantly changing amount of solvents from two or three separate reservoirs. The mixing may be done either before the high pressure pump, or after it. The low pressure mixing method uses low pressure metering pumps to deliver the components of the solvent mixture to the inlet of the high pressure pump. High pressure gradient systems use separate high pressure pumps for each component, feeding into a small mixing chamber just before the injector. These are usually controlled by a microprocessor, which gradually increases the speed of one pump, while slowing the other. This is the most commonly used method, although it is somewhat more expensive. a-delivers variable mobile phase composition and b-can be used to mix and deliver an isocratic mobile phase or a gradient mobile phase. Binary gradient pump delivers two solvents whereas Quaternary gradient pump delivers four solvents [2].

Gradient Elution

The mobile phase composition does not have to remain constant. A separation in which the mobile phase composition is changed during the separation process is described as a **gradient elution [3]**. One example is a gradient starting at 10 per cent methanol and ending at 90 per cent methanol after 20 minutes. The two components of the mobile phase are typically termed "A" and "B"; **A** is the "weak" solvent which allows the solute to elute only slowly, while **B** is the "strong" solvent which rapidly elutes the solutes from the column. In reverse-phase chromatography, solvent **A** is often water or an aqueous buffer, while **B** is an organic solvent miscible with water, such as acetonitrile, methanol, or isopropanol.

Gradient elution decreases the retention of the later-eluting components so that they elute faster, giving narrower (and taller) peaks for most components. This also improves the peak shape for tailed peaks, as the increasing concentration of the organic eluent pushes the tailing part of a peak forward. This also increases the peak height, which is important in trace analysis. The gradient program may include sudden "step" increases in the percentage of the organic component, or different slopes at different times-all according to the desire for optimum separation in minimum time. In gradient elution, the elution order may change as the dimensions or flow rate change. The driving force in reversed phase chromatography originates in the high order of the water structure. The role of the **organic component of the mobile phase** is to reduce this high order and thus **reduce the retarding strength of the aqueous component.**

Comparative Characteristics of Isocratic and Gradient Pumping Conditions

The details are tabulated below [2]:

Sl.No.	*Isocratic Condition*	*Gradient Condition*
1	Isocratic mobile phase solvent composition remains constant with time.	Gradient mobile phase solvent ("B") composition increases with time. it is faster and easier to make successive runs.
2	Best for simple separations	Best for the analysis of complex samples
3	Often used in quality control applications and are in close proxymity to a manufacturing process.	Often used in method development for unknown mixtures. It has the ability to change the elution mixture easily.
4	-	Linear gradients are most popular (for example, the "gradient" shown at right). In a laboratory with several instruments available, the gradient instrument should be used to set up methods, even for isocratic systems.

3. *Sample Injector [1,2]*

The requirement for an injector in HPLC is the same as it is for other types of chromatography. It must put a very narrow plug of sample into the eluent stream.

(*i*) The use of a syringe is impractical because the eluent is under high pressure.

(*ii*) The usual method is the use of sampling valve, containing a small sampling loop, with a volume of a few microliters. An excess of dissolved sample is flushed through the sampling loop, to fill it completely, with the excess passing out to waste. When the valve is rotated, the eluent flow is diverted through the loop, picking up the sample and moving it on to the column [2].

The sample injectors should have the following characteristics [1]:

(i) The injector serves to introduce the liquid sample in to the flow stream of the mobile phase.

(ii) Typical sample volumes are 5-to 20-microliters (µL).

(iii) The injector must also be able to with stand the high pressures of the liquid system.

(iv) An autosampler is the automatic version for when the user has many samples to analyze or when manual injection is not practical.

Types of Injector Module [1]

Manual Injector

User manually loads sample into the injector using a syringe and then turns the handle to inject sample into the flowing mobile phase, which transports the sample into the beginning (head) of the column, which is at high pressure.

Autosampler

(i) User loads vials filled with sample solution into the autosampler tray (100 samples).

(ii) The autosampler automatically measures the appropriate sample volume, injects the sample, then flushes the injector to be ready for the next sample, *etc.*, until all sample vials are processed.

4. Column

The analytical column is expensive and can be damaged by particulate material depositing at the head of the column, as well as by attack of the eluent on the packing itself. Eluents with pH outside the 2 to 7 range may dissolve the silica support of the column. To lengthen the useful column life, guard and pre columns are used. **Precolumns** are short segments of tubing packed roughly with similar material to that used in the column. The precolumn will pick up any particulates which are present at the exit of the pump. These particles may arise from poorly filtered eluents, or from wear fragments from the pump. More importantly, if the eluent is aggressive, and is dissolving the silica backbone of the packing, the precolumn serves to saturate the eluent with silica. Another source of particulate matter which may damage columns is the sample. To protect the column from materials in the sample which may deposit at the entrance of the column or which may be irreversibly adsorbed on the column packing, a **guard column** may be used. Guard columns can contribute to loss of separation efficiency because the sample passes through them. These Precolumns and Guard columns must be packed as carefully as the analytical column [2]. The some important points of main analytical column are given below:

(i) Considered the "heart of the chromatograph" the column's stationary phase separates the sample components of interest using various physical and chemical parameters.

(ii) The small particles inside the column are what cause the high backpressure at normal flow rates.

(iii) The pump must push hard to move the mobile phase through the column and this resistance causes a high pressure within the chromatograph [1].

Note-Proper choice of column is critical for success in HPLC.

Types of Columns in HPLC [1]

The following types of main columns are being used in HPLC:

(i) **Analytical** [internal diameter (i.d.) 1.0 -4.6-mm; lengths 15 –250 mm].

(ii) **Preparative** (i.d. > 4.6 mm; lengths 50 –250 mm).

(iii) **Capillary** (i.d. 0.1 -1.0 mm; various lengths).

(iv) **Nano** (i.d. < 0.1 mm, or sometimes stated as < 100 μm).

Materials of Construction for the Tubing [1,2]

The main Columns are usually constructed of stainless steel (the most popular; gives high pressure capabilities) tubing with inner diameters of 4 or 5 mm. Microbore columns, with diameters of 1 and 2 mm are also available, as are columns with inside diameters above 10 mm, which are mainly used for preparative scale work. The end fittings on the column contain a frit to hold the packing in place, and a flow distributing plate which spreads the flow from the pump over the end of the column, thus helping to achieve constant flow through the entire cross section of the column. In addition to the stainless steel, main columns are also of Glass (mostly for biomolecules) and PEEK polymer (biocompatible and chemically inert to most solvents).

HPLC Columns Packing Materials [1]

(i) Columns are packed with small diameter porous particles. The most popular sizes are: 5-μm, 3.5-μm and 1.8-μm.

(ii) Columns are packed using high-pressure to ensure that they are stable during use. Most users purchase pre-packed columns to use in their liquid chromatographs

(iii) These porous particles in the column usually have a chemically bonded phase on their surface which interacts with the sample components to separate them from one another. For example, C18 is a popular bonded phase.

(iv) The process of retention of the sample components (often called analytes) is determined by the choice of column packing and the selection of the mobile phase to push the analytes through the packed column.

Temperature Control in HPLC

The Temperature Control in HPLC is needed for the followings:

Reproducibility

(*i*) Retention in HPLC is temperature-dependent.

(*ii*) If temperature varies, then it is difficult to assign "peaks" to specific compounds in the chromatogram and the peak areas/heights may vary

Solubility

(*i*) Certain chemical compounds may have low solubility in the HPLC mobile phase.

(*ii*) If they are injected into the flow stream they may precipitate or other difficulties may arise.

Stability

(*i*) Certain chemical compounds, especially biological compounds such as enzymes or proteins, may not be stable at room temperature or higher.

(*ii*) The temperature needs to be much lower down to 4°C.

Ways for Temperature Control

There are following three ways by which the temperature of a column could be controlled in HPLC [4]:

1. Oven.
2. Heater Block.
3. Water bath.

5. Detector [1,2]

There is no sensitive universal detector available for use in HPLC. The only really universal, bulk property HPLC detector is the refractive index detector, which cannot be used with gradient elution, requires excellent temperature control, and is as much as 10^3 times less sensitive than other detectors [2]. The detectors normally have the following properties:

(*i*) The detector can see (detect) the individual molecules that come out (elute) from the column.

(*ii*) A detector serves to measure the amount of those molecules so that the chemist can quantitatively analyze the sample components.

(*iii*) The detector provides an output to a recorder or computer that results in the liquid chromatogram (*i.e.*, the graph of the detector response).

In HPLC, there are many detection principles used to detect the compounds eluting from an HPLC column. The most common are:

(*i*) Spectroscopic Detection.

(*ii*) Refractive Index Detection.

(*iii*) Fluorescence Detection

i. Spectroscopic Detection

a. Ultraviolet (UV) Absorption

Ultra violet detectors are fairly general in application, since most organic compounds absorb some wavelengths in the UV spectrum. However, the spectral region of wavelengths below 210 is usually not useable for analysis because most solvents which would be used as eluents would also absorb in these areas. The response of this detector depends on Beer's Law, and therefore gives a linear response over four to five orders of magnitude. The detection limits vary widely, depending on the sample component and its extinction coefficient at the wavelength being used. In the most favorable cases, 1 ng or less of a compound may be detected [2]. It is working as per the followings:

(i) An ultraviolet light beam is directed through a flow cell and a sensor measures the light passing through the cell.

(ii) If a compound elutes from the column that absorbs this light energy, it will change the amount of light energy falling on the sensor.

(iii) The resulting change in this electrical signal is amplified and directed to a recorder or data system.

(iv) A UV spectrum is sometimes also obtained which may aid in the identification of a compound or series of compounds.

Fixed wavelength detectors, using filters to isolate a single band of radiation, are inexpensive and stable. Generally, these are single beam instruments, but dual beam systems are possible. They lack versatility, since the only compounds which can be analyzed are those which absorb at the fixed wavelength. Variable wavelength detectors, are, however, much more versatile.

b. Mass Spectroscopy (MS) Detector [1]

(i) An MS detector senses a compound eluting from the HPLC column first by ionizing it then by measuring it's mass and/or fragmenting the molecule into smaller pieces that are unique to the compound.

(ii) The MS detector can sometimes identify the compound directly since its mass spectrum is like a fingerprint and is quite unique to that compound.

Here is a mass spectrum of a simple chemical compound, **toluene**. The pattern of lines is very unique to this compound. The largest peak in the spectrum occurs at a mass of 91, which is a fragment ion generated by loss of a hydrogen atom.

ii. Refractive Index (RI) Detection [1]

(i) The ability of a compound or solvent to deflect light provides a way to detect it.

(ii) The RI is a measure of molecule's ability to deflect light in a flowing mobile phase in a flow cell relative to a static mobile phase contained in a reference flow cell.

(iii) The amount of deflection is proportional to concentration.

(iv) The RI detector is considered to be a universal detector but it is not very sensitive.

iii. Fluorescence Detection

Fluorescence Detectors [2]

These are among the most sensitive available. These are most suitable for, but are not limited to, compounds which fluoresce. Non-fluorescent compounds may be derivatized by adding a reagent after the column, which supplies a fluorescent tag to the sample molecules. Alternatively, the eluent may be made fluorescent and the sample peaks detected by the decrease of fluorescence as the peak elutes. This is known as **"vacancy chromatography".** Fluorescence detectors need an intense high energy source, either line or continuous, to excite the fluorescence. Mercury lamps are used for line source excitation, and deuterium or xenon arc sources for continuum source. A monochromator is used to select the wavelength for excitation and for emission. The wavelength selection can also be done with filters, at the expense of versatility and sensitivity. The wavelength at the absorbance peak may not be available in a filter instrument, so that the highest sensitivity cannot be achieved. An photomultiplier is used to capture and amplify the weak emission from the fluorescent molecules. This type of detectors has the following important characteristics:

(i) Compared to UV-Vis detectors fluorescence detectors offer a higher sensitivity and selectivity that allows to quantify and identify compounds and impurities in complex matrices at extremely low concentration levels (trace level analysis).

(ii) Fluorescence detectors sense only those substances that fluorescence.

6. Computer

Frequently called the data system, the computer not only controls all the modules of the HPLC instrument but it takes the signal from the detector and uses it to determine the time of elution (retention time) of the sample components (Qualitative analysis) and the amount of sample (Quantitative analysis).

Source

1. *Wikipedia.*

2. *F: \Instrumental Analysis\Chromatography.htm.*

3. *A recent book provides a comprehensive treatment of the theory of high-performance gradient chromatography: Lloyd R. Snyder and John W. Dolan (2006). High-Performance Gradient Elution: The Practical Application of the Linear-Solvent-Strength Model. Wiley Interscience. ISBN 0471706469.*

4. *Lee, M.L. Nollet Lee, (Ed.) (1996). Handbook of Food analysis. Vol. I. Physical Characterization and Nutrient Analysis.*

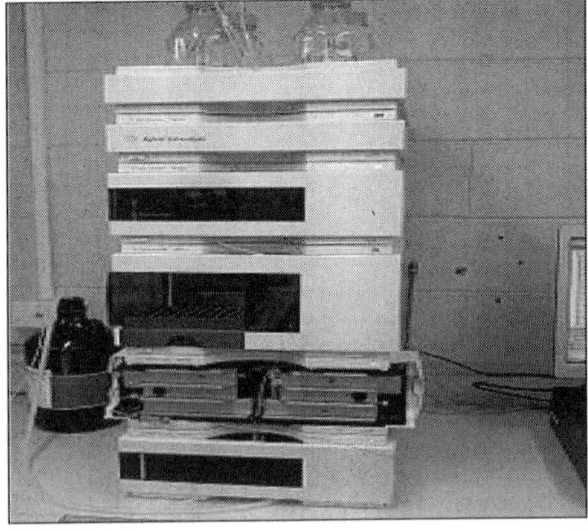

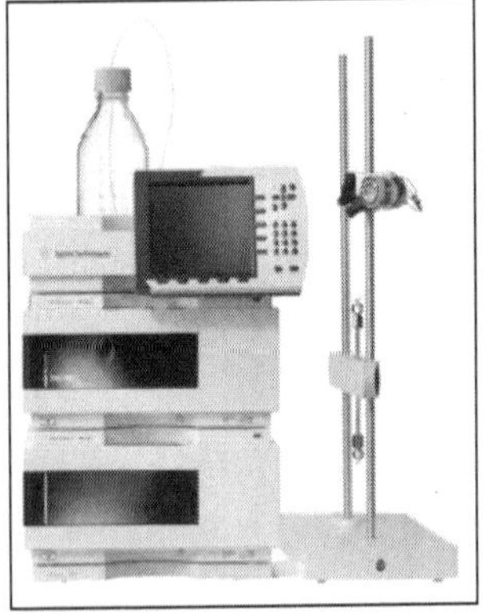

Modular HPLC System–Basic Configuration with Isocratic Pump, Manual Injector, Variable Wavelength Detector, and Hand-HeldController.

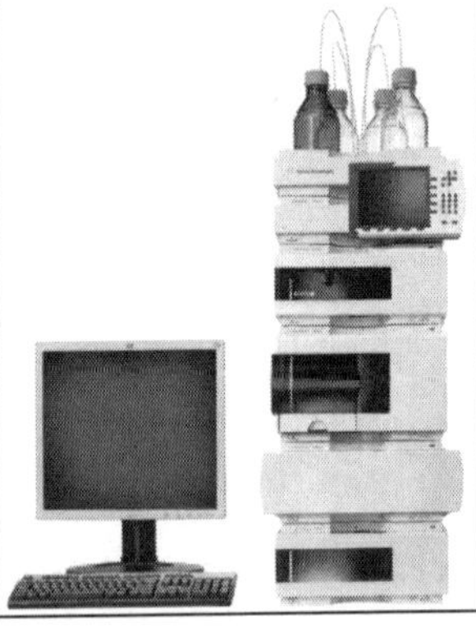

Modular HPLC System–High-End Configuration with Quaternary Pump, Autosampler, Column Thermostat, Diode Array Detector, and Computer with Control and Data Analysis SW.

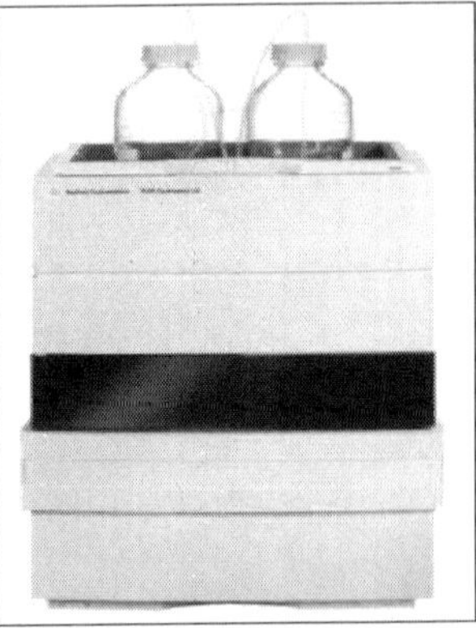

Integrated HPLC System "All parts in one box"-different Configurations Possible, here with Gradient Pump, Auto-Sampler, Column Oven, VWD, and Computer with Control and Data Analysis SW (not shown on picture).

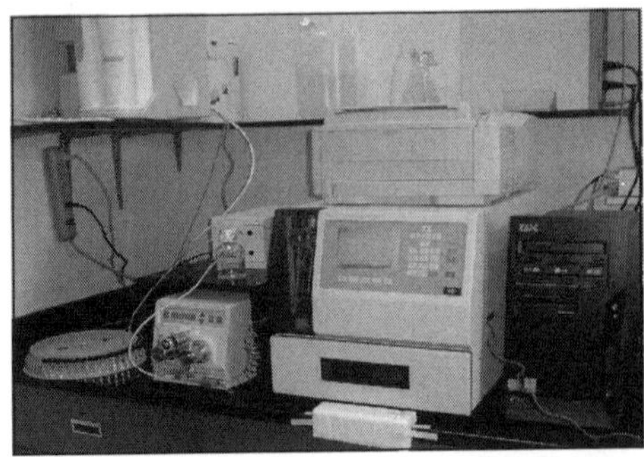

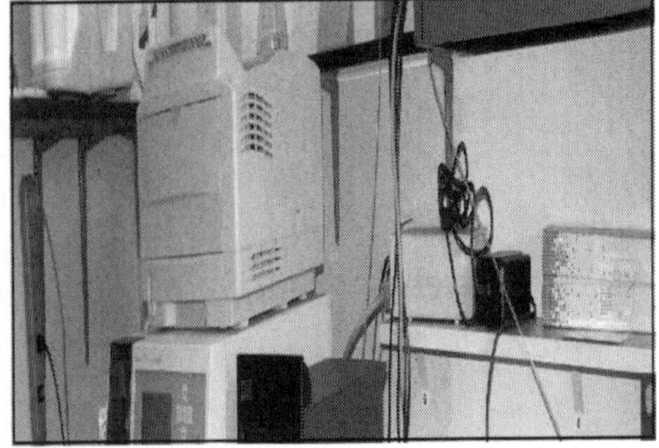

Different Models of HPLC (*Source*: Wikipedia).

Operational Procedure [1,2,3,4]

(i) **Sample Preparation [1]**: The success or failure of HPLC methods dependent upon the adequacy of sample preparation. The methodologies, such as extraction, ultra filtration, and precipitation are generally needed to remove interfering materials prior to chromatography. Variables such as extraction efficiency, analyte stability and consistency of chemical or enzymatic pretreatment must be considered during each sample preparation step. The use of HPLC to analyze foods for pesticide, mycotoxin and drug residues also requires fastidious sample preparation and data evaluation.

(ii) The sample to be determined, is introduced, in small volumes, into the stream of mobile phase. The solution moved through the column is slowed by specific chemical or physical interactions with the stationary phase present within the column. The velocity of the solution depends on the nature of the sample and on the compositions of the stationary (column) phase. The time at which a specific sample elutes is called the retention time; the retention time under particular conditions is considered an identifying characteristic of a given sample. The use of smaller particle size column packing (which creates higher back pressure) increases the linear velocity giving the components less time to diffuse within the column, improving the chromatogram resolution. Common solvents used include any miscible combination of water or various organic liquids (the most common are methanol and acetonitrile) [2,3,4].

Water may contain buffers or salts to assist in the separation of the sample components, or compounds such as trifluoroacetic acid which acts as an ion pairing agent. A further refinement of HPLC is to vary the mobile phase composition during the analysis; **gradient elution**. A normal gradient for reversed phase chromatography might start at 5 per cent methanol and progress linearly to 50 per cent methanol over 25 minutes; the gradient depends on how hydrophobic the sample is. The gradient separates the sample mixtures as a function of the affinity.

Source

1. *Rajesh Kumar Bajaj and Priyanka Singh Rao, HPLC and its Application in Food Analysis, Dairy Chemistry Division, Compendium of Lectures on Advanced Techniques and Novel Approaches for Quality and Safety Evaluation of Dairy Foods- 2015, NDRI, Karnal.*
2. *A recent book provides a comprehensive treatment of the theory of high-performance gradient chromatography: Lloyd R. Snyder and John W. Dolan (2006). High-Performance Gradient Elution: The Practical Application of the Linear-Solvent-Strength Model. Wiley Interscience. ISBN 0471706469.*
3. *Agilent Technologies, Inc.*
4. *Wikipedia.*

Parameters Affecting the Detection Sensitivity and Separation Selectivity

Some parameters *e.g.* Internal diameter of columns, Particle size, Pore size, Pump pressure, are important and responsible parameters which are affecting the HPLC working and sensitivity. The details are tabulated below:

Sl.No.	*Parameters*	*Details of effects*
1	Internal diameter (ID) of HPLC columns	ID is an important parameter that influences the detection sensitivity and separation selectivity in gradient elution. It also determines the quantity of analyte. Larger columns are usually seen in industrial applications.Low-ID columns have improved sensitivity and lower solvent consumption at the expense of loading capacity. *(i)* Larger ID columns (over 10 mm) are used to purify usable amounts of material. *(ii)* Analytical scale columns (4.6 mm) have been the most common type of columns. They are used in traditional quantitative analysis of samples and often use a UV-Vis absorbance detector. *(iii)* Narrow-bore columns (1-2 mm) are used for applications when more sensitivity is desired either with special UV-vis detectors, fluorescence detection or with other detection methods. *(iv)* Capillary columns (under 0.3 mm) are used almost exclusively with alternative detection means such as mass spectrometry. They are usually made from fused silica capillaries, rather than the stainless steel tubing that larger columns employ.
2	Particle size	The traditional HPLC is performed with the stationary phase attached to the outside of small spherical silica particles (very small beads). These particles come in a variety of sizes with 5 μm beads being the most common. *(i)* Smaller particles generally provide more surface area and better separations, but the pressure required for optimum linear velocity increases by the inverse of the particle diameter squared [2,3,4]. *(ii)* Larger particles are used in preparative HPLC (column diameters 5 cm up to >30 cm) and for non-HPLC applications such as solid-phase extraction.
3	Pore size	Stationary phases are porous to provide greater surface area. Small pores provide greater surface area while larger pore size has better kinetics, especially for larger analytes.
4	Pump pressure	Pumps vary in pressure capacity, but their performance is measured on their ability to yield a consistent and reproducible flow rate. *(i)* Pressure may reach as high as 40 MPa (6000 lbf/in^2), or about 400 atmospheres. *(ii)* Modern HPLC systems have been improved to work at much higher pressures, and therefore are able to use much smaller particle sizes in the columns (<2 μm). *(iii)* These "Ultra High Performance Liquid Chromatography" systems (UPLC- a trademark of the Waters Corporation) or RSLC/UHPLCs can work at up to 100 MPa (15,000 lbf/in^2), or about 1000 atmospheres.

Source

1. *Wikipedia.*
2. *Majors, Ronald E. (2010-09-07) Fast and Ultrafast HPLC on sub-2 μm Porous Particles -Where Do We Go From Here? -LC-GC Europe. Lcgceurope.com.*
3. *Xiang, Y.; Liu Y. and Lee M.L. (2006). "Ultrahigh pressure liquid chromatography using elevated temperature". Journal of Chromatography A 1104 (1-2): 198–202.*
4. *Horváth, Cs.; Preiss B. A. and Lipsky S.R. (1967). "Fast liquid chromatography. Investigation of operating parameters and the separation of nucleotides on pellicular ion exchangers". Analytical Chemistry 39 (12): 1422–1428.*

Application of HPLC [1,2,3]

1. General Applications

HPLC is being used for Qualitative, Quantitative analysis, Preparation of Pure Compound (s) and Trace analysis.

Qualitative Analysis

The identification (ID) of individual compounds in the sample:

(i) The most common parameter for compound ID is its retention time (the time it takes for that specific compound to elute from the column after injection).

(ii) Depending on the detector used, compound ID is also based on the chemical structure, molecular weight or some other molecular parameter.

Quantitative Analysis

The measurement of the amount of a compound in a sample (concentration); meaning, how much is there?. There are two main ways to interpret a chromatogram (*i.e.* perform quantification):

(i) Determination of the peak height of a chromatographic peak as measured from the baseline.

(ii) Determination of the peak area. In order to make a quantitative assessment of the compound, a sample with a known amount of the compound of interest is injected and its peak height or peak area is measured. In many cases, there is a linear relationship between the height or area and the amount of sample.

Preparation of Pure Compound(s)

By collecting the chromatographic peaks at the exit of the detector and concentrating the compound (analyte) by removing/evaporating the solvent. A pure substance can be prepared for later use (*e.g.* organic synthesis, clinical studies, toxicology studies, *etc.*). This methodology is called **Preparative chromatography.**

Trace Analysis

A trace compound is a compound that is of interest to the analyst but it's concentration is very low, usually less than 1 per cent by weight, often parts per million (ppm) or lower.

The determination of trace compounds is very important in pharmaceutical, biological, toxicology, and environmental studies since even a trace substance can be harmful or poisonous. In a chromatogram trace substances can be difficult to separate or detect. High resolution separations and very sensitive detectors are required.

Source

1. *HS. Lee, HPLC Analysis of Phenolic Compounds, in Nollet MLL Food Analysis by HPLC, New York, Marcel Dekker,2000; 804.*
2. *Marini D. HPLC of Lipids in Nollet LML. Food Analysis by HPLC, New York, Marcel Dekker, 2000; 231-232.*
3. *Wikipedia.*

2. Applications in Manufacturing [4]

HPLC is a common technique used in pharmaceutical development, as it is a dependable way to obtain and ensure product purity [1]. While HPLC can produce extremely high quality (pure) products, it is not always the primary method used in the production of bulk drug materials [2]. HPLC is used in only 15.5 per cent of syntheses [3]. However, it plays a role in 44 per cent of syntheses in the United States pharmacopoeia [4].

Source

1. *Siddiqui, Masoom Raza; AlOthman, Zeid A.; Rahman, Nafisur (2013). "Analytical techniques in pharmaceutical analysis: A review". Arabian Journal of Chemistry. 10: S1409.*
2. *The European Pharmacopoeia, 2002. fourth ed., Council of Europe, Strasbourg.*
3. *United States Pharmacopoeia, 2004. 27th ed. The USP Convention Inc., Rockville, MD.*
4. *Pesce, Amadeo; Rosenthal, Murray; West, Robert; West, Cameron; Crews, Bridgit; Mikel, Charles; Almazan, Perla; Latyshev, Sergey (2010-06-01). "An evaluation of the diagnostic accuracy of liquid chromatography-tandem mass spectrometry versus immunoassay drug testing in pain patients". Pain Physician. 13 (3): 273–281.*

3. Legal Applications [1]

This technique is also used for detection of illicit drugs in urine. As HPLC is a method of determining (and possibly increasing) purity, using HPLC alone in evaluating concentrations of drugs is some what insufficient. With this, HPLC in this context is often performed in conjunction with mass spectrometry [2]. This technique has been used to detect a variety of agents like doping agents, drug metabolites, glucuronide conjugates, amphetamines, opioids, cocaine, BZDs, ketamine, LSD, cannabis, and pesticides [3,4]. Performing HPLC in conjunction with Mass spectrometry reduces the absolute need for standardizing HPLC experimental runs.

Source

1. *Wikipedia.*
2. *Weinmann, W.; Renz, M.; Vogt, S.; Pollak, S. (2000-01-01)."Automated solid-phase extraction and two-step derivatisation for simultaneous analysis of basic illicit drugs in serum by GC/MS". International Journal of Legal Medicine. 113 (4): 229–235.*
3. *Pelander, Anna; Ojanperä, Ilkka; Laks, Suvi; Rasanen, Ilpo; Vuori, Erkki (2003-11-01). "Toxicological screening with formula-based metabolite identification by liquid chromatography/time-of-flight mass spectrometry". Analytical Chemistry. 75 (21): 5710–5718.*
4. *Nobilis, Milan; Pour, Milan; Senel, Petr; Pavlík, Jan; Kunes, Jirí; Voprsalová, Marie; Kolárová, Lenka; Holcapek, Michal (2007-06-15). "Metabolic profiling of a potential antifungal drug, 3- (4-bromophenyl)-5-acetoxymethyl-2,5-dihydrofuran-2-one, in mouse urine using high-performance liquid chromatography with UV photodiode-array and mass spectrometric detection". Journal of Chromatography B. 853 (1–2): 10–19.*

4. Applications in Research [1]

Similar assays can be performed for research purposes. This technique is useful in observing multiple species in collected samples, as well, but requires the use of standard solutions when information about species identity is sought out. It is used as a method to confirm results of synthesis reactions, as purity is essential in this type of research.

5. Medical Applications [1]

Medical use of HPLC can include drug analysis. While urine is the most common medium for analyzing drug concentrations, blood serum is the sample collected for most medical analyses with HPLC [2].

Source

1. *Wikipedia.*
2. *Zahedi Rad, Maliheh; Neyestani, Tirang Reza; Nikooyeh, Bahareh; Shariatzadeh, Nastaran; Kalayi, Ali; Khalaji, Niloufar; Gharavi, Azam (2015-01-01). "Competitive Protein-binding assay-based Enzyme-immunoassay Method, Compared to High-pressure Liquid Chromatography, Has a Very Lower Diagnostic Value to Detect Vitamin D Deficiency in 9–12 Years Children". International Journal of Preventive Medicine.* **6**: *67.*

6. Application in Food Industry

i. Analysis of Food Ingredients [1,8]

The ingredients of Foods are generally grouped as Major and Minor depending upon the quantity present in food. Fat, Protein and Carbohydrates are the major ingredients of food where as Vitamins, Minerals, Food additives are the minor ingredients. The details related to the application of HPLC in analysis of food ingredients are given below:

Sl.No.	*Types of Ingredients*	*Food Ingredients*	*Details related to HPLC application*
1	Major	Fat	Fat characterization is inmportant for food industry. *(i)* The reverse phase C18 column allowing the separation of individual components from the simplest natural mixtures of triacylglycerols. *(ii)* The triacylglycerols composition of soyabean, peanut, olive, palm, avocado, corn palm, oils, cocoa butter, lard and tallow were thus determined. *(iii)* The greatest value of HPLC is for volatile components (short chain fatty acids),for preparative scale separations or for studying isotopically labelled fatty acids. Further more, positional and conformational isomers are more easily separated by HPLC than GLC. *(iv)* A number of methods have been devised for the separation of trans isomers of unsaturated fattyacids (*e.g.* Conjugated linoleic acid -18:2n-6-CLA *etc.*).
		Protein	HPLC shows a growing importance for the analysis of specific protein components in foods and food products. The following three basic separating principles can be discerned- *(i)* Ion-exchange chromatography (IE-HPLC)-Proteins are separated on the basis of their net charge. Modern IE supports are mainly based on silica or hydrophilic organic polymers. *(ii)* Size exclusion chromatography (SE-HPLC)- Proteins are separated on the basis of their molecular size. Protein detection is usually achieved in the far (220nm) or near (280nm) UV region. SE-HPLC has also been proposed as a fast and non destructive method for the separation of unreduced flour proteins into glutenin, gliadin and albumin globulin components. *(iii)* Reverse phase chromatography (RP-HPLC)-Protein separation is based on difference in surface hydrophobicity between the protein molecules. Hydrophobic amino acid residues of proteins are bounded through hydrophobic interaction to stationary matrix composed of a non polar surface (usually C2-C18 alkyl chains) and silica based.
		Carbohydrates	HPLC analysis using an amino-based stationary phase is the most popular technique for the routine analysis of simple sugars. This analysis uses isocratic elution (*e.g.*, acetonitrile: water,75:25) and a refractive index detector (RID). *(i)* In N-phase HPLC, the elution order of carbohydrates, monosaccharides and polyhydric alcohols then di saccharides followed by oligosaccharides. *(ii)* It is successfully used to analyse the carbohydrate content of beverages, breakfast,honey,cakes, malts, cereals, ice creams, vegetables, meat infant foods, fruits, *etc.*

Contd.

Sl.No.	Types of Ingredients	Food Ingredients	Details related to HPLC application
			(iii) Amino bonded silica gel has a the tendency for glycosylamine formation between reducing carbohydrates and amino groups which results in deterioration of column performance. *(iv)* Reverse phase chromatography has been used to analyse the sucrose, raffinose and stachylose content of soybeans/soy bean products and juices, syrups, beverages and wort have been analysed for invert sugar, sucrose, maltose and malto triose.
2	Minor	Vitamins	Due to relative simplicity, accuracy and precision, the chromatographic methods using HPLC are preferred over other available methods. *(i)* C18 HPLC column features a retentive, high-purity packing that is ideal for separating a range of fat-soluble vitamins(Vitamins A -retinol, E -alpha tocopherol, D, and K). *(ii)* The very polar compounds are difficult to retain by reversed phase HPLC, and many methods use ion-pairing reagents to improve retention.
		Minerals (Ions and ionic components)	HPLC - Ion pair chromatography opens up another field of ion chromatography not only useful in analyzing inorganic anions or cations but also for analysis of organic ions, *e.g.*, organic acids. *(i)* Other applications like - In beverages-organic acid, transition metals, sulphite, p hydroxybenzoic acid ester, sorbic acid, benzoic acid. In meat products-nitrate and nitrite, sulphanilamide and sulfaguanidine, free amino acid. In milk and milk products- nitrate and bromide. In cheese- iodide; In milk chocolate-mono valent cations,In blue cheese- bromate,In bread dough- phosphate,fluoride and monofluorophosphate.
		Food Additives	Food additives can be separated into several groups: preservatives, fungicides, artificial sweeteners, anti- oxidants, food dyes and flavour components. *(i)* Propionic acid, sorbic acid, benzoic acid p-hydroxy benzoic acid, methyl ester,natamycin can be detected with C-18 columns at 40°C with gradient mobile aqueous phase and organic phase using UV detector at 230,240 and 278 nm. *(ii)* Sweeteners - asparatame,saccharin, dulrin, cyclamate and sucrose can be detected by RP-HPLC with indirect photometric detection. *(iii)* Flavor compounds in vanilla extract and artificial vanilla flavor are analyzed using HPLC. The analytes are separated on a C8 column and quantitated by comparing their UV absorbance at 254nm to an external standard. *(iv)* The antioxidents can be determined by HPLC with UV and fluorescence detection.

Source

1. *Rajesh Kumar Bajaj and Priyanka Singh Rao, HPLC and its Application in Food Analysis, Dairy Chemistry Division, Compendium of Lectures on Advanced Techniques and Novel Approaches for Quality and Safety Evaluation of Dairy Foods -2015, NDRI, Karnal.*
2. *Lee, M.L. Nollet (Ed.) (1996). Handbook of Food analysis. Vol. I. Physical Characterization and Nutrient Analysis.Marcel Dekker, Inc., New York, Basel, Hong Kong, pp. 197-1051.*
3. *Matissek, R. and Wittkowski, R. High Performance Liquid Chromatography in Food Control and Research. Technomic Publishing Co. Inc., Lancaster-Basel, pp. 87-296.*
4. *Nielsen, S.S. (1994) Introduction to Chemical analysis of foods, Jones and Bertlett Publishers International, London, Part IV.*
5. *Oliver, RWA (1989) HPLC of macromolecules-a practical approach, IRL Press, Oxford.*
6. *Pomeranz, Y. and Meloan, C.F. (1994) Food Analysis-Theory and Practice, Chaman and Hall, NY.*
7. *Lee HS. HPLC Analysis of Phenolic Compounds, in Nollet MLL Food Analysis by HPLC, New York, Marcel Dekker,2000; 804.*
8. *Marini D. HPLC of Lipids in Nollet LML. Food Analysis by HPLC, New York, Marcel Dekker, 2000; 231-232.*

ii. Analysis of Mycotoxin

Determination of Aflatoxins B1, B2, G1 and G2, in Spices, Tea, Coffee, Nutmeg and Rice by HPLC-MSMS

Principle of LC/MS

LC/MS is a hyphenated technique, combining the separation power of HPLC, with the detection power of mass spectrometry. Even with a very sophisticated MS instrument, HPLC is still useful to remove the interferences from the sample that would impact the ionisation. Interface that will eliminate the solvent and generate gas phase ions, then transferred to the optics of the mass spectrometer.

Instruments and Apparatus

Centrifuge, Sonicator, Blender, Horizontal shaker, HPLC complete setup with a Mass Spectrometry detector (triple Quadrupole), variable volume micropipette, 150ml beaker, glass funnels, 10ml pipettes, 50ml graduated cylinders with ground glass stoppers, 50ml Vol. Flask, Para film, Aflatest pump stand, fluted filter papers and glass fiber filter paper and Immuno affinity/Afla test columns.

Chemicals/Reagents

Sodium chloride AR/GR grade, Methanol (MeOH), HPLC grade, Tween-20, Acetonitrile HPLC, formic acid AR grade, Ammonium acetate AR/GR grade.

Reference Standards

(*i*) Aflatoxin mixed std. (B1, B2, G1, G2) of 20 ppm concentration from Fluka.

(*ii*) Intermediate standard solution of 120 ppb and the working standards solutions of 1.0,2.5,5.0,7.5,10 and 12.5ppb concentration are prepared as follows:

Aflatoxin Standard dilution Procedure–a. Stock Solution (1ml) in Acetonitrile- 20mg/ml (20ppm), b. Add 250µl of the stock diluted to 50ml (50: 50 V/V, MeOH: H2O) Vol.flask–100ppb (Afla-Mix)

The following dilutions should be done in Methanol: Water mix for calibration dilutions.

10 µl of 100 ppb (Afla-mix) +990 µl of MeOH: H2O (50: 50)-1.0ppb

-25 µl of 100ppb (Afla-mix)+975 µl of MeOH: H2O (50: 50)-2.5ppb

-50 µl of 100ppb (Afla-Mix)+950 µl of MeOH: H2O (50: 50)-5.0ppb

-75 µl of 100ppb (Afla-Mix)+925 µl of MeOH: H2O (50: 50)-7.5ppb

-100 µl of 100ppb (Afla-Mix) +900 µl of MeOH: H2O (50: 50)-10ppb

-125 µl of 100ppb (Afla-Mix) +875 µl of MeOH: H2O (50: 50)-12.5ppb

The stability of Stock, intermediate and working standards have to be ensured in the lab before they are put to use.

Instrumental Condition

HPLC Mobile Phase

A- 10mM Ammonium acetate with 0.1 per cent formic acid and B-Methanol and Acetonitrile (V/V 50: 50) with 0.1 per cent formic acid.

Injection volume-10 µL

Mobile Phase Gradient

Time	*Mobile Phase-A*	*Mobile Phase-B*
0.0	50	50
4.0	10	90
7.0	10	90
7.2	50	50
9.0	50	50

Instrument

Triple quadrupole HPLC-MSMS and Analytical Column RP-18 end-capped, 150x4.6 mm, 5µm particle size or its equivalent, Flow rate: 0.3-1.0 ml per min. Depending on column ID and length, Run time: 9-12 min.

MSMS-Conditions: ES -Positive MRM of

Aflatoxin-G2-331>285 and 245

Aflatoxin-G1-329>283 and 243

Aflatoxin-B2-315>287 and 259

Aflatoxin-B1-313>285 and 241

The resolution, Dwell time, Collision energy *etc.* those are specific to model and make of MS system need to be standardized by the analyst.

Test Procedure-Sample Preparation

Weigh 25g of sample into a 250ml Conical flask, to this add 5g of AR/GR grade Sodium Chloride and 100mL of Aflatoxin extraction solution (containing 70 per cent methanol and 30 per cent water mix in case of Rice and incase of coffee and oleoresin the extraction solution shall be 80 per cent methanol and 20 per cent water mix). Cap and seal the conical flask with Para film, Shake at about 140rpm on a horizontal shaker for 30 minutes. After shaking keep the flasks standing for 5 minutes for settling of suspended particles if any. Filter all the extraction solution through a fluted filter into a 150ml Beaker. From the beaker take 15ml of the filtrate into 50ml graduated measuring cylinder and to this add 30ml of H_2O.

If the sample is nutmeg, Oregano or black pepper, from the 150ml beaker take 10ml of the filtrate into 50ml graduated measuring cylinder add 40mL of 20 per cent Tween-20 solution. If the sample is not one of the above products mentioned, add 40ml of De-ionized water to the cylinder and mix.

Filter the contents of the graduated cylinder through a glass fiber filter into a150mL beaker. This filtrate will be used in the Immuno Affinity column/Aflatest column.

Immuno-Affinity Column Clean-up

Attach an Afla test-column to the pump stand. Pipette 10mL of filtrate on the column and allow it to absorb on column. Once the entire filtrate has passed through the column, rinse the column with 10mL of De-ionized water, Repeat Deionized water rinse. Place a 2 or 4 ml vial under the tip of the column and add 1mL of methanol to the column collect the eluate and again add 1ml H_2O to the column and collect the eluate in the same 2 or 4ml vial. This sample is now ready for injection into the HPLC-MSMS.

Injection-Sequence

- *(a)* Inject calibration standard (s).
- *(b)* Inject the recovery sample.
- *(c)* Inject the blank sample and verify the absence of analytes above 5 per cent of the recovery or sample concentration (s).
- *(d)* Inject sample extract (s).
- *(e)* Re-inject the calibration standard at the appropriate level at least after every 20 injections and at the end of the run to verify instrument response.

Note: If significant carry over is detected, inject wash solution as needed until it is reduced to an acceptable level.

Calculations

1. **For Quantitation of each compound of interest-** a-Review the chromatograms to verify that the analyte peaks are within the retention time windows and that the peaks are integrated correctly, b- Calculate the normalized peak for each component of interest by dividing the component response by the internal standard response, if any:

   $$\text{Normalized Response Component 1} = \frac{\text{Response of Component 1}}{\text{Response of Internal Standard}}$$

   c-Generate a linear curve fit to each analyte in standard curve using normalized response to concentration in tissue (µg/g or ppm), d- Standard curve must have a correlation coefficient greater than or equal to 0.995. and e- Blank must exhibit a response of less than 5 per cent of the recovery used contemporaneously in the set.

2. **For Confirmation**-a-Choose a standard or recovery containing the analyte of interest, b- Identify 2 product ion peaks in the sample and verify that their peaks are present with a signal to noise ratio ≥ 3. Auxiliary ions may be used if necessary, c-Identify the retention time of the two product ion peaks in the standard or recovery and in the sample of interest. The sample peak retention times must be within ± 5 per cent of the standard or recovery retention times, d- Calculate the ratio of the response of product ion #2 to product ion #1 in the standard or recovery for the analyte of interest: Ratio=Product ion#2/Product ion#1.

   Note-Ion ratio should be less than1. If not, then invert the ratio.

   e-Ion ratios determined for each analyte shall be within tolerance limits as described in the EU document 2002/657/EC incase of positive samples. Suggested tolerances are based on EU guidelines and range from ± 20 per cent for peaks greater than 50 per cent of the base peak and to ± 50 per cent for those less than or equal to 10 per cent of the base peak.

   Software provided in the instrument can be used for auto Quantitation by using linear regression (y=mx+b), where y=peak area/height, x= Analyte concentration in ppb/µg/kg, m=slope of curve, and b= intercept of y) for samples taking in to account dilution factor, if any.

Source

1. *Official analytical Methods of the ASTA,4th Edition, 1997 Method no.24.2.*
2. *Commission Decision 2002/657/EC of 12 August 2002, implementing Council directive 96/23/EC concerning the performance of analytical methods and interpretation of results.*
3. *AOAC Official Methods of Analysis (2005), Ch.49.2.18 Method,991.31*

iii. Estimation of Oxalic Acid in Solvent Extracted Sesame Flour by HPLC-UV Method

Apparatus

The analytical HPLC system equipped with a Waters 996 photodiode array detector. The separation was carried out using a Waters Atlantis dC18 steel cartridge (150 x 4.6 mm i.d.) with a particle diameter of 3lm, using a Waters Atlantis (20 x 4.6 mm) dC18 guard column to protect the analytical column.

Reagents

1. Methanol of HPLC gradient grade.
2. Sodium dihydrogen phosphate (NaH_2PO_4) and phosphoric acid.
3. Ethanol.
4. Standards of cis-aconitic, fumaric, quinic, l- (+)-tartaric acids.
5. Sodium pyruvate.
6. Ultrapure water.

Procedure

Homogenize the samples (approximately 100 g) till the formation of a fifth and homogeneous flour. Take five subsamples of fresh homogenised wheat to measure the pH and acidity. Determine the pH (2 g of flour with 50 mL of ultrapure water) by potentiometric measurement at T =200C with pH meter (AOAC, 1990). Determine the acidity by means of titration with NaOH 0.1m until pH 8.1 and the results are expressed in grams of anhydrous citric acid per kilogram (AOAC,1990).

Determination of Organic Acids by HPLC coupled to a diode array detector according to the method (Hernandez Suarez *et al.*, 2008) previously optimised. For each accession, weigh about 1 g of homogenized flour directly in poly-propylene tubes and mix with 2 mL of 80 per cent ethanol. Afterwards, put the tubes into an ultrasound bath at 50°C for 5 min and then centrifuge for 5 min at1090g. Carefully recover the supernatant to prevent the contamination with the dried pellet. Then, add another 2 mL of 80 per cent ethanol to the pellet and place in an ultrasound bath and centrifuge as described above. Recover the two supernatants in the same tube. Concentrate this liquid phase with nitrogen stream until the elimination of all ethanol; adjust the residue to 5 mL with ultrapure water and store at T=80°C in the freezer. Pass this dissolution through a 0.45-µm filter and through a strong anion-exchanger, which was previously preconditioned with 3 mL of ultrapure water. Elute the compounds by washing with two fractions of 1 mL of sodium dihydrogen phosphate (NaH_2PO_4 =20mm=to pH=1). Perform duplicate injections and use average peak areas for the quantification.

- ☆ Mobile phase Sodium dihydrogen Phosphate (NaH_2PO_4 =20mm to pH= 2.7).
- ☆ The injection volumes 10µL of both the standard solutions and sample extracts,

☆ Flow rate: 0.7 mL min^{-1}.

☆ Wavelength: 210 nm.

Identify the HPLC peaks by comparing the retention times and spectral data obtained from standards.

Source

1. *International Journal of Food Science and Technology 2012, 47, 627–632.*
2. *Nour, V.et al./Not. Bot. Hort. Agrobot. Cluj. HPLC Organic Acid Analysis inDifferent Citrus Juices under Reversed Phase Conditions.38 (1) 2010,44-48.*
3. *Vanda Pereira et al., HPLC-DAD methodology for the quantification of organic acids, furans and polyphenols by direct injection of wine samples. J. Sep. Sci.2010, 33, 1204–1215.*
4. *Manual of Methods of analysis of Foods (Cereals and Cereal Products), Lab Manual-3, FSSAI, New Delhi.*

iv. Determination of Free and Total Gossypolin Solvent Extracted Cotton Seed Flour by HPLC–UV Method

For Free Gossypol

(i) HPLC - dual-wavelength absorbance detector.

(ii) Zorbax Eclipse XDB-C18 column (4.6 mm × 250 mm, 5 μm particle).

(iii) C18 pre-column (4.6 mm × 20 mm, 5 μm) (Wang *et al.*, 1985).

(iv) The mobile phase - 90: 10 (v/v) methanol-0.5 per cent acetic acid aqueous.

(v) Solution flow rate of 0.8 ml/min..

(vi) The wavelength for UV detection was 254 nm.

(vii) Injection volume: 5 μl.

(viii) The assays can be performed at room temperature.

Source

1. *Yingfanetal.J.Biosci.Vol.29No.1, March,67-71,2004.*
2. *Manual of Methods of analysis of Foods (Cereals and Cereal Products), Lab Manual-3, FSSAI, New Delhi.*

v. Determination of Coffein Contents in Coffee-HPLC Method

Dissolve 1.0 gm of sample in100ml hotwater. Filter 20 ml through a Millipore filter (0.45μm) under vacuum and apply to a Bond Elut C 18 cartridge or equivalent under vacuum. Elute the caffeine with 5ml of mobile phase (0.005M Sodium acetate: tetrahydrofuran–95: 5 at pH5).Collect in a 10ml flask and make upto volume. Inject 20ul in to a Spherisorb ODS, C18, 5um packed column 25cm long x4 mm internal dia.Elute with the mobile phase at 1ml/min, observe absorbance at 280nm, calibrate with standard Caffeine solution, 0-1 mg Caffeine in 10 ml mobile phase. For routine purposes the HPLC step can be eliminated and the absorbance of eluant from the cartridge measured at 280nm in a spectro-photometer.

Source

1. *Pearson's Composition and Analysis of Foods 9th edn,1991, page 373.*
2. *Manual of Methods of analysis of Foods (Beverages - Coffee, Tea, Sugar, Sugar Products etc.), Lab Manual-4, FSSAI, New Delhi.*

7. Application of Chromatographic Techniques in Dairy

The details related to some of the application of Chromatographic techniques in Dairy are given below:

A. HPLC

i. Analysis of Melamine and Related Compounds

Melamine is a type of organic nitrogen compound with a triazine ring at the center of its structure. Melamine and formaldehyde are the main raw materials used to make melamine resin. HPLC is commonly used for the screening analysis and quantitative analysis of melamine. GC/MS or LC/MS is used for qualitative checks and for batch analysis of other reaction by-products. HPLC-PDA-A liquid medium flows through the flow channels of a High-Performance Liquid Chromatograph (HPLC). When a sample is introduced into the instrument, the components of the sample are separated in a column and the amount of each component is measured by a detector. HPLC-PDA can rapidly analyze melamine while separating out the impurities in the sample. The PDA detector permits quantitative analysis of melamine while confirming that no impurity peaks overlap.

The analysis of a commercial dairy product that was similarly spiked with melamine to a concentration of 10 mg/L, can be done. The commercial dairy product was dissolved ten times with water and then diluted an additional two times with mobile phase before centrifuging (12000 rpm for 5 minutes) and filtering through a membrane filter to produce the analysis sample. To ensure separation from impurities, the analysis conditions were changed from the analysis conditions for the melamine standard solution (A/B=90/10). The 0.5 mg/L melamine after dilution could be accurately detected. A good melamine recovery rate of 99 per cent was achieved for this sample.

ii. HPLC-MS Analysis of Chloramphenicol Residues in Milk and Powdered Milk Products

Chloramphenicol (CAP) is a broad-spectrum antibiotic with bacteriostatic action but also has toxic properties, which is why its presence in food and feed is prohibited in Croatia and the European Union. In the aim of consumer protection it is essential to develop a sensitive analytical method for detection of CAP fractions lower than w= 0.3 $\mu g\ kg^{-1}$. For the efficient control and monitoring of CAP, a rapid, sensitive, and selective method for its identification and quantification, using high performance liquid chromatography in combination with mass spectrometry LC-MS, has been developed. The cleaning procedure was based on the AOAC official method 993.32. HPLC-MS analysis used the ODS Hypersile column and the water/acetonitrile gradient. Electrospray negative ionization (neg ESI) was used before single ion monitoring (SIM) detection of three m/z 321, 323 and 325. As additional criteria, the ratio between these masses in real and spiked milk samples was also investigated in accordance with theoretical values of the isotope pattern for 2 chlorine atoms present in the analyte. The detection limit of 0.1 µg kg^{-1}- was achieved. The mean value of recovery was 94 per cent, the correlation coefficient of the calibration curves calculated for 2 m/z values was higher than 0.99.Fourty samples of milk and milk products were tested with the HPLC-MS method, and obtained results showed that samples had CAP 0.37, 0.29, 0.39 µg kg^{-1}-, respectively. All the other analysed samples contained CAP concentrations below the detection limit.

Source

1. *Chemistry in Industry, Vol.56 No.2 Feb.2007.*

iii. Milk and Milk Products – Determination of Lactose Content by HPLC (Reference Method)- IDF 198/ISO 22662 (FDIS)

Principle and Method

It involves the following steps:

(i) Sample + Internal Std[1] —Biggs Solution[2]→ Precipitation of fat and protein

(ii) Filtration of the sample

(iii) Injection through an analytical column

(iv) Detection by a differential refractometer

(v) Quantification

Internal Std: Melezitose[1] (Glu-Fru-Fru) and Zinc acetate + Glacial acetic acid+ Phosphotungstic acid[2]

Chromatographic Conditions

(i) Mobile phase: degassed HPLC grade water.

(ii) Flow rate: 0.6ml/min.

(iii) Internal detector temperature: 35°C.

(iv) Column temperature: 85° C.

(v) Volume to be injected: 20 µl.

(vi) Run time: 15 min.

(vii) Column type-Styrene divinylbenzene resin.

Source

1. *IDF International Symposium -Lactose and it derivatives.*

2. *Sbrons, C, and Olieman, C, 1983. Study of the high performance liquid chromatography separation of reducing sugars applied to the determination of lactose in milk, J. Chromato., 259, 79.*

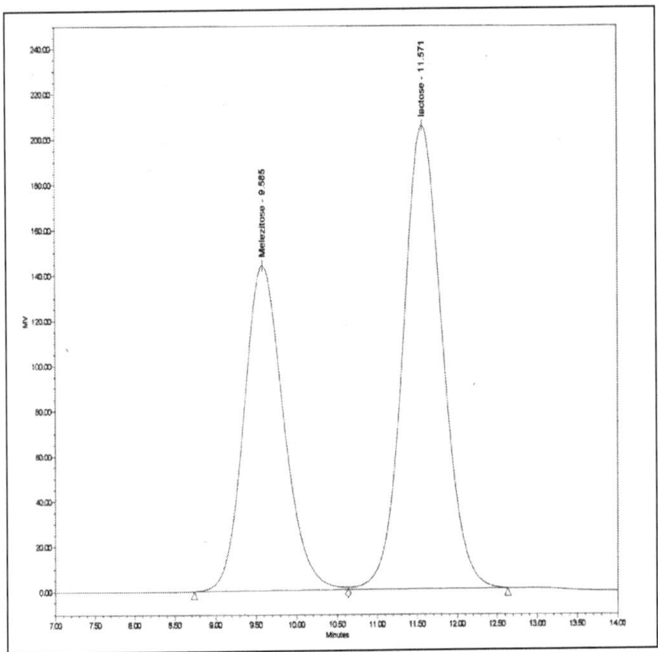

Chromatogram of Raw Milk Sample.

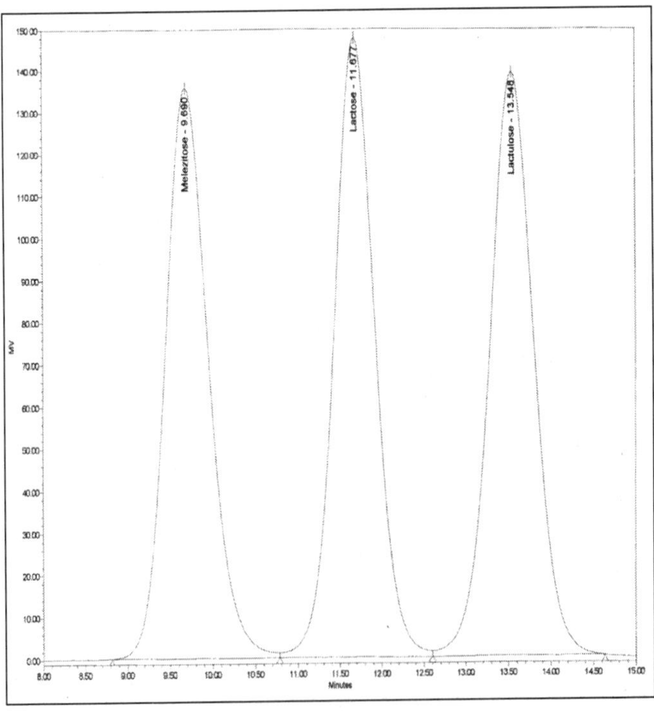

Chromatogram of Raw Milk Sample with Added Lactulose.

3. *IDF 141C: 2000 -Whole Milk- Determination of milkfat, protein and lactose content, Guidance on the operation of Mid-infrared instruments.*
4. *IDF 79-1: 2002%ISO 5765-1- Dried milk, dried ice-mixes and processed cheese-Determination of lactose content- Part 1: Enzymatic methodutilizing the glucose moiety of the lactose.*
5. *IDF 79-2: 2002/ISO 5765-2- Dried milk, dried ice-mixes and processed cheese-Determination of lactose content- Part 2: Enzymatic method utilizing the galactose moiety of the lactose.*
6. *IDF 106: 2004%ISO 5548- Caseins and caseinates- Determination of lactose content-Photometric method.*
7. *Official Methods of Analysis, 16th Ed., 4th Revision, 1998, AOAC International, Gaithersburg, MD, methods 896.01, 984.15, 930.28 and 972.16.*
8. *Nickerson, T. A., Vujicic, I. F., and Lin, A. Y, 1976. Colorimetric estimation of lactose and its hydrolytic products. J. Dairy Sci., 59, 386.*

iv. Milk and Milk Products – Determination of Aflatoxin M_1 Content

Clean -up by immunoaffinity and determination by HPLC- IDF 171/ISO 14501. The details related to the operation of HPLC and quantitative estimation of Aflatoxin $M_1$are given in chapter-09 of author's another book entitled Hand Book on Analysis of Milk.

v. Determinaton of Denatured Whey Protein Content in Milk Powder

HPLC based IDF method is available for quantifying the denatured whey protein content in milk powder under IDF (2002), Dried Milks –Assessment of Heat treatment Intensity- HPLC Method, IDF Standard162.

vi. Determination of Antibiotic/Veterinary Drugs Residues and Pesticide Residues

The details are given in chapter-09 of author's another book entitled Hand Book on Analysis of Milk. In addition to these applications, HPLC is also being used for determination of followings [1]:

Sl.No.	*Pesticide Group*	*HPLC Analytical Technique*
1	Benzimidazoles (Carbendazim, Thiabendazole, Albendazole, Fenbendazole).	HPLC with UV and fluorescence detection. GC not suitable but still necessary for confirmation of the HPLC analysis.
2	Carbamates (Aldicarb, Carbaryl, Methomyl, Methiocarb, Carbofuran)	HPLC with post column modification/ derivatization to form fluorescent compounds. GCnot commonly used.
3	Alkyl Phenolic Acid Herbicides (2,4-D [2,4-Dichlorophenoxy acetic acid]).	HPLC with UV detector. GCpossible after esterification.

Source

1. *Jeon, I.J. and Ikins, W.G. (1995). Analyzing Food for Nutrition Labeling and Hazardous Contaminants. Marcel Dekker, Inc. USA, pp. 333-362.*
2. *Technews Issue No.47 (Nov- Dec 2003), NDDB.*

vii. Determination of Natamycin Content in Cheese, Cheese Rind and Processed Cheese

IDF has prescribed a High Performance Liquid chromatographic method under IDF 140-2/ISO 9233-2 for determination of Natamycin content.

viii. Testing Adulteration of Milk [1,2,3,4]

(a) The milk may be adulterated by the addition of soybean proteins and hence several testing methods are available to detect soy proteins in unheated milk.

(b) Various methods have also been developed to detect the presence of bovine, ovine, and caprine (goat) proteins in mixtures of milk or cheeses. RP-HPLC-UV methods have been developed, optimised and validated for the separation and quantification of α-, β-, and k- caseins with the idea of using such methods for the detection of the adulteration of milk.

(c) RP-HPLC has been used to find alternative β-lactoglobulin markers.

(d) An SE-HPLC (size exclusion HPLC) detection of the marker glycomacropeptide (GMP) have been developed to detect adulteration of skim milk powder with solids from whey or buttermilk.

ix. Estimation of Vitamin A in Fortified Milk [6]

Vitamin A deficiency (VAD) exists in more than 60 countries. It is the second largest cause of global blindness next to cataracts. Vitamin A fortification is an effective strategy to control this Vitamin A deficiency and Vitamin A content in the fortified samples can be estimated by HPLC. The details related to the principle and method for determination of Vitamin A by HPLC are given below:

a. Principle

The principle of separation is adsorption. Sample (Mixture of component) is introduced into the column. Various chemical and/or physical interactions take place between the sample molecules and the particles of the column packing and they travel according to their relative affinities towards the stationary phase (More affinity – travel slower and vice versa). The separated components are detected by a flow-through device (detector) that measures their amount. The output from the detector is called as liquid chromatogram.

b. Method

The stepwise details are given below:

Preparation of vitamin A standard (all trans retinol) curve:

Stock Solution

13.2 mg retinol was accurately weighed and transferred into 10 mL low actinic volumetric flask and dissolved in HPLC grade methanol and volume was made up to 10 mL. This had vitamin A concentration of 4356000IU/L.

Standard Solutions

The stock solution was diluted to such an extent that the resulting solutions were having concentration of 100, 1000, 2000, 3000 and 4000 IU/L. The standard solutions were stored in a low actinic flask (amber) which remained stable for at least one and four weeks at 4 and -20°C, respectively.

Sample Preparation

Milk samples were pasteurised at 63°C for 30 minutes in high precision water bath in air tight glass bottles. The samples were immediately cooled to 4°C. After 2 hrs of storage at 4°C, milk samples were analysed for vitamin A content.

Extraction of Vitamin A from Milk

2 mL milk was taken in centrifuge tube. 1 mL of 2.5 per cent ammonium hydroxide solution was added to it, the contents were sonicated for 10 minutes and then vortexed. 1 mL 0.1 M HCl was added, vortexed and sonicated for 10 minute again. 1 mL of aqueous potassium hydroxide (60 per cent, w/v) and 100 µl of alcoholic pyrogallol (30 per cent, w/v) were added. The contents were flushed with ultrapure nitrogen gas and then kept for saponification at 70°C for 30 minutes in water bath. The contents were then cooled to room temperature in an ice bath and vortexed. 2 mL hexane was added, contents were vortexed and a few drops of ethanol were added and then centrifuged at 1800 rpm for 10 minutes. Upper hexane layer was transferred to another centrifuge tube with the help of a pipette. This extraction step was repeated twice by adding 2 mL hexane each time and all the extracts were collected in centrifuge tubes. The collected extracts were dried under nitrogen. The dried material was reconstituted to 1 mL of mobile phase, transferred to a 2 mL vial and centrifuged in refrigerated centrifuge at 3500 rpm for 15 minutes at 0°C. The supernatant was then filtered through 0.22 µm syringe filter and 20 µl of filtrate was then injected in HPLC system.

Instrumentation

As mentioned earlier.

Chromatographic Conditions

Reverse phase HPLC analysis procedure was standardised for vitamin A, selecting acetonitrile: chloroform: methanol (88: 4: 8) as mobile phase, with flow rate 1mL/minute, in C18 column, temperature 40°C and PDA detector (lmax 325 nm).

Source

1. *Rajesh Kumar Bajaj and Priyanka Singh Rao, HPLC and its Application in Food Analysis, Dairy Chemistry Division, Compendium of Lectures on Advanced Techniques and Novel Approaches for Quality and Safety Evaluation of Dairy Foods-2015, NDRI, Karnal.*

2. *Settle, F. (1997) Hand Book of Instrumental techniques for analytical chemistry, Pretice – Hall International, NJ.*

3. *Swadesh, J. (1997) HPLC–Practical and Industrial Applications, CRC, Press, Boca Raton, FL.*

4. R.Matissek, and R.Wittkowski, High Performance Liquid Chromatography in Food Control and Research. Technomic Publishing Co. Inc., Lancaster-Basel, pp. 87-296.

5. S.S. Nielsen, (1994) Introduction to Chemical analysis of foods, Jones and Bertlett Publishers International, London, PartIV.Oliver, RWA (1989) HPLC of macromolecules-a practical approach, IRL Press, Oxford.

6. Sumit Arora, Chitra Gupta, Apurva Sharma and Prince Chawla, Estimation of Vitamin A in Fortified Milk, Dairy Chemistry Division, NDRI, Karnal.

International Standards for Analysis of Milk Products by HPLC [1]

(a) Some important international standards that have been specified for HPLC analysis of specific components of milk and milk products are given in following table:

Sl.No.	Component	Product	International Standard
1	Vitamin A	Dried skimmed milk	IDF 142:1990
2	Vitamin D	Dried skimmed milk	IDF 177:2002
3	Aflatoxin M1	Milk and milk powder	ISO 14501/IDF 171:2007 and ISO 14674/IDF 190:2005
4	Benzoic and sorbic acid	Milk, dried milk, yoghurt and other fermented milks	IDF 139:1987
5	Natamycin	Cheese, cheese rind and processed cheese	ISO 9233-2/IDF 140-2:2007
6	Lactose	Milk and milk products	ISO 22662/IDF 198:2007
7	Lactulose	Heat treated milk	ISO 11868/IDF 147:2007
8	Heat treatment intensity	Dried milk	ISO 11814/IDF 162:2002
9	Acid soluble β-lactoglobulin	Liquid milk	ISO 13875/IDF 178:2005

Source

1. Technews Issue No.74 (May-June 2008), NDDB.

(b) Some important international standards that have been specified for HPLC/GC analysis of Aflatoxin M_1 in milk and milk products [1,2,3] are given below:

Sl.No.	Component	Product	GC.TLC/HPLC	International Standard
1	Aflatoxin M_1	Milk and Milk Products	TLC	AOAC 980.21
		Milk and Milk Products	Immunoaffinity column and HPLC	IDF 171: 1995 ISO 14 501: 1998
		Milk and Milk Products	Immunoaffinity column and LC-applicable for determination at > 0.02 ppb (ppb=μg/kg)	AOAC 986.16.AOAC 2000.08
2	Antibiotic	Milk and Milk Products	LC, applicable to tetracycline, chlortetracycline and oxytetracycline; validated detection level are lower than Codex MRLs	AOAC 995.04

Source

1. *Technews Issue No.47 (Nov-Dec2003), NDDB.*
2. *Codex (1999). Methods of Analysis and Sampling. Standard no. XOT* **16**. *Codex Alimentarius Commission, Rome.*
3. *19. IDF (2000). Inventory of IDF/ISO/AOAC International Adopted Methods of Analysis and Sampling for Milk and Milk Products. Bulletin no. 350.International Dairy Federation, Brussels, pp. 4-21.*

B. Gas Chromatography Mass Spectrometry (GC/MS) [1,2,3]

It is an analytical method that combines the features of gas-chromatography and mass spectrometry to identify different substances within a test sample [1,2]. A Gas Chromatograph Mass Spectrometer (GC/MS) uses a mass spectrometer as the detector. This technique is used for the screening of target compounds, due to its excellent ability to confirm that a peak really is the target component (qualitative capacity) that permits the measurement of the masses of compounds and fragments. It is a method that combines the features of gas-liquid chromatography and mass spectrometry to identify different substances within a test sample.

History and Year-wise Development of GCMS [1]

The details related to the history and year wise development of GCMS are tabulated below:

Sl.No.	*Year*	*Developments*
1	1959	The first on-line coupling of gas chromatography to a mass spectrometer was reported [2,3,4].
2	1964	Electronic Associates, Inc. (EAI), a leading U.S. supplier of analog computers, began development of a computer controlled quadrupole mass spectrometer under the direction of Robert E. Finnigan [5].
3	1966	Finnigan and collaborator Mike Uthe's EAI division had sold over 500 quadrupole residual gas-analyzer instruments [6].
4	1968	The first prototype quadrupole GC/MS instruments was delivered to Stanford and Purdue University [5].
5	1996	The top-of-the-line high-speed GC-MS units completed analysis of fire accelerants in less than 90 seconds.
6	By 2000	Computerized GC/MS instruments using quadrupole technology had become both essential to chemical research. Now computerized GC/MS instruments are widely used in environmental monitoring of water, air, and soil; in the regulation of agriculture and food safety; and in the discovery and production of medicine.

Source

1. *Wikipedia.*
2. *Gohlke, R. S. (1959). "Time-of-Flight Mass Spectrometry and Gas-Liquid Partition*

Chromatography". Analytical Chemistry. **31** *(4): 535–541. ISSN 0003-2700.*

3. *Gohlke, Roland S.; McLafferty, Fred W. (1993). "Early gas chromatography/mass spectrometry". Journal of the American Society for Mass Spectrometry. 4 (5): 367–371. ISSN 1044-0305.*
4. *Hites, Ronald A. (2016). "Development of Gas Chromatographic Mass Spectrometry". Analytical Chemistry. 88 (14): 6955–6961. ISSN 0003-2700.*
5. *Brock, David C. (2011). "A Measure of Success". Chemical Heritage Magazine. 29 (1). Retrieved 22 April 2014.*
6. *Webb-Halpern, Leah (2008). "Detecting Success". Chemical Heritage Magazine.* **26** *(2): 31.*

Instrumentation and Working [1,2,3]

The GC-MS is composed of a gas chromatograph and a mass spectrometer.

i. Gas Chromatograph

It utilizes a capillary column which depends on the column's dimensions and phase properties. The difference in the chemical properties between different molecules in a mixture will separate the molecules as the sample travels the length of the column. The molecules take different amounts of time (called the retention time) to come out of (elute from) the gas chromatograph, and this allows the mass spectrometer downstream to capture, ionize, accelerate, deflect, and detect the ionized molecules separately.

ii. Mass Spectrometer

It does this by breaking each molecule into ionized fragments and detecting these fragments using their mass to charge ratio. The mass spectrometry process normally requires a very pure sample while gas chromatography using a traditional detector (*e.g.* Flame Ionization Detector-FID) detects multiple molecules that happen to take the same amount of time to travel through the column which results in two or more molecules to co-elute.

Combining the two processes reduces the possibility of error, as it is extremely unlikely that two different molecules will behave in the same way in both a gas chromatograph and a mass spectrometer.

In GC-MS, the separated compounds from GC are detected and further analyzed by mass spectrometric technique. In mass spectrometry, a compound is ionized and the resulting ions are separated according to their mass-to-charge ratio, and the ion abundances are measured. By using GC/MS, we can identify the following:

(i) What compounds are contained in a sample.

(ii) Determine whether certain substances/compounds are present in a sample.

(iii) Determine the amount (per cent weight, per cent composition) of specified

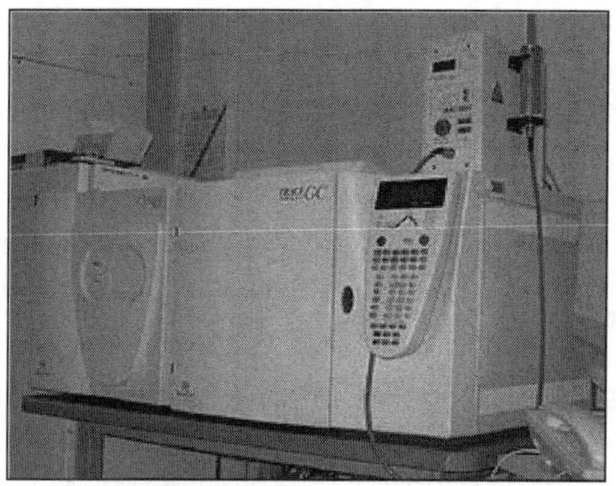

GC-MS instrument (Wikipedia).

The insides of the GC-MS, with the column of the gas chromatograph in the oven on the right (Wikipedia).

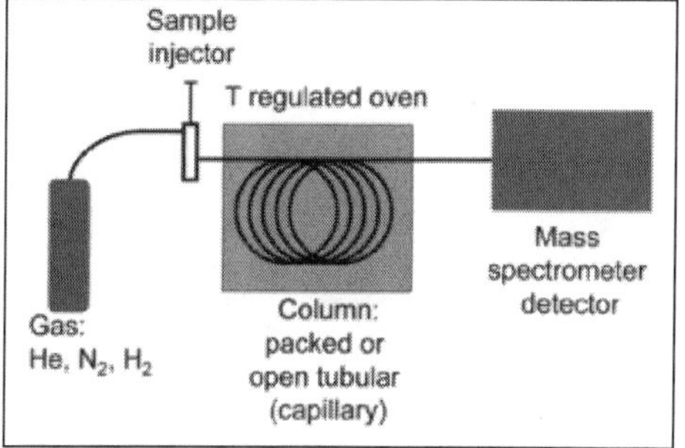

GC-MS Schematic (Wikipedia).

GCMS Thermofisher Scientific.

Examples of some GC-MS Instrument.

compounds in a sample.

The details of Gas chromatogreaph and Mass spectrometer are given in respective section of the concerned chapters. Some special requirements for combined GCMS, are given below:

Purge and Trap GC-MS

For the analysis of volatile compounds, a purge and trap (P and T) concentrator system may be used to introduce samples. The volatile compounds are drawn along a heated line onto a 'trap'. The trap is a column of adsorbent material at ambient temperature that holds the compounds by returning them to the liquid phase. The trap is then heated and the sample compounds are introduced to the GC-MS column via a volatiles interface, which is a split inlet system. P and T GC-MS is particularly suited to volatile organic compounds (VOCs) and BTEX compounds (aromatic compounds associated with petroleum) [3,4].

Mass Spectrometer Detectors

The most common type of mass spectrometer (MS) associated with a gas chromatograph (GC) is the quadrupole mass spectrometer, (trade name of HP/ Agilent -"Mass Selective Detector" -MSD). Another relatively common detector is the ion trap mass spectrometer.

Ionization

After the molecules travel the length of the column, pass through the transfer line and enter into the mass spectrometer they are ionized by various methods. Once the sample is fragmented it will then be detected, usually by an electron multiplier diode, which essentially turns the ionized mass fragment into an electrical signal that is then detected. Several ionisation techniques *e.g.* Electron ionization, Cold electron ionization, Chemical ionization *etc.* are being used [3]. The details of some of the ionisation techniques are discussed under Mass spectrometry in chapter 3.

Analysis [3]

A mass spectrometer is typically utilized in one of following two ways:

a. Full Scan

A "full spectrum" analysis considers all the "peaks" within a spectrum. Full scan is useful in determining unknown compounds in a sample. It provides more information than SIM when it comes to confirming or resolving compounds in a sample.

b. Or Selective Ion Monitoring (SIM)

Conversely, selective ion monitoring (SIM) only monitors selected ions associated with a specific substance. SIM analysis allows for a smaller quantity of a compound to be detected and measured, but the degree of certainty about the identity of that compound is reduced.

The typical GC-MS instrument is capable of performing both functions either individually or concomitantly, depending on the setup of the particular instrument [3]. Two kinds of analysis are possible, comparative and original.

- *(i)* **Comparative Analysis**: It essentially compares the given spectrum to a spectrum library to see if its characteristics are present for some sample in the library. This is best performed by a **computer** because there are a myriad of visual distortions that can take place due to variations in scale.
- *(ii)* Another method of analysis measures the peaks in relation to one another. In this method, the tallest peak is assigned 100 per cent of the value, and the other peaks being assigned proportionate values. Total mass of the unknown compound is normally indicated by the parent peak.

Source

1. *O. David Sparkman; Zelda Penton; Fulton G. Kitson (17 May 2011). Gas Chromatography and Mass Spectrometry: A Practical Guide. Academic Press. ISBN 978-0-08-092015-3.*

2. *Nicole E. Rabaud, Susan E.Ebeler, Lowell L. Ashbaugh, and Robert G. Flocchini, Crocker Nuclear Laboratory and Department of Viticulture and Enology, University of California, One Shields Avenue, Davis, California 95616, J. Agric. Food Chem.,2002, 50 (18), pp 5139–5145.*

3. *Wikipedia.*

4. *Optimizing the Analysis of Volatile Organic Compounds – Technical Guide" Restek Corporation, Lit. Cat. 59887A.*

Application of GC-MS

Typical applications of GC/MS is Pesticide residues and pollutants in water agricultural products foodstuff, organic solvents in packaging materials, ink, *etc.*, Drugs of abuse in urine, blood, tablets fatty acid contents in edible oils fat, Essential oil characterization, Allergens in fragrance, polymer characterization, *etc.*

A. Multiple Applications in different Fields

GC-MS are being successfully used in drug detection, fire investigation, environmental analysis, explosives investigation, airport security to detect substances in luggage or on human beings and identification of unknown samples [1]. It is also being used in the following fields:

(i) Environmental monitoring and cleanup.
(ii) Criminal forensics.
(iii) Law enforcement.
(iv) Sports anti-doping analysis.
(v) Security.
(vi) Chemical warfare agent detection.
(vii) Chemical engineering.
(viii) Astrochemistry.
(ix) Medicine.

B. Food, Beverage Analysis [1]

Foods and beverages contain numerous aromatic compounds (some naturally present in the raw materials and some forming during processing). GC-MS is extensively used for the analysis of these compounds which include esters, fatty acids, alcohols, aldehydes, terpenes *etc.* Gas chromatography-mass spectrometry (GC-MS) is an important technique for qualitative and quantitative analysis of food contaminants and residues. It is fast and sensitive, provides a high peak capacity and allows determination of thermally stable and volatile compounds [2]. More sophisticated GC-MS instruments are also used in many other areas of food analysis, especially in the identification of toxic residues (*e.g.*, dioxins) [3].

C. In Dairy Industry

In recent years, Gas Chromatography Mass Spectrometry (GC/MS) has been widely adopted in the field of food analysis. In addition to the aforesaid applications, it is now being used in analysis of Dairy Products/Food [1]. Some examples are given below:

(i) **Thermal Desorption GC/MS with Simultaneous Olfactory Evaluation for the Characterization and Quantification of Odour Compounds from a Dairy**-Samples were analyzed by thermal desorption into a cryotrap and sub-sequent gas chromatographic separation, followed by simultaneous olfactometry and mass spectrometry. Twenty odorous and nonodorous compounds were identified and quantified, including straight-chain and aromatic hydrocarbons, chlorinated compounds, alcohols, ketones, aldehydes, and organic acids, at air concentrations of 0.55–320.20 $\mu g/m^3$. Compound peaks were characterized by odors ranging from offensive to pleasant, demonstrating the integrative nature of olfaction. This method could be useful in studying many kinds of odors in air.

Source

1. *Wikipedia.*
2. *P. Vazquez-Roig, Y. Pico, in Chemical Contaminants and Residues in Food, 2012.*
3. *F. A. Mellon, in Encyclopedia of Food Sciences and Nutrition (Second Edition), 2003.*

ii. Analysis of Melamine in Dairy Products

The analysis of milk and dairy products for melamine requires the use of SPE to remove interferences caused by the high fat and sugar content of the matrix. A method including strong cation exchange, SPE, derivatization, and GC/MS analysis has demonstrated that melamine in dairy product scan successfully be analyzed by GC/MS well below the limits required by regulation.

The batch analysis of melamine and related compounds using a GCMS-QP2010 Plus instrument was done. Analysis was conducted on four types of melamine and related compounds and on TMS (trimethylsilyl) derivative spiked with benzoguanamine as the internal standard. Satisfactory linearity was obtained in the range from 5 ng to 50 ng.

Source

1. *Screening for Melamine Adulteration in Protein-Based Foods by GC/MS" James Neal-Kababick, Flora Research Laboratories, William Goodman, PerkinElmer (007969A_01).*

C. Liquid Chromatography Mass Spectrometry (LC/MS)

In recent years, Liquid Chromatography Mass Spectrometry (LC/MS) has been widely adopted in the field of food analysis. LC/MS is an analytical technique which combines the physical separation capabilities of liquid chromatography (or HPLC) with the mass analysis capabilities of mass spectrometry. Liquid chromatography separates mixtures with multiple components, mass spectrometry provides structural identity of the individual components with high molecular specificity and detection sensitivity. LC-MS is a powerful technique used for many applications which has very high sensitivity and selectivity [1].

Advantages of LCMS

(i) Chromatography offers great resolution, but accurately qualifying and quantitating substances can be difficult if multiple components elute approximately at the same time, such as during simultaneous multianalyte analysis. In contrast, mass spectrometry (MS) offers a highly sensitive detection technique that ionizes the sample components using various methods, then separates the resulting ions in vacuum based on their mass-to-charge ratios and measures the intensity of each ion. Since the mass spectra provided by MS can indicate the concentration level of ions that have a given mass, it is extremely helpful for qualitative analysis. That is because mass is information particular to specific molecules and MS enables obtaining that information directly. Therefore, LC-MS systems combine the outstanding separation resolution of liquid chromatography with the outstanding qualitative capabilities of mass spectrometry. In addition, SIM (selected ion monitoring) measurements detect substances based on mass, which is a highly-selective parameter. This enables quantitative analysis that avoids the effects of impurities even when separation by LC is inadequate. In terms of providing both broad applicability for a wide range of substances and high selectivity, mass spectrometers offer excellent characteristics as an LC detector. [1].

(ii) **The advantage of LC/MS over GC/MS- a**-Mass spectrometers have been used as a detector for gas chromatography (GC) systems since the relatively early days of mass spectrometry and their advantages are widely recognized. GCMS provides a very effective means for separating and qualifying substances, but its applicability is limited to gases or volatile compounds with relatively low molecular mass, and samples with high thermal stability. In contrast, if the substance is dissolved in a mobile phase (liquid), then LC is able to analyze even least volatile or thermally-unstable compounds that are difficult to analyze using GCMS. In other words, one of the advantages of LC-MS is its broad applicability [1]. **b**-In comparison with HPLC and the UV detection method in particular, LC/MS offers unsurpassed selectivity and superior qualitative and quantitative analysis. It was found that LC/MS achieved superior selectivity and sensitivity than HPLC [2].

Source.

1. *Introduction to LC-MS Part1- Shimadzu Corporation.*

2. *Wikipedia.*

History and Year-wise Development of LCMS [1]

The details related to the history and year wise development of LCMS are tabulated on next page:

Sl.No.	Year	Developments
1	1950s	The coupling of chromatography with MS is a well developed chemical analysis strategy.
2	1952	Gas chromatography (GC) - MS was originally introduced.
3	1970	LC-MS systems were first commercialized [2]. The development of LC-MS systems took longer than GC-MS and was directly related to the development of proper interfaces. V.L. Tal'roze and collaborators started the development of LC-MS in the early 1970s, when they first used capillaries to connect LC columns and MS ion sources [3,4].
4	1973	A similar strategy was investigated by McLafferty and collaborators. This was the first and most obvious way of coupling LC with MS, and was known as the capillary inlet interface.

Source

1. *Wikipedia.*
2. *Pitt, James J (2017-03-12). "Principles and Applications of Liquid Chromatography-Mass Spectrometry in Clinical Biochemistry". The Clinical Biochemist Reviews. 30 (1): 19–34. ISSN 0159-8090.*
3. *Niessen, Wilfried M. A (2006). Liquid Chromatography-Mass Spectrometry, Third Edition. Boca Raton: CRC Taylor and Francis. pp. 50–90. ISBN 9780824740825.*
4. *Tal'roze, V.L.; Gorodetskii, I.G.; Zolotoy, N.B; Karpov, G.V.; Skurat, V.E.; Maslennikova, V.Ya. (1978). "Capillary system for continuous introducing of volatile liquids into analytical MS and its application". Adv. Mass Spectrom. 7: 858.*

Instrumentation

Mass spectrometer systems include a device for introducing samples (such as an HPLC or GC unit), an interface for connecting such device, an ion source that ionizes samples, an electrostatic lens that efficiently introduces the generated ions, a mass analyzer unit that separates ions based on their mass-to-charge (m/z) ratio, and a detector unit that detects the separated ions. LCMS system broadly consists of the following parts:

(i) **Liquid chromatography-** Details given in respective section of chapter 4.

(ii) **Mass spectrometer-** Details given in respective section of chapter 3.

(iii) **Interface-**LC-MS system contains an interface that efficiently transfers the separated components from the LC column into the MS ion source [2,3]. It is necessary because the LC and MS devices are fundamentally incompatible. While the mobile phase in a LC system is a pressurized liquid, the MS analyzers commonly operate under vacuum (around 10^{-6} torr). Thus, it is not possible to directly pump the eluate from the LC column into the MS source. The interface is a mechanically simple part of

the LC-MS system that transfers the maximum amount of analyte, removes a significant portion of the mobile phase used in LC and preserves the chemical identity of the chromatography products (chemically inert). Now-a-days, most extensively applied LC-MS interfaces are based on atmospheric pressure ionization (API) strategies like electrospray ionization (ESI), atmospheric pressure chemical ionization (APCI), and atmospheric pressure photo-ionization (APPI). These interfaces became available in the 1990s after a two decade long research and development process [3,4].

Currently, the most common LC-MS interfaces are electrospray ionization (ESI), atmospheric pressure chemical ionization (APCI), and atmospheric pressure photo-ionization (APPI). These are newer MS ion sources that facilitate the transition from a high pressure environment (HPLC) to high vacuum conditions needed at the MS analyzer [3,5]. These can also be commercially available as dual ESI/APCI, ESI/APPI, or APCI/APPI ion sources [4]. The following different types of interfaces are also being used:

(i) Moving-belt interface.

(ii) Direct liquid introduction interface.

(iii) Thermo- spray interface.

(iv) FAB Based Interfaces.

The atmospheric ionization unit is based on electrospray ionization (ESI), atmospheric chemical ionization (APCI), or other ionization methods, and serves as the ion source and interface with the HPLC system. Ions generated in this unit are stripped of solvent, then focused into a beam using an octupole or other means, then delivered to a quadrupole. Both a direct current and high frequency alternating current are applied to the quadrupole, so that only the ions with the target m/z ratio successfully pass through the quadrupole. The quantity of ions that reach the detector is converted to a signal and output to a computer [6].

Agilent, Brucker, Perkin Elmer, SCIEX, Shimadzu, Thermofisher and Water corporation are the major manufacturers of LCMS.

Source

1. *Wikipedia.*

2. *Dass, Chhabil (2007-01-01).Fundamentals of Contemporary Mass Spectrometry.John Wiley and Sons, Inc.pp. 151-194.ISBN 97804701 18498.*

3. *Pitt, James J (2017-03-12). "Principles and Applications of Liquid Chromatography-Mass Spectrometry in Clinical Biochemistry". The Clinical Biochemist Reviews. 30 (1): 19–34. ISSN 0159-8090.*

4. *Niessen, Wilfried M. A (2006). Liquid Chromatography-Mass Spectrometry, Third Edition. Boca Raton: CRC Taylor and Francis. pp. 50–90. ISBN 9780824740825.*

5. *Arpino, Patrick (1992)."Combined liquid chromatography mass spectrometry. Part III. Applications of thermospray". Mass Spectrometry Reviews. 11: 3.*

Agilent Technology LCMS.

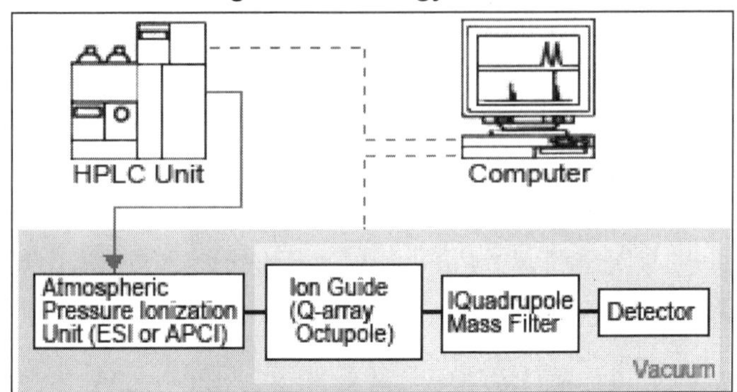

Components of an LC-MS System
(*Source*: Introduction to LC-MS Part1- Shimadzu Corporation).

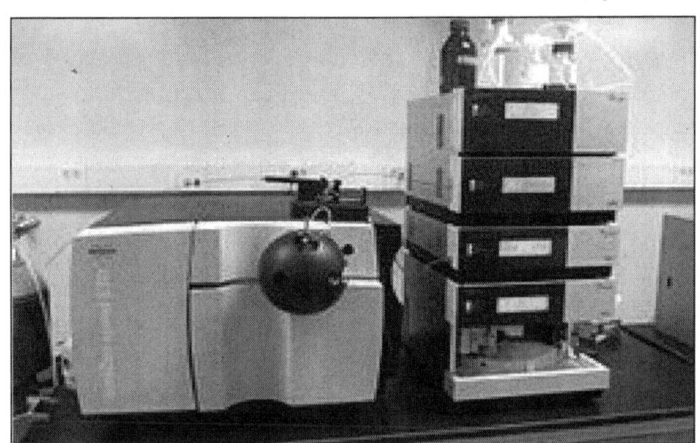

Ion Trap LCMS System with ESI Interface (Wikipedia).

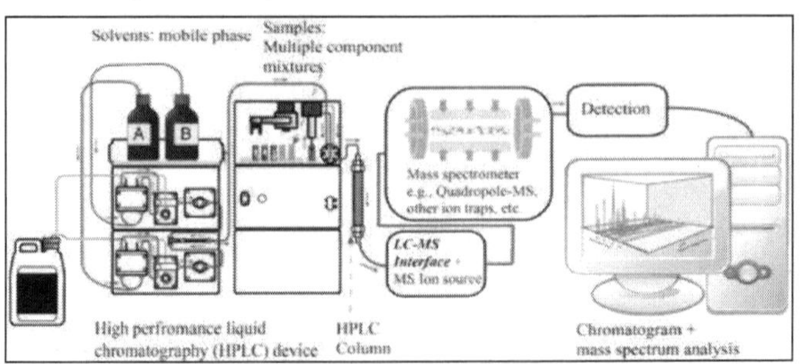

Diagram of an LC-MS System (Wikipedia).

6. *Introduction to LC-MS Part1- Shimadzu Corporation.*

Applications of LCMS

LC-MS may be applied in a wide range of sectors including biotechnology, environment monitoring, food processing, and pharmaceutical, agrochemical, and cosmetic industries [1,2,3]. The coupling of MS with LC systems is attractive because liquid chromatography can separate delicate and complex natural mixtures, which chemical composition needs to be well established. Now-a-days, LC-MS has become one of the most widely used chemical analysis techniques because more than 85 per cent of natural chemical compounds are polar and thermally labile and GC-MS cannot process these samples. LCMS is being used in the following fields:

i. Pharmacokinetics [1]

Pharmacokinetic studies are needed to determine how quickly a drug will be cleared from the body organs and the hepatic blood flow. LC-MS is widely used in the field of bioanalysis and is specially involved in pharmacokinetic studies of pharmaceuticals. Because of their shorter analysis time, and higher sensitivity and specificity compared to UV detectors, MS analyzers are useful in these studies.

ii. Proteomics [1,5]

LCMS is used in proteomics as a method to detect and identify the components of a complex mixture. This approach generally involves protease digestion and denaturation using trypsin as a protease, urea to denature the tertiary structure, and iodoacetamide to modify the cysteine residues. After digestion, LC-MS is used for. peptide mass fingerprinting, or LC-MS/MS (tandem MS) is used to derive sequence of individual peptides [5]. LC-MS/MS is most commonly used for proteomic analysis of complex samples where peptide masses may overlap even with a high-resolution mass spectrometry.

iii. Metabolomics

LC-MS is also used for the analysis of natural products and the profiling of secondary metabolites in plants [6]. In this regard, MS-based systems are useful to acquire more detailed information about the wide spectrum of compounds from a complex biological plant sample. The first application of LC-MS in plant metabolomics was the detection of a wide range of highly polar metabolites, oligosaccharides, amino acids, amino sugars, and sugar nucleotides from Cucurbita maxima phloem tissues [7]. Another example of LC-MS in plant metabolomics is the efficient separation and identification of glucose, sucrose, raffinose, stachyose, and verbascose from leaf extracts of Arabidopsis thaliana **[8]**.

iv. Drug Development

LC-MS applications for drug development are highly automated methods used for peptide mapping, glycoprotein mapping, natural products dereplication, bioaffinity screening, *in vivo* drug screening, metabolic stability screening, metabolite identification, impurity identification, quantitative bioanalysis, and quality control

[9].

v. Other Important Applications of LC-MS

These application include the analysis of food, pesticides, and plant phenols [4].

a. Analysis of Foods [10]

Food products are complex mixtures containing both organic and inorganic constituents. The analysis of food products is generally directed towards the assessment of food safety and authenticity, the control of a technological process, the determination of nutritional values and the detection of molecules with a possible beneficial or toxic effect on human health. Liquid chromatography–mass spectrometry (LC–MS) is an increasingly valuable tool in food analysis and has been widely applied to the analysis of many food products. Recent achievements in instrumentation and data processing have allowed LC-MS to play a central role in food-related analysis. The existing recent applications of LC–MS-based techniques in food analysis include the analysis of MS analysis of LC-amenable food compounds, namely triacylglycerols, carotenoids and polyphenols.

b. Analysis of Pesticides in Foods [11]

LC/MS/MS is successfully being used for the detection of Analysis of Pesiticides, in foods. The 3/4 of 800 used pesticides are detectable by HPLC/MS/MS.

Method

It involves the following steps:

- *(i)* Sample Preparation and HPLC-Add water to 5g or 10g of dry sample (sum 10mL).
- *(ii)* Homogenize with 20mL methanol; if necessary filtration or centrifugation.
- *(iii)* Add NaCl solution to an aliquot and soak ChemElut.
- *(iv)* Elution with dichloromethane; evaporate to dryness.
- *(v)* Reconstitute in methanol/water (final sample concentration 1g/mL).
- *(vi)* Pass through a 0.45µm syringe filter.
- *(vii)* HPLC on 50x2mm Aqua RP-18 column with H_2O/CH_3OH+5mM NH_4form
- *(viii)* MS/MS detection in positive ESI withAPI 3200®LC/MS/MS system.

The details of Pesticide Classes Analyzed by LC/MS/MS are given below [11]:

Sl.No.	*Class of Compounds*	*Example Compounds*
1	Triazines	Atrazine, cyanazine, Sebutylazine, Simazine, Terbutylazine, Desmetryn, Prometryn, Terbutryn, Terbumeton, DEA, DIA.
2	Triazones	Hexazinone.
3	Ureas	Chlorotoluron, Diuron, Neburon, Dimefuron,Isoproturon, Monoluron, Linuron, Metoxuron, Methabenzthiazuron.
4	Amide herbicides	Dimethenamide.
5	Dinitroanliline herbicides	Pendiméthaline.

Contd.

Sl.No.	Class of Compounds	Example Compounds
6	Chloroacetanilide herbicides	Acetochlor, Alachlor, Metazachlor, Metolachlor, Propachlor.
7	Carbamates	Carbaryl, Carbofuran, Benfuracarb, Pirimicarb, Methomyl, Aldicarb, Thiodicarb, Methiocarb.
8	Conazoles	Cyproconazole, Difenoconazole, Epoxyconazole, Hexaconazole, Propiconazole, Tebuconazole, Tetraconazole.
9	Strobines	Azoxystrobine.
10	Organophosphorous Pesticides	Dimethoate,Acephate, Omethoate, Dicrotophos, Mevinphos, Phosphamidon, Dichlorvos.

c. Analysis of Antibiotics (Nitrofuran and Chloramphenicol) and Mycotoxins in Foods [11]

Analysis of Nitrofuran in Chiken Meat

The Nitrofuran is a class of synthetic antibiotics. It is mutagenic and genotoxic and banned in some countries (EU). It is used for treatment of bacterial and protozoal infections in animals, used as feed additives in animal husbandry. It is metabolized rapidly into reduced forms and inexpensive and very effective.

Analytical Method

It involves the following steps:

i. Sample Preparation

1g of a homogenized sample + ISTD (AMOZ-D5) HCl 0.125 and 2-Nitrobenzaldehyde. Homogenise and heat at 37°C in water bath for 16 hrs. Adjust the pH - Value to 7.4 and centrifuge. Extract the supermatant with EtOAC on an extrelut column. Evaporate and reconstitute in 500 µL mobile phase.

HPLC and MS Conditions

Mobile Phase- A- H2O + 1 mM ammonium acetate, B- Methanol, Gradient mode,

Flow rate- 200 µL/min, **Injection Volume-**50 µL, **Column-**LUNA C18, 3 x 150, 3 µm @ 25 °C, **API 3200™System** with TurboIonSpray®Source, Nitrofuran Metabolites monitored with 2 MRM transitions (1 for quantitation, 1 for confirmation):

- ☆ AOZ 236.1/134.0 + 236.1/192.1
- ☆ AMOZ 335.1/291.1 + 335.1/262.1 (D5-AMOZ: 340.1/291.1)
- ☆ SC 209.1/192.1 + 209.1/166.0
- ☆ AD 249.1/134.0 + 249.1/178.1

Analysis of Chloramphenicol in Honey

Chloramphenicol is Broad spectrum antibiotic. It is used in treatment prophylactically in food-producing animals and it is inexpensive and effective. It is potential carcinogenicity and genotoxicity, aplastic anemia, leukemia, causing antimicrobial resistance. A safe dose in human unknown, the use of the contaminated food over a long period would increase the risk (CFIA).

Analytical Method

It involves the following steps:

Sample Preparation

Prepare sample for analysis of above antibiotic as per instruction.

HPLC and MS Conditions

Phenomenex Aqua 50x2mm, Eluent A: H_2O + 0.1 per cent formic acid + 5mM NH_4ac, Eluent B: CH_3OH + 0.1 per cent formic acid + 5mM NH_4ac, Flow rate = 400µL/min,

Gradient: 0 min 90/10

5 min 10/90

7.5 min 10/90 + equilibration

Injection of 25µL, Column oven 25°C, API 3200™LC/MS/MS system with Turbo V™Source, Probe (500°C) negative polarity, CAP: 320.9/151.9 + 320.9/259.9 and D5-CAP: 325.9/156.9.

Analysis of Mycotoxins in Foods [11]

Mycotoxins are secondary metabolites produced by various molds (for instance fusarium species).

(i) **Fumonisines**- metabolites of molds which grow on senescent or stressed plants,

(ii) **DON and NIV**-metabolites of plant pathogens.

(iii) **OTA and Aflatoxins**- metabolites of molds that colonise the plant before harvest.

(iv) Some common mycotoxins - Fumonisines, Deoxinivalenol (DON), Zearalenon (ZON), Ochratoxin A, Aflatoxins, Trichothecenes and Patulin.

Analytical Method

It involves the following steps:

Sample Preparation

Take 20g of a homogenized sample.Extract with 100mL (ACN/H_2O vv 85/15) Shake (1h), filter and go for following clean up processes:

(i) SPE-Clean-up Active coal/Al_2O_3/Celite.

(ii) IAC Clean-up AOZ-column VICAM.

Inject 20 µL of aforesaid treated sample.

HPLC and MS conditions

HPLC Method Conditions–Column- SB-RP18-Zorbax, 150x3, 3.5 um Thermo

Aquasil C18, 150 x 4.6, Column oven: 50°C, Integrated Valco valve, Injection volume: 20 uL, Eluents: Water (A) and MeOH (B)

Time	*Flow (uL/min)*	*Eluent A H_2O*	*Eluent B MeOH*
0	250	30	70
10	250	100	0
14	250	100	0
15	250	30	70
25	250	30	70

MS method conditions and Mass Spectrometer Parameters for Selected Mycotoxins:

RT, Min	*Com- pound*	*Parent, m/z*	*Production*		*Ratio**	*Linearity (R2)*	*LOQ, ppb*
			Primary	*Secondary*			
3.80	NIV	311	281	205	10	0.998	10
5.25	DON	295	265	138	1.4	0.998	10
9.19	VOL	265	125	143	1.6	0.995	10
9.7	OTA	402	358	167	1.6	-	1
13.95	ZAN	319	187	136	1.3	0.991	10
14.09	ZON	317	175	187	1.2	0.990	10

* - Ratio of primary to secondary production. Run on an API 2000™ System in ESI mode

d. *Analysis of Dicyandiamide and Melamine in Milk Powders by LC/MS/MS Method [12]*

Melamine was found to be used as a protein-rich adulterant first in pet-food in 2007, and then in infant formula in 2008. Recently, low levels of dicyandiamide (DCD) residues were found in milk products from New Zealand. Instead of addition directly, the DCD present in the milk products was explained to be due to that cows eating the grass "contaminated by DCD" may produce milk containing traces of DCD residues. Dicyandiamide is a toxic agrichemical compound and could be used to promote the growth of pastures where cows graze. LC/MS/MS is successfully used for the detection of Dicyandiamide and Melamine in Milk Powders.

e. *Analysis of Melamine and Cyanuric Acid in Milk by LC-MS/MS [13]*

Melamine is an organic base and a trimer of cyanamide, with a 1,3,5-triazine skeleton. Melamine is combined with formaldehyde to produce melamine resin, a very durable thermosetting plastic, and melamine foam, a polymeric cleaning product. Melamine is nontoxic in low doses, but when combined with cyanuric acid it can cause fatal kidney stones due to the formation of an insoluble melamine cyanurate. The most common techniques to detect melamine are liquid chromatography (LC) or combined with tandem mass spectrometry (LC/MS/MS). Alternatively, gas chromatography combined with tandem mass spectrometry (GC/MS/MS) can also be used. Chinese authorities have just released the National

Standard of Melamine test. The requested detection limit for the LC assay is 2 mg/kg, for LC/MS/MS 0.01 mg/kg and for GC/MS/MS 0.005 mg/kg (the General Administration of Quality Supervision, the National Standardization Committee, GB/T22388-2008, "Raw milk and dairy products in the detection of melamine").

Analytical Method

It involves the following steps:

Pretreatment of Milk Sample

Milk + (water/acetonitrile = 20/80) = 10 + 90 (v + v) → Mixture, Ultra-centrifugation (5,000 rpm, 5 min) → Filtration (0.5 μm pore size).

Chromatographic Conditions

Column-	TSKgel Amide-80 3 μm 2.0 mm ID x 15 cm L
Eluent-	A: 0.05 per cent formic acid in water
	B: 0.05 per cent formic acid in actetonitrile, 75 per cent B
Flow rate-	0.2 ml/min
Injection vol-	5 μl
Temperature-	40°C
Detection-	LC-ESI-MS/MS, QTrap (MDS SCIEX),
	127/85+ (Melamine); 128/42- (Cyanuric acid)

Source

1. *Wikipedia.*
2. *Chaimbault, Patrick (2014-01-01). Jacob, Claus; Kirsch, Gilbert; Slusarenko, Alan; Winyard, Paul G.; Burkholz, Torsten, eds. Recent Advances in Redox Active Plant and Microbial Products. Springer Netherlands. pp. 31–94. ISBN 9789401789523.*
3. *Dass, Chhabil (2007-01-01). Fundamentals of Contemporary Mass Spectrometry. John Wiley and Sons, Inc. pp. 151-194. ISBN 9780470118498.*
4. *Niessen, Wilfried M. A (2006). Liquid Chromatography - Mass Spectrometry, Third Edition. Boca Raton: CRC Taylor and Francis. pp. 50–90. ISBN 9780824740825.*
5. *Wysocki VH, Resing KA, Zhang Q, Cheng G; Resing; Zhang; Cheng (2005). "Mass spectrometry of peptides and proteins". Methods. 35 (3): 211–22.*
6. *Stobiecki, M.; Skirycz, A.; Kerhoas, L.; Kachlicki, P.; Muth, D.; Einhorn, J.; Mueller-Roeber, B. (2006). "Profiling of phenolic glycosidic conjugates in leaves of Arabidopsis thaliana using LC/MS". Metabolomics. 2 (4): 197.*
7. *Tolstikov, Vladimir V.; Fiehn, Oliver. "Analysis of Highly Polar Compounds of Plant Origin: Combination of Hydrophilic Interaction Chromatography and Electrospray Ion Trap Mass Spectrometry". Analytical Biochemistry. 301 (2): 298–307.*
8. *Antonio, Carla; Larson, Tony; Gilday, Alison; Graham, Ian; Bergström, Ed; Thomas-Oates, Jane. "Hydrophilic interaction chromatography/electrospray mass spectrometry analysis of carbohydrate-related metabolites from Arabidopsis thaliana leaf tissue".*

Rapid Communications in Mass Spectrometry. 22 (9): 1399–1407.

9. *Lee, Mike S.; Kerns, Edward H. (1999). "LC/MS applications in drug development". Mass Spectrometry Reviews. 18 (3–4): 187–279.*

10. *Luigi Mondello, Paola Donato, Paola Dugo, Francesco Cacciola, Marco Beccaria, Advances in LC–MS for Food Analysis, LCGC chromatography.com.*

11. *Overview of Food Analysis by LC/MS/MS Applied Markets Division: HID, BioSecurity, Quality and Safety Testing.*

12. *F:/LCMS3/Analysis of Dicyandiamide and Melamine in Milk Powders by LC_MS_MS Method _ SHIMADZU (Shimadzu Corporation).html.*

13. *Analysis of Melamine and Cyanuric Acid in Milk by LC-MS/MS, Agilent Technology, TOSOH and JSB international.*

Chapter 5

Polarimetry

Polarization by reflection was discovered in 1808 by **Etienne-Louis Malus** (1775–1812) [1]. Polarimetry was discovered by Étienne-Louis Malus, a French engineer who was studying reflective glass. Several years later another Frenchman, Jean-Baptiste Biot, found that molecules such as sugar could rotate polarized light as well. It was not until 1874 that Dutch chemist Jacobus Henricus van't Hoff proposed that carbon's tetrahedral structure was responsible for the optical activity (the ability to rotate plane-polarized light)of many organic compounds [10].

A nicol prism or a polaroid lens can be used to polarize the light. Polarimetry measures the rotation of polarized light as it passes through an optically active fluid. The measured rotation can be used to calculate the value of solution concentrations; especially substances such as sugars, peptides and volatile oils.

Optical activity is a property unique to chiral substances, for example 2-butanol, which possess a chiral center (one carbon bound to four different ligands). 2-butanol exists as two mirror-image isomers, or ***enantiomers.*** The atomic connectivity in the *S*-isomer is identical to that of its mirror-image *R*-isomer, except that two of the groups attached to carbon were interchanged [10]. The ratio, the purity, and the concentration of two **enantiomers** can be measured via polarimetry. Enantiomers are characterized by their property to rotate the plane of **linear polarized** light. Therefore, those compounds are called optically active and their property is referred to as **optical rotation**. Light sources such as a light bulb, a light-emitting diode (LED), or the sun emit electromagnetic waves at the frequency of visible light. Their electric field oscillates in all possible planes relative to their direction of propagation. In contrast to that, the waves of linear-polarized light oscillate in parallel planes [6].

The *R* and *S* designations are based on the Cahn-Ingold-Prelog rules for assigning priority to substituent groups. Visualize the smallest substituent pointing downward below the plane of the paper or computer screen. The three remaining groups are ranked by molecular weight. If the direction of the heaviest, next heaviest, and lightest group is clockwise the molecule is designated as *R*; if counter-clockwise the designation is *S*. When the *d* and *l* isomers are present in exactly equal concentrations they are still chiral, but their rotations cancel out and the sample is referred to as a racemate or racemic mixture [10].

Definition

Polarimetry is a sensitive, nondestructive technique for measuring the optical activity exhibited by inorganic and organic compounds. A compound is considered to be optically active if linearly polarized light is rotated when passing through it. The amount of optical rotation is determined by the molecular structure and concentration of chiral molecules in the substance [3]. Specific rotation, [α], is a fundamental property of chiral substances that is expressed as the angle to which the material causes polarized light to rotate at a particular temperature, wavelength, and concentration [10]. Each optically active substance has its own specific rotation as defined in Biots law [3,10]:

$$[\alpha]_{\lambda}^{T} = \frac{\alpha_{\lambda}^{T}}{c \cdot l}$$

[α] = specific rotation, *l* = optical pathlength in dm;
λ = wavelength, T = temperature, α = optical rotation,
c = concentration in g/100ml

Specific rotation may also be expressed as degrees per mole of the substance where the conditions of measurement (*i.e.* solvent, light source, and path length) are also specified [10].

Working/Measuring Principle

Normal monochromatic light emerging from a light bulb consists of an infinite number of oscillating waves in all possible planes perpendicular to the line of propagation. A polarizer is a special type of slit or opening that allows light propagating on one plane to pass through. When this light interacts with a chiral substance it speeds up or slows down, the net effect being an apparent rotation in the plane of polarized light.

There is no correlation exists between the absolute configuration of the molecule (*e.g. R*- or *S*-) and the direction in which it rotates polarized light. Molecules that shift the angle clockwise are known as ***dextrarotatory*** (right-turning), *d* or (+), while those that shift the angle counter-clockwise are called ***levarotatory*** (left-rotating), *l*, or (-). It is difficult to predict the precise rotation of a molecule having more than one chiral center since both chiral centers contribute to optical rotation [10].

If light encounters a polarizer, only the part of the light that oscillates in the defined plane of the polarizer may pass through. That plane is called the plane of polarization. The plane of polarization is turned by optically active compounds. According to the direction in which the light is rotated, the enantiomer is referred to as dextro-rotatory or levo-rotatory.

The optical activity of enantiomers is additive. If different enantiomers exist together in one solution, their optical activity adds up. That is why racemates are optically inactive, as they nullify their clockwise and counter clockwise optical activities. The optical rotation is proportional to the concentration of the optically active substances in solution. Polarimeters may therefore be applied for concentration measurements of enantiomer-pure samples. With a known concentration of a sample, polarimeters may also be applied to determine the specific rotation when characterizing a new substance. The specific rotation $[\alpha]^{T}\gamma$ is a physical property and defined as the optical rotation α at a path length l of 1 dm, a concentration c of 1g/100 mL, a temperature T (usually 20 °C) and a light wavelength λ (usually sodium D line at 589.3 nm) [7]:

$$[\alpha]^{T}\gamma = \frac{100 \text{ X } \alpha}{1 \text{ X } c}$$

This tells us how much the plane of polarization is rotated when the ray of light passes through a specific amount of optically active molecules of a sample. Therefore, the optical rotation depends on temperature, concentration, wavelength, path length, and the substance being analyzed [8].

Polarimeters

Historically, polarimetry was performed using an instrument where the extent of optical rotation is estimated by visual matching of the intensity of split fields. For this reason, the D-line of the sodium lamp at the visible wavelength of 589nm was most often employed. Specific rotation determined at the D-line is expressed by the symbol [1].

[α] {25} [D] [α] {20} [D]

A polarimeter consists of a polarized light source, an analyzer, a graduated circle to measure the rotation angle, and sample tubes. The polarized light passes through the sample tube and exhibits angular rotation to the left (-) or right (+). On the side opposite the polarizer is the analyzer. Using optics, visual fields are manually adjusted by the user to measure the optical rotation angle [2].

Regular (unpolarised) light includes waves that are inclined in any direction of space, polarised light is inclined at one defined angle.Polarised light is 'created' by eliminating all light other than that which is inclined at one angle. To do this, a very fine meshed grating – a polarisation filter – is used to filter out all waves from the regular light that do not have the same inclination as the filter's grating. If the polarised light beam were then directed through a second grating, set at exactly 90 degrees to the first grating, no light would pass through it.Optically active substances change the inclination of the light wave. So if an optically active substance is placed between the two filters, then the polarised beam would be rotated. A component of the light would then be able to pass through the second filter. Examples of optically active fluids are sugars, lactic acid, tartaric acid, and some biologically-active substances [9].

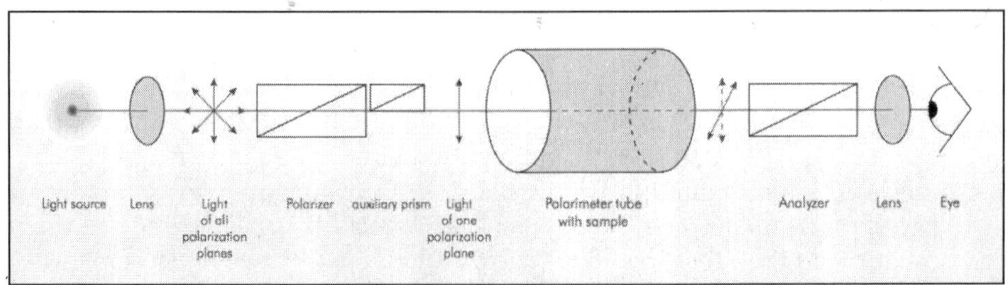

Outlay of Polarimeter.

A **polarimeter** is a scientific instrument used to measure the angle of rotation caused by passing polarized light through an optically active substance. Depending on the design of the device, the second filter is then rotated (manually or automatically) until no light falls on the detector. The amount of rotation and its direction (clockwise or anticlockwise) provides a powerful technique for measuring features of the optically active substance.

Some chemical substances are optically active, and polarized (uni-directional) light will rotate either to the left (counter-clockwise) or right (clockwise) when passed through these substances. The amount by which the light is rotated is known as the angle of rotation. The angle of rotation is basically known as observed angle [1,4,5].

Polarimetry will provide information about the molecular structure, the concentration of the substance and sometimes information about the solvent used. The optical activity is also influenced by temperature, the wavelength of the light, and the length of the optical path - the longer the path of the light through an optically active substance, the larger the angle of rotation. These variables may be used to control the sensitivity of the instrument [9].

Wavelengths lower than 589 nm are available with mercury and deuterium lamps isolated through filters and may provide greater sensitivity. Generally, the observed optical rotation at 436 nm is approximately double and at 365nm about three times that at 589 nm. Reduction in the concentration of the solute required for measurement may sometimes be accomplished by conversion of the substance under test to one that has a significantly higher optical rotation. Optical rotation is also affected by the solvent used for the measurement, and this is always specified [1,10].

It is now common practice to use other light sources, such as xenon or tungsten halogen, with appropriate filters, because these light sources offer advantages of cost, long life [1,10], and broad wavelength emission range over traditional light sources [9,10].

Polarimeters offer high accuracies where precision is critical in determining the concentration of samples. Manual polarimeters where you look through a viewing scope to read values on a vernier scale, and semiautomatic polarimeters that have a digital display are available. Polarimeters can measure in angle of rotation, International Sugar Scale (°Z), or both [2].

The polarimeter is made up of two Nicol prisms (the polarizer and analyzer). The polarizer is fixed and the analyzer can be rotated. The prisms may be compared to as slits S1 and S2. The light waves may be considered to correspond to waves in the string. The polarizer S1 allows only those light waves which move in a single plane. This causes the light to become plane polarized. When the analyzer is also placed in a similar position it allows the light waves coming from the polarizer to pass through it.

When it is rotated through the right angle no waves can pass through the right angle and the field appears to be dark. If now a glass tube containing an optically active solution is placed between the polarizer and analyzer the light now rotates through the plane of polarization through a certain angle, the analyzer will have to be rotated in same angle.

Polarimeters measure the observed rotation designated by the Greek lower case letter α. From this value, and knowledge of the specific rotation, one can easily calculate the concentrations of both isomers of a pure substance. For example it is possible to determine the conversion of an achiral material to a chiral substance, or the relative concentrations of optical isomers, known as ***enantiomeric excess.***

Polarimetry provides an additional check on a pure substance before it is added to an expensive batch to determine the ingredient's concentration or purity. For example a 25 per cent glucose syrup will have an observed rotation that is five-sixths that of a 30 per cent syrup.

Similarly the optical rotation of a mixed-component ingredient, intermediate, or finished product will have a characteristic optical rotation that may arise from the presence of several chiral compounds. Once a standard is determined for the composite observed rotation, one can establish quality criteria based on optical rotation. In these situations the polarimeter measurement becomes a type of screen for further testing to determine which ingredient is out of specification.

Operation

Polarimeters measure this by passing **monochromatic light** through the first of two polarising plates, creating a polarized beam. This first plate is known as the polarizer. This beam is then rotated as it passes through the sample. After passing through the sample, a second polarizer, known as the analyzer, rotates either via manual rotation or automatic detection of the angle.

Types of Polarimeters [5]

The following types of Polarimeters are available:

i. Laurent's Half-Shade Polarimeter

When plane-polarised light passes through some crystals, the velocity of left-polarised light is different from that of the right-polarised light thus the crystals are said to have two refractive indices *i.e.* **Double refracting Construction**-It consists of a monochromatic source S which is placed at focal point of a convex lens L. Just after the convex lens there is a Nicol Prism P which acts as a polariser. H is a half shade device which divides the field of polarised light emerging out of the Nicol P into two halves generally of unequal brightness. T is a glass tube in which optically active solution is filled. The light after passing through T is allowed to fall on the analyzing Nicol A which can be rotated about the axis of the tube. The rotation of analyser can be measured with the help of a scale C. Laurent's half-shade polarimeter is shown in following figure:

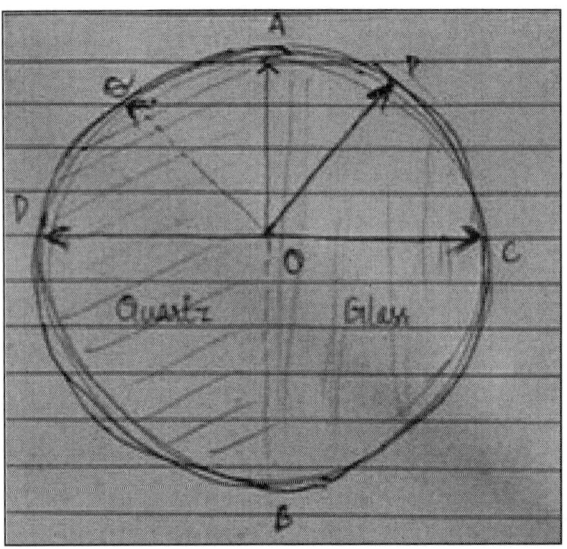

Working

In order to understand the need of a half-shade device, let us suppose that half-shade device is not present. The position of the analyzer is so adjusted that the field of view is dark when tube is empty. The position of the analyzer is noted on circular scale. Now the tube is filled with optically active solution and it is set in its proper position. The optically active solution rotates the plane of polarization of the light emerging out of the polariser P by some angle. So the light is transmitted by analyzer A and the field of view of telescope becomes bright. Now the analyzer is rotated by a finite angle so that the field of view of telescope again become dark. This will happen only when the analyzer is rotated by the same angle by which plane of polarization of light is rotated by optically active solution.

The position of analyzer is again noted. The difference of the two readings will give you angle of rotation of plane of polarization (8).

A difficulty is faced in the above procedure that when analyzer is rotated for the total darkness, then it is attained gradually and hence it is difficult to find the exact position correctly for which complete darkness is obtained. To overcome above difficulty half-shade device is introduced between polariser P and glass tube T.

Half Shade Device

It consists of two semicircular plates ACB and ADC. One half ACB is made of glass while other half is made of quartz.Both the halves are cemented together. The quartz is cut parallel to the optic axis. Thickness of the quartz is selected in such a way that it introduces a path difference of A/2 between ordinary and extraordinary ray. The thickness of the glass is selected in such a way that it absorbs the same amount of light as is absorbed by quartz half.

Let us consider that the vibration of polarisation is along OP. On passing through the glass half the vibrations remain along OP. But on passing through quartz half these vibrations will split into 0- and £-components. The £-components are parallel to the optic axis while O- component is perpendicular to optic axis. The O-component travels faster in quartz and hence an emergence 0-component will be along OD instead of along OC. Thus components OA and OO will combine to form a resultant vibration along OQ which makes same angle with optic axis as OP. Now if the Principal plane of the analyzing Nicol is parallel to OP then the light will pass through glass half unobstructed. Hence glass half will be brighter than quartz half or we can say that glass half will be bright and the quartz half will be dark. Similarly if principal plane of analyzing Nicol is parallel to OQ then quartz half will be bright and glass half will be dark.

When the principal plane of analyzer is along AOB then both halves will be equally bright. On the other hand, if the principal plane of analyzer is along DOC. then both the halves will be equally dark.

Thus it is clear that if the analyzing Nicol is slightly disturbed from DOC then one half becomes brighter than the other. Hence by using half shade device, one can measure angle of rotation more accurately.

Determination of Specific Rotation

In order to determine specific rotation of an optically active substance (say sugar) the polarimeter tube T is first filled with pure water and analyzer is adjusted for equal darkness (Both the halves should be equally dark) point. The position of the analyzer is noted with the help of scale. Now the polarimeter tube is filled with sugar solution of known concentration and again the analyser is adjusted in such a way that again equally dark point is achieved. The position of the analyzer is again noted. The difference of the two readings will give you angle of rotation θ. Hence specific rotation S is determined by using the relation.

$$[S]\ t\ \lambda = \theta / LC$$

ii. Biquartz Polarimeter

In biquartz polarimeters, a biquartz plate is used. Biquartz plate consists of two semi circular plates of quartz each of thickness 3.75mm. One half consists of right-handed optically active quartz, while the other is left-handed optically active quartz.

iii. Lippich Polarimeter

iv. X-Ray Polarimeter

v. Quartz-Wedge Polarimeter

vi. Manual

The earliest polarimeters, which date back to the 1830s, required the user to physically rotate one polarizing element (the analyzer) whilst viewing through another static element (the detector). The detector was positioned at the opposite end of a tube containing the optically active sample, and the user used his/her eye to judge the "alignment" when least light was observed. The angle of rotation was then read from a simple fixed to the moving polariser to within a degree or so.

Although most manual polarimeters produced today still adopt this basic principle, the many developments applied to the original opto-mechanical design over the years have significantly improved measurement performance. The introduction of a half-wave plate increased "distinction sensitivity", whilst a precision glass scale with vernier drum facilitated the final reading to within ca. ±0.05°. Most modern manual polarimeters also incorporate a long-life yellow LED in place of the more costly sodium arc lamp as a light source.

vii. Semi-automatic

Today, semi-automatic polarimeters are available. The operator views the image via a digital display adjusts the analyzer angle with electronic controls.

viii. Fully Automatic

Fully automatic polarimeters are now available and simply require the user to press a button and wait for a digital readout. Fast automatic digital polarimeters yield an accurate result within a second, regardless of the rotation angle of the sample. In addition, they provide continuous measurement, facilitating High-performance liquid chromatography and other kinetic investigations.

Another feature of modern polarimeters is the Faraday modulator. The Faraday modulator creates an alternating current magnetic field. It oscillates the plane of polarization to enhance the detection accuracy by allowing the point of maximal darkness to be passed through again and again and thus be determined with even more accuracy.

As the temperature of the sample has a significant influence on the optical rotation of the sample, modern polarimeters have already included Peltier Elements to actively control the temperature. Special techniques like a temperature controlled sample tube reduce measuring errors and ease operation. Results can directly be transferred to computers or networks for automatic processing. Traditionally, accurate filling of the sample cell had to be checked outside the instrument, as an appropriate control from within the device was not possible. Now-a-days a camera system allows accurate monitoring of the sample and filling conditions in the sample cell from inside the instrument. A telecentric camera gives a sharp image over the complete length of any sample cell placed within modern instruments. The online monitoring of the filling process ensures that no bubbles or particles obstruct the measurement. A picture can be saved together with the recorded data. Any temperature gradients, in homogeneous sample distributions or air bubbles can immediately be recognized before measurement, so that potential errors caused by bubbles or particles are no longer an issue.

Manual V/S Automatic Polarimeter [9]

Polarimetry is a powerful technique for the analysis of optically-active fluids such as sugars, lactic acid, tartaric acid, and biologically-active substances. The method provides valuable information on the chemical structure, chirality, and concentration of a sample by measuring the angle through which a ray of polarised light is diverted.

In a traditional manual polarimeter, this requires a skilled operator and takes a variable length of time. A modern automatic polarimeter, on the other hand, makes the process much quicker – an accurate reading is provided in just one second. Operator-induced errors are eliminated and measurement consistency is greatly improved – an important consideration for regulated laboratories. Touch screen controls and a clear digital readout make the analysis fast, clean, and reliable. Automatic polarimetry is compatible with Good Laboratory Practice (GLP) and US FDA 21 CFR part 11.

The increased speed means that lab productivity is greatly improved, and the investment in an automated instrument pays real dividends. Manual polarimetry still has its place, for low-throughput labs and for training, but for many modern industries such as pharmaceutical, chemical, sugar, and food processing, an automatic polarimeter is the better choice. Lab productivity is greatly improved, and the investment pays real dividends.

Features of Automatic Polarimeters

(*i*) Compatible with Good Laboratory Practice (GLP) and US FDA 21 CFR part 11.

(*ii*) Accurate reading is provided in a few seconds or less.

(*iii*) Operator-induced errors are eliminated.

(*iv*) Measurement consistency is greatly improved – important for regulated laboratories.

Source

1. *Polarimeter.kenyon.edu.*
2. *www.coleparmer. In/Tech Library Article/606.*
3. *rudolphresearch. com/products/polarimeters/polarimetry-definitions.*
4. *Polarimeter. Princeton WordNet.*
5. *Wikipedia.*
6. *Hart, C. (2002), Organische Chemie, Wiley-VCH.*
7. *F. A. Carey; R.J.Sundberg (2007). Advanced Organic Chemistry, Part A: Structure and Mechanisms (Fifth ed.). Springer.*
8. *http://www.anton-paar.com/corp-en/products/group/polarimeter/and.*
9. *A.Kruss Optronic, Germany.*
10. *Angelo DePalma, PHD, Rudolph research analytical.*

Technical Specification of Some Models of Polarimeters

1. Anton Paar

Parameters	*MCP150*	*MCP100*
Measuring scales	°Optical Rotation, °Specific Rotation, per cent Concentration, user-defined scales	°Optical Rotation, °Specific Rotation, per cent Concentration
Measuring range	±89.9°	±89.9°
Resolution	0.001°	0.001°
Accuracy	±0.005°	±0.01°
Repeatability	±0.005°	±0.01°
Response time	max. 12 sec.	12 sec.
Wavelength	589 nm	589 nm
Light source	LED, with average lifetime of 100 000 hours	LED, with average lifetime of 100.000 hours
Sensitivity	Light intensity control compensates attenuation up to Optical Density (OD) 2.0	A light intensity control compensates attenuation up to Optical Density (OD) 2.0
Temperature sensor	Built-in PT 100 sensor	Built-in PT 100 sensor
Temperature resolution	0.1 °C	0.1 °C
Temperature accuracy	±0.1 °C	±0.2 °C
Temperature control	Automatic Peltier temperature control between 15 °C and 35 °C, or without temp. control	Automatic Peltier tempe- rature control at 20°C and 25°C, or without temp. control
Sample cell/quartz control plate	**Toolmaster™** automatic sample cell and quartz control plate Identification; sample cell path length of 2.5 mm to 100 mm	**Toolmaster™** - Automatic sample cell and quartz control plate identification; sample cell path length of 2.5 mm to 100 mm
Display	TFT touch screen 5.7" (145 mm), 640 x 480 Pixels.	TFT touch screen 5.7" (145 mm), 640 x 480 Pixels
Interfaces	3 USB, RS232, Ethernet, CAN bus, 1 USB typ "B" for service Easy connection of keyboard, mouse, printer, bar code. Network capability.	3 USB, RS232, Ethernet, CAN bus, easy connection of keyboard, mouse, printer, bar code. Network capability 370 mm x 320 mm x 130 mm (14.6 in x 12.6 in x 5.1 in)
Dimensions (LxWxH)	370 mm x 320 mm x 130 mm (14.6 in x 12.6 in x 5.1 in)	370 mm x 320 mm x 130 mm (14.6 in x 12.6 in x 5.1 in)
Power supply/ consumption	AC 100 to 240 VAC, 50/60Hz/70VA, 120 VA max.	AC 100 to 240 VAC, 50/60Hz/70VA, 120 VA max.

Parameters	MCP 200	MCP 300	MCP 500
Measuring range	± 89.9°	± 89.9°	± 89.9°
Resolution	0.001° (0.0001° optional)	0.001° (0.0001° optional)	0.0001°
Accuracy*	± 0.002° (most precise measuring mode) ± 0.005° (fast measuring mode)	± 0.003°	better than ± 0.002°
Repeatability	± 0.002°	± 0.002°	± 0.001°
Response time	12 seconds	12 seconds	15 seconds
Wavelength	589 nm and optionally up to eight wavelengths. Standard spectral (365, 405, 436, 546, 578, 633, 880 nm), customer-specific wavelengths on request.		
Sensitivity	A light intensity control compensates attenuation up to an Optical Density (OD) of 4.0.		
Light source	Standard: LED light source with 100 000 hours life time Optionally for multi wavelength instruments: Tungsten halogen lamp, 6 V, 20 W, with average lifetime of 2000 h.		
Temperature interface			
Resolution	0.1 °C	0.1 °C	0.01 °C
Accuracy**	± 0.2 °C	± 0.1 °C	± 0.03 °C
Temperature control	Peltier system for automatic temperature control between 10 °C and 45 °C		
Temperature control range***	20°C + 25°C (optional 10 °C to 45 °C)	10 °C to 45 °C	-
Sample cell	ToolMaster™: Wireless automatic identification of sample cells via RFID, sample cell path length from 2.5 mm to 200 mm		

2. Rudolph Research Analytical

a. General Specification

Autopol I	Autopo II	Autopol III
The Autopol I is our entry-level polarimeter model designed for: University Education • Natural Product Research,Starch Analysis, Invert Sugar Lactose in Milk, Vitamins. Flavors Fragrance, Chemicals.	The Autopol II offers the same standard and optional features as the Autopol I, but allows greater wavelength flexibility.	The Autopol III is Rudolph's entry level pharmaceutical solution: Accuracy: 0.002°C, 0.2 per cent above 1° Standard. Wavelengths: 589nm and 546nm.
Standard Features-0.01° Arc Accuracy, 589nm fixed wavelength, built in thermo-probe for temperature measurement	**Standard Wavelengths**-589nm and 546nm Optional Wavelengths: 365nm, 405nm, 436nm, 578nm, 633nm	Complete Accessory Package with IQOQPQ documentation, 2 user selectable cells and 1 NIST traceable calibration standard with NIST Certificate.

Contd.

Autopol I	*Autopo II*	*Autopol III*
Windows Embedded 7 for direct connection to the network server and flexible USB printing.	A total of 2-4 extra wavelengths may be ordered at time of purchase or added later when needed.	
Optional Features-TempTrol™ heating and cooling: 15°C – 40°C ±.2°C.AP Accuracy Option: 0.004° Arc Optical Rotation,AP Resolution Option: 0.01, 0.001°Arc Selectable.	**Optional Features**-100mm or 200mm TempTrol™ heating and cooling: 15°C – 40°C ±.2°C.AP Accuracy Option: 0.004° Arc Optical Rotation. AP Resolution Option: 0.01, 0.001°Arc Selectable	**Optional Features-**100mm or 200mm TempTrol heating and cooling:15°C - 40°C ±2°C Optional Wavelengths:365nm, 405nm, 436nm,578nm,633nm, (up to 4 wave lengths may be selected or added later. AP Accuracy Option: ±0.002° up to 10° Arc, ±0.004° 10°- 89° Arc. AP Resolution Option: 0.01, 0.001, 0.0001°, Arc Sele

b. Technical Specification

Parameters	*Autopol I*	*Autopol II*	*Autopol III*
Measuring Mode	Optical Rotation, Specific Rotation, Concentration, Sugar Degrees, °Z (ISS)	Optical Rotation, Specific Rotation, Concentration, Sugar Degrees, °Z (ISS)	Optical Rotation, Specific Rotation, Concentration, Sugar Degrees, °Z (ISS)
Measuring Scale	Degrees Arc Optical Rotation	Degrees Arc Optical Rotation	Degrees Arc Optical Rotation
Resolution	0.01° Arc Optical Rotation, 0.01 per cent Concentration. 0.01 Specific Rotation	0.01° Arc Optical Rotation, 0.01 per cent Concen-tration, 0.01 Specific Rotation	0.001° Arc Optical Rotation,0.001 per cent Concentration, 0.001 Specific Rotation
Accuracy	0.01° Arc Optical Rotation, 0.03°Z (ISS) Sugar Degrees	0.01° Arc Optical Rotation, 0.03°Z (ISS) Sugar Degrees	0.002° Arc up to 1°, 0.2 per cent above 1°,0.01° Z (ISS)
AP Accuracy Option	Resolution: 0.001° Arc Optical Rotation Reprodu cibility: 0.002° Arc Optical Rotation, Accuracy: ±0.004° Arc Optical Rotation	Resolution: 0.001° ArcOptical Rotation Reproducibility: 0.002° Arc Optical Rotation, Accuracy: ±0.004° Arc Optical Rotation	Accuracy: ±0.002° up to 10°, ±0.004° 10° - 89° Arc Optical Rotation
AP Resolution Option	Selectable Resolution: 0.01, 0.001° Arc	Selectable Resolution: 0.01, 0.001° Arc	Selectable Resolution: 0.01, 0.001,0.0001°Arc
Reproducibility	0.01° Arc Optical Rotation	0.01° Arc Optical Rotation	0.002° Arc Optical Rotation
Measuring Range	± 89.9° Arc Optical Rotation, ± 999.99° Arc Specific Rotation, 0-99.9 per cent Concentration	± 89.9° Arc Optical Rotation, ± 999.99° Arc Specific Rotation, 0-99.9 per cent Concentration	± 89.9° Arc Optical Rotation, ± 999.99° Arc Specific Rotation, 0-99.9 per cent Concentration
Prism	Glan Thompson Calcite	Glan Thompson Calcite	Glan Thompson Calcite
Optical Wavelengths	589nm	589nm, 546nm standard Optional wavelengths: 365nm, 405nm, 436nm, 578nm	589nm, 546nm standard Optional wavelengths: 365nm, 405nm, 436nm, 578nm

Contd.

Parameters	*Autopol I*	*Autopol II*	*Autopol III*
Wavelength Selection	Fixed	Touchscreen Selectable: 2 standard, 4 optional	Touchscreen Selectable: 2 standard, 4 optional
21 CFR Part 11 compliant Software	Yes with optional PC software	Yes with optional PC software	Yes with optional PC software
Temperature Control	By external water bath (standard)TempTrolTM Automatic Electric Heating and Cooling 15°-40°C±0.2°C (optional)	By external water bath (standard)TempTrolTM Automatic Electric Heating and Cooling 15°-40°C±0.2°C (optional)	By external water bath (standard)TempTrolTM Automatic Electric Heating and Cooling 15°-40°C±0.2°C (optional)
Temp.Probe Range	10° - 100°C	10° - 100°C	10° - 100°C
Temp. Probe Accuracy	±0.1°C	±0.1°C	±0.1°C
Measurement Time	5 measurements in less than 25 seconds (avg.)	5 measurements in less than 25 seconds (avg.)	5 measurements in less than 25 seconds (avg.)
Light Source	Tungsten-Halogen 6V, 20W, avg. 2,000 hour life	Tungsten-Halogen 6V, 20W, avg. 2,000 hour life	Tungsten-Halogen 6V, 20W, avg. 2,000 hour life
Sample Chamber	Accepts sample tubes up to 200 mm	Accepts sample tubes up to 200 mm	Accepts sample tubes up to 200 mm
Data Storage	8 GB Non-removable Compact Flash	8 GB Non-removable Compact Flash	8 GB Non-removable Compact Flash
Communication Interface	3 – USB Ports, 2 – RS232 Ports, Ethernet Port for Network Connection	3 – USB Ports, 2 – RS232 Ports, Ethernet Port for Network Connection	3 – USB Ports, 2 – RS232 Ports, Ethernet Port for Network Connection
Calibration	Automatic calibration via touchscreen	Automatic calibration via touchscreen	Automatic calibration via touchscreen
Display	8" color, 800 x 600 pixel resolution with 400 nits of brightness	8" color, 800 x 600 pixel resolution with 400 nits of brightness	8" color, 800 x 600 pixel resolution with 400 nits of brightness
User Interface	Touchscreen	Touchscreen	Touchscreen
Automatic Sensitivity Control	Measures samples with transmittance as low as 0.01 per cent (up to O.D. 4.0)	Measures samples with transmittance as low as 0.01 per cent (up to O.D. 4.0)	Measures samples with transmittance as low as 0.01 per cent (up to O.D. 4.0)
Input Power	100 - 240V, 50/60 Hz	100 - 240V, 50/60 Hz	100 - 240 V, 50/60 Hz

Applications in Industries

Optical rotation is an indispensable quality and identity assay for a wide range of critical industries. Research organic chemists use polarimetry to test the effectiveness of catalysts and asymmetric synthetic processes. Food, drug, and flavors industries utilize polarimetry as a quality attribute for raw ingredients and finished products [1]. For mixtures of chiral organic molecules in pure or diluted form, polarimetry provides a rapid, reliable, quality check that eliminates conventional analysis like liquid chromatography which can take an hour to do

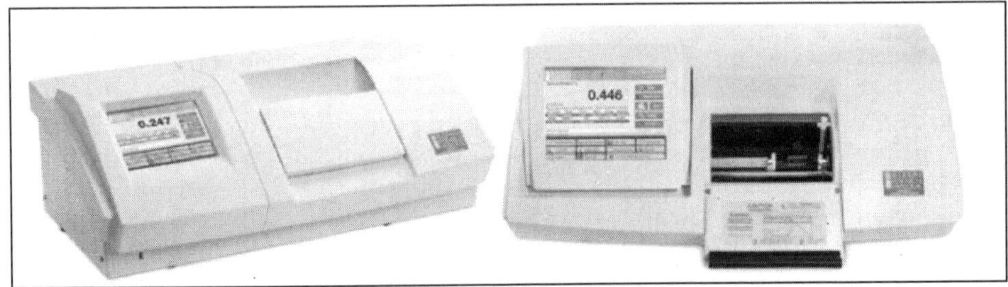

Automatic Polarimeter Rudolph Research Analytical.

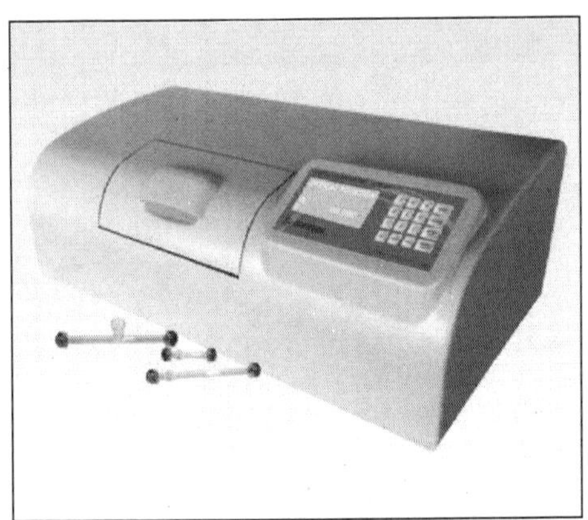

Digital Polarimeter CDP - 001
CONTECH Instruments Ltd.

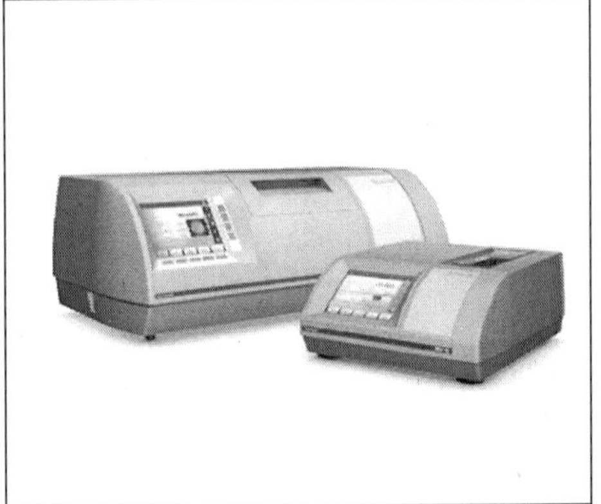

MCP Polarimeter, Anton Paar.

Model D7 Polarimeter - Bellingham + Stanley is Part of Xylem Analytics.

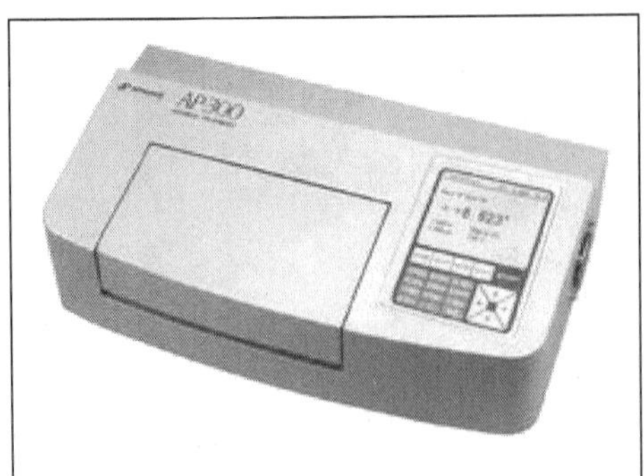

ATAGO™ AP-300
Automatic Polarimeter

what the polarimeter accomplishes in minutes. Research chemists use polarimetry to test the effectiveness of catalysts and asymmetric synthetic processes [2].

Source

1. *Angelo DePalma, PHD, Rudolph research analytical.*
2. *A.Kruss Optronic, Germany.*

A. Application in Food/Dairy Industry

1. Sucrose Determination

a. Polarimetric Determination of Sucrose Content in Sweetened Condensed Milk (BIS Method-IS 11764: 2005/1S02911: 2004):

Principle (BIS and ISO)

The dairy industry needs to be able to accurately determine the sucrose content of milk and milk products. The **sucrose content** is the total amount of unaltered sucrose in the milk product, and is determined by the procedure specified in the International Standard ISO 2911 and IS11764: 2005. It is expressed as a percentage by mass.

A test sample is treated with ammonium hydroxide, so as to bring mutarotation of lactose to final equilibrium. It is neutralized and then clarified by successive additions of zinc acetate and potassium hexacyanoferrate (II), followed by filtration, The optical rotation is determined on a portion of the filtrate. On another portion of the filtrate, inversion is induced (based on the Clerget principle) by mild acid hydrolysis of the sucrose, leaving lactose and other sugars virtually unaffected. The optical rotation is determined after inversion.

Reagents (BIS)

(i) Zinc acetate solution, 1,0mol/1 (Dissolve in water 21,9 g of zinc acetate dihydrate [Zn $(C_2H_30_2)2H_2O$] and add 3 ml of glacial acetic acid. Mix and dilute to 100 ml).

(ii) Potassium hexacyanoferrate (ll) solution, 0,25 mol/1 (Dissolve in water 10,6 g of potassium hexacyanoferrate (ll) trihydrate [$K_4Fe\ (CN)_6 3H_20$] and dilute to 100 ml).

(iii) Dilute hydrochloric acid, HCI = (6,35 ± 0,20) mol/1 [20 per cent to 22 per cent (mass fraction)].

(iv) Ammonium hydroxide solution, NH_4OH- (2,0 ±02) mol/1 [3,5 per cent (mass fraction).

(v) Dilute acetic acid, CH_3COOH - (2,0 ± 0,2) mol/1 [12 per cent (mass fraction)], of exactly known concentration,

Apparatus (BIS)

(i) Analytical balance, capable of weighing to the nearest 0,01 g.

(ii) Glass beaker, of capacity 100 ml.

(iii) Volumetric flasks, of capacities 200 ml and 50 ml, conforming to class A of ISO 1042.

(iv) Pipette, either 20 ml, conforming to class A of ISO 648, or 40 ml, of corresponding accuracy.

(v) Graduated measuring cylinders, of capacity 25 ml.

(vi) Graduated pipettes, of capacity 10 ml.

(vii) Filter funnel diameter 8 cm to 10 cm, and folded medium-grade filter papers, of diameter 15 cm.

(viii) Polarimeter tube, exactly 2 dm long.

(ix) Polarimeter -using sodium light or mercury green light (mercury vapour lamp with prism or the special Wratten screen No. 77A), capable of being read to an accuracy of at least 0,05 angular degrees.

(x) saccharimeter with international sugar scale, using white light passing through a filter of 15 mm depth of a 6 per cent solution of potassium bichromate, or sodium light, capable of being read to an accuracy of atleast 0,1 international sugar scale degrees.

(xi) Water baths, capable of being maintained at about 40°C and at (60±1) °C respectively.

Sampling

A representative sample should have been sent to the laboratory, It should not have been damaged or changed during transport or storage. Sampling should be done as per method specified in ISO 707.

Procedure

Preparation of Test Sample

As per BIS Method

Samples of recently "manufactured products in which no appreciable separation of components may be expected

Open the container, transfer all material adhering to the lid into the container and thoroughly mix by an up and down movement of a spoon, in such a way that the top layers and the contents of the lower corners are moved and mixed. When the product is in a can, transfer the contents to a jar with a well-fitting lid, When the product is in a collapsible tube, transfer as much as possible of the contents to a jar with a well-fitting lid, then cut open the tube, scrape out all material adhering to the interior and transfer this also to the jar. Mix the contents of the jar as described above,

Samples of older products and samples in which separation of components may be expected

Heat in a water bath at about 40 °C until the sample has nearly reached this temperature. Open the container and proceed as described above. When the product is in a can or tube, transfer the contents to a jar, scrape out all material adhering to the walls (in the case of a collapsible tube, after cutting open the tube) and continue the mixing until the whole mass is homogeneous, reducing the size of any large crystals –by crushing them with a glass rod. Close the jar with a well-fitting lid. Allow to cool.

As per ISO Method

(i) Shake and invert the container of the sample of milk several times.

(ii) Open the container and stir the sample carefully using a glass rod.

(iii) Transfer the sample to a glass bottle and close it,

If the sample is too old or too thick-Transfer the sample to another glass bottle. The sample must be warmed up to 40°C in a water bath for about 2 hours. Every 15 minutes during the warming time, take the sample out and shake it well. Cool it down to room temperature.

Check Test (BIS)

In order to check the procedure, the reagents and the apparatus, make a check test as described below in duplicate, on a mixture of 100 g of whole milk (or 110 g of skimmed milk) and 18,00 g of pure sucrose, This mixture corresponds to 40,00 g of condensed milk containing 45.0 **per cent** of sucrose. Calculate the sugar content by means of the equations given under method of calculation, using in Equation (2) form, F and P respectively the quantity of milk weighed and the fat and protein content of this milk, and in Equation (1) for m, the value **40,00**

The mean of the values found shall be within the range (45±0,1) per cent (mass fraction).

Determination (BIS and ISO)

(i) Weigh, to the nearest 0,01 g, a test portion of approximately 40 g of the well-mixed test sample into the glass beaker. Add 50 ml of hot water (80 °C to 90 °C) and mix well.

(ii) Transfer the mixture quantitatively to the 200-ml volumetric flask, rinsing the beaker with successive quantities of water at 60 °C, until the total volume is between 120 ml and 150 ml. Mix and cool to **(20 ± 2) °C.**

(iii) Add 5 ml of the ammonium hydroxide solution. Mix again and then allow to stand for 15 min at **(20 ±2) °C.**

(iv) Neutralize the ammonium hydroxide by adding the stoichiometrically equivalent quantity of the diluteacetic acid. Mix

(v) Add, mixing gently by rotating the tilted flask, 12,5ml of the zinc acetate solution.

(vi) In the same manner as for the addition of the zinc acetate solution, add 12,5 ml of the potassiumhexacyanoferrate (ll) solution.

(vii) Bring the contents of the flask to 20°C and dilute to the mark with water at 20°C.

(viii) Up to this stage, all additions of water or reagents shall be made in such a manner as to avoid the formation of air bubbles and, with the same object in view, all mixing shall be carried out by rotation of the flask rather than by shaking. If air bubbles are found to be present before completion of the dilution to 200 ml, they may be removed by temporarily connecting the flask to a vacuum pump, and rotating the flask.

(ix) Close the flask with a dry stopper and mix thoroughly by vigorous shaking

(x) Allow the precipitate to settle for a few minutes and then filter the solution through a dry filter paper, rejecting the first 25 ml of filtrate **(In ISO, it is called Solution A).**

Direct Polarization

Determine the optical rotation of the filtrate at (20 ±2) °C.

Inversion

Pipette 40 ml (two 20 ml portions if a 40 ml pipette is unavailable) of the filtrate into the 50 ml volumetric flask. Add 6,0 ml of the dilute hydrochloric acid. Place the flask in a water bath set at 60°C for 15 min, the flask being immersed to the base of the neck. Mix by rotating the flask during the first 5 min, in which time the contents of the flask should have reached the temperature of the water bath. Cool to 20°C and dilute to the mark with water at 20°C. Mix and allow to standfor 1 h at this temperature **(In ISO, it is called Solution B).**

Invert Polarization

Determine the optical rotation of the inverted solution at (20±2)°C. If the temperature of the liquid in the polarization tube differs by more than 0,2 °C from 20 °C during the measurement, the temperature correction [Equation (4)] given under method of calculation shall be applied.

Expression of Results

Method of Calculation and Equations (BIS and ISO)

The sucrose content, *ws,* of the sample, expressed as a percentage by mass, is given by:

$$ws = \frac{A - I{,}25B}{Q} \times \frac{I - \Delta}{1} \times \frac{IV}{L \times m}$$

where,

m- the mass of the test portion in grams, A-the direct polarimeter reading before inversion, B- the polarimeter reading after inversion, L-the length, in decimetres, of the polarimeter tube, Q- the inversion division factor (the values of which-are given below), I'- the volume, in millilitres, to which the sample is diluted before filtration,Δl'- the correction, in millilitres, for the volume of the precipitate formed during the clarification.

For example, using the 200 mm tube, the formula above for per cent sucrose simplified to:

$$S = \frac{A-1.25B}{Q} X \frac{200-\Delta V}{2xm}$$

$$\text{BIS and ISO } \Delta I' = \frac{m}{100} (1{,}08F+1{,}55P) \qquad \textit{Equation 2}$$

where,

m- the mass of the test portion in grams, F-the mass fraction, as a percentage, of fat in the sample (determined in accordance with ISO1737), P-the mass fraction, as a percentage, of protein (6,38 times the nitrogen content) in the sample,

Note: When exactly 40.00 g of condensed milk are weighed and a polarimeter with sodium light, angular degrees and a 2 dm polanmeter tube at (20,0 ± 0,1)°C are used. the sucrose content of normal condensed milk [*i.e.* when C, defined below, is 9 per cent (mass fraction)] can be calculated from the following equation:

ws = (A -1,25 B) (2,833 -0,00612 F– 0,00878 P) *Equation 3*

If the invert polarization is measured at a temperature, t, other than (2O±0,2) °C, the value B shall be multiplied by the correction factor

BIS and ISO 1 + 0,0037 (t - 20) *Equation 4*

and this corrected value shall be used in the calculation.

Values of the Inversion Division Factor, Q

The following equations give accurate values for Q, for various sources of light with corrections, where necessary, for concentration and temperature.

Sodium light and polarimeter with scale in angular degrees:

BIS Q = 0,8825 + 0,0006 ((C– 9) -0,0033 (t–20) *Equation 5*

ISO Q = 0.8825 + 0.0006 (C-9) – 0.0033 (t-30)

Mercury green light and polarimeter with scale in angular degrees:

Q = 1,0392 + 0,0.007 (C- 9)– 0,0039 (t- 20) *Equation 6*

White light with bichromate filter or sodium light and saccharimeter with scale in international sugar scaledegrees:

Q = 2,549+ 0,0017 (C-9) - 0,0095 (t– 20) *Equation 7*

where,

C-the mass fraction, as a percentage, of total sugars in the inverted solution according to the polarimetric reading-, t-the temperature, in degrees Celsius, of the inverted solution during the polarimetric reading.

Note: 1-The mass fraction of total sugars, C, in the inverted solution may be calculated from the direct reading and the change on inversion in the usual manner, using the usual values for the specific rotations of sucrose, lactose and invert sugar.

The corrections 0,0006 (C–9), *etc.*, in Equations (5), (6) and (7) are only accurate when C is approximately 9; for normal condensed milk, this correction can be neglected, C being very close to 9,

2. Variations in temperature from 20°C make little difference in the direct reading, but variations of more than 0.2°C in the invert reading necessitate a correction. The correction factor given in Equation (4) given above, is only accurate between 18 °C and 22°C.

Source

1. *IS11764: 2005.*
2. *ISO 2911: 2004.*
3. *Reichert Technologies- Analytical Instruments.*
4. *Ametek Ultra Precision Technologies.*

b. Polarimetric Method for Sugar Determination (Before and After inversion with HCl)

a. Direct Reading

Pipette 50 ml Pb free filtrate into 100 ml volumetric flask, add 2.315 g NaCl and 25 ml water. Dilute to volume with water at 20°C and polarize in 200 mm tube at 20° C. Multiply reading by 2 to obtain direct reading.

b. Invert Reading

Pipette 50 ml aliquot Pb free filtrate into 100 ml volumetric flask and add 20 ml water. Add little by little while rotating flask 10 ml HCl (sp. gr 1.109). Heat water bath and adjust heater to keep bath at 60°C. Place flask in water bath, agitate continuosly for 3 minutes and leave flask in bath exactly 7 minutes longer. Place flask at once in water at 20° C. When contents cool to 35°C dilute almost to mark. Leave flask in bath at 20°C at least 30 minutes longer and finally dilute to mark.Mix well and polarize in 200 mm tube provided with lateral branch and water jacket keeping temperature at 20°C. Multiply by 2 obtain invert reading

Calculate

Sugar per cent as follows

$$S = \frac{100\,(P - I)}{132.66 - 0.0794\,(13\text{-}m) - 0.53\,(t - 20)}$$

where,

P- direct reading, normal solution, I- Invert reading, normal solution, t- Temperature at which readings are made and m - gm of total solids from original sample in 100 ml inverted solution

Source

1. *A.O. A.C 17th edn, 2000 Official Method 925.48 Sucrose in sugars and Syrups).*
2. *Manual of Methods of analysis of Foods (Beverages - Cofffee, Tea, Sugar, Sugar Products etc.), Lab Manual-3, FSSAI, New Delhi.*

c. Determination of Sucrose in Cane Sugar and Refined Sugar

Principle

An aqueous solution of the sugar sample (26 gm, *i.e.* the normal weight of sucrose in 100 ml water.) is polarized by means of a saccharimeter which is calibrated to read 1000 S on the' International Sugar Scale' under specified conditions.

Apparatus

Saccharimeter, calibrated with quartz plates. Basis of calibration of 100°S points on international sugar scale is polarization of normal solution of pure sucrose (26.000 gm/100 ml) at 20°C in 200 mm tube using white light and dichromate filter defined by the commission. This solution polarized at 20° C must give saccharimeter reading of exactly 100 ° S. Temperature of sugar solution during polarization must be kept constant at 20° C.

Following rotations hold for normal quartz plate of international sugar scale:

Normal Quartz Plate =100°S = 40.690 ± 0.002 (´= 546.1 nm) at 20°C. In general make all polarizations at 20° C. For countries where mean temperature is above 20° C. Saccaharimeters may be adjusted at 30 °C or any other suitable temperature, provided sugar solution is diluted to final volume and polarized at this temperature.

Procedure

In determining polarization use whole normal weight (26± 0.002 gm) for 100 ml or multiple for any corresponding volume. Bring solution exactly to mark at proper temperature and after wiping out the neck of the flask with filter paper add minimum amount of dry basic lead acetate, shake to dissolve. Repeating addition till precipitation is complete. Pour all clarified sugar solution on rapid air dry filter. Cover funnel at start of filteration. Reject first 25 ml filtrate and use remainder (must be perfectly clear) for polarization. In no case return whole solution or any part to filter. To remove excess lead used in clarification add anhydrous Pot or Sod Oxalate to clarified filtrate in small amounts until test for lead in filtrate is negative, then refilter. Polarise in 200 mm tube. Other permissible clarifying and decolorizing agents are alumina cream or conc alum solution. Temperature correction for polarization of sugars. Polarisation when made at temperatures other than 20°C may be calculated to polarization at 20°C by the following formula:

$$P\,20 = P\,t\,[1 + 0.0003\,(t - 20)]$$

where,

Pt= polarization at temperature read.

2. Determination of Lactose by Polarimetry

a. AOAC Method (AOAC 896.01)

Principle

(*i*) Precipatation of fat and protein.

(ii) Measurement of the specific rotation of the polarized light due to the asymmetric carbon of lactose.

Advantages

Costs.

Disadvantages

(i) Interference with optically active components.

(ii) No differentiation between carbohydrates.

(iii) Empirical calculation

Calculation

$$C\,[g/100ml] = \frac{a\,[Vm + Vr - 0.01145Vm\,(F+P)]}{Vm\,(b)}$$

where,

a- observed reading, Vm: - Volume of milk used, Vr- Volume of reagent used, b-rotation in a 40 cm tube of 100 ml solution containing 1 gram of lactose at t 0C (b=2.096at 25 0C), F- per cent fat and P- per cent protein.

Source

1. *IDF International Symposium -Lactose and its Derivatives.*
2. *Rachid Kouaouci, Ing., M.Sc., Chemist, Analytical Methods for Lactose, Quantification in Milk and Milk Products, Valacta, Ste-Anne de Bellevue, Québec, Canada.*

b. Determination of Lactose Content in Milk

One of the easiest methods of determining the lactose concentration in milk is through polarimetry. Normal milk contains 45-50 g lactose per liter. For the lactose measurement to be successful, all optically-active compounds except for the lactose (for instance, proteins) must be removed, and the solution must be transparent enough for polarized light to pass through readily. Setting the polarimeter to the correct angle for the optical rotation value is tricky and must be done carefully to avoid erroneous values.

The **specific rotation** of a substance is defined as the optical rotation of a solution containing 1 g/ml in a 100 mm polarimeter tube. It is affected by temperature (20°C reference temperature) and wavelength (usually the sodium D line, 589 nm, is used), and can be measured using the following formula:

$$[\alpha]^{20}_{D} = \frac{\alpha}{c \cdot d}$$

where, α is the measured optical rotation; *d* the light path in dm, and *c* the concentration as g/ml. The specific rotation, $[\alpha]$, of lactose in solution at equilibrium is +52.3°.

Procedure

It involves the following steps:

(i) **Removal of the fat and protein from the sample-**The first step in preparing the milk sample requires the removal of all fat and protein. For this, take 50.0ml of milk to a 125 ml Erlenmeyer flask. Add 7.5 ml of concentrated sulfuric acid to the milk. Mix it and add 7.5 ml of a mercury iodine/ potassium iodine solution to this mixture. Filter the solution to remove the precipitate. Rinse the Erlenmeyer and wash the protein solids with water to ensure that the dissolved lactose is transferred completely. Transfer the filtrate to a 100 ml volumetric flask and carefully dilute the solution to 100 ml total volume. Discard the solidified proteins.

(ii) **Measurement of the optical rotation-**Fill a clean, 100 mm-long polarimeter tube with the diluted milk solution at 20°C. Place the filled sample tube in the polarimeter and record the measured value (angular degree).

(iii) **Calculation of the concentration of lactose -**Calculate the, concentration of the diluted milk by using the following formula:

$$[\alpha] = \frac{a}{c \times l}$$

Where, *c*-is the concentration in grams per 100 ml, *l*-is the path length in dm, [α] - the specific rotation, and *a*-is measured value.

Use this concentration value to calculate the number of grams of lactose present in the diluted milk (which is the same as grams in the undiluted milk). In calculating the concentration of lactose, a correction should be used for the concentration of fat and protein in the precipitate.

Source

1. *Reichert Technologies- Analytical Instruments.*
2. *Ametek Ultra Precision Technologies.*

B. Research Applications for Polarimetery [1]

Research applications for polarimetry are found in industry, research institutes and universities as a means of:

(i) Isolating and identifying unknowns crystallized from various solvents or separated by high performance liquid chromatography (HPLC).

(ii) Evaluating and characterizing optically active compounds by measuring their specific rotation and comparing this value with the theoretical values found in literature.

(iii) Investigating kinetic reactions by measuring optical rotation as a function of time.

(iv) Monitoring changes in concentration of an optically active component in a reaction mixture, as in enzymatic cleavage

(*v*) Analyzing molecular structure by plotting optical rotatory dispersion curves over a wide range of wavelengths.

(*vi*) Distinguishing between optical isomers.

C. Using Polarimetry in Quality and Process Control Applications [1]

Quality and process control applications, both in the laboratory or on-line in the factory, are found throughout the pharmaceutical, essential oil, flavor, food and chemical industries. A few examples are listed below.

i. Pharmaceutical Industry

Analysis of pharmaceuticals complies with Pharmacopoeia, DAB and other national and international standards. It helps in [3]:

(*i*) Monitoring chemical processes.

(*ii*) Purity control and determination of concentrations.

(*iii*) Characterisation of new synthetic substances.

Polarimetery determines product purity by measuring specific rotation and optical rotation of:

(*i*) Amino Acids, Antibiotics, Dextrose. Steroids, Amino Sugars, Cocaine, Diuretics, Tranquilizers, Analgesics, Codeine, Serums, Vitamins.

ii. Polarimetry for Flavor, Fragrance, and Essential Oil Industry

Utilizes polarimetry for incoming raw materials inspection of: Camphors, Gums, Orange oil, Citric acid, Lavender oil, Spearmint oil, Glygeric acid, Lemon oil.

iii. Polarimetery for Food Industry Applications

Polarimetery ensures product quality by measuring the concentration and purity of the following compounds in sugar based foods, cereals, syrups and essential Oils: Carbohydrates, Lactose, Raffinose, Various starches, Fructose, Levulose, Sucrose, Natural monosaccharides, Glucose, Maltrose, Xylose.

It helps in [3]:

(*i*) Determination of concentration.

(*ii*) Purity control.

(*iii*) Quality control.

Polarimeters are used in the **sugar industry** for determining quality of both juice from sugar cane and the refined sucrose. Often, the sugar refineries use a modified polarimeter with a flow cell (and used in conjunction with a refractometer) called a saccharimeter. These instruments use the International Sugar Scale, as defined by the International Commission for Uniform Methods of Sugar Analysis (ICUMSA) [2]. It helps in [3]:

(*i*) Quality control of original and end product.

(*ii*) Determination of fructose and glucose.

(iii) Sugar concentrations in refined beet and cane sugar, molasses and beet pulp.

iv. Polarimetery in the Chemical Industry

Polarimetery analyzes optical rotation as a means of identifying and characterizing-Biopolymers, Natural polymers, Synthetic polymers. It helps in [3]:

(i) Purity control and determination of concentrations.

(ii) Analysis of optically-active components (qualitative and quantitative).

(iii) Determination of changes in the configuration.

(iv) Monitoring chemical processes

Source

1. *Rudolphresearch.com/products/polarimeters/polarimetry-definitions.*
2. *Wikipedia.*
3. *A.Kruss Optronic, Germany.*

Section C

Instrumentation in Food Safety and Microbiological Analysis

Chapter 6

Techniques for Food Safety and Microbiological Analysis

Food borne pathogens are a growing concern for human illness and death. Pathogens are virtually present everywhere, reaching every aspect of life. There is increasing demands to ensure safe food supply. Current trends in nutrition and food technology are increasing the demands on food microbiologist to ensure a safe food supply. Bacterial pathogens encountered to human illness in the last decades are through consumption of undercooked or minimally processed ready-to-eat meats (hotdogs, sliced luncheon meats and salami), dairy products (soft cheeses made with unpasteurized milk, ice cream, butter, *etc.*), or fruits (apple cider, strawberries, cantaloupe, *etc.*) and vegetables [1-5]. The application of untreated manure onto farmland may contaminate soil or water and eventually transmits microbes to fruits or vegetables [6,7]. Seafoods are another potential source of pathogens, such as Vibrio, Listeria, Yersinia, Salmonella, Shigella, Clostridium, Campylobacter and Hepatitis A [8,9]. Because of the aforesaid reasons and more awareness about food safety, there are urgent need to implement programmes such as HACCP as a part of Good Manufacturing Practices (GMP) and Sanitary and Phytosanitary measures (SPS) to monitor the quality of the products produced for the presence of the pathogens and microbial toxins (APHA, 1987). The analysis of food for the presence of both pathogenic and spoilage bacteria is a standard practice for ensuring **food safety** and quality [10]. There is continuous development of methods for the rapid and reliable detection of food borne pathogens. Advent of biotechnology has greatly altered food testing methods. Improvements in the field of immunology, molecular biology, automation and computer technology continue to have a positive effect on the development of faster, more sensitive and more convenient methods in food microbiology.Further, development of on-line microbiology, including ATP bioluminescence and cell counting methods, is important for rapid monitoring of cleanliness in HACCP programs. One of the most challenging problems is sample preparation. More research is needed on techniques for separating microorganisms from the food matrix and for concentrating them before detection to ensure **food safety**, by immunological or nucleic acid-based assays [11].

The possibilities of combining different rapid methods, including immunological and DNA based methods should be further exploited. Further developments in immunoassays and PCR protocols should result in quantitative detection of microorganisms and the simultaneous detection of more than one pathogen or toxin. Lastly, technology continuing to advance at a great pace, the next generation of assays currently being developed potentially has the capability for near real time and online monitoring of multiple pathogens. Modern methods are based on molecular biology techniques like PCR, RFLP, DNA microarray assay, immunological techniques like ELISA, biophysical and biochemical principles with the application of biosensers like bioluminescence sensor, bio-analytical sensors utilizing enzymes, electrical impedometry and flow cytometry [11].

Conventional/Traditional Methods for Microbiological Analysis

The Conventional microbiological methods have served microbiologists well over the past century and have helped to ensure the production of microbiologically safe products. For example, a wide range of microbiological methods have been successfully verified using plate count methods to enumerate and identify microorganisms within an accepted margin of error [11a]. Conventional bacterial testing methods rely on specific media to enumerate and isolate viable bacterial cells in food. These methods are very sensitive, inexpensive and can give both qualitative and quantitative information on the number and the nature of microorganisms present in the food sample [10]. Traditional methods for the detection of bacteria involve the following basic steps: pre-enrichment, selective enrichment, selective plating, biochemical screening and serological confirmation [12]. Hence, a complete series of tests is often required before any identification can be confirmed [13]. The conventional methods for detecting the foodborne bacteria specially pathogens present in food are based on culturing the microorganisms on agar plates followed by standard biochemical identifications. Conventional methods are usually inexpensive and simple but these methods can be time consuming as they depend on the ability of the microorganisms to grow in different culture media such as pre-enrichment media, selective enrichment media and selective plating media. Usually conventional methods require 2 to 3 days for preliminary identification and more than a week for confirmation of the species of the pathogens. Conventional methods are laborious as they require the preparation of culture media, inoculation of plates and colony counting. Further more, conventional methods may be limited by their low sensitivity. False negative results may occur due to viable but non-culturable (VBNC) pathogens. The failure to detect foodborne pathogens would increase the transmission risk of pathogens.

However, conventional methods have limitations. These limitations include the time taken to produce a result and the inability of many methods to recover all of the microorganisms that might be present in a sample. These conventional methods require several days to give results because they rely on the ability of the organisms to multiply to visible colonies [14]. Conventional methods generally regarded as the golden standard often takes days to complete the identification of viable pathogens. In Dairy and Food Industries following traditional microbiological tests are being carried out to find out the microbiologial quality of food/dairy products [12e]:

Sl. No.	*Traditional Microbiologial Tests*	*Dairy/Food Products*					
		Milk and Milk Products	*Sea-foods*	*Spices*	*Tomato products*	*Jam/ Marm alade/Fruit Jelly/Fruit Chutney*	*Meat and Fish*
1	Total Plate Count	√	√	√	Tomato ket-chup and Tomato Sauce	-	Total Viable Count (TPC)/ Aerobic Plate Count (APC)
2	Coliform Count	√	-	√	-	-	Faecal
3	*E. coli*	√	√	√	-	-	√
4	*Salmonella*	√	√ and Shigella	√ and Shigella			
5	*Staph. aureus* (coagulase positive)	√	√	√	-	-	√
6	Yeast and mold count	√ (Butter, Cheeses, yoghurt, Dahi, Khoa *etc.*)	-	√	Tomato juices and soups-Mould Tomato puree and paste-Yeast, Tomato ketchup and TomatoSauce-√	√	√
7	Spore Count Aerobic (*B.cereus*)	√	-	√	Tomato puree and paste -√ Tomato ketch up and Tomato Sauce-√	√	-
	Anaerobic (*Clostridium perfringern*)	√	√	-	-	-	√
8	Listeria Monocytogenes	√	-	√	-	-	-
9	Vibro Choleare	-	√	-	-	-	-
10	Vibro para-haemolyticus	-	√	-	-	-	-

National and International Standards for Microbiological Test Methods

The following microbiological test methods are prescribed in National and International Standards and these are being used successfully for knowing the microbiological quality of food products:

Sl. No.	*Microbiological Methods for Enumeration*	*National Standards*		*Internatioal Standards*	
		FSSAI Test Lab manual 14	*BIS Standards*	*ISO, Geneva*	*AOAC Official Methods*
1	Standard Plate counts	Chapter 01-1	IS 5402:2002 Methods for detection and estimation of Standard Plate Count in food stuff	*(i)* ISO. (2003c).Yogurt – Enumeration of Characteristic Micro organisms –Colony-Count Technique at 37º C. *(ii)* ISO:4833 : 1991 Reaffirmed 2007.	-
2	Coliform counts	Chapter 01-5	*(i)* IS:5404, 1995. *(ii)* IS5401(Part 1): 2002 or IS 5401(Part 2):2002. *(iii)* IS 5887(Part 1):1976 Reafirmed 2009- E.coli.	*(i)* /ISO 4832:1991 Reaffir- med 2007. *(ii)* ISO 4831:1991 Reaffirmed 2007.	AOAC, International (1995),16th edition, published by AOAC, International, USA,Test No. 17.2.02, page - 4-5.
3	Yeast and mould counts	Chapter 01-14	IS 5403:1999 Reaffirmed 2005.	ISO 7954:1987 Reaffirmed 2009.	-
4	Pathogens	Chapter 01	-	*ISO (2005a.b,2006a,b,c,2011 a.b,2012,2013) Microbiology of Food and Animal Feeding Stuffs.Polymerase Chain Reaction (PCR) for the Detection of FoodBorne pathogens.*	-
	i. Listeria mono-cytogenes	Chapter 01-14	IS 14988(Part 1):2001 Reaffirmed 2007 or IS:14988(Part 2): 2002 Reaffirmed 2007. ISO:11290-2 :1998	*(i)* ISO 10560, 1999. *(ii)* ISO 11290-1 :1996	*(i)* 993.09.2000. *Listeria* in dairy products, seafoods, and meats. Colorimetric deoxyribonucleic acid hybriddization method (GENE-TRAK *Listeria* Assay).

Contd.

Sl. No.	Microbiological Methods for Enumeration	National Standards		International Standards	
		FSSAI Test Lab manual 14	BIS Standards	ISO, Geneva	AOAC Official Methods
					(ii) 993.12. 2000. *Listeria monocytogenes* in Milk and Dairy Products, Selective Enrichment and Isolation Method (IDF Method). *(iii)* 994.03. 2000. *Listeria monocytogenes* in dairy products, seafoods, and meats. Colorimetric monoclonal enzyme-linked immunosorbent assay method (*Listeria* -Tek). *(iv)* *995.22. 2000. Listeria* in foods. Colorimetric polyclonal enzyme immunoassay screening method
	ii. Staphylo-coccus aureus	Chapter 01-11	*(i)* IS 5887(Part 8/Sec1): 2002. Reaffirmed 2007 or IS 5887 (Part 8/Sec2):2002 Reaffirmed 2007.	*(i)* Draft ISO/DIS 6888, 1983. *(ii)* ISO 6888-1 :1999. *(iii)* ISO 6888-2 :1999.	*(i)* AOAC INTERNATION AL.1995.Official Methods of Analysis, 15th ed., sec. 987.09-Most Probable Number Methodfor *Staphylococcus* spp(less than 100). *(ii)* AOAC INTERNATIO-NAL.1995.Official Methods of Analysis, 16th ed.,sec.975.55-Direct Plate Count Method for *Staphylococcus spp*(more than 100).

Contd.

Sl. No.	*Microbiological Methods for Enumeration*	*National Standards*		*Internatioal Standards*	
		FSSAI Test Lab manual 14	*BIS Standards*	*ISO, Geneva*	*AOAC Official Methods*
5	Salmonella	Chapter 01-9	IS 5887(Part 3):1999 Reaffirmed 2009.	ISO 6579:1993	*(i)* 978.24 (API 20E, Enterotube II, and *Enterobacteriaceae* II), and Method 991.13 (Vitek GNI)Enterotube II, *Enterobacteriaceae* II, MICRO-ID, or Vitek GNI) for rapid presumeptive generic identification of *Salmonella*. *(ii)* AOAC, International (1995),16 th edition, published by AOAC, International, USA,Test No. 17.9.01
6	Shigella	Chapter 01-15	IS: 5887(Part-3)-1999	-	-
7	Vibrios (V.cholerae V. parahae-molyticus)	Chapter 01-14	-	AOAC International (1995). 16th Edn.(Edited by Patricia Cuniff. Publis-hed by AOAC, USA), TestNo. 17. 11.01 p.108-110	-
8	Spore formers	Chapter 01-3 and 4	*(i)* IS 5887(Part 6):1999.*B.cereus.* *(ii)* IS:5887 Part IV:1999 Reaffirmed 2009.*C.perfringens, C.botulinum.*	*(i)* ISO 7932: 1993 Reaffirmed 2007. B.cereus.	*(i)* AOAC, International (1995),16 th edition, published by AOAC, International,USA,Test No. 17.7.02-C.perfringens. *(ii)* AOAC, International (1995),16 th edition, published by AOAC, International, USA,Test No. 17.8.01, page - 52-54

Contd.

Sl. No.	Microbiological Methods for Enumeration	National Standards		Internatioal Standards	
		FSSAI Test Lab manual 14	BIS Standards	ISO, Geneva	AOAC Official Methods
9	Others	Chapter 01-1,13 and 15	IS:12899-1989 Special Lactic acid bacterial Count in Yoghurt	*(i)* ISO. (1998)- Microbiology of Food and Animal Feeding Stuffs - Horizontal Method for the Enumeration of Mesophilic Lactic Acid Bacteria-Colony Count Technique at 30 °C. *(ii)* ISO.(2003a)- Microbiology of Food and Animal Feeding Stuffs -Protocol for the Valida tion of Alternative Methods. *(iii)* ISO (2003b). Yogurt-Identifi cation of Characteristic Micro organisms (Lactobacillus delb rueckii subsp. bulgaricus and Streptococcus thermophilus).	-

The details of the some of the traditional microbiological test methods for milk and milk products prescribed under FSSAI and BIS are given in author's another books [12f,12g],

Constraints in Food Analysis

Microbiological analysis of food, especially for particular pathogenic species remains a challenging task for virtually all assays and technologies. The problems may be due to the following facts [10]:

i-Bacteria are not uniformly distributed in the food, ii- Heterogenicity of food matrices (Ingredients such as proteins, carbohydrates, fats, oil, chemicals, preservatives *etc.*), Physical form of food (powder, liquid, gel, semisolid or other forms) and Difference in viscosity due to fats and oils, which may interfere in proper mixing) and iii-Presence of indigenous microbes which do not cause health risk but their presence often interferes with the selective identification and isolation of specific pathogens, which are usually found in low numbers.

Need for rapid method [10c,11] -Any modification in Traditional methods that reduces the analysis time can technically be called rapid method. Recently, different rapid methods with high sensitivity and specificity have been developed to overcome the limitations of conventional methods for the detection and identification of food borne pathogens. The effective testing of bacteria requires methods of analysis that can meet a number of challenging criterions. Time and sensitivity of analysis are the most important limitation related to the usefulness of microbial testing. Time, sensitivity, specificity *etc.* of some rapid methods and comparative status on charcteristics of TCM and rapid methods are given in following tables:

Characteristics of some Alternative and Rapid Methods

Sl.No.	*Method*	*Detection Limit (cfu mL _1or g*	*Time before Result*	*Specificity*
1	Plate techniques	1	1-3 days	Good
2	Bioluminescence	10^4	1/2 h	No
3	Flow cytometry	10^2 10^3	1/2 h	Good
4	DEFT	10^3 10^4	1/2 h	No
5	Impedimetry	1	6-24 h	Moderate/Good
6	Immunological methods	10^4	1-2 h	Moderate/Good
7	Nucleic acid based assays	10^2	6-12 h	Excellant

Requirements for Rapid Microbiological Methods [11]

There are several following factors which must be considered before adapting a new alternative or rapid method:

i. Accuracy

False-positive and false-negative results must be minimal or preferably zero. The Method must be as sensitive as possible and the detection limit as low as

Methodological Properties of Selected Rapid Methods in Food Microbiology [11I]

Sl.No.	Methods	Purpose			Detection Limit Cells (per ml or g)	Rapidity
		Quali.	Quant.	Char.		
1	Direct Methods - Epifluorescence microscopy					
	Flow cytometry	+	+	-	10^4	<0.5h
	DEFT	-	+	-	10^4	<1h
	Ab-DEFT	+	+	+	10^3	<1h
	MMCF	-	+	-	10^3	≥7h
2	Nucleic acid–based methods					
	PCR	+	(+)	+	100 to direct in food 10^4	<2days
	Colony hybridization	+	+	+	10^1	<2days
	Dot Blot	+	(+)	+	10^4	<2days

Note: i- qual.- qualitative result (presence/absence), ii-quant.- quantitative result (enumeration), iii-char.- microbiological characterization, iv- + - applicable, v- (+) - applicability restricted, vi- - - not applicable. a If special selective media are available.

possible. ln many cases, the demand is, less than one cell per 25 g of food, as small numbers of some pathogens may cause disease. Analytical tests for these agents need only be qualitative (presence/absence). For rapid screening methods, a higher false positive frequency may be acceptable, as positive screening tests are followed by confirmation tests.

ii. Speed

Rapid tests for the detection of bacteria/pathogens or toxins should give an accurate result within hours or at the utmost one day.

iii. Automaton and Computerization

The ability to test many samples at the same time. Many systems utilizing the microliter plate format can handle 96 samples at one time.

iv. Sample Matrix

New systems should give a good performance of the matrices to be tested. Baseline extinction values may depend on the type of food being tested. Background flora, natural substances or debris can interfere with the lest method and invalidate the test result.

v. Validation

The alternative Rapid test should be validated against standard tests and evaluated by collaborative studies. Results obtained with samples containing a low contamination level should be emphasized, since there is sufficient evidence that in most cases high numbers of target cells will lead to positive lest results.

vi. Costs

Purchasing reagents, supply, operational costs, up-keeping. The initial financial investment for rapid methods may be high, because, many systems require expensive instruments. Operating costs of many commercial rapid test kits are also high.

vii. Simplicity

Methods should be user-friendly, which means easy to operate and manipulate.

viii. Rapid Availability of Reagents and Supply

The required reagents and other supply should be easily and rapidly available.

ix. Training, Technical Service and Company Support

These are essential and should also be easily and rapidly available.

x. Space Requirements

Instruments are preferably compact and small.

The comparative status on charcteristics of TCM and rapid methods are given in following table:

Comparative Status of Traditional Culture Methods (TCM) and Rapid Methods (PCR, ATP Bioluminescence Technique *etc.*)

Sl. No.	*Characteristics**	*Traditional Culture Methods**	*PCR**	*ATP Bioluminescence Technique***
1	Skill Level Required	*(i)* Different activities -making culture, plating, counting *etc.*, needs expertise. Bacterial identification is an art form that can take years to perfect. *(ii)* High Labour requirement.	*(i)* Real-Time PCR requires a seasoned eye with the technical skills to interpret results but now latest development (New Instrument Hunter®) greatly simplify the process and can consistently and automatically interpret the results. *(ii)* Low Labour requirement.	*(i)* Requires skilled manpower. *(ii)* Low Labour requirement
2	Time to Result (Speed of Test)	TCM can take from 3 to 7 days for the most common food contaminants.	Real-Time PCR-including primary enrichment-can often be run in 10-24 hours	Quick method. The ATP monitoring system helps to quickly identify contaminated areas so that corrective actions can be taken in a timely manner.

Contd.

Sl. No.	Characteristics*	Traditional Culture Methods*	PCR*	ATP Bioluminescence Technique**
3	Accuracy	There is always the possibility that a pathogenic organism could be missed even by the most seasoned technician.	Real-Time PCR, on the other hand, usually hunts for highly conserved genes or essential genes. Hence there is no chance of missing. In many of the new instruments of Real Time PCR,there is provision for automatic electronic report-generating.	Accurate
4 ***	Sensitivity	Often very low	High- Viable and non culturable (VBNC_/Stressed organism	High
5	Waste Generated	The amount of waste gene-rated per test per sample doesn't seem like an important concern but when detection of one micro-organism requires plating in different media, then it becomes matter of concern. Disposal of waste is another concern as it requires special treatment *e.g.* Autoclaving *etc.*	It can hunt for the correct DNA segment without all the trial and error,there is substantially less waste to be considered,typically only a few small tubes beyond the primary enrichment.	-
6	Cost	*(i)* Cost of reagents is perhaps less but as this method is time consuming and requires skill set hence overall cost per sample is more than PCR. *(ii)* Release of Products after getting negative results, delayed and hence reducing the shelf life and increasing the warehousing cost. *(iii)* Upfront Cost - Low initial and ongoing cost.	*(i)* PCR is 38 per cent less expensive to run than TCM. *(ii)* Real-Time PCR means faster release of product, which decreases these types of cost. *(iii)* Upfront Cost - Often very high.	-
7	Diversity of Available Tests	The Diversity of Available Tests is better than PCR.	There are still some very rare and emerging pathogens for which Real-Time PCR tests are not available.	-
8	***Time to Return on investment	Short - Several months	Significant - years	-

Source: *-Technical Literature of Lauren Bambusch, Instant Labs,**-Zubair M. Azizkhan.Comparison between ATP Bioluminescence Technique and Traditional Microbiological Method to Detect Contamination within Food Facilities, Journal of Public Health Frontier Mar. 2014, Vol.3 Iss.1, PP.11-18 and***Technical Literature of Wickham Laboratories.

It is evident from the aforesaid tables that the rapid detection methods are generally more sensitive, specific, time-efficient, labor-saving, and reliable than conventional methods. The food industry is in need of more rapid methods which are sensitive for the following reasons [12]:

(i) To provide immediate information on the possible presence of pathogen in raw material and finished products.

(ii) Low numbers of **pathogenic bacteria** are often present in complex biological environment along with many other non-pathogenic organisms.

(iii) The presence of even a single pathogenic organism in the food may be a infectious dose.

(iv) For monitoring of process control, cleaning and hygienic practices during manufacture.

(v) To reduce human errors and to save time and labor cost.

Rapid microbiological method technologies aim to provide more sensitive, accurate, precise, and reproducible test results when compared with conventional, growth-based methods. Rapid methods normally involve some form of automation, and the methods often capture data electronically [12a].

Advantages and Limitations/Disadvantages of Rapid Methods

Rapid microbiological methods (RMMs) have the following advantages and disadvantages.

Advantages

Rapid microbiological methods (RMMs) have the following advantages [12a] and because of these advantages, RMMs attract interest from microbiologists and are areas of considerable investment by vendors:

(i) These methods are essentially used as alternatives to the following four major types of conventional microbiological determinations [12b,12c,12d]:

(a) Qualitative tests for the presence or absence of microorganisms (*e.g.*, enrichment turbidity measurements of growth). For example, to determine if *Escherichia coli* is in a sample of water.

(b) Quantitative tests for enumeration of microorganisms (*e.g.*, plate count methods to determine the bioburden of a sample).

(c) Quantitative tests for potency or toxicity (*e.g.*, what level of endotoxin is in the sample?).

(d) Identification tests (*e.g.*, biochemical and morphological characterization).

(ii) Another advantage afforded by rapid methods, aside from the time-to-result, is throughput. Most rapid systems allow for higher volumes than the traditional method. In environments with considerable volumes of raw ingredients, in-process batches, and final products to test, a

high throughput can confer an important advantage for maintaining manufacturing uptime and moving inventory as quickly as possible.

(iii) These methods (RMMs) are helpful in following activities:

(a) Designing more robust processes that could reduce the opportunities for contamination (fitting in with some quality-by-design objectives).

(b) Developing a more efficient corrective and preventative action process.

(c) Confirming that the process is in a continuous state of microbiological control through "real-time" monitoring (that meets some process analytical test objectives).

(d) Assisting with continuous process and product improvement.

(iv) Labour efficiency and error reduction are the another important advantages. Reducing errors is one of the greatest potential benefits of rapid enumeration. Microbial counting, incubation changeovers, and data entry can all become far more reliable because of the right equipment.

(v) RMMs enable a proactive approach to be taken to instances of microbial contamination, especially in relation to out-of-specification results. Here, RMMs enable quicker responses to out-of-trend situations through providing real-time or near real-time results. This allows for corrective actions to be taken earlier.

(vi) The new RMM must offer a higher level of quality assurance. There needs to be a clear and demonstrable benefit in adopting the alternative method. Examples of this include:

(a) The ability to make critical business decisions more quickly.

(b) The prevention of recalls through greater method sensitivity to microorganisms.

(c) The detection of "objectionable" microorganisms.

(d) Recovery of higher or more accurate microbial numbers.

(e) Potential reduced stock holding through faster release times.

(f) Improvement in manufacturing efficiency.

(g) A more proactive, rather than reactive, decision making.

Limitations/Disadvantages [11]

Rapid microbiological methods (RMMs) have the following Limitations/ disadvantages:

Limitations

(i) AOAC international approved rapid methods are mostly designed for preliminary screening, negative results are regarded definitive, but positive results are considered presumptive and must be confirmed.

(ii) Evaluation of rapid methods shows that same methods may perform better in some foods. This may be attributed mostly to interference by normal microbiota or inhibitors in food.

(iii) In case of an illness investigation, the food implicated may be suspected to contain a particular pathogen based on clear symptoms, but the actual pathogen is unknown. In these situations, when multiple pathogen analysis may be needed it makes the procedure complex and costly.

Disadvantages

(i) The most rapid methods need damaging of the cells and therefore, viable cells for confirmation and further characterization can only be obtained by repeat analysis using standard cultural procedures. This is a major disadvantages of these methods over traditional methods.

(ii) The rapid methods usually detect only one specific pathogen, while cultural methods may simultaneously detect and isolate many pathogens by including several types of numerous microbiological examinations or samples, selective media in the analysis.

(iii) The use of several rapid assays to do multi pathogen analyses on a food makes this analysis unacceptably expensive.

Techniques for Separation and Concentration of Bacteria/Pathogens [11]

The efficient pathogen separation and concentration techniques need to be evolved for specific detection of pathogens (Discrimination of target pathogen from other cells) and to avoid false-negative results as food samples are highly complex consisting of fats, proteins, minerals and even sometimes contains antimicrobial preservatives [10] and pathogens are numerically very low.

Several antibody-based and physical- and chemical-based separation and concentration methods have been developed for separation and concentration of pathogens from various sample matrices [10a,10b,10c]. This subsequently generates large quantity of material of which only a portion is used for further analysis unless a concentration step is also used. By concentrating the target organisms the detection period can be shortened and more efficient. The following methods are generally being used for concentration and separation:

1. Concentration by Membrane Filtration-Direct Epiflourescent Technique (DEFT) [11]

Membranes can be made from nitrocellulose, cellulose acetate esters, nylon, polyvinyl chloride and polyesters. Membrane filters are used in modified conventional methods for a variety of purposes. The details are given below:

(i) To concentrate target organisms from a large volume to improve detection limit.

(ii) To remove growth inhibitors.

(iii) To transfer organisms between growth media without physical injury through resuspension.

This (DEFT) is a direct method used for enumeration of microbes based on binding properties of flurochrome acridine orange. In this food samples are pretreated with detergents and proteolytic enzymes, filtered on to a polycarbonate membrane stained with acridine orange and examined under fluorescent microscope. The number of viable cells is determined based on the count of orange cells on the filter and can be performed in 10 min.

Application of DEFT in Analysis of Milk [11l]

It uses membrane filtration and epifluorescence microscopy. It takes less than 30 min to complete analysis and it is suitable for milk containing 5 x 103 to 5 x 108 bacteria per ml and does not suffer from many of the disadvantages of other microscopic methods.

Pre-treatment of the milk with proteolytic enzyme and surfactant at 50°C lyses somatic cells and makes fat globules sufficiently fluid to enable 2 ml of milk to be filtered routinely through a 25 mm diameter membrane filter. Filtration concentrates and distributes the bacteria in a manner that makes counting easier and the techniques 100 times more sensitive than the other likes- Breed smear.

Procedure

It involves the following steps:

(i) Whole milk, enzyme and surfactant are incubated at 50°C for 10 min in a sterile capped test tube.

(ii) A polycarbonate membrane filter (Nucleopore, 0.6-µm pore size) is mounted, shiny side uppermost, in a membrane filtration unit of a manifold connected to a vacuum line.

(iii) The filter is warmed by filtration of surfactant at 50°C before the treated milk is filtered. Surfactant at 50°C is used to rinse the test tube and filter tower after filtration of the treated milk. After filtration of the treated milk and rinse, the membrane is overlaid with stain.

(iv) After 2 min the vacuum is reapplied and the membrane is rinsed with buffer followed by isopropanol. The stained membrane is air dried and then mounted in non-fluorescent immersion oil on a slide beneath a cover slip.

(v) The mounted membrane filter is then examined by means of an epifluorescence microscope suitable for use with acridine orange.

(vi) **In the DEFT, because of the higher RNA content, metabolically active bacteria fluoresce orange whereas inactive bacteria fluoresce green.**

(vii) Clumps of orange fluorescing bacteria are counted in microscopic fields of view taken at random. Depending on the number of clumps per field, bacteria in 2-15 fields are counted.

The DEFT count per millilitre is estimated from the mean clump count per field and the microscopic factor (MF). The MF depends upon the amount of sample filtered through the given area of the membrane and the area of the microscopic field. The MF calculated as follows:

$$\mathbf{MF} = \frac{\text{Area of membrane through which sample is filtered (mm}^2\text{)}}{\text{Microscopic field area (mm}^2\text{) X sample volume (ml)}}$$

2. Separation by Immunomagnetic Separation (IMS) Method [11,11l]

The isolation/separation stage can be shortened by replacing a selective enrichment stage with non growth related procedures. IMS uses super-paramagnetic particles, which are coated with antibodies against the target organisms to selectively isolate the organisms from a mixed population. IMS is analogues to selective cultural enrichment. In selective cultural enrichment, the growth of other bacteria is suppressed while the pathogen of interest is allowed to grow [11]. **This technique is approved by AOAC international to perform [11l].** The separation process consists of following two fundamental steps [11]:

i. First Step

The suspension containing target cells is mixed with immunomagnetic particles for incubation no longer than 60 min and finally, they are separated using an appropriate magnetic separator.

ii. Second Step

In the second step, the magnetic complex is washed repeatedly to remove unwanted contaminants and the target cells with attached magnetic particles can be used for the further experiments.

The immune magnetic separation may be employed either directly or indirectly. However, in selective enrichment stage separation, chemical reagents (antibiotics) are used to select pathogens. Since reagents can be harsh and may cause cells stress or injury, LMS is a milder alternative to enrichment; also the elimination of selective enrichment step shortens analysis time. The major drawbacks of the IMS-based assays are the requirement of enrichment and a sample clean up step.

Advantages of IMS [11k,11l]

It has following merits:

(i) This technique enables to scale-up the application of IMS from the usual 1 ml to 400 ml volumes.

(ii) By recirculating the sample over a capture phase, comprised of immobilized antibody coated magnetic beads, the sensitivity of the capture is increased and thus accomplishes potential significant reduction of time to detection.

(iii) A high-volume wash also enables the efficient removal of the sample matrices, nonspecific microorganisms, and PCR inhibitors [11l].

(iv) The detection limit for IMS with PCR was1 cfu/1-25 g of sample following enrichment for *L. monocytogenes* [11k].

Application of IMS

(i) Polystyrene beads coated with iron oxide and antibodies [11b] are the most common magnetic carriers used for concentration and separation of selected microorganism from foods [11c]. Monosized super paramagnetic polymer particles known as 'Dynabeads' are available commercially from Invitrogen-Dynal. Pathatrix, an automated system, is a patented recirculating immunomagnetic separation technology [11l].

(ii) The Immunomagnetic beads have been used for capture of E. coli O157: H7 [11d], Salmonella [11e] and Listeria [11f]. In the **ISO 16654** detection methods for E. coli O157: H7 from foods, IMS is incorporated in order to pick up selectively the O157 serotype [11].

(iii) In recent years, applications of IMS coupled with PCR assays are showing very promising results for the detection of E. coli O157: H7 [11g], Salmonella enteric [11h] and *Listeria monocytogenes* [11i,11j].

Types of Rapid Microbiological Methods [11,12a,15]

Rapid methods micro-biological analysis can be categorized in multiple means. It can also be categorised on the basis of technology or application.

A. Classification on the Basis of Technology [11,12a]

The RMMS can be grouped into the following categories:

1. Growth-based Methods

Growth-based methods are those where a detectable signal is usually achieved following a period of subculture (*e.g.*, electrochemical methods). These methods generally involve the measurement of biochemical or physiological parameters that reflect the growth of microorganisms. These methods aim to decrease the time at which once can detect actively growing microorganisms. The methods continue to use conventional liquid or agar media. These methods include:

(i) **Impedance microbiology** (measurable electrical threshold during microbial growth-**Impedimetry**).

(ii) The detection of carbon dioxide (CO_2).

(iii) The utilization of biochemical and carbohydrate substrates.

(iv) The use of digital imaging and auto-fluorescence for the rapid detection and counting of micro-colonies.

(v) Fluorescent staining and enumeration of micro-colonies by laser excitation.

(vi) Selective media for the rapid detection of specific microorganisms.

2. Direct Measurement

Direct measurement is where individual cells are differentiated and visualized (*e.g.*, flow cytometry). These methods generally use viability stains and laser excitation for the detection and quantification of microorganisms without the need for cellular growth. These methods include:

(*i*) Demonstration of direct labeling of individual cells with viability stains or fluorescent markers with no requirement for cellular growth.

(*ii*) **Flow cytometry** (individual particles are counted as they pass through a laser beam).

(*iii*) **Solid-phase cytometry** (staining and laser excitation method).

3. Cell Component Analysis

Cell component analysis is where the expression of specific cell components offers an indirect measure of microbial presence (*e.g.*, genotypic methods). These methods generally involve the detection and analysis of specific portions of the microbial cell, including ATP, endotoxin, proteins, and surface macromolecules. The methods include:

(*i*) **ATP bioluminescence** (the generation of light by a biological process).

(*ii*) Endotoxin testing (LAL).

(*iii*) Fatty acid analysis (methods that utilize fatty acid profiles to provide a fingerprint for microorganism identification).

(*iv*) Matrix-Assisted Laser Desorption Ionization-Time of Flight (MALDI-TOF) mass spectrometry (microbial identification).

4. Optical Spectroscopy

Optical spectroscopy methods utilize light scattering and other optical techniques to detect, enumerate, and identify microorganisms (*e.g.*, "real time" airborne particle counters). These methods include:

(*i*) Real-time and continuous detection, sizing, and enumeration of airborne microorganisms and total particles. These methods are applied to the monitoring of clean rooms.

5. Nucleic Acid Based Assays/Nucleic Acid Amplification

Nucleic acid amplification technologies are those such as **PCR**-DNA amplification, RNA-based reverse-transcriptase amplification, 16S rRNA typing, gene sequencing, and other novel techniques. These methods include:

(*i*) Riboprinting: 16S sequence of rRNA is highly conserved at the genus and species level.

(*ii*) **PCR methods** for targeting specific microorganisms (millions of copies of the target DNA in a short period of time).

(*iii*) Gene sequencing (specific dye labeling).

Nucleic acid-based methods such as PCR, mPCR, qPCR, and DNA microarray have high sensitivity and they are widely used for the detection of foodborne pathogens, but these methods require trained personnel and specialized instruments. Alternative nucleic acid-based methods such as NASBA and LAMP are available for the detection of foodborne pathogens and their toxins. NASBA and LAMP are relatively sensitive, specific and cost efficient [15a].

6. Micro-Electrical-Mechanical Systems

Micro-Electrical-Mechanical Systems (MEMS) utilize microarrays, biosensors, Bioluminescence biosensor, Piezoelectric biosensors and nanotechnology to provide miniaturized technology platforms. These methods include:

(i) **Microarrays** (DNA chips, Gene chip technolog), evolved from Southern Blot technology, to measure gene expression (*e.g.*, mycoplasma detection).

7. Modified and Automated Conventional Methods

8. DNA Hybridization

In short, Rapid methods can be classified into the following broad categories:

(i) Modified and automated conventional methods.

(ii) Biosensors [Bioluminescence biosensor, Impedimetry (electrical impedance) Piezoelectric biosensors, Flow cytometry, Solid phase cytometry, Electronic nose].

(iii) Immunological methods.

(iv) Nucleic acid based assays (and DNA hybridization, Polymerase chain reaction, DNA micro assay (Gene chip technology).

B. Classification on the Basis of Application [15]

RMM can be classified in to the following three categorries:

1. Qualitative Methods

These RMMs detect the presence of microorganisms in products. Impedimetry, Polymerase chain reaction (PCR), Flow cytometry, ATP measurement and endotoxin tests (Limulus ameobocyte assay test) can be used to more rapidly detect presence of microorganisms [16].

2. Quantitative Methods

These RMMs are capable of providing a numerical value on the microorganisms present in a known sample unit. This is useful across the spectrum of microbiological tests. PCR (RT-PCR), ATP bioluminescence and auto fluorescence methods. Flow cytometry and Raman spectroscopy are coming in quantitative **RMMs [16,17,18].**

3. Identification Methods

Traditional biochemical tests can still be found in microbiological identification. An alternative to the more traditional microbiological tests is matrix-assisted laser desorption ionisation time of flight mass spectrometry (MALDI-ToF). PCR or nucleic acid-based methods can be used to identify organisms based on the 16s rRNA gene (or 16sDNA) in bacteria or the 26s rRNA gene, ITS or D2 region in fungi [18].

Application of Different Rapid Microbiological Methods in Food Analysis

The application of some Rapid Microbiological Methods (Biosensors, Electronic nose, Elisa, etc) in microbiological quality analysis of dairy products and other food products are already given in respective chapters. The application of some another important rapid microbiologial methods *e.g.* PCR, Flow cytometry *etc.* are given in following chapters. Here applications of rapid test using Petrifilms and modified/ automated conventional methods are described:

A. Modified and Automated Conventional Methods [11]

Modified and automated conventional methods are also very useful for rapid microbiological analysis These methods used traditionally for microbial analysis are regarded as gold standards except for time delay and labor involved. The following modifications in sample preparation, plating techniques, counting and identification systems have made these conventional methods faster, convenient, easier, user friendly and to reduce the cost of material and labor [10,12].

i. Sample Preparation

Gravimetric diluters-automatically adds the correct amount of diluents to the test sample before homogenization.

ii. Stomacher

Massages samples in a sterile disposable bag eliminating need to sterilize and to use blender cups.

iii. Pulsifier

This apparatus beats the outside of a sterile disposal bag at high frequency (3500 rpm) producing a combination of shock waves and intense stirring which drives the microbes into suspension.

iv. Plating Technique

There are several methods of adding sample homogenate to the agar plates.

a. Spiral Plater

This deposits a small volume on to the surface of the agar in a spiral fashion such that there is a dilution ratio of 10^4 from the centre to edge of the plate. The colonies appearing along the spiral pathway can be counted either manually or electronically. As the volume dispensed at any point is known, this technique eliminates the need for serial dilution before plating and less time requited for colony counting.

b. Dip Slides

The agar slides containing selective or non selective media are pressed on to the surface to be examined and replaced within a sterile sleeve.

c. Use of Flurogenic and Chromogenic Substrate

In selective media detection, enumeration and identification. This eliminates the use of subculture media and further biochemical tests. These compounds yield bright color fluorescent products when reacting with specific bacterial enzymes or metabolites. Flurogenic enzyme substrates are derived from coumarin, such as 4-

methylnubelliferone, while chromogenic enzymes compounds arc mainly phenol derivatives.

v. Hydrophobic Grid Membrane Filter (HGMF)

Works by confining colony growths 10 a set of 1600 grid cells. These techniques has the advantage of removing inhibitors or unwanted nutrients, concentrating organisms, as well as three log unit range. The food samples obtained are homogenized and pre filtered in nitrocellulose membrane filters, which trap food particles larger than micrometers. This filtrate is then filtered through HGMF, which traps target microbes. The inoculated HGMF are placed in suitable agars and colonies are counted after incubation.

vi. Colony Counting

This process is time consuming and several attempts have been made to automate this last step in enumeration of microbes to improve efficiency and labor cost. Image analysis systems have been shown to be useful and cost effective. An image of the plate is stored and can be viewed, printed or imported to other programs. The user can set variables such as top or bottom lighting and colony size limit to exclude spreaders or background particulars.

vii. Identification Systems

The morphological, physiological and biochemical tests are used for identification of microorganisms in conventional methods. Now several commercially available kits have been developed to simplify and automate the identification of individual organisms, the result of which is comparable to that of conventional identification systems.

B. Rapid Petrifilm Test Method [19,11l]

Petrifilms (3M) are alternative to agar poured plates for microbiological testing of food and beverages and these are ready-to-use products, which are composed of rehydratable films coated with standard nutrients, a cold water soluble gelling agent, and indicators that facilitate colony enumeration. Petrifilms eliminate the need for preparation of traditional media bringing many advantages over traditional enumeration techniques.One milliliter of liquid sample is placed on the centre of film system and the rehydrated growth of microorganism. After incubation, the colonies can be counted directly from the film system as in conventional plates.Rapid Petrifilm Test Method was sucessfully used for enumeration of aerobes. It is simple to use, is small in size and has a long shelf life over one year in cold storage. doesn't require preparing agar, and has results that are easy to read. Dry rehydratable films are marketed by 3M® for enumeration of aerobic count plate, Enterobacteriaceae (including Salmonella, Shigella, and Yersinia), coliform and Escherichia coli, yeast and mold, selective *E. coli* count plate, high-sensitivity coliform count plate, rapid coliform count plate, staph express count plate and environmental Listeria count plate. **Enumeration by using Petrifilms got International recognitions, certificates and validations.**

Procedure

It involves the following steps [111]:

(i) To obtain a viable cell count, the protective top layer is lifted, 1 ml of liquid sample is introduced to the center of the unit, and then the cover is replaced.

(ii) A plastic template is placed on the cover to make a uniform distribution of sample and round mold.

(iii) The rehydrated medium will support the growth of microorganisms after suitable incubation time and temperature.

(iv) The colonies are directly counted in the unit. Recently the company also introduced a Petrifilm.

(v) Counter, the unit automatically count and record the viable cell count in the computer.

Applications

Enumeration of Aerobic Microorganisms, Coliforms and E. coli in Food by Rapid Petrifilm Test Method [19]

These petrifilm products are available for yeast and mould counts, TVC, coliforms and E. coli, OI57. The Aerobic Plate Count (APC), Coliform Plate Count (CPC) and E. coli obtained using Rapid Petrifilm test method were compared with the corresponding standard methods using samples of milk powder, poultry and ready to serve drink (RTS). The counts obtained using rapid Petrifilm method were not significantly ($p<0.05$) different from the counts obtained using the conventional methods.

The corresponding correlation coefficients for Coliform counts were 0.9891, 0.9940 and 0.9990 while the corresponding correlation coefficients for E. coli were 0.9817, 0.9969 and 0.9983, respectively. The study shows that the Rapid Petrifilm Test method can effectively be used for APC and CPC for foods. On the basis of aforesaid results, it is concluded that the rapid Petrifilm technique can effectively be applied for routine laboratory analysis of regular food samples as a convenient alternative to conventional method for enumeration of aerobic microorganisms, coliforms and E. coli.

C. Compact Dry Plates [111]

These plates also have a dedicated user-friendly small plastic dish format that contains sterilised dehydrated nutrients and differentiating components. Similar to the petrifilm, Compact Dry plates are also thin, light, and convenient to handle sample diffuses automatically and evenly in to the plates.

The presence of chromogenic substrates and redox indicators stain the grown colonies on the incubated plates with different colors, facilitating interpretation of the plates. The results are validated with the conventional methods. Compact Dry Total counts, Coliform, E coli, and Yeast and molds have AOAC PTM (Performance Tested Methods) certificate, and Compact Total Count has also MicroVal certificate

(European validation and certification alternative methods according to ISO 16140).

D. Hydrophobic Grid Membrane Filter (HGMF) [11l,11m]

It is a new approach to microbiological enumeration, it consists of a micro porous membrane filter of 0.45μm pore size on which a grid pattern has been imposed. The working of HGMF involves the following steps:

- *(i)* A conventional membrane filter is placed on the surface of an agar medium, water containing dissolved nutrients from the medium is carried by capillary action up through the filters pores to feed the microorganisms, which are retained on the top surface during filtration.
- *(ii)* This produces a microscopic film of moisture on the top surface of the membrane filter through which motile organisms can travel.
- *(iii)* The hydrophobic grid pattern of an HGMF interrupts this film of moisture, preventing motile organisms from travelling beyond the boundaries of grid squares in which they have landed.
- *(iv)* Also, wherever the hydrophobic materials has been deposited on the filter, the pores are locked and water is prevented from bringing nutrients to any stray microorganisms, which might be pushed on to a grid line as colony increase in size and fills its square.
- *(v)* The analyst can then count the squares as individual colonies. Automatic instruments are also available to count these square colonies in seconds.

Procedure and Application

HGMF method has been used to test a great variety of foods in the past. This method is also approved by AOAC official method. **ISO-GRID/NEO-GRID** (Neogen, Lansing, MI) tests are based on the concept of hydrophobic grid membrane filtration.

- *(i)* The filter membrane is embossed with hydrophobic ink to produce a grid containing 1600 individual squares.
- *(ii)* The ink's hydrophobic nature contains the growth of an organism isolated during the filtration process within the square of capture.
- *(iii)* Samples are first diluted in a sterile buffer. The ISO-GRID and/or NEO-GRID filtration unit is then placed onto a manifold connected to a vacuum source, and a sterile filter membrane is placed in the ISO-GRID filtration unit (NEO-GRID already contains a sterile filter membrane).
- *(iv)* Using the vacuum source, a portion of the diluted sample is filtered through the filter membrane.
- *(v)* The membrane is then transferred to an agar plate prepared with a medium formulated to promote/differentiate the growth of the target organism.

Calculation

Following incubation, the plate's squares containing presumptively positive colonies are counted to yield a most probable number (MPN) result by using following formula.

$$\mathbf{MPN} = N\ X \log c\ [(N/N\text{-}X)]$$

where,

N = total number of grid squares on the filter; and **X** =number of positive grid squares.

Application-HGMF in Food Microbiology

The HGMF has formed the basis for a wide range of quantitative food microbiology applications, including total bacterial counts, coliform and Escherichia coli, fluorescent pseudomonads, lactic acid bacteria, yeasts and molds, Aeromonas, E. coli O157: H7, Vibrio para-haemolyticus, fecal streptococci, and Staphylococcus aureus. It also has been used for rapid detection of Salmonella and Yersinia, as well as for disinfectant efficacy tests.

E. Bio Control System (SimPlate®-Bio-Control, Bellevue, WA) [111]

The method is simple to use with minimum amount of preparation. Using different media, the unit can also make counts of total coliforms and Escherichia coli counts, yeast and mold counts, and even Campylobacter. The detection and enumeration of microorganisms by the SimPlate® methods rely on a binary detection technology. It uses Multiple Enzyme TechnologyTM (IDEXX's METTM) to detect bacteria in food and in water. Visible colour changes occur as a result of bacterial enzyme reactions with substance present in the liquid culture medium.

Procedure

It involves the following steps:

(i) The SimPlate system has 84 wells (A 198-well unit is also available for samples with high counts) imprinted in a round plastic plate.

(ii) After the lid is removed, a diluted food sample (1ml) is dispensed onto the centre landing pad, and 10 ml of rehydrated nutrient liquid provided by the manufacturer is poured onto the landing pad.

(iii) The mixture (food sample and nutrient liquid) is distributed evenly into the wells by swirling the SimPlate in a gentle, circular motion. Excessive liquid is absorbed by a pad housed in the unit.

(iv) After 24 h of incubation at 35°C, the plate is placed under UV light. Positive fluorescent wells are counted, and the number is converted in the most probable number (MPN) table to determine the number of bacteria present in the SimPlate.

Source

1. *Bhunia, A.K., 2007. Biosensors and bio-based methods for the separation and detection of foodborne pathogens. Adv. Food Nutr. Res., 54: 1-44.*
2. *Altekruse, S.F., N.Bauer, A.Chanlongbutra, R. DeSagun and A.Naugle et al.,2006. Salmonella Enteritidis in broiler chickens, United States, 2000-2005. Emerg. Infect. Dis., 12: 1848-1852.*
3. *CDC.,2006.Update on Multi-State Outbreak of E. coli O157: H7 Infections from Fresh Spinach E. coli O157: H7 Outbreak in Spinach. CDC., New York, New Jersey.*
4. *Doyle, M.P. and M.C. Erickson, 2006. Reducing the carriage of foodborne pathogens in livestock and poultry. Poult. Sci., 85: 960-973.*
5. *Lynch, M., J.Painter, R.Woodruff, C.Braden and Centers for Disease Control and Prevention 2006. Surveillance for foodborne disease outbreaks: United States, 1998-2002. MMWR. Surveill. Summ. Rep., 55: 1-42.*
6. *Brandl, M. T., 2006. Fitness of human enteric pathogens on plants and implications for food safety. Annu. Rev. Phytopathol., 44: 367-392.*
7. *Solomon, E.B., S. Yaron and K.R. Mathews, 2002. Transmission of Escherichia coli 0157: H7 from contaminated manure and irrigation water to lettuce plant tissue and its subsequent internalization. Applied Environ. Microbiol.,68: 397-400.*
8. *Carter, M.J., 2005. Enterically infecting viruses: Pathogenicity, transmission and significance for food and water borne infection. J. Applied Microbiol., 98: 1354-1380.*
9. *Feldhusen, F., 2000. The role of seafood in bacterial foodborne diseases. Microbes Infect., 2: 1651-1660.*
10. *Doyle, M.P.,2001. Food Microbiology: Fundamentals and Frontiers. 2nd Edn., ASM Press, Washillgton.*

10a. *Chen, W. T., R.L. Hendrickson, C.P. Huang, D. Sherman, T. Geng, A.K. Bhunia and M.R. Ladisch, 2005. Mechanistic study of membrane concentration and recovery of Listeria monocytogenes. Biotechnol. Bioeng., 89: 263-273.*

10b. *Stevens, K. A. and L. A. Jaykus, 2004. Bacterial separation and concentration from complex sample matrices: A review. Crit. Rev. Microbiol., 30: 7-24.*

10c. *Vasavada PC, Rapid methods and automation in dairy microbiology, Journal of dairy science- Oct 1993.*

11. *Mandal, P.K., Biswas, A.K., Choi, K and Pal, U.K., Methods for Rapid Detection of Foodborne Pathogens: An Overview, American Journal of Food Technology, 6: 87-102.*

11a. *S. Sutton, "Accuracy of Plate Counts,"Journal of Validation Technology17 (3),42–46, 2011.*

11b. *DynabeadsÒ, Dynal, Inc., Oslo, Norway.*

11c. *Skjerve, E., L.M. Rorvik and O. Olsvik, 1990. Detection of Listeria monocytogenes in foods by immunomagnetic separation. Applied Environ. Microbiol., 56: 3478-3481.*

11d. *Chapman, P. A. and R. Ashton, 2003. An evaluation of rapid methods for detecting Escherichia coli O157 on beef carcasses. Int. J. Food Microbiol., 87: 279-285.*

11e. Jordan, D., T. Vancov, A. Chowdhury, L.M. Andersen, K. Jury, A.E. Stevenson and S.G. Morris, 2004. The relationship between concentration of a dual marker strain of Salmonella typhimurium in bovine faeces and its probability of detection by immunomagnetic separation and culture. J. Applied Microbiol., 97: 1054-1062.

11f. Kaclikova, E., T. Kuchta, H. Kay and D. Gray, 2001. Separation of Listeria from cheese and enrichment media using antibody-coated microbeads and centrifugation. J. Microbiol. Methods, 46: 63-67.

11g. Fu, Z., S. Rogelj and T.L. Kieft, 2005. Rapid detection of Escherichia coli O157: H7 by immune-magnetic separation and real-time PCR. Int. J. Food Microbiol., 99: 47-57.

11h. Mercanoglu, B. and M.W. Griffiths, 2005. Combination of immunomagnetic separation with real-time pcr for rapid detection of Salmonella in milk, ground beef and alfalfa sprouts. J. Food Prot., 68: 557-561.

11i. Amag liani, G., E. Omiccioli, A. Campo, I.J.Bruce, G.Brandi and M. Magnani, 2006. Development of a magnetic capture hybridization-PCR assay for Listeria monocytogenes direct detection in milk samples. J. Applied Microbiol., 100: 375-383.

11j. Ueda, S., T.Maruyama and Y.Kuwabara, 2006. Detection of Listeria monocytogenes from food samples by PCR after IMS-plating. Biocontrol Sci., 11: 129-134.

11k. Hudson, J. A., R.J. Lake, P. Savill, P. Scholes and R.E. Mc-Connick, 2001. Rapid detection of L. monceytogenes in harn sample using irnmunornagnetic seperation followed by PCR. J. Applied Microbiol., 90: 614-621.

11l. Technews, NDDB, Technological Advancement in Methods of Microbiology in Milk Analysis- Part 1 (Direct methods) Issue no.91, October-December 2016.

11m. A. N. Sharpe and G. L. Michaud. Hydrophobic Grid-Membrane Filters: New Approach to Microbiological Enumeration, Applied Microbiology, Vol. 28, No. 2, 1974, p 223-225.

12. Vunrcrzant, C. and D.F.pllustoesser,1987.Compendium of Methods for Microbiological Examination of Food. 3rd Edn., American Public Health Association, New York.

12a. Tim Sandle, Bio Products Laboratory Limited, Approaching the Selection of Rapid Microbiological Methods.

12b. J.Moldenhauer,"Overview of Rapid Microbiological Methods," in M. Zourob, S. Elwary, A. and Turner, eds. Principles of Bacterial Detection: Biosensors, Recognition Receptors and Microsystems, pp 49-79,2008.

12c. M.J.Miller, Encyclopedia of Rapid Microbiological Methods. Parenteral Drug Association and Davis, Healthcare International Publishing, LLC, p.103–135,2005.

12d. R. T. Noble, S.B. and Weisberg, "A Review of Technologies for Rapid Detection of Bacteria," Journal of Water and Health 3, 381–391, 2005.

12e. FSS (Products specification and Food Additives)Regulation-2011 and FSSAI Testing manual for Fish and Meat products.

12f. M.K.Srivastava, Handbook on analysis of milk, CBS Publisher, ISBN-978812392668.

12g. M.K.Srivastava, Handbook on analysis of milk products, Write and print publishers, ISBN-9789384649258.

13. *Invitski, D., L. A. Harnid, P. Atanasov and E. Wilkins, l999. Biosensors for detection of pathogenic bacteria. Biosensors Bioelectronics.*

14. *599-624,14-Biswas, A.K., 2005. Evaluation of export buffalo meat for some chemical residues and microbial contaminants. Ph.D. Thesis, Indian Veterinary Research Institute, Izatnagar, Bareilly.*

15. *Technical Literature of Wickham Laboratories.*

15a. *Jodi Woan-Fei Law, Nurul-Syakima Ab Mutalib, Kok-Gan Chan, and Learn-Han Lee, Rapid methods for the detection of foodborne bacterial pathogens: principles, applications, advantages and limitations.*

16. *PHSS. Bio-contamination Technical monograph No. 20 Bio-contamination characterisation, control, monitoring and deviation management in controlled/GMP classified areas. (2014).*

17. *Jeanne Moldenhauer. Rapid Sterility Testing. (2011).*

18. *http://rapid micro methods.com/files/matrix. php, 2015.*

19. *N.K. A. Lakmini and T. Madhujith, Comparison of Performance of Rapid Petrifilm Test Method and Standard Test Method for Enumeration of Aerobic Microorganisms, Coliforms and E.coli in Food, Tropical Agricultural Research Vol. 23 (4): 363 – 369 (2012).*

Part 2

Section A

Instrumentation in Analysis of Taste of Food/Milk Products

Chapter 1

Analysis of Taste of Food/Milk Products

Electronic Tongue

Electronic tongues (also called as artificial tongue taste sensor) are an emerging and promising field in modern chemical sensor science. The electronic tongue proved to be a valuable tool for assessment, comparison and prediction of the taste of Food, Dairy Products pharmaceuticals and related products. The **electronic tongue** is an instrument that measures and compares tastes. It was invented by **Professor Fredrik Winquist** of Linkoping University, Sweden [13]. The electronic tongue is based on electrochemical sensors combined with multivariate data analysis. The Chemical compounds that responsible for taste, are perceived by human taste receptors, and the seven sensors of electronic instruments (Electronic Tongue).Like human receptors, each sensor has a spectrum of reactions different from the other. The information given by each sensor is complementary and the combination of all sensors results generates a unique fingerprint. Most of the detection thresholds of sensors are similar or better than those of human receptors [12]. The electronic tongue is capable of discriminating between substances with different taste modalities and can also distinguish different substances eliciting same taste. Electronic taste instrument have been developed and optimized to reduces human sensory test panels, to precise measurement of taste for formulation development time and cost [1,2,3]. In the biological mechanism, taste signals are transducted by nerves in the brain into electric signals. E-tongue sensors process is similar: they generate electric signals as potentiometric variations. Taste quality perception and recognition is based on building or recognition of activated sensory nerve patterns by the brain and on the taste fingerprint of the product. This step is achieved by the **e-tongue's statistical software** which interprets the sensor data into taste patterns [12].

A system (The first multi-sensor system termed as **taste sensor system** [4] for liquid analysis), mimicking the sense of taste, was presented by Toko and co-workers [5]. For this concept of transformation of taste quality information (sweetness, bitterness, sourness, saltiness or umami) into an electric signal, the terms **electronic tongue (e-tongue)** or **taste sensor** have been coined [5,5a,5b]. Later the instrument was named as **Electronic tongue**. The multi-sensor array system [5,6] or electronic tongue shows the clear correlation of the instrument output with human perception for various substances. It can be stated that Electronic tongue as "the system for automatic analysis of liquid including an array of non-specific chemical sensors with partial specificity for different component in liquid samples and an appropriate pattern recognition capable of recognizing the qualitative and quantitative composition of sample and complex solutions". Generally speaking, an electronic tongue or a taste sensor functions in a manner similar to that of an electronic nose, that is to say, the sensor array produces signals which are not necessarily specific to a particular species, but do form a pattern that can be correlated to certain features or qualities of the sample [6a,6b].

An electronic tongue is generally tested for its capability of recognising the following standard tastes: sweet, acid, bitter, salty, and umami. An electronic tongue uses a range of sensors that respond to salts, acids, sugars, bitter compounds, *etc.* Interpretation of complex data sets generated by electronic tongue signals is accomplished by use of multivariate statistics including PCA, LDA, discriminant function analysis (DFA), hierarchical cluster analysis (HCA), soft independent modelling of class analogy (SIMCA) and partial least squares (PLS). For non-linear responses, ANN can be used for data modelling (Baldwin *et al.*, 2011). The main purpose of electronic tongues is qualitative analysis, such as recognition, classification or identification of samples, which depends on the composition of the sensor array and the mathematical procedure chosen for data processing [7].

Mechanism/Procedure for Recognition of Taste [3]

Mechanism for taste recognition in human and electronic tongue shows the following three same levels:

1. **The receptor level (taste buds in humans, lipid membrane of sensors in the electronic tongues):** At this level, to detect dissolved organic and inorganic compounds the sensor-probe assembly is used by the electronic tongue. Each probe shows cross-sensitivity and selectivity so that each sensor could concurrently contribute to the detection of most substances found in the liquid matrix. These sensors are composed of an organic coating sensitive to the species to analyze in the samples and a transducer that allows converting the response of the membrane into signals that will be analyzed.
2. **The circuit level (neural transmission in humans, transducers in the electronic tongues):** At this level, the sample is quantified, digitized and results are recorded potentiometrically.
3. **The perceptual level (cognition in the thalamus in humans, statistical analysis by software in the electronic tongues):** At this level, process

of taste perception or sensation that occurs in the computer where the electronic tongue systems statistical software interprets the sensor data into taste patterns

Principle [3]

Taste is a survival mechanism, alerting us to potential harmful or potential nutritious substances. Approximately 10, 000 chemoreceptors or taste buds reside on the tongue. In the biological mechanism, taste signals are transducted by nerves in the brain into electric signals. E-tongue sensors process is similar and they generate electric signals as potentiometric variations. Taste quality perception and recognition is based on building or recognition of activated sensory nerve patterns by the brain and on the taste fingerprint of the product. This step is achieved by the e-tongue's **statistical software** which interprets the sensor data into taste patterns. The intensity of human response to taste stimuli depends logarithmically on the concentration of the stimuli [7a].

Response of potentiometric chemical sensor is also logarithmic in relation to the analyte concentration as follows the **Nernst or Nikolsky-Eisenmann equation** [8,9]. Thus the principle of the electronic tongue [10,11] is based on utilization of non-specific or low-selective potentiometric chemical sensors with enhanced cross-sensitivity to as many different components in solution as possible.Cross-sensitivity means that the sensor responds not to a single analyte but to several substances simultaneously present in the analyzed media. Different sensors should exhibit different sensitivity and selectivity pattern that might be partly overlapping or even completely distinct. The responses of the sensors are defined by the interaction with ionic and redox species, both inorganic and organic at the membrane-solution interface.

The electronic tongue uses taste sensors to receive information from chemicals on the tongue and send it to a pattern recognition system. The result is the detection of the tastes that compose the human palate. The types of taste that is generated is divided into five categories sourness, saltiness, bitterness, sweetness, and umami (deliciousness). Sourness, which includes HCl, acetic acid, and citric acid is created by hydrogen ions.Saltiness is registered as NaCl, sweetness by sugars, bitterness, which includes chemicals such as quinine and caffeine is detected through MgCl (2), and umami by monosodium glumate from seaweed, disodium in meat/fish/ mushrooms [12].

Source

1. *Legin, A., Rudnitskaya, A. and Vlasov, Y., In; Alegret, S., Eds.*
2. *Integrated analytical systems, comprehensive anal. Chem., Elsevier, 2003, 39, 437.*
3. *P. D. Chaudhari, S.P.Chaudhari, A.P.Chaudhari, N.S.Barhate and C. J mistry, Department of Pharmaceutics, Padm. Dr.D.Y.Patil Institute of Pharmaceutical Sciences and Research, Pimpri, Pune-411018, (MS), India.*
4. *Hung-Weichen., Ren-Jang Wu., Horn-How Chen. and Yih-Lu Sun., ISNCGS., 2004, 31.*

5. *Hayashi, K., Yamanaka, M., Toko, K., Yamafuji, K. (1990): Multichannel taste sensor using lipid membranes. Sensors and Actuators B: Chemical 2 (3), 205-213.*

5a. *Toko, K., Meas. Sci. Tech., 1998, 9, 1919.*

5b. *Tian, S.Y., Deng, S.P., Chen, Z.-X. (2007): Multifrequency large amplitude pulse voltammetry: A novel electrochemical method for electronic tongue. Sensors and ActuatorsB 123, 1049-1056.*

6. *Legin, A., Rudnitskaya, A. and Vlasov, Y., In; Analytical and Bioanalytical chemistry, Springerlink-Verlag GmbH., 2002, 373, 136.*

6a. *Winquist, F., Wide, P., Lundstrom, I. (1997): An electronic tongue based on voltammetry. Analytica Chimica Acta 357, 21-31.*

6b. *Winquist, F., Krantz-Rülcker, C., Lundström, I. (2002): Electronic tongues and combinations of artificial senses. Sensors Update 11 (1), 279-306*

7. *Dias, L. A., Peres, A.M., Veloso, A.C. A., Reis, F.S., Vilas- Boas, M., Machado, A. A.S.C. (2009): An electronic tongue taste evaluation: Identification of goat milk adulteration with bovine milk. Sensors and ActuatorsB 136, 209-217.*

7a. *Indow, T., Percept. Psychophysio., 1969, 5, 347.*

8. *Nikolsky, B., Zurn. Anal. Chem., 1937, 10, 495.*

9. *Eisenmann, G., Rudin, G. and Casb, J., Science., 1957, 126, 831.*

10. *Legin, A., Rudnitskaya, A. and Vlasov, Y., In; Alegret, S., Eds.*

11. *Integrated analytical systems, comprehensive anal. Chem., Elsevier, 2003, 39, 437.*

12. *Wikipedia.*

13. *"Fredrik Winquist". www.ifm.liu.se. Retrieved2015-09-11.*

Apparatus/Instrumentation

Depending upon the application of various recognition techniques (potentiometry, voltammetry, conductometry *etc.*) in electronic tongues, these are classified as follows:

A. Potentiometric Electronic Tongues [10]

Potentiometry represents a direct application of the Nernst equation through the measurement of potentials of non-polarized electrodes lacking current flow. These potentials depend on the nature and concentration of ions present in the solution, as well as on the medium and the type of electrodes employed [1]. It is using lipid/polymer membranes have a concept of global selectivity which implies the ability to classify vast kinds of chemical substances into several groups, as really found in the taste reception in biological systems [1a]. The measuring principle is based on the potential of electrodes across an ion-sensitive membrane with zero current flow [2]. A potentiometry-based electronic tongue was first presented by Toko and co-workers [3,10]. It was composed of several kinds of lipid/polymer membranes used for the transformation of taste quality information into an electric signal. The output of the system was not the amount of specific taste substances but rather the taste quality and intensity, because different output electric patterns are obtained for chemical substances producing different taste qualities [1a,3,10]. Another type of presented potentiometric electronic tongue configured by several non-specific sensors based on chalcogenide glasses as transducers [3a].

The equipment necessary for potentiometric studies includes following [10]:

(i) An ion-selective electrode

(ii) A reference electrode

(iii) A voltage-measuring device.

Commonly used reference electrode is a silver/silver chloride (Ag/AgCl) electrode. The electrode consists of a silver wire coated with silver chloride placed into the chloride ions solution. A porous plug serves as a voltage bridge to the outer solution. An ion-selective electrode has a similar configuration, but makes use of an ion-selective membrane rather than a voltage bridge. This membrane should be non-porous, water insoluble and mechanically stable. It should have an affinity for the selected ion that is high and selective. Due to the binding of ions, a membrane potential will develop [5b].

B. Voltammetric Electronic Tongues [10]

In voltammetric techniques, a potential is applied on the working electrode, followed by the measurement of the resulting current between the working electrode and the reference electrode. As a result, an electrochemical redox reaction occurs at the electrodes' surface and gives rise to the measured current. All particles in the measured liquid, electrochemically active below the applied voltage, will contribute to the redox response [3b]. The use of voltammetry as a sensing principle, under which an electronic tongue operates, appears to have several advantages; the technique is commonly used in analytical chemistry due to its very high sensitivity,

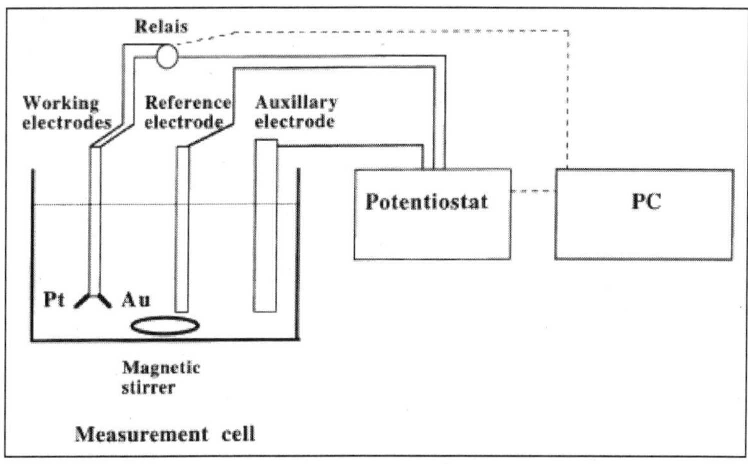

Schematic Diagram of the Experimental Setup of the First Voltammetry-Based Electronic Tongue [5].

versatility, simplicity and robustness. Besides, the technique offers the possibility of using and combining different analytical principles such as cyclic, stripping or pulsed voltammetry.

Voltammetric electronic tongues are based on metal electrodes with different selectivity and sensitivity patterns [4]. One sensor system using this technique is the voltametric electronic tongue which was developed at Linkoping University [5]. This sensor system can be used to measure foodstuffs in order to try to mimic the human tongue. The first voltammetry-based electronic tongue was presented in 1997 [5], the technique was chosen due to its high sensitivity, simplicity and robustness. The first sensor system had a three electrode-configuration with an Ag/AgCl reference electrode, a stainless steel counter. The experimental set up of this system is given in figure on previous page. Pulse voltammetry was used in order to obtain information on the diffusion coefficient of charged species [3b,5].

The electronic tongue consists typically of four working electrodes made of the metals gold, iridium, platinum and rhodium, an Ag/AgCl reference electrode and a stainless steel counter electrode. A relay box is used, enabling the working electrodes to be connected consecutively to form four standard three-electrode configurations. The potential pulses/steps are applied by a potentiostat which is controlled by a PC. The PC is used to set and control the pulses, measure and store current responses and to operate the relay box. The set-up is given in following Figure.

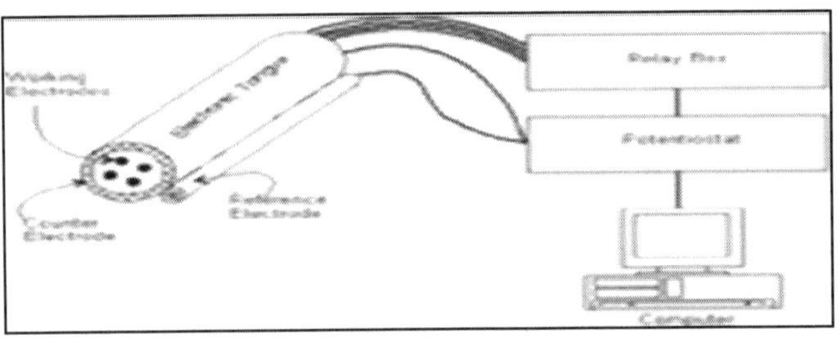

Schematic of a Voltammetric Electronic Tongue.

Among various techniques, **pulse voltammetry** is of special interest due to its advantages in terms of greater sensitivity and resolution. Two types of pulse voltammetry are commonly used:

(i) Large amplitude pulse voltammetry (LAPV)

(ii) Small amplitude pulse voltammetry (SAPV) [5a].

Depending on the technique, various kinds or aspects of information can be obtained from the measured solution. A number of possibilities are available for obtaining information using voltammetry: choice of the potential applied (size, sweep rates, pulses *etc.*), type of working electrode or stripping techniques. When using voltammetry in complex media consisting of several redox active compounds and various ions, the selectivity of the system is often insufficient for species analysis of single component, because the steps in the voltammogram are too close to be individually discriminated. Thus, rather complicated "spectra" are obtained and the interpretation of data may be difficult due to their complexity. Still, the voltammograms contain a large amount of information and to extract this, multivariate analysis methods have found increasing interest and use in the field [5].

Voltammetric electronic tongue has been further developed; the concept of voltammetric electronic tongue has been advanced throughout the recent years and proven valuable in many applications. Studies have shown that voltammetric electronic tongue could be used to monitor quality changes in milk [5,5c].

In measurements with the voltametric electronic tongue, data are collected over the whole pulse and not only at the end of the pulse, as in traditional electrochemistry. This is done since it has been found that extra information is also found at the beginning of the pulse (mainly conductivity and mobility).

Voltametric methods provide high sensitivity, a wide linear range and simple, robust instrumentation. In voltametric measurements a current is measured between the metal working electrode and the counter-electrode when a voltage pulse is applied over the working electrode and the reference electrode. A set of pulses can be put together to form a pulse train in order to extract as much information as possible from the solution. When the potential is applied, electroactive compounds that react to that potential will be reduced or oxidized and a current, that can be measured, will arise.

C. Conductometric Electronic Tongues

It is based on Conductometry (direct measurement of the conductance between two inert identical electrodes), so that most specific effects due to electrodes are eliminated [6].

D. Impedance Spectroscopy-Based Electronic Tongue [10]

Recently, electrochemical impedance spectroscopy (EIS) has started to be employed into qualitative applications of electronic tongues [7]. Impedance spectroscopy-based electronic tongue was first described [8]. The sensors were constructed from pure and composite nano-structured films of conducting polymers. Impedance measurement is often used for preliminary characterization of complex circuits [8,1,9a,5c]. In comparison with potentiometry and especially with voltammetry, impedance measurements are advantageous because of their simplicity and shorter response times [7].

Source

1. *Pioggia, G., Di Francesco, F., Marchetti, A., Ferro, M., Ahluwalia, A. (2007a): A composite sensor array impedentiometric electronic tongue Part I. Characterization. Biosensors and Bioelectronics 22, 2618-2623.*

1a. *Toko, K. (1998): Electronic tongue.Biosensensors and Bioelectronics 13, 701-709.*

2. *Winquist, F., Krantz-Rülcker, C., Lündstrom, I. (2004): Electronic tongues. MRS Bulletin 29, 1-6.*

3. *Hayashi, K., Yamanaka, M., Toko, K., Yamafuji, K. (1990): Multichannel taste sensor using lipid membranes. Sensors and Actuators B: Chemical 2 (3), 205-213.*

3a. *Legin, A.V., Vlasov, Y.G., Rudnitskaya, A.M., Bychkov, E. A. (1996): Cross-sensitivity of chalcogenide glass sensors in solutions of heavy metal ions. Sensors and ActuatorsB 34, 456-461.*

3b. *Iliev, B., Lindquist, M., Robertsson, L., Wide, P. (2006): A fuzzy technique for food and water quality assessment with an electronic tongue. Fuzzy Sets and Systems 157, 1155-1168.*

4. *Söderström, C., Winquist, F., Krantz-Rülcker, C. (2003): Recognition of six microbial species with an electronic tongue.Sensors and Actuators B 89, 248-255.*

5. *Winquist, F., Wide, P., Lundstrom, I. (1997): An electronic tongue based on voltammetry. Analytica Chimica Acta 357, 21-31.*

5b. *Winquist, F., Krantz-Rülcker, C., Lundström, I. (2002): Electronic tongues and combinations of artificial senses. Sensors Update 11 (1), 279-306.*

5c. *Wei, Z., Wang, J., Zhang, X. (2013): Monitoring of quality and storage time of unsealed pasteurized milk by voltammetric electronic tongue. Electrochimica Acta 88, 231-239.*

6. *Pioggia, G., Di Francesco, F., Marchetti, A., Ferro M., Ahluwalia, A. (2007): A composite sensor array impedentiometric electronic tongue Part I. Characterization. Biosensors and Bioelectronics 22, 2618-2623.*

7. *Cortina-Puig, M., Muñoz-Berbel, X., Alonso-Lo-millo, M. A., Muñoz-Pascual, F.J., del Valle, M. (2007): EIS multianalyte sensing with an automated SIA system-An electronic tongue employing the impedimetric signal. Talanta 72, 774-779.*

8. *Riul, A., Malmegrim, R.R., Fonseca, F.J., Mattoso, L.H.C. (2003): An artificial taste sensor based on conducting polymers. Biosensors and Bioelectronics, 18, 1365-1369.*

9a. *Pioggia, G., Di Francesco, F., Marchetti, A., Ferro, M., Leardi, R., Ahluwalia, A. (2007b): A composite sensor array impedentiometric electronic tongue Part II. Discrimination of basic tastes. Biosensors and Bioelectronics 22, 2624-2628.*

10. *M. Tudor Kalit et al.: Electronic nose and electronic tongue in the dairy industry, Mljekarstvo 64 (4), 228-244 (2014).*

Different Units of Electronic Tongue [5]

It comprises of the following units:

1. Liquid Autosampler

The autosampler allows running automatically upto 15 different liquid samples plus one extra for sensor cleaning purpose. A programmed sample agitator, a robotic sensor head and individual connections for each sensor are present [1]. Recently, liquid autosampler by ALPHA M.O.S. consist of 16 beakers (80-ml capacity) and 48 beakers (20 ml capacity) [2].

2. Chemical Sensor or Sensor Array

Chemical sensors have been widely used in critical care, safety, industrial hygiene, process controls, product quality controls, human comfort controls, emission monitoring, automotive, clinical diagnostics and home safety alarms.

Chemical sensors can be defined as small devices that result in transformation of chemical or biochemical information of a qualitative or quantitative type into an analytical useful signal by the process of chemical interaction between sample and sensors devices [2,3]. Same as with electronic noses, different sensing principles can be used with electronic tongues or taste sensors. A wide variety of chemical sensors can be employed into the design of electronic tongue. Some examples are as follows:

i. Electrochemical Sensors

The use of electrochemical measurements for analytical purposes has found a vast range of applications. All commercially available systems are based on potentiometric measurements, especially Ion-Selective Electrodes (ISEs). Besides ISEs, Ion-Sensitive Field-Effect Transistors (ISFETs) are now-a-days also gaining greater interest as sensors forming electronic tongue sensor arrays [6,7]. The features of electronic tongue sensors are different from those of traditional chemical sensors: instead of high selectivity in substance detection, electronic tongue sensors have an overall selectivity that provides for global information on the analysed solution. That global information, later applied as a digital fingerprint of gustatory compounds, is analyzed using the appropriate pattern recognition tools [8].

Currently these technologies are based on electrochemical sensors. The oxidation and reduction of chemical species on a conducting electrode can be observed by measuring the movement of charge. There are following primary methods of sensing electrochemical reactions.

a. Potentiometric Sensors

These sensors [2] can be used to measure the equilibrium potential established between the electrode material and the solution, a potential that is dependent on the chemistry involved. Potentiometry represents a direct application of the Nernst equation through the measurement of potentials of non-polarized electrodes lacking current flow. These potentials depend on the nature and concentration of ions present in the solution, as well as on the medium and the type of electrodes employed [9]. The equipment necessary for potentiometric studies includes an ion-selective electrode, a reference electrode and a voltage-measuring device. Commonly used reference electrode is a silver/silver chloride (Ag/AgCl) electrode. The electrode consists of a silver wire coated with silver chloride placed into the chloride ions solution. A porous plug serves as a voltage bridge to the outer solution. An ion-selective electrode has a similar configuration, but makes use of an ion-selective membrane rather than a voltage bridge. This membrane should be non-porous, water insoluble and mechanically stable. It should have an affinity for the selected ion that is high and selective. Due to the binding of ions, a membrane potential will develop [6].

Potentiometric electrodes are the most widely used sensors in electronic tongue systems. The sensing mechanism of the most of potentiometric sensors is based on a membrane made of inorganic (polycrystalline and chalcogenide glass) or organic (plasticized polymeric matrix doped with membrane active components, Langmuir-Blodgett or electropolymerised films, *etc.*) materials. The membrane is the main active component, but in the same time, also one of the sensor's weak points, because most drawbacks, such as reproducibility and long-term stability, depend on the membrane. Therefore, most of the research in this field is devoted to the development of suitable electrodes [9,10].

b. Amperometric Sensors

These measure the current generated by a reaction and thus give a measure of reaction rates.

ii. Voltammetric Sensors

By controlling the potential of the electrode relative to the solution and measuring the charge flow induced the presence of specific ions can be determined by observing the potential at which they undergo oxidation or reduction. This process is called a voltammetry.

iii. Optical Sensors-Optical or Enzymatic Sensors (Biosensors)

Almost all electronic tongues or taste sensors developed in so far are based on potentiometry or voltammetry. However, some other techniques such as optical techniques, enzymatic sensors (biosensors) or techniques based on mass-sensitive devices are also interesting to use; their special features make them useful for electronic tongues.

a. An Electronic Tongue Based on Optical Techniques [12]

Optical techniques are based on light absorption at specific wavelengths ranging from ultraviolet (UV) across the visible light to near infrared (NIR) and infrared (IR) spectrum. Many compounds have distinct absorption spectra, so that scanning at a certain wavelength may yield the spectrum specific to the tested sample. Optical methods offer advantages of high reproducibility and good long-term stability [6].

b. An Electronic Tongue Based on Biosensors [12]

Biosensors used with electronic tongues are systems equipped with a biochemical transducer, an enzyme and a solid electrode positioned in an intimate proximity. Enzymes are oxidases that consume oxygen and produce hydrogen peroxide or the reduced form of β-nicotinamide adenine dinucleotide (phosphate) (NAD (P)H) as a dehydrogenase. Improvement in performance of metal sensors, conductive polymers or biosensors, is linked to their downsizing to nano-dimensions, which increases the surface-to-volume ratio of the sensors, lowering thereby the detection limits [11].

c. An Electronic Tongue Based on Mass-Sensitive Devices [12]

Mass sensitive devices based on piezoelectric crystals can also be useful. A quartz crystal resonator is operated at given frequency influenced by the absorption of certain compounds onto the crystal surface. As for a surface acoustic wave-based device, a surface wave propagates across the surface of the device. Due to the adsorption of compounds taking place underway, the properties of the surface wave can be changed. These types of devices give overall rather than specific information and can therefore be used to detect a number of different compounds [6].

Source

1. *Legin, A., In lecture; Artificial Sensing Techniques for Evaluating Quality., 1999.*
2. *ALPHA M. O. S. Newsletter, 2002.*
3. *Technical Note T- SAS- 01. ALPHA M. O. S., 2002.*
4. *Ciosek, P., Electronic Tongue/Electronic Nose, Acquired from www.electronictongue.com.*
5. *P. D. Chaudhari, S.P. Chaudhari, A.P. Chaudhari, N. S. Barhate AND C. J mistry, Department of Pharmaceutics, Padm. Dr. D. Y. Patil Institute of Pharmaceutical Sciences and Research, Pimpri, Pune-411018, (MS), India.*
6. *Winquist, F., Krantz-Rülcker, C., Lundström, I. (2002): Electronic tongues and combinations of artificial senses. Sensors Update11 (1), 279-306.*
7. *Ciosek, P., Wróblewski, W. (2011): Potentiometric electronic tongues for foodstuff and biosample recognition-an overview. Sensors 11, 4688-4701.*
8. *Wei, Z., Wang, J., Zhang, X. (2013): Monitoring of quality and storage time of unsealed pasteurized milk by voltammetric electronic tongue. Electrochimica Acta 88, 231-239.*
9. *Pioggia, G., Di Francesco, F., Marchetti, A., Ferro, M., Ahluwalia, A. (2007a): A composite sensor array im-pedentiometric electronic tongue Part I. Characterization. Biosensors and Bioelectronics 22, 2618-2623.*
10. *Lvova, L., Martinelli, E., Mazzone, E., Pede, A., Paolesse, R., Di Natale, C., D' Amico, A. (2006): Electronic tongue based on an array of metallic potentiometric sensors. Talanta 70, 833-839.*
11. *Baldwin, E. A., Bai, J., Plotto, A., Dea, S. (2011): Electronic noses and tongues: Applications for the food and pharmaceutical industries, Sensors 11, 4744-4766.*
12. *M. Tudor Kalit et al.: Electronic nose and electronic tongue in the dairy industry, Mljekarstvo 64 (4), 228-244 (2014).*

iv. Biosensors

Biosensors [1] use biomolecules and structures to measures something with biological importance or bioactivity. They are also called the subsets of chemical sensors, as the transduction mechanism is similar to those of chemical sensors.

The sensors are made up of silicon transistors with organic coating that are necessary for the sensitivity and selectivity of each individual sensor to ensure good repeatability, sensitivity and selectivity, The detecting sensor part consist of lipid/polymer membranes. These lipids membrane are made up of different types of lipids, which are pasted on the opening of the sensors [3]. Each membrane possessed an electric charge due to each lipid. These are the working electrodes. The reference electrodes are made of an Ag wire whose surface was plated with Ag/AgCl, with an internal cavity filled with 3 M KCl solution and opening tube filled with 100 mM KCl and Agar.

3. Acquisition System [2]

This system establishes the communication between the sensor array and data processing system. The electrical potential responses produced at the sensor array are received by this system and it converts these electrical signals to digital signals with less distortion. An analog signal input devices, a filter mechanism and an analog to digital converter (ADC) should be present in the system as the speed of the sensor system is very slow, a low speed and high resolutions delta sigma ADC is enough to this electronic tongue system [4]. The filter mechanism is necessary to provide a clean and no distortion transmission channel between the sensors and ADC.

Source

1. *Ciosek, P., Electronic Tongue/Electronic Nose, Acquired from www. electronictongue. com*
2. *P.D.Chaudhari, S.P.Chaudhari, A.P.Chaudhari, N. S. Barhate and C. J mistry, Department of Pharmaceutics, Padm. Dr. D. Y. Patil Institute of Pharmaceutical Sciences and Research, Pimpri, Pune-411018, (MS), India.*
3. *Technical Note T- SAS- 01. ALPHA M. O. S., 2002.*
4. *Hung-Weichen., Ren-Jang Wu., Horn-How Chen. and Yih-Lu Sun., ISNCGS.2004, 31.*

4. Data Processing System [3]

Interpretation of data with multivariable required the use of statistical interpretation methods. Chemometric techniques a type of multivariate statistic used in the analytical field provide data processing, which consist of recognition, classification and identification and multivariate calibration [1]. Various recognition techniques are applied in electronic tongues, potentiometry, voltammetry and conductometry being most commonly used [2]. The data obtained from the electronic tongue is processed by multivariate data analysis (MVDA). Recognition, identification and classification is mainly done by Principle Component Analysis (PCA), Discrimination factorial analysis (DFA) or by Partial least square (PLS) while multivariable calibration is undertaken with the help of PLS [2,3].

Source

1. *Lenin, A., Rudnitskaya, A., Clapham, D., Selezney, B., Lord, K. and Vlasov, Y., In; Analytical and Bioanalytical chemistry., Springerlink-Verlag GmbH., 2004, 380 (1), 136.*
2. *Winquist, F., Krantz-Rülcker, C., Lündstrom, I. (2004): Electronic tongues. MRS Bulletin 29, 1-6.*
3. *P. D. Chaudhari, S.P. Chaudhari, A.P. Chaudhari, N. S. Barhate and C. J mistry, Department of Pharmaceutics, Padm. Dr. D. Y. Patil Institute of Pharmaceutical Sciences and Research, Pimpri, Pune-411018, (MS), India.*

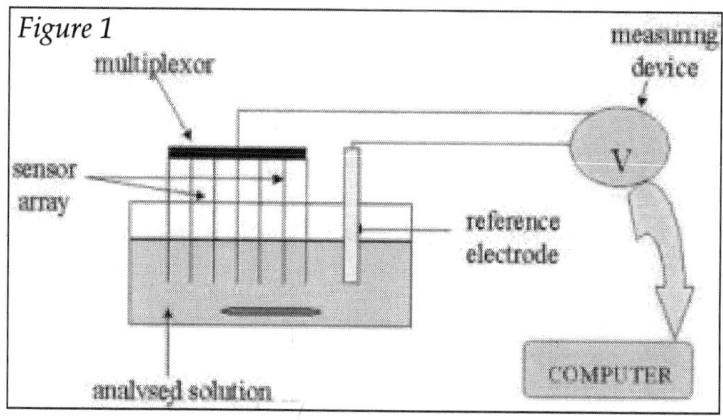

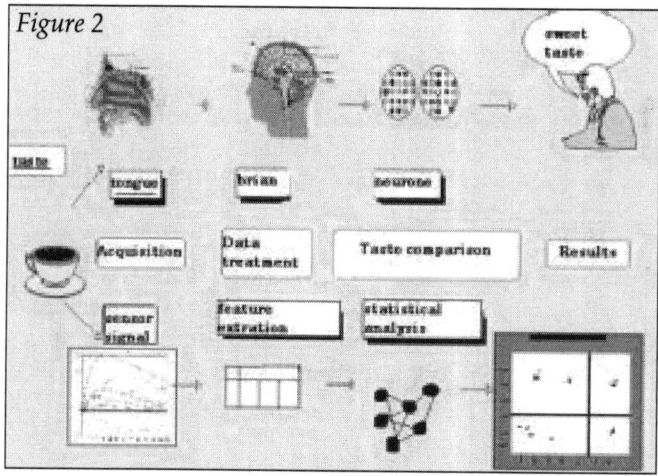

Schematic Diagram of Electronic Tongue.

Source

Figure 1: Winquist, F., Krantz-Rulckerc, andLundstrom, I., In: Pearce, T., Schiffman, S., Nagle., Gardner., Eds., Handbook of Machine Olfaction., Wiley, Winheim, 2003, 267.

Figure 2: Technical Note T- P- 01. ALPHA M. O. S., 2002.

Operation of Electronic Tongue [1]

Liquid samples are directly analyzed without any preparation, whereas solids require a preliminary dissolution before measurement. Reference electrode and sensors are dipped in a beaker containing a test solution for 120 seconds. A potentiometric difference between each sensor and a reference electrode is measured and recorded by the E-Tongue software. These data represent the input for mathematical treatment that will deliver results.

Source

1. *Wikipedia.*

Applications

Electronic Tongues have several applications in various industrial areas (*e.g.* Dairy, food and beverage, pharmaceutical industry), sector, *etc.* It can be used to [6]:

(a) analyze flavor ageing in beverages (for instance fruit juice, alcoholic or non alcoholic drinks, flavored milks).

(b) Quantify bitterness or "spicy level" of drinks or dissolved compounds (*e.g.* bitterness measurement and prediction of teas).

(c) Quantify taste masking efficiency of formulations (tablets, syrups, powders, capsules, lozenges.),

(d) Analyze medicines stability in terms of taste and e-benchmark target products. Electronic tongue systems seem to be very useful for process monitoring and as a quality-control tool in the food industry, in clinical analysis, and in research laboratories [6].

Taste quantification and foodstuff recognition are the main area of application of the taste sensor. The details related to Industry wise application of Electronic Tongue [1] are given below:

A. Dairy Industry

The main purpose of electronic tongues is qualitative analysis, and thus several works on the application of electronic tongue devices in recognition, classification or identification of milk and fermented milk samples have been reported. A hybrid electronic tongue based on voltammetric, potentiometric and conductometric measurements was successfully used to classify different fermented milk samples [7].

1. Potentiometric Electronic Tongue

Dias *et al.* [2] developed a potentiometric electronic tongue and used it in the detection of goat milk adulteration with bovine milk. In dairy, potentiometric electronic tongue (α-ASTREE, Alpha M.O.S. Co., Toulouse, France) for differentiation of different milk and yoghurt samples was used by Mirjana Hruskar [4]. It included the automatic sampling system, the sensor array with the reference electrode, the signal processing unit and a personal computer with the Software Astree 3.0.1. installed. Summary of the aforesaid studies is given below:

a. To Differentiate (Classification) Milk Samples

Taste sensors were found capable in discriminating between fresh and spoiled milk and to follow the deterioration of the milk quality when it was stored at room temperature. Two packaged commercial milk; the ultra high temperature (UHT) and the pasteurized milk were tested [3]. The electronic tongue α-ASTREE (Alpha M.O.S) has successfully distinguished five brands of milk and differentiated among various products from one dairy producer [4].

b. To Differentiate Yoghurt Samples

The electronic tongue α-ASTREE (Alpha M.O.S) has also successfully distinguished five brands of yoghurt and differentiated among various products from one dairy producer [4].

c. To Differentiate Different Dairy Products

Kefir samples, sour cream *etc.* were also differentiated [4].

The usefulness of potentiometric sensor array in the classification of different types of probiotic fermented milk during the entire shelf-life was emphasized [5]. The sensor array successfully reported changes in composition that occurred during the storage of probiotic fermented milk and was capable of tracking different rates of degradation of such milk stored at two different temperatures. Measurements done by potentiometric sensor array achieved a high correlation with the results of human sensory panel [5,7].

Source

1. *Ciosek, P., Electronic Tongue/Electronic Nose, Acquired from www.electronictongue.com.*
2. *Dias, L. A., Peres, A.M., Veloso, A.C. A., Reis, F.S., Vilas-Boasa, M., Machado, A. A.S.C. (2009): An electronic tongue taste evaluation: Identification of goat milk adulteration with bovine milk. Sensors and ActuatorsB 136, 209-217.*
3. *Kikkawa, Y., Toko, K. and Yamafuji, K., Sens Mater., 1993, 5, 83.*
4. *Mirjana Hruskar, Nikola Major, Marina Krpan, Ines Panjkota Krbavcic, Goran saric, Ksenija Markoviæ, Nada Vahcic, Department of Food Quality Control and Nutrition, Faculty of Food Technology and Biotechnology, University of Zagreb, Pierottijeva 6, Zagreb Received - Prispjelo: 01.04.2009. Accepted - Prihvaceno: 09.07.2009.*
5. *Hruskar, M., Major, N., Krpan, M. (2010a): Application of a potentiometric sensor array as a technique in sensory analysis. Talanta 81, 398-403.*
6. *Wikipedia.*
7. *M. Tudor Kalit et al.: Electronic nose and electronic tongue in the dairy industry, Mljekarstvo 64 (4), 228-244 (2014).*

d. Determination of Compounds Responsible for Aroma Development [1]

Complex data sets from electronic tongue signals and standard analytical measurements employing prediction models have been used for the prediction of different compounds' concentrations. The performance of potentiometry-based electronic tongue in rapid determination of compounds responsible for aroma development in probiotic fermented milk such are ethanol, acetaldehyde, diacetyl, lactic acid, acetic acid and citric acid was evaluated [2]. Different types of probiotic fermented milk samples were analysed both using standard analytical methods and an electronic tongue. ANN models were used for rapid estimation of aroma compounds' content in probiotic fermented milk samples. The best prediction capability with a low error of prediction was obtained with the estimation of ethanol content. Good prediction capability with slightly higher prediction errors was exhibited by ANN models suitable for acetic acid, citric acid, lactic acid and diacetyl concentration determination. Only the model intended for acetaldehyde determination exhibited low accuracy of prediction. The obtained results indicate that potentiometric electronic tongues coupled with ANN models can be applied as a rapid method of determination of aroma compounds in probiotic fermented milk [2].

Source

1. *M. Tudor Kalit et al.: Electronic nose and electronic tongue in the dairy industry, Mljekarstvo 64 (4), 228-244 (2014).*
2. *Hruškar, M., Major, N., Krpan, M., Vahèiæ, N. (2010b): Simultaneous determination of fermented milk aroma compounds by a potentiometric sensor array. Talanta 82, 1292-1297.*

e. Mastitis Detection [1]

An experiment to determine whether an electronic tongue multi-sensor system based on potentiometric chemical sensors and suitable data processing could be used to discriminate between milk secretions from infected and healthy glands, was conducted [2]. It was demonstrated that a multi-sensor system can distinguish between the control and clinically mastitic milk. The results showed that a multi-sensor system offers a novel method for mastitis detection and discrimination between milk samples coming from cows suffering from clinical mastitis and those coming from healthy animals. An array of fifteen chemical sensors (an electronic tongue) proved itself as the best detection method above an array of three ion-selective electrodes and conductivity [2].

Source

1. *M. Tudor Kalit et al.: Electronic nose and electronic tongue in the dairy industry, Mljekarstvo 64 (4), 228-244 (2014).*
2. *Mottram, T., Rudnitskaya, A., Legin, A., Fitzpatrick, J.L., Eckersall, P.D. (2007): Evaluation of a novel chemical sensor system to detect clinical mastitis in bovine milk. Biosensors and Bioelectronics 22, 2689-2693.*

f. Milk Adulteration [1]

There is an open field for the development of sensors to be used for the characterization of milk adulteration. An all-solid-state potentiometric electronic tongue constructed with two units of 20 polymeric membranes each (18 different and 2 selected membranes' replicates), working in parallel, in which the polymeric mixtures were applied on solid conducting silver-epoxy supports to use the taste recognition ability of the electronic tongue to detect raw goat milk adulterations with raw cow milk by means of LDA [2]. The results showed that a multi-sensor device applied to study caprine milk adulterations with bovine milk, which poses a problem in the dairy industry, was able to give different signal profiles associated with specific sensory characteristics of each skimmed milk sample, allowing the discrimination between goat, cow and goat/cow raw skimmed milks with satisfactory sensibilities and specificities. This methodology can be used as a fast and economic procedure to evaluate possible adulterations of goat raw milk with cow raw milk in a real time frame [1]. A disposable electronic tongue with gold CD-R and copper sheet substrates and golden sensing elements, the copper and gold surface thereby being modified with a Prussian Blue layer, was also constructed [2,3]. The performance, the electrode materials and the capability of the device to differentiate samples were evaluated using a taste substances model, that is to say, milk obtained using different pasteurization processes (homogenized/ pasteurized, ultra high temperature (UHT)- pasteurized and low fat analysed cases, a good separation between different samples was noticed in PCA score plots. The researchers highlighted the importance of fabrication of disposable sensors in the particular case of milk analysis, because high contents of organic materials which adsorb onto the electrode surface may change the voltammetric response [3].

Source

1. *M. Tudor Kalit et al.: Electronic nose and electronic tongue in the dairy industry, Mljekarstvo 64 (4), 228-244 (2014).*
2. *Dias, L. A., Peres, A.M., Veloso, A.C. A., Reis, F.S., Vilas-Boas, M., Machado, A. A.S.C. (2009): An electronic tongue taste evaluation: Identification of goat milk adulteration with bovine milk. Sensors and ActuatorsB 136, 209-217.*
3. *Paixao, T.R.L.C., Bertotti, M. (2009): Fabrication of disposable voltammetric electronic tongues by using Prussian Blue films electro deposited onto CD-R gold surfaces and recognition of milk adulteration. Sensors and ActuatorsB 137, 266-273.*

2. Voltametric Electronic Tongue

Extensive research has been conducted on the voltammetric electronic tongue. The following are some examples of application of Voltametric Electronic Tongue in dairy Industry:

a. Classification of Milk

A voltammetric electronic tongue is specially designed for use in the dairy industry which consisted of four working electrodes (gold, platinum, rhodium and stainless steel), embedded in PEEK (TM). It was mounted in a housing of stainless steel, which was inserted in the process line for direct online measurements. The voltammetric electronic tongue was used to follow different sources of milk coming into the process and to monitor the cleaning process.

It was started to use the voltammetric electronic tongue to monitor the deterioration of milk quality due to microbial growth [1]. A specially designed voltammetric electronic tongue inserted directly online in dairy industry process line was also used [2]. The results of the study showed that milk from different sources and thus also of different quality, can be mutually separated. Thus, the possibility of monitoring off-flavours in the incoming milk was opened, indicating that an electronic tongue could be a valuable security measure to prevent economic loss suffered due to the contamination of a large volume of raw milk with off-flavour milk entering a storage silo in the dairy plant [2]. A disposable voltammetric electronic tongue was developed and it was used to differentiate among milks with different pasteurization processes and adulterated milk with hydrogen peroxide [3]. A voltammetric electronic tongue to monitor yoghurt production in a pilot facility was used [4].

Source

1. *Winquist, F., Krantz-Rülcker, C., Wide, P., Lundstrom, I. (1998): Monitoring of freshness of milk by an electronic tongue on the basis of voltammetry. Measurement Science and Technology 9, 1937-1946.*
2. *Winquist, F., Bjorklund, R., Krantz-Rülcker, C., Lundstrom, I., Ostergren, K., Skoglund, T. (2005): An electronic tongue in the dairy industry. Sensors and Actuators B111-112, 299-304.*
3. *Paixao, T.R.L.C., Bertotti, M. (2009): Fabrication of disposable voltammetric electronic tongues by using Prussian Blue films electrodeposited onto CD-R gold surfaces and recognition of milk adulteration. Sensors and Actuators B137, 266-273.*
4. *Bjorklund, R.B., Magnusson, C., Martensson, P., Winquist, F., Krantz-Rülcker, C. (2009): Continuous monitoring of yoghurt fermentation using a noble metal electrode array. International Journal of Food Science and Technology 44 (3), 635-640.*

b. Microbial Quality [1]

The electronic tongue has been used in monitoring of microbial properties of milk that makes use of prediction models and pattern recognition techniques employed to the goal of bacterial count determination. An electronic tongue measurement principle based on pulsed voltammetry was described, in which successive voltage pulses of gradually changing amplitudes were applied to the working electrodes connected in a standard three-electrode configuration [2]. An electronic tongue was used to follow the deterioration of quality of the milk stored at the room temperature, arising on the grounds of microbial growth. The obtained data were subject to PCA, while the deterioration process could be clearly followed in the diagrams. To compile prediction models, projections to the latent structure and ANN were used, both of them being proven to satisfactorily predict the course of bacterial growth in milk samples [2]. A self-developed voltammetric electronic tongue based on multi-frequency rectangle pulse voltammetry (MRPV) and multi-frequency staircase pulse voltammetry (MSPV) was used to monitor the storage time of unsealed pasteurized milk and to explore the relation between physicochemical and microbiological properties of the pasteurized milk and voltammetric data gathered during storage [3]. The unsealed pasteurized milk samples were measured at seven different time points spread over 72 h, and classified using two pattern recognition techniques, *i.e.* PCA and cluster analysis (CA). Additionally, total bacterial count, acidity and viscosity were analyzed immediately after voltammetric electronic tongue experiments, employing two pattern recognition techniques (PLSR and LS-SVM) for total bacterial count and viscosity forecast. The milk samples stored for a different length of time were clearly classified using a voltammetric electronic tongue based on MRPV and MSPV. PLSR and LS-SVM showed high correlation between the measured and the predicted values both when it comes to viscosity and prediction of total bacterial count [3].

Source

1. *M. Tudor Kalit et al.: Electronic nose and electronic tongue in the dairy industry, Mljekarstvo 64 (4), 228-244 (2014).*

2. *Winquist, F., Krantz-Rülcker, C., Wide, P., Lundström, I. (1998): Monitoring of freshness of milk by an electronic tongue on the basis of voltammetry. Measurement Science and Technology 9 (12), 1937-1946.*
3. *Wei, Z., Wang, J., Zhang, X. (2013): Monitoring of quality and storage time of unsealed pasteurized milk by voltammetric electronic tongue. Electrochimica Acta 88, 231-239.*

c. Determination of Antibiotic Residues [1]

A voltammetric electronic tongue used for the detection of antibiotic residues in bovine milk was developed [2]. Six antibiotics (Chloramphenicol, Erythromycin, Kanamycin sulphate, Neomycin sulphate, Streptomycin sulphate and Tetracycline HCl), present at four concentration levels (0.5, 1, 1.5 and 2 maximum residue limits (MRLs)), were measured using a voltammetric electronic tongue, while the obtained response signals were analysed using four pattern recognition techniques: PCA, DFA, PLS regression and least squares-support vector machines (LS-SVM). The voltammetric electronic tongue was composed of five working electrodes (gold, silver, platinum, palladium, and titanium) positioned in a standard three-electrode configuration. Multi-frequency large amplitude pulse voltammetry (MLAPV) consisting of four segments was applied as a potential waveform. At the MRLs, the six antibiotics could not be completely separated from bovine milk using PCA, but were clearly demarcated by DFA. Three regression models - PCR, PLSR, and LS-SVM - were used for the prediction of antibiotic concentrations; all the regression models performed well, PCR thereby having the most stable results [2].

Source

1. *M. Tudor Kalit et al.: Electronic nose and electronic tongue in the dairy industry, Mljekarstvo 64 (4), 228-244 (2014).*
2. *Wei, Z., Wang, J. (2011): Detection of antibiotic residues in bovine milk by a voltammetric electronic tongue system. Analytica Chimica Acta 694, 46-56.*

B. Food Stuffs Industry

(i) Food quality control during processing and storage (water, wine, coffee, juices).

(ii) Optimalization of bioreactors.

(iii) Control of ageing process of whiskey.

Some other examples of application are as follows:

a. Mineral Water

Different brands of commercial mineral water were classified by using the taste sensor [1]. Mineral water was classified on the basis of the hardness of water. The tastes of 20 bottled nutritive drinks, all commercially available were evaluated both in human sensory test and by using electronic tongue. The electronic tongue was able to differentiate between low price group products, middle price group products, high price group products and played important role in evaluating the palatability of bottled nutritive drinks [2].

b. Quality of Tomatoes

Different varieties of tomato were recognized using taste sensor by measuring the crushed tomatoes [3]. The taste sensor were first calibrated in canned tomato juice to which four basic taste substances, NaCl, citric acid, monosodium glutamate, and glucose were added, for quantification of tomato taste.

c. Quality of Coffee

Ten brands of coffee of different origin (one of which was used as standard) were measured at 60°C by using taste sensors [4]. Oleic acid contained in sensor was correlated with coffee acidity as perceived by tasters with a correlation coefficient of 0.98. The correlation between coffee bitterness and the response of the sensor contained dioctyl phosphate (DOP)and Trioctylmethyl ammonium chloride (TOMA) was found to be 0.94.

d. Voltammetric tongue was used to investigate the quality of juices [5].

e. Voltammetric tongue was used to investigate the quality of tea [6].

Source

1. *Ishizaka, T., Miyanaga, Y., Mukai, J., Asaka, K., Nakai., Tsuji, E. and Uchida, T., Chem. Pharm. Bull., 2004, 52, 943.*
2. *Toko, K., Iiyama, S. and Yahiro, M., Sens Mater., 1995, 7, 191.*
3. *Kataoka, M., Miyanaga, Y., Tsuji, E. and Uchida, T., Int.J. Pharm., 2004, 279, 107.*
4. *Sim, M. Y. M., Shya, T.J., Ahmad, M. N., Shakaff, A.V.M., Othman, A.R. and Hitam, M. S., Sensors., 2003, 3, 340.*
5. *Winquist, F., Wide, P., Lundstrom, I. (1997): An electronic tongue based on voltammetry. Analytica Chimica Acta 357, 21-31.*
6. *Ivarsson, P., Holmin, S., Höjer, N.-E., Krantz-Rülcker, C. and Winquist, F. (2001) Discrimination of tea by means of a voltametric electronic tongue and different applied waveforms. Sensors Actuators B, 76, 449–454.*

C. Medicine

(i) Non-invasive diagnostics (patient's breath, analysis of urine, sweat, skin, odor).

(ii) Clinical monitoring in vivo.

(iii) Identification of unpleasant taste of pharmaceuticals. Some examples are given below:

(a) The components of medical liquids- dialysis solutions for an artificial kidney that contained Ca^{2+}, HCO_3^-, $H^2PO_4^-$, Na^+, K^+, Cl^-, Mg^{2+}, pH were determined by quantitative performance of the electronic tongue [1]. Measurement procedures were developed such that the system detects with precision of 2 per cent - 4 per cent acceptance for clinical analysis.

(*b*) Bitterness of 18 different antibiotics and antiviral drug formulation for pediatric use were evaluated as suspension in water and in an acidic sport drink [2]. Bitterness intensities of suspension in an acidic sport drink and in water were compared using the taste sensors. Suspension in an acidic sport drink would enhance or reduce the bitter intensity of the pediatric drug formulation compared with suspension in water; taste sensors were able to predict it.

Source

1. *Puech, J.L., Prida, A., Isz, S. and Radoux, C., ASEV 53rd Annual meeting, 2002, 42.*
2. *Iiyama, S., Suzuki, Y., Ezaki, S., Arikawa, Y. and Toko, K., Mater. Sci.Eng.C., 1996, 4, 45.*

D. Safety [2]

(*i*) Searching for chemical/biological weapon.
(*ii*) Searching for drugs, explosives.
(*iii*) Friend-or-foe identification.

E. Environmental Pollution Monitoring [2]

(*i*) Monitoring of agricultural and industrial pollution of air and water.
(*ii*) Identification of toxic substances.
(*iii*) Leak detection.

F. Chemical Industry [2]

(*i*) Products purity.
(*ii*) In the future-detection of functional groups, chiral distinction. Some examples are given below:
  (*a*) An electronic tongue made up of micro-sensor array of three enzyme sensors had been developed for determination of glucose, urea, and triglyceride (triolein) [1].
  (*b*) Voltammetric tongue was used to investigate the quality of spirit (Comina *et al.*, unpublished data).

Source

1. *Fukunaga, T., Toko, K., Mori, S., Nakabayashi, Y. and Kanda, M., Sens. mater., 1996, 8, 47.*
2. *P.D. Chaudhari. Electronic Tongue: A Review.*

G. Quality control of air in buildings, closed accommodation (*i.e.*space station, control of ventilation systems).

H. **Legal protection of inventions** – digital "fingerprints" of taste and odors.

I. **Others**-Voltammetric Electronic Tongue- However, most of the work has focused on supervising processes, including water cleaning processes [1], dishwasher processes [2], the wet-end process of a paper machine [3] and the process of mold growth [4].

Source

1. *Krantz-Rülcker, C., Stenberg, M., Winquist, F. and Lundström, I. (2001) Electronic tongue for environmental monotoring based on sensor arrays and pattern recognition: a review. Anal.Chim. Acta, 426, 217–226.*
2. *Ivarsson, P., Hojer, N.-E., Krantz-Rülcker, C. and Winquist, F. (2003) Process supervision of a dishwasher and washing machine with a voltametric electronic tongue. Manuscript, S-SENSE and Laboratory of AppliedPhysics.*
3. *Carlsson, A., Krantz-Rülcker, C. and Winquist, F. (2001) An electronic tongue as a tool for wet-end monitoring. Nordic Pulp Paper Res. J., 16, 319–326.*
4. *Soderström, C., Borén, H., Winquist, F. and Krantz-Rulcker, C. (2003) Use of an electronic tongue to analyze mold growth in liquid media. Int.J.Food Microbiol., 83, 253–261.*

Advantages of Electronic Tongue [1]

Electronic Tongue shows the following advantages:

(i) Requires small sample volume.
(ii) Decreased measurement time
(iii) Objectivity compared to sensory panel.
(iv) Small size of sensors.
(v) Easily operated by unskilled personnel.
(vi) Amenability to fully automatic long-term routine application [1].

The use of voltammetry as a sensing principle, under which an electronic tongue operates, appears to have several advantages; the technique is commonly used in analytical chemistry due to its very high sensitivity, versatility, simplicity and robustness [2].

Source

1. *Technical Note T- P- 01. ALPHA M. O. S., 2002.*
2. *M. Tudor Kalit et al.: Electronic nose and electronic tongue in the dairy industry, Mljekarstvo 64 (4), 228-244 (2014).*

Combination of Electronic Nose and Electronic Tongue and Application

It is also possible to combine the electronic tongue with other sensor systems. An electronic tongue based on the combination of potentiometry, conductivity and voltametry techniques has been constructed [1]. Additionally, a combination of the voltametric electronic tongue and the electronic nose [2] has been investigated. However, an electronic nose, as well as an electronic tongue, does not provide information on the nature of constituting compounds, but only give a digital fingerprint of the food that can be investigated by means of multivariate statistical analysis [3]. This is why the use of electronic nose and electronic tongue cannot become more than just a method complementary to sensory analysis [4,5]. Combined with pattern recognition techniques, an electronic nose represents a rapid and efficient tool for the classification of volatile compounds with no need for previous separation procedures [6], unlike traditional analytical methods, this technique is comparison- rather than separation-based [7].

Source

1. *Winquist, F., Holmin, S., Krantz-Rulcker, C., Wide, P. and Lundstrom, I. (2000) A hybrid electronic tongue. Anal. Chim. Acta, 406, 147-157.*
2. *Winquist, F., Wide, P. and Lundström, I. (1999)The combination of an electronic tongue and an electronic nose. Sensors Actuators B, 58, 512–517.*
3. *Cosio, M.S., Ballabio, D., Benedetti, S., Gigliotti, C. (2006): Geographical origin and authentication of extra virgin olive oils by an electronic nose in combination with artificial neural networks. Analytica Chimica Acta 567, 202-210.*
4. *Hodgins, D. (1997): The electronic nose: Sensor array-based instruments that emulate the human nose. In: Techniques for Analysing Food Aroma (R. Marsili, ed.), Marcel Dekker Inc., New York, 331-371.*
5. *Gutierrez-Mendez, N., Vallejo-Cordoba, B., Gonzalez- Cordova, A.F., Nevarez-Moorillón, G. V., Rivera- Chavira, B. (2008): Evaluation of Aroma Generation of Lactococcus lactis with an Electronic Nose and Sensory Analysis. Journal of Dairy Science 91, 49-57.*
6. *Cerrato Oliveros, M.C., Perez Pavon, J.L., Garca Pinto, C., Fernandez Laespada, M.E., Moreno Cordero, B., Forina, M. (2002): Electronic nose based on metal oxide semiconductor sensors as a fast alternative for the detection of adulteration of virgin olive oils. Analytica Chimica Acta 459, 219-228.*
7. *O'Riordan, J.P., Delahunty, C.M. (2003): Characterisation of commercial Cheddar cheese flavour. 1: traditional and electronic nose approach to quality assesment and market classification. International Dairy Journal 13, 355-370.*

Advantages and Disadvantages of Electronic Nose and Electronic Tongue [1]

Although human olfactory and taste systems and sensory assessment of food cannot be replaced by any device. Numerous works have shown that electronic nose and electronic tongue pose as excellent non-destructive methods for the determination of volatile and non-volatile food compounds. Their wide application can be found in food industry, where they can be employed into the identification and classification of products, monitoring of the ripening process (especially in cheese industry) and determination of the optimal ripening time (especially when it comes to fruit), monitoring of food spoilage (milk spoilage determination thereby presenting a showcase), food shelf-life determination and food adulteration detection.

Advantages of Electronic Nose and Electronic Tongue

Their major advantage is the possibility of continuous monitoring of product quality from the very beginning (*i.e.* from the field) across all processing stages down to the final product. Compared with the electronic nose and electronic tongue, classical instrumental analysis employed for the determination of volatile and non-volatile compounds (*i.e.* chromatography) is much more expensive and time-consuming. Electronic nose and electronic tongue can be calibrated to be reliably consistent and can give objective data for quality and safety control. These instruments can also test samples unfit for human consumption. Although the electronic nose and electronic tongue are considered to be a revolutionary discovery in the field of sensory evaluation, some shortfalls are still witnessed.

Disadvantages of Electronic Nose and Electronic Tongue

A disadvantage of electronic nose and electronic tongue systems is that they are affected by environmental conditions including temperature (both electronic nose and electronic tongue) and moisture (electronic nose), which can cause sensor drift, although calibration systems and built-in algorithms help compensate for this.

In addition, it is necessary to conduct a more extensive research so as to come up with the type of sensors best fit for the detection of a particular type of volatile and taste components of certain dairy products, as well as with the best statistical method for processing of data obtained by an electronic nose and electronic tongue, so as to refine their practical applications in the dairy industry to a reliable level.

From an instrumental point of view, it would be interesting to combine an electronic nose and an electronic tongue to obtain complementary information on the same sample by means of a hybrid array.

Source

1. *M. Tudor Kalit et al.: Electronic nose and electronic tongue in the dairy industry, Mljekarstvo 64 (4), 228-244 (2014).*

Section B

Instrumentation in Chemical Analysis

Chapter 2

Immunoassay (Immuno-chemical) Techniques

An **immunoassay** is a biochemical test that measures the presence or concentration of a macromolecule or a small molecule in a solution through the use of an antibody (usually) or an antigen (sometimes) [1]. Immunoassays can be defined as quantitative binding reaction between antibodies and target antigens. In this technique, the application of a specific antigen or antibody, capable of binding to the analyte, to identify and quantify substances, is used [2,3]. Immunochemical methods are based on the reaction of an antigen with an antibody-a reaction of antigenic determinants with the antibody linking site.**"Immuno"**refers to an immune response that causes the body to generate antibodie and **"assay"** refers to a test. Thus, an immunoassay is a test that utilizes immune complexing when antibodies and antigens are brought together. Immunoassays are used to quantify molecules of biological interest based on the specificity and selectivity of antibody reagents generated.

(i) **Radio immunoassay**: The antibody can be linked to a radioisotope then it is called radio immunoassay.

(ii) **Enzyme-linked immunosorbent assay**: The antibody can be linked to an enzyme which catalyses an easily monitored reaction then it is called enzyme-linked immunosorbent assay.

(iii) **Immunofluorescene**: The antibody can be linked to a highly fluorescent compound by which the location of an antigen can be visualized, then it is called immunofluorescene.

The measurement of trace amounts of physiologically active small molecules (for example, lipids, drugs, other synthetic chemicals and metals) is critical for Food, clinical and environmental analyses. Most small molecules can be analyzed using highly sophisticated analytical techniques, including high pressure liquid chromatography (HPLC), gas chromatography (GC), and inductively coupled plasma atomic emission spectroscopy (ICPAES). However, these methods require extensive purification, experienced technicians, and expensive instruments and reagents. Immunoassays offer an alternative to these instrument-intensive methods. Immunoassays rely on an antibody (Ab), or mixture of antibodies, for recognition of the molecule being analyzed (the analyte). Immunoassays are frequently applied to the analysis of both low molecular ligands and macromolecular drugs, and are also applied in such important areas as the quantitation of biomarkers that indicate disease progression and immunogenicity of therapeutic drug candidates. The performance of immunoassays is critically dependent on the binding properties of the antibody used in the analysis, and identification of suitable antibodies is often a major hurdle in assay development. Recombinant antibodies will play a major role in future immunoassay development [5]. Immunoassays are different from other types of laboratory tests, such as colorimetric tests, because they use antibody: antigen complexes to generate a signal that can be measured.

Immunoassay techniques provide simple, powerful and inexpensive methods for the measurement of small ligands. However, the progress of the development of new immunoassays and related immuno technologies is still limited by the availability of antibodies with the desired affinities and specificities for given applications. Advances in molecular biology have led to the ability to synthesize antibodies in vitro, completely without the use of animals. Recombinant molecular technology that can generate variability, combined with high-throughput screening methodologies, can be used to produce engineered antibody-like molecules and novel antibody-mimic domains on scaffold proteins. The rAbs fused with other functional proteins can enhance the sensitivity of antibody-based assays and reduce the cost and labor involved in chemically synthesizing conjugates. Antibody engineering had already matured into a technology available to the general scientific community. Further advances will lead to better binding proteins that will permit the development of novel, high-throughput sensing systems for low molecular weight ligands [5].

Development of Imunnoassays [5a]

Immunoassays are used when an unknown concentration of an analyte within a sample needs to be quantified. To obtain the most accurate determination of the unknown concentration, an immunoassay must be developed based not only on the usual assay development criteria (standard deviation or optimal signal window) but also on how well the immunoassay can predict the value of an unknown sample. First, one needs to establish the assay critical success factors. Then the immunoassay needs to be developed, which establishes proof of concept. During the optimization phase, the quantifiable range of the immunoassay method is determined by calculating a precision profile in the matrix in which the experimental samples will be measured. A spiked recovery is then performed by spiking the analyte into the matrix and determining the per cent recovery of the analyte in the matrix. If the precision profile is within the desired working range, then assaying spiked recovery samples over several days completes the validation of the immunoassay.

If the precision profile limits are not within the desired working range, further optimization of the immunoassay is required prior to validation.

a. Basic Steps for Developing and Running an Immunoassay [5a]

These are as follows:

(i) Establish assay critical success factors (*i.e.* sensitivity required).

(ii) Ensure appropriate antibody and antigen reagents are available.

(iii) Adsorb antigen or capture antibody to a solid surface.

(iv) Wash off unbound reagents.

(v) Block nonspecific binding sites to reduce background.

(vi) Incubate the secondary antibody with the sample.

(vii) Wash off unbound reagents.

(viii) Incubate secondary antibody-conjugate with sample.

(ix) Wash off unbound reagents.

(x) Incubate substrate to generate signal.

(xi) Calibration curve fitting, data analysis and quantitation by non-linear regression.

b. Imunnoassay Parameters [5a]

To begins the development, optimization and validation of an immunoassay, it is important to define the following relevant immunoassay parameters:

(i) Analyte (hapten or antigen) to be measured.

(ii) Sample matrices in which measurements will be made (serum, plasma, cell lysates, culture media, *etc.*).

(iii) Source of antibody, analyte standards and detection reagents (labeled antibody, enzyme substrates, *etc.*). Availability of these reagents is a critical requirement.

(iv) Detection mode (colorimetric, fluorescence or chemiluminescence) and appropriate plate readers.

(v) Type of immunoassay to develop: Sandwich, competitive or antigen-down formats.

(vi) Expected analyte concentration ranges to be measured: pg/ml, ng/ml or µg/ml in the sample matrix of choice. This would determine the detection limits and the measurable range that should be achieved in a validated assay.

(vii) Data analysis models and format for reporting results.

(viii) Validation and optimization criteria using statistical experimental design tools.

(ix) Recovery, accuracy and precision expected at the limits of quantification and the measurable range.

(x) Sample throughput, frequency of use, automation and the number of laboratories that would run the assay.

(xi) Control samples that would be used for optimization, validation and quality control runs.

c. Reagents [5a]

Reagents are a critical piece of any assay development process. This refers to all of the reagents that will be used in the assay. There are following certain items that need to be considered when obtaining reagents:

(i) Quality of standards and antibodies.

(ii) Quantity of standards and antibodies.

(iii) Purity of standards and antibodies (when possible antibodies are affinity purified).

(iv) Selectivity and specificity of antibodies.

d. Buffers [5a]

The following types of buffers are being used for immunoassay:

Coating Buffers	*Blocking Buffers*	*Wash Buffers*	*Antobody Diluents Buffers*
(i) 50 mM sodium bicarbonate, pH 9.6 (ii) 0.2 M sodium bicarbonate, pH 9.4 (iii) PBS - 50 mM Phosphate, pH 8.0, 0.15 M NaCl (iv) Carbonate-bicarbonate (v) Phosphate Buffer- 1.7 mM NaH_2PO_4, 98 mM $Na_2HPO_4{\cdot}7H_2O$, 0.1 per cent NaN_3, pH 8.5 (vi) TBS - 50 mM TRIS, pH 8.0, 0.15 M NaCl.	(i) 1 per cent BSA or 10 per cent host serum in TBS, or TBS with 0.05 per cent Tween-20 (ii) Phosphate Buffer-73 mM Sucrose, 1.7 mM NaH_2PO_4, 98 mM $Na_2HPO_4{\cdot}7H_2O$, 0.1 per cent NaN_3, pH 8.5 (iii) 1 per cent HSA in PBS (iv) Casein Buffer-Pierce Blocker (v) Protein Free Block-Pierce (vi) Pierce has many blocking buffers that are available in their catalog (vii) Heterophilic Blocking Reagent (HBR): Scantibodies Laboratory, Inc. (viii) Scantibodies has many other blocking reagents that are available in their catalog.	(i) PBST, 0.05 per cent Tween-20 (ii) TBST, 0.05 per cent Tween-20	(i) 1 per cent BSA or 10 per cent host serum in TBS, or TBS with 0.05 per cent Tween-20 (ii) 1 per cent BSA or 10 per cent host serum in PBS, or PBS with 0.05 per cent Tween-20 (iii) 50 mM HEPES, 0.1 M NaCl, 1 per cent BSA, pH 7.4 (iv) Blocking buffer

e. Enzymes and Substrates

i. Horseradish Peroxidase (HRP) Substrates

(i) TMB: 3,3,5, 5′-tetramethyl benzidine (colorimetric).

(ii) OPD: o-phenylenediamine (colorimetric).

(iii) ABTS: 2,2′-azino-bis (3-ethylbenzthiazoline-6-sulfonicacid) (colorimetric).

(iv) Pierce Super signal (chemiluminescent).

(v) QuantaBlu (chemifluorescent).

(vi) Pierce Quanta Red (chemifluorescent).

(vii) Pierce has other substrates that provide strong signal and sensitivity with HRP enzyme conjugates that are available in their catalog.

B. Alkaline Phosphatase Substrate

(i) pNpp (p-Nitrophenyl Phosphate)

Principle [1]

Immunoassays rely on the ability of an antibody to recognize and bind a specific macromolecule in what might be a complex mixture of macro-molecules. In immunology the particular macromolecule bound by an antibody is referred to as an antigen and the area on an antigen to which the antibody binds is called an epitope. In some cases an immunoassay may use an antigen to detect for the presence of antibodies, which recognize that antigen, in a solution. In other words, in some immunoassays, the analyte may be an antibody rather than an antigen.

In addition to the binding of an antibody to its antigen, the other key feature of all immunoassays is a means to produce a measurable signal in response to the binding. Most, though not all, immunoassays involve chemically linking antibodies or antigens with some kind of detectable label. A large number of labels exist in modern immunoassays, and they allow for detection through different means. Many labels are detectable because they either emit radiation, produce a color change in a solution, fluoresce under light, or because they can be induced to emit light.

Instrumentation [5a]

Immunoassays include following three major elements:

(i) Preparation of the target-specific antibody.

(ii) Antigen-antibody recognition based on a specific carrier.

(iii) Acquisition of the detection signals.

The details of antibody and antigen are given under reagents. For acquisition of the detection signals, the majority of **plate readers** are being used. These employ **UV-Vis Absorbance, fluorescence or chemiluminescence signals** as the measured response, because the products of enzyme labels are chromophores, fluorophores or emit luminescent signals. The instrument used to read the output of the immunoassay should be tested initially for both linearity and performance.

Instrument performance should be regularly calibrated according to manufacturer's specifications. Linearity in response to the specific enzyme product of an enzyme-linked immunoassay (ELISA) should be checked at the appropriate wavelengths and instrument settings. The details of some plate readers are given below:

i. Colorimetric/Spectrophotometric Plate Readers

Lamp sources and Photo-multiplier Tubes (PMT) vary in quality and performance in many plate readers. The linear range of many plate readers is generally between 0-3 Absorbance Units (AU), but other instruments have a linear range up to 4.0 AU. A malfunctioning lamp source or photomultiplier tube can significantly affect the linear response range.

ii. Fluorescence Plate Readers

These readers employ excitation and emission filter sets in addition to excitation lamp sources and PMTs. In addition to the lamps and PMTs, the filter sets also vary in quality, light throughput and band width. Fluorescence signals are generally in Relative Fluorescence Units (RFU) and linearity should be verified with appropriate filter sets for the fluorophore employed according to instrument specifications.

iii. Chemiluminescence Readers

These type of instruments have sensitive photo-multipliers to detect light emitted from a chemical reaction hence Lamp sources are not necessary. These readers usually have a much larger dynamic range, thus allowing for the increase in sensitivity. Signals or responses are measured in Relative Light Units (RLU) and can be significantly different depending on the instrument design.

Reagents of Immunoassays [5]

Antibody and Antigen are the major reagents of immunoassays.

Antobodies and Antibody Substitues

The antibody is the key reagent of an immunoassay and it can be produced by animal immunization, hybridoma technology, and/or recombinant techniques. Most, but not all, production methods require immunization of an animal with an antigen. An antibody is a protein that is produced by the body in response to an "invading" (foreign) substance. Antibodies are produced as part of the body's immune response to protect itself. For instance, some immunoassays test for the presence of antibodies to cancer molecules. Thus, if the antibodies are present, it means invading cancer cells are, too.

Antibodies are proteins, specifically immunoglobulins, formed by the host animal in response to the invasion by antigens. There are several distinct classes of natural antibodies (IgG, IgM, IgA, and IgE) that provide key defenses against pathogenic organisms and toxins. Most immunoassay systems rely upon IgG as the immuno-globulin of choice. IgG is bivalent, and its ability to bind to two antigenic sites greatly increases its functional affinity and confers high retention time on cell surface receptors. Animmunoglobulin (IgG) molecule contains two identical light (L)and two identical heavy (H) polypeptidic chains mutually linked with disulphide bonds. One of the Ig molecules always contains one light and one heavy chain type. The light chain determine the immunoglobulin molecule type. The heavy chains are decisive in classifying into the immunoglobulin class. C-ending sections of the chain make up their constant area, N-endings of chain are denoted as the variable sections and they represent the section of the molecule binding to the antigen-aparatope. Immunochemical methods make use of monoclonal or polyclonal antibodies.

Antibodies possess high specificity and affinity for a specific antigen. It is the specific binding of an antibody to an antigen that allows the detection of analytes by a variety of immunoassay methods.

Types of Antibodies and Antibody Substitutes

In addition to the continuous improvements in traditional antibody production, the discovery and production of novel antibodies or antibody substitutes with specific properties has been reported including single-chain antibody fragments [5b], aptamers [5c], receptor proteins [5d], and molecularly-imprinted polymers [5e].

i. Traditional Antibodies [9]

Polyclonal and monoclonal antibodies are generated from immunized animals (rabbits, mice, or goats). Polyclonal antibodies can recognize multiple epitopes on any one antigen, and are inexpensive to produce but they are not as popular as monoclonal antibodies because they have high batch-to-batch variability and non-specificity. Monoclonal antibody technology (invented by British scientists) [5f] is based on secreted antibodies from a hybridoma amplified in vitro. Traditional antibodies are inferior in stability to many of their alternatives considered below because of the long time required for preparation and ethical concerns associated with the use of animals.

ii. Single-chain Antibodies [9]

To overcome the bottle-neck of screening of hybridomas, recombinant antibodies with improved specificity and affinity were generated.Single chain antibody fragments (scFv) are an important class of recombinant antibodies. The antibodies are usually expressed in **E.coli** by genetic engineering and are composed of 15-20 short peptides from the variable regions of the heavy chain and light chain of the monoclonal antibody. They are only 1/6 of the molecular weight of the whole IgG. Versus monoclonal antibodies, scFv have higher affinity for the antigen, smaller size, better tissue penetration, shorter preparation time and easier labeling.

iii. Receptor Proteins [9]

Most receptor proteins are membrane-associated proteins. Studies have indicated that after some drugs enter an organism, they can function by binding to receptor proteins on the cell membrane and hence the receptor protein acts as an "antibody".

iv. Aptamers [9]

Aptamers are a single-stranded oligonucleotide sequence of DNA or RNA that consist of 20-60 nucleotides and can specifically bind to a transfer ligand.

The specific affinity of the aptamers and binding ligands are based on a diverse conformation of single-stranded nucleic acids. Some aptamers have already been screened to identify specific proteins, veterinary drugs, nucleotides, heavy metal ions and small molecule toxins. Aptamers offer a rapid and efficient identification strategy for clinical diagnosis and food safety inspection with significant prospects in practical applications.

In comparison to antibody, aptamers have following advantages:

(i) Preparation is based on simple chemical synthesis and is cost efficient.

(ii) Good stability with highly reversible denaturation.

(iii) The single nucleic acids are usually composed of no more than 30 bases with relatively small molar amounts and molecular sizes. The low steric hindrance is suitable for binding with nano-materials for the generation of high-precision sensors.

(iv) There is a stronger affinity and specificity between the aptamer and the target molecules but with lower dissociation coefficients.

Imprinted Polymer

Imprinted polymers are generated by molecular imprinting. They are also known as plastic antibodies and have selective adsorption or specific recognition of target molecules and their structural analogs. Imprinted polymers have stable chemical properties and higher susceptibility to acids and bases than biological antibodies [9].

Antigen

An antigen is a molecule that can be recognized by the immune system (immunogenicity) and that can be bound specifically to an antibody (reactogenicity). Chemically, antigens are polymers-proteins, polypeptides, polysaccharides, or nucleoproteins. They have two fundamental properties – theyincite a specific immune response, and they react specifically with the products of this response (with antibodies and immune complementary cells). An antigen is the substance that the body is trying to "fight off" (eliminate or reduce) by mounting an immune response. Some immunoassays test for antigens directly, rather than looking for the antibodies. In a test to measure the concentration of a therapeutic drug, for example, the drug is the antigen that binds to the antibody. There is complete and incomplete antigens.

(i) **Complete antigens:** Molecules with both immunogenicity and reactogenicity are called "complete antigens".Both properties can be found in the complete antigen animmunogen which is made of a macromolecular carrier and antigenic determinants-epitopes.

(ii) **Incomplete antigens**: Molecules that possess only reactogenicity are called "incomplete antigens". A low molecular substance which itself can not incite the antibody production but which reacts specifically with the products of the immune responseis termed as incomplete

antigens. Incomplete antigens, also called **haptens**, encompass a wide variety of molecules, including drugs, explosives, pesticides, herbicides, polycyclicaromatic hydrocarbons, and metal ions. These haptens can induce the immune system to produce antibodies only when they are covalently conjugated to a larger carrier molecule such as a protein.

Analyte

An analyte is anything measured by a laboratory test. In immunoassay testing, the analyte may be either an antibody, or an antigen. The molecule detected by the immunoassay is often referred to as an "analyte" and is in many cases a protein, although it may be other kinds of molecules, of different size and types, as long as the proper antibodies that have the adequate properties for the assay are developed. Analytes in biological liquids such as serum or urine are frequently measured using immunoassays for medical and research purposes [1,4].

Immunoassays utilize one or more select antibodies to detect analytes of interest. The analytes being measured may be those that are naturally present in the body (such as a thyroid hormone), those that the body produces but are not typically present (such as a cancer antigen), or those that do not naturally occur in the body (such as an abused drug).

Modern immune analytical methods accomplish an increase in the sensitivity by labelling one of the reagents- the antigen or the antibody. The label can be a radio-isotope, enzyme, fluorescence or chemical scintillation agent [6,7,8].

Nonisotopic immunoassays such as ELISA (enzyme linked immunosorbent assay), FPIA (fluorescence polarisation immunoassay), PCIA (particle-concentration immunoassay), PCFIA (particle-concentration fluorescence immunoassay), and monoclonal-based immunoassays will, in all likelihood, play an increasingly important role in antibiotics screening immunoassay determinations [7].

Types of Immunoassay Techniques [2,3]

These techniques are named on the basis of linkage of antibody to radioisotope (RadioImmunoassay), or to an enzyme which catalyses an easily monitored reaction (Enzyme linked Immunosorbent assay) or to a highly fluorescent compound (Immuno fluorescene) for visualising the location of an antigen. The following immunoanalytical techniques are being used in the analysis of food components, Chemical contaminants in food and Food borne pathogens:

(i) Enzyme immunoassay (EIA).

(ii) Enzyme linked Immunosorbent assay (ELISA).

(iii) Enzyme linked Coagulation assay (ELCA).

(iv) Radio immuno-assay (RIA).

(v) Immunosensor with multiple amperometric detection (ISA).

(vi) Immunosensor based on antibodies attached to quartz crystal (ISC).

Types of Immunoassay Carriers [9]

The most immunoassays are based on immobilizing the antigen or antibodies on the support surface of the carrier. The following types of carriers are generally used in immunoassay:

i. Microplate-Based Carriers

The most currently used microplates for immuno-assays are polystyrene which is a linear polymerized styrene monomer linked via a covalent radical reaction. The weak polarity and strong hydrophobicity easily makes the antigens or antibodies adsorb to the microplate surface by hydrophobic interactions. The preparation of microplates is also cost efficient. Microplates with 96, 384, and even larger numbers of wells are available, but plates with 96-well are the most common. The plates can be of following different colours:

(i) Transparent plates (for ELISA).

(ii) White opaque plates (for CIA).

(iii) Black opaque plates (for NFIA).

(iv) Black opaque plates with low non-specific binding surface (for FPIA).

This method has following advantages and disadvantages:

Advantages	*Disadvantages*
High throughput (one plate can detect 40 to 80 samples simultaneously)	More analysis time (>two hours)
Semi-quantitative results	Have need for a microplate reader or chemilumine-scence detector.
High sensitivity and specificity	—

ii. Membrane-Based Carrier

Immunoassays based on membrane carriers are common format for the rapid detection. Its initial development is flow-through immunoassay, which is known as immuno-filtration assay. The test principle involves a flow of fluid containing the analyte through a porous membrane and into an absorbent pad. It is a rapid (within 5 min) and simple immunoassay for qualitative screening, with the advantage of large volumes of samples.

Lateral flow tests are also known as lateral flow immunochromatographic assays. When a sample solution is applied to the end of a test strip, it moves forward by capillary effects; when it reaches the conjugation pad, it dissolves and reacts with the colloidal gold-labeled reagents on the conjugation pad. It then further moves to an area on the test strip with antigens or antibodies attached. If the compound formed between samples and the gold-labeled reagents specifically reacts with the antigens or antibodies, they will accumulate in this zone. This causes a color change visible to the naked eye.

Labeling Molecules [9]

The different labels are being used in immunoassays.

(a) The most commonly used label in immunochromatographic assay is colloidal gold. It is negative under weak basic conditions and this is attractive to positively charged protein molecules. They bind strongly without changing the biological properties of the proteins. Colloidal gold as a label is used because it is easy to prepare, inexpensive, stable, and yields a bright color. It is observed from reports that it is used in milk for detection including aminoglycoside antibiotics, cephalosporins, quinolones, sulfonamides and aflatoxin.

(b) Latex particles are another common type of labeling molecule in immunoassay. Latex particles bind antibodies with strong covalent bonds. By mixing with the dye molecules, the latex particles show a variety of colors. However, the preparation of latex particles is relatively complicated and is not as extensively studied as colloidal gold.

(c) The fluorescent labeling materials including nanosilver, fluorescent silica particles, liposomes, colloidal 245 carbon, upconverting phosphors, quantum dots, fluorescent biological markers and LMW fluorescent dyes are also being used in immunoassays. Fluorescent microspheres and quantum dots have gained the most attention.

iii. Chip-based Carriers [9]

Chip is another carrier for the detection of different antibiotic classes in milk. Biochips technologies use various strategies to immobilize biological molecules (oligonucleotides, DNA, cDNA, polypeptides, antigens, antibodies, *etc.*) on solid supports such as silicon slides, glass slides (beads), plastic sheets (beads), and nylon films. According to the different coating compounds, common biochips could be divided into the following categories:

(i) DNA microassay.

(ii) Protein-chip.

(iii) Microfluid

The system can capture targets from the sample with signal capture facilitated by CCD cameras or laser scanner. Digital processing offers qualitative and quantitative analyses.Biochip-based immunoassays are still being improved. The synthesis of the biochip base involves complicated chemical processing, and the capture of chemiluminescent signals requires extremely high sensitivity CCD detectors for one-time acquisition. On the other hand, this strategy uses small sample volumes and requires specialists for such sophisticated operation. However, the high cost will likely limit its utility.

iv. Gel-based Carriers [9]

Sephrose 4B gel is commonly used as a carrier in protein purification. It can bind antibodies through covalent bonds after being activated by CNBr.

Conventionally, sepharose 4B gel has been used for sample cleaning in pre-processing, *i.e.* an immunoaffinity column. Recently, researchers have introduced it into immunoaffinity assays.Gel-ELISA is one of the immunoassays and it was firstly reported in 2008.Gel-ELISA is mostly used to detect toxins and organic molecules in liquid samples. As a novel test strategy, Gel-ELISA has shown its following unique advantages and brought application potentials:

(i) The range of sample volumes is large. In Gel-ELISA, the pore size governs sample size in contrast to microplate approaches. The applied sample volume is usually 50μL, and the total reaction volume is less than 100μL. In a gel column, up to five column bed volumes can be used and eluted with gravity. Practically, sample volumes of 1 -3 mL are usually applied, but that may theoretically contain up to 20-fold the amount of target actually as in DcELISA-this effectively enhances sensitivity.

(ii) The matrix interferences are small and sample pre-treatment is minimal. Gel-ELISA has strong tolerance to sample matrix effects. After protein precipitation, the supernatant can be completely transferred to the column-there is no sample dilution. In contrast, to eliminate sample matrix interferences in DcELISA, the supernatant sample from protein precipitation still requires dilution to reduce the matrix background. This reduces detection limits.

(iii) Gel-ELISA offers rapid assay times. Because it uses natural gravity flow, the method does not need any incubation or washing steps with total assay times of 15-20 min. This saves up to 80 per cent of the time versus the 90-120 min of ELISA.

Application of Immunoasay Techniques in Dairy Industry

Immunoassays are increasingly important in the detection of chemical contaminants in milk because of developments in material techniques, antibody techniques, bioconjugate techniques, and signal acquisition techniques. Immunoassay technology has been applied successfully to dairy quality control due to its simple operation, short analysis times, and low cost. The most mature method is microplate-based ELISA that uses small sample volumes and simple pretreatment steps. It has a high capacity without the need for complex instrumentation. It can achieve detection limits similar to HPLC or LC-MS/MS. The test efficiency is 10-fold better than instrument-dependent measurements [9]. These are used to detect bacterial and fungal cells and Veterinary drugs, pesticides, toxins *etc.* in milk and milk products.

(i) The various immunochemical techniques are capable of detecting residues of antibiotics, toxins and other undesirable compounds in milk with high degree of sensitivity and accuracy.

(ii) The ELISA combined with an enrichment step substantially reduces the time in mould and bacterial detection in food/Dairy products as compared to traditional/conventional microbiological methods.

(iii) ELISA with antibodies specific for native α-lactalbumin and β-lactoglobulin may be used to differentiate between raw, pasteurised and ultra high temperature (UHT) milk.

(iv) ELISA based test methods can indirectly identify recombinant chymosin from E.coli, K.lactis and A.niger, which are used for cheese manufacturing.

(v) AOAC has specified several test methods based on immunoassays for screening of Listeria, Staphylococcal toxins, Salmonella *etc.* in food/dairy products.

The test strip method using a NC membrane as the carrier can complete a test in 5-10 min, and the results are visible to the naked eyes. The utility of this rapidly developed method has extended from the field of medical diagnostics to the another important field of food safety. The following test methods based on these techniques, are developed and used for milk with specified detection limits:

Test Methods for Detetion of Contaminants/Microorganism in Milk [3,3a]

Sl.No.	*Methods*	*Analytes*	*Detection Limits*
1	EIA	Aflatoxin M_1	0.25 ppb
		Chloramphenicol	0.5-2 ppb
		β-Lactuam antibodies	0.05-1 ppb
		Ochratoxin A	-
		Suphamethazine	2-10 ppb
		Fragment of κ́-casein macropeptide to detect proteolysis	100 ppb
2	ELISA	Clostridium perfringens enterotoxins A	5 ppb
		Chymosin	0.25 ppb
		Heatstable proteinase from Pseudomonos	0.08 ppb
		Staphylococcus spp.	10^4-10^5 cells/ml
3	EIA or RIA	Staphylococcal enterotoxins A,B,C,D,E.	0.1-1 ppb
		Pesticides and Herbicides	0.2-1 ppb
4	ELCA	*Clostridium botulinum* toxins A,B,C.	0.01 ppb
5	ISA	Listeria monocytogenes	$2x10^2$ cells/ml
6	ISC	Listeria monocytogenes	$2.5x10^5$cells/ml

1. Immunoassay of Chemical Contaminants in Milk (Food Safety Evaluation of Milk) [9]

Immunoassays have recently been used to detect contaminants [veterinary medicines, antibiotics, pesticides, heavy metals, mycotoxins, and persistent organic pollutants (POPs)] in milk because of their simple operation, high speed and low cost. The application of the four major substrates used for immunoassays (microplates, membranes, gels, and chips) as well as their use in the detection of milk contaminants have been successfully established. The source of these aforesaid contaminants are farmland contaminated with medication, forage pollution, drinking water contamination, illegal additions as well as transportation and processing pollution.

Examples of Microplate-Immunoassay for Milk Analysis [9]

Sl.No.	*Species*	*Analyte*	*Method*	*Signal*	*Reference*
1	Antibiotics	Gentamicin	CIA	Luminol	Li *et al.*, 2012
		Macrolide antibiotics	ELISA	TMB	Zhang *et al.*, 2013
		Cephalexin/Cefadroxil	FPIA	FITC	Zhang *et al.*, 2014
		Tetracycline	ELISA	TMB	Wang *et al.*, 2014
2	Veterinary drug	Chloramphenicol	ELISA/CIA	TMB/ Luminol	Liu *et al.*, 2014, Tao *et al.*, 2013
		Fluoroquinolones/ sulfonamides	FPIA	FITC	Chen *et al.*, 2014
		Sulfonamides	ELISA/ FLISA	TMB	Liang *et al.*, 2014, Shen *et al.*, 2007
		Fluoroquinolone	ELISA/ TRFIA	TMB/ Europium	Suryoprabowo *et al.*, 2014, Leivo *et al.*, 2013
		Sulfonamides/ melamine/Quinolones	FLISA	Quantum dot	Zhu *et al.*, 2011
		Orbifloxacin	FPIA	FITC	Mi *et al.*, 2014
3	Pesticides	Atrazine	ELISA	TMB	Barchanska *et al.*, 2012
		Avermectins	ELISA	TMB	Wang *et al.*, 2012
4	Toxin	Aflatoxin B-1 and Aflatoxin M-1	ELISA	TMB	Jiang *et al.*, 2013
		Abrin	ELISA	TMB	Zhou *et al.*, 2012
5	Heavy metal	Mercury(II) ion	ELISA	TMB	Wang *et al.*, 2012
6	POPs (Persistent organic pollutants)	tris-(2,3-dibromopropyl) isocyanurate	ELISA	TMB	Feng *et al.*, 2014
7	Others	Melamine	ELISA/ FLISA	Quantum dot	Gong *et al.*, 2014, Wu *et al.*, 2013
		5-hydroxymethyl-2-furfural	ELISA	TMB	Guan *et al.*, 2013

The different countries (EU, U.S., China, and others) have enacted many validated methods for monitoring milk safety including Gas chromatography [10,11], High-performance liquid chromatography [12], Gas chromatography-mass spectrometry [13,14], Liquid chromatography tandem mass spectrometry [15,16,17], Differential pulse anodic stripping voltammetry [18] *etc.* These technologies are highly sensitive, specific, and reliable, but their application was limited by costly instrumentation and complicated sample preparation.

In contrast, immunoassays based on specific antibody recognition are a rapid method of screening samples [19]. Unlike chromatography-based methods, they are simple, specific, easy to use, high-throughput, and do not require costly instrumentation. This makes them suitable for on-site detection. Therefore, a combination strategy with fast immunoassay screening followed by instrument-based confirmation has been widely considered. Moreover, immunoassay is also suitable for the analysis of milk. Because milk is a fluid, milk samples can be analyzed after only simple pretreatment such as dilution or protein precipitation. They can even be analyzed without sample preparation.

(i) Several companies, such as Biopharm (Germany), Idexx (U.S.), and Randox (U.K.), have produced rapid screening products (kits and strips) for the assessment of milk contamination using monoclonal antibodies.

The β-lactam antibiotics including penicillins and cephalosporins are widely used in dairy cows and are thus the main contaminants in milk. Receptor proteins are being used to determine said antibiotics.Currently, drug receptor proteins expressed in vitro include sulfonamide receptor proteins and tetracycline receptor proteins.

(ii) Molecularly-imprinted polymers (MIPs) are used in dairy quality control. Analytes measured with These polymers are used to determine different analytes include bisphenol A, estrogen, sulfonamide, quinolones, oxytetracycline, amphenicols, and selenium.

(iii) Competitive ELISA is mostly used in dairy studies because milk contaminants are mostly small-molecule compounds. The enzymes used in the assay are the horseradish peroxidase (HRP). Quantification of the target analytes is based on TMB color change. ELISA is simple and low cost, but is also limited by low sensitivity of the photometric measurement and poor stability of the labeled enzymes.

(iv) Immuno-chromatographic methods are widely applied in dairy quality control. It is most commonly used with a nitrocellulose membrane (NC) because of the following reasons:

- *(a)* They have high affinity to target across a wide range and at low ion strength buffer. Here, most negatively charged proteins will be associated with the nitrocellulose membrane due to hydrophobicity.
- *(b)* The binding site is easily blocked, which reduces nonspecific binding.
- *(c)* It is suitable for mass production and is stable with low cost.

(v) IDEXX Laboratories (Idexx 2015) and Charm Sciences (Charm 2015) use latex particles for dairy assays including β-lactam antibiotics, tetracyclines, amino-glycosides, and sulfonamides with many latex color choices available depending on the analyte.

(vi) Most quantum dot-or fluorescent microsphere-based immuno-chromatography techniques in the dairy field were reported.Detection of cephalexin in milk, serum, and animal tissues based on fluorescent microsphere strips in 2013, sulfamethazine in milk with a fluorescent microsphere strip method in 2013, lincomycin in milk, honey, cattle, and swine urine with a fluorescent microsphere strip method in 2014 and detection of chloramphenicol in milk with a quantum dot-based immuno chromatographic method in 2013 were reported.

(vii) **Gel-ELISA** was used to detect chloramphenicol in animal tissues, milk, honey, fish, and shrimp with a LOD of 1.0 μg/L. 286 Versus the lateral flow strip test of CAP with the same antibody, the Gel-ELISA provided a reliable assay with 100-fold sensitivity improvement. In 2014, use of Gel-ELISA was also reported to detect apramycin in milk, muscle, and liver, to measure 14 sulfonamides and 13 quinolones in milk.

(viii) The biochip method has huge potential in the detection of antibiotic residues in milk, which uses low sample volumes, requires no sample pre-treatment, and is high-throughput and rapid. In these methods, the analytes (chemical contaminants in milk) were coupled to the chip surface with or without linking agents, and competitive immunoassay format was used to implement the detection.

Milk studies using bio-chip technologies to detect chemical contaminants in milk mainly focus on antibiotics and drug residues.

(a) In 2008- A microplate array-based SMM-FIA for the simultaneous detection of sulfamethazine, streptomycin, and tylosin in milk was reported.

(b) In 2009-**i**-A label-free multiplex detection system for streptomycin, gentamicin, neomycin, kanamycin, sulfamethazine, enrofloxacin, and chloramphenicol residues in milk using a surface plasmon resonance-based immunosensor was reported. **ii.** A hapten microarray designed for the parallel analysis of 13 different antibiotics in milk was reported.

(c) In 2012 -A biochip array kit for multiplex screening of residues for more than 20 anthelmintic drugs in milk and animal tissues was developed.

Source

1. *Wikipedia.*

2. *Stepaniak, L., Sorhaug, T. andJedrychowski, L. (2002), Immunochemical, Encyclopedia of dairy Sciences (Volume-1), pp. 62-67, ElsevierScience Limited.*

3. *Technews, No.74, May-June-2008, NDDB.*

3a. *M.K.srivastava, Hand book on analysis of milk, IBDC, Luknow.*

4. *Yetisen A. K. (2013)."Paper-based microfluidic point-of-care diagnostic devices".Lab on a Chip.13 (12): 2210-2251.*

5. *Bhupal Ban and Diane A.Blake, Recombinant Antibodies and Non-Antibody Scaffolds for Immunoassays, Department of Biochemistry and Molecular Biology, Tulane University School of Medicine, New Orleans, Louisiana, USA.*

5a. *Immunoassay Methods-Assay Guidance Manual, Sittampalam GS, Coussens NP, Brimacombe K, et al., editors. Assay Guidance Manual [Internet]. Bethesda (MD): Eli Lilly and Company and the National Center for Advancing Translational Sciences; 2004.*

5b. *Ahmad Z A, Yeap S K, Ali A M, Ho W Y, Alitheen N B M, Hamid M. 2012. ScFv antibody: 369 Principles and clinical application. Clinical and Developmental Immunology, 2012, 1-15.*

5c. *Toh S Y, Citartan M, Gopinath S C, Tang T. 2015. Aptamers as a replacement for antibodies in 648 enzyme-linked immunosorbent assay. Biosensors and Bioelectronics, 64, 392-403.*

5d. *Beltrán M C, Borràs M, Nagel O, Althaus R L, Molina M P. 2014. Validation of receptor-binding 390 assays to detect antibiotics in goat's milk. Journal of Food Protection, 77, 308-313.*

5e. *Song X, Xu S, Chen L, Wei Y, Xiong H. 2014. Recent advances in molecularly imprinted polymers 626 in food analysis. Journal of Applied Polymer Science, 131, 1-18.*

5f. *Köhler G, Milstein C. 1975. Continuous cultures of fused cells secreting antibody of predefined 522 specificity. Nature, 256, 495-497.*

6. *Mitchell J.M., Griffiths M.W., McEwen S. A., McNabW.B., Yee A.J. (1998): Antimicrobial drug residues in milk and meat: cause, concerns, prevalence, regulations, tests, and test performance. Journal of Food Protection, 61: 742–756.*

7. *Roeder A.M., Roeder M. (2000): Antibiotics in food of animal origin. In: Francis F.: Encyclopedia of Food Science and Technology. Vol. 1, 2nd Ed. John Wiley and Sons Inc., New York: 54–63.*

8. *Stepaniak L., SorhaugT., JedrychowskiL. (2003): Immuno-chemical.In: Roginski H.: Encyclopedia of Dairy Sciences. Vol. 1. Elsevier Science Ltd., London: 62-66.*

9. *XU Fei, Ren Kang, Yang Yu-ze, Guo Jiang-peng, MA Guang-peng, LIU Yi-ming, LU Yong-qiang and LI Xiu-bo, Food Safety Special Issue: Immunoassay of Chemical Contaminants1 in Milk: A Review, Journal of Integrative Agriculture Advance Online Publication 2015.*

10. *Hernandes T, Dores E, Ribeiro M L, Rossignoli P A, Malm O.2014. Simple method to determine residual cypermethrin and deltamethrin in bovine milk. Journal of the Brazilian Chemical Society, 25, 1656-1661.*

11. *Meneghini L Z, Rubensam G, Bica V C, Ceccon A, Barreto F, Ferrao M F, Bergold A M. 2014. Multivariate optimization for extraction of pyrethroids in milk and validation for GC-ECD and CG-MS/MS analysis. International Journal of Environmental Research and Public Health, 11, 11421-11437.*

12. *Karami-Osboo R, Shojaee M H, Miri R, Kobarfard F, Javidnia K. 2014. Simultaneous determination of six fluoroquinolones in milk by validated QuEChERS-DLLME HPLC-FLD. Analytical Methods, 6, 5632-5638.*

13. *Liu L Q, Luo L J, Suryoprabowo S, Peng J, Kuang H, Xu C L. 2014. Development of an immunochromatographic strip test for rapid detection of ciprofloxacin in milk samples. Sensors, 14, 16785-16798.*

14. *Zhang J, Wang Z H, Mi T J, Wenren L Q, Wen K.2014. A homogeneous fluorescence polarization immunoassay for the determination of cephalexin and cefadroxil in milk. Food Analytical Methods, 7, 879-886.*

15. *Freitas S, Paim A, Silva P. 2014. Development of a LC-IT-TOF MS procedure to quantify veterinary drug residues in milk employing a QuEChERS Approach. Food Analytical Methods, 7, 39-46.*

16. *Sniegocki T, Posyniak A, Gbylik-Sikorska M, Zmudzki J. 2014. Determination of chloramphenicol in milk using a QuEChERS-Based on liquid chromatography tandem mass spectrometry Method. Analytical Letters, 47, 568-578.*

17. *Young M S, Tran K V, Goh E, Shia J C. 2014. A rapid spe-based analytical method for UPLC/MS/MS determination of aminoglycoside antibiotic residues in bovine milk, muscle, and kidney. Journal of Aoac International, 97, 1737-1741.*

18. *Sadeghi N, Oveisi M R, Jannat B, Hajimahmoodi M, Behfar A, Behzad M, Norouzi N, Oveisi M, Jannat B. 2014. Simultaneous measurement of zinc, copper, lead and cadmium in baby weaning food and powder milk by DPASV. Iranian Journal of Pharmaceutical Research, 13, 345-349.*

19. *Wild D.2013. TheImmunoassay Handbook: Theory and Applications of Ligand Binding, ELISA and Related Techniques. Sockton Press, NewYork.*

2. Determination of Veterinary Drug Residues in Raw Cow Milk [1]

The most frequently used immunochemical method for rapid diagnostics of veterinary drug residues is enzyme immunoanalysis (EIA). As an enzyme label, horseradish peroxidase, alkaline phosphatase, glucoseoxidase, pyruvate dehydrogenase and recombinant β-galactosidase are used. These enzymes catalyse the reactions that cause the substrates degradation and form coloured products that can be read spectrophotometrically or visually [2].

LacTec test (β-lactam) is based on the immunobinding principles of the enzyme-linked immunosorbent assay system [2,3,4]. Among other methods is used *e.g.* the FluorophosBeta Screen E.U.test-qualitative enzyme immunoassay based on fluorescence detection for the determination of six β-lactam antibiotics-amoxicillin, ampicillin, cloxacillin, dicloxacillin, oxacillin, penicillin G in raw milk [5]. Parallux (IDEXX Laboratories Inc., USA) can be used to detect different antimicrobial agents. The method is based on competitive fluorescent immuno-assays in glass capillary tubes [6].

Most of immunoassays are normally performed in a laboratory and therefore it takes quite a long time before the outcome of the test is known. For this reason, there is a growing need for on-site screening at the beginning of the food chain using fast and easy to perform test methods. A number of immunochemical tests are commercially available in a kit format for many drugs (card format, one step strip test). A promising approach in the rapid antibiotics detection is the use of a dipstick format (lateral flow devices).

Source

1. *Pavlína Navrátilová, Department of Milk Hygiene and Technology, Faculty of Veterinary Hygiene and Ecology, University of Veterinary and Pharmaceutical Sciences in Brno, Brno, Czech Republic.Vol. 26, No. 6: 393–401 Czech J.Food Sci.*

2. *Botsoglou N. A., Fletouris D.J. (2001): Drug Residues in Foods, Pharmacology, Food Safety, and Analysis.Marcel Dekker, New York.*

3. *Cullor J.S., Van Eenennaam A., Dellinger J., PeraniL., Smith W., Jensen L. (1992): Antibiotic residue assays: Can they be used to test milk from individual cows? Veterinary Medicine, 87: 477-494.*

4. *Cullor J.S. (1993): Antibiotic Residue Tests for Mammary Gland secretions. The Veterinary Clinics of NorthAmerica. Food Animal Practice, 9: 609-620.*

5. *Sternesjo A., Johnsson G. A. (1998): Novel rapid enzyme immunoassay (Fluorophos Beta Screen) for detection of β-lactam residues in ex-farm raw milk. Journal of Food Protection, 61: 808–811.*

6. *Gustavsson E., Sternesjo A. (2004): Biosensor analysis of β-lactams in milk: comparison with microbiological, immunological and receptor-based screening methods.Journal of AOACInternational, 87: 614–620.*

3. Rapid Detection of Foodborne Pathogens [1]

The increased use of immunoassay for rapid detection of microbes is due to the following reasons:

(i) Development of new and highly sensitive essays.

(ii) Mechanical devices to automate tedious steps.

(iii) Techniques to construct predetermined antibodies of specificity (Hybridoma technology).

Polyclonal antibodies contain a collection of antibodies having different cellular origin and therefore somewhat different specificity. The development of Monoclonal antibodies greatly enhanced the field of immunology by providing a constant and reliable source of characterized antibodies. Both Homogenous immunoassay (Marker free) and Heterogeneous assays are being used to detect pathogens.ELISA for pathogens have detection limits ranging from 10^3-$1O^5$ cfu mL^{-1} for whole bacterial cells and few ng mL^{-1} for toxins/protein. Therefore, direct detection of pathogens in food is not possible and enrichment is required for at least 16-24 h.

Source

1. *P.K. Mandal, A.K. Biswas, K. Choi and U.K. Pal, 2011. Methods for Rapid Detection of Foodborne Pathogens: An Overview. American Journal of Food Technology, 6: 87-102.*

1. EIA [1]

In spite of the great variety of enzyme immunoassays (EIA) they can be classified into two groups'analyte-observed' and 'reagent-observed' assays, depending on their reaction principle. The latter are favored by use of monoclonal antibodies and are characterized by a greater sensitivity, a larger measuring range, a lower susceptibility to disturbing influences. They can be used only for detection of macromolecules. For heterogeneous EIAs to be used on laboratory seale, simple adsorption of antigens and antibodies is still recommendable though affinity constants decrease bv at least one order of magnitude and antibody density at the solid phase and analyte binding capacity are not parallel due to increasing steric hindrance. For this reason, the antibody with the higher affinity constant should therefore always be used as solid-phase antibody. Microparticles used as solid phase for heterogeneous assays, due to their very high binding capacity for the analyte and extremely short diffusion distances, guarantee 'one step' assays of only a few minutes. Of the limited number of enzymes suitable as markers in immunoassays, horseradish peroxidase is the enzyme of choice followed by alkaline phosphatase. Although enzyme and enzyme-labelled reagents are detectable by fluorogenic product measuring with a sensitivity, which is 10-1000 times higher than using chrornogenic substrates, the sensitivity of the assays can be inereased only by factor 2-10.Labelling enzymes cannot only be covalently bound to the antibody, but also via anti-enzyme antibodies. Different EIA variants to detect specific antibodies are reviewed. Among them only capture EIAs permit precise isotype analysis of antibodies of a distinct idiotype. Homogeneous EIAs are widely spread for hapten determination but even variants based on proximal linkage are no alternatives to heterogeneous EIAs for determination of macromolecules.Different parameters are defined which permit to assess the quality of an immunoassay and which should be used in routine assays as internal controls in the laboratory.

Immunoassays can be generally divided into unlabelled and labelled ones. Most unlabelled techniques are based on secondary immune reactions, as, *e.g.*, precipitation and agglutination. They are measured by light scattering or partied counting methods. Labelled immunoassays are based on the primary immune reaction. There are two types, depending on the reaction principle:

1. Type I'Reagent-observed'.
2. Type II'Analyte-observed' [2].

Whereas in the type I assay there is excess of antibody as binder, it limits the reaction in the type II assay with excess of analyte. Type II assays have got a maximum theoretical sensitivity of about 10- 14 moljl (Ekins, 1981b) and are therefore one to two orders of magnitude less sensitive than type I assays, The maximum sensitivity of type I assays, which are also called 'two site assay' or 'sandwich assay', ranges between10- 15 and 10- 16 moljl [3].

Unlike type II assays, which are used to determine both haptens and substances of high molecular weight, type I assays require substances to be used as analyte with at least two different epitopesd early distinct from each other in order to permit epitope-different labelled and unlabelled antibodies to bind without steric hindrance. Type I assays are thus unsuitable for hapten quantification.

The different characteristics of type I and type II assays are listed in following Table. The decision to use one or the other assay principle is mainly influenced by the following aspects:

Characteristics of Type I 'Reagent-Observed' and Type II 'Analyte-Observed' [4]

Type I 'Reagent-Observed' (Excess antibody)	Type II' Analyte Observed' (Excess analyte)
(i) Maximal sensitivity is attained as amount of antibody approaches infinity,	*(i)* Maximal sensitivity is attained when antibody concentration approaehes zero.
(ii) Theoretical sensitivity of assay is one moleeule of analyte,	*(ii)* This saturation assay is regulated by the equilibrium constant of the reaction between analyte and antibody.
(iii) Cross-reaction antigens will be equipotent with exeess antibody system.	*(iii)* Sensitivity of the assay is dependent upon the affinity constant of the antibody.
(iv) Antigen-antibody reaction is less influenced by substances such as salt and urea.	*(iv)* Cross-reactive antigen will demonstrate a relative potency dependent upon the rate of the equilibriurn constants of the analyte and cross-reactive antigen.
(v) Assay time is relatively rapid with labelled antibody procedures.	*(v)* Assay reaction is slow since equilibrium must be reached.

Source

1. *T.Porstmann and S. T.Kiessig, Enzyme immunoassay techniques, An overview, Journal of Immunological Methods, 150 (1992) 5-21.*
2. *Ekins, R. (1981a) toward immunoassays of greater sensitivity, specificity and speed: an overview.in: a.albertini and R.ekins (eds.), monoclonal antibodies and developrnents in Immunoassays, Elsevier Amsterdam, p.3.*
3. *Jackson, T., Marshall, N. and Ekins, R. (1983) Optimization of immunoradiometric (labelled antibody) assays, In: W.M.Hunter and J.E. T. Corrie (Eds.), Immuno-assays for Clinical Chemistry Churchill Livingstone, Edinburgh, 557.*
4. *Nakamura, R.M., Voller, A. and Bidwell, D.E., 1986.*

2. ELISA [16]

The **enzyme-linked immunosorbent assay (ELISA)** is a test that uses antibodies and color change to identify a substance. **ELISA** (Enzyme linked immunosorbent assay) is a technique in which a substance (antigen, antibody) is detected by its ligand conjugated to an enzyme which produces a color reaction by acting on a chromogenic substrate. It is a popular format of "wet-lab" type analytic biochemistry assay that uses a **solid-phase enzyme immuno-assay** (**EIA**) to detect the presence of a substance, usually an antigen, in a liquid sample or wet sample. An immunoassay is a test that uses antibody and antigen complexes as a means of generating a measurable result. An antibody: antigen complex is also known as an immuno-complex. The ELISA has been used as a diagnostic tool in medicine and plant pathology, as well as a quality-control check in various industries including food industries. Today many ELISA tests are available as robotized automated systems not only to reduce the hands-on time but also to improve the reproducibility and standardization of each step of the assay.

Antigens from the sample are attached to a surface. Then, a further specific antibody is applied over the surface so it can bind to the antigen. This antibody is linked to an enzyme, and, in the final step, a substance containing the enzyme's substrate is added. The subsequent reaction produces a detectable signal, most commonly a color change in the substrate.

Performing an ELISA involves at least one antibody with specificity for a particular antigen. The sample with an unknown amount of antigen is immobilized on a solid support (usually a polystyrene microtiter plate) either non-specifically (via adsorption to the surface) or specifically (via capture by another antibody specific to the same antigen, in a "sandwich" ELISA). After the antigen is immobilized, the detection antibody is added, forming a complex with the antigen. The detection antibody can be covalently linked to an enzyme, or can itself be detected by a secondary antibody that is linked to an enzyme through bioconjugation. Between each step, the plate is typically washed with a mild detergent solution to remove any proteins or antibodies that are non-specifically bound. After the final wash step, the plate is developed by adding an enzymatic substrate to produce a visible signal, which indicates the quantity of antigen in the sample.

ELISA can perform other forms of ligand binding assays instead of strictly "immuno" assays, though the name carried the original "immuno" because of the common use and history of development of this method. The technique essentially requires any ligating reagent that can be immobilized on the solid phase along with a detection reagent that will bind specifically and use an enzyme to generate a signal that can be properly quantified. In between the washes, only the ligand and its specific binding counter parts remain specifically bound or "immunosorbed" by antigen-antibody interactions to the solid phase, while the nonspecific or unbound components are washed away. **Unlike other spectrophotometric wet lab assay formats where the same reaction well (*e.g.* a cuvette) can be reused after washing, the ELISA plates have the reaction products immunosorbed on the solid phase which is part of the plate, and so are not easily reusable.**

Principle [16]

As an analytic biochemistry assay, ELISA involves detection of an "analyte" (*i.e.* the specific substance whose presence is being quantitatively or qualitatively analyzed) in a liquid sample by a method that continues to use liquid reagents during the "analysis" (*i.e.* controlled sequence of biochemical reactions that will generate a signal which can be easily quantified and interpreted as a measure of the amount of analyte in the sample) that stays liquid and remains inside a reaction chamber or well needed to keep the reactants contained; It is opposed to "dry lab" that can use dry strips-and even if the sample is liquid (*e.g.* a measured small drop), the final detection step in "dry" analysis involves reading of a dried strip by methods such as reflectometry and does not need a reaction containment chamber to prevent spillover or mixing between samples.

As a heterogenous assay, ELISA separates some component of the analytical reaction mixture by adsorbing certain components onto a solid phase which is physically immobilized. In ELISA, a liquid sample is added onto a stationary solid phase with special binding properties and is followed by multiple liquid reagents that are sequentially added, incubated and washed followed by some optical change (*e.g.* color development by the product of an enzymatic reaction) in the final liquid in the well from which the quantity of the analyte is measured. The qualitative "reading" usually based on detection of intensity of transmitted light by spectrophotometry, which involves quantitation of transmission of some specific wavelength of light through the liquid (as well as the transparent bottom of the well in the multiple-well plate format). The sensitivity of detection depends on amplification of the signal during the analytic reactions. Since enzyme reactions are very well known amplification processes, the signal is generated by enzymes which are linked to the detection reagents in fixed proportions to allow accurate quantification-thus the name "enzyme linked".

The analyte is also called the ligand because it will specifically bind or ligate to a detection reagent, thus ELISA falls under the bigger category of ligand binding assays. The ligand-specific binding reagent is "immobilized", *i.e.*, usually coated and dried onto the transparent bottom and sometimes also side wall of a well (the stationary "solid phase'/"solid substrate" here as opposed to solid microparticle/ beads that can be washed away), which is usually constructed as a multiple-well plate known as the "ELISA plate". Conventionally, like other forms of immunoassays, the specificity of antigen-antibody type reaction is used because it is easy to raise an antibody specifically against an antigen in bulk as a reagent. Alternatively, if the analyte itself is an antibody, its target antigen can be used as the binding reagent.

History [16]

Before the development of the ELISA, the only option for conducting an immunoassay was radioimmunoassay, a technique using radioactively labeled antigens or antibodies. In radioimmunoassay, the radioactivity provides the signal, which indicates whether a specific antigen or antibody is present in the sample. Radioimmunoassay was first described in a scientific paper by Rosalyn Sussman Yalow and Solomon Berson published in 1960 [1].

Because radioactivity poses a potential health threat, a safer alternative was sought. A suitable alternative to radioimmunoassay would substitute a nonradioactive signal in place of the radioactive signal.When enzymes (such as horseradish peroxi-dase) react with appropriate substrates (such as ABTS or TMB), a change in color occurs, which is used as a signal. However, the signal has to be associated with the presence of antibody or antigen, which is why the enzyme has to be linked to an appropriate antibody. This linking process was independently developed by Stratis Avrameas and G.B.Pierce [2]. In the end of the 60s, **Stratis Avrameas** (National Scientific Research Centre) and **G.B Pierce** have developed a reaction **analysis technique between antigens and antibodies, using markers enzymes**. This technique **enzyme immunoassay** is safer than the radioimmunoassay method that uses radioactivity to cause a chemical reaction [18]. Since it is necessary to remove any unbound antibody or antigen by washing, the antibody or antigen has to be fixed to the surface of the container; *i.e.*, the immunosorbent must be prepared. A technique to accomplish this was published by Wide and Jerker Porath in 1966 [3].

In 1971, **Peter Perlmann and Eva Engvall** at Stockholm University in Sweden, and Anton Schuurs and Bauke van Weemen in the Netherlands independently published papers that synthesized this knowledge into methods to perform EIA/ ELISA [4,5].

Advantages of ELISA [18]

This test has been used worldwide. This has following many advantages [18]:

(i) Its accessibility for all biologists.

(ii) A more specific detection through the use of Monoclonal Antibodies.

(iii) Achieving range in parallel which allows quantification.

(iv) The technique is sensitive due to the use of antibodies.

(v) A specialized apparatus is not necessary for the detection of a signal.

Types of ELISA [16, 18]

Direct, Indirect, Sandwich and Competitive ELISA are generally being used for different purposes including food analysis. The details of aforesaid types of ELISA are given below:

1. Direct ELISA [16]

The steps of direct ELISA follows the mechanism given below:

(i) A buffered solution of the antigen to be tested for is added to each well of a microtiter plate, where it is given time to adhere to the plastic through charge interactions.

(ii) A solution of nonreacting protein, such as bovine serum albumin or casein, is added to well (usually 96-well plates) in order to cover any plastic surface in the well which remains uncoated by the antigen.

(iii) The primary antibody with an attached (conjugated) enzyme is added, which binds specifically to the test antigen coating the well.

(iv) A substrate for this enzyme is then added. Often, this substrate changes color upon reaction with the enzyme and v-The higher the concentration of the primary antibody present in the serum, the stronger the color change. Often, a spectrometer is used to give quantitative values for color strength.

The enzyme acts as an amplifier; even if only few enzyme-linked antibodies remain bound, the enzyme molecules will produce many signal molecules. Within common-sense limitations, the enzyme can go on producing color indefinitely, but the more antibody is bound, the faster the color will develop. A major disadvantage of the direct ELISA is the method of antigen immobilization is not specific; when serum is used as the source of test antigen, all proteins in the sample may stick to the microtiter plate well, so small concentrations of analyte in serum must compete with other serum proteins when binding to the well surface. The sandwich or indirect ELISA provides a solution to this problem, by using a "capture" antibody specific for the test antigen to pull it out of the serum's molecular mixture.

ELISA may be run in a qualitative or quantitative format. Qualitative results provide a simple positive or negative result (yes or no) for a sample. The cutoff between positive and negative is determined by the analyst and may be statistical. Two or three times the standard deviation (error inherent in a test) is often used to distinguish positive from negative samples. In quantitative ELISA, the optical density (OD) of the sample is compared to a standard curve, which is typically a serial dilution of a known-concentration solution of the target molecule. For example, if a test sample returns an OD of 1.0, the point on the standard curve that gave OD = 1.0 must be of the same analyte concentration as the sample.

The use and meaning of the names "direct ELISA" and "indirect ELISA" differs in the literature and on web sites depending on the context of the experiment. When the presence of an antigen is analyzed, the name "direct ELISA" refers to an ELISA in which only a labelled primary antibody is used, and the term "indirect ELISA" refers to an ELISA in which the antigen is bound by the primary antibody which then is detected by a labeled secondary antibody. In the latter case a sandwich ELISA is clearly distinct from an indirect ELISA. When the "primary" antibody is of interest, *e.g.* in the case of immunization analyses, this antibody is directly detected by the secondary antibody and the term "direct ELISA" applies to a setting with two antibodies.

2. Indirect ELISA [18]

The indirect ELISA test is used when searching for particular antibodies. Thereof may be made quantitative sight (for the various protein assay) or qualitative (indicating the presence or absence of an antigen in the sample) [18].

The assay is usually carried out in two stages. Like Direct ELISA, the antigen is coated to a polystyrene multiwell plate. In the first stage, unlabeled primary antibody is introduced into the well which is specific to an antigen. In the second stage, enzyme labeled secondary antibody (often polyclonal antibody) is introduced to the well [19].

Advantages [18a]

- *(i)* Increased sensitivity, since more than one labeled antibody is bound per primary antibody.
- *(ii)* wide variety of labeled secondary antibodies are available commercially.
- *(iii)* Maximum immunoreactivity of the primary antibody is retained because it is not labeled.
- *(iv)* Versatile because many primary antibodies can be made in one species and the same labeled secondary antibody can be used for detection.
- *(v)* Flexibility, since different primary detection antibodies can be used with a single labeled secondary antibody.
- *(vi)* Cost savings, since fewer labeled antibodies are required.
- *(vii)* Different visualization markers can be used with the same primary antibody.
- *(viii)* The experimental procedure can be made flexible as per the demands of the study given the stages in Indirect ELISA [18].

Disadvantages [18a]

(i) Cross-reactivity might occur with the secondary antibody, resulting in nonspecific signal.

(ii) An extra incubation step is required in the procedure.

Procedure

It involves the following steps [19]:

(i) The antigen is diluted to a concentration of 20 μg/ml in PBS. The wells of a PVC microtiter plate are coated with the antigen by pipetting 50 μl of the antigen dilution in the top wells of the plate.

(ii) 200 μl blocking buffer is added to block the remaining protein-binding sites in the coated wells with 5 per cent non-fat dry milk.

(iii) The plate is covered with adhesive plastic and incubated for at least 2 hr at room temperature.

(iv) The plate is washed twice with PBS.

(v) 100 μl of diluted primary antibody is added to each well.

(vi) The plate is covered with an adhesive plastic and incubated for 2 hours at room temperature.

(vii) The plates should be washed with PBS four times.

(viii) 100 μl of conjugated secondary antibody diluted at optimal concentration in blocking buffer is added.

(ix) The plate is covered with adhesive plastic and incubated for 1-2 hours at room temperature

(x) Again, the plates are washed with PBS four times.

3. Sandwich ELISA [16,18]

ELISA is a biochemical technique that couples an immunoassay with an enzyme assay. In most of the alternative methods, a sandwich ELISA is used. The assay is highly efficient in sample antigen detection and quantification. Antigen can be detected by sandwich ELISA. In this technique, antibody is coated on the microtiter well. A sample containing antigen is added to the well and allowed to react with the antibody attached to the well, forming antigen-antibody complex. After the well is washed, a second enzyme-linked antibody specific for a different epitope on the antigen is added and allowed to react with the bound antigen. Then after unbound secondary antibody is removed by washing. Finally substrate is added to the plate which is hydrolyzed by enzyme to form colored products] 18a].

The assay is suitable for antigens that contain at least two antigenic epitopes capable of binding to antibody [19]. This highly specific technique involves isolating a protein to be assayed between two antibodies against it but recognizing different epitopes. Concretely, the first antibody is fixed on a support, followed by incubation of the protein to be assayed. Finally, the second antibody comes recognizing the protein/first antibody, then we say that the protein is sandwiched between the two antibodies, hence the name of this method. An increasingly strong staining indicates an increasing concentration of antigen [18].

Both monoclonal and polyclonal antibodies are widely used as capturing and detection antibodies in Sandwich ELISA systems. The quantification of antigens happens between two layers of antibodies and therefore the method is termed as sandwich ELISA [19].

Advantages [18a]

(i) High specificity, since two antibodies are used the antigen is specifically captured and detected. The assay is more sensitive and specific than direct or Indirect ELISA *i.e.* up to 2 to 5 times.

(ii) Suitable for complex samples, since the antigen does not require purification prior to measurement..

(iii) Flexibility and sensitivity, since both direct and indirect detection methods can be used.

Difficulty

The difficulty in sandwich ELISA lies in optimization and selecting tested pair antibodies. This ensures antibodies chosen are detecting different epitopes on the target protein without interfering with other antibody binding.

Procedure [16]

In a sandwich ELISA:

(i) Plate is coated with a capture high precision assay, the sample need not be purified before analysis [19], antibody.

(ii) Sample is added, and any antigen present binds to capture antibody.

(iii) Detecting antibody is added, and binds to antigen.

(iv) Enzyme-linked secondary antibody is added, and binds to detecting antibody.

(v) Substrate is added, and is converted by enzyme to detectable form.

A "sandwich" ELISA is used to detect sample antigen [9]. The following steps are needed to perform the test:

(i) A surface is prepared to which a known quantity of capture antibody is bound.

(ii) Any nonspecific binding sites on the surface are blocked.

(iii) The antigen-containing sample is applied to the plate, and captured by antibody.

(iv) The plate is washed to remove unbound antigen..

(v) A specific antibody is added, and binds to antigen (hence the 'sandwich': the Ag is stuck between two antibodies). This primary antibody could also be in the serum of a donor to be tested for reactivity towards the antigen.

(vi) Enzyme-linked secondary antibodies are applied as detection antibodies that also bind specifically to the antibody's Fc region (nonspecific).

(vii) The plate is washed to remove the unbound antibody-enzyme conjugates.

(viii) A chemical is added to be converted by the enzyme into a color or fluorescent or electrochemical signal.

(ix) The absorbance or fluorescence or electrochemical signal (*e.g.*, current) of the plate wells is measured to determine the presence and quantity of antigen.

The image to the right includes the use of a secondary antibody conjugated to an enzyme, though, in the technical sense, this is not necessary if the primary antibody is conjugated to an enzyme (which would be direct ELISA). However, the use of a secondary-antibody conjugate avoids the expensive process of creating enzyme-linked antibodies for every antigen one might want to detect. By using an enzyme-linked antibody that binds the Fc region of other antibodies, this same enzyme-linked antibody can be used in a variety of situations. Without the first layer of "capture" antibody, any proteins in the sample (including serum proteins) may competitively adsorb to the plate surface, lowering the quantity of antigen immobilized. Use of the purified specific antibody to attach the antigen to the plastic eliminates a need to purify the antigen from complicated mixtures before the measurement, simplifying the assay, and increasing the specificity and the sensitivity of the assay. A sandwich ELISA used for research often need validation because of the risk of false positive results [10].

4. Competitive ELISA [18]

The competition takes place between the labeled antigens (in known quantity) and unmarked (in an amount to be determined) for binding to antibodies, which are in default. Thus, the more there are antigens to measure, the more their proportion amid the antigens identified by the antibody is, the more the signal will be low. Conversely, if the initial concentration of the antigen is low, the signal will be strong [18].

This test is used to measure the concentration of an antigen in a sample. In this test, antibody is first incubated in solution with a sample containing antigen. The antigen-antibody mixture is then added to the microtitre well which is coated with antigen. The more the antigen present in the sample, the less free antibody will be available to bind to the antigen-coated well. After the well is washed, enzyme conjugated secondary antibody specific for isotype of the primary antibody is added to determine the amount of primary antibody bound to the well. The higher the concentration of antigen in the sample, the lower the absorbance [18a]

Procedure [16]

A third use of ELISA is through competitive binding. The steps for this ELISA are some what different from the first two examples of the ELISA. It involves the following steps:

(i) Unlabeled antibody is incubated in the presence of its antigen (sample). These bound antibody/antigen complexes are then added to an antigen-coated well.

(ii) The plate is washed, so unbound antibodies are removed. (The more antigen in the sample, the more Ag-Ab complexes are formed and so there are less unbound antibodies available to bind to the antigen in the well, hence "competition".).

(iii) The secondary antibody, specific to the primary antibody, is added. This second antibody is coupled to the enzyme.

(iv) A substrate is added, and remaining enzymes elicit a chromogenic or fluorescent signal.

(v) The reaction is stopped to prevent eventual saturation of the signal.

Some competitive ELISA kits include enzyme-linked antigen rather than enzyme-linked antibody. The labeled antigen competes for primary antibody binding sites with the sample antigen (unlabeled). The less antigen in the sample, the more labeled antigen is retained in the well and the stronger the signal .Commonly, the antigen is not first positioned in the well.

For the detection of HIV antibodies, the wells of microtiter plate are coated with the HIV antigen. Two specific antibodies are used, one conjugated with enzyme and the other present in serum (if serum is positive for the antibody).Cumulative competition occurs between the two antibodies for the same antigen, causing a stronger signal to be seen. Sera to be tested are added to these wells and incubated at 37 °C, and then washed. If antibodies are present, the antigen-antibody reaction occurs. No antigen is left for the enzyme-labelled specific HIV antibodies. These antibodies remain free upon addition and are washed off during washing. Substrate is added, but there is no enzyme to act on it, so a positive result shows no color change.

The stepwise details related to the procedures of Incomplete, Sandwich and Competitive ELISA are given below [18a]:

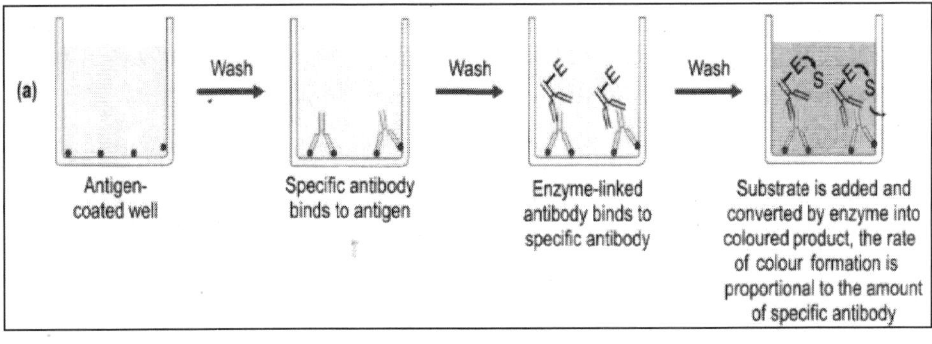

Indirect ELISA

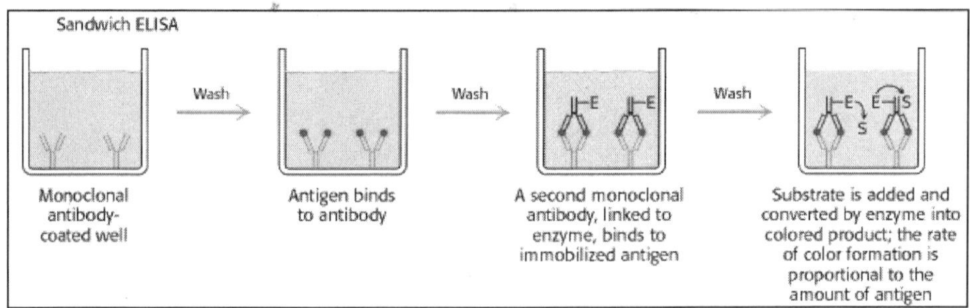

Sandwich Elisa

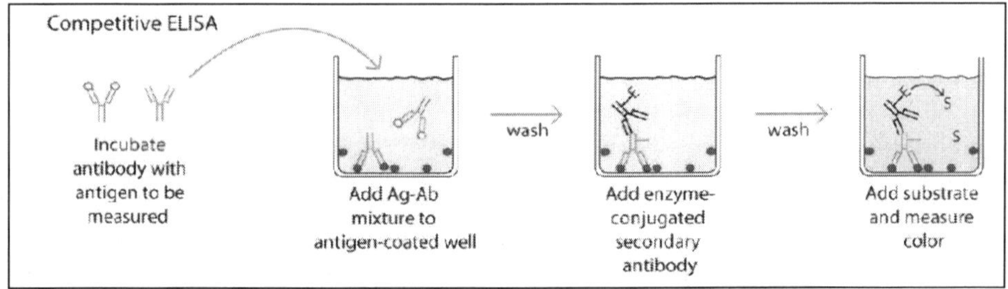

Competative Elisa

Detection [19]

ELISA (as a biochemical technique) is mainly used to detect the presence of an antigen or antibody in a sample. Therefore, detection occupies an important role in ELISA. In simple terms, the final step in ELISA always involves adding a signaling agent to the enzyme to produce a visible signal *i.e.* color change. Based on detection strategies, ELISAs are classified into the following three groups:

i. Chromogenic Assays

Chromogenic assays result in colored reaction product that absorbs in light in the visible spectrum and therefore visible to naked eyes. The antigen-antibody complex reacts with the substrate of choice giving rise to colored reaction product. Like Chemifluorescent assays, the color changes are directly proportional to the amount of the analyte. Some absorbance Readout are:

(i) HRP TMB: 450 nm.

(ii) OPD: 490 nm.

(iii) ABTS: 405 nm.

ii. Chemifluorescent Assays

Some enzymes on addition to the substrate produces a reaction that fluoresces when light particles are emitted at a specific wavelength. When these enzymes find utility in ELISAs, those assays are called Fluorescent immunoassays. Fluorescence units detected are directly proportional to the quantity of analyte in sample. Some Fluorescent Readout are:

(*i*) QuantaBlu 420/325.

(*ii*) QuantaRed 585/570.

(*iii*) FITC 518/494.

iii. Chemiluminescent Assays

An immunoassay technique where the antibody or the antigen is labeled with a molecule capable of emitting light during a chemical reaction. For detection, a luminescent plate reader is required (available from PerkinElmer).

Advancement in ELISA

Traditional ELISA typically involves chromogenic reporters and substrates that produce some kind of observable color change to indicate the presence of antigen or analyte. Newer ELISA-like techniques use fluorogenic, electrochemi-luminescent, and quantitative PCR reporters to create quantifiable signals. These new reporters can have various advantages, including higher sensitivities and multiplexing [6,7]. In technical terms, newer assays of this type are not strictly ELISAs, as they are not "enzyme-linked", but are instead linked to some nonenzymatic reporter. However, given that the general principles in these assays are largely similar, they are often grouped in the same category as ELISAs [16].

To further enhance assay sensitivity, the following new methods have been developed on the basis of ELISA **[7a]:**

(*i*) Chemiluminescence assays (CIA) are efficient non-radioactive detection methods using a luminescent substrate to replace the ELISA chromogenic substrate. This method has high sensitivity, stable luminescent markers, a wide detection range and high automation. Therefore, CIA has been widely used in food safety tests. Fluorescent labels can also be used to increase the sensitivity of the methods.

(*ii*) The major methods via fluorescence include time-resolved immunoassay method (TRFIA), fluorescence polarization analysis (FPIA), liposome immunoassay (LIA) and fluorescent-linked immunosorbent assay (FLISA). Of these, it is vital to select a tracer that is efficient, highly sensitive, and easy to label. Fluorescent markers are mostly organic fluorescent dyes such as fluorescein, cyanine dyes, fluorescent marker proteins *e.g.* phycoerythrin, green fluorescent proteins, metal complex markers such as Eu3+, Tb3+, noble metal ion complexes, *etc.*, nano-particles (*e.g.* quantum dots), fluorescent microsphere and carbon nanodots. These fluorescent markers not only enhance the sensitivity of the current detection methods, but also provide unparalleled flexibility for labeling.

(*iii*) In 2012 an ultrasensitive, enzyme-based ELISA test using nanoparticles as a chromogenic reporter was able to give a naked-eye colour signal from the detection of mere attograms of analyte. A blue color appears for positive results and red color for negative. Note that this detection only can confirm the presence or the absence of analyte not the actual concentration [8].

Applications [16]

ELISA can be performed for the following analysis:

A. Toxicological/Diseases and Diagonostic Analysis

i. Determination of Antibody Concentrations [16]

Because the ELISA can be performed to evaluate either the presence of antigen or the presence of antibody in a sample, it is a useful tool for determining serum antibody concentrations (such as with the HIV test [11] or West Nile virus). The ELISA was the first screening test widely used for HIV because of its high sensitivity. In an ELISA, a person's serum is diluted 400 times and applied to a plate to which HIV antigens are attached. If antibodies to HIV are present in the serum, they may bind to these HIV antigens. The plate is then washed to remove all other components of the serum. A specially prepared "secondary antibody" (an antibody that binds to other antibodies) is then applied to the plate, followed by another wash. This secondary antibody is chemically linked in advance to an enzyme.

Thus, the plate will contain enzyme in proportion to the amount of secondary antibody bound to the plate. A substrate for the enzyme is applied, and catalysis by the enzyme leads to a change in color or fluorescence. ELISA results are reported as a number; the most controversial aspect of this test is determining the "cut-off" point between a positive and a negative result. A cut-off point may be determined by comparing it with a known standard. If an ELISA test is used for drug screening at workplace, a cut-off concentration, 50 ng/ml, for example, is established, and a sample containing the standard concentration of analyte will be prepared. Unknowns that generate a stronger signal than the known sample are "positive." Those that generate weaker signal are "negative".

ii. Toxicological Analysis [16]

ELISA can also be used in toxicology as a rapid presumptive screen for certain classes of drugs.

iii. Detection of Diseases

Dr Dennis E Bidwell and Alister Voller created the ELISA test to detect various kind of diseases, such as Malaria, Chagas disease, and Johne's disease [15].

iv. Diagnostics in Medical Laboratories [16]

ELISA tests also are used as in in vitro diagnostics in medical laboratories. The other uses of ELISA include:

- *(i)* Detection of Mycobacterium antibodies in tuberculosis.
- *(ii)* Detection of rotavirus in feces.
- *(iii)* Detection of hepatitis B markers in serum.
- *(iv)* Detection of enterotoxin of E. coli in feces.
- *(v)* Detection of HIV antibodies in blood samples and vi-as serological blood test for coeliac disease [13,14].

B. Food Analysis

It has also found applications in the **food industry** in detecting potential food allergens, such as milk, peanuts, walnuts, almonds, and eggs [12].

Source

1. *Yalow, Rosalyn S.; Berson, Solomon A. (1960). "Immunoassay of endogenous plasma insulin in man". The Journal of Clinical Investigation. 39: 1157–75.*
2. *Lequin, R. M. (2005). "Enzyme Immunoassay (EIA)/Enzyme-Linked Immunosorbent Assay (ELISA)". Clinical Chemistry. 51 (12): 2415-8.*
3. *Wide, Leif; Porath, Jerker (1966)."Radioimmunoassay of proteins with the use of Sephadex-coupled antibodies". Biochimica et Biophysica Acta. 130 (1): 257–60.*
4. *Engvall, Eva; Perlmann, Peter (1971)."Enzyme-linked immunosorbent assay (ELISA) quantitative assay of immunoglobulin G". Immunochemistry. 8 (9): 871-4.*
5. *Van Weemen, B.K.; Schuurs, A.H.W.M. (1971)."Immunoassay using antigen-enzyme conjugates".FEBS Letters. 15 (3): 232–236.*
6. *Leng, S. X.; McElhaney, J. E.; Walston, J.D.; Xie, D.; Fedarko, N.S.; Kuchel, G. A. (2008)."ELISA and Multiplex Technologies for Cytokine Measurement in Inflammation and Aging Research". The Journals of Gerontology Series A: Biological Sciences and Medical Sciences. 63 (8): 879–84.*
7. *Adler, Michael; Schulz, Sven; Spengler, Mark (2009). "Cytokine Quantification in Drug Development: A comparison of sensitive immunoassay platforms". Chimera Biotech.*

7a. *XU Fei, Ren Kang, Yang Yu-ze, Guo Jiang-peng, MA Guang-peng, LIU Yi-ming, LU Yong-qiang and LI Xiu-bo, Food Safety Special Issue: Immunoassay of Chemical Contaminants1 in Milk: A Review, Journal of Integrative Agriculture Advance Online Publication 2015.*

8. *de la Rica, Roberto; Stevens, Molly M. (2012)."Plasmonic ELISA for the ultra-sensitive detection of disease biomarkers with the naked eye".Nature Nanotechnology.7 (12): 821–4.*
9. *Schmidt, SD; Mazzella, MJ; Nixon, RA; Mathews, PM (2012). "Aβ measurement by enzyme-linked immunosorbent assay". Methods in Molecular Biology. 849: 507–27.*
10. *Kragstrup, Tue W; Vorup-Jensen, Thomas; Deleuran, Bent; Hvid, Malene (2013). "A simple set of validation steps identifies and removes false results in a sandwich enzyme-linked immunosorbent assay caused by anti-animal IgG antibodies in plasma from arthritis patients".SpringerPlus.2 (1): 263.*
11. *Medline Plus Encyclopedia ELISA/Western blot tests for HIV.*
12. *"Food Allergen Partnership " (Press release). FDA. January 2001. Retrieved August 20, 2015.*
13. *Sblattero, D.; Berti, I.; Trevisiol, C.; Marzari, R.; Tommasini, A.; Bradbury, A.; Fasano, A.; Ventura, A.; Not, T. (2000)."Human recombinant tissue transglutaminase*

ELISA: an innovative diagnostic assay for celiac disease". The American Journal of Gastroenterology. 95 (5): 1253–7.

14. *Porcelli, Brunetta, Ferretti, Fabio, Vindigni, Carla, Terzuoli, Lucia (2014)."Assessment of a Test for the Screening and Diagnosis of Celiac Disease". Journal of Clinical Laboratory Analysis.*
15. *Griffin, J.F. T., Spittle, E.; Rodgers, C.R., Liggett, S., Cooper, M.; Bakker, D. Bannantine, J.P. (2005)."Immunoglobulin G1 Enzyme-Linked Immunosorbent Assay for Diagnosis of Johne's Disease in Red Deer (Cervus elaphus)".Clinical and Vaccine Immuno-logy. 12 (12): 1401–9.*
16. *Wikipedia.*
17. *H.J.Lee, M.H.Lee and In K. Han, College of Veterinary Medicine, Seoul National University, Suwon, Korea.*
18. *Interchim -Blog-Bio Chromatography-ELISA Kits: The solution to immunoassay methods.*

18a. *ELISA- Principle, Types and Applications 4.61/5 (354), https://www.thermofisher.com.*

19. *Elisa testing and applications.My Biosource.com.*

C. Application of ELISA Methods in Dairy Industry

1. Detection of Antibiotics in Milk and in Animal Tissues

a. Detetion of Oxytetraycline Residues in Animals [1]

Oxytetraycline has been widely used to control pneumonia, footrot shipping fever, bacterial enteritis and uterine infection in cattle. Excess and Extensive application of antibiotics in treatment of milch animals has resulted in residues in tissue and bacterial resistsnce to antibiotics. To check the drug residue in milk (human food chain), extensive control measures have been established by Government authorites and Food Industries. A increasing demand for reliable, simple, sensitive, rapid and low cost method for determination of antibiotic residues in food (milk) is found to operate the aforesaid control measures effectively. In the modified ELISA method, the rcommended therapeutic dose of Oxytetraycline (Withdrawal time,14 days) was administered to cattle, blood samples were collected before drug administration and during withdrawal period, concentration of and oxytetraycline in plasma is determined by semi quantitative ELISA method and result s were compared with internal standard (10ppb). The absorbane ratio of internal standard to sample (B/Bs) was employed as an index to determine whether the residues in cattle tissues were negative or positive.If B/Bs ratio is less than 1, then positive and if it is more than 1 then negative.

Source

1. *H.J.Lee, M.H.Lee and In K.Han, Appliation of Elisa for detection of Oxy-tetracyline residues in Live animals, College of Veterinary medine, Seoul National University, Seoul. Asian- Aus. J. Anim. Sci.2000.Vol.13, No, 12, 1775-1778.*

b. Detection of other Antibiotics in Milk [1]

The 'traditional' tests for antibiotics in milk, known as **'microbial inhibitor'** tests, involve incubating a susceptible organism in the presence of the milk sample. The Delvotest (Gist-brocades BV, The Netherlands) is the best known microbial inhibitor test The first version to be developed, in the 1970s, was the Delvotest P, designed to detect ß-lactams but recently Delvotest SP (capable of detecting a wider spectrum of substances, notably sulphonamides), Delvotest Cow Test (introduced into the UK in 1994) for testing individual animals and Delvotest MCS test are being used.

The following **immunoassay methods** are also available:

(i) The most widely used immune receptor test is the LacTek test for ß-lactams (Idexx Laboratories, USA).

(ii) The Delvo X-Press ß-lactam II test (Gist-brocades BV, The Netherlands employs the same principle as the LacTek test.

(iii) The SNAP ß-L test (Idexx Laboratories, USA) was the first of a growing number of tests to employ capillarity to draw the milk sample and test reagents over an immobilised antibody. It consists of a test tube and a disposable plastic unit or 'SNAP' device that contains the test reagents.

(iv) The Beta Screen test (Advanced Instruments, USA) is an 'add-on' to the Fluorophos test forphosphatase used in many dairy laboratories and employs a fluorescent end-point. The milk sample and enzyme conjugate are first incubated in an antibody-coated test tube, placed in a heating block. After washing the tube, fluorescence is developed by the addition of reagents and further incubation; finally, the fluorescence intensity is measured in the Fluorophos fluorimeter and compared with that of a penicillin 'standard'. The Beta Screen test takes 10 min.; it is highly specific for penicillins and does not detect cephalosporins.

(v) The Charm II assay (Charm Sciences Inc., USA) is not a single test but a family of separate tests for specific groups of antibiotics, notably ß-lactams, sulphonamides, tetracyclines, novobiocin, aminoglycosides and macrolides, as well as various other substances such as chloramphenicol.

(vi) The Parallux test (Idexx Laboratories, USA) is designed to detect a range of antimicrobials as well as other substances of interest to the dairy industry.

Source

1. *Paul Neaves, Williams and Neaves, The Food Microbiologists, "Moleview", 28, Randalls Road, Leatherhead, Surrey, KT22 7TQ, Monitoring antibiotics in milk - The Changing world of test methods.*

2. Detetion of Allergens [1]

Regulatory authority has required manufacturers to label their products with regards to eight specific allergens-Milk, Peanuts, Wheat, Soy-beans, Tree Nuts, Eggs, Fish and Shellfish. These allergens are responsible for over 90 per cent of the documented food allergen-related cases. Milk is a source of food allergies and must be excluded from the diet of susceptible individuals. Milk protein is comprised of casein and whey proteins. These proteins are present together in whole milk but they may also be present individually in some products. Companies producing non- dairy products should always ensure their raw materials, finished products and environmental surfaces have not been accidentally contaminated with milk during manufacture. The following Elisa methods are being used for detection of milk allergen:

i In food industry, the two most common and preferred methods for the detection of allergens are **ELISA and PCR**. These techniques can detect levels of these contaminants at concentrations in the low parts per million (ppm) range. These techniques detect the food allergen at the molecular level and provide a quick and definitive result that allows manufacturers to dispose of or re-label contaminated products before they are released. It also alerts them to areas of their processing facilities that need to be decontaminated or to production lines that need to be used for other products.

ELISA methods detect the actual allergen protein molecule by binding antibodies to the allergen and then using an enzyme-linked conjugate to create a colorimetric change that can be measured. There are certain instances though, that ELISA methods should not be used. Some matrices can interfere with the ELISA method, such as chocolate, or can cause cross reactivity as seen between different types of nuts. This method is also not the most suitable for cooked or heated products because the protein molecules are denatured or broken down and the allergen is no longer detectable, but may still cause problems to sensitive individuals.

Source

1 *Sino Biological Inc., Elisa encyclopedia.*

ii. Elisa Systems Total Milk Kit

The new kits are available to detects both alpha-S casein as well as Beta-lactoglobulin to make screening easy. Elisa Systems total milk kit is a sensitive, yet simple to use, kit for the detection of unwanted milk which can prevent the possible allergic response in susceptible individuals. The Elisa Systems Total Milk Residue kit is a rapid test which detects both casein and whey proteins. It can be used to detect contamination of foods by milk, casein or whey. The Elisa systems Total Milk Residue assay is an enzyme-linked immuno-sorbent assay (ELISA) that may be used to screen food products and environmental surfaces for the presence of milk proteins. The assay detects αS-Casein as a marker of casein and β-lactoglobulin as a marker of whey protein. Time required for testing extracted samples: approximately 45 minutes.Results are expressed in terms of ppm milk protein or equivalents.

Source

1. *ELISA Systems at sales@elisasystems.net.*

3. Detection of Melamine Residuals in Milk (High Sensitivity ELISA Assays [1]

Melamine is an industrial chemical most commonly used as a fire retardant in dry powder extinguishers due to its release of nitrogen gas when burned or charred. Melamine has been used illegally by adding it to food products in order to increase the apparent protein content. Standard tests such as the Kjeldahl and Dumas tests estimate protein content by measuring the nitrogen content, therefore nitrogen rich melamine increases the nitrogen content leading to over estimation of the protein content.

Melamine can cause fatal kidney stones, especially when combined with cyanuric acid due to the formation of an insoluble melamine cyanurate. It has also been explained that when melamine and cyanuric acid are absorbed into the blood stream, they concentrate and interact in the urine-filled renal microtubules, then crystallize and form large numbers of round yellow crystals, which in turn block and damage the renal cells that line the tubes, causing the kidneys to malfunction.

Chromatographic methods are rather expensive, require special instrumentation and are laborous. Therefore a simple high throughput screening assay for melamine with reasonable costs is urgently needed. Classic ELISA type immuno-assays are very good and easy solutions for this need. They can be performed in microplates where large amount of samples can be simultaneously analyzed and require low instrumentation costs and affordable running prices.

Even if there has not been a vital need for melamine testing in milk products, there are a couple of commercial ELISA kit manufacturers that have melamine assay kits in their portfolio. Melamine testing has been needed in some earlier cases where melamine has been added to animal food. Melamine is also a metabolite of cyromazine (a cyclopropyl derivative of melamine that is used as pesticide). Melamine is therefore formed in the body of mammals who have ingested cyromazine and melamine testing can be used to detect cyromazine residuals.

Principle

This melamine ELISA assay study was performed with two ELISA assay kits from two different manufacturers. Both kits use the same competitive ELISA operational principle. These ELISA assay kits include melamine antibody coated microplates, horse radish peroxidase (HRP) conjugated melamine and chromogenic HRP substrate. Both unknown samples and HRP conjugated melamine are added into the microplate well coated with a melamine antibody. Then the HRP conjugated and free melamine are bound to the antibody according to the competition principle: the binding ratio is the same as the concentration ratio. As a result, the amount of bound HRP conjugated melamine is conversely dependent on the amount of the free melamine. After the binding phase, unbound material is removed with the microplate washer and chromogenic HRP substrate is added. HRP enzyme activity in this step is directly proportional to the amount of bound HRP conjugated melamine (and therefore conversely dependent on the melamine concentration in the sample). As a last step, the HRP enzyme reaction is stopped after a certain incubation time and the amount of the coloured dye formed in the HRP enzyme reaction is measured with the microplate photometer using 450nm measurement wavelength.

Kits and Instrumentation

Two following different commercially available melamine ELISA assay kits were used.

(i) **Agra QuantR Melamine Kit**-Sensitive Assay, Romer Labs Singapore Pte Ltd, Singapore.

(ii) **Abraxis Melamine Plate Kit**- Abraxis LLC, US.

Instrumentation

(i) Centrifuge Machine – for centrifugation of the melamine spiked milk samples during sample preparation was performed with a ThermoScientific centrifuge at 10° C temperature.

(ii) Multidrop R Combimicroplate dispenser-for addition of HRP (Horseradish peroxidase)-melamine conjugate, chromogenic HRP substrate andstop solution additions.

(iii) Well wash R AC microplate washer- for wahing of microplate.

(iv) Different microplatephotometer or multitechnology Readers Multiskan FC, Multiskan EX, Multiskan Ascent-allmicroplate photometer, Multiskan Spectrum-microplate spectrophotometer, Appliskan-filter-based multi-technology reader and Varioskan-Flash (monochromator-based multitechnology reader) for Photometric absorbance readings.

Samples and Sample Preparation

Milk samples were spiked with melamine using pure melamine (Sigma-Aldhrich, cat no. M 2659) and different milk products from the local grocery store as melamine contaminated milk samples were not available. Three different types of milk products-Natural full milk, fat free milk and natural milk powder were used. Artificial milk used in these tests was prepared from the milk powder by mixing 20 g of milk powder with 100 ml of distilled water. A group of spiked milk samples was prepared by mixing melamine stock solution (2 mg/ml dissolved in distilled water) with different types of milk products. With all spiked milk products (full milk, fat free milk, and artificial milk from milk powder) two different samples were analyzed-Milkspiked with 20 and 100 µg/l melamine in the final assay well. In addition, milk samples spiked with 500 and 1,000 µg/l melamine in the assay well were used with full and fat free milk. All these spiked samples were measured with an Abraxis Melamine Platekit. Only spiked full and fat free milk samples were measured with the AgraQuant Sensitive Assay kit because all results with the Abraxis kit showed no differences whether milk or milk powder was used.

Assay Procedure

Both ELISA assay kits were used according to the instructions from the manufacturer. The following procedural steps were taken to determine the melamine:

(i) Either 100 (Abraxis kit) or150µl (AgraQuant kit) of melamine standard or spiked sample was added into the antibody-coatedwell*.

(ii) HRP-melamine conjugate was added (50 µl) to each well with Multidrop Combi.

(iii) The plate was mixed for a short time and incubated for 30 minutes at room temperature.

(iv) All unbound sample was removed by washing the plate with distilled water. Washing was repeated four times with 300 µl volume using WellwashAC.

(v) An aliquot of 100 µl of the HRP-substrate was added into each well with another MultidropCombi.

(vi) The plate was incubated at room temperature for 20 (Agra Quantkit) or 30 (Abraxis kit) minutes.

(vii) The enzyme reaction was stopped by adding 100 µl of stop solution into each well using a third Multidrop Combi.

(viii) The absorbance was read with a microplate photometer using 450 nm wavelength.

Note*

These kits include only calibrators of 20, 100 and 1000µg/l but for this study additional calibrators with concentrations of 2, 5 and 10 µg/l were prepared by diluting the highest calibrator from the kits with water.

Microplate Reader Measurement and Calculation

Appropriate measurement and calculation protocols were created with either Thermo ScientificSkanIt Software or ThermoScientific Ascent Software, depending on the microplate photometer model used. The example below shows how the protocol was created for Multiskan FC. A similar protocol was used for all other models.

(i) A microplate layout was created to define different sample types, calibrators, controls and unknown samples. A sample without any melamine was defined to be the zero level control (giving the highest absorbance).

(ii) A measurement step using a 450nm filter was added to measure the absorbances.

(iii) Calculations of the results can be performed with SkanItSoftware. The calculation is based on generating the calibration curve and determining the concentrations of the unknowns based on the fitted curve. The calibration curve can be calculated based on either measured absorbances or normalized values where absorbance of the zero control has been set to 100 per cent. The type curve fitting used was a linear regression using Log-Lintrans formation and scaling.

Results

The results show that with both ELISA kits it is possible to easily detect below 10 µg/l melamine concentrations (the required sensitivity for LC/MS/MS in the Chinese melamine testing standard). This clearly shows the power of ELISA testing for purposes such as melamine screening.Both of these melamine ELISA kits are well suitable for screening dairy products for melamine residuals. Assay sensitivity is about equal to mass spectrometric analyses but assays with these ELISA kits are easier to perform.

Source

1. *Jorma Lampinen, Arto Perälä, and Reija-Riitta Harinen, Appliation note-AP-MIB-MSFCO3-1108, Thermo Fisher Scientific, Vantaa, Finland.*

4. Detection of Aflatoxin M1 in Milk [2]

The available rapid methods for AFM1detection are mostly based on the following formats:

(i) Competitive immunoassay formats

(ii) Surface based formats

The competitor (AFM1-protein conjugate) or the receptor (antibody or aptamer) is immobilized onto a surface (microtiter plate, membrane, electrode). The ELISA method involves the following steps:

(i) Coating of Ab onto a microtiter.

(ii) Add sample extract.

(iii) Add AFM1-enzyme conjugate.

(iv) Competition between free and anzime-conjugate AFM1.

(v) Add enzyme substrate for colour development.

(vi) Observe the signal which is inversely proportional to AFM1 concentration. Earlier Aflatxin M1in milk is being determined by conventional method but some rapid metrhods are also available. Elisa method and mycotoxin florometer method are among them. The comparative flow chart for conventional method (AOAC Method) and Elisa method are given in following table:

Detetion of Aflatoxin M1-Flow Chart for Rapid vs Conventional Methods

Sl.No.	*Conventional Methods (AOAC 2000.08)*	*Rapid/Emerging Methods*
1	Sample Preparation	Sample Preparation
	Centrifugation to separate the fat layer	(not always needed)
2	Immunoaffinity Column	
	CLEAN UP	
3	HPLC determination with	Detection
	Fluorescence Detection	(immunoassays - sensors)
4	Time of analysis: 4 h/sample	Time of analysis: 5 - 20 min

The details of different Elisa kits available for detection of Aflatxin M1 in milk and milk products are given in following table:

Different ELISA Kits Available for Detection of Aflatoxin M1 [1]

Sl.No.	Manufaturers	Test Kit	Test Kit Range	Inubation Time (min.)	Intended Matries
1	Beacon Analytical Systems Inc.	Aflatoxin M1 Plate Kit	0.3 - 300 ng/ kg	75	Milk and Milk powder
2	Bioo Scientific Corporation	Max Signal Aflatoxin M1 ELISA Test Kit	>0.1 μg/kg	30	Milk
3	EuroProxima	Aflatoxin M1 ELISA	>6 ng/kg	60	Milk, milk powder, cheese and butter
4	MP Biomedicals	Fast TOX Aflatoxin M1 ELISA Kit	N/A	N/A	Milk and Milk powder
5	Neogen Corporation	Veratox for Aflatoxin M1	5-100 ng/kg	45	Milk, milk powder, butter and cheese
6	R-Biopharm	RIDASCREEN AflatoxinM1	> 5 ng/kg	75	Milk, milk powder and Cheese
7	R- Biopharm	RIDASCREEN FAST Aflatoxin M1	>125 ng/kg	15	Milk and milk powder
8	Reagen LLC	Aflatoxin M1 ELISA Test Kit	>0.05 μg/kg	30	Milk and milk powder
9	Romer Labs	Agra Quant Aflatoxin M1 Sensitive	25-500 ng/kg	80	Milk, milk powder and cheese
10	Romer Labs	Agra Quant Aflatoxin M1 fast	100-2000 ng/ kg	30	Milk, milk powder and cheese

Source

1. *Wei et al., World Mycotoxin Journal (2014) 7 (4): 417-430.*
2. *Veronica M. T. Lattanzio and Michelangelo Pascale, Institute of Sciences of Food Production (ISPA), National Research Council (CNR) of Italy. The milk day -February 4th, 2015.*

5. Detection of Undeclared Milk in other Species Milk [1]

The presence of undeclared milk in other species milk or cheese, in principle, can be detected through using two basic ELISA methods: sandwich ELISA and indirect ELISA, including their different variations.

ELISA is able to detect cows' and goats' milk in milk mixtures by polyclonal and monoclonal antibodies produced to combat whey proteins, caseins or short-string peptides from milk proteins. The caseins which represent the main part of the protein fraction, feature advantage in being more or less stable under high temperature conditions. Therefore they can be successfully used as the main antigens in the heat treatment (pasteurization, UHT) of milk and dairy products. Their major disadvantage is weak immunogenicity and higher sensitivity to protheolytic degradation. Whey proteins, contrary to casein, are much better immunogens and they are protheolytically degradable only in minimal quantity. In respect of high temperatures, whey proteins are less resistant. At present, there are a small number of ELISA tests with really sufficient sensitivity for detection of additives in the heat treated milk [2].

An indirect enzyme-linked immunosorbent assay (ELISA) was developed for the detection and quantification of bovine milk adulteration in goat'smilk. The polyclonal antibodies have been modified by mixing with goat's milk for the assay purposes. The absorbance at 450 nm in indirect ELISA revealed a linear relationship with the concentration of adulterated bovine milk at the range of 4 per cent - 50 per cent. Detection limit was 4 per cent for mixed milk samples. The assay was characteristics of high reproducibility with intra- and inter-assay variation coefficients less than 5 per cent. Therefore, the ELISA can be successfully used to determine the adulteration of milk samples and it is suitable for developing a kit in routine inspection of milk [3]. An approach to estimate the proportion of goat'smilk in sheep milk is made [2] where focus on laboratory testing and evaluation of quality parameters of the ELISA tests, was made. The detection and quantification of goat milk was based on the presence of the specific immunoglobulins (IgG).

Source

1. *Peter Zachar, Michal Šoltés, Radovan Kasarda, Jaroslav Novotný, Miroslava Novikmecovѡá, Dana Marcinèákovѡá, Review on Analytical methods for the species identification of milk and milk products published as Identification of milk and milk products in Mljekarstvo 61 (3), 199-207 (2011).*
2. *Zeleòákovѡá, L., Golian, J. (2008): Application of ELISA tests for detection of milk and cheese adulteration, Sci.Monograph. SPU Nitra, 98. ISBN 978-80-552-0075-0.*
3. *Xue, H., Hu, W., Son, H., Han, Y., Yang, Z. (2010): Indirect ELISA for detection and quantification of bovinemilk in goat milk. J. Food Sci. Tech. 31 (24), 370-373.*

Lateral Flow Devices [1,2]

Lateral flow tests [9] also known as **lateral flow immunochromatographic assays**, are simple devices intended to detect the presence (or absence) of a target analyte in sample (matrix) without the need for specialized and costly equipment, though many lab based applications exist that are supported by reading equipment [10].

Food quality and Food safety are the important quality aspects. The consumers are inadvertently getting exposed to numerous contaminants (antibiotics, pesticides, heavy metals, pathogenic microorganisms, aflatoxins and many other harmful agents) via food chain. In the livestock management practice, many of these contagion enters to milk. If appropriate measures are not taken, then milk supply gets contaminated. To safeguard the consumer, there is an increased demand for ensuring the safe quality milk for human consumption. Screening for these contaminants, especially for antibiotics in milk, at farm level is very essential. Numbers of test methods are available for detection or estimation of antibiotics in milk. While rapid screening tests are commonly preferred to detect presence or absence of antibiotics, lateral flow assay (LFA) is one of the emerging rapid test method [1].

It's an immuno assay method, An LFD generally comprises a porous membrane, typically nitrocellulose, with an immobilized capture protein for the target analyte, forming a visible line in a viewing window, due to nanoparticles of gold or colored latex particles, after contact with the specified analyte. For the test to be valid, a control line should form in a second viewing window. In most devices, an antibody is commonly used as capture protein, which specifically binds and captures a particular antigen if present in the sample. In most cases a sandwich assay is used. The test is fast (reading in terms of minutes) and simple both in use and in interpretation [2].

Point of care (POC) testing has become the most famous way of diagnosis in clinical analysis, food safety and environment. Lateral flow assay (LFA) based POC devices are among very rapidly growing strategies for qualitative and quantitative analysis. LFA is performed over a strip, different parts of which are assembled on a plastic backing. These parts are sample application pad, conjugate pad, nitrocellulose membrane and adsorption pad. Nitrocellulose membrane is further divided into test and control lines. Pre-immobilized reagents at different parts of the strip become active upon flow of liquid sample. LFA combines unique advantages of biorecognition probes and chromatography. LFA based strips have different detection formats. Drawbacks associated with conventional clinical technique, enzyme linked immunosorbent assay (ELISA), were flabbergasted by LFA.

Lateral Flow Immunoassay has gained wide acceptance and **recognized by WHO** as point of care diagnostics. It gets importance because of the following characteristics/advantages:

(i) The simplicity of the test design.

(ii) Rapid result.

(iii) Long-term stability over a wide range of weather conditions.

(iv) Relatively low manufacturing costs.

(v) User friendly format.

(vi) Rapidity and one step analysis.

(vii) Low operational cost, simple instrumentation.

(viii) Less or no interferences due to chromatographic separation, high specificity, better sensitivity.

(ix) Portability of the device are unique advantages related to LFA strips [2a].

Principle

Lateral flow assay (LFA) is one of the emerging test methods. Traditionally, LFA is an immune-chromatographic assay. This method derives the basic principles from the subjects of biology, chemistry, physics and engineering fields.

The fundamental principle underlying in LFA is the movement of very minute quantity of liquid sample containing the target analyte (s) along a strip of polymeric material by the action of capillary flow across various reactive zones, where the reorganization molecules have been attached, that exert more or less specific interactions with the analytes [3,4].

Instrument/Components of LFA Strips [1]

Lateral flow assay basically combines a number of variants such as formats, biorecognition molecules, labels, detection systems and applications. This test device is functionally divided into the following segments:

i. Sample Pad

Main function of sample pad is to promote, even distribution of the sample to the conjugate pad and to releases the analyte with high efficiency.

ii. Conjugate Pad

It is also known as reagent pad. The role of this pad is to accept the conjugate, hold it, stabilize it over its entire pad, and release it efficiently and reproducibly when the assay is run.

iii. Testing Zone (Capture Zone)/Membrane or Analytical Region

The purpose of the analytical region in a LFA is to bind proteins at the test and control areas and to maintain their stability and activity over the shelf-life of the product. When the strip is run, it must accept the conjugate and sample from the conjugate pad, flow them consistently to the reaction area, allow the reaction at the test and control lines to happen, and allow excess fluids, label, and reactants to exit without binding. Backing material are the most common and flexible part of the strip system.

iv. Wicking Pad

The wick is the engine of the strip. It is also known as absorbent pad or reservoir. Mainpurpose of this pad is to pull out all of the fluid added to the strip and to hold it for the duration of the assay [3].

Typically, all these segments are made up from different porous and flexible membranes, which serve one or more functions. These different parts overlap onto one another and are pasted on a backing card using a pressure sensitive adhesive. LFA can be used as point of care (POC) testing device, and has become the most famous way of diagnosis [5,6].

v. Other Components

Labels

These are the essential part of the lateral flow test, because the attachment of the antibody and antigen is difficult to visualize without them. Colloidal gold is the most used label in lateral flow assays but coloured latex particles and other coloured particles such as carbon and selenium are also used. Fluorescence and chemiluminescence emitting dyes along with the enzyme labels are also tested. The use of paramagnetic particles is a relatively new invention. Gold nanoparticles, Magnetic particles and aggregates, Fluorescent and luminescent materials,

Enzymes, Colloidal carbon are being used as labels.

Antibodies

These are also called immunoglobulin, are proteins which are part of the hormonal, adaptive immune system. A basic property of antibodies is that they can recognize specific substances efficiently with the antigen binding sites and that is the reason why they are applied in many recognition systems. Basically following three types of antibodies can be used:

(i) Monoclonal.

(ii) Polyclonal

(iii) Recombinant antibodies.

vi. Detection Systems

Selection of detector is mainly determined by the label employed in analysis.

(a) **In case of gold nanoparticles or other color producing labels-** qualitative or semi-quantitative analysis can be done by visual inspection of colors at test and control lines. But for quantification, optical strip readers are employed for measurement of the intensity of colors produced at test and control lines of strip. This is achieved by inserting the strips into a strip reader and intensities are recorded simultaneously by imaging softwares.

(b) **In case of fluorescent labels**- a fluorescence strip reader is used to record fluorescence intensity of test and control lines.

(c) A photoelectric sensor was also used for detection in LFA where colloidal gold is exposed to light emitting diode and resulting photoelectrons are recorded.

(d) Chemiluminescence which results from reaction of enzyme and substrate is measured as a response to amount of target analyte.

(e) Magnetic strip readers and electrochemical detectors are also reported as detection systems in LFTS.

Working of Lateral Flow Assay for Antibiotics Detection [1]

Following are the main steps involved in the design of LFA:

(i) Antibody development against target analyte (antibiotics).

(ii) Preparation and characterization of gold nanoparticles or suitable label material.

(iii) Preparation of conjugates of target antibody and label material, analyte with a carrier protein and

(iv) Construction of lateral flow strip and analyte detection.

Formats of LFA [6]

Different formats are adopted in LFA. The details of some of these formats are given below:

i. Sandwich Format

In a typical format, label (Enzymes or nanoparticles or fluorescence dyes) coated antibody or aptamer is immobilized at conjugate pad. This is a temporary adsorption which can be flushed away by flow of any buffer solution. A primary antibody or aptamer against target analyte is immobilized over test line. A secondary antibody or probe against labeled conjugate antibody/aptamer is immobilized at control zone.

Sample containing the analyte is applied to the sample application pad and it subsequently migrates to the other parts of strip. At conjugate pad, target analyte is captured by the immobilized labeled antibody or aptamer conjugate and results in the formation of labeled antibody conjugate/analyte complex. This complex now reaches at nitrocellulose membrane and moves under capillary action. At test line, label antibody conjugate/analyte complex is captured by another antibody which is primary to the analyte. Analyte becomes sandwiched between labeled and primary antibodies forming labeled antibody conjugate/analyte/primary antibody complex. Excess labeled antibody conjugate will be captured at control zone by secondary antibody. Buffer or excess solution goes to absorption pad. Intensity of color at test line corresponds to the amount of target analyte and is measured with an optical strip reader or visually inspected. Appearance of color at control line ensures that a strip is functioning properly.

ii. Competitive Format

This format suits best for low molecular weight compounds which cannot bind two antibodies simultaneously. Absence of color at test line is an indication for the presence of analyte while appearance of color both at test and control lines indicates a negative result. Competitive format has two layouts.

(a) **In the First Layout**-solution containing target analyte is applied onto the sample application pad and prefixed labeled biomolecule (antibody/aptamer) conjugate gets hydrated and starts flowing with moving liquid. Test line contains pre-immobilized antigen (same analyte to be detected) which binds specifically to label conjugate. Control line contains pre-immobilized secondary antibody which has the ability to bind with labeled antibody conjugate. When liquid sample reaches at the test line, pre-immobilized antigen will bind to the labeled conjugate in case target analyte in sample solution is absent or present in such a low quantity that some sites of labeled antibody conjugate were vacant. Antigen in the sample solution and the one which is immobilized at test line of strip compete to bind with labeled conjugate [12].

(b) **In another layout**- labeled analyte conjugate is dispensed at conjugate pad while a primary antibody to analyte is dispensed at test line. After application of analyte solution a competition takes place between analyte and labeled analyte to bind with primary antibody at test line.

iii. Multiplex Detection Format

Multiplex detection format is used for detection of more than one target species and assay is performed over the strip containing test lines equal to number of target species to be analyzed. It is highly desirable to analyze multiple analytes simultaneously under same set of conditions [13] . Miniaturized versions of LFA based on microarrays for multiplex detection of DNA sequences have been reported to have several advantages such as less consumption of test reagents, requirement of lesser sample volume and better sensitivity [14]. Multiplex detection format resulted in LODs in the range of 0.05–3.0 ppb for a series of mycotoxins in cereal samples and results were in good agreement with LC-MS/MS [15].

Applications [1]

LFA has following types of applications:

(i) LFA has gained popularity because of pregnancy detection kit which detects human chorionic gonadotropin (hCG), a chemical marker associated with pregnancy [7].

(ii) There after successful application (Commercialization) of LFA has expanded to many analytical fields. The applications of LFA are summarized [5] below:

- *(a)* It is used to detect presence of toxins in food samples and environmental samples.
- *(b)* To detect the use of illegal drugs.
- *(c)* in the field of biowarfare.
- *(d)* agricultural applications such as soil borne plant pathogenic fungus, plant viruses and bacterial pathogens.
- *(e)* for veterinary drugs used in animal production and veterinary drug residue detection.
- *(f)* Food and feed product quality and **g-** in molecular diagnostics.

(iii) **Antibiotics (small molecules; lower Mol wt.: <1000 Da) Detection –** Because of some technical reason, competitive format will be used in detection of antibiotics by LFA. In competitive design, antigens will be coated over test line and coating of antigen on test line is facilitated through coupling with protein. The assay is based on competitive reaction concept wherein free analyte in sample and analyte coated at test line compete for limited amount of colloidal gold-labeled target antibody. If analyte is absent in the sample or present below detection limit, the target antibiotic-protein conjugate coated at test line would bind gold labeled target antibodies and thus starts accumulation of gold will create a red line. The intensity of the red color at test line is inversely proportional to the concentration of target analyte in the sample *i.e.* the more analyte in the test sample, the weaker the test signal and vice-versa. If the target antibiotics concentration in the sample is above the detection limit, whole of colloidal gold-labeled target antibody would bind to the analyte which is present in the test sample and thus would not be available for target

antibiotic-protein at test line and thus no colour line is formed. **Thus, in competitive format when test is positive; only one line will appear and in absence of analyte two lines will appear.**

(iv) **Detection of veterinary drug residues in milk by LFA-** A lateral flow test comprises five different basic components: sample filter (a paper like material with two functions-filtering out a solid material and buffering the sample after extraction/pretreatment); conjugate pad (gold pad) is a fibreglass-like pad which can either be sprayed by, or bathed in gold-, latex- or carbon-conjugated solutions; membrane (membrane is composed of nylon or nitrocellulose and glued upon a backing material) are made to display a maze structure with various pore sizes; reservoir-absorption pad (has one purpose-to-adsorb the liquid at the end of the strip); test line and control lines (lines are sprayed onto the nitrocellulose/nylon membrane). The sample is applied to the sample filter, housed inside the device casing. The sample runs through the sample filter and conjugate pad. This conjugate pad contains a labelled antibody, specific for the analysis under test. Antibiotics, if present, will form a complex with the conjugate and migrate further slowly to the membrane in the test zone. This zone contains an immobilised antibody, specific for the analysis, but preferably not competing with the conjugated antibody for the same or adjacent epitopes. The test line will thus capture the migrating analyse conjugate complex. The intensity of the test line correlates well with the amount of analyte in the sample [8]. Immunoassay can be sensitive, class specific, accurate, and can provide a means for a rapid screening of samplesfor antibiotics.

(v) **Detection of Micotoxins-** A brief review of the LFA based test strips for detection of mycotoxins in food and feed is given by Rudolf and Alexandra [11].

(vi) **Other [6] -** The details related to the application area, analyte, labels *etc.* are tabulated below:

Sl. No.	*Application Area*	*Analyte*	*Label*	*Sample Type*	*Detection limit*	*Detection Time*
1	Pesticides	Carbofuran	Gold nanoparticles	Water	32 μg/L	8–10 min
		Triazophos	Gold nanoparticles	Water	4 μg/L	8–10 min
		Carbaryl	Gold nanoparticles	Agricultural products	100 μg/L	-
		Dichloro-diphenyl-trichloroethane	Gold nanoparticles	Food	27 ng/mL	-
2	Metal Ions	Hg^{2+}	Gold nanoparticles	Water	0.1 nM	–
		Cd^{2+}	Gold nanoparticles	Drinking and tap water	0.4 ppb	-
3	Pharma-ceuticals and drugs	Enrofloxacin	Gold nanoparticles	Chicken muscles	0.138 μg/kg	5–10 min
		Chloramphenicol	Gold nanoparticles	Milk	10 ng/mL	10 min
		Ofloxacin	Gold nanoparticles	Milk, chicken and pork meat	30 ng/mL	10 min

Source

1. *Laxmana Naik, Rajan Sharma, Y.S. Rajput and Bimlesh Mann, Lateral Flow Assay - its Application in Assessing the Quality of Milk, Compendium on Advanced Techniques and Novel Approaches for Quality and Safety Evaluation of Dairy Foods-2015, NDRI, Karnal.*

2. *Technews, NDDB, Technological Advancement in Methods of Microbiology in Milk Analysis- Part 1 (Direct methods), Issue no.91, October-December 2016.*

2a. *Z. Li, Y. Wang, J. Wang, Z. Tang, J.G. Pounds, Y. Lin Rapid and sensitive detection of protein biomarker using a portable fluorescence biosensor based on quantum dots and a lateral flow test strip Anal. Chem., 82 (2010), pp. 7008-7014.*

3. *O'Farrell, B. (2009). Evolution in Lateral Flow–Based Immunoassay Systems. In: Lateral Flow Immunoassay, Humana Press. Pp. 1-33.*

4. *Anfossi, L., Baggiani. C., Giovannoli, C., Giraudi, G. (2013). Lateral Flow Immunoassays for Aflatoxins B and G and for Aflatoxin M1. In: Aflatoxins, Rrecent Advances and Future Prospects, Pub: InTech Chapters. 315.*

5. *Lata, K., Naik, L., Sharma, R., Mann, B., and Rajput, Y.S. (2013). Lateral Flow Assay - Concept and its Applications in Food Analysis. Indian Food Industry Mag. 32 (5): 22-32.*

6. *Sajid, M., Kawde, A. N., and Daud, M. (2014). Designs, formats and applications of lateral flow assay: A literature review. J Saudi Chemical Society.*

7. *Butler, S. A., Khanlian, S. A., Cole, L. A. (2001). Detection of early pregnancy forms of human chorionic gonadotropin by home pregnancy test devices. Clin. Chem 47: 2131-2136*

8. *Van Herwijnen R. (2006): The use of lateral flow device to detect food allergens. In: Koppelman J, Helfe S.L.: Detecting Allergens in Food. Woodhead Publishing, Cambridge: 175-181.*

9. *Concurrent Engineering for Lateral-Flow Diagnostics (IVDT archive, Nov 99) Archived 2014-04-15 at the Wayback Machine.*

10. *Yetisen A. K. (2013). "Paper-based microfluidic point-of-care diagnostic devices". Lab on a Chip. 13 (12): 2210–2251,*

11. *R. Krska, A. Molinelli, Rapid test strips for analysis of mycotoxins in food and feed, Anal. Bioanal. Chem., 393 (2009), pp. 67-71.*

12. *C. Liu, Q. Jia, C. Yang, R. Qiao, L. Jing, L. Wang, et al. Lateral flow immunochromatographic assay for sensitive pesticide detection by using Fe_3O_4 nanoparticle aggregates as color reagents, Anal. Chem., 83 (2011), pp. 6778-6784.*

13. *B.R. Panhotra, Z.U. Hassan, C.S. Joshi, A. Bahrani, Visual detection of multiple viral amplicons by dipstick assay: its application in screening of blood donors a welcome tool for the limited resource settings, J. Clin. Microbiol., 43 (2005), pp. 6218-6219.*

14. *D.J. Carter, R.B. Cary. Lateral flow microarrays: a novel platform for rapid nucleic acid detection based on miniaturized lateral flow chromatography, Nucleic Acids Res., 35 (2007), p. e74.*

15. *N. Song, Z. Liu, E. Zhao, E. Njumbe Ediage, S. Wu, C. Sun, et al.Multiplex lateral flow immunoassay for mycotoxin determination, Anal. Chem. (2014).*

Chapter 3

Biosensors Techniques

A biosensor is an analytical device which converts a biological response into an electrical signal (Measurable signal). Biosensors can also be defined as analytical devices that combine bio specific recognition systems with physical or electrochemical signalling. Biosensors are devices that can combine a biochemical molecule with a physical signal that can be translated into an indication of safety or quality of the food. Biosensors help carrying out the procedures that are sensitive, selective, rapid, cost effective and portable. A biosensor is for the detection of an analyte that combines a biological component with a physicochemical detector component [1,2]. It involves the combination of biological entities like DNA, RNA, and proteins/enzymes to the electrochemical transducers in order to detect and observe certain biological analytes like antibody-antigen interaction. The sensors use biological material to make measurement, are called Biosensors. Biosensors use specific biochemical reactions mediated by isolated enzymes, immunosystem, organelles or whole cells, tissues to detect chemical compounds usually by thermal, electrical or optical signals. The term 'biosensor' is often used to cover sensor devices used in order to determine the concentration of substances and other parameters of biological interest even where they do not utilise a biological system directly. The biologically sensitive elements can also be created by biological engineering. The transducer or the detector element (works in a physicochemical way; optical, piezoelectric, electrochemical, *etc.*) transforms the signal resulting from the interaction of the analyte with the biological element into another signal (*i.e.*, transduces) that can be more easily measured and quantified. It is an interdisciplinary technology and involves the collaborative efforts of engineering, microbiology, physics, chemistry, biology, biotechnology and so on [2a]. A great majority of biosensprs have immobilized enzymes.

The biosensor reader device with the associated electronics or signal processors that are primarily responsible for the display of the results in a user-friendly way

[3]. Biosensors are a potentially powerful and important tool for quality control testing of food/dairy products, product composition, the control of production and processing and the management of dairy animals. They could detect the presence of pesticides, antibiotics, toxins, pathogens and chemicals in food products. Bio-sensors that use antibodies are called immunobiosensors. The performance of the biosensors is mostly dependent on the specificity and sensitivity of the biological reaction.

The Biosensors are evolving as excellent substitutes for the existing conventional techniques. But still operation of biosensors is a challenging task for their utility owing to the cost and shelf life of bio-recognition molecule [3a].

E–Nose and E-Tongue (described in chapter-1) are trending biosensors. E-Noses are being used in detection of odd volatile compounds in tea, wine, coffee, spices industry whereas E. Tongue (colorimetric sensor array) is being used in liquid food industry.

Definition of Biosensor

As per **IUPAC 1998,** biosensor is defined as:

Self contained integrated device that is capable of providing specific qualitative or semi qualitative analytical information using a biological recognition element which is in direct spatial contact with transduction element.

A Clark oxygen electrode was invented by Leyland C.Clark (Father of Biosensors) and it was a pivotal device that allows real time monitoring of patient"s blood oxygen level.

The general aim of the design of a biosensor is to enable quick, convenient testing at the point of concern or care where the sample was procured [4]. A common example of a commercial biosensor is the blood glucose biosensor, which uses the enzyme glucose oxidase to break blood glucose down. In doing so it first oxidizes glucose and uses two electrons to reduce the FAD (a component of the enzyme) to FADH2. This in turn is oxidized by the electrode (accepting two electrons from the electrode) in a number of steps. The resulting current is a measure of the concentration of glucose. In this case, the electrode is the transducer and the enzyme is the biologically active component.

Biosensor System Apparatus

A biosensor typically consists of following three important units:

(i) A bio-recognition component (Biologically Active Material).

(ii) Biotransducer component (Detector Element).

(iii) Electronic system which include a signal amplifier, processor, and display (Signal Processor).

i. A Recognition Sensitive Element (Biological Receptor)

The recognition component, often called a bioreceptor, uses biomolecules from organisms or receptors modeled after biological systems to interact with the analyte of interest. The types of bioreceptor could be:

(*i*) Non catalytic (cell receptor or antibodies) or catalytic (microorganism, plant, enzymes or mammalian tissue, organelles etc.)

(*ii*) A biologically derived material or biomimic).

Enzymes, whole cells (bacterial, fungal, animal or plant), organelles, tissues, receptors, antibodies and nucleic acids can all be used in biosensors as biorecognition elements [1]. Specificity is the main reason for the widespread use of enzymes, of these, oxidases are by far the most commonly used. Enzymes with a high specific activity may be immobilized on a water insoluble membrane and placed on the sensing area of the transducer. This method offers the advantage of simple preparation and linearity over a greater range of analyte concentrations. Enzymes can also be covalently immobilized onto water-soluble particles and held between two selectively permeable membranes.

The attachment of the biological elements (small molecules/protein/cells) to the surface of the sensor (be it metal, polymer or glass) is an important part of the biosensors. The simplest way is to functionalize the surface in order to coat it with the biological elements. This can be done by polylysine, aminosilane, epoxysilane or nitrocellulose in the case of silicon chips/silica glass. Subsequently the bound biological agent may be for example fixed by Layer by layer depositation of alternatively charged polymer coatings [7]. The most commonly used hydrogel is sol-gel, a glassy silica generated by polymerization of silicate monomers. Another group of hydrogels, which set under conditions suitable for cells or protein, are acrylate hydrogel, which polymerize upon radical initiation. Smart materials that mimic the biological components of a sensor can also be classified as biosensors using only the active or catalytic site or analogous configurations of a biomolecule [8].

ii. Signal Conversion Unit (Transducer or the Detector element)

It works in a physicochemical way; optical, piezoelectric, electrochemical, *etc.*, that transforms the signal resulting from the interaction of the analyte with the biological element into another signal that can be more easily measured and quantified. The key part of a biosensor is the transducer, which makes use of a physical change accompanying the reaction. This may be:

(*i*) The heat output (or absorbed) by the reaction (calorimetric biosensors).

(*ii*) Changes in the distribution of charges causing an electrical potential to be produced (potentiometric biosensors).

(*iii*) Movement of electrons produced in a redox reaction (amperometric biosensors).

(*iv*) Light output during the reaction or a light absorbance difference between the reactants and products (optical biosensors), or effects due to the mass of the reactants or products (piezo-electric biosensors) [15].

Types of Tranducers

Several types of transducer, including amperometric transducers, ion-selective electrodes (ISEs), ion-sensitive field effect transistors (ISFETs), conductometric electrodes, calorimetric devices, optical transducers, and piezoelectric crystals, have been used in biosensors.

Amperometric Tranducers

In general, amperometric electrodes based on oxygen or hydrogen peroxide detection together with an oxidase system or an aerobic microorganism, dominate food and drink applications. A comparative details of different tranducers used in biosensors are given in following table **[9]:**

Sl.No.	*Transducers*	*Advantages*	*Disadvantages*
1	Amperometric	Simple, high selectivity, wide linear range, high sensitivity.	Interference resulting for redox active compounds.
2	Conductometric	Microfabrication	Influence of pH
3	Optical systems	Remote sensing, low cost, can be miniaturized, free from electrical interference.	Interference form ambient light, requires high energy sources, only applicable to a narrow concentration range.
4	Piezoelectric	Fast response, simple, stable output signal, low cost for readout device, no special sample handling	Low sensitivity in liquid applications, interference resulting for non-specific binding.
5	Calorimetric	Versatility, free from optical interferences such as colour or turbidity.	Expensive, cumbersome, requires a large amount of enzyme.
6	ISEs	Simple, reliable	Sluggish response, require a stable reference electrode, susceptible to electronic noise.
7	ISFETs	Low cost, mass product- ion, stable output,requires very small amount of biological material,can monitor several analytes simultaneously	Temperature sensitive, fabrication of diferent layers on the gate is not perfected

Amperometric biosensor systems with enzyme electrode received the greatest attention. Of these, oxidase enzyme mounted sensors with following reaction mechanism, have proved to be the most popular.

$$\text{Substrate} + O_2 \xrightarrow{\text{Oxidase}} \text{product} + H_2O_2$$

Several biosensors that use amperometric transducers have been developed for analytes that may be of interest to the food and beverage industries. The details of some of these are given in table on next page [9]:

Sl.No.	Target Compound (Analyte)	Biosensing Material
1	Glucose	Glucose oxidase, glucose dehydrogenase, Pseudomonas fluorescences.
2	Sulphite	Sulphite oxidase.
	Xanthine	Xanthine oxidase.
	Cholesterol	Cholesterol oxidase.
	Acetic acid	Trichosporon brassicae.
3	Aspartame	Peptidase, aspartate amino transferase, glutamate oxidase.
4	Glutamine	Glutaminase, glutaate oxidase Lactate oxidase, lactate dehydrogenase.
	Lactate	
5	Urea	Urease. Tyrosinase.
	Phenols	Malate dehydrogenase.
	Malate	

Amperometric biosensor technology suffered from following drawbacks:

(i) Showed short operational and storage stability.

(ii) Poor reproducibility and a very little versatility of the systems when they were used to test real samples.

(iii) To undergo scaling up in industrial production.

Solid Binding Matrix (SBM)

To solve aforesaid problems, the concept of **Solid binding matrix (SBM)** has been developed [10-13] and now, this innovative technology hold an important position among industrialized amperometric (mediated transducers) biosensor systems presently available and have already found a wide commercial market. Solid binding matrices are compounds having a hydrophobic skeleton, mostly substituted by polar groups.SBMs are solid at room temperature, insoluble in water and in most organic solvents. They show a good compatibility with biocatalysts, co-factors, redox mediators, and they have a good mechanical properties, especially compactness, plasticity, polishability, with good resistance against progressive erosion of the electrode surface, a very common occurrence in the case of carbon paste biosensors.

The commercial SBM biosensing system dedicated to determination of glucose, fructose, ethanol, malate and lactate in wine samples (industrial product-'Perbacco 200X' series licensed by POLYtech SCrl, Trieste, produced by Biofutura Srl, Gorizia) is already marketed. After proper interface the POLYtech biosensor system delivers a fast and precise determination of glucose, fructose, reducing sugars, lactate and malate in grapes, must and wine. Analytical applications of the POLYtech biosensor systems are under development in several fields, such as fruit juices, meat, fish, milk and dairy products, oil and fats, employing variable transducers set-ups (tip-type, niddle type, screen printed type) [13]. The details related to the potential application of the SBMs based POLYtech Biosensor Systems in food and drink industries are given in following table [9,13]:

Sl.No.	*Samples*	*Analytes*
1	Milk and Diary products	Lactose, lipids, casein, Lactic acid, citric acid, phosphates, glycerol, urea, acetone, lactulose, residual pesticides, bio-organic amines, antibiotics, phosphatase activity, hydrogen peroxide.
2	Meat and meat extracts	Lactose, Glucose, Lactate, Creatinine, starch sulphites.
3	Fats and oils	Peroxides, antioxidants, fatty acids, polyphenols, cholesterol.
4	Honey	fructose, glucose, maltose, sucrose, reducing sugars.
5	Feeds	Urea, Sucrose, starch, ammonia.
6	Soft drinks and concentrated fruit juices	Ethanol, ascorbate, caffeine, citrate, glucose, fructose, malic acid, methanol, sucrose, reducing sugars.
7	Fruit and vegetables	Sucrose, starch ethanol, methanol, citric acid, lactic acid, malic acid, reducing sugars, glucose, fructose, residual pesticides.
8	Fish	Biogenic amines, ATP, Hipoxanthine, ammonia, ethanol.
9	Beer	Ethanol, glycerol, total acidity, reducing sugars, dextrins, diacetyl, starch, biogenic, amines, polyphenols, sulphites. a-amylase.
	Wine	Glucose, fructose, ethanol, lactate, malate, sulphites, carbon dioxide, aldehydes, citrate, glycerol, sucrose, reducing sugars.
	Liquors	Ethanol, Sucrose, aldehyde, glycerol, methanol.

iii. Associated Electronics or Signal Processors

Primarily responsible for the display of the results in a user-friendly way. It is most expensive part of the sensor device, however it is possible to generate a user friendly display that includes transducer and sensitive element.

Transducers and electronics can be combined, *e.g.*, in CMOS-based microsensor systems [5,6].

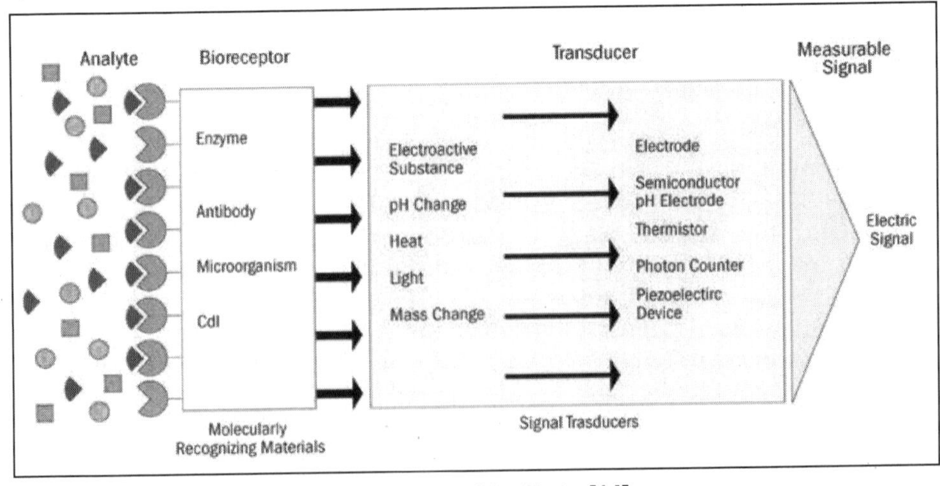

Biosensor and its Parts [14]

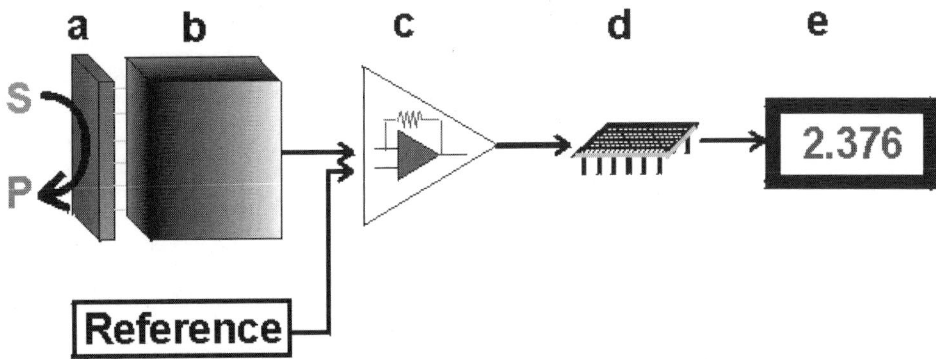

Schematic Diagram Showing the Main Components of a Biosensor [15].

(a) The biocatalyst converts the substrate to product; (b) the transducer determine the said reaction and converts it to an electrical signal; (c) Amplifier -the output from the transducer is amplified; (d) Processsor-the amplified output from the transducer is processed; (e) Display where amplified and processed output is displayed.

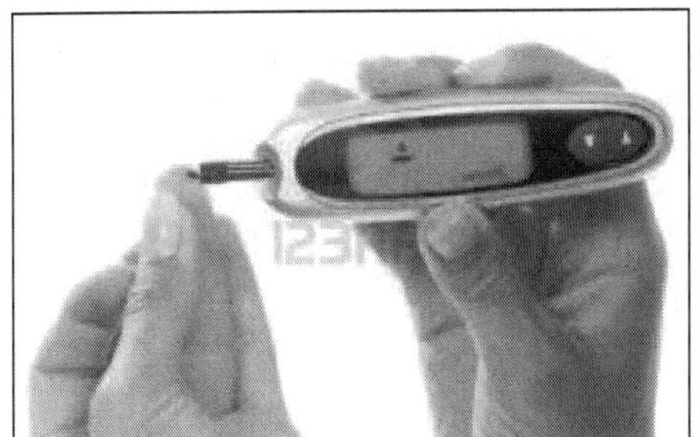

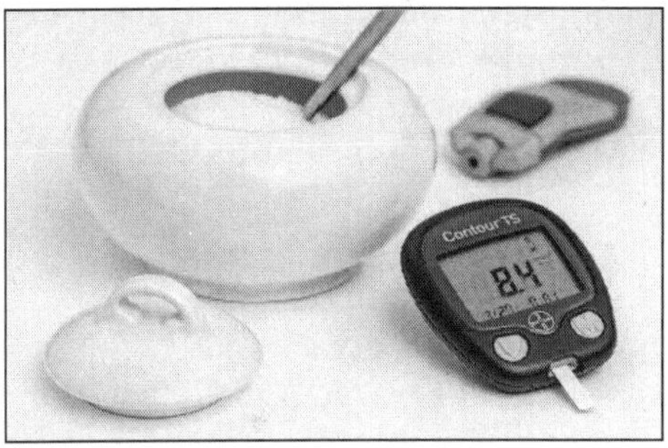

Biosensors in Health Care and other Fields.

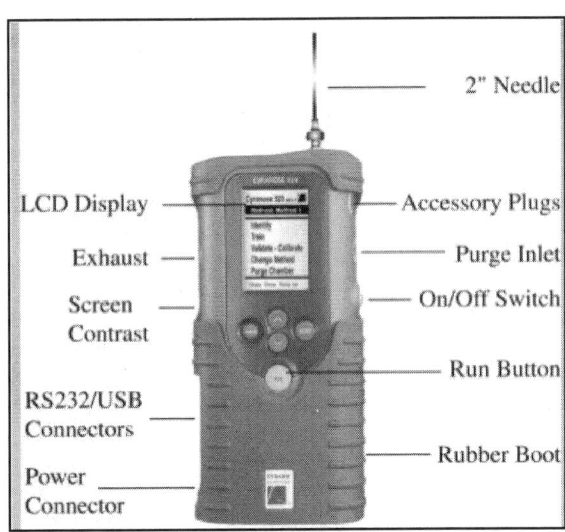

A Typical Biosensor.

Source

1. *Turner, Anthony; Wilson, George; Kaube, Isao (1987).Biosensors: Fundamentals and Applications. Oxford, UK: Oxford University Press. p.770.*

2. *Banica, Florinel-Gabriel (2012). Chemical Sensors and Biosensors: Fundamentals and Applications. Chichester, UK: John Wiley and Sons. p. 576.*

2a. *Hinze S (1994) Bibliographical cartography of an emerging interdisciplinary discipline: The case of bioelectronics. Scientometrics 29: 353-376.*

3. *Cavalcanti A, Shirinzadeh B, Zhang M, Kretly LC (2008)."Nanorobot Hardware Architecture for Medical Defense" (PDF).Sensors.8 (5): 2932-2958.*

3a. *Kumar N., Thakur G., Raghu H. V., Singh N., Sharma P. K., Singh V.K., Khan A., Balhara M., Avinash, Lawaniya R., Kouser S., Tehri N., Gopaul R. and Arora S. (2013) Bacterial Spore Based Biosensor for Detection of Contaminants in Milk. J. Food. Process. Technol. 4: 277.*

4. *"Biosensors Primer" Retrieved 28 Jan. 2013.*

5. *A.Hierlemann, O.Brand, C.Hagleitner, H.Baltes, Micro-fabrication techniques for chemical/biosensors, Proceedings of the IEEE91 (6), 2003, 839-863.*

6. *A. Hierlemann, H.Baltes, "CMOS-based chemical micro-sensors, The Analyst 128 (1), 2003, pp.15-28.*

7. *Pickup JC, Zhi ZL, Khan F, Saxl T, Birch DJ. Nanomedicine and its potential in diabetes research and practice.Diabetes Metab Res Rev.2008, Nov-Dec; 24 (8): 604-10.*

8. *http://www.technologyreview.com/biomed icine/21603.*

9. *Andrea Pizzariello, Miroslav Stredansky and Stanislav Miertus, Biosensors applications in food analysis, Science and Technology.*

10. *J. Svorc, S. Miertus, M.Stredansky, PCT WO97/02359.*

11. *J.Svorc, M. Stredansky et al., Italian Pat. MI97A00216.*

12. *A.Pizzariello, M. Stredansky et al., Italian Pat.MI99000210.*

13. *S. Miertus, J. Katrlik et al., Biosensors and Bioelectronics, 998, 13, 911.*

14. *Sarvesh Rustagi, and Pravesh Kumar, Department of Food Technology and Zoology, Baba Farid Institute of Science and Research, Dehradun, Uttrakhand, Biosensor and It's Application in Food Industry, Advances in Bio-research, Adv. Biores., Vol. 4 (2) June 2013: 168-170.*

15. *www1.lsbu.ac.uk/water/enztech/biosensors.html*

Classification of Biosensor [1]

For classifications of biosensors, following approaches can be utilized [2,3].

(*a*) Depending upon the used transduction principle, biosensors could be distributed into groups of electrochemical, mass dependent, optical, radiation sensitive, *etc.* [4].

(*b*) Enzyme, nucleic acid, proteins, saccharides, oligonucleotides, ligands *etc.* are the various sets of biosensors which could be acquired if bioelement is considered as the basis of categorization [5].

Biosensor is broadly classified into two classes:

i. On the Basis of Biological Element

(*a*) **Enzyme Biosensor**: enzyme because of its high selectivity and activity towards substrate are the best candidate to be used as biologically active materials. Most of the enzymes used in biosensor are oxidizes but there are certain limitations as their activity is susceptible to pH, temperature, ionic strength *etc.*

(*b*) **Microbial Biosensor**-it uses microorganism as biological element. The metabolism of microorganism is used as basis of their activity. They are cheaper and versatile as compared to other sensorelements.

(*c*) **Antibody Based Biosensor-**It uses antibodies as sensor. They are immobilized on the surfaces by covalent/Ionic/Vander wall attachment in conjugation with amino-carboxyl groups *etc.*

ii. On the Basis of Transducing Element

(*a*) **Electrochemical transducers**–The two common types of electrochemical transducers are amperometric and potentiometric transducer. In the Amperometric transducers the potential between the two electrodes is set and current produced by oxidation –reduction of electro active species is measured and correlated with the concentration of the analyte of the interest [6]. The Potentiometric measures the potential of electrochemical cell with very low current value.

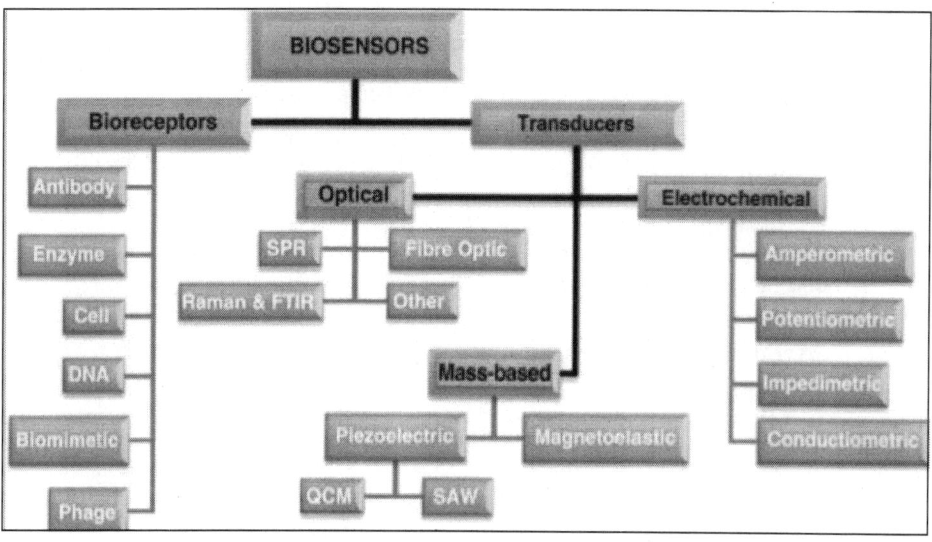

Classification of Biosensors.

(b) **Optical Transducer**–They are used for determining the concentration of analyte on the basis of change in optical density at appropriate wavelength. *I.e.* Total Internal Reflection is measured with a Photodetector as a function of incident angle.

(c) **Calorimetric Transducers**–it is used for calculating the heat of biochemical reactions bymeasuring the temperature difference between the reaction vessel and isothermal heat sink surrounding it.

(d) **Piezoelectric sensors**–These sensors utilise crystals which undergo an elastic deformation when an electrical potential is applied to them. An alternating potential (A.C.) produces a standing wave in the crystal at a characteristic frequency.

Source

1. *Sarvesh Rustagi, and Pravesh Kumar, Department of Food Technology and Zoology, Baba Farid Institute of Science and Research, Dehradun, Uttrakhand, Biosensor and It's Application in Food Industry, Advances in Bio-research, Adv. Biores., Vol. 4 (2) June 2013: 168-170.*
2. *Buenger D, Topuz F, Groll J (2012) Hydrogels in sensing applications. Progress in Polymer Science 37: 1678-1719.*
3. *Monošík R, Stre?anský M, Šturdík E (2012) Biosensors-classification, characterization and new trends. Acta Chimica Slovaca 5: 109-120.*
4. *Thévenot DR, Toth K, Durst RA, Wilson GS (2001) Electrochemical biosensors: Recommended definitions and classification. Biosensors and Bioelectronics 16: 121-131.*
5. *Mohanty SP, Kougianos E (2006) Biosensors: a tutorial review. IEEE Potentials 25: 35-40.*
6. *Mamas I. Prodromidis and Miltiades I.Karayannis (2001).Enzyme Based Amperometric Biosensors for Food Analysis.*

Advantages over other Methods of Analysis

Application of biosensors for analysis has following potential advantages over other methods of analysis:

(i) Many biosensors allow analysis of larger volumes of liquids in a continuous flow.

(ii) Fast and real time detection enable Food technologists or manufacturers to take corrective action before a product is further processed or delivered to consumers for consumption.

(iii) Biosensors can be integrated into HACCP programme, enabling critical examination of the entire manufacturing process.

(iv) Biosensors can be miniaturized and can be integrated into dairy equipment.

(v) Biosensors can also be integrated with on-line process monitoring schemes and hence enable us for better control and automation of milk and dairy facilities.

Beneficial Characteristics of Biosensors [1]

A successful biosensor must possess at least some of the following beneficial characteristic/features:

(i) The biocatalyst must be highly specific for the purpose of the analyses, be stable under normal storage conditions and show good stability over a large number of assays.

(ii) The reaction should be as independent of such physical parameters as stirring, pH and temperature as is manageable.

(iii) The response should be accurate, precise, reproducible and linear over the useful analytical range, without dilution or concentration and it should also be free from electrical noise.

(iv) The probe must be tiny and biocompatible, having no toxic or antigenic effects or sterilisable as per the requirements. The biosensor should not be prone to fouling or proteolysis.

(v) The complete biosensor should be cost effective, small, portable and capable of being used by semi-skilled operators.

Source

1. *www1.lsbu.ac.uk/water/enztech/biosensors.html*

Applications of Biosensors

1. Application in Dairy Industry/Food Industry

The food industry needs rapid and affordable techniques, both for compounds that have not previously been monitored, and to replace existing but complex and/ or expensive procedures. To establish a niche in this sector, biosensors must be inexpensive and reliable, and be robust enough to operate under conditions that may be hostile. Other requirements include increased speed in sample detection (instantaneous or just in time determinations), short sample preparation time, decreased requirement for skilled labour, measurements *in situ*, and a departure from the use of dangerous or 'environmentally unfriendly' organic reagents [1]. Biosensors are defined as indicators of biological compound that can be as simple as temperature sensitive paints or as complex as DNA-RNA probes. The science of biosensor is a multidisciplinary area. The potential application of biosensor technology to food testing offers several attractive features. Many of the system are portable and hence can be used for field testing or on the spot analysis and are rapid test which are capable of testing multiple analysis simultaneously.

There are many potential applications of biosensors of various types. The main requirements for a biosensor approach to be valuable in terms of research and commercial applications are the identification of a target molecule, availability of a suitable biological recognition element, and the potential for disposable portable detection systems to be preferred to sensitive laboratory-based techniques in some situations. There are several applications of biosensors in food analysis. In food industry optic coated with antibodies are commonly used to detect pathogens and food toxins. The light system in these biosensors has been fluorescence, since this type of optical measurement can greatly amplify the signal. Different immuno- and legend-binding assays for the detection and measurement of small molecules *e.g.* water-soluble vitamins and drug residues (chemical contaminants) *e.g.* sulfonamides and Beta-agonists have been developed for use on **SPR** based sensor systems (often adapted from existing **ELISA** or other assay).

A. Dairy Industry [2,3,4,5]

Biosensors are generally being used for the detection of DNA of the cow, Pathgogen testing in milk, micro-organism identification in milk, test for preservation and pasteurization. The details related to the various applications of biosensors based analytical techniques in dairy are given below:

I. Bacterial/Pathogens Analysis

Biosensing methods for pathogen detection are centered on four basic physiological or genetic properties of microorganisms-a-metabolic patterns of substrate utilization, b-phenotypic expression analysis of signature molecules by antibodies, c- nucleic acid analysis and d-the analysis of the interaction of pathogens with eukaryotic cells.

Many of today's popular commercially available rapid methods use culture-based methods coupled with automated or semi-automated **nucleic acids,** antibody, or substrate utilization-based methods to obtain results in 24-72 h. Interestingly, many of the modern-day biosensor-based methods are developed utilizing one of the above four principles or combinations of some sort. However, antibody-based methods are the most popular because of their versatility, convenience and relative ease in interpretation of the data. It is interesting to note that a majority of biosensors use antibody for capture and detection of the target analyte [6].

The detection of pathogenic bacteria can be done in following two ways [7,5]:

i. Direct Detection

These biosensors are those in which Biospecific reactions are directly measured in real time by measuring the physical changes induced by the complex formation.

a. Optical Biosensor

These biosensors are used for direct detection of bacteria. These sensors are able to detect the small change in refractive indices, when cell bind to receptor which are immobilized on the transducer [8]. The common optical biosensor which are used now a days are Elapsometric, Ewascent wave interferometer, resonant mirror, and piezoelectric biosensor.Fiber optic biosensor is one of the first commercially available optical biosensors, marketed by Research International (Monroe, WA) for the detection of foodborne pathogens. The basic principle of the fiber optic sensor is that when light propagates through the core of the optical fiber *i.e.* wave guide, it generates an evanescent field outside the surface of the waveguide. The waveguides are generally made up of polystyrene fibers or glass slides. When fluorescent labeled analytes such as pathogens or toxins bound to the surface of the waveguide, are excited by the evanescent wave generated by a laser (635 nm) and emit fluorescent signal [9,10], the signal travels back through the waveguide in high order mode to be detected by a fluorescence detector in real time.

b. Bioluminescence Biosensor

The use of photons as a byproduct of reaction for Bio analytical sensor led to the so called bioluminescence biosensor which may be use to detect the presence orphysiological state of cell. This phenonmenon is known as bioluminescence. This technique is giving results in a short time. The following two distinct areas of bioluminescence are of use in food industry [5]:

i. ATP Bioluminescence

All living cells contain the molecule ATP. This molecule may be analyzed simply using an enzyme and coenzyme complex (Luciferase- Luciferin) found in the tail of fire fly (Photinus pyralis). The total light output of the sample is directly proportional to the amount of ATP present and can be quantified by luminometers. Because of rapid response time for obtaining results, this system is very suitable for on-line monitoring of HACCP programs but this system lacks specificity. This technique has a detection limit of 1 pg ATP which is equivalent to 1000 bacterial cells and at least 10^4 cells are required to produce a signal. ATP is present in both non-microbial and microbial cells. To determine microbial ATP selective extraction is used. First, non-microbial ATP is extracted with non-ionic detergents and then destroyed with high levels of potato ATPase for 5 minutes. Subsequently, microbial ATP is extracted using either trichloro-acetic acid (5 per cent) or an organic solvent (ethanol, acetone or chloroform).

ii. Bacterial Bioluminescence

The gene responsible for bacterial bioluminescence (lux gene) has been identified and cloned. The DNA carrying this gene can be introduced into host specific phages. On transfer of lux gene to the host bacterium during infection results in light emission that can be easily detected by luminometers. This technique can detect 1×10^2 cells in 60 min. The specificity of this assay depends on phage specificity *e.g.*: Bacteriophage p22 is specific for Salmonella typhimurium.

c. Electrical Impedence Biosensor

Impedance microbiology detects microbes either directly due to production of ions from metabolic end products or indirectly from liberation of CO_2. In this biosensor microbial metabolism bring increase in both conductance and capacitance and thus cause a decrease in impedance [5, 11]. The biosensor is based on impedance measure of adherently growing cells on inter digitized electrode structure. *i.e.* Cell density, growth of cells on electrode changes the impedance of biosensor. This method is well suited for monitoring of quality and detect specific food pathogens and for detection of bacteria in clinical samples. Typically most impedance analysis of food samples can be completed in 24 h. This technique is not suited for testing samples with low number of microorganisms and that the food matrix may interfere with the analysis.In this method, a population of microbes is provided with nutrients (non-electrolyte) like lactose and microbes may utilize that nutrient and convert it to lactic acid (ionic form) thus changing the impedance. This impedance is measured over a period of 20 h after inoculation in specific media. This system is capable of analyzing hundreds of sample at the same time since the instrument (Bactometer) is computer driven and automated to enable continuous monitoring [5]. **Impedance-based biochip sensor** is also used to detect pathogens. Impedance is based on the changes in conductance in a medium due to the microbial breakdown of inert substrates into electrically charged ionic compounds and acidic by-products. The threshold value for the detection of target pathogens is mainly depends on initial inoculums and the physiological state of the cells. In media-based impedance methods, bacterial metabolism results in increased conductance and capacitance, with decreased impedance [12]. The major advantage of this system is that it allows the detection of only the viable cells, which is the major concern in **food safety** [5].

ii. Indirect Detection

These biosensors are those in which a preliminary bio chemical reaction takes place and the product of the reaction are detected by a sensor.

a. Fluorescence Labeled Biosensor

Microorganism have protein and polysaccharides in their outer coat which permit development of bioassays for bacterial detection. In Fluoro immunoassay, Fluor chrome molecules are used to label immunoglobulin. These Fluor chrome molecules absorb short wavelength light and emit higher wavelength light which detect by the fluorescent microscopy. FITC, RITC, BSA are most useful Fluorochrome used to tag the antibody *e.g.* Using an Antibody to the protective protein express with the anthrax toxin for use of detection of Anthrax.

b. Microbial Metabolism Based Biosensors

Micro organism is able to transducer their metabolic Redox reaction in to quantify electric signals by using Oxido Reductase reaction and mediator.

c. Flow Immune Sensors

Now days, many of the assays for microbes are based on Elisa using micro titration plates on completion of Chromogenic reaction, the quantitative determination is done using Elisa reader *e.g.* E.coli detection.

iii. Others

a. Piezoelectric Biosensors [5]

This system is very attractive and offers a real time output, simplicity of use and cost effectiveness. The general principle is based on coating the surface of piezoelectric sensor with a selective binding substance for example antibodies to bacteria and then placing it in a solution containing bacteria. The bacteria will bind to the antibodies and the mass of the crystal will increase while the resonance frequency of oscillation will decrease proportionally.

b. Surface Plasmon Resonance (SPR) Sensor [5]

SPR is a phenomenon that occurs during optical illumination of a metal surface and it can be used for biomolecular interaction analysis. Receptors or antibodies immobilized on the surface of a thin film of a precious metal (gold) deposited on the reflecting surface of an optically transparent waveguide are used to capture the target analyte. The sensing surface is located above or below a high index-resonant layer and a low index coupling layer. When a visible or near-infrared radiation (IR) is passed through the waveguide in such a way, it causes an internal total reflection on the surface of the waveguide. At a certain wavelength in the red or near-IR region, the light interacts with a plasma or cloud of electrons on the high-index metal surface and the resonance effect causes a strong absorbance. The exact wavelength of this absorption depends on the angle of incidence, the metal, the amount of capture molecules immobilized on the surface and the surrounding material. The presence of ligands or antigens interacting with the receptor or antibody causes a shift in the resonance to longer wavelengths and the amount of shift can be related to the concentration of the bound molecules.

SPR-based sensors are governed by two basic principles:

(i) Wavelength interrogation.

(ii) Angle interrogation.

Wavelength interrogation uses a fixed angle of incidence but measures spectral changes, while in angle interrogation, a fixed wavelength is used but the angle of reflectance is monitored. Most of the commercial SPR systems are operated based on the angle interrogation mode. SPR-based sensors allow real-time or near real-time detection of binding events between two molecules. The detection system is label free, thus eliminating the need for additional reagents, assay steps and time. The sensor can be reused for the same analyte repeatedly. It is highly sensitive and it can detect molecules in the femtomolar range [9,13].

c. Cell based Sensor [5]

Cell-based assays (CBAs) continue to serve as a reliable method for detection of pathogens in food samples. The CBA systems can report perturbations in the normal physiological activities of mammalian cells as a result of exposure to an external or environmental challenge. For this, mammalian cells are used as electrical capacitors. Electrical Impedance (EI) uses the inherent electrical properties of cells to measure the parameters related to the tissue environment. The mechanical contact between cell-cell and cell-substrates is measured via conductivity or EI. The cell can be equated to a simple circuit since it is nothing more than conductive fluid encapsulated by a membrane surrounded by another conductive fluid. The conductive fluids make up the resistance elements of the circuit, while the membrane acts as a capacitor. Changes in impedance were able to detect changes in cell density, growth, or cellular behavior. These biosensors are able to provide detailed information about the growth characteristics of the **tissue culture,** including information on spreading, attachment and cellular morphology. Mammalian cells have been widely used for the analysis of the pathogenic potential of foodborne bacteria [14,15].

The details of the biosensors developed for the analysis of microbial contamination in milk and milk products are given in following table [2,3,4]:

Sl.No.	*Biosensor Techniques*	*Microorganism*	*Detection Limit (cells/ml)*
1	Electrochemical	*Listeria monocytogenes, Staphylococcus aureus, Salmonella typhimurium, E.coli* O157H7.	10^3-10^5
2	Resonant mirror	*Staphylococcus aureus, Salmonella* spp., *E.coli*	10^5-10^7
3	Light addressable potentiometric sensor	*Salmonella typhimurium, E.coli* O157H7	10^2-10^4
4	Fluorescent fibre optic	*E.coli* O157H7, *Salmonella typhimurium*	10^2-10^5
5	Peizoelectric	*Salmonella* spp., *E.coli, Vibrio cholerae*	10^5-10^8

II. Detection of Antibiotic/Drug Residues

Several Biosensors have been developed and available for detection of Antibiotic residues in milk and milk products. The use of antibiotics for the treatment of food-producing animals generates the risk to human health, as these compounds and their metabolites can be transferred into milk. Rapid testing of the presence of antibiotics in raw milk to grant its quality has become a major task for farmers and dairy industry. The conventional analytical methods are either too slow or do not enable quantitative detection of antibiotic residues, so alternative methods *e.g.* Biosensing [16] *etc.* which are rapid, cost effective, and easy to perform, are being used to detect amtibiotic residues in milk.

III. Analysis of Effectiveness of Milk Heat Treatment

Biosensors techniques are used to monitor the Lactulose produced during UHT process and generally present in UHT milk (around 500 ppm)to assess the intensity of thermal process.

IV. Detection of Microbial Toxin/Toxic Metabolites (Mycotoxins)

The real time analysis of Staphylococcal enterotoxins in milk has been done by Optical biosensors. Aflatoxin M_1 in milk is being determined by many biosensors. The following table shows the detection limits and detection times for some biosensors developed for detrection of various microbial toxins in milk and milk products:

Sl.No.	*Toxin*	*Detection Time*	*Detection Limit*
1	AflatoxinM_1	< 2min	0.1ppb
2	AflatoxinM_1	18 min	0.05ppb
3	Staphylococcal enterotoxin A	< 5min	10ppb
4	Staphylococcal enterotoxin B	< 2min	5ppb

V. Detection of Mastitis

The detection of Mastitis can be done by an electric conductivity biosensors.

VI. Analysis of Components of Milk Solids

The on- line measurement of major components of milk solids (Fat, Protein, Lactose and Calcium *etc.*) can be done by biosensors.

VII. Integration with HACCP Programme

The screening of antibiotic residues, assaying for pathogens and assaying the effectiveness of milk heat treatments in dairy plants, are the possible uses of biosensors in HACCP monitoring.

VIII. Application in Quality Control

The level of L- and D- amino acids in milk (depends upon the type of processing and duration and condition of storage) can be detected by biosensors and would give information about the effect of milk ageing.

Source

1. *Andrea Pizzariello, Miroslav Stredansky and Stanislav Miertus, Biosensors applications in food analysis, Science and Technology.*
2. *Technews, No.74, May-June-2008, NDDB.*
3. *Rasooley, A. (2007), Bio-sensors, Encyclopedia of Dairy Sciences (Voln.-1), pp-85-93, Elsevier Science Limited, Holroyd, A. (2007), Rapid methods for the chemical and microbiological testing of dairy products, IDF bulletin 424/2008, pp-3-9, IDF, Brussels.*
4. *Pohanka M, Skladal P, Kroca M (2007)."Biosensors for biological warfare agent detection". Def.Sci. J. 57 (3): 185-93.*
5. *P.K. Mandal, A.K. Biswas, K. Choi and U.K. Pal, Methods for Rapid Detection of Foodborne Pathogens: An Overview, Journal of Food Technology, 6: 87-102.*
6. *Ritcher, E.R., 1993. Biosensors: Application for dairy food industry. J. Dairy Sci., 76: 3114-3117.*
7. *Sarvesh Rustagi, and Pravesh Kumar, Department of Food Technology and Zoology, Baba Farid Institute of Science and Research, Dehradun, Uttrakhand, Biosensor and It's Application in Food Industry, Advances in Bio-research, Adv. Biores., Vol. 4 (2) June 2013: 168-170.*
8. *P.J. O'Connell, C. K.O'Sullivan, G.G. (2000). Biosensors for Food Analysis: Guilbault Source: Irish Journal of Agricultural and food Research, Vol. 39, No. 2, Special Issue on Food Safety.*
9. *Bhunia, A.K., 2007. Biosensors and bio-based methods for the separation and detection of food borne pathogens. Adv. Food Nutr. Res., 54: 1-44.*
10. *Taitt, C.R., G.P. Anderson and E.S. Ligler, 2005. Evanescent wave fluorescence biosensors. Biosensors Bioelectronics, 20: 2470-2487.*

11. *Dmitri Ivnitski, Ihab Abdel-Hamid, Plamen Atanasov 1999. Biosensors for detection of pathogenic bacteria, Ebtisam Wilkins USA.*

12. *Invitski, D., L. A. Harnid, P. Atanasov and E. Wilkins, 1999. Biosensors for detection of pathogenic bacteria. Biosensors Bioelectronics, 14: 599-624.*

13. *Rasooly, A. and K.E. Herold, 2006. Biosensors for the analysis of food and waterborne pathogens and their toxins. J. AOAC. Int., 89: 873-883.*

14. *Bhunia, A.K. and J.L. Wampler, 2005. Animal and Cell Culture Models for Foodborne Bacterial Pathogens. In: Foodborne Pathogens: Microbiology and Molecular Biology, Fratamico, P., A.K. Bhunia and J.L. Smith (Eds.). Caister Academic Press, Norfolk, UK.*

15. *Gray, K.M., 2004. Cytotoxicity and cell-based sensors for detection of Listeria monocytogenes and Bacillus species. Ph.D. Thesis, Purdue University, West Lafayette.*

16. *Kairi Kivirand, Margarita Kagan and Toonika Rinken, Institute of Chemistry, University of Tartu, Tartu, Estonia, Biosensors for the Detection of Antibiotic Residues in Milk,*

B. Food Industry

In order to ensure the safety of processed foodstuffs, specific methods have been adopted by the food industries to sort out the problems leading to food spoilage and detection and destruction of such chemicals or biological agents that are responsible for the spread of some serious health related problems [1,2]. Biosensors being target specific, highly sensitive and quickly responsive would be helpful in that regard to enable us to determine the cause of chemical activities that leads to the food spoilage, not only the canned and industrial food products. There are several applications of biosensors in food analysis but mainly enzyme-based biosensors and immunosensors are employed.In food industry optic coated with antibodies are commonly used to detect pathogens and food toxins. The light system in these biosensors has been fluorescence, since this type of optical measurement can greatly amplify the signal.

A range of immuno- and ligand-binding assays for the detection and measurement of small molecules such as water-soluble vitamins and chemical contaminants (drug residues) such as sulfonamides and Beta-agonists have been developed for use on SPR based sensor systems, often adapted from existing ELISA or other immunological assay. These are in widespread use across the food industry.

(i) Detection of Heavy metals [3a].

(ii) **Analysis of Carbohydrates** -Enzyme Biosensor are used in industries [3] as Wine, Beer Soft drink *etc.* for detecting or measuring the carbohydrates from Alcohol, Amino acids, Amines, Amides, Phenol *etc.* [3,4].

(iii) **Determination of Vitamins-**Routine analytical measurement of folic acid, biotin, vitamin B_{12} and panto thenic acid as an alternative to microbiological assay.

(iv) Detectin of residual agro chemicals [3a].

Sl.No.	Food Components	Enzyme Used
1	Glucose	Glucose oxidase
2	Fructose	Fructose -5-dehydrogenase
3	Sucrose	Glucose Oxidase, Mutarotase,Invertase
4	Lactose	Galactose oxidase and Peroxidase
5	Glutamate	Glutamate Oxidase
6	Malate	Malate Dehydrogenase, Diaphorase
7	Glycerol	Glycerol Dehydrogenase
8	Cholesterol	Cholesterol Oxidase
9	Essential fatty acids	Lipoxygenase
10	Ethanol	Alcohol Dehydrogenase
11	Choline	Choline Oxidase

(v) Detection of Toxic metals [3a].

(vi) **Detection of Microorganism and fungi in foods**-Biosensors are successfully being used for the detection of microorganism and fungi.

a. Bacterial Monitoring

E.coli strain 0157: H7, *Listeria monocytogenes,* campylo- bacter and salmonella are common food spoiling bacteria and these bacteria are problems for the food industries as they reduce the consumer demands of the food if the food provided by company gets contaminated with these food spoiling biological entities.

Salmonella

It is the major cause of food poisoning. Food industries had been striving to get rid of the cause of food poisoning by timely detection and removal of this bacterium. The piezoelectric biosensors are commonly used for monitoring Salmonella and these can detect monoclonal antigen-antibody complex quickly and easily but are poor detectors of polyclonal antibody complexes, which is its main drawback and needs to be addressed for upgradations [5].

Listeria monocytogenes

This bacterium is commonly found in processed meat. It can also be found in raw milk or its products. They are the cause of a common disease in humans named as listeriosis. The fibre-optic biosensors have been commonly used to monitor the presence of L. monocytogenes [6]. Surface Plasmon Resonance (SPR) biosensors along with fibre-optic biosensors have proven to be best for L. monocytogenes detection [5].

Campylobacter jejuni

It is also the common cause of food poisoning worldwide. Optical SPR can also be used for quick detection of campylobacter [5].

E. coli strain 0157: H7

It leads to the cause of bloody diarrhea and sometimes it may cause kidney failure. It is mostly present in poorly cooked foods, cheese, raw milk, juices and also in contaminated water sources *etc.* Amperometric biosensors based on bienzymatic system had been successfully used for the detection of E. coli 0157: H7 [7]. Before that optical sensors had also showed precise monitoring of 0157: H7.

b. Fungal Pathogens Detection

The fungi are also the common cause of food spoilage and causing severe health related problems that may prove to be life threatening in most cases. Fungi that cause food contamination are commonly *Botrytis* sp., *Aspergillus, Colletotrichum* and many other fungal species like that. Due to the remarkable specificity, reduced costs and easy and quick monitoring through biosensors, fungal toxins can be also detected using optical SPR biosensors [8,9,10,11] . After their detection, steps to remove them from food products can be taken within the time. The details related to the bacteria, types of biosensors used, foods, detection range *etc.* are given in following table:

Sl.No.	*Bacteria*	*Types of Biosensors*	*Found in Foods*	*Detection Range/ Limit*	*Refere-nces*
1	*Salmonella*	Piezoelectric	Eggs, meat, raw milk and dairy products.	1.7×10^{2} cells/mL	[9]
2	*L. mono-cytogenes* (pure culture)	Fibre-optic SPR sensor	Raw milk, meat, milk products.	10^{7} to 10^{8} CFU/mL low concentrations upto 10^{5} CFU/mL	[6] [10]
3	*C. jejuni* (pure culture)	SPR sensor	Raw milk, uncooked chicken, contaminated water.	10^{1} to 10^{8} CFU/mL	[11]
4	*E. coli* O157:H7	Amperometric	Raw milk, milk products, juices, cheese *etc.*	1.0×10^{7} cells/mL	[7]

(vii) Assure Foods Safety and as indicators of product acceptability [3a].

(viii) Sensory analysis [3a].

(ix) Detection of unpermitted chemicals [3a].

Source

1. *Wei N (2011) A novel hydrogen peroxide biosensor based on the immobilization of hemoglobin on three-dimensionally ordered macroporous (3DOM) gold-nanoparticle-doped titanium dioxide (GTD) film. Biosensors and Bioelectronics 26: 3602-3607.*

2. *Villalonga R (2011) Wiring horseradish peroxidase on gold nanoparticles-based nanostructured polymeric network for the construction of mediatorless hydrogen peroxide biosensor. Electrochimica Acta 56: 4672-4677.*

3. *F.Scheller, F.Schubert, (1992).Biosensors, Elsevier, Amsterdam.*

3a. *Muruga-boopathi G., Parathasarathi V, Chellaram C., Anand T.P. and Vinurajkumar S (2013) Application of Biosemsors in food industry, Biosciences Biotechnolgy Reasearch Asia 10 (2), 707-711.*

4. *Sarvesh Rustagi, and Pravesh Kumar, Department of Food Technology and Zoology, Baba Farid Institute of Science and Research, Dehradun, Uttrakhand, Biosensor and It's Application in Food Industry, Advances in Bio-research, Adv. Biores., Vol. 4 (2) June 2013: 168-170.*

5. *Serna-Cock L, Perenguez-Verdugo JG (2011) Biosensors applications in agri-food industry. INTECH Open Access Publisher.*

6. *Geng T, Morgan MT, Bhunia AK (2004) Detection of low levels of Listeria monocytogenes cells by using a fiber-optic immunosensor. Applied and Environmental Microbiology 70: 6138-6146.*

7. *Tang H (2006) A new amperometric method for rapid detection of Escherichia coli density using a self-assembled monolayer-based bienzyme biosensor. Analytica Chimica Acta 562: 190-196.*

8. *Li Y, Liu X, Lin Z (2012) Recent developments and applications of surface plasmon resonance biosensors for the detection of mycotoxins in foodstuffs. Food Chemistry 132: 1549-1554.*

9. *Fung Y, Wong Y (2001) Self-assembled monolayers as the coating in a quartz piezoelectric crystal immunosensor to detect Salmonella in aqueous solution. Analytical Chemistry 73: 5302-5309.*

10. *Leonard P Hearty S, Quinn J, O'Kennedy R (2004) A generic approach for the detection of whole Listeria monocytogenes cells in contaminated samples using surface plasmon resonance. Biosensors and Bioelectronics 19: 1331-1335.*

11. *Wei D (2007) Development of a surface plasmon resonance biosensor for the identification of Campylobacter jejuni. Journal of Microbiological Methods 69: 78-85.*

2. Applications of Biosensors in Enviromenrtal Studies

Different types of biosensors based on enzymatic inhibition and those based on DNA with high specificity to certain DNA, RNA or proteins that they can encounter have been used for the removal of contaminants like, pesticides, other toxins and heavy metals from soil, air and water including many other versatile types of biosensors. The major pollutant that can be successively detected and removed using biosensors includes heavy metals, polychlorinated biphenyls, pesticides, Biochemical Oxygen Demand (BOD), nitrogenous compounds and various pathogens (including many viruses and bacteria).

a. Heavy Metals

Heavy metals pose maximum threat to Human health. Bacteria-based cell biosensors requires the use of genes that resist certain types of heavy metals like copper, mercury, tin cobalt *etc.* [1,2]. However, they can act when the heavy metals interact with their cytoplasm, dependence of these sensors is based on the conjugation of some luminescent proteins like luciferin, with those genes that resist heavy metals [3]. Enzyme-based biosensors have also provided promising results in that regard, like fibre-optic biosensors have been used for the detection of the toxic levels of various heavy metals like lead, cadmium, mercury, copper, nickel, cobalt *etc.* Amperometric biosensors were used for the successful detection of inhibition of mercury ions (Hg+2) by urease enzyme action [4]. Inhibition of cobalt, nickel, mercury, gold and lead with same urease enzyme lead to the monitoring of the toxic levels using fibreoptic sensors [5].

b. Polychlorinated Biphenyls (PCBs)

Polychlorinated biphenyls are non-biodegradable agents used in various types of herbicides and insecticides that though prove to be good for pest control but also lead to the accumulation of these PCBs in the soil which are then taken up by the crops and in turn they enter human body and causes serious health problems. Biosensors have been used in precise detection of these organic compounds in foods and soils, using immunological biosensors that monitors antigen-antibody interaction using piezoelectric transducers [6,7].

c. Pesticides

Organophosphates being commonly used as insecticides (pesticides) pose changes to the soil fertility, thus damaging many beneficial insects and microbes in soil and leads to the loss of biodiversity. Another type of nanotechnological sensors have been used recently to measure toxic levels of these pesticides in soils and in water. Based on nanotechnology, enzymatic biosensors have been modified by allowing them to be immobilized. The common example is of acetylcholinesterase (enzyme) sensors which work by inhibiting acetylcholine esterase activity in order to detect organophosphates, where acetylcholinesterase activity is constantly monitored [8].

The details related to the various biosensors for environmental pollutants are given in following table:

Sl.No.	*Pollutants*	*Biosensors*	*Detectability source*	*References*
1	Heavy metals (Hg, Cd, Ni, Co, Zn, Cu, Pb, Au, Pb, Cu)	Bacterial-based sensors	Soil, sewage water	[1]
		Amperometric (Ureases enzyme-based) for Hg		[4]
		Optical sensor		[5]
2	PCBs	Piezoelectric immuno-sensor	Soil, water and various foods (fruits,vegetables)	[6]
3	Organophosphates (Pesticides)	Enzymatic sensors (Acetylcholinesterase based)	Soils, water sources	[8]

Source

1. *Magrisso S, Erel Y, Belkin S (2008) Microbial reporters of metal bioavailability. Microbial Biotechnology 1: 320-330.*

2. *Rathnayake I (2009) Tolerance of heavy metals by gram positive soil bacteria. World Academy of Science Engineering and Technology.*

3. *da Costa Silva LM, Melo AF, Salgado AM (2004) Biosensors for Environmental Applications. Pure Appl.Chem 76: 723–752.*

4. *Domínguez-RenedoO (2009) Development of urease based amperometric biosensors for the inhibitive determination of Hg (II). Talanta 79: 1306-1310.*

5. *Kuswandi B (2003) Simple optical fibre biosensor based on immobilised enzyme for monitoring of trace heavy metal ions. Analytical and Bioanalytical Chemistry 376: 1104-1110.*

6. *Hepel M, Skládal P (2006) Piezoelectric immunosensors for polychlorinated biphenyls operating in aqueous and organic phases. Sensors and Actuators B: Chemical 113: 900-910.*

7. *Ivask A, Virta M, Kahru A (2002) Construction and use of specific luminescent recombinant bacterial sensors for the assessment of bioavailable fraction of cadmium, zinc, mercury and chromium in the soil. Soil Biology and Biochemistry 34: 1439-1447.*

8. *Zhang W (2014) Nanomaterial-based biosensors for environmental and biological monitoring of organophosphorus pesticides and nerve agents. TRAC 54: 1-10.*

Spore Based Biosensors [1]

The members of genera Clostridium and Bacillus have the aptitude to form endospores during stress and starvation conditions. The living phase of bacterial spores revolves around following two phases:

- *(i)* **The dormant state:** The dormant spore state is very much stable allowing them to survive even for millions of years. The dormant bacterial spores having unique ability to sense environmental changes in response to specific "germinant".
- *(ii)* The metabolically active vegetative state.

The conversion from one phase (dormant state) to another phase (metabolically active vegetative state) is only possible if spore senses favorable conditions in the environment. The presence or absence of microbial or nonmicrobial contaminants directly or indirectly affect this conversion. The spore germination concepts involved the release of DPA/or marker enzymes and their action on specific chromogenic/ or fluorogenic substrates as a mean for detection of contaminants in dairy foods has been explored successfully. This phenomenon can therefore be used to develop spore based biosensor systems which sense the presence of contaminants in milk. These biosensing systems are superior over existing methods in terms of better sensitivity, low cost and help on site rapid analysis of milk and milk products. The spore based biosensor is a novel strategy being exploited to ensure safe and healthy milk to each consumers. A number of spore based sensing system have been developed to detect aflatoxins [2,3], antibiotics and microbial pathogens in milk. The details of some of these biosensors are tabulated on next page:

Sl.No.	*Principle/ Process etc.*	*Detection of antibiotic residues in milk*	*Detection of pesticides*
1	**Principle**	Working on principle of spore germination and its inhibition in presence of antibiotic residues in milk [4]. If antibiotic residues are absent in milk, the calcium dipicolinic acid (DPA) is released during germination which change the color of the bromocresol purple (BCP) indicator from purple to yellow. If antibiotic residues are present in milk, the spore germination process is inhibited at ≥ MRL level of contaminants and no change in color of BCP indicates the presence of drug residues in milk.	Spore based assay which exploit the principle of **'spore germination and enzyme inhibitio**n' for detection of pesticide residues was successfully developed.
2	**Process**	Take milk sample, add bromocresol purple solution and incubate at 64°C for 2.50-3.00 hrs. Observe thew colour [4].	The assay protocol involves the following three steps- i-spore germination,ii-pesticide exposure and iii-enzyme-substrate reaction. Using this protocol all marker enzymes were screened for their inhibition by pesticide. The optimized 3 step assay protocol was evaluated for LOD after spiking pesticides in acetonitrile and LOD for organophosphorous (OP), carbamate (CM) and organo-chlorine (OC) were observed at 10 ppb-10 ppm, 1-10 ppm and 100 ppm respectively. The protocol developed in tube was further attempted to miniaturize on paper strip for exploring its potential for field application.
3	**Others**	i. Cost effective ii. Semi-quantitative detection at Codex MRL iii. Validated with AOAC approved Charm 6602 Assay iv. Minimal false positive/ negative results v. No interference of inhibitors other than antibiotic residues.	The specificity of said assay was evaluated by checking for cross reactivity with other inhibitors, however, marker enzyme activity was found unaffected by antibiotics, afla-toxin,detergents, heavy metals, preservatives and sanitizers. Subsequently, a protocol for extraction of pesticides from food sample *i.e.*spiked reconst-ituted milk samples was optimized. "Three step Spore based assay" miniaturised on 96 well plate/paper strip was found useful in detection of OP and CM group of pesticides employing spores as an inexpensive source of marker enzyme(s) alternative to acetyl cholinesterase enzyme which is present only in eukaryoticsystem.

Source

1. *Naresh K., Raghu H. V., Sharma. P K., Ankita K., Nimisha T., Birjesh K., Spurthi, M, Spores/Enzymes based Intervention for Rapid Detection of Contaminants in Milk, Microbial Biosensor, Food Safety and Quality Assurance Lab, Compendium on Advanced Techniques and Novel Approaches for Quality and Safety Evaluation of Dairy Foods-2015, NDRI, Karnal.*
2. *Prandini A., Tansini G., Sigolo S., Filippi L., Laporta M. and Piva G. (2009). On the occurrence of Aflatoxin M1 in milk and dairy products. Food. Chem. Toxicol. 47: 984-991.*
3. *Shundo L., Navas S. A., Lamardo L.C. A., Ruvieri V. and Sabino M. (2009) Estimate of Aflatoxin M1 exposure in milk and occurrence in Brazil. Food Control 20: 655-657.*
4. *Kumar N., Patil G. R., Rane S., Malik R. K. (2014). A Novel Process of Sporulation, Activation and Germination in Thermophilic Bacteria for Rapid Detection of Antibiotic Residues in Milk. Indian Patent No. 264145 dated 9.12.2014.*

Chapter 4

Electrophoresis

Electrophoresis is the separation of charged molecules in an applied electric field. It is the motion of dispersed articles relative to a fluid under the influence of a spatially uniform electric field [1]. This electro-kinetic phenomenon was observed for the first time in 1807 by Ferdinand Frederic Reuss [2] who noticed clay particles dispersed in water to migrate on applying constant electric field for the first time. It is ultimately caused by the presence of a charged interface between the particle surface and the surrounding fluid. It is the basis for a number of analytical techniques used in biochemistry for separating molecules by size, charge, or binding affinity.

Because many biological compounds have charges (Positive or negative electrical charges) or ionizable groups (*e.g.*, DNA and proteins), electrophoresis is frequently utilized in biochemical and medical researches. When these biomolecules are placed in an electric field, charged biomolecules move towards the electrode of opposite charge (positively charged ions in the sample will move toward the

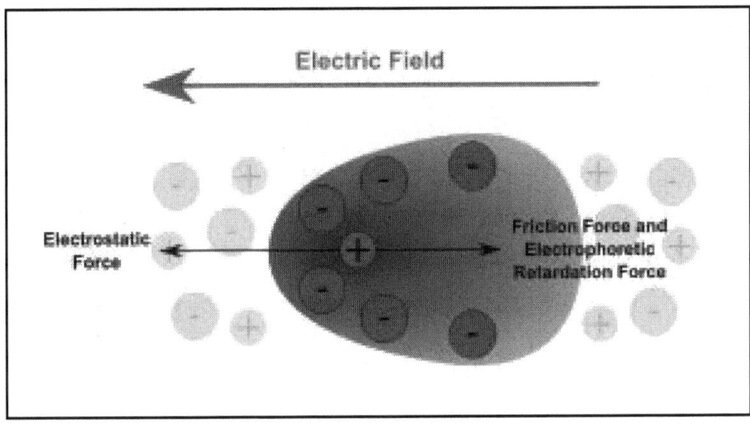

Illustration of Electrophoresis.

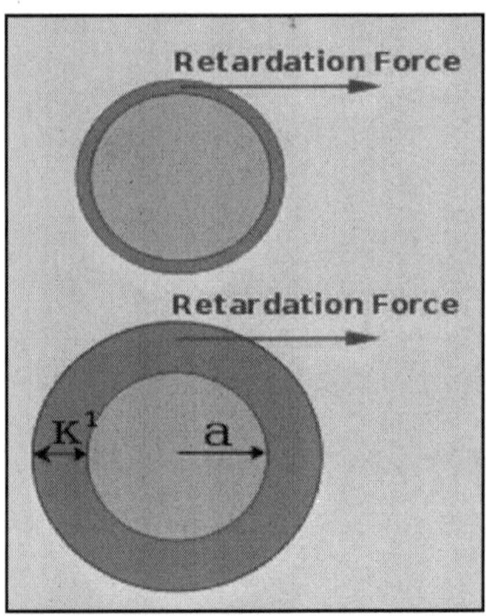

Illustration of Electrophoresis Retardation.

negative electrode -**the cathode**, while negatively charged ions will move toward the positive electrode -**the anode**, due to the phenomenon of electrostatic attraction. The relative mobility of individual molecules depends on several factors. The most important of which are net charge, charge/mass ratio, molecular shape and the temperature, porosity and viscosity of the matrix through which the molecule migrates [3-7]. Electrophoresis of positively charged particles (cations) is called **cataphoresis**, while electrophoresis of negatively charged particles (anions) is called **anaphoresis**.

Source

1. *Wikipedia.*
2. *Ferdinand Frederic Reuss, Electrophoresis.*
3. *Pulimamidi Rabindra Reddy and Nomula Raju, Gel-Electrophoresis and Its Applications, Department of Chemistry, Osmania University, Hyderabad, India.*
4. *M.L. Riekkola, J. A. Jonsson, and R. M. Smith, "Terminology for Analytical Capillary Electromigration Techniques (IUPAC Recommendations 2003), "Pure and Applied Chemistry, 76 (2004) 443–451*
5. *T. Blanc, D. E. Schaufelberger, and N. A. Guzman, "Capillary Electrophoresis." In Analytical Instrumentation Handbook, 2nd ed., G. W. Ewing, Ed., Marcel Dekker, NewYork, 1997, Chapter 25.*
6. *J. Inczedy, T. Lengyel, and A. M. Ure, Compendium of Analytical Nomenclature, 3rd ed., Blackwell Science, Malden, MA, 1997.*
7. *B. L. Karger, L. R. Snyder, and C. Hovath, An Introduction to Separation Science, Wiley, New York, 1973, Chapter 17.*

Developmental Steps/History of Electrophoresis [6]

The details related to the development of electrophoresis are given in following table:

Sl.No.	Year	Developments
1	1930s	First reports of the use of sucrose for gel electrophoresis
2	1955	Introduction of starch gels, mediocre separation
3	1959	Introduction of acrylamide gels; disc electrophoresis (Ornstein and Davis); accurate control of parameters such as pore size and stability; and (Raymond and Weintraub)
4	1966	A gels [1].
5	1967	Electrophoretic separation of molecules in a glass tube and subsequent detection of the separated compounds by ultraviolet absorption was first described by Hjerten [1a].
6	1969	Introduction of denaturing agents especially SDS separation of protein subunit [2].
7	1970	Laemmli separated 28 components of T4 phage using a stacking gel and SDS.
8	1972	Agarose gels with ethidium bromide stain [3].
9	1975	2-dimensional gels ; isoelectric focusing then SDS gel electrophoresis
10	1977	Sequencing gels
11	1981	electrophoresis in a tube did not become popular until 1981, when Jorgenson and Lucaks demonstrated the high-resolution power of capillary zone electrophoresis(CZE) [3a].
12	1983	Pulsed field gel electrophoresis enables separation of large DNA molecules
13	1983	Introduction of capillary electrophoresis
14	others	i-A 1959 book on electrophoresis by Milan Bier cites references from the 1800s [4]. ii-Oliver Smithies made significant contributions [5].

Source

1. *Thorne HV (1966). "Electrophoretic separation of polyoma virus DNA from host cell DNA". Virology 29 (2): 234–9.*

1a. *Hjerten, S. (1967). Free zone electrophoresis. Chromatogr. Rev. 9, 122–219.*

2. *Weber, K; Osborn, M (1969). "The reliability of molecular weight determinations by dodecyl sulfate-polyacrylamide gel electrophoresis". The Journal of Biological Chemistry 244 (16): 4406–12.*

3. *Aaij C, Borst P (1972)."The gel electrophoresis of DNA".Biochim Biophys Acta 269 (2): 192-200.*

3a. *Jorgenson, J. W., and Lukacs, K. D. (1981). Free-zone electrophoresis in glass capillaries. Clin. Chem. 27, 1551–1553.*

4. *Milan Bier (ed.) (1959). Electrohoresis. Theory, Methods and Applications (3rd printing ed.). Academic Press. p. 225.*

5. *Minde DP (2012)."Determining biophysical protein stability in lysates by a fast proteolysis assay, FASTpp". PLOS ONE 7 (10): e 46147.*

6. *Wikipedia.*

Theory and Principle [1,2]

Suspended particles have an electric surface charge, strongly affected by surface adsorbed species [3], on which an external electric field exerts an electrostatic Coulomb force.

i. Double Layer Theory

According to this theory, all surface charges in fluids are screened by a diffuse layer of ions, which has the same absolute charge but opposite sign with respect to that of the surface charge. The electric field also exerts a force on the ions in the diffuse layer which has direction opposite to that acting on the surface charge. This latter force is not actually applied to the particle, but to the ions in the diffuse layer located at some distance from the particle surface, and part of it is transferred all the way to the particle surface through viscous stress. This part of the force is also called electrophoretic retardation force. When the electric field is applied and the charged particle to be analyzed is at steady movement through the diffuse layer, the total resulting force is zero:

$$F_{tot} = 0 = F_{el} + F_f + F_{ret}$$

Considering the drag on the moving particles due to the viscosity of the dispersant, in the case of low Reynolds number and moderate electric field strength E, the drift velocity of a dispersed particle v is simply proportional to the applied field, which leaves the electrophoretic mobility μ_e defined as:

$$\mu_e = \frac{v}{E}.$$

ii. Smoluchowski Theory

The most well known and widely used theory of electrophoresis was developed by Smoluchowski [4] in 1903, which can be better represented by the following formula:

$$\mu_e = \frac{\varepsilon_r \varepsilon_0 \zeta}{\eta}$$

where,

ε_r - the dielectric constant of the dispersion medium, ε_0 - the permittivity of free space ($C^2 N^{-1} m^{-2}$), η -dynamic viscosity of the dispersion medium (Pa s), and ζ - zeta potential (*i.e.*, the electrokinetic potential of the slipping plane in the double layer).

Merits and Limitations

Merit

This theory is very powerful because it works for dispersed particles of any shape at any concentration.

Limitations

(a) It has limitations on its validity. It follows, for instance, because it does not include Debye length κ^{-1}. However, Debye length must be important for electrophoresis. Increasing thickness of the double layer (DL) leads to removing the point of retardation force further from the particle surface. The thicker the DL, the smaller the retardation force must be.

Detailed theoretical analysis proved that the Smoluchowski theory is valid only for sufficiently thin DL, when particle radius a is much greater than the Debye length:

$$a\kappa \gg 1$$

This model of "thin double layer" offers tremendous simplifications not only for electrophoresis theory but for many other electrokinetic theories. This model is valid for most aqueous systems, where the Debye length is usually only a few nanometers. It only breaks for nano-colloids in solution with ionic strength close to water.

(b) This theory also neglects the contributions from surface conductivity. This is expressed in modern theory as condition of small Dukhin number:

$$Du \ll 1$$

In the effort of expanding the range of validity of electrophoretic theories, the opposite asymptotic case was considered, when Debye length is larger than particle radius:

$$a\kappa < 1$$

Under this condition of a "thick double layer", Hückel [5] predicted the following relation for electrophoretic mobility:

$$\mu_e = \frac{2\varepsilon_r \varepsilon_0 \zeta}{3\eta}$$

This model can be useful for some nanoparticles and non-polar fluids, where Debye length is much larger than in the usual cases.

iii. Other Theories

There are several analytical theories that incorporate surface conductivity and eliminate the restriction of a small Dukhin number, pioneered by Overbeek [6] and Booth [7]. Modern, rigorous theories valid for any Zeta potential and often any $a\kappa$ stem mostly from Dukhin–Semenikhin theory [8]. In the thin double layer limit, these theories confirm the numerical solution to the problem provided by O'Brien and White [9].

Factors Affecting Analyte Migration [9a]

a. Electrophoretic Mobility

Electrophoresis is similar to chromatography in that both involve the separation of compounds by differential migration. **Chromatography brings about differential**

migration through chemical interactions between analytes with the stationary phase and mobile phase. In **electrophoresis, differential migration is produced by the movement of analytes in an electric field,** where their rate of migration will depend on their size and charge.

Electrophoresis is the process of migration of charged molecules through solutions in an applied electric field. Electrophoresis is often classified according to the presence or absence of a solid supporting medium or matrix through which the charged molecules move in the electrophoretic system. The overall rate of travel of a charged solute in electrophoresis will depend on the following two opposing forces:

(i) The first of these forces (*F*+) is the attraction of a charged solute toward the electrode of opposite charge. This force depends on the strength of the applied electric field (*E*, units of volts per distance) and the charge on the solute (*z*).

(ii) The second force acting on the solute is resistance to its movement, as created by the surrounding medium. The force of this resistance (*F*–) depends on the "size" of the solute (as described by its solvated radius *r*), the viscosity of the medium (η) and the solute's velocity of migration (v, in units of distance per time).

When an electric field is applied, a solute will accelerate toward the electrode of opposite charge until the forces *F*+ and *F*– become equal in size (although opposite in direction) [9b,9c]. At this point a steady-state situation is produced in which the solute begins to move at a constant velocity. This velocity can be found by setting the expressions for *F*+ and *F*– equal to each other and rearranging the resulting equation in terms of v.

$$6\pi r\, v = E\, z \quad \text{or} \quad v = \frac{E\, z}{6\pi\eta}$$

$$v = \mu E$$

where

μ= z/6πη. This new combination of terms is known as the **electrophoretic mobility**, which is represented by the symbol **μ.**

So we can say, that all types of electrophoresis are governed by the single set of general principles illustrated by the following equation [11]:

$$\text{Mobility of a molecule} = \frac{\text{(applied voltage) (net charge on the molecule)}}{\text{(friction of the molecule)}}$$

The rate of migration of particle depends on the followings [12]:

(i) Strength of the field.

(ii) On the net charge size and shape of the molecules.

(iii) On the ionic strength.

(iv) Viscosity.

(v) Temperature of medium in which the molecules are moving.

The mobility, or rate of migration, of a molecule increases with increased applied voltage and increased net charge on the molecule. Conversely, the mobility of a molecule decreases with increased molecular friction, or resistance to flow through the viscous medium, caused by molecular size and shape. Total actual movement of the molecules increases with increased time, since mobility is defined as the rate of migration.

b. Secondary Interactions

To obtain good separations in electrophoresis, it is often necessary to adjust the conditions of this method to change the electrophoretic mobility of a solute.

(i) This goal can be achieved by using secondary reactions that alter the charge or apparent size of the solute. If an analyte is a weak acid or weak base, for example, its net charge can be varied by changing the pH.

(ii) It is also possible to use side reactions to change the effective size or charge of the analyte. This effect occurs in a method known as sodium dodecyl sulfate polyacrylamide gel electrophoresis (SDS-PAGE), which is a technique for separating proteins according to their size.

(iii) Another approach for altering the apparent electrophoretic mobility of an analyte is to use a solubility equilibrium.

c. Electroosmosis

It is also possible for the running buffer to move in such a field. This phenomenon can occur if there are any fixed charges present in the system, such as on the interior surface of an electrophoretic system or on a support within this system. The presence of these fixed charges attracts ions of opposite charge from the running buffer and creates an electrical double layer at the surface of the support. In the presence of an electric field, this double layer acts like a piston that causes a net movement of the buffer toward the electrode of opposite charge versus the fixed ionic groups. This process is known as **electroosmosis** and results in a net flow of the buffer and its contents through the system [12].

As an analytical tool, electrophoresis is simple, rapid and highly sensitive. It is used analytically to study the properties of a single charged species and as a separation technique. It provides the basis for a number of analytical techniques used for separating molecules by size, charge, or binding affinity. DNA Gel electrophoresis is usually performed for analytical purposes, often after amplification of DNA via PCR, but may be used as a preparative technique prior to use of other methods such as mass spectrometry, RFLP, PCR, cloning, DNA sequencing, or Southern blotting for further characterization [13].

Source

1. *Lyklema, J. (1995). Fundamentals of Interface and Colloid Science. vol. 2.p.3.208.*
2. *Reuss, F.F. (1809), "Sur un nouvel effet de l'electricite galvanique", Mémoires de la Societe Impériale des Naturalistes de Moscou. 2: 327–337.*

3. *Hanaor, D. A.H.; Michelazzi, M.; Leonelli, C.; Sorrell, C.C. (2012)."The effects of carboxylic acids on the aqueous dispersion and electrophoretic deposition of* ZrO_2*". Journal of the European Ceramic Society. 32 (1): 235-244.*
4. *Von Smolu-chowski, M. (1903)."Contribution a la theorie de l'endosmose électrique et de quelques phénomènes corrélatifs". Bull. Int. Acad. Sci. Cracovie.184.*
5. *Hückel, E. (1924). "Die kataphorese der kugel".Physik.Z. 25: 204.*
6. *Overbeek, J. Th. G (1943). "Theory of electrophoresis-The relaxation effect".Koll. Bith.: 287.*
7. *Booth, F. (1948), Theory of Electrokinetic Effects", Nature.161 (4081): 83-86. Bibcode: 1948 Natur.161.83B.*
8. *Dukhin, S.S.; N.M. Semenikhin (1970). Koll. Zhur. 32: 366.*
9. *O'Brien, R.W.; L.R. White (1978). "Electrophoretic mobility of a spherical colloidal particle". J. Chem. Soc. Faraday Trans. 2 (74): 1607.*

9a. *Electrophoresis pdf, Chapter 23, 96943_23_ch23_p571-596 1/8/10.*

9b. *B. L. Karger, L. R. Snyder, and C. Hovath, An Introduction to Separation Science, Wiley, New York, 1973, Chapter 17.*

9c. *J. P. Landers, Handbook of Capillary Electrophoresis, CRC Press, Boca Raton, FL, 1992.*

10. *Wikipedia.*
11. *Basic Techniques of Experimental Biochemistry- Experiment-4, Electrophoresis.*
12. *M.-L. Riekkola, J. A. Jonsson, and R. M. Smith, "Terminology for Analytical Capillary Electromigration Techniques (IUPAC Recommendations 2003), " Pure and Applied Chemistry, 76 (2004) 443–451.*
13. *Electrophoresis modules. pdf.*

Types of Electrophoresis

Electrophoresis can be performed in a variety of formats. Some are explained below:

(i) One format is to apply small amounts of a sample to a support (usually a gel) and allow the analytes in this sample to travel in a running buffer through the support when an electric field is applied. This approach is known as **gel electrophoresis [1,2,3].**

(ii) In second format, it is also possible to separate the components of a sample by using a narrow capillary that is filled with a running buffer and placed into an electric field. This second format is called **capillary electrophoresis [1,3,4].**

Depending on the type of electrophoresis being used, the resulting separation can be viewed in one of two ways.

Gel Electrophoresis	*Capillary Electrophoresis*
In the case of gel electrophoresis, the separation is stopped before analytes have traveled off the support.The result is a series of bands where the **migration distance (*d*m)** characterizes the extent to which each analyte has interacted with the electric field. This approach is similar to that used to characterize the retention of analytes **in thin-layer chromatography and paper chromatography**. Because the migration distance of an analyte through a gel for electrophoresis will depend on the exact voltage and time used for the separation, it is common to include standard samples on the same support as the sample to help in analyte identification.The intensity of the analyte band is then used to measure the amount of this substance in the sample.	In capillary electrophoresis all analytes travel the same distance, from the point of injection to the opposite end where a detector is located. The analytes will differ, however, in the time it takes them to travel this distance, in a manner similar to what occurs in the chromatographic methods of **gas chromatography (GC) and high-performance liquid chromatography (HPLC).**In this situation the **migration time (*t*m)** for each analyte is measured and recorded [5]. The resulting plot of detector response versus migration time is called an **electropherogram**.

Source

1. *J. C. Giddings, Unified Separation Science, Wiley, New York, 1991, Chapter 8.*
2. *L. Stryer, Biochemistry, Freeman, New York, 1988, Chapter 3.*
3. *D. S. Hage, "Chromatography and Electrophoresis, " In Contemporary Practice in Clinical Chemistry, W. Clarke and D. R.Dufour, Eds., AACC Press, Washington, DC, 2006, Chapter 7.*
4. *J. M. Anderson and G. A. Tetrault, "Electrophoresis, " In Laboratory Instrumentation, 4th ed., M. C. Haven, G. A. Tetrault, and J. R. Schenken, Eds., Van Nostrand Reinhold, New York, 1995, Chapter 12.*
5. *M.-L. Riekkola, J. A. Jonsson, and R. M. Smith, "Terminology for Analytical Capillary Electromigration Techniques (IUPAC Recommendations 2003), " Pure andApplied Chemistry, 76 (2004) 443–451.*

The following types of electrophoresis are being applied for different analysis purposes:

1. Affinity electrophoresis
2. Immuno electrophoresis
3. Capillary electrophoresis
4. Paper electrophoresis
5. Gel Electyrophoresis [Pulsed field gel electrophoresis, Sodium Dodecyl Sulfate-Polyacrylamide (SDS-PAGE), Native Gels, Electrofocusing Gels, DNA Agarose Gels].

1. Affinity Electrophoresis [1]

The methods are based on changes in the electrophoretic pattern of molecules (mainly macromolecules) through biospecific interaction or complex formation. The interaction or binding of a molecule, charged or uncharged, will normally change the electrophoretic properties of a molecule. Membrane proteins may be identified by a shift in mobility induced by a charged detergent.

Applications

(i) Nucleic acids or nucleic acid fragments may be characterized by their affinity to other molecules. The methods has been used for estimation of binding constants, as for instance in lectin affinity electrophoresis or characterization of molecules with specific features like glycan content or ligand binding.

(ii) For enzymes and other ligand-binding proteins, one dimensional electrophoresis similar to counter electrophoresis or to "rocket immune electrophoresis", affinity electrophoresis may be used as an alternative quantification of the protein. The methods include the followings:

   (a) Mobility shift electrophoresis.

   (b) Charge shift electrophoresis

   (c) Affinity capillary electrophoresis.

Source

1. *Electrophoresis modules. pdf.*

2. Immuno Electrophoresis [1]

It is a general name for a number of biochemical methods for separation and characterization of proteins based on electrophoresis and reaction with antibodies. All types of immuno electrophoresis require immunoglobulins (also known as antibodies) reacting with the proteins to be separated or characterized. The following types of Immuno-electrophoresis are being applied in food analysis:

(a) **Rocket immunoelectrophoresis-**It is one-dimensional quantitative immuno electrophoresis.

(b) **Fused rocket immunoelectrophoresis-** It is a modification of one-dimensional quantitative immunoelectrophorsis used for detailed measurement of proteins in fractions from protein separation experiments.

(c) **Affinity immunoelectrophoresis-**This type of immuno-electrophoresis is based on changes in the electrophoretic pattern of proteins through specific interaction or complex formation with other macromolecules or ligands.

Source

1. *Electrophoresis modules.pdf.*

3. Capillary Electrophoresis (CE) [1]

It can be used to separate ionic species by their charge and frictional forces and hydrodynamic radius. In traditional electrophoresis, electrically charged analytes move in a conductive liquid medium under the influence of an electric field. Capillary electrophoresis (CE) was designed to separate species based on their size to charge ratio in the interior of a small **capillary** filled with an electrolyte. This method was first reported in the late 1970s and early 1980s and is sometimes known as **Capillary zone electrophoresis** [2,3].

CE offers high-resolution separations at a minimal cost in terms of staff time and reagent use. However, in comparison to other fields of study, the application of CE for food analysis has been limited [4].

Source

1. *Electrophoresis modules.pdf.*
2. *J. W. Jorgenson and K. D. Lukacs, "Zone Electrophoresis in Open-Tubular Glass Capillaries, " Analytical Chemistry, 53 (1981) 1298–1302.*
3. *D. E. Garfin, "Two-Dimensional Gel Electrophoresis: An Overview, " Trends in Analytical Chemistry, 22 (2003) 263–272.*
4. *Richard A. Frazier, Review on Recent advances in capillary electrophoresis methods for food analysis, Electrophoresis 2001, 22, 4197–4206.*

Principle and Instrumentation [1]

a. Principle

CE is based on the movement of molecules in an electric field. Initially simple buffer solutions were employed to separate ionic water-soluble solutes. However, different than classical electrophoresis, CE is not restricted to the separation of large molecules based on size and charge. Several modifications of CE have been developed to separate molecules with low molecular weight or neutral charge.

The current form of CE is typically conducted in capillaries with inner diameters of 20–100 μm and lengths of 20–100 cm [2]. The use of these narrow-bore tubes provides efficient removal of Joule heating by allowing this heat to be quickly dissipated to the surrounding environment [3,4,5]. This removal of heat helps to decrease band-broadening and provides much more efficient and faster separations than gel electrophoresis.

b. Instrumentation

In principle, all CE techniques are carried out using the same following equipment:

(i) A high voltage power supply.
(ii) The anode and cathode buffer reservoirs with corresponding electrodes.
(iii) The separation chamber (the capillary tube).
(iv) The injection system
(v) The detector.

The arrangement of these equipment to perform analysis by CE, is well shown on next page:

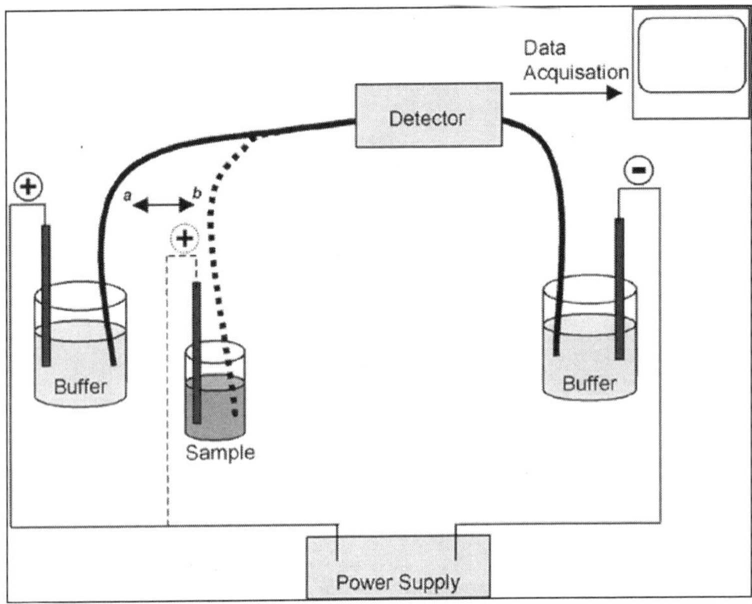

General Design of a Capillary Electrophoresis System.

Commercial Instrument for CE (Beckman Instrument).

A capillary tube is placed between the two buffer reservoirs, and an electric field is applied by means of a high power supply. Then, a defined volume of the analysate is introduced into the capillary by replacing one buffer reservoir with the sample vial. The result of the electrophoretic separation is measured by a detector. A detector is located at the opposite side of the capillary as the injection site. The sample data are collected and stored by computer, and analyzed using dedicated software.

(i) **The capillary**- In a CE system, capillary is typically made of fused silica. This capillary can be used directly or it can be modified to place various coatings on its interior surface.

(ii) **Injection Techniques**-There are following two techniques that make it possible to inject small sample volumes onto a CE system:

(a) **Hydrodynamic injection** –This is a first technique, which uses a difference in pressure to deliver a sample to the capillary. This method can be carried out by placing one end of the capillary into the sample in an enclosed chamber and applying a pressure to this chamber for a fixed period of time, where the amount of injected sample will depend on the size of the pressure difference and the amount of time that this pressure is applied. Once the sample has entered the capillary, the separation is begun after the capillary has been put back into contact with the running buffer and electrodes.

(b) **Electrokinetic injection** -This is a second technique that allows the injection of small sample volumes. This method again begins by placing the capillary into the sample, but an electrode is also now in contact with the sample. When an electric field is applied across the capillary, electroosmostic flow and the electrophoretic mobility of the analytes cause them to enter the capillary. The amount of each analyte that is injected in this method will depend on the analyte's electrophoretic mobility, the electric field, and the time over which this field is applied [3].

Sample Stacking

There are various methods for concentrating samples and providing narrow analyte bands in CE. Sample stacking is one of such method [6].

Sample stacking occurs when the ionic strength (and therefore the conductivity) of the sample is less than that of the running buffer. When an electric field is applied to such a system, analytes will migrate quickly through the sample matrix until they come to the boundary between the sample and running buffer. Because the running buffer has a higher ionic strength than the sample, the rate of analyte migration decreases at this boundary. This decrease in migration rate causes the analytes to concentrate into a narrower band as they enter the running buffer. **The overall effect is similar to what occurs when using stacking gels in traditional electrophoresis.**

Detection Methods

The different detection methods are used for capillary electrophoresis. Many of these methods are also used in liquid chromatography [3,6]. Selective monitoring methods that work well for this purpose are electrochemical and fluorescence detection. Ultraviolet-visible (UV-vis) absorbance, conductance, and mass spectrometry detection are also often employed in CE. The details of some of the detection techniques are tabulated on next page:

Sl.No.	Types of Detectors	Name of Detectors	Detection Limits	Compounds Detected
1	General detectors	Ultraviolet-visible (UV/vis) absorbance detector	10^{-13}–10^{-16} mol	Compounds with chromophores
2	Selective detectors	*(i)* Fluorescence detector	10^{-15}–10^{-17} mol	Fluorescent compounds
		(ii) Laser-induced fluorescence detector	10^{-18}–10^{-20} mol	Fluorescent compounds
		(iii) Conductivity detector	10^{-15}–10^{-16} mol	Ionic compounds
		(iv) Electrochemical detector	10^{-18}–10^{-19} mol	Electrochemically active compounds
3	Structure-specific detectors.	Mass spectrometry.	10^{-16}–10^{-17} mol	Compounds forming gas-phase ions

Due to the possibility for automation and parallelization, CE allows fast, large scale and high-throughput analysis. Sophisticated CE autosamplers are available that allow quantitative injection and analysis of a large number of samples in an unattended sequence.

Differences between CE and LC Detection

The followings two differences are reported:

(i) An important difference between detection in LC and CE is the need in CE for methods that can work with very small sample sizes. This need is a result of the small injection volumes that are required in capillary electrophoresis to avoid excessive band-broadening.

(ii) Another difference between detection in LC and CE concerns how their signals vary with analyte retention or migration. In LC, all analytes pass at the same flow rate through the detector and spend the same amount of time in this device. This effect makes it possible to directly compare the peak areas of two analytes with different retention times. In capillary electrophoresis analytes with different migration times also spend different amounts of time in the detector.

Advantages of CE over other Electrophoretic Techniques [1]

CE has following advantages over classical techniques like slab polyacrylamide gel electrophoresis (PAGE), southern blotting, sequencing, or conventional gel electrophoresis:

(i) CE has not only the ability to analyze nonideal tissue samples such as archival paraffin embedded formalin fixed tissue, but also automated digital imaging capabilities that can use either peak height or peak area in a semiquantitative or quantitative manner.

(ii) One reason capillary electrophoresis is more efficient than gel electrophoresis is that Joule heating is greatly reduced as a source of band-broadening.

(iii) CE is often used with no gel or support present, which eliminates eddy diffusion and secondary interactions with the support.

(iv) Besides being faster and more efficient than gel electrophoresis, capillary electrophoresis is easier to perform as part of an instrumental system.

Special Types of Capillary Electrophoresis [2]

The main capillary electrophoresis method is zone electrophoresis. It is possible to include other chemical and physical interactions in a CE system to give additional types of separations.Some special types of CE are explained here:

(i) **Capillary Sieving Electrophoresis.** One useful feature of gel electrophoresis is the ability of some supports to separate analytes based on size, as occurs for proteins in SDS-PAGE. The same effect can be obtained in capillary electrophoresis by including an agent in the CE system that **"sieves"** the analytes, or separates them based on size. This approach is known as **capillary sieving electrophoresis** (**CSE**) [2]. The capillary sieving electrophoresis can be performed by several ways:

(a) **Capillary gel electrophoresis (CGE)-**The first way is to place a porous gel in the capillary, like the polyacrylamide gels employed in SDS-PAGE. This method is called **CGE** [2]. One problem with these gels is they are not always stable in the high electric fields used in capillary electrophoresis and must frequently be replaced.

(b) A second approach, and the one now used in DNAsequencing by CE, is to add to the running buffer a large polymer that can entangle with analytes and alter their rate of migration. This approach provides a system with better reproducibility and stability than those using gels, because the polymer is continuously renewed as the running buffer passes through the capillary.

(ii) **Electrokinetic Chromatography-** Ordinary capillary electrophoresis works well for separating cations and anions, but it cannot be used to separate neutral substances from each other. Instead, these substances migrate as a single peak that travels with the electroosmotic flow. It is possible to employ CE with such compounds if we place in the running buffer a charged agent that can interact with these substances. This approach is called **electrokinetic chromatography**.

(iii) **Other Methods-** There are many other types of CE that have been explored for use in chemical analysis and separations.

(a) Isoelectric focusing can be carried out in a capillary, creating the method **of capillary isoelectric focusing (CIEF) [2,6].**

(b) **Affinity capillary electrophoresis (ACE)-** This is the another type of capillary electrophoresis occurs when biologically related agents are placed as additives in the running buffer. As analytes travel through this buffer, their overall mobility will be affected by their binding to these agents. One common use of ACE is in the separation of chiral analytes through the use of binding agents like cyclodextrins or proteins.

Source

1. *Anja Bosserhoff and Claus Hellerbrand, Capillary Electrophoresis, University of Regensburg, Regensburg, Germany.*
2. *M.L. Riekkola, J. A. Jonsson and R. M. Smith, "Terminology for Analytical Capillary Electromigration Techniques (IUPAC Recommendations 2003), " Pure and Applied Chemistry, 76 (2004) 443-451.*
3. *T. Blanc, D. E. Schaufelberger, and N. A. Guzman, "Capillary Electrophoresis." In Analytical Instrumentation Handbook, 2nd ed., G. W. Ewing, Ed., Marcel Dekker, New York, 1997, Chapter 25.*
4. *J. C. Giddings, Unified Separation Science, Wiley, New York, 1991, Chapter 8.*
5. *J. W. Jorgenson and K. D. Lukacs, "Zone Electrophoresis in Open-Tubular Glass Capillaries, " Analytical Chemistry, 53 (1981) 1298–1302.*
6. *J. P. Landers, Handbook of Capillary Electrophoresis, CRC Press, Boca Raton, FL, 1992.*

Subtypes of the CE Methodology [1]

On the basis of the methodologies used, CE is subdivided in to the following categories:

Sl.No.	*Subtypes of CE*	*Details*
1	Capillary Zone Elect rophoresis CZE	It is the simplest form of CE. A capillary is filled with a relatively low viscosity buffer and the analysates migrate from one end of the capillary to the other with velocities determined by the charge-to-mass ratio to form discrete peaks (zones).
2	Capillary Gel Elect rophoresis CGE	It separates analytes,such as DNA or proteins, based on molecular weight. A variety of compounds, such as bispoly acrylamide, agarose,or methylcellulose, have been used to act as a molecular sieve to cause separation.
3	Micellar Electrokin- etic Chromatoraphy MEKC	MEKC was introduced for the separation of neutral compounds by the addition of a micelle sodium dodecyl sulfate (SDS) to the buffer solution.
4	Capillary Isoelectric Focusing CIEF	Similar to classic isoelectric focusing, in CIEF a pH gradient is created inside a capillary to separate peptides and proteins on the basis of their isoelectric point.
5	Capillary Isotacho- phoresis CITP	CITP, a combination of two buffers with different mobilities is used causing the analytes of interest to be concentrated in zones between the leading and terminating constituents.
6	Constant Denaturant Capillary Electrophoresis CDCE	CDCE permits high-resolution separation of single-base variations occurring in an approximately 100 bp isomelting DNA sequence based on their differential melting temperatures [2]. It requires different degrees of partial denaturetion at a given temperature in the presence of denaturing agents [3].

Source

1. *Anja Bosserhoff and Claus Hellerbrand, Capillary Electrophoresis, University of Regensburg, Regensburg, Germany.*
2. *Li-Sucholeiki, X.C., Khrapko, K., Andre, P. C., Marcelino, L. A., Karger, B. L., and Thilly, W. G. (1999). Applications of constant denaturant capillary electrophoresis/ high fidelity polymerase chain reaction to human genetic analysis.Electrophoresis 20, 1224-1232.*
3. *Khrapko, K., Hanekamp, J. S., Thilly, W. G., Belenkii, A., Foret, F., and Karger, B. L. (1994). Constant denaturant capillary electrophoresis (CDCE): a high resolution approach to mutational analysis. Nucleic Acids Res. 22, 364–369.*

Application of Capillary Electrophoresis [1,1a]

The speed, resolution and simplicity of CE, combined with low operating costs, make the technique an attractive option for the development of improved methods of food analysis. CE is used for the analysis of pharmaceuticals, DNA, chiral compounds, proteins, peptides, clinical and forensic samples, metal ions, and inorganic anions. In addition to the aforesaid applications, CE finds wide use in the analysis of pesticides, food content and composition, and pollutants in water and soil samples.

A. Molecular Diagnostics/DNA Studies/Specific Diagnostic Applications [1]

Capillary Electrophoresis is now successfully being used in Molecular Diagnostics/DNA Studies (DNA Sequencing, Analyses of DNA Fragment Length or Restriction Patterns and Microsatellites, Analyses of Single-strand Conformation Polymorphism, DNA amplification), Specific Diagnostic Applications (Diagnosis of Neoplastic Disorders, Diagnosis of Hereditary Diseases and Prenatal Testing, Diagnosis of Infectious Diseases) and Identity Testing and Forensic Science Applications.

Source

1. *Anja Bosserhoff and Claus Hellerbrand, Capillary Electrophoresis, University of Regensburg, Regensburg, Germany.*

B. Food Industry

CE has found application to a number of important areas of food analysis, including quantitative chemical analysis of food additives, biochemical analysis of protein composition, and others. The Capillary Electrophoresis is being applied for the analysis of following components in foods:

1. Food-Related Proteins

The analysis of proteins and peptides is undoubtedly the most common and multidisciplinary CE application area [2,3]. CE offers an attractive alternative to more traditional gel electrophoresis methods, providing analogous separation mechanisms combined with a high degree of automation. For human nutrition studies, methodology has been developed that employs CE for the quantification and characterization of lipoproteins in human plasma [4].

Maillard reaction has important implications in terms of flavor, aroma and color characteristics in foods and involves the interaction between carbonyl containing compounds (primarily reducing sugars) and amino groups, such as those of the lysine residue of proteins.CE was reported as tool to study the Maillard reaction [5].

a. Milk Proteins

CE was used to study the evolution of casein throughout the ripening process of ewe's milk cheese and to assess the affect of the type of rennet used in its hydrolysis [6,7]. The detection of adulteration of dairy products, particularly those of goat's and ewe's milk with cows milk, has seen many recent developments, such as the use of isoelectric buffers to enable the application of high-voltage gradients [8].

b. Cereal Proteins

New developments in CE methodology and applications for separating and characterizing cereal proteins have been recently reviewed [9]. SDS-CE is an alternative approach to the separation of cereal proteins that is analogous to SDS-PAGE. Separation is based on size as in SDS-PAGE, but SDS-CE presents potential advantages in terms of automation and ease of quantitation.

2. Amino Acids in Foods

a. Free Amino Acids

CE has recently gained recognition as a useful approach to the analysis of amino acids in foods [10,11]. The analysis of nonderivatised free amino acids in foods by CE, usually by limiting the analysis to amino acids having a chromophore (*e.g.*, tryptophan) [12], or by utilizing an indirect UV detection strategy [13] was repored.

b. Biogenic Amines

Biogenic amines are bacterial degradation products of amino acids, which can have aliphatic, aromatic or heterocyclic structures. They are normal constituents of fermented foods, such as cheeses, wines and beers, but excessive dietary intake can lead to toxic effects. The application of CE as a rapid screening technique to determine levels of biogenic amines in foods was reported [14-18]. CE has also been applied to study the formation of biogenic amines through the action of proteinases [19]. A rapid CZE method was developed for the simultaneous determination of four heterocyclic aromatic amines (HAAs) in fried beef steak, meat extract, and baked salmon after solid-phase extraction to isolate the HAAs [14].

3. Vitamins

CE, by its very nature, is more suited to the analysis of water-soluble vitamins, although it is still possible to separate fat-soluble vitamins by use of modes of separation such as MEKC, or those based on interactions with a microemulsion or hydrophobic pseudostationary phase [5]. The major CE applications in vitamin analysis are for determinations of vitamin C in fruits and beverages, niacin in a range of foods, and thiamin in meat and milk.

4. Carbohydrates and Gelling Agents

CE has been used widely for the analysis of carbohydrates and their polymers (polysaccharides) [20]. Common problems with carbohydrate analysis by CE are that the majority of carbohydrates are neutral and have no chromophore for UV detection. Recently, both issues were overcome in a simple and reproducible CE method with indirect UV detection for the determination of carbohydrates in food samples [21].

5. Phenolic Compounds

Phenolic compounds occur naturally in many plant-derived foods and beverages, and are primarily associated with flavor and color and because of this, there has been intense interest in the analysis of phenolic compounds in foods. Tea, most popular beverage in the world, is rich in phenolic antioxidants, including catechins and theaflavins. CE analysis is used for analysis of these tea components.CE is able to resolve these analytes within shorter analysis times than HPLC [22,23,24]. Several CE methods have been developed for the analysis of other phenolic compounds in foods. For example, CZE at alkaline pH has been applied to the analysis of enzymatically glucosylated flavonoids, which are used as natural food additives [25]. In addition, CZE has been applied to the analysis of anthocyanins, which are water-soluble polyphenols responsible for red, purple and blue colors in flowers and fruits.

6. Food Additives

CE is used to the analysis of food additives, primarily antioxidants, preservatives, colors and sweeteners, [26]. CE is also used to the analysis of natural food pigments extracted from commercial food samples [27]. Nonaqueous CE using pure organic solvents has been investigated for the separation of positional isomers of dinitrobenzoic acid and isomeric pairs of artificial food colors [28].

7. Toxins and Residues

a. Toxins

CE with LIF detection has been applied to the determination of the products of a solution-phase immunoassay to quantify total glycoalkaloids in potatoes [29]. CZE, using 100 mM ammonium acetate (pH 9.0) as buffer can be used for the analysis of fluorescent whitening agents extracted from paper samples using a standard hot-water extraction [30].

b. Agricultural and Veterinary Residues

CE is used to the analysis of pesticide [31,32] and veterinary antibiotic residues [33]. Recently, a method was developed for the multiresidue analysis of eight commonly used fungicides by MEKC [34].

Source

1. *Richard A.Frazier, Review on Recent advances in capillary electro phoresis methods for food analysis, Electrophoresis 2001, 22, 4197-4206.*

1a. *Anja Bosserhoff and Claus Hellerbrand, Capillary Electrophoresis, University of Regensburg, Regensburg, Germany.*

2. *Kasicka, V., Electrophoresis 1999, 20, 3084-3105.*
3. *Dolník, V., Electrophoresis 1999, 20, 3106–3115.*
4. *Cruzado, I. D., Song, S., Crouse, S. F., O'Brien, B. C., Macfar-lane, R. D., Anal. Biochem. 1996, 243, 100-109.*
5. *Frazier, R. A., Ames, J. M., Nursten, H. E., Electrophoresis, 1999, 20, 3156-3180.*
6. *Recio, I., Ramos, M., Lopez-Fandino, R., Electrophoresis 2001, 22, 1489–1502.*
7. *Irigoyen, A., Izco, J. M., Ibanez, F. C., Torre, P., J. Chromatogr. A 2000, 881, 59-67.*
8. *Herrero-Mart-inez, J. M., Simo-Alfonso, E. F., Ramis-Ramos, G., Gelfi, C., Righetti, P. G., J. Chromatogr. A 2000, 878, 261-271.*
9. *Bean, S. R., Lookhart, G. L., Electrophoresis 2001, 22, 1503-1509.*
10. *Zamora, R., Alaiz, M., Hidalgo, F. J., J. Agric. Food Chem.1999, 47, 1942-1947.*
11. *Molnar-PerI, I., J. Chromatogr. A 2000, 891, 1-32.*
12. *Cancalon, P. F., J. AOAC Int. 1999, 82, 95-106.*
13. *Soga, T., Ross, G. A., J.Chromatogr. A1999, 837, 231–239.*
14. *Mardones, C., Arce, L., Rios, A., Valcarcel, M., Chromatographia1998, 48, 700–706.*
15. *Mendonsa, S. D., Hurtubise, R. J., J. Liq. Chromatogr. Relat. Technol. 1999, 22, 1027-1040.*
16. *Liao, W., Paek, H., Mabuni, C., Angold, S., Soliman, M., J. Chromatogr. A 1999, 853, 541-544.*
17. *Nakashima, M., Sugiyama, A., J. Food Hyg. Soc. Jpn. 1999, 40, 285–290.*
18. *Pais, P., Knize, M.G., J.Chromatogr.B 2000, 747, 139–169.*
19. *Femandez- Garcia, E., Tomillo, J., Nunez, M., Int. J. Food Microbiol. 1999, 52, 189-196.*
20. *El Rassi, Z., Electrophoresis 1999, 20, 3134–3144.*
21. *Soga, T., Serwe, M., Food Chem. 2000, 69, 339–344.*
22. *Aucamp, J. P., Hara, Y., Apostolides, Z., J. Chromatogr. A 2000, 876, 235–242.*
23. *Horie, H., Kohata, K., J. Chromatogr. A 2000, 881, 425–438.*
24. *Lee, B. L, Ong, C. N., J. Chromatogr. A 2000, 881, 439-447.*
25. *Akiyama, T., Yamada, T., Maitani, T., J. Chromatogr. A 2000, 895, 279-283.*
26. *Boyce, M. C., Electrophoresis 2001, 22, 1447–1459.*
27. *Watanabe, T., Terabe, S., J. Chromatogr. A 2000, 880, 311-322.*
28. *Razee, S., Tamura, A., Masujima, T., J. AOAC Int.999, 82, 1577-1586.*
29. *Driedger, D. R., LeBlanc, R. J., LeBlanc, E. L., Sporns, P., J. Agric. Food Chem. 2000, 48, 4079–4082.*

30. *Damant, A. P., Castle, L., J. Microcol. Sep. 1999, 77, 259-262.*

31. *Eash, D. T., Bushway, R. J., J. Chromatogr. A 2000, 880, 281–294.*

32. *Menzinger, F., Schmitt-Kopplin, P., Freitag, D., Kettrup, A., J. Chromatogr. A 2000, 891, 45–67.*

33. *Oka, H., Ito, Y., Matsumoto, H., J. Chromatogr. A 2000, 882, 109-133.*

34. *Molina, M., Silva, M., Electrophoresis 2000, 27, 3625–3633.*

4. Paper Electrophoresis [1]

Paper Electrophoresis is one of the zone electrophoresis.

This is very important method in all laboratories.

Principle

"The charge carried by a molecule depends on the pH of the medium. Electrophoresis at low voltage is not usually to separate low molecular weight compounds because of diffusion, but it is easier to illustrate the relationship between charge and pH with amino acids than with proteins (or) other macromolecules".

Apparatus and Materials

(i) **Filter paper-**Paper of good quality should contain at least 95 per cent α-cellulose and should have only a very slight adsorption capacity.

(ii) **Apparatus-**The equipment for this electrophoresis consists basically of following two items:

(a) **Power Pack-**Power pack provides a stabilized direct current and has controls for both voltage and current out put, which have an out put of 0 to 500V and 0 to 150 mA are available.

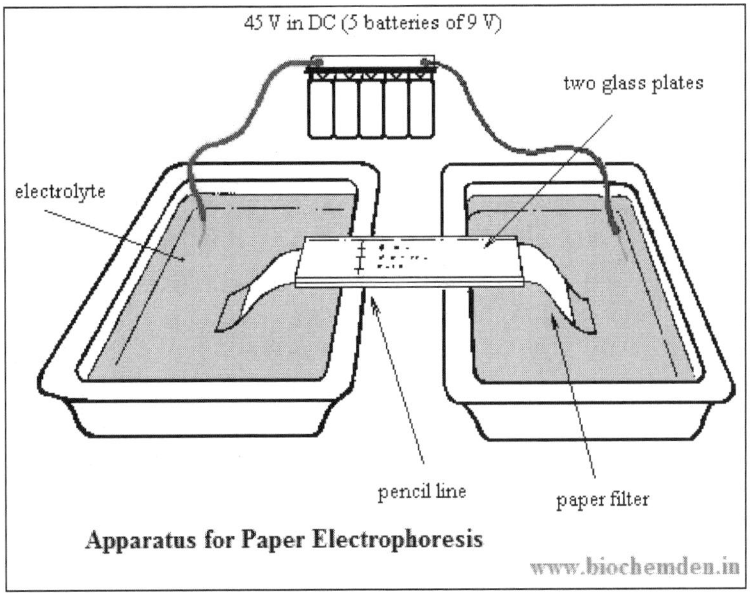

Apparatus for Paper Electrophoresis

(*b*) **Electrophoretic Cell-**The Electrophoretic cell contains the electrodes, buffer reservoirs, a support for paper and a supporting transparent insulating cover. The electrodes are usually made of platinum.

Arrangement of Filter Paper Strips

The two arrangements (horizontal and vertical arrangements) of the filter strips are commonly used and both arrangements are equally viable and the choice usually depends upon personal preferences.

Sample Application and Method

The details related to the sample application and testing method are given below:

i. Sample Application

The sample may be applied as a spot (about 0.5cm in diameter) or as a narrow uniform streak by the available commercial special devices. The sample can be applied before the paper has been equilibrated with buffer (or) after it.

Procedure

After the sample has been applied to the paper and the paper has been equilibrated with the buffer. The current is switched on. The details of the commonly used buffers are given in following table:

Sl.No.	*Indented Separation*	*Buffer*	*pH*	*Ionic Strength*	*Composition liter*
1	Proteins	Barbital	8.6	0.05	10.3g Sodium barbiturate
					1.84 g Barbital
		Phosphate	7.4	-	0.6g $NaH_2PO_4.H_2O$
					2.2g Na_2HPO_4
2	Nucleoproteins	Acetate	4.5	0.1	3.51g NaCl
					3.28g Sodium acetate
					(pH adjusted with HCl to 4.5)
		Citrate	4.5	0.13	28.46g Sodiun citrate
					20.6g Citric acid
3	Aminoacids	Phosphate	4.6	0.15	20.4g KH_2PO_4
		Michalies	8.6	0.1	9.8 g Sodium barbiturate

The device providing stable voltage (or) current is available. Frequent observation is necessary to run electrophoretic apparatus. Overheating can be avoided by placing the entire equipment in the cold room. The process does not take longer than two hours. After 2 hours switched off the power and paper is removed. Once removed, the paper is dried in hot oven at 110°C.

Detection and Quantitative Assay

To identify the unknown electrophorogram, compare the this electogram with standard electrogram under standard conditions. Individual compounds are usually identified by physical properties by the following methods:

i. Fluorescence

a-Staining with "Ethidium bromide" and subsequent visualization of the electrophoreticgram under UV light makes DNA and RNA fluoresce and thus facilitates their detection, **b**-lourescamine staining is utilized for detecting amino acids, amino acid derivatives, peptides and proteins and **c**-DANSYl chloride may be used in place of fluorescamine.

ii. UV absorption

Proteins, Peptides and nucleic acids absorb in the range of 260 to 280nm, this property these can be detected.

iii. Staining

Sl.No.	*Compound*	*Dye*	*Comments*
1	Proteins	Bromophenol blue in acetic acid	Visual, Quantitaive
		DANSYl Chloride	Fluorescent, Quantitaive
		Fluorescamine	Fluorescent,Very sensitive
2	Nucleic acids	Methyl green- pyronine	DNA- Blue, RNA-Red, Sensitive
		Ethedium bromide	Fluorescent,Very sensitive
3	Polysaccharide	Iodine	Visual, Sensitive
4	Lipoprotein	Sudan Black in 60 per cent ethanol	Visual, Sensitive
5	Glycoprotein	Alcian blue	Visual, Sensitive

iv. Detection of Enzymes In situ

if the component to be separated is an enzyme, special techniques may be used to detect it. The paper strip, which have separated enzyme, is impregnated with the substrate for the enzyme desired to be separated. The paper strip is now placed in a suitable buffer along with electrophoretogram. The color bands will appear which indicates the position of enzyme.

v. Quantitative Estimation

The color density of the zone may be multiplied with the area of the zone and the resulting value would be a rough estimate of the concentration of the component.

Applications of Paper Electrophoresis

It is being used for the following purposes:

1. Serum analysis for diagnostic purpose is routinely carried about by paper electrophoresis.
2. Muscle proteins, egg white proteins, milk proteins and snake, insect venom analysis done by this technique.

Source

2. *Biotechnique Den.*

5. Gel Electrophoresis [1,31]

One of the most common types of electrophoresis is the method of **gel electrophoresis**. This technique is an electrophoretic method that is performed by applying a sample to a gel support that is then placed into an electric field. Hydrated gel networks have many desirable properties for electrophoresis. They allow a wide variety of mechanically stable experimental formats such as horizontal/vertical electrophoresis in slab gels or electrophoresis in tubes or capillaries. The mechanical stability also facilitates post electrophoretic manipulation making further experimentation possible such as blotting, electro-elution or MS identification/finger printing of intact proteins or of proteins digested in gel slices [1].

Gel electrophoresis is a method for separation and analysis of macro-molecules (DNA, RNA and proteins) and their fragments, based on their size, charge and conformation. It is used in clinical chemistry to separate proteins by charge and/or size and in biochemistry and molecular biology to separate a mixed population of DNA and RNA fragments by length, to estimate the size of DNA and RNA fragments or to separate proteins by charge [1a]. Proteins can have either a net positive or net negative charge and nucleic acids have a consistent negative charge imparted by their phosphate backbone, and migrate toward the anode.

(i) Nucleic acid molecules are separated by applying an electric field to move the negatively charged molecules through an agarose matrix. Shorter molecules move faster and migrate farther than longer ones because shorter molecules migrate more easily through the pores of the gel. This phenomenon is called sieving [2].

(ii) Proteins are separated by charge in agarose because the pores of the gel are too large to sieve proteins.

(iii) Gel electrophoresis can also be used for separation of nanoparticles.

(iv) DNA Gel electrophoresis is usually performed for analytical purposes, often after amplification of DNA via PCR, but may be used as a preparative technique prior to use of other methods such as mass spectrometry, RFLP, PCR, cloning, DNA sequencing, or Southern blotting for further characterization.

Principle of Gel Electrophoresis

Electrophoresis is a process which enables the sorting of molecules based on size. Using an electric field, molecules (such as DNA) can be made to move through a gel made of **agar** or **polyacrylamide**. The electric field consists of a negative charge at one end which pushes the molecules through the gel, and a positive charge at the other end that pulls the molecules through the gel. The molecules being sorted are dispensed into a well in the gel material. The gel is placed in an electrophoresis chamber, which is then connected to a power source. When the electric current is applied, the larger molecules move more slowly through the gel while the smaller molecules move faster. The different sized molecules form distinct bands on the gel.

Electrophoresis refers to the **electromotive force** (EMF) that is used to move the molecules through the gel matrix. By placing the molecules in wells in the gel and applying an electric field, the molecules will move through the matrix at different rates, determined largely by their mass when the charge to mass ratio (Z) of all species is uniform. However when charges are not all uniform then, the electrical field generated by the electrophoresis procedure will affect the species that have different charges and therefore will attract the species according to their charges being the opposite. Species that are positively charged (cations) will migrate towards the **cathode** which is negatively charged. If the species are negatively charged (anions) they will migrate towards the positively charged anode [3].

If several samples have been loaded into adjacent wells in the gel, they will run parallel in individual lanes. Depending on the number of different molecules, each lane shows separation of the components from the original mixture as one or more distinct bands, one band per component. Incomplete separation of the components can lead to overlapping bands, or to indistinguishable smears representing multiple unresolved components.Bands in different lanes that end up at the same distance from the top contain molecules that passed through the gel with the same speed, which usually means they are approximately the same size.

There are **molecular weight size markers** available that contain a mixture of molecules of known sizes. If such a marker was run on one lane in the gel parallel to the unknown samples, the bands observed can be compared to those of the unknown in order to determine their size. The distance a band travels is approximately inversely proportional to the logarithm of the size of the molecule.

Gel Electrophoresis Apparatus and Supports [1]

(a) Some typical systems for carrying out gel electrophoresis are shown in following figure. These systems may have a support that is held in either a vertical or horizontal position. This support contains a running buffer with ions that carry a current through the support when an electric field is applied. To replenish this buffer and its components as they move through the support or evaporate, the ends of the support are placed in contact with two reservoirs that contain the same buffer solution and the electrodes. Once samples have been placed on the support, the electrodes are connected to a power supply and used to apply a voltage across the support. This electric field is passed through the system for a given amount of time, causing the sample components to migrate. After the electric field has been turned off, the gel is removed and examined to locate the analyte bands. The type of support we use in such a system will depend on our analytes and samples [3a,3b]. Cellulose acetate, filter paper, and starch are useful supports for work with relatively small molecules, like amino acids and **nucleotides.**

(*b*) An agarose gel is placed in this buffer-filled box and electrical field is applied via the power supply to the rear. The negative terminal is at the far end (black wire), so DNA migrates toward the anode (red wire). Gel electrophoresis separates DNA fragments by size in a solid support

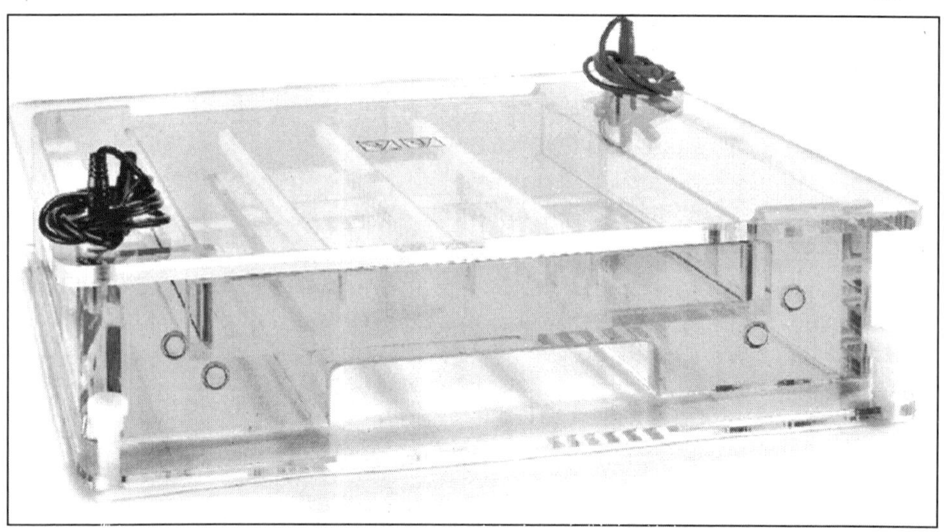

Horizontal Gel Electrophoresis System.

Vertical Gel Electrophoresis System.

***Source*: Thermo Fisher Scientific.**

medium (an agarose gel). DNA samples are pipetted into the sample wells. Application of an electric current at the top (anodal, negative) end causes the negatively-charged DNA to migrate (electrophorese) towards the bottom (cathodal, positive) end. Some models of gel electrophoresis units are given in following figures:

Gel Electrophoresis Tank with Power Unit.

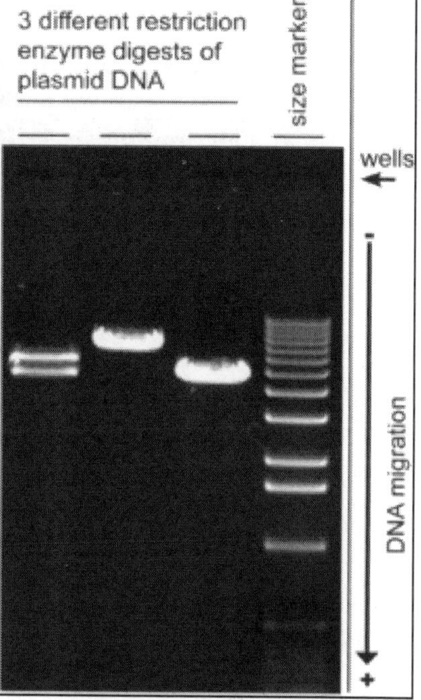

Digital Image of 3 Plasmid Restriction Digests Run on a 1 per cent w/v Agarose Gel.

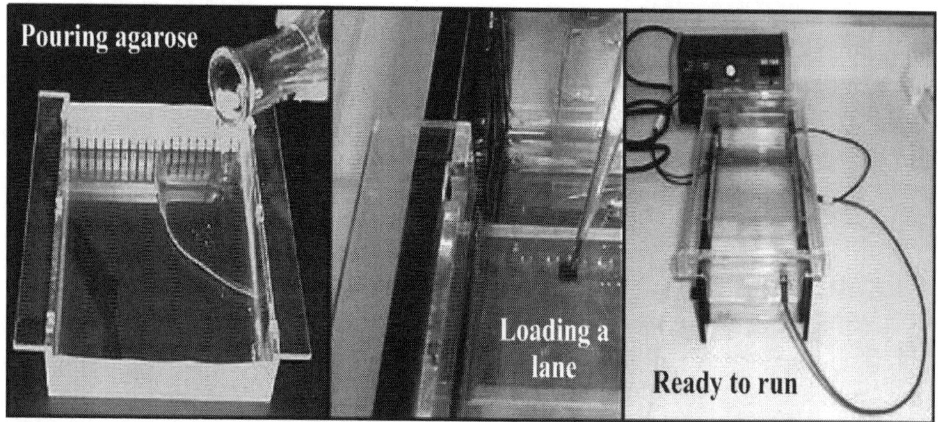

Gel Electrophoresis Unit [21].

(c) The equipment and supplies necessary for conducting agarose gel electrophoresis are relatively simple and include the followings:

(i) **An electrophoresis chamber and power supply.**

(ii) **Gel casting trays**, which are available in a variety of sizes and composed of UVtransparent plastic. The open ends of the trays are closed with tape while the gel is being cast, then removed prior to electrophoresis.

(iii) **Sample combs**, around which molten medium is poured to form sample wells in the gel.

(iv) **Electrophoresis buffer**, usually Tris-acetate-EDTA (TAE) or Tris-borate-EDTA (TBE).

(v) **Loading buffer**, which contains something dense (*e.g.* glycerol) to allow the sample to "fall" into the sample wells, and one or two tracking dyes, which migrate in the gel and allow visual monitoring or how far the electrophoresis has proceeded.

(vi) **Staining.**

(vii) **Transilluminator** (an ultraviolet light box), which is used to visualize stained DNA in gels.

Gel

Gel electrophoresis uses a gel as an anticonvective medium and/or sieving medium during electrophoresis, the movement of a charged particle in an electrical field.Gels suppress the thermal convection caused by application of the electric field, and can also act as a sieving medium, retarding the passage of molecules; gels can also simply serve to maintain the finished separation, so that a post electrophoresis stain can be applied [4].

In general the macromolecules solution is electrophoresed through some kind of matrix. The matrix acts as a molecular sieve to aid in the separation of molecules on the basis of size. The kind of supporting matrix used depends on the type of molecules to be separated and on the desired basis for separation: charge, molecular weight or both.

The term gel refers to the matrix used to contain, then separate the target molecules. In most cases, the gel is a crosslinked polymer whose composition and porosity is chosen based on the specific weight and composition of the target to be analyzed.

(i) **When separating proteins or small nucleic acids (DNA, RNA, or oligonucl- eotides)**-The gel is usually composed of different concentrations of **acrylamide** and a **cross-linker**, producing different sized mesh networks of **polyacrylamide**. **Acrylamide**, in contrast to **polyacrylamide**, is a **neurotoxin** and must be handled using appropriate safety precautions to avoid poisoning.

(ii) **When separating larger nucleic acids (greater than a few hundred bases)-** The preferred matrix is purified **agarose**. **Agarose** is composed of long unbranched chains of uncharged carbohydrate without cross links resulting in a gel with large pores allowing for the separation of macromolecules and **macromolecular complexes**.

There are limits to electrophoretic techniques. Since passing current through a gel causes heating, gels may melt during electrophoresis. Electrophoresis is performed in buffer solutions to reduce pH changes due to the electric field, which is important because the charge of DNA and RNA depends on pH, but running for too long can exhaust the buffering capacity of the solution. Further, different preparations of genetic material may not migrate consistently with each other, for morphological or other reasons.

Types of Gel [31]

The most commonly used materials for the separation of nucleic acids and proteins are agarose and acrylamide [4a].

Sl.No.	*Medium*	*Conditions*	*Principal Uses*
1	Starch	Cast in tubes or slabs	Proteins
2	Agarose gel	Cast in tubes or slabs. No cross-linking	Very large proteins, nucleic acids, nucleo-proteins *etc.*
3	Acrylamide gel	Cast in tubes or slabs	Proteins and nucleic acids

In both cases, the gel forms a solid, yet porous matrix. Each type of gel is well-suited to different types and sizes of analyte. Agarose and Polyacrylamide gels differ in following ways:

(i) **Polyacrylamide gels are run in a vertical configuration** while **agarose gels are typically run horizontally** in a submarine mode.

(ii) They also differ in their casting methodology, as **agarose sets thermally**, while **polyacrylamide forms in a chemical polymerization reaction**.

1. Agarose [31]

The most widely used polysaccharide gel matrix now-a-days is that formed with agarose. This is a polymer composed of a repeating disaccharide unit called agarobiose which consists of galactose and 3,6-anhydrogalactose. Agarose is extracted from seaweed. Agarose gels are easily cast and handled compared to other matrices, because the gel setting is a physical rather than chemical change. Samples are also easily recovered. After the experiment is finished, the resulting gel can be stored in a plastic bag in a refrigerator. Agarose gels have following characteristics:

(i) Agarose gels on the other hand **have lower resolving power for DNA but have greater range of separation,** and are therefore used for DNA fragments of usually 50-20,000 bp in size, but resolution of over 6 Mb is possible with **pulsed field gel electrophoresis** (PFGE) [5].

(ii) Agarose gels do not have a uniform pore size, but are optimal for electrophoresis of proteins that are larger than 200 kDa [6].

(iii) Agarose gel electrophoresis can also be used for the separation of DNA fragments ranging from 50 base pair to several megabases (millions of bases), the largest of which require specialized apparatus.

(iv) The distance between DNA bands of different lengths is influenced by the per cent agarose in the gel, with higher percentages requiring longer run times, sometimes days. Instead high percentage agarose gels should be run with a pulsed field electrophoresis (PFE), or field inversion electrophoresis.

(v) Most agarose gels are made with between 0.7 per cent (good separation or resolution of large 5-10kb DNA fragments) and 2 per cent (good resolution for small 0.2-1kb fragments) agarose dissolved in electrophoresis buffer. Up to 3 per cent can be used for separating very tiny fragments but a vertical polyacrylamide gel is more appropriate in this case.

(vi) Low percentage gels are very weak and may break when try to lift them. High percentage gels are often brittle and do not set evenly. The higher the agarose concentration, stiffer the gel. Agarose gels are extremely easy to prepare, simply mix agarose powder with buffer solution, melt it by heating, and pour the gel. It is also non-toxic.

(vii) 1 per cent gels are common for many applications [7].

(viii) Most modern DNA separation methods now use agarose gels, except for particularly small DNA fragments. It is currently most often used in the field of immunology and protein analysis, often used to separate different proteins or isoforms of the same protein into separate bands. These can be transferred onto a nitrocellulose or PVDF membrane to be probed with antibodies and corresponding markers, such as in a western blot.

2. Polyacrylamide

It is a cross-linked polymer of acrylamide. A far stronger gel suitable for electrophoretic separation of both proteins and nucleic acids may be formed by the polymerization of acrylamide.Polyacrylamide gels are usually used for proteins, and have very high resolving power for small fragments of DNA (5-500 bp). Poly-acrylamide gel electrophoresis (PAGE) is used for separating proteins ranging in size from 5 to 2,000 kDa due to the uniform pore size provided by the polyacrylamide gel. The length of the polymer chains is dictated by the concentration of acrylamide used, which is typically between 3.5 and 20 per cent. Polyacrylamide gels are significantly more **annoying to prepare than agarose gels**. Because oxygen inhibits the polymerization process, they must be poured between glass plates (or cylinders).Polyacrylamide gels have following characteristics:

(i) Pore size is controlled by controlling the concentrations of acrylamide and bis-acrylamide powder used in creating a gel.

(ii) Polyacrylamide gels have a rather small range of separation, but very high resolving power.

(iii) Traditional **DNA sequencing** techniques such as **Maxam-Gilbert** or **Sanger** methods used polyacrylamide gels to separate DNA fragments differing by a single base-pair in length so the sequence could be read.

(iv) **Acrylamide is a potent neurotoxin and should be handled with care.** Wear disposable gloves when handling solutions of acrylamide, and a mask when weighing out powder. Polyacrylamide is considered to be non-toxic, but polyacrylamide gels should also be handled with gloves due to the possible presence of free acrylamide

Typically resolving gels are made in 6 per cent, 8 per cent, 10 per cent, 12 per cent or 15 per cent. Stacking gel (5 per cent) is poured on top of the resolving gel and a gel comb (which forms the wells and defines the lanes where proteins, sample buffer and ladders will be placed) is inserted. The percentage chosen depends on the size of the protein that one wishes to identify or probe in the sample. The smaller the known weight, the higher the percentage that should be used. Changes on the buffer system of the gel can help to further resolve proteins of very small sizes [8].

3. Starch [31]

Partially hydrolysed potato starch makes for another non-toxic medium for protein electrophoresis. The gels are slightly more opaque than acrylamide or agarose. Non-denatured proteins can be separated according to charge and size. They are visualised using Napthal Black or Amido Black staining. Typical starch gel concentrations are 5 per cent to 10 per cent [9,10,11].

Advantages and Disadvantages of using Agarose Gels and Polyacrylamide Gels

The advantages and disadvantages of using both gels for DNA electrophoresis are tabulated below [11a]

Sl.No.	*Gels*	*Advantages*	*Disadvantages*
1	Agarose gels	Nontoxic gel medium.	High cost of agarose.
		Gels are quick and easy to cast.	Fuzzy bands.
		Good for separating large DNA molecules.	Poor separation of low molecular weightsamples
		Can recover samples by melting the gel,digesting with enzyme agarose or treating with chaotropic salts	-
2	Polyacrylamide gels	Stable chemically cross-linked gel.	Toxic monomers.
		Sharp bands.	Gels are tedious to prepare and often leak.
		Good for separation of low molecular weight Fragments.	Need new gel for each experiment.
		Stable chemically cross-linked gel.	-

Gel Conditions [31]

Denaturing and Non**denaturing** are the two conditions being used in gel electrophoresis. The details of aforesaid conditions are given below:

i. Denaturing

Denaturing gels are **run under conditions that disrupt the natural structure**

of the analyte, causing it to unfold into a linear chain. Thus, the mobility of each **macromolecule** depends only on its linear length and its mass-to-charge ratio. Thus, the secondary, tertiary, and quaternary levels of **biomolecular structure** are disrupted, leaving only the primary structure to be analyzed.

Nucleic acids are often denatured by including urea **in the buffer, while proteins are denatured using** sodium dodecyl sulfate**, usually as part of the** SDS-PAGE **process.** For full denaturation of proteins, it is also necessary to reduce the covalent disulfide bonds that stabilize their tertiary and quaternary structure, a method called reducing PAGE. Reducing conditions are usually maintained by the addition of beta-mercaptoethanol or dithiothreitol. For general analysis of protein samples, reducing PAGE is the most common form of protein electrophoresis.

Denaturing conditions are necessary for proper estimation of molecular weight of RNA. RNA is able to form more intramolecular interactions than DNA which may result in change of its electrophoretic mobility. Urea, DMSO **and** glyoxal **are the most often used denaturing agents to disrupt RNA structure.** Originally, highly toxic methylmercury hydroxide was often used in denaturing RNA electrophoresis [12], but it may be method of choice for some samples [13].

Denaturing gel electrophoresis is used in the DNA and RNA banding pattern-based methods DGGE **(denaturing gradient gel electrophoresis) [14],** TGGE **(temperature gradient gel electrophoresis), and TTGE (temporal temperature gradient electrophoresis) [15].**

ii. Non-denaturing

Native gels are run **in non-denaturing conditions, so that the analyte's natural structure is maintained.** This allows the physical size of the folded or assembled complex to affect the mobility, allowing for analysis of all four levels of the biomolecular structure. For biological samples, detergents are used only to the extent that they are necessary to lyse lipid membranes in the cell.Complexes remain-for the most part-associated and folded as they would be in the cell.

Unlike denaturing methods, native gel electrophoresis does not use a charged denaturing agent. The molecules being separated (usually proteins or nucleic acids) therefore differ not only in molecular mass and intrinsic charge, but also the cross-sectional area, and thus experience different electrophoretic forces dependent on the shape of the overall structure. For proteins, since they remain in the native state they may be visualised not only by general protein staining reagents but also by specific enzyme-linked staining.

Native **gel electrophoresis is typically used in** proteomics **and** metallomics **[16]. However, native SDS-PAGE is also used to scan genes (DNA) for unknown mutations as in** Single-strand conformation polymorphism.

Preparation and Running of Standard Agarose Gels [1]

To prepare gel, agarose powder is mixed with electrophoresis buffer to the desired concentration, and heated in a microwave oven to melt it. Ethidium bromide is added to the gel (final concentration 0.5 ug/ml) to facilitate visualization of DNA after electrophoresis.

After cooling the solution to about 60°C, it is poured into a casting tray containing a sample comb and allowed to solidify at room temperature. After the gel has solidified, the comb is removed, taking care not to rip the bottom of the wells. The gel, still in plastic tray, is inserted horizontally into the electrophoresis chamber and is covered with buffer. Samples containing DNA mixed with loading buffer are then pipetted into the sample wells, the lid and power leads are placed on the apparatus and a current is applied. The current flow can be confirmed by observing bubbles coming off the electrodes. DNA will migrate towards the positive electrode, which is usually colored red, in view of its negative charge. The distance DNA has migrated in the gel can be judged by visually monitoring migration of the tracking dyes like bromophenol blue and xylene cyanol dyes.

Sample Application

The samples in gel electrophoresis are applied to small "wells" that are made in the gel during its preparation. A sample volume of 10-100 μL is then placed into one of these wells by using a micropipette. These sample volumes help provide a sufficient amount of analyte for later detection and collection, but they also create a danger of introducing band-broadening by creating a large sample band at the beginning of the separation.

A common approach to create narrow sample bands is to employ two types of gels in the system- **a "stacking gel" and a "running gel"** [3b]. The running gel is the support used for the electrophoretic separation of substances in the sample. In a vertical gel electrophoresis system, this gel is formed first and is located throughout the middle and lower section of the system. The stacking gel has a lower degree of crosslinking (giving it larger pores) and is located on top of the running gel. The stacking gel is also the section of the support in which the sample wells are located. After a sample has been placed in the wells and an electric field has been applied, analytes will travel quickly through the stacking gel until they reach its boundary with the running gel.

Buffers

The buffers in gel electrophoresis are used to provide ions that carry a current and to maintain the pH at a relatively constant value. There are a number of buffers used for electrophoresis.

(i) The most common being, for nucleic acids **Tris/Acetate/EDTA** (TAE), **Tris/Borate/EDTA** (TBE).

(ii) Many other buffers have been proposed, *e.g.* **lithium borate**, which is almost never used, iso electric histidine, pK matched goods buffers, *etc.*; in most cases the purported rationale is lower current (less heat) and or matched ion mobilities, which leads to longer buffer life.

(iii) Most SDS-PAGE protein separations are performed using a "discontinuous" (or DISC) buffer system that significantly enhances the sharpness of the bands within the gel. During electrophoresis in a discontinuous gel system, an ion gradient is formed in the early stage of electrophoresis that causes all of the proteins to focus into a single sharp band in a process called

isotachophoresis. Separation of the proteins by size is achieved in the lower, "resolving" region of the gel. The resolving gel typically has a much smaller pore size, which leads to a sieving effect that now determines the electrophoretic mobility of the proteins.

Merits and Demerits of different Buffers

(i) Borate is problematic; Borate can polymerize, and/or interact with cis diols such as those found in RNA. LB is relatively new and is ineffective in resolving fragments larger than 5 kbp; However, with its low conductivity, a much higher voltage could be used (up to 35 V/cm), which means a shorter analysis time for routine electrophoresis. As low as one base pair size difference could be resolved in 3 per cent agarose gel with an extremely low conductivity medium (1 mM Lithium borate) [17].

ii TAE has the lowest buffering capacity but provides the best resolution for larger DNA. This means a lower voltage and more time, but a better product.

Detection Methods and Visualization

a. Detection Methods

The analytes can be detected in gel electrophoresis by several aways. Analyte bands can be examined directly on the gel or they can be transferred to a different support for detection. Direct detection can sometimes be performed visually (when dealing with intensely colored proteins like hemoglobin) or by using absorbance measurements and a scanning device known as a **densitometer [17a,17b].**

b. Visualization/Staining [1,31]

The most common approach for detection in gel electrophoresis is to treat the support with a stain or reagent that makes it easier to see the analyte bands. After the electrophoresis is complete, the molecules in the gel can be stained to make them visible.

(i) DNA may be visualized using ethidium bromide which, when intercalated into DNA, fluoresce under ultraviolet light. Alternatively, nucleic acids can be stained after electrophoretic separation by soaking the gel in a solution of ethidium bromide. When intercalated into double stranded DNA, fluorescence of this molecule increases greatly. It is also possible to detect DNA with the extrinsic fluor 1-anilino 8-naphthalene sulphonate.

Note: Ethidium bromide is a known mutagen and should be handled as a hazardous chemical. Need wearing gloves while handling.

(ii) While protein may be visualised using silver stain or Coomassie Brilliant Blue dye and amido black.

Silver Staining [1]: The most sensitive staining for protein is silver staining. This involves soaking the gel in Ag NO_3 which results in precipitation of metallic silver (Ag0) at the location of protein or DNA forming a black deposit in a process similar to that used in black-and-white photography.

(a) Steps involving formaldehyde solutions should be performed in the fume hood.

(b) Chill the developer solution to 4°C. Prepare the developer fresh before each use.

(c) Be sure to save the fix/stop solution from the first step in the silver staining to add to the developer solution once the bands are visible and d- The 10 second deionized water rinse must not exceed this time frame. If it does, the deposited silver may be rinsed away and the staining must be done again.

(iii) Other methods may also be used to visualize the separation of the mixture's components on the gel.

(iv) If the molecules to be separated contain radioactivity, for example in DNA sequencing gel, an autoradiogram can be recorded of the gel.Photographs can be taken of gels, often using Gel Doc.

The details related to the commonly used stains for biopolymers after electrophoretic separation in agarose or polyacrylamide gels., uses and detection limits are given in following table [1]:

Sl.No.	*Stain*	*Detection limits (ng)*	*Uses*
1	Amido black	400	Proteins
2	Bis-1-anilino-8-Naphthalene sulphonate	150	Proteins
3	Coomassie blue	200	Proteins
4	Ethidium bromide	10	DNA/RNA
5	Fluorescamine (protein treated prior to electroph-oresis)	1	Proteins
6	Nile red	20	Proteins (reversible)
7	Ponceau red	200	Proteins (reversible)
8	Silver chloride	1	Proteins/DNA
9	SYPRO orange	10	Proteins
10	SYPRO red	0.5	Proteins

c. Blotting

Another possible approach for detection in gel electrophoresis is to transfer a portion of the analyte bands to a second support (such as nitrocelluose), where they are reacted with a labeled agent. This approach is known as blotting [17b]. There are several blotting methods.

Southern Blot

One such method is a **Southern blot** which is named after its discoverer Edwin Southern] 17c].

Western Blotting [26]

(i) This technique refers to identification of specific proteins, which are first separated on acrylamide gel using electrophoresis and then subsequently transferred to nitrocellulose membrane and identified.

(ii) The identification of protein (antigen) is carried out by performing antigen-antibody (first antibody) reactions on the membrane itself.

(iii) Second antibody-enzyme conjugates were then allowed to interact with immobilized first antibody and then using appropriate substrate, protein bands are detected. Although, antigen-antibody interactions are widely employed in Western blot, other kind of interactions such as glycoprotein-lactin and biotin-avidin have allowed research workers to employ this technique for other applications including carbohydrate staining of glycoprotein, protein sequencing *etc.*.

(iv) Many workers described this technique simultaneously during 1979-80, but the method described by Towbin (1979) is most cited.

(v) The technique of Western blotting involves following two distinct techniques:

(a) SDS-PAGE.

(b) Electrophoretic transfer of protein from gel to membrane and immuno-detection of proteins.

Downstream Processing

After separation, an additional separation method may then be used, such as **isoelectric focusing** or **SDS-PAGE**. The gel will then be physically cut, and the protein complexes extracted from each portion separately. Each extract may then be analysed, such as by **peptide mass fingerprinting** or **de novo sequencing** after **in-gel digestion**. This can provide a great deal of information about the identities of the proteins in a complex.

Applications [31]

(i) Estimation of the size of DNA molecules following restriction enzyme digestion, *e.g.* in **restriction mapping** of cloned DNA.

(ii) Analysis of **PCR** products, *e.g.* in molecular **genetic diagnosis** or **genetic fingerprinting**.

(iii) Separation of restricted genomic DNA prior to **Southern transfer**, or of RNA prior to **Northern transfer**.

Gel electrophoresis is used in forensics, molecular biology, genetics, microbiology and biochemistry. The results can be analyzed quantitatively by visualizing the gel with UV light and a gel imaging device. The image is recorded with a computer operated camera, and the intensity of the band or spot of interest is measured and compared against standard or markers loaded on the same gel. The measurement and analysis are mostly done with specialized software.

Depending on the type of analysis being performed, other techniques are often implemented in conjunction with the results of gel electrophoresis, providing a wide range of field-specific applications.

1. Nucleic Acids [31]

In the case of nucleic acids, the direction of migration, from negative to positive electrodes, is due to the naturally occurring negative charge carried by their sugar-phosphate backbone [18].

Double-stranded DNA fragments naturally behave as long rods, so their migration through the gel is relative to their size or, for cyclic fragments, their radius of gyration. Circular DNA such as plasmids, however, may show multiple bands, the speed of migration may depend on whether it is relaxed or supercoiled. Single-stranded DNA or RNA tend to fold up into molecules with complex shapes and migrate through the gel in a complicated manner based on their tertiary structure. Therefore, agents that disrupt the hydrogen bonds, such as sodium hydroxide or formamide, are used to denature the nucleic acids and cause them to behave as long rods again [19].

Gel electrophoresis of large DNA or RNA is usually done by agarose gel electrophoresis.Characterization through ligand interaction of nucleic acids or fragments may be performed by mobility shift affinity electrophoresis.

Electrophoresis of RNA samples can be used to check for genomic DNA contamination and also for RNA degradation. RNA from eukaryotic organisms shows distinct bands of 28s and 18s rRNA, the 28s band being approximately twice as intense as the 18s band. Degraded RNA has less sharpely defined bands, has a smeared appearance, and intensity ratio is less than 2: 1.

Gel electrophoresis separates DNA fragments by size in a solid support medium (an agarose gel). DNA samples are pipetted into the sample wells, seen as dark slots at the top of the picture. Application of an electric current at the top (anodal, negative) end causes the negatively-charged DNA [remember it's an acid] to migrate (electrophorese) towards the bottom (cathodal, positive) end. The rate of migration is proportional to size - smaller fragments move more quickly, and wind up at the bottom of the ge [20].

DNA is visualized by including in the gel an intercalating dye, ethidium bromide. DNA fragments take up the dye as they migrate through the gel. Illumination with ultraviolet light causes the intercalated dye to fluoresce with a pale pink colour [20].

The larger fragments fluoresce more intensely. Although each of the fragments of a single class of molecule are present in equimolar proportions, the smaller fragments include less mass of DNA, take up less dye, and therefore fluoresce less intensely. This is most evident in the lane at the extreme right, which shows a "**ladder**" set of **DNA** fragments of known size that can be used to estimate the sizes of the other unknown fragments [20].

2. Proteins [31]

Proteins, unlike nucleic acids, can have varying charges and complex shapes, therefore they may not migrate into the polyacrylamide gel at similar rates, or at all, when placing a negative to positive EMF on the sample. Proteins therefore, are usually **denatured** in the presence of a **detergent** such as **sodium dodecyl sulfate** (SDS)that coats the proteins with a negative charge [3]. Generally, the amount of SDS bound is relative to the size of the protein (usually 1.4g SDS per gram of protein), so that the resulting denatured proteins have an overall negative charge, and all the proteins have a similar charge to mass ratio.

Since denatured proteins act like long rods instead of having a complex tertiary shape, the rate at which the resulting SDS coated proteins migrate in the gel is relative only to its size and not its charge or shape [3].

Proteins are usually analyzed by sodium dodecyl sulfate polyacrylamide gel electrophoresis (SDS-PAGE), by native gel electrophoresis, by quantitative preparative native continuous polyacrylamide gel electrophoresis (QPNC-PAGE), or by 2-D electrophoresis.

Characterization through ligand interaction may be performed by electroblotting or by affinity electrophoresis in agarose or by capillary electrophoresis as for estimation of binding constants and determination of structural features like glycan content through lectin binding.

Types of Gel Electrophoresis

PFGE, SDS-PAGE Isoelectric Focusing, 2-Dimensional Electrophoresis are some examples of Gel Electrophoresis.Details of these gel electrophretic techniques are given below:

Pulsed Field Gel Electrophoresis [21]

It is a technique used for the separation of large deoxyribonucleic acid (DNA) molecules by applying to a gel matrix an electric field that periodically changes direction.While in general small fragments can find their way through the gel matrix more easily than large DNA fragments, a threshold length exists above 30–50 kb where all large fragments will run at the same rate, and appear in a gel as a single large diffuse band. However, with periodic changing of field direction, the various lengths of DNA react to the change at differing rates. That is, larger pieces of DNA will be slower to realign their charge when field direction is changed, while smaller pieces will be quicker. Over the course of time with the consistent changing of directions, each band will begin to separate more and more even at very large lengths. Thus separation of very large DNA pieces using PFGE is made possible.

Sodium Dodecyl Sulfate - Polyacrylamide Gel Electrophoresis (SDS-PAGE) [21,22]

It is one of the most common means of analyzing proteins by electrophoresis and is the most widely used method for quantative analysis of protein mixture. It is particularly useful for monitoring protein purification and determination of the relative molecular mass of proteins.

Principle

Sodium Dodecyl Sulfate SDS (CH_3- (CH_2)10-CH_2OSO_3-Na+) is an anionic detergent which denatures proteins when proteins are treated with SDS in presence of a reducing agent like b-mercaptoethanol or dithiothreitol, SDS binds to hydrophobic regions of protein molecule and provides net negative charge on protein molecule. The SDS is negatively charged and thus becomes the dominant charge of the complex. The binding of SDS to per-unit length of protein molecules is almost constant for large number of different proteins and this brings charge to–mass ratio almost constant for most proteins. The number of SDS molecules that bind is simply proportional to the size of the protein. Therefore the charge to mass ratio should not change with size [22]. In solution (water), in principle all different sized proteins covered with SDS would run at about the same mobility. **However, the proteins are not run through water. Instead they are run through an inert polymer, polyacrylamide.** The density and pore size of this polymer can be varied by just how you make it (concentration of monomer and of cross-linking agent). Thus, the size of molecules that can pass through the matrix can be varied. This determines in what molecular weight range the gel will have the highest resolving power [21].

The electrophoretic movement of protein in acrylamide gel is determined by molecular weight of proteins. Lower molecular weight proteins move faster than high molecular weight proteins. Laemmli method [23] is widely used. In this method, discontinuous buffer system is employed.

A **continuous buffer system** has only single separating gel and uses same buffer in the tanks and gel. In **discontinuous buffer system**, large pore gel (stacking gel) is layered over small pore gel (separating or running gel). For preparation of stacking gel and separating gel, different buffers are used and also tank buffers are different from gel buffers. When electrophoresis is started, ions from stacking gel (leading ion), ions from buffer tank (trailing ion) and proteins start moving into stacking gel. In stacking gel, protein moves between leading ion and trailing ion and this leads to concentration of protein in a thin zone referred as stack. The protein molecules continue to move in the stack until they reach the separating gel.

Crosslinked polyacrylamide gels are formed from the polymerization of acrylamide monomer in the presence of smaller amounts of N, N′-methylene-bis-acrylamide (normally referred to as "bis-acrylamide"). For both **SDS-PAGE** and **Native-PAGE**, samples may be applied directly to the top of the gel in which protein separation is to occur **(the separating gel).** To get better resolution by polymerizing a short **stacking gel** on top of the separating gel. The purpose of this stacking gel is to concentrate the protein sample into a sharp band before it enters the main separating gel thus giving sharper protein bands in the separating gel. This modification allows relatively large sample volumes to be applied to the gel without any loss of resolution. The stacking gel has a very large pore size (4 per cent acrylamide) which allows the proteins to move freely and concentrate, or stack under the effect of the electric field [22].

Test Procedures [22,26]

SDS-PAGE is basically of following two type:

(i) **Glycine** - Glycine is used for separation of proteins which have molecular weight more than 20 kDa.

(ii) **Tricine**-Tricine is used for low molecular weight protein (<20 kDa).

1. Method for Glycine-SDS-PAGE [22,24,25]

It also known as **Laemmli-SDS-PAGE** and It mainly involes electrophoretic separation and staining of protein, Electrophoretic Transfer of Proteins from gel to Membrane, Detection of Transferred Protein on Nitrocellulose Membrane and Visualization of Molecular Weight Markers and Protein (Antigen).

A. Electrophoretic Separation and Staining of Protein [22]

Equipments-Mini-vertical gel electrophoresis dual model with glass plates, spacer, comb and power-supply, orbital shaker; 1ml glass syringe with 2″22G needle.

Reagents and Chemicals

Acrylamide, N, N1 methylene bisacrylamide, ammonium persulfate, b-mercaptoethanol, sodium dodecyl sulfate, molecular weight markers, coomassie brilliant blue R-250, TEMED, tris, glycine, dithiothreitol.

Stock Solutions

(i) Acrylamide/Bisacrylamide (30 per cent)- 29.2 g acrylamide and 0.8 g bisacrylamide are dissolved in distilled water and total volume was made to 100 ml. The solution is filtered and filtered solution can be stored at 4°C in dark bottle up to 3 months.

(ii) **4 x Running Gel Buffer (1.5 M Tris-HCl, pH 8.8)** - 18.15 g Tris is dissolved in about 80 ml distilled water. pH is adjusted to 8.8 with 1 N HCl and total volume is made to 100 ml with distilled water. Prepared buffer can be stored up to 3 months at 4°C in dark bottle..

(iii) **4 x Stacking Gel Buffer (0.5M Tris-HCl, pH 6.8)**- 3.0 g Tris is dissolved in about 40 ml distilled water. pH is adjusted to 6.8 with 1 N HCl and total volume is made to 50 ml with distilled water. Prepared buffer can be stored up to 3 months at 4°C in dark bottle.

(iv) **10 per cent SDS-** 10 g sodium dodecyl (lauryl) sulfate is dissolved in distilled water and total volume is made to 100 ml with distilled water. Prepared solution can be stored at room temperature.

(v) **5 x Electrode Buffer (125 mM Tris, 960 mM Glycine, 0.5 per cent SDS, pH 8.3)**-15 g Tris,72 g glycine and 5 g SDS are dissolved in distilled water and total volume is made to 1 litre with distilled water. The pH of buffer should be 8.3± 0.2. Stock electrode buffer is diluted five times with distilled water before use. The stock buffer can be stored at room temperature up to 1 month. The diluted stock buffer is 25 mM tris, 192 mM glycine, and 0.1 per cent SDS.

(vi) **10 per cent Ammonium Persulfate**-100 mg ammonium persulfate is dissolved in 1.0 ml distilled water. The solution is always prepared fresh.

(vii) **2 x Sample Buffer (0.125 M Tris, 4 per cent SDS, 20 per cent glycenol 0.2 M DTT, 0.02 per cent bromophenol blue, pH 6.8)**- 2 x sample buffer is prepared by mixing following solutions/chemical:

a- 4 x stacking gel buffer - 2.5 ml, b- glycerol - 2.0 ml, c-10 per cent SDS - 4.0 ml, d- Bromophenol blue - 2.0 mg, e-Dithiothreitol (DTT) - 0.31 g and f-Distilled water - 1.5 ml..

(viii) 2 x sample buffer can be stored in small aliquots at - 20°C up to 6 months. Instead of DTT, 1.0 ml of b-mercaptoethanol can be used but the volume of water is reduced to 0.5 ml.

(ix) **Overlay Buffer (0.375 M Tris, 0.1 per cent, SDS, pH 8.8)**- Overlay buffer is prepared by mixing 25 ml running gel buffer, 1 ml 10 per cent SDS and 74 ml distilled water. This buffer can be stored up to 3 months at 4°C in the dark bottle.

Procedure

This involvers the following steps:

a. Glass Sandwich

(i) One notched glass plate is placed on a flat surface.

(ii) One spacer (1.0 mm) each is then placed along the each of two edges so that spacer aligns with the notch.

(iii) Subsequently, rectangular glassplate is placed over it.

(iv) The sandwich is held firmly between thumb and fingers.

(v) Side-ways of both spacers were sealed with appropriate tape to overcome any possible gel-leak during gel plate preparation.

(vi) There is always the possibility of leakage at the bottom of the plate. This is taken care by placing molted agar (1 per cent in water) up to 5 mm height in trough of gel-casting uni.

(vii) The plate in standing position is then quickly placed in casting unit and screws are finger tightened.

b. Preparation of Running Gel

i. Final Gel Concentrations

The running gel of desired concentration is prepared by mixing appropriate volumes of solutions as shown in following table:

Sl.No.	*Solutions*	*7.5 per cent*	*10 per cent*	*12.5 per cent*	*15 per cent*
1	Acrylamide/bisacrylamide (30 per cent)	5.0 ml	6.7 ml	8.3 ml	10.0 ml
2	4 x Running gel buffer	5.0 ml	5.0 ml	5.0 ml	5.0 ml
3	Distilled water	9.7 ml	8.0 ml	6.4 ml	4.7 ml
4	SDS 10 per cent	0.2 ml	0.2 ml	0.2 ml	0.2 ml
5	Ammonium persulfate 10 per cent	0.1 ml	0.1 ml	0.1 ml	0.1 ml
6	TEMED	6.7 µl	6.7 µl	6.7 µl	6.7 µl

ii. Mixing

Acrylamide/bisacrylamide, running gel buffer, SDS and distilled water are added to conical flask and degassed and then ammonium persulfate and TEMED are added and contents mixed gently. With the help of glass pipette, the running gel solution is delivered to sandwich to a level about 3 cm below the top of rectangular plate. There should not be air trapping while filling sandwich with running gel solution. A small volume of water or overlay-buffer (~ 200 µl) is layered over gel solution by glass syringe with 22 G needle to prevents exposure to oxygen.

c. Preparation of Stacking Gel

4 per cent Stacking gel is prepared by mixing appropriate volumes of solutions as tabulated below:

Sl.No.	Solutions	Volume
1	Acrylamide/bisacrylamide (30 per cent)	1.3 ml
2	4 x Running gel buffer	2.5 ml
3	Distilled water	6.1 ml
4	SDS 10 per cent	0.1 ml
5	Ammonium persulfate 10 per cent	50 µl
6	TEMED	10 µl

The stacking-gel solution preparation is similar to running-gel solution preparation. After removal of water or overlay buffer, stacking-gel solution is layered over running gel. Appropriate comb is inserted into the stacking gel to make wells for sample application and which is removed after polymerization of gel.

d. Sample Preparation

(i) Protein samples (1 mg/ml) are centrifuged (10,000 g, 5 min.) to remove any insoluble material and are mixed with equal volume of 2 X sample buffer.

(ii) The resultant solution is boiled for 3 to 5 min.

(iii) Molecular weight markers are also prepared in a similar way.

e. Electrophoresis

(i) Gel plates are then tightly attached to electrophoresis unit.

(ii) Stock electrode buffer is five times diluted with cold water.

(iii) Anode and cathode chambers are filled with buffer. 5 to 20 µl of sample is applied to each well.

(iv) Electrophoresis is carried out at constant voltage of 50V till sample crosses stacking gel.

(v) When sample enters running gel, voltage is increased to 100V.

(vi) Complete electrophoretic run takes around 2.5 to 3.0 h. During electrophoresis, temperature is kept low by circulating water in electrophoretic assembly.

(vii) After electrophoretic run, stacking gel is removed.

(viii) Small cut on top left side in running gel is made to remember the orientation of the gel.

f. Staining of Proteins in Gel

It involves the following steps:

(i) The gel is placed in glass tray containing coomassie brilliant blue solution (0.25 per cent) prepared in methanol: acetic acid: water (40: 7: 53) mixture.

(ii) Glass tray is then placed on orbital shaker for 4 h at room temperature.

(iii) After staining for 4 h, the gel is transferred to the destaining solution I (methanol, acetic acid and water mixture in ratio of (40: 7: 53) for 30 min.

(iv) Subsequently gel is placed in destaining solution II (methanol, acetic acid and water mixture in ratio of 7: 5: 88) till bands become visible against light background. During staining and destaining, gel should float free in glass tray.

B. Electrophoretic Transfer of Proteins From Gel to Membrane [22,24,25]

a. Equipments

Mini-trans blot assembly (Bio-rad), power pack, orbital shaker, tris, glycine, methanol, nitrocellulose membrane, Whatman No. 3 paper.

b. Chemicals

Transfer buffer (25 mM tris, 192 mM glycine, 20 per cent methanol, pH 8.3)- 3.03 g tris and 14.4 g glycine are dissolved in distilled water. 200 ml methanol is then added. Volume is made up to 1 litre with distilled water. The pH of buffer will range from 8.1 to 8.4 depending on quality of tris, glycine and methanol. Methanol should be analytical grade as metallic contaminants in low grade methanol will plate on the electrode. The pH of buffer is not adjusted with acid or base.

c. Procedure

(i) Membrane and gel are handled only after wearing gloves.

(ii) After completion of SDS-PAGE, spacer gel is removed. A small cut on top left side in running gel is made to remember the orientation of gel.

(iii) The running gel is equilibrated with transfer buffer for 30 min. to remove salts and SDS. Transfer buffer is changed at least once during equilibration. Membrane of appropriate size is cut from sheet. A small cut on top left side of membrane (glossy side facing worker) is made to remember orientation of the membrane.

(iv) While the gel is equilibrating, nitrocellulose membrane is activated by placing it in transfer buffer at an angle of 45°. Also, fiber pads and pre-cut filter papers (Whatman No. 3) are immersed in transfer buffer. Air bubbles trapped in fiber pads and filter papers are removed.

(v) Gel holder cassete is opened and placed in glass vessel so that the gray panel is flat on the bottom of the vessel and clear panel rests at an angle against wall of the vessel.

(vi) Gel holder cassette is assembled in following sequence: gray panel (cathode), fiber pad, filter paper, gel, nitrocellulose membrane (glossy side facing the gel), filter paper, fiber pad, clear panel (anode). This arrangement allows transfer of proteins on membrane where well position remains the same as that in acrylamide gel. The care should be taken not to allow trapping of air-bubbles during assembling. Nearly adhesive contact is essential between the membrane and gel otherwise swirled or missing transfer patterns and overall high background will be observed.

(vi) Buffer tank is filled with transfer buffer (4°C). Bio-freeze cooling unit containing ice is placed in buffer tank.

(vii) Gel holder cassette is closed and placed in the buffer tank such that gray panel of the cassette faces the gray cathode electrode panel.

(viii) The whole of blotting assembly is then placed over the magnetic stirrer.

(ix) Electrophoretic transfer is carried out at constant voltage of 30 V overnight at 4°C. The starting current should be around 40 mA. At the end of transfer, the current should be 90 mA (if current is less than 90 mA, a constant voltage of 100 V is additionally applied for 1 h).

(x) After run, nitrocellulose membrane is stained with different reagents for visualization of proteins or antigens. For ascertaining transfer of proteins from gel, the gel is also stained with coomassie brilliant blue.

C. Detection of Transferred Protein on Nitrocellulose Membrane [22,24,25]

In Western blot, molecular weight markers and protein (antigen) samples are loaded in separate lanes in SDS-PAGE. Whereas, methods used for staining of molecular weight markers are based on non-specific reaction of dye with protein, antigenic proteins are detected employing antigen-antibody interaction. Therefore, after electrophoretic transfer, the membrane-portion containing molecular weight marker is cut from rest of membrane containing protein antigens.

D. Visualization of Molecular Weight Markers and Proteins [22,24,25]

The molecular weight markers can be stained by Ponceau S or congo-red dye. The proteins (antigens) are stained using primary antibody and secondary antibody-enzyme conjugates.

i. Visualization of Molecular Weight Markers [22,24,25]

The details related to the staining of molecular weight markers are tabulated on next page:

Sl.No.	*Procedural Steps*	*Ponceau S Staining*	*Congo-red Staining*
1	Dye preparation	*(i)* Stock dye solution is prepared by dissolving 200 mg Ponceau S in 10 ml of 3 per cent trichloroacetic acid. It can be stored at room temperature. *(ii)* The stock solution is diluted ten fold with distilled water before use.	*(i)* Stock dye solution is prepared by dissolving 1 g Congo-red in 100 ml distilled water. This solution is stable at room temperature. *(ii)* The working congo-red solution is prepared just before use by diluting 1 ml of stock dye solution with 9 ml of 0.2 M acetate buffer, pH 3.5.
2	Staining	*(i)* The membrane is added slowly to vessel containing diluted dye solution so that membrane absorbs dye uniformly. *(ii)* The membrane is then submerged for 5 to 10 min with mild shaking *(iii)* After staining, the membrane is then rinsed with water or PBS until a clear contrast between the bands (pink) and background (white) is observed. *(iv)* Staining of proteins with Ponceau S is reversible.	*(i)* The membrane is submerged in working congo-red solution for 5 min. at room temperature. *(ii)* The destaining is carried out by immersing the membrane in distilled water until brown bands become visible against light pink background. *(iii)* During staining and destaining, mild shaking is employed.

ii. Visualization of Protein (Antigen) [22,24,25]

The details related to Visualization of Protein are tabulated below:

Sl.No.	*Visualization of Antigen using Secondary Antibody Peroxidase Conjugate*	*Visualization of Antigen using Secondary Antibody-Alkaline Phosphatase Conjugate*
1	Reagents- Primary antibodies directed against antigen and raised in rabbit, secondary antibody enzyme conjugates such as goat anti rabbit immunoglobulin-peroxidase goat, anti rabbit immunoglobulin-alkaline phosphatase, diamino benzedine, hydrogen peroxide, bovine serum albumin, nitro blue tetrazolium (NBT) bromochloroindolyl phosphate (BCIP), dimethyl formamide (DMF).	
2	All steps are carried out at room temperature.	The method is similar to the method described using secondary antibody-peroxidase conjugate except the followings.
3	Membrane is washed with PBS (3x10 min.)	Appropriately diluted secondary antibody-alkaline phosphatase conjugate is used instead of antibody-peroxidase conjugate.
4	The membrane is treated with blocking solution (3 per cent BSA prepared in PBS) for 1 h.	Instead of DAB-H2O2, the enzyme substrate used is BCIP-NBT. Stock solutions of nitroblue tetrazolium (NBT) and bromo chloro indolyl phosphate (BCIP) are prepared and stored at -20°C.

Contd.

Sl.No.	Visualization of Antigen using Secondary Antibody Peroxidase Conjugate	Visualization of Antigen using Secondary Antibody-Alkaline Phosphatase Conjugate
5	The membrane is treated with diluted rabbit antiserum for 1 h. The antiserum in rabbit is raised against the antigen. The dilution is decided by antibody litre in immune serum and is carried out in 1 per cent BSA – PBS-(0.05 per cent) Tween – 20. The membrane is washed with PBS -.05 per cent Tween 20 (3x10min.).	Stock NBT is prepared by dissolving 30 mg NBT in 1 ml of 70 per cent DMF.
6	The membrane is treated with goat anti-rabbit immunoglobulin - peroxidase conjugate (1:1000 diluted with 1 per cent BSA – PBS – Tween 20) for 1 h. The dilution of conjugate is done as per instruction from manufacturer.	Stock BCIP is prepared by dissolving 15 mg BCIP in 1 ml of DMF. The working substrate solution is prepared by addition of 200 µl of stock NBT and 200 µl of stock BCIP to 20 ml of 100 mM Tris-HCl, pH 9.5 containing 100 mM NaCl and 5 mM MgCl2. When membrane is treated with enzyme substrate, light violet colour blots become visible against light background.
7	The membrane is washed with PBS (4 x 10 min.).	-
8	The membrane is immersed in enzyme substrate DAB - H2O2 (6 mg diaminobenzidine in 10 ml of 0.05 M Tris-HCl buffer, pH 7.6 containing 100 µl of 3 per cent H2O2) till brown bands become visible. The membrane at that stage is washed with distilled water and air-dried.	-
9	Membrane strips containing molecular weight markers and proteins (antigens) are aligned and photographed.	-

2. Procedure for Tricine-SDS-PAGE [27]

In this system, Tricine, used as the trailing ion, allows a resolution of small proteins at lower acrylamide concentrations than in glycine-SDS-PAGE systems. Tricine-SDS-PAGE is commonly used to separate proteins in the mass range 1-100 kDa but it is the preferred system for the resolution of proteins smaller than 30 kDa. A superior resolution of proteins, especially in the range between 5 and 20 kDa, is achieved without the necessity to use urea.

Apparatus and Reagents [27]

A. Apparatus

Mini vertical gel electrophoresis unit.

B. Reagents

The following solutions are required for proper running of Tricine-SDS-PAGE:

(i) **Acrylamide Solution (49.5 per cent T, 3 per cent C)**-Acrylamide - 48 g, N, N′- Methylene-bis-acrylamide –1.5 g, Dissolve in distilled water to a

final volume of 100 ml. Filter the solution and refrigerate (7-10°C).Gentle warming may be required for complete dissolution after refrigeration, per cent T= Total acrylamide percentage of both monomers (acrylamide and the crosslinker bisacrylamide), per cent C = Percentage concentration of the crosslinker relative to total concentration.

(ii) **Gel Buffer (3X)**-Tris -36.34 g, SDS - 0.3 g, HCl, Dissolve in 60 ml water. Adjust the pH to 8.45 with concentrated HCl. Make up the final volume upto 100 ml with water. Store at 20 – 25°C.

(iii) **Cathode Buffer (1X)**-Tris - 12.11 g, Tricine -17.92 g, SDS -1 g.Dissolve in distilled water and make volume upto 1 L. The pH of the solution should be approx 8.2. Do not correct the pH.

(iv) **Anode Buffer**-Tris 6.05 g.Dissolve in 50 ml. Adjust pH upto 8.9 with conc. HCl/1N HCl and then makeup volume up to 500 ml with distilled water.

(v) **Sample Buffer (4x)**-SDS (12 per cent, w/v) -12 ml of 20 per cent SDS, Glycerol (30 per cent, w/v) - 6 g, Mercaptoethanol (6 per cent.v/v) -1.2 ml, Coommassie blue G 250 (0.05 per cent) -0.01g, Tris/HCl (pH 7) (150 mM) - 3 ml of 1 M Tris-HCl.Make up the volume upto 20 ml.

(vi) **Marker (2.5µl/well)**-Take 3.5µl marker and add 10µl sample buffer (1.33x). For 1.33x sample buffer, dilute 1ml of 4x sample buffer with 2 ml distilled water.

(vii) **Ammonium persulfate**-Dissolve 100 mg of ammonium persulfate in 1 ml of distilled water immediately before use.

viii **Separating gel solution (16 per cent)**-Acrylamide solution - 5 ml, Gel buffer - 5 ml, Glycerol - 1.5 ml, Water -.5 ml, APS - 30 µl (last to be added), TEMED (N, N, N′, N′-Tetramethylethylenediamine) -10 µl.

(ix) **Stacking gel solution (4 per cent)**-Acrylamide solution -1 ml, Gel buffer - 3 ml, Water - 8 ml, APS -60 µl, TEMED - 10 µl.

(x) **Fixing Solution**-10 per cent TCA (Trichloroacetic acid)

(xi) **Staining Solution** (0.025 per cent Coommassie brilliant blue G250 in 10 per cent acetic acid)-Dissolve 25 mg of the dye in 100 ml of 10 per cent acetic acid.

(xii) **Destaining Solution**-10 per cent Acetic acid. Destain the gel twice. Each incubation should last for 15-60 min. Then transfer the gel to distilled water.

Method [27]

It involves–Gel Preparation, Sample Preparation and Electrophoresis. The details related to these steps are givcn below:

Gel Preparation

(i) Prepare gel plates of the size 8 x 10 cm in casting stand for gel electrophoresis.

(ii) Prepare the separating gel (16.5 per cent), pipette it down into each of the gel cassettes to a height of 4 cm.

(iii) Overlay the gel mix with water or butanol to cuy oxygen action and to give a flat gel surface for flat sample bands.

(iv) After polymerization carefully remove water with the help of blotting paper.

(v) Then overlay stacking gel solution over it.

(vi) Insert preparative comb into the stacking gel solution to make troughs and wells and keep the whole system undisturbed (1 -2 hours) for the setting of gel.

Sample Preparation

(i) **Solution A** -Dissolve 3 mg sample in 500 µl of Tris buffer (150 mM, pH 7.0).

(ii) Take 3 x 15 µl of Solution A and 3 x 5 µl of 4x buffer and load 10 µl in each well. Each well would contain 45 µg of protein.

(iii) Then take 100 µl of the Solution A and mix with equal volume of Tris buffer (150 mM, pH 7.0). Now again take 3 x 15 µl of this sol in 3 x 5µl of 4x buffer and load 10 µl in each well. Each well would contain 22.5 µg of protein.

(iv) Incubate samples and marker at 37°C for 15 min.

(v) The concentration of protein in sample must be adjusted according to sensitivity of the stain. The protein concentration in loaded sample is given below.

Sl.No.	*Type of Stain*	*Protein Concentration (µg/µl)*	*Sensitivity*
1	Sliver-nitrate	2 – 3	Silver nitrate staining is highly sensitive followed by coomassiee brilliant blue (CBB) and periodic acid Schiff 's (PAS) staining respectively.
2	CBB	6 – 8	
3	PAS	20 – 25	

Electrophoresis [27]

(i) Place the gel sandwich after removing the comb in the mini vertical gel electrophoresis unit.

(ii) Clamp the sandwich in place.

(iii) Fill the lower buffer chamber with anode buffer. Check that the lower electrode is completely submerged and after then fill the upper buffer chamber with cathode buffer.

(iv) Before loading sample gel will be pre-run (50 volts, 30 min) in refrigerator at 10°C, for smooth movement of protein sample through gel, which can help in better and uniform separation of protein according to their respective molecular weight.

(v) Load 15-20 µl of protein samples and molecular weight markers carefully to these wells and place the safety lid on the unit.

(vi) Run the experiment at 10°C by keeping the whole assembly in the refrigerator.

(vii) Carried out electrophoresis at constant current of 20 mA till the sample crosses the stacking gel. Then the increase the current to 25 mA and maintain throughout the remainder of the run until the marker dye was within 1 cm of the anodic end of the gel.

(viii) Remove the gel carefully and then transfer it to the fixative solution. Keep it over an orbital shaker for 60 min.

(ix) Stain the gel with staining solution for twice the length of time used for fixing again using orbital shaker.

(x) Transfer the gel to the destaining solution till bands become visible against light background renewing the solution every 30 min.

Cautionary Notes for Problems and Probable Reasons

(a) **If loaded sample does not move downwards**–i–Check the electric cord which may be connected in reverse order or there is possibility of electrode buffer has been filled in different order and ii–It is also possible that electrode wire broken from somewhere.

(b) **If protein sample floats over buffer during loading**- i-Stacking gel has not polymerized completely (high probability) or glycerol concentration in sample buffer is very low (high probability) and ii-Other possible reason for this might be pH of electrode buffer is not adjusted properly.

(c) **If protein bands appear uneven in gel after staining-**i-it may be due to overheating during the run which results in melting of gel and ii- It could also be due to improper formation (polymerized) of separating gel. The air bubbles entrapped in the bottom of gel can also be cause of disturbance in uniformity of current flow.

2-Dimensional Electrophoresis [28]

Another way gel electrophoresis can be utilized is in **two-dimensional** (or **2-D) electrophoresis**, which is a high-resolution technique used to look at complex protein mixtures [3b,29]. In this method, two different types of electrophoresis are conducted on a single sample. The first of these separations is usually based on a isoelectric point, as accomplished by using isoelectric focusing. A typical 2-D electrophoresis method involves the followings:

(i) First, a small band of sample is applied to the top of a support for use in isoelectric focusing. The support used in this case is typically agarose or a polyacrylamide gel with large pores.

(ii) After this first separation has been finished, some proteins will have been separated based on their pI values, but there may still be many proteins with similar isoelectric points and overlapping bands.

(iii) After a 2-D separation has been finished, the protein bands can be detected using the methods discussed earlier. Staining with Coomassie blue or silver nitrate is often used in the location and measurement of these bands. Analysis by mass spectrometry is another option.

Isoelectric Focusing [28]

Another type of electrophoresis that often employs supports is **isoelectric focusing (IEF)** [30]. It is a method used to separate zwitterions (substances with both acidic and basic groups). Zwitterions are separated in IEF based on their isoelectric points by having these compounds migrate in an electric field across a pH gradient. In this pH gradient, each zwitterion will migrate until it reaches a region where the pH is equal to its isoelectric point. At this point, the zwitterion will no longer have any net charge and its electrophoretic mobility will become zero, causing the analyte to stop migrating.

The reason isoelectric focusing produces tight bands for these analytes is that even if a zwitterions momentarily diffuses out of the region where the pH is equal to its pI, the system will tend to "focus" the zwitterions back into this. This focusing occurs because of the way the pH gradient is aligned with the electric field. High pH's occur toward the negative electrode, so as solutes diffuse out of their band and toward this region they will take on a more negative charge and be attracted back to the positive electrode. At the same time, zwitterions that move toward the positive electrode and region of lower pH will acquire a more positive charge and be attracted back toward the negative electrode. **It is this focusing property that makes it possible for IEF to separate zwitterions with only very small differences in their pI values.** To obtain a separation in IEF, it is necessary to have a stable pH gradient. This pH gradient is produced by placing in the electric field a mixture of small reagent zwitterions known as **ampholytes**.

Applications of IEF

(i) IEF is a valuable tool for separating proteins or other compounds that contain both positive and negative charges which include some drugs, as well as bacteria, viruses, and cells.

(ii) Applications of this method range from biotechnology and biochemistry to forensic analysis and paternity testing.

(iii) IEF is particularly useful in providing high-resolution separations between different forms of enzymes or cell products.

Source

1. *Pulimamidi Rabindra Reddy and Nomula Raju, Gel-Electrophoresis and Its Applications, Department of Chemistry, Osmania University, Hyderabad, India.*

1a. *Kryndushkin DS, Alexandrov IM, Ter-Avanesyan MD, Kushnirov VV (2003). "Yeast [PSI+] prion aggregates are formed by small Sup35 polymers fragmented by Hsp104". Journal of Biological Chemistry 278 (49): 49636–43.*

2. *Sambrook J, Russel DW (2001). Molecular Cloning: A Laboratory Manual 3rd Ed. Cold Spring Harbor Laboratory Press. Cold Spring Harbor, NY.*

3. *Robyt, John F. and White, Bernard J. (1990). Biochemical Techniques Theory and Practice. Waveland Press. ISBN 0-88133-556-8.*

3a. *J. C. Giddings, Unified Separation Science, Wiley, New York, 1991, Chapter 8.*

3b. *D. S. Hage, "Chromatography and Electrophoresis, " In Contemporary Practice in Clinical Chemistry, W. Clarke and D. R.Dufour, Eds., AACC Press, Washington, DC, 2006, Chapter 7.*

4. *Berg JM, Tymoczko JL Stryer L (2002). Biochemistry (5th ed.). WH Freeman. ISBN 0-7167-4955-6.*

4a. *Van Holde, K. E.; Johnson, W. C. and Shing Ho, P. (1998). Principles of Physical Biochemistry, Prentice-Hall Inc. ISBN: 10: 0137204590, New Jersey, pp. 213-233.*

5. *Tom Maniatis, E. F. Fritsch, and Joseph Sambrook. "Chapter 5, protocol 1". Molecular Cloning - A Laboratory Manual 1 (3rd ed.). p.5.2–5.3. ISBN 978-0879691363.*

6. *Smisek, D. L.; Hoagland, D. A. (1989). "Agarose gel electrophoresis of high molecular weight, synthetic poly electrolytes". Macro-molecules 22 (5): 2270.*

7. *"Agarose gel electrophoresis (basic method)". Biological Protocols. Retrieved 23 August 2011.*

8. *Schägger, Hermann (2006). "Tricine–SDS-PAGE". Nature protocols 1 (1): 16–22.*

9. *Gordon, A.H. (1975). Electrophoresis of proteins in polyacrylamide and starch gels. New York: American Elsevier Publishing Company, Inc.*

10. *Smithies, O. (1955). "Zone electrophoresis in starch gels: group variations in the serum proteins of normal adults".Biochem. J. 61 (4): 629–641.*

11. *Wraxall, B.G.D.; Culliford, B.J. (1968). "A thin-layer starch gel method for enzyme typing of bloodstains". J. Forensic Sci. Soc.8 (2): 81–82.*

11a. *Stellwagen, N.C. 1998. DNA gel electrophoresis. Nucleic acid electrophoresis laboratory manual. (D Tietz, Ed.). Springer Verlag. Berlin-Heidelberg-New York.*

12. *Buell, GN; Wickens, MP; Payvar, F; Schimke, RT (Apr 10, 1978). "Synthesis of full length cDNAs from four partially purified oviduct mRNAs.". The Journal of Biological Chemistry 253 (7): 2471-82.*

13. *Schelp, C; Kaaden, OR (May 1989). "Enhanced full-length transcription of Sindbis virus RNA by effective denaturation with methylmercury hydroxide.". Acta virologica 33 (3): 297–302.*

14. *Cell. 1979 Jan; 16 (1): 191-200. Length-independent separation of DNA restriction fragments in two-dimensional gel electrophoresis. Fischer SG, Lerman LS.*

15. *Fromin N., Hamelin J., Tarnawski S., Roesti D., Jourdain-Miserez K., Forestier N., Teyssier-Cuvelle S., Gillet F., Aragno M. and Rossi P. (2002) Statistical analysis of denaturing gel electrophoresis (DGE) fingerprinting patterns. E+nviron Microbiol 4: 634-643.*

16. *Preparative gel electrophoresis of native metalloproteins.*

17. *Brody JR, Kern SE (October 2004). "History and principles of conductive media for standard DNA electrophoresis". Anal. Biochem. 333 (1): 1-13.*

17a. *J. Inczedy, T. Lengyel, and A. M. Ure, Compendium of Analytical Nomenclature, 3rd ed., Blackwell Science, Malden, MA, 1997.*

17b. *J. M. Anderson and G. A. Tetrault, "Electrophoresis, " In Laboratory Instrumentation, 4th ed., M. C. Haven, G. A. Tetrault, and J. R. Schenken, Eds., Van Nostrand Reinhold, New York, 1995, Chapter 12.*

17c. *E. M. Southern, "Detection of Specific Sequences among DNA Fragments Separated by Gel Electrophoresis, " Journal of Molecular Biology, 98 (1975) 503–517.*

18. *Lodish H, Berk A, Matsudaira P (2004). Molecular Cell Biology (5th ed.). WH Freeman: New York, NY. ISBN 978-0-7167-4366-8.*

19. *Trouble shooting DNA agarose gel electrophoresis. Focus 19: 3 p.66 (1997).*

20. *Steven M. Carr.*

21. *Electrophoresis modules. pdf.*

22. *Rajan Sharma and Y.S. Rajput, SDS-PAGE-Principle and Applications, Compendium of Lectures on Advanced Techniques and Novel Approaches for Quality and Safety Evaluation of Dairy Foods 2015, NDRI, Karnal.*

23. *Laemmli, U.K. (1970) Cleavage of structural proteins during the assembly of the head of bacteriophage T4. Nature, 227L680-685.*

24. *Towbin, H., Staehelin, T. and Gordon, J. (1979). Electrophoretic transfer of proteins from polyacrylamide gels to nitrocellulose sheets: Procedure and some applications. Pro Natl Aca. Sci USA. 76: 4350-4354.*

25. *Protein electrophoresis- applications guide (1994). Hoefer Scientific Instruments, USA.*

26. *Y.S. Rajput, Rajan Sharma, Dhiraj K. Nanda and Satyanagalakshmi Karri, Demonstration and Self-learning of Western Blotting, Compendium of Lectures on Advanced Techniques and Novel Approaches for Quality and Safety Evaluation of Dairy Foods 2015, NDRI, Karnal.*

27. *Vikrant Narwal, Rajan Sharma and Y.S. Rajput, Separation and identification of low molecular weight proteins using Tricine SDS-PAGE, Compendium of Lectures on Advanced Techniques and Novel Approaches for Quality and Safety Evaluation of Dairy Foods 2015, NDRI, Karnal.*

28. *Electrophoresis, Chapter 23, p 571-596, 2010.*

29. *D. E. Garfin, "Two-Dimensional Gel Electro phoresis: An Overview, " Trends in Analytical Chemistry, 22 (2003) 263–272.*

30. *B. L. Karger, L. R. Snyder, and C. Hovath, An Introduction to Separation Science, Wiley, New York, 1973, Chapter 17.*

31. *Wikipedia.*

Application of Different Electrophioretic Techniques

A. Application in Dairying

The details of applications of Gel Electrophoresis are given below:

1. Use of Pulsed-Field Gel Electrophoresis for Detecting differences in *Staphylococcus aureus* Strain Populations between Dairy Herds with different Cattle Importation Practices

The hypothesis tested was that dairy herds which import cattle for replacement or expansion have a higher prevalence of Staphylococcus aureus mastitis and a greater number of new Staphylococcus aureus strains enter their herds than closed herds. Fifteen commercial dairy herds were divided into four groups based on cattle importation practices. Composite foremilk samples were collected at 4-monthly intervals for 1 year from all lactating cattle. Additionally, foremilk samples were collected from cattle at parturition and skin swabs were taken from the udder of primiparous heifers. All samples were cultured for Staphylococcus aureus and isolates were strain-typed using pulsed-field gel electrophoresis. Herds that purchased replacement heifers had a higher prevalence of Staphylococcus aureus mastitis than herds that purchased lactating cattle for expansion ($P = 0.02$). Herds that purchased replacement heifers had more total strains of Staphylococcus aureus ($P = 0.01$) and more new strains ($P = 0.04$) enter the herd than closed herds.

Source

1. *Middleton JR, Fox LK, Gay JM, Tyler JW, Besser TE., Department of Veterinary Medicine and Surgery, University of Missouri, Columbia, MO 65211, USA.*

2. Density-Gradient Electrophoresis of Native and of Rennet-Treated Casein Micelles

Analytical electrophoresis of native and of rennet-treated casein micelles was carried out at 26 and at 5°C in a sucrose density gradient in a medium of the same ionic composition as milk. Under normal circumstances, the micelles were negatively charged and showed little heterogeneity in the electric field, the fastest moving ones having at 26°C a mobility not more than 30 per cent greater than that of the slowest. Native micelles were more highly charged at 5 than at 26°C; at each temperature, the mobility was approximately halved by rennet treatment. The results suggest that the effects of rennet treatment and of alterations in temperature on the tendency of the micelles to clot can be entirely explained as being caused by changes in charge.

Source

1. *Margaret L. Green and G. Crutchfield. National Institute for Research in Dairying, Shinfield, Reading RG2 9AT, Journal of Dairy Research, Volume 38/Issue 02/June 1971, pp 151-164.*

3. Molecular Fingerprinting of Dairy Microbial Ecosystems by Use of Temporal Temperature and Denaturing Gradient Gel Electrophoresis

Numerous micro-organisms, including bacteria, yeasts, and molds, constitute the complex ecosystem present in milk and fermented dairy products. The aim was to describe the bacterial ecosystem of various cheeses that differ by production technology and therefore by their bacterial content. For this purpose, researchers developed a rapid, semi systematic approach based on genetic profiling by temporal temperature gradient electrophoresis (TTGE) for bacteria with low-G+C-content genomes and denaturing gradient gel electrophoresis (DGGE) for those with medium- and high-G+C-content genomes. Bacteria in the unknown ecosystems were assigned an identity by comparison with a comprehensive bacterial reference database of approximately 150 species that included useful dairy microorganisms (lactic acid bacteria), spoilage bacteria (*e.g.*, Pseudomonas and Entero-bacteriaceae), and pathogenic bacteria **(*e.g.*, Listeria monocytogenes and Staphylococcus aureus).**

The analyses provide a high resolution of bacteria comprising the ecosystems of different commercial cheeses and identify species that could not be discerned by conventional methods; at least two species, belonging to the Halomonas and Pseudoalteromonas genera, are identified for the first time in a dairy ecosystem. The analyses also reveal a surprising difference in ecosystems of the cheese surface versus those of the interior; the aerobic surface bacteria are generally G+C rich and represent diverse species, while the cheese interior comprises fewer species that are generally low in G+C content. TTGE and DGGE have proven here to be powerful methods to rapidly identify a broad range of bacterial species.

Source

1. *Ogier JC, Lafarge V, Girard V, Rault A, Maladen V, Gruss A, Leveau JY, Delacroix-Buchet A, Unité de Recherches Laitières et Génétique Appliquée, Institut National de la Recherche Agronomique, Domaine de Vilvert, 78352 Jouy-en-Josas, France. ogier@jouy.inra.fr, Appl Environ Microbiol. 2004 Sep; 70 (9): 5628-43.*

4. Genotyping of *Staphylococcus aureus* Isolated from Various Sites on Farms with Dairy Sheep using Pulsed-Field Gel Electrophoresis

It was investigated the genetic diversity of 179 Staphylococcus aureus isolates recovered from various sites in 10 farms producing cheeses manufactured with raw ewe's milk. Isolates were collected from handcrafted cheeses, bulk tank milk, milk from half-udders, skin abscesses on the udder if present, hands and anterior nares of farmers, and air of the milking area. The isolates were typed using pulsed-field gel electrophoresis (PFGE) of DNA SmaI digests and compared to other isolates of S. aureus isolated in different hosts or in different locations. The results showed that nine farms were contaminated by S. aureus isolates with identical banding patterns (named OV) or by genetically related isolates (named OV′). These dominant banding patterns were found in a variable proportion of the samples from each farm (range: 11-100 per cent). Most of the strains isolated from nasal carriage or strains isolated from other regions or from other animal species had different PFGE patterns to OV or OV′, except for three strains. These results show that a single clone of S. aureus is widely distributed both in infected mammary glands and in cheese produced from raw milk. This study confirms that infected mammary glands are the main source of the contamination of dairy products in sheep.

Source

1. *Vautor E, Abadie G, Guibert JM, Huard C, Pépin M. AFSSA (Agence Française de Sécurité Sanitaire des Aliments), Unité de Pathologie des Petits Ruminants, BP 111, 06902 Cedex, Sophia-Antipolis, France. e.vautor @sophia. afssa.fr, Vet Microbiol. 2003 Oct 8; 96 (1): 69-79.*

5. Use of Transfer RNA-Intergenic Spacer PCR Combined with Capillary Electrophoresis to Identify Coagulase-Negative *Staphylococcus* Species Priginating from Bovine Milk and Teat Apices

Coagulase-negative staphylococci (CNS) are the most frequently isolated bacteria in milk samples from cows with and without mastitis.Elucidating their relevance in bovine udder health is hampered because identification at the species level, if done at all, used to be performed based on phenotypic features. To provide a rapid, cheap, and easy-to-use genotypic technique that can be used to identify CNS species from milk and teat apices from cows, the performance of transfer RNA-intergenic spacer PCR (tDNA-PCR) in combination with capillary electrophoresis was evaluated. After updating the tDNA library with CNS reference strains, 288 field isolates were identified with tDNA-PCR and gene sequencing, and the latter was used as the reference method. The field isolates were divided in 2 groups of 144. Isolates of the first group were identified with tDNA-PCR with a typeability of 81.9 per cent and an accuracy of 94.1 per cent.Peak patterns of these isolates were then added to the tDNA library with species identity as determined by DNA sequencing. The second group was identified with the updated tDNA library, resulting in 91.0 per cent typeability and 99.2 per cent accuracy. This study showed that the updated tDNA-PCR in combination with capillary electrophoresis was almost as accurate as gene sequencing but faster and cheaper and is a useful tool in observational studies concerning the epidemiology of bovine CNS species.

Source

1. *Supré K, De Vliegher S, Sampimon OC, Zadoks RN, Vaneechoutte M, Baele M, De Graef E, Piepers S, Haesebrouck F, Department of Reproduction, Obstetrics, and Herd Health, Faculty of Veterinary Medicine, Ghent University, Belgium. Karlien.Supre@ UGent.be.J Dairy Sci. 2009 Jul; 92 (7): 3204-10.*

6. Identification of the Bacterial Microflora in Dairy Products by Temporal Temperature Gradient Gel Electrophoresis [1]

Numerous microorganisms, including bacteria, yeasts, and molds, are present in cheeses, forming a complex ecosystem. As bacteria are responsible for most of the physicochemical and aromatic transformations that are intrinsic to the cheese making process hence the identification of the bacteria that constitute the cheese ecosystem is essential for understanding their individual contributions to cheese production.

The transformation of milk to cheese involves a complex and dynamic microbial ecosystem in which numerous biochemical reactions occur. The following two main groups of bacteria are involved in cheese manufacture and ripening:

(i) **The first group**: It consists of starter bacteria (mainly *Lactococcus*) that are added to milk during cheese manufacture. About 10^9 CFU of starter bacteria per g is present in the final product.

(ii) **The second group**: It consists of adventitious microorganisms (secondary microflora) from the environment which contaminate the milk or cheese curd during manufacture and ripening [2,3]. This group includes numerous species of lactic acid bacteria (*Lactobacillus, Pediococcus, Enterococcus,* and *Leuco-nostoc*) and surface cheese bacteria (*Micrococcus* and *Staphylococcus*).

Temporal temperature gradient gel electrophoresis (TTGE) was used to identify different bacterial species present in several dairy products, including members of the genera *Lactobacillus, Lactococcus, Leuconostoc, Enterococcus, Pediococcus, Streptococcus,* and *Staphylococcus.* The TTGE technique is based on electrophoretic separation of 16S ribosomal DNA (rDNA) fragments by using a temperature gradient. It was optimized to reveal differences in the 16S rDNA V3 regions of bacteria with low-G+C-content genomes. Using multiple control strains, a species database was established first in which each species (or group of species) was characterized by a specific TTGE fingerprint. TTGE was then applied to controlled dairy ecosystems with defined compositions, including liquid (starter), semisolid (home-made fermented milk), and solid (miniature cheese models) matrices. Finally, the potential of TTGE to describe the bacterial microflora of unknown ecosystems was tested with various commercial dairy products. Subspecies, species, or groups of species of lactic acid bacteria were distinguished in dairy samples. TTGE has proven to be a powerful and simple method for identifying species of bacteria in both liquid and solid dairy food environments. This method allowed differentiation generally at the species level and sometimes at the subspecies level. The capacity of TTGE to differentiate among *Lactobacillus acidophilus, Lactobacillus delbrueckii* ***subsp.*** *bulgaricus,* and *Lactobacillus delbrueckii subsp. lactis* species is of particular interest for the dairy industry. TTGE was shown to distinguish bacterial species in vitro, as well as in both liquid and solid dairy products.

Source

1. *Jean-Claude Ogier, Olivier Son, Alexandra Gruss, Patrick Tailliez, and Agnes Delacroix-Buchet, Identification of the Bacterial Microflora in Dairy Products by Temporal Temperature Gradient Gel Electrophoresis, Appl Environ Microbiol. 2002 Aug; 68 (8): 3691-3701.*

2. *H. R. Chapman, and M. S. Sharpe, 1990. Microbiology of cheese, p. 203-289. In R. K. Robinson (ed.), Dairy microbiology, the microbiology of milk products, 2nd ed., vol. 2. Elsevier, London, United Kingdom.*

3. *P.L.H.Mesweeney, M.Walsh, P.F.Fox, T.M. Cogan, F. D. Drinan, and M. Castelo-Gonzales, 1995. A procedure for the manufacture of Cheddar cheese under controlled*

bacteriological conditions and the effect of adjunct lactobacilli in cheese quality. Ir. J. Agric. Food Res. 33: 183-192.

7. Detection of Adulterated Protein Ingredient in Milk by Capillary Gel Electrophoresis Method

Capillary gel electrophoresis method was applied to detect extrogeneous protein in milk fast. After optimization of experimental conditions, CGE for milk was established by which soybean hydrolyzed protein specific elute peak was observed. The peak area-concentration linear relationships of pure soybean hydrolyzed protein solution and milk-soybean mixture were measured respectively.CGE method was further compared with SDS-PAGE and protein chip electrophoresis to verify the result as well as learn about the advantage and disadvantage of different methods.

Source

1. *WU Ya-jun, LIU Ming-chang, ZHAO Gui-ming HAN Jian-xun, WANG Bin, CHEN Ying (Agro-product Research Center, Chinese Academy of Inspection and Quarantine, Beijing, China Dairy Industry, 2013-05.*

B. Application in Food Industry

1. Detection of Kesari Dal Powder (*Lathyrus sativus*) in Besan by Electrophoresis

I. Electrophoresis Method

Principle

The presence of kesari dal powder is detected on the basis of the presence of an unusual amino acid namely beta–N–oxalyl L amino alanine which is not present in the seeds of other legumes.

Apparatus

(*i*) Centrifuge.

(*ii*) Capillary Electrophoresis HPLC System- CZE system.

(*iii*) Filter

(*iv*) Filter paper

Reagents

(*i*) Ethanol.

(*ii*) Ultra distilled waterand iii- NaOH

Procedure

Weigh approximately 0.5 g of L. sativus seeds. Soak in 50 ml ethanol- water (30: 70, v/v) solution and shake for 2 h (in ice). After centrifugation (3500 rpm at 15 min), filter the upper clear solution with 0.45 μm filter paper. Dilute the clear solution with ultra distilled water (1: 1) and inject directly into the CZE (capillary zone electrophoresis) system for 40 s at 50 mbar. Capillar was 55 cm (44 cm effective length) x 50 m. Perform the analyses at a constant voltage of 20 kV at 20°C in an electrolyte of 75 mM (H_3BO_3) buffer at pH 7.5. Adjust the required pH of the buffer by adding NaOH. Use all chemicals of analytical- reagent grade. Determine the β-ODAP using capillary zone electrophoresis (CZE- modified -Zhao *et al.*, 1999) and UV detection at 195 nm.

Source

1. *UgurBasaranet al. African Journal of Biotechnology Vol. 10 (20), pp.4072-4080, 16 May, 2011.*
2. *Manual of Methods of Foods (Cereals and Cereal Products) Lab Manual No.-3, 2012, FSSAI, GOI, New Delhi.*

II. Second Method (Capillary Electrophoretric Method)

Apparatus

(i) Capillary electrophoresis instrument equippes with a diode array detector and to operate via chemstation software.

(ii) Capillary of uncoated fused silica with a dimension of 48.5 cmx50μm and length of 40 cm.

Reagents

i Double Glass distilled water.

(ii) AR grade 3 – (N-Oxalyl, 1,23-diamino propnoic acid (β-ODAP).

(iii) Spectrophotometry.

Seed extracts were analyzed for α- and β-ODAP according to the spectrophotometric method of Rao (1974) as modified by C. C. Campbell (1993. Agriculture Canada, personal communication), for each sample solutions.

Procedure

Prepare the hydrolysates by boiling 2 ml. of seed extract extract in 4 ml. of 3 M KOH for 30 minutes. Centrifuge the hydrolysates (0.25ml).Pipette in to 0.75 ml. of water. Add 2 ml. of OPA reagent (100 mg. of OPA, 200 μl of mercaptoethanol, 1 ml. of ab solute ethanol, 99 ml. of 0.05 M $Na_2B_4O_7$, pH 9.9). Add 2 ml. of 0.05 M $Na_2B_4O_7$ pH 9.9 for the A4 solution. Prepare another pair of tubes with unhydrolyzed extract to correct for any background absorbance (A1 with OPA reagent, A2 without OPA reagent). Keep the solution at 40°C for 2 h and read the absorbance at 425 nm. Perform on-column UV detection at 195 nm. The analyses were performed at a constant voltage of 25kV at 40 °C in an electrolyte of 20 mM Na_2HPO_4 buffer at pH 7.8. Condition the capillary prior to each run by flushing it with 0.1 M NaOH for 2 min and with the electrolyte for 3 min. The electrolyte was replenished every third run.

Inject the filtered seed extracts for 4-12 s at 25 mbar depending on the concentration of the extract. Ascribe the formula (A3-A4)- (A1–A2)/3 for the calculation of the absorbance change to ODAP. Use A standard curve based on dilutions of a DAP solution (2.89 mM in ethanol-water,6: 4) for the quantification purpose.

Source

1. *Arenfotet al. J. Agric. Food Chem., Vol. 43, No. 4, 1995.*
2. *Manual of Methods of Foods (Cereals and Cereal Products) Lab Manual No.-3, 2012, FSSAI, GOI, New Delhi.*

Chapter 5

Polarography and Voltammetry

Electrochemical techniques are powerful and versatile analytical techniques that offer high sensitivity, accuracy, and precision as well as large linear dynamic range, with relatively low-cost instrumentation.Electrochemical techniques are inevitable tools in almost every chemical and biochemical research laboratory [1]. Some of the most useful electroanalytical techniques are based on the concept of continuously changing the applied potentials to the electrode-solution interface and the resulting measured current. Among the electrochemical techniques applied in food analysis, the principal ones are polarographic and voltammetric techniques [2]. Currently, polarographic techniques have almost been completely excluded from research laboratories, and are being replaced by the more sophisticated voltammetric techniques [3].

Chromatographic and spectroscopic techniques are traditionally highly desirable for the analysis of food and agricultural products. The recent data related to the study about instrumental methods of food analysis revealed that new analytical techniques such as electrophoresis, flow injection analysis, bioassays and stripping voltammetric technique had been increasingly employed in foodstuffs analysis over the last two decades [4].

Electro-Analytical Techniques (Polarography and Voltammetry)is being used for Qualitative, Quantitative and In one step analysis metals, other inorganics and organic compounds. The nature of sample is practically any types. The level of detection (LOD) is $10^{-5} \rightarrow 10^{-15}$M and μg → pg [5].

The beginning of voltammetry was facilitated by the discovery of polarography in 1922 by the Nobel Prize–winning chemist **Jaroslav Heyrovský.** Early voltammetric techniques had many problems, limiting their viability for everyday use in analytical chemistry. In polarography, these problems included the fact that Mercury is oxidized at a potential that is more positive than +0.2, which makes it harder to analyze that results for the analytes in the positive region of the potential. Another problem included the residual current obtained from the charging of the large capacitance of the electrode surface [6]. **Jaroslav Heyrovský** was also an important part of the introduction of dropping mercury electrode as a modern-day tool [7]. In 1942 Archie Hickling built the first three electrodes potentiostat, which was an advancement for the field of electrochemistry [8]. He used this potentiostat to control the voltage of an electrode. In the meantime, the late 1940s, Kenneth Cole invented an electronic circuit which he called a voltage clamp. The voltage clamp was used to analyze the ionic conduction in nerves.

The 1960s and 1970s saw many advances in the theory, instrumentation, and the introduction of computer added and controlled systems. Modern polarographic and voltammetric methods on mercury electrodes came about in three sections [10].

First Section

Development of the following mercury electrodes:

(a) Dropping mercury electrode
(b) Mercury steaming electrode
(c) Hanging mercury drop electrode
(d) Static mercury drop electrode
(e) Mercury film electrode
(f) Mercury amalgam electrodes.
(g) Mercury microelectrodes
(h) Chemically modified mercury electrodes
(i) Controlled growth mercury electrodes
(j) Contractible mercury drop electrodes

Second Section

An advancement of the following measuring techniques used:

(a) Classical DC polarography
(b) Oscillopolarography
(c) Kaloussek's switcher
(d) AC polarography
(e) Tast polarography
(f) Normal pulse polarography
(g) Differential pulse polarography
(h) Square-wave voltammetry
(i) Cyclic voltammetry
(j) Anodic stripping voltammetry
(k) Convolution techniques
(l) Elimination methods.

Third Section

Lastly, there was also an advancement of preconcentration techniques that produced an increased the sensitivity of the mercury electrodes. This came about through the development of anodic stripping voltammetry, cathodic stripping voltammetry and adsorptive stripping voltammetry [9].

These advancements improved sensitivity and created new analytical methods, which prompted the industry to respond with the production of cheaper potentiostat, electrodes, and cells that could be effectively used in routine analytical work.

Source

1. *Rubin Gulaboski and Carlos M. Pereira, Electroanalytical Techniques and Instrumentation in Food Analysis, Handbook of Food Analysis Instruments 45660_ C017.*
2. *Brainina, Kh.Z., Malakhova, N. A., and Stojko, N.Y., Stripping voltammetry in environmental and food analysis, Fresenius J. Anal. Chem., 368, 307, 2000.*
3. *Bard, A.J. and Faulkner, L., Electrochemical Methods, Wiley, New York, 2001.*
4. *Ahmad H. Alghamdi, Applications of stripping voltammetric techniques in food analysis.*
5. *Electrochemical Methods, Fundamentals and Applications, Allen J. Bard and Larry R. Falkner (Wiley)*
6. *Kosuke Izutsu "Electro-chemistry in Nonaqueous Solutions" Second Edition Wiley-VCH, Weinheim. 2009.*
7. *Geiger, W. E.; Barriere, F., "Organometallic Electrochemistry Based on Electrolytes Containing Weakly-Coordinating Fluoroarylborate Anions", Accounts of Chemical Research 2010, volume 43, pp. 1030-1039.*
8. *http://www.abc.chemistry.bsu.by/vi/*
9. *Nicholson, R.S. (1965). "Theory and Application of Cyclic Voltammetry for Measurement of Electrode Reaction Kinetics". Anal. Chem. 37: 1351–1355.*
10. *Wkikpedia.*

Voltammetry

Voltammetry (abbreviation of volt-amper-metry) is a branch of electrochemistry that was developed by the discovery of polarography. The term voltammetry refers to a class of electroanalytical techniques, and it is used to designate the current-voltage measurement obtained at a given electrode. It is an analytical technique based on the measure of the current flowing through an electrode dipped in a solution containing electro-active compounds, while a potential scanning is imposed upon it or voltammetry is the study of current as a function of applied potential. It is concerned with study of voltage-current-time relationships during electrolysis carrid out in a cell [1,2]. Voltammetry experiments investigate the half cell reactivity of an analyte. The voltammetric techniques commonly involves studying the influence of changes in applied voltage on the current flowing in the cell but in some circumstances, the variation in current with time also studied [2].

The common characteristic of all voltammetric techniques is that they involve the application of a potential (E) to an electrode and the monitoring of the resulting current (i) flowing through the electrochemical cell. In many cases the applied potential is varied or the current is monitored over a period of time (*t*). Thus, all voltammetric techniques can be described as some function of E, *i*, and *t*. They are considered active techniques (as opposed to passive techniques such as potentiometry) because the applied potential forces a change in the concentration of an electroactive species at the electrode surface by electrochemically reducing or oxidizing it.

The analytical advantages of the various voltammetric techniques include excellent sensitivity with a very large useful linear concentration range for both inorganic and organic species (10–12 to 10–1 *M*), a large number of useful solvents and electrolytes, a wide range of temperatures, rapid analysis times (seconds), simultaneous determination of several analytes, the ability to determine kinetic and mechanistic parameters, a well-developed theory and thus the ability to reasonably estimate the values of unknown parameters, and the ease with which different potential wave forms can be generated and small currents measured.

This relationship could be explained when potential (E) is applied to the working electrode and the resulting current (i) flowing through the electrochemical cell will be recorded. The applied potential could be changed or the resulting current will be recorded over a period of time (t). The potential applied to the working electrode serves as driving force for the reaction; it is controlled the parameter that causes the chemical species present in solution to be electrolyzed (reduced or oxidized) at the electrode surface. The resulting current-voltage plot is called voltamogram where current is displayed in vertical axis and potential in horizontal axis.

The potential is varied arbitrarily either step by step or continuously, and the actual current value is measured as the dependent variable. The opposite, *i.e.*, amperometry, is also possible but not common. The shape of the curves depends on the speed of potential variation (nature of driving force) and on whether the solution is stirred or quiescent (mass transfer). Most experiments control the potential (volts) of an electrode in contact with the analyte while measuring the resulting current (amperes) [3].

Voltammetric methods can also easily be hyphenated with other techniques, *e.g.*, their use as detectors in chromatographic separations [4]. Another advantage of voltammetric techniques in food analysis is the various possible modes and methodologies that can be used for the determination of both electrochemically active and electrochemically inactive compounds [5].

Source

1. *Vogel's Textbook of quantitative Chemical Analysis, Fifth Edition, ELBS, Longman, UK.*

2. *AMEL Electrochemistry, Introduction to Modern Voltammetric and Polarographic Analysis Techniques IV Edition, 2001.*
3. *Bard, Allen J., Larry R. Faulkner (2000-12-18). Electrochemical Methods: Fundamentals and Applications (2 ed.). Wiley.*
4. *Delgado-Zamarreño, M.M. et al., Pressurized liquid extraction prior to liquid chromatography withelectrochemical detection for the analysis of vitamin E isomers in seeds and nuts, J.Chromatogr. A, 1056, 249, 2004.*
5. *Xu, Q. et al., Application of a single electrode, modified with polydiphenyl-amine and dodecyl sulfate, for the simultaneous amperometric determination of electro-inactive anions and cations in ion chromatography, J.Chromatogr. A, 997, 65, 2003.*

Principles of Voltammetric Techniques

Voltammetric experiments are usually carried out in simple electrochemical cells, similar to that shown in Figure. A common electrochemical cell consists of a working electrode, a reference electrode, and usually an auxiliary (counter) electrode [1]. Generally system contains two or three electrodes. There are many systems which have more electrodes.

a. System with Two and Three Electrodes

To conduct such an electrochemical experiment one requires at least **two electrodes** but procedure normally involves the use of cell with assembly of following three electrodes:

(i) A Working Electrode.
(ii) A Reference Electrode.
(iii) An Axillary Electrode.

i. Working Electrode

It makes contact with the analyte, must apply the desired potential in a controlled way and facilitate the transfer of charge to and from the analyte.Electrolysis under investigation takes place at this electrode. The working electrode is an electron conductor at which the reaction or transfer of interest takes place. In practice it can be very important to have a working electrode with known dimensions and surface characteristics. As a result, it is common to clean and polish working electrodes regularly. The working electrode could be made with several materials. Usually, it has a very little surface in order to assume quickly and accurately the potential imposed by the electrical circuit. The electrode can be solid (gold, platinum or glassy carbon) or formed by a drop of mercury hanging from a tip of a capillary [2].

ii. Reference Electrode

In all voltammetric experiments, it is necessary to keep one of the electrodes at a constant potential. This electrode, designed to have a constant (reversible) potential, is called a reference electrode. It is used to measure the potential of working electrode. A second electrode acts as the other half of the cell. This second electrode must have a known potential with which to gauge the potential of the working electrode, further more it must balance the charge added or removed by the working electrode. While this is a viable setup, it has a number of short comings. Most significantly, it is extremely difficult for an electrode to maintain a constant potential while passing current to counter redox events at the working electrode. To solve this problem, the roles of supplying electrons and providing a reference potential are divided between two separate electrodes. The reference electrode is a half cell with a known reduction potential. Its only role is to act as reference in measuring and controlling the working electrode's potential and at no point does it pass any current.

The reference is the most complex of the three electrodes; there are a variety of standards used and it is worth investigating else where. For non-aqueous work, IUPAC recommends the use of the ferrocene/ferrocenium couple as an internal standard [4].

iii. Auxiliary Electrode

The auxiliary electrode is one at which a counter reaction to that at the working electrode takes place, for the sake of balancing the total charge in the system. It passes all the current needed to balance the current observed at the working electrode. To achieve this current, the auxiliary will often swing to extreme potentials at the edges of the solvent window, where it oxidizes or reduces the solvent or supporting electrolyte. The auxiliary electrode can be almost anything as long as it doesn't react with the bulk of the analyte solution and conducts well.

These electrodes, the working, reference, and auxiliary make up the modern three electrode system.

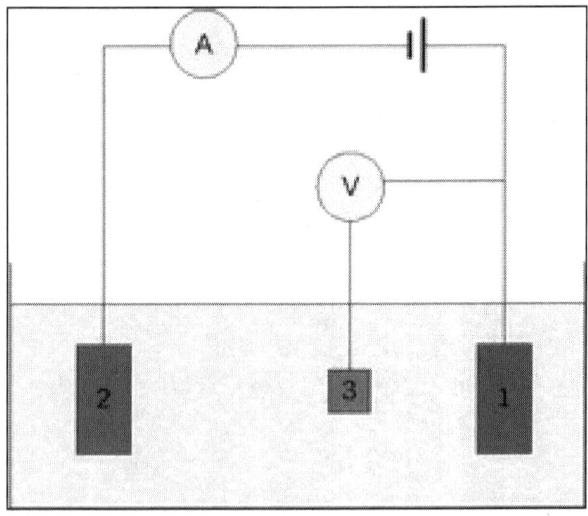

Three Electrode Setup.

(1) Working electrode; (2) Counter electrode; (3) Reference electrode.

b. System with more Electrodes

There are many systems which have more electrodes, but their design principles are generally the same as the three electrode system.Some examples are as follows:

(i) **An Overall Four Electrode (Rotating ring-disk electrode) Design:** The system with the rotating ring-disk electrode as working electrode is good example of this type of system. The rotating ring-disk electrode has two distinct and separate working electrodes, a disk and a ring, which can be used to scan or hold potentials independently of each other. Both of these electrodes are balanced by a single reference and auxiliary combination for an overall four electrode design.

(ii) More complicated experiments may add working electrodes as required and at times reference or auxiliary electrodes.

In most voltammetry experiments, a bulk electrolyte (also known as a supporting electrolyte) is used to minimize solution resistance. It is possible to run an experiment without a bulk electrolyte, but the added resistance greatly reduces the accuracy of the results. With room temperature ionic liquids, the solvent can act as the electrolyte.

In the presence of electroactive species in the electrochemical working cell, the applied potential will provoke a change in the concentration of the monitored electroactive species at the electrode surface by electrochemically reducing or oxidizing them. Changing the concentration of any electroactive participant at the working electrode surface will cause mass transport toward the electrode, and current will flow through the electrodes that is directly proportional to the analyte concentration. This simple dependence between measured current and analyte concentration makes voltammetric techniques to be routinely used for the quantitative determination of a variety of inorganic and organic compounds. Although there are many voltammetric techniques they are all based on the same electrochemical theory.

Theory of Voltammetry

Data analysis requires the consideration of kinetics in addition to thermodynamics, due to the temporal component of voltammetry. Idealized theoretical electrochemical thermodynamic relationships such as the **Nernst** equation are modeled without a time component.While these models are insufficient alone to describe the dynamic aspects of voltammetry, models like the Tafel equation and **Butler-Volmer equation** lay the ground work for the modified voltammetry relationships that relate theory to observed results [5].

Consider the simplest electrochemical reaction of the type

$$O_x + ne^- \underset{\rightarrow}{\leftarrow} Red \quad (I)$$

where O_x refers to the oxidized form of an electroactive substance initially present in the electrode cell, while Red is its reduced form.

At least two well known laws can be applied for expressing the interdependence between the applied potential and the surface concentrations of O_x and Red. For a thermodynamically reversible electrochemical reaction (*i.e.*, a fast reaction where equilibrium is always re established as changes in electrode potential are made), the following type of **Nernst equation** always holds **[1]:**

$$E = E^{\circ} + \frac{RTc\,(Ox)_{x=0}}{nFc\,(Red)_{x=0}} \ln \text{———}$$

where,

R is the gas constant (8.3144 J/mol K), **T** is the absolute temperature (K),

n is the number of electrons exchanged, **F** is the Faraday constant (96,485 C/mol)

E° is the standard redox potential for the couple Ox/Red.

For many electrochemical systems, especially for kinetics-controlled equations, the **Butler–Volmer equation** is often used **[1]:**

$$\frac{I}{nFA} = k_s e^{-\alpha\varphi}\,[c\,(Ox)_{x=0} - e^{\varphi}c\,(Red)_{x=0}]$$

where,

$\varphi = nF\,(E - E^{\circ})/RT$, **ks** is the standard rate constant of electron transfer, α is the electron transfer coefficient, **A** is the exposed (active) area of the working electrode.

This equation is useful for estimating the heterogeneous standard rate constant of electron transfers. When performing electrochemical experiments, applying a potential difference between the working and reference electrodes would alter the surface concentrations of both forms of the redox couple. This will cause a mass transfer toward (or from) the working electrode and current will flow through the cell. The current resulting from the electrochemical transformation of the electrochemically active species is known as faradaic current, and it is related to the material flux at the electrode–solution interface, as described by well-known **Fick's laws [1,5].** The instrumental output that is obtained in the voltammetric techniques is known as a voltammogram, which is a current–potential or I–E curve. The features of the obtained voltammograms depend on the type of the mass transfer phenomena, the number of exchanged electrons, the timescale of the measurement, as well as on the nature of the various coupled chemical reactions and surface phenomena that can happen in the electrochemical cell, and on the voltammetric technique used, in particular **[1,5].** For analytical purposes, one is mainly interested to know the magnitude of the faradaic current of the voltammograms, which is a quantitative measure of concentration and the kinetics of the redox transformation of given electroactive species at the electrode surface. The magnitude of the faradaic current is a function of the analyte concentration, but it is also affected by additional factors, such as the size, shape, and material of the electrode, the solution resistance, the cell volume and the number of electrons transferred in the electrode reaction **[1,5].**

Mechanism for Voltammetric Technique [2]

It involves the following steps:

1. Discharge Process at a Constant Potential Electrode

For better understanding and for a complete comprehension of the mechanism on which the voltammetric technique is based on, following simple model can be considered:

Suppose that a working electrode is dipped in a solution containing an electro-active compound, Aox, that can be reducible (or is able to gain an electron at the electrode) accordingly to the following reaction:

$$A_{ox} + ne \rightarrow A_{red}$$

Example- $$Pb^{2+} + 2e \rightarrow Pb$$

and that a reducible (meaning a lower or more negative) potential is imposed to the electrode respect to the reduction potential of the compound A_{ox}.

In this way the conditions for the **discharge** of A_{ox} to the surface of the electrode are fulfilled.

Finally, we can suppose to measure the electric current flowing through the electrode, while Aox is discharging itself.

In order that the discharge (and than the flowing of the current through the electrode) can occur, is necessary that A_{ox} can reach the electrode starting from the bulk solution and also that A_{ox} can accept the electrons from the electrode. The process then proceeds under the influence of following two kinetic factors:

(i) The speed (vd) of the chemical compound that reach the electrode from the bulk solution.

(ii) The speed (ve) of the electronic exchange between the electrode and the solution.

In analytical Voltammetry is necessary to use redox processes or, at least, to realise operating conditions, for which the second speed is higher than the first **(ve>>vd),** or, that is the same, that the discharge of A_{ox} on the electrode is practically **instantaneous**. In this way the discharging current can depend only on vd and than on the modality with which A_{ox} reaches the electrode.

Motion of the Particles in a Solution

There are principally following three processes allowing the motion of a chemical compound (Ion transportation during electrolysis) in solution:

(a) Convection

(b) Migration

(c) Diffusion

a. Convection

It is transport of ions towards electrode due to agitation, vibration and temperature gradients. The **convection,** is realised when, independently by the discharge process, a solution is stirred or when in the solution is present a temperature or a density gradient. In this case the molecules of the solvent and the analyte move themselves with a more or less troublesome motion, but that become more laminar in the vicinityof the electrode surface. The layer of solution closer to the electrode surface is practically stationary.

b. Migration

The **migration** is the moving due to the **attraction force** of the electric field generated by the electrode toward every ion having opposite charge and also due to the contemporary **repulsion force** of every ion having the same charge of the electrode. It is movement of oppositely charged ions towards electrode due to electrostatic attractions.

c. Diffusion

The diffusion is the spontaneous movement of those chemical compounds subjected to a concentration gradient that means a situation in which a zone of the solution is poorer than another. Diffusion is movement of ions from region of higher concentration (bulk) to region of lower concentration (near the electrode surface). With the process of diffusion the system tries to destabilise its homogeneity.

During the discharging process the solution closer to the electrode becomes ever more poor in Aox in respect to the bulk solution. The growing concentration gradient recalls other electroactive compound, Aox, from bulk solution toward the electrode.

The diffusion speed is directly proportional to the concentration gradient and then to the concentration of the electro-active compound in the solution.

Among the three described phenomena only diffusion can be related to the concentration of the electro-active compound, called also **depolarizer**. This is way in a voltammetric analysis is necessary to take adequate steps in order to secure that an electro-active compound can move itself prevalently by diffusion during a discharge process.

Supporting Electrolyte

Limitation of migration is achieved by screening the electrode using a **supporting electrolyte,** meaning a solution which ions do not discharge themselves at the electrode in the experimental conditions. This electrolyte is added at high concentration to the sample and could be a simple salt, or acid, or base or also a buffer solution or a chelating reagent. The supporting electrolyte surrounds the electrode with ions having the same charge of the depolarising agent, reducing in this way the electrostatic attraction toward the latter.

Relating to the behaviour of the depolarizer, the choice of the supporting electrolyte has to be made on the basis of the following characteristics:

(*a*) Be chemically not reactive.

(*b*) Do not interfere with diffusion and with the electrons exchange on the electrode surface

(*c*) Have a different discharge potential (at least 100 – 200 mV).

(*d*) Have an high ionic conductivity and guarantee a low electrical resistance

Faradic Current and Capacitive Current

The electric current flowing through the working electrode has two components:

(*a*) **The faradic current,** follows the Faraday laws and is due to the discharge of the electro active compound (A_{ox}),

(*b*) **The capacitive current**, is produced by the growth of adouble electric layer on the interface between the electrode and the solution. This double layer is due to the high concentration of the supporting electrolyte in the solution and acts as a condenser with high capacity. The total current flowing through the electrode is finally due to the sum of the charging current (capacitive current) of this condenser and the faradic current.

The capacitive current acts as a non specific background interference of the faradic current, and sometimes can be higher than the latter, when the depolarizer is present at low concentration in solution. In this cases the measure of the faradic current is difficult and some electronic adjustment has to be used. That's why Polarography (and Voltammetry) is growth, as analytical technique, only after the progress in the electronic field: so we can now affirm that the development of this technique is strictly linked to the tentative to electronically overcome problems due to capacitive current.

Diffusion Layer

Near the electrode surface, during the discharge, a thin layer of solution, called **diffusion layer**, is formed, either the solution is stirred or not. This layer has a thickness of about 10-100 μm if the solution is stirred, or about 500 μm if the solution is quiescent and the motion of the particles in it is due only to the diffusion. For this reason the quantity of analyte that reaches the electrode is directly proportional to its concentration in the solution and the intensity of the discharging (faradic) current gets to a limit value due to the velocity with which the electro active compound reaches the diffusion layer starting from bulk solution.

2. Discharging Process – II (Potential Scanning)

We can suppose now a model in which, instead of keeping constant the electrode potential, it is varying during the time, to make or in other words, a **scanning** of the potential in increasing or decreasing way.

If we consider the kinetics aspects, when an electro active compound is submitted to a potential scanning, following three possibilities could take place:

(*a*) The exchange of electrons between the working electrode and the solution is faster than the potential variation and hence than the velocity of the diffusion (vs > vd). In this way the potential of the working electrode corresponds, in every moment, to the one expected on the basis of **Nernst law**. The electron exchange is than due to a rapid (or reversible) redox couple, or, anyway the over potential is absent and the discharging process is due only to the diffusion.

(*b*) The velocity of the electron transfer is lower than the diffusion velocity (vs< vd). The redox system, in this case, is called slow or irreversible, or is characterised by high over potential phenomenon: its potential is, in fact, "inlate" respect on the one expected on the basis of Nernst law.

(*c*) Both the velocity are similar. The redox couple has intermediate characteristics and the discharging process is ruled by both the diffusion and the velocity of the electron transfer.

Different Steps in Voltammetric Analysis [2]

The following steps are required to be performed in voltammetric analysis:

i. Sample Treatment

For a voltammetric analysis the sample has to be a solution. If the sample is solid, it has to be dissolved in a proper way. Liquids with complex matrix and low soluble solids are to be digested or extracted. Rarely, it is not necessary to digest complex matrices, *i.e.* the analysis of ascorbic acid in fruit juice can be performed without any treatment.

At the end of any treatment the obtained solution has to have the following characteristics:

(*a*) No suspended solids. If present they have to be eliminated by filtering or centrifuging. The separate solid can be analysed separately.

(*b*) No colloids. Colloids compete with electrochemical processes, sometime stopping them or keeping them not reproducible. Often a simple oxidizing treatment with UV lamp is sufficient to overcome this interference.

(*c*) No surfactants, because they also can stop the electrochemical processes or decreasing the sensitivity of the method. Also in this case a simple oxidising treatment with UV lamp is sufficient to overcome this interference.

(*d*) pH between 1 and 10 (also 12 in NaOH is not present) and atmosphere either too much oxidising nor too much reducing, otherwise the mercury electrode can react or give raise to a very noisy signal. This latter phenomena effect also solids electrode. Digested solutions have to be accurately neutralised or buffered. It is necessary also to boil the digested solution until the nitrous or hydrogen peroxide not react or other vapours are removed from solution.

(*e*) If the analysis is performed in a solvent different from water, when possible, an electrolyte has to be added increasing the electrical conductivity of the solution.

ii. Addition of the Supporting Electrolyte

Every analysis has to be performed using a typical supporting electrolyte.

iii. Bubbling of the Solution with Nitrogen

Oxygen is always present in solution and give rise to 2 voltammetric peaks; the first at 0 V and the other one at 1 V. This gas has to be eliminated from the solution before the analysis, by a prolonged bubbling of nitrogen. Lower is the analyte concentration to be found, greater has to be the bubbling time.

iv. Pause (Delay)

Solution has to be leave quiescent stopping after the bubbling.

v. Electrode Cleaning

Mercury electrode-some drops are discharged, Solid electrode- an inverse scanning has to be performed.

vi. Scanning and registration of the sample voltammogram.

vii. Addition of standard solution of analyte.

viii. Scanning and registration of the voltammogram after the addition (the process is repeated 2-8 times).

ix. Measure of the all peak (or wave) heights.

x. Drawing the graph.

xi. Reading of the unknown concentration.

Precautions for Trace Analysis of Metals by Voltammetry [2]

1. Cleaning Procedure for Glassware and Working Area

Trace analysis need sophisticated and very sensitive instruments. Results cannot reach the same reproducibility as for high concentration analysis. In fact, typical percentages of standard deviation of results in trace analysis are in the range of 15-20 per cent.

Lead, Zinc, Iron and Sodium

These are environmental pollutants. These metals are very wide spread and it is very difficult to find any material completely lacking in these metals. Distilled and deionized water, acids and chemicals normally used for the preparation of solutions or for the sample treatment are anyway polluted with these metals.

For these reasons, it is mandatory to adopt strict cleaning procedure, especially for the analysis of lead, zinc, iron and sodium. The best strategy to avoid errors consist in:

(a) The cleaning of the glassware for the analysis.

(b) The analysis of the blank solutions.

(c) The control of the analytical data.

Mercury

The analysis of mercury has to be performed with dedicated glassware to avoid any pollution. (don't use the same glassware used for the analysis with mercury drop or film electrode). Stock gold electrode and cells far from mercury and accurately clean Pt and reference electrodes, magnetic rod and the top of the cell.Obviously analyse Pt using a glassy carbon reference electrode (instead a Pt one).

2. Glassware Cleaning

Stock the glassware (volumetric flasks, pipette and bottles) dedicated to the preparation of standard solutions far from the one for the sample treatments.Clean the glassware with copious distilled water, then with 1-2 per cent HNO_3 and finally, again with copious distilled water.

Dip the cells in 1-5 per cent HNO_3 at the end of the analysis till a new analytical session. Don't use the same cell for the following analysis. Accurately clean the instrument and the working area (pay attention to the powder falling down from old fume hood).

Copiously wash the electrodes and the top of the cell at the end of the analysis. Preferably start with low concentration sample analysis and continue with high concentration ones.

Maintenance and Cleaning of Electrodes and Cell [2]

A. Electrodes

The maintenance and cleaning of different electrodes are to be done as per the following:

1. Mercury Drop Electrode

The following are the ways for maintenance and cleaning of this electrode after at the end of each analysis, after at the end of a working day *etc.*:

- *(a)* **At the end of each analysis,** wash the capillary with distilled water; take care to avoid to aim the water flow directly on the tip of the capillary. Clean while the mercury valve (function n. 6 of the supporting block: Start to open mercury valve). Pay attention to avoid the water flows into the capillary and to avoid the scattering of mercury drops on the working area.
- *(b)* **At the end of a working day,** accurately wash the capillary, dry it and leave the capillary with a drop dropping from the tip (use function n. 6 of the supporting block: Start to open mercury valve) in a clean and dry cell. Check that there are not salt deposits nearby the tip of the capillary.
- *(c)* **During long periods of inactivity** it is better to make a scanning with an indifferent supporting electrolyte (*i.e.* 0.1 M KCl) every week (or two) avoiding the stagnation of mercury in the capillary.
- *(d)* **If the capillary is obstructed,** or if the current don't flow in the circuit, or if the noise of the signal is very high, try to set it free applying the vacuum

pump at the end of the capillary. When the maximum of the vacuum is reached, open the mercury valve (use function n. 6 of the supporting block: Start to open mercury valve). Aspirate some mercury drop and disconnect the tube of the vacuum pump (remember: don't close the water tap before disconnecting the tube). If the problem cannot to be solved, change the capillary.

(*e*) **Stock the exhausted mercury** in a strictly closed polythene bottles, under a water layer. Don't leave the bottle open.

(*f*) **Mercury waste** can occur during the maintenance procedure or during the replacement of the capillary. Due to the mercury toxicity and vapour pressure the scattered drops must be recovered using a vacuum pump and the working area must be cleaned using proper devices (cleaning kits are available).

2. *Gold, Platinum and Glassy Carbon Electrodes*

Main sources of surface pollution for solid electrodes are fat (or hydrophobic compounds), arising typically from the fingers of the operator. Clean solid electrodes carefully rubbing the active surface with soft paper imbibed with acetone or thin abrasive paste (*i.e.* 0.3 μ alumina powder in distilled water). Rinse finally with distilled water and check that the electrode surface is mirrored using a lens.

Deep incrusted electrodes in 2 M HNO_3 and rinse with distilled water. After the cleaning and, anyway, every time a new electrode is used, make a cathodic cleaning at a potential which can discharge any metallic impurity. It's better to make the cathodic cleaning also before each potential scanning, during an analysis.

Store clean and dry electrodes at the air, without any further particular care. Quality of the electrode surface can be controlled analysing a blank or a reference solution.

3. *Auxiliary Platinum Electrode*

The auxiliary platinum electrode has to be cleaned and stored just like the other solid electrodes. The cathodic cleaning is no necessary.

4. *Reference Electrode*

Reference electrode contains a concentrated solution of a salt (normally KCl) which ions have the same mobility to diminish the junction. Those solution is, of course, saturated of AgCl. During analysis, it is mandatory to check that the ions of the sample donot interfere with the ones present in the internal reference electrode solution. It could happen in the following situations:

(*a*) Analysis of traces of silver.

(*b*) Analysis of chloride.

(*c*) Analysis of sulphide, bromide and iodide.

These can react with Ag developing in soluble salts and obstructing the diaphragm and, at the end, increasing the junction potential. These problems can be overcome using a double junction reference electrode, containing a KNO_3 solution as a final junction.

Moreover the level of the internal solution of the reference electrode has to be higher than the level of the external sample solution in the cell assuring a very low flow of solution from the inner toward the outer (and not vice and versa). In this way the pollution of the internal solution of the reference electrode is reduced and the potential readings are more reproducible.

When a diaphragm is obstructed (that means that the peak position fluctuate among different scanning) it is better to change completely the glassy body of the electrode.

After periods of inactivity the electrode cover itself by a white layer of KCl. Simply wash the electrode with distilled water and if necessary add more 3 M KCl internal solution.

5. *Film Electrodes*

(a) **Mercury film electrode on glassy carbon-**Store the mercury film electrode dipped in distilled water. When in use, leave it on air only for short periods. When the film is clearly scratched or striped, clean it with soft paper, rinse with distilled water and depose e new film.

(b) **Gold film electrode-**Store the gold film electrode dipped in distilled water. When in use, leave it on air only for short periods.When the film is clearly scratched or striped, clean it with very thin abrasive paste on soft paper, rinse with distilled water and depose e new film.

B. Cells

Clean glass and Teflon cells using distilled water or acid solution.Dip the cells in 5-10 per cent HNO_3 for 2-3 days before the trace analyse of metals.

Preparation of Film Electrodes [2]

The preparation of the different Film electrodes are to be done by the following ways:

1. Mercury Film on Glassy Carbon Electrode

Dip the clean glassy carbon electrode in a 800 mg/l in 1.3 M HC1 Hg^{2+} solution, or in a 1000 mg/l Hg^{2+} solution for AA Spectrophotometry.

(a) Bring the electrode potential at -1100 mV for 2 minutes.

(b) Rinse with distilled water.

(c) Never leave the electrode to dry.

(d) Don't touch the Hg film for any reason film.

(e) Check if the surface of the film is uniform, grey and opaque before the use. If not, clean the electrode using soft paper, rinse with water and depose another film.

(f) Store the Hg film electrode in distilled water until next analysis. Before starting a new series of analysis, check the electrode performances.

(g) After the first time it is necessary to run a series of scanning before the electrode rise to a complete settlement.

2. Gold Film Electrode on Glassy Carbon

Dip the clean glassy carbon electrode in a Au^{3+}solution (1000 mg/l in 1 M HC1). Bring the electrode potential at -400 mV for 2 minutes. Rinse with distilled water. Never leave the electrode to dry. Donot touch the Au film for any reason film.Check if the surface of the film is uniform, gilded before the use. If not, clean the electrode using soft paper, rinse with water and depose another film.Store the Au film electrode in distilled water until next analysis. Before starting a new series of analysis, check the electrode performances.

Reagents for Trace Analysis [2]

The quality of reagents plays important role. Hence these are of required quality.Use only reagents for trace analysis grade. If this is not possible analyse blank solution before sample analysis.

1. Preparing Standard Solutions

Buy directly the concentrated standard solution of metals (1 g/l of theanalyte) from manufacturers. The solution for AAS are generally suitable.

Prepare standard solutions anions or organic compounds starting from a pure, analytical grade substance.Dry the substance in a oven at 110°C eliminating the humidity if this procedure do not compromise the stability of the compound.

Prepare diluted standard solutions starting from the concentrated ones. Add 1-2 per cent HNO_3 unless it is differently specified. Throw away the standard solutions less than 100 mg/l at the end of an analytical session.

2. Preparing Supporting Electrolytes

Prepare supporting electrolytes using pure, analytical grade reagents. It is better to check the presence of the analyte in the blank, before starting the analysis. The most common supporting electrolyte and buffer are the following:

(i) 0.1 M EDTA solution-Dissolve 37.2 g of EDTA-Na_2 in 1 l of distilled water, in a volumetric flask.

(ii) 0.1 or 1 M HCl solution-Dilute 8.2 (or 82) ml of 37 per cent HCl in 1 l of distilled water, in a volumetric flask.

(iii) 0.1 or 1 M KCl solution-Dissolve 7.5 (or 75) g of KCl in 1 l of distilled water, in a volumetric flask.

(iv) 0.1 M KCNS solution-Dissolve 9.72 g of KCNS in 1 l of distilled water, in a volumetric flask.

(v) 0.1 o 1 M KNO_3 solution-Dissolve 10 (o 100) g of KNO_3 in 1 l of distilled water, in a volumetric flask.

(vi) 0.1 M LiCl/LiOH solution-Dissolve 2.4 g of LiOH and 4.3 g of LiCl in 1 l of distilled water, in a volumetricflask.

(vii) 1 M NaF solution-Dissolve 42 g of NaF in 1 l of distilled water, in a volumetric flask. Heat if necessary.

(viii) 0.1 M acetate buffer solution at pH 4.5 (CH_3COONa/CH_3COOH)-Dissolve 8.2 g of anhydrous CH_3COONa (or 13.6 g of CH_3COONa and 3H2O) in 800 mlof distilled water. Add 5.75 ml of glacial CH_3COOH. Check the pH. Bring to mark volume in a 1 l volumetric flask with distilled water.

(ix) 0.1 M ammonia buffer solution at pH 9.4 (NH_4Cl/NH_3)-Dissolve 5.4 g of NH_4Cl in 900 ml of water. Add 6.9 ml of 26 per cent NH_3. Check the pH. Bring to mark in a 1 l volumetric flask with distilled water.

(x) 0.1 M borate buffer solution at pH 9.5 (NaH_2BO_3/H_3BO_3)-Dissolve 5.1 g of H_3BO_3 in 900 ml of water. Add 2 g of NaOH. Check the pH. Bring to mark in a 1 l volumetric flask with distilled water.

(xi) Britton Robins on buffer solution at various pH:

(a) Common solution-Mix- 100 ml of 0.04 M H_3BO_3 (2.04 g/100 ml) aqueous solution, 100 ml of 0.04 M CH_3COOH (2.3 ml of glacial CH_3COOH/100 ml) aqueous solution and 100 ml of 0.04 M H_3PO_4 (2.8 ml of 85 per cent H_3PO_4/100 ml) aqueous solution.

(b) Solutions at desired pH-Bring the preceding solution at the desired pH Using a 0.2 M NaOH solution.

(xii) 0.2 M ammonia citrate buffer solution at pH 3 (ammonia citrate/citric acid)-Dissolve 42.5 g of mono hydrate citric acid in 800 ml of water. Add 26 per cent NH_3 until pH 3. Bring to mark in a 1 l volumetric flask with distilled water.

(xiii) 0.2 M sodium citrate buffer solution at pH 3 (sodium citrate/citric acid)-Dissolve 42.5 g of mono hydrate sodium citrate in 800 ml of water. Add 20 per cent NaOH until pH 3. Bring to mark in a 1 l volumetric flask with distilled water.

(xiv) 0.2 M phosphate buffer solution at pH 6.8-Dissolve 24 g of NaH_2PO_4 and H_2O in 800 ml of water. Add 85 per cent H_3PO_4 until pH 6.8.Bring to mark in a 1 l volumetric flask with distilled water.

(xv) 0.2 M ammonia tartrate buffer solution at pH 9 (ammonia tartrate/tartaric acid)-Dissolve 300 g of tartaric acid in 800 ml of water. Add 26 per cent NH_3 until pH 9. Bring to mark in a 1 l volumetric flask with distilled water.

(xvi) 0.2 M sodium tartrate buffer solution at pH 9 (sodium tartrate/tartaric acid)-Dissolve 300 g of tartaric acid in 800 ml of water. Add 20 per cent NaOH until pH 9. Bring to mark in a 1 l volumetric flask with distilled water.

(xvii) 0.3 M TEA/0.1 M KOH buffer solution-Dissolve 45 g of TEA (triethanolamine) and 5.6 g of KOH in 1 l of distilled water, in a volumetric flask.

(xviii) 0.3 M TEA/0.1 M NaOH buffer solution-Dissolve 45 g of TEA (triethanolamine) and 8 g of NaOH in 1 l of distilled water, in a volumetric flask.

Checking the Polarograph Performances

The following KCl test is to be done to check the polarograph performance:

0.1 M KCl Test

The following test allows to check if the instrument (particularly the capillary) works correctly and maintains the performances for a long the time. Activate this procedure immediately, just before the instrument reach laboratory, collecting in this way an "historical" data series as reference.

Reagents

(i) 10 mg/l Cadmium standard solution. Prepare a fresh solution before each analysis, starting from a 1000 mg/l standard solution.

(ii) 0.1 M KCl checking solution. Dissolve 15 g of KCl in 2 litres of 1 per cent HNO3 (20 ml of 65 per cent HNO3 in 2 l of distilled water), in a volumetric flask, add 100 µl of Cd standard solution (1 g/l). Store this solution in a polythene bottle. Use the same solid KCl, stored in a polythene bottle, if prolonged check are needed.

Procedure

Every time an analytical series start (it means each day, or week) pour 10 ml of checking solution in the cell. Bubble nitrogen for 5 minutes. Analyse Cadmium using single (quicker) or multiple (longer) standard addition (100 µl of 10 mg/1 Cd each addition). Prepare a control chart after 10-15 analysis.

Results Control

(i) The peak position has to be constant, into a specific confidential range. Use statistics and control chart for this purpose.If the peak position significantly varies check the conditions of the reference electrode.

(ii) During the time, the results of the analysis have to agree to each other, into a confidential range. Use statistics and control chart for this purpose. If the results do not agree to each other verify the following:

(a) Quality of the solutions (KCl and Cd standard solutions)

(b) The conditions of the capillary.

(c) The cleaning procedure for the glassware, the instrument and the operator

Types of Voltammetry

There are following two major divisions in voltammetry:

(i) Voltammetry under diffusion controlled mass (ion) transfer- Polarography, LSV, CV, NPP, DPP, *etc.*

(ii) Voltammetry under convection controlled mass (ion) transfer:

(a) Movement of electrode in a still solution which promote "convection"-RDE, RRDE.

(b) Movement of solution past the stationary electrode-electrochemical detection for LC where, flow cells, channel electrode (wall jet electrode) are used.

The following types of voltammetry are in application:

i. Linear Sweep Voltammetry LSV (Rapid scan Voltammetry)

Rapid scan Voltammetry is the simplest technique. At the working electrode is applied a rapid potential scanning that varies linearly (20 – 100 mV/s). The scanning starts before the discharging potential and stops afterwards, Capacitive current increases when the velocity of scanning is increased and cannot be electronically compensated. Thus the performance of this technique are strongly restricted. Detection limits range at mg/l levels [2].

ii. Cyclic Voltammetry (CV)

A voltammetric method that can be used to determine diffusion coefficients and half cell reduction potentials [14]. Cyclic voltammetry is one of the most exploited techniques in electrochemical studies. Its primary advantage comes from the fact that it gives insight into both the half-reactions taking place at the working electrode, providing at the same time information about the chemical or physical phenomena coupled to the studied electrochemical reaction.Cyclic voltammetry is often considered as electrochemical spectroscopy [13]. The elaboration of the principles of cyclic voltammetry is important, since very electroanalytical study almost inevitably commences with this technique. In cyclic voltammetry, starting from an initial potential Ei, a staircase potential sweep (or linear sweep in older potentiostats) is applied to the working electrode. After reaching a switching potential Ef, the sweep is reversed and the potential returns to its initial value. The main instrumental parameter in the cyclic voltammetry is the scan rate (v=dE/t), since it controls the timescale of the voltammetric experiment. The useful scan rates range from 1 to 1000 mV=s, although scan rates of over 10V=s are technically achievable. The instrumental output in cyclic voltammetric techniques is acurrent–potential curve, a **cyclic voltammogram.** The main features of the cyclic voltammogram are the cathodic and anodic peak potentials, the cathodic and anodic peak currents, and the formal (or half-peak) potential. While the half-peak potential (defined simply as a median between the cathodic and the anodic peak potentials) provides mainly thermodynamics information, the magnitudes of the peak currents reveal the kinetics involved in the electrochemical reaction. The shape of the cyclic voltammogram gives information about the type of the electrode reaction, the number of electrons involved in the elementary step of electrochemical transformation, as well as about the additional phenomena coupled to the electrochemical reaction of interest, like those for coupled chemical reactions or adsorption and crystallization [5,13].

Cyclic Voltammetry is a technique devote to the theoretical study of the behaviour of redox couples. Cyclic Voltammetry is a particular LSV that perform

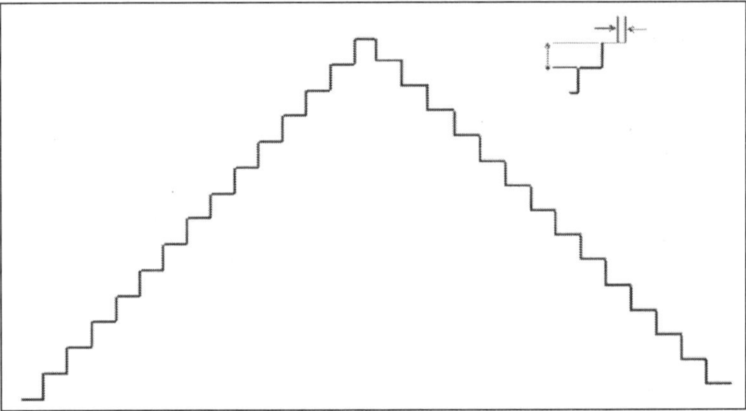

Staircase Potential Ramp Used in Cyclic Voltammetry.

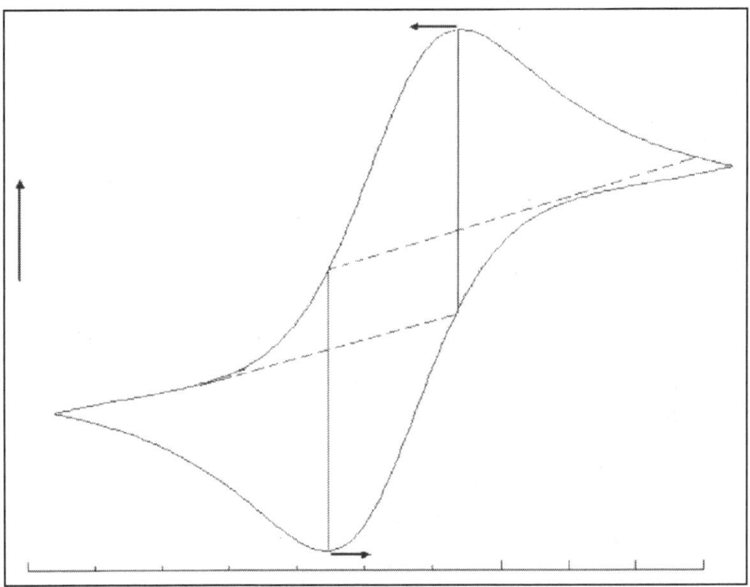

A Cyclic Voltammogram.

a triangular shaped scanning at the working electrode. In this way a redox couple in solution is exposed before to an oxidation and afterwards to a reduction or vice and versa. The cyclic scanning has to be ruled on the basis of few simply basic rules:

(i) A scanning starts in the cathodic sense has to be stopped at a potential lower than 100 mV (at least) respect on the peak potential.

(ii) A scanning starts in the anodic sense has to be stopped at a potential higher than 100 mV (at least) respect on the Epa peak potential.

(iii) Alternatively, a pause of 1 minute can be set between the go and backscanning.

The plot of a cyclic voltammetry consist on a close curve: reversible redox couples show both as cathodic and anodic peak, while irreversible redox systems show only one peak.

The following relations can be useful to establish the standard potential of a reversible redox couple and the number of electrons involved in the discharge process

$$E_o = E_{½} = \frac{\sim E_{pa} + E_{pc}}{2}$$

and

$$(E_{pa} - E_{pc}) = \frac{\sim 0.1984.\ T}{N}$$

where,

Epa: Anodic peak potential, in mVe Epc: Cathodic peak potential.

Because of its poor sensitivity [2], generally this technique (CV) is not used for quantitative analysis hence its usage is relatively minimal in quantitative food analysis.

Cyclic voltammetry is an electrolytic method that uses microelectrodes and an unstirred solution so that the measured current is limited by analyte diffusion at the electrode surface. The electrode potential is ramped linearly to a more negative potential, and then ramped in reverse back to the starting voltage. The forward scan produces a current peak for any analytes that can be reduced through the range of the potential scan. The current will increase as the potential reaches the reduction potential of the analyte, but then falls off as the concentration of the analyte is depleted close to the electrode surface. As the applied potential is reversed, it will reach a potential that will reoxidize the product formed in the first reduction reaction, and produce a current of reverse polarity from the forward scan. This oxidation peak will usually have a similar shape to the reduction peak.

Cyclic Voltammetry (CV) is a versatile electroanalytical technique for the study of electroactive species. CV consists of linearly cycling the potential of an electrode immersed in an unstirred solution while measuring the resulting current. Thus, a voltammogram is a display of current versus potential. The most useful aspect of this technique is its application to the qualitative diagnosis of electrode reactions. Such as voltammetry of a redox couple (*e.g.* ferro-/ferricyanide or Fe2+/Fe3+ in aqueous solution) using a rotating disc electrode (Pt or Au) or cyclic voltammetry of a redox system. A simple potential wave form that is often used in electro-chemical experiments is the linear wave form *i.e.*, the potential is continuously changed as a linear function of time. The rate of change of potential with time is referred to as the scan rate. Cyclic voltammetry can be used as a technique to:

(i) Reveal surface contamination.

(ii) Estimate relative surface area and roughness.

(iii) Evaluate electrolyte leakage at electrode-insulator interfaces.

(iv) "Fingerprint" electrochemical reactions for bench marking and quality control, estimate potentials at which reduction-oxidation reactions occur.

(v) Determine charge storage capacity

iii. Staircase Voltammetry (STCV)

A different variant of the LSV technique consist in a regular potential step scanning. The current is sampled just before the subsequent step. Thus the signal is less influenced by the capacitive current [2].

iv. Squarewave Voltammetry

This technique represents a further development of the preceding one. A rapid step scanning of potential is applied to the electrode and, moreover on each step is superimposed an high frequency square wave (20- 100 Hz). The current is sampled two times at the end of the two half waves. If the amplitude of the wave is very little and the redox system is reversible, during the first half wave the electro active compound can be reduced (or oxidised), while, in the second half wave, at the contrary, it can be oxidised (or reduced). The two current are then summed up and so, the sensitivity is increased. The sensitivity of this technique can be increased by enhancing the amplitude of the square wave or the frequency. The limits of the enhancing is strictly related to the kinetics aspects of the redox system: it has not to be slower than the velocity of the scanning of potential. The interference due to capacitive current are lowered to minimum because the current is sampled just at the end of the half waves, when the current of the double electrical layer is the least. Detection limits range from 5–50 μg/l [2].

v. Stripping Voltammetry (SV)

There are following two types of Stripping Voltammetry:

(a) Anodic Stripping Voltammetry (ASV)

(b) Cathodic Stripping Voltammetry (CSV).

Some metals form an amalgam with the working electrode and some anions form insoluble salts with it. The stripping voltammetry is based upon these properties.

The technique follows two main steps:

(a) A pre-concentration of the analyte on to the electrode

(b) The successive stripping of the cumulate compound onto the electrode toward the solution.

The stripping takes place during a scanning with the usual techniques (linear or differential pulsed or square wave scanning).

The usual working electrode is a dropping (or a film) mercury and the most common technique is the **anodic stripping**, meaning that a negative potential is applied to the electrode and the cations are discharged as metallic atoms into the mercury (amalgam). Successively the metal atoms are oxidised again, during ananodic scanning of potential. During the scanning of the potential, the current is measured and plotted, so the resulting voltammogram is a peak shaped graphic. The position and the height of this peak are related, respectively, to the type and to the concentration of the analyte.

The analysis of anions on mercury takes place following a inverse way- first mercury oxidise itself and forms a insoluble salt with the anion and after, by applying a **cathodic scanning**, mercury reduce itself and the anion come back in solution.

Example

$$2Hg + 2Cl\text{-}\ Hg_2Cl_2 + 2e \textbf{ pre-concentration}$$

$$Hg_2Cl_2 + 2e - 2Hg + 2Cl\text{-} \textbf{ reducing (cathodic) scanning}$$

This technique allows to considerably enhance the sensitivity because during the pre-concentration step a great quantity of analyte in a small volume of electrode (the mercury drop) and the measured stripping currents are greater than the ones obtained using a non accumulative voltammetric techniques.In this way detection limits below µg/l are achieved [2].

(*a*) **Anodic stripping voltammetry-**A quantitative, analytical method for trace analysis of metal cations. The analyte is deposited (electroplated) onto the working electrode during a deposition step, and then oxidized during the stripping step. The current is measured during the stripping step [13].

(*b*) **Cathodic stripping voltammetry-**A quantitative, analytical method for trace analysis of anions. A positive potential is applied, oxidizing the mercury electrode and forming insoluble precipitates of the anions. A negative potential then reduces (strips) the deposited film into solution [13],

vi. Adsorptive Voltammetry (AdV) or Adsorptive Stripping Voltammetry (AdSV)

A quantitative, analytical method for trace analysis. The analyte is deposited simply by adsorption on the electrode surface (*i.e.*no electrolysis), then electrolyzed to give the analytical signal.Chemically modified electrodes are often used [13].

Some metallic complex and some organic substances are adsorbed onto the surface of the electrode at specific potential. This phenomenon can be useful exploited when metals do not form amalgam onto the mercury, but form complex that can be adsorbed onto the latter and analysed with a direct scanning or a stripping technique. The detection limits are below the µg/l, while the linear range is of tenshort [2].

vii. Alternating Current Voltammetry

viii. Polarography

A subclass of voltammetry where the working electrode is a dropping mercury electrode (DME), useful for its wide cathodic range and renewable surface [13]. Classical Polarography, developed by Heyrowsky, is now an obsolete technique.

a. Rapid DC Polarography

The mercury drops fall down, rhythmically, from the capillary with a imposed rhythm, while a linear scanning is imposed to the electrode. The obtained polarogram is a wave characterised by strong oscillations due to the rhythmic falling of the drop (that means a rhythmic interruption of the electrical circuit) [2].

b. Sampled (or Test) DC Polarography-Staircase Polarography

The interference of capacitive current in the previous technique is strong. Capacitive current reach its maximum in the first moments of life of the drop and goes to the minimum just before the drop fall down.In this technique, the potential is increased (or decreased) at the beginning by a step and maintained constant during the life of each drop. The current is measured in the last moment of life of the drop. A stepped wave polarogram, with a smaller noise, is thus obtained [2].

ix. Hydrodynamic Voltammetry (Rotating Disk Voltammetry, RDV - Rotating Ring-Disk Voltammetry, RRDV)

In the hydrodynamic techniques the scanning of the potential takes place while the solution is in motion toward the electrode. In this way the solution closed to the Nernst (practically quiet) layer is continuously renewed. The motion is actuated in following three different ways:

- *(a)* The electrode is stationary and the solution is stirred. This technique has low reproducibility and is practically not used.
- *(b)* The solid electrode rotates around the own axis, the solution is than stirred. The electrode is a disk or a disk coupled with a ring.
- *(c)* The solution flows into a tubular electrode at constant velocity. This system is used as voltammetric detector in HPLC [2].

Hydradynamic voltammetry is voltammetry under convection control where rate of convection is made faster.Diffusion also occurs and there is no migration.

Convection >> Diffusion >> Migration

Rotated electrode voltammetry is a hydrodynamic technique in which the working electrode, usually a rotating disk electrode (RDE) or rotating ring-disk electrode (RRDE), is rotated at a very high rate. This technique is useful for studying the kinetics and electrochemical reaction mechanism for a half reaction [13].

x. Normal Pulse Voltammetry/Normal Pulse Polarography, NPP

Pulsed techniques allow to reduce the capacitive current. In the Normal Pulse Polarography a potential pulse is imposed to the drop during its last moment of life. The pulse height is linearly increased during the time. In the Differential Pulse Polarography the potential pulse is constant but is over imposed to a linear scanning of the potential. The polarogram is a wave in NPP and a peak in DPP and the height of both are related to the concentration of the analyte.DPP was been, for a long time, the most used technique and it has allowed to develop the first procedures and is the basis for the further developments of the analytical voltammetric techniques [2].

xi. Differential Pulse Voltammetry (DPV)

If a series of periodical constant pulse of potential is superimposed to a linear scanning, a consistent enhancement of the signal is achieved. Moreover, if the difference between the current just before and at the end of the pulse is measured, a reading less influenced by the capacitive current can be performed. In this way this differential reading of the current generates a peak shaped voltammogram. The technique is very sensitive and detection limits range near 10 – 100 μg/l. [2].

xii. Chronoamperometry

xiii. Tast Polarography

In this, voltage variations is same as classical polarography and Current measurement only over the last few ms of the drop life. (Just before it detached). The advantage of this is the improvement of precision and accuracy.

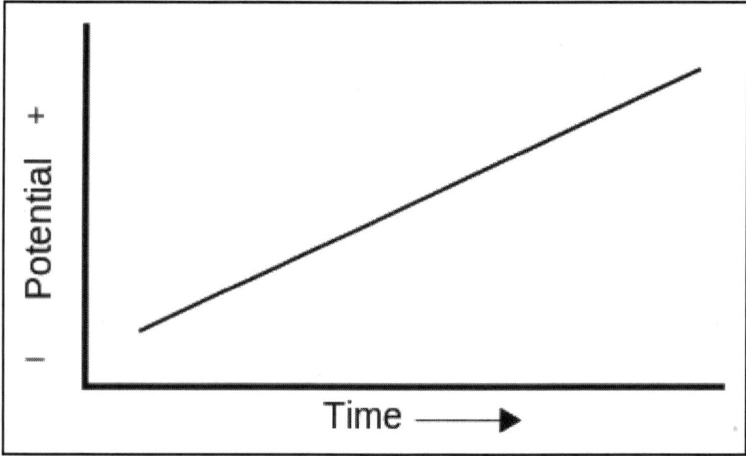

Linear Potential Sweep.

Applications

i. Voltammetric Sensors

A number of voltammetric systems are produced commercially for the determination of specific species that are of interest in industry and research. These devices are sometimes called electrodes but are, in fact, complete voltammetric cells and are better referred to as sensors. These sensors can be employed for the analysis of various organic and inorganic analytes in various matrices [10].

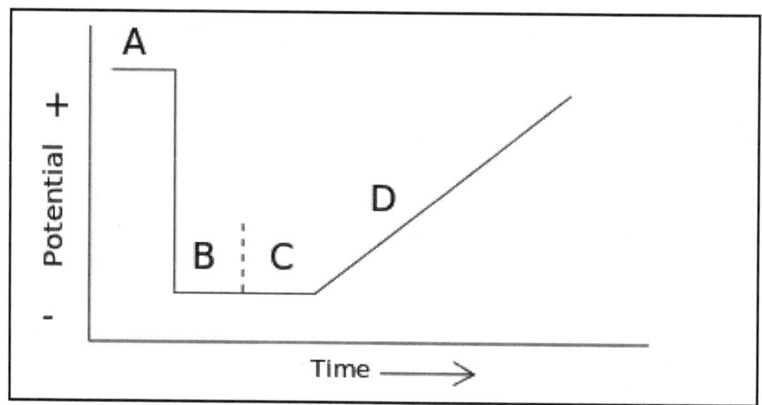

Potential as a Function of Time for Anodic Stripping Voltammetry.

ii. The Oxygen Electrode

The determination of dissolved oxygen in a variety of aqueous environments, such as sea water, blood, sewage, effluents from chemical plants, and soils is of tremendous importance to industry, biomedical and environmental research, and clinical medicine. One of the most common and convenient methods for making such measurements is with the Clark oxygen sensor, which was patented by L.C. Clark, Jr. in 1956.

iii. Research Purposes [2]

Voltammetry is a versatile technique for research purposes, it allow to search into several aspects of the electrochemical reactions, namely those reactions in which electrons exchanges are involved between reagents and products. For those reactions it is possible to investigate on the laws governing the dependence of the current by the potential imposed on an electrode dipped into the reaction environment.

Generally those laws are very complicated, just like the redox reactions and the environment in which they take place are.

The use of the voltammetric techniques is the basis of the comprehension ofthe laws concerning several electrochemical phenomena and has a great importance in several thecnological fields, like:

(a) Research of corrosion proof materials (corrosion is a consequence of a series of electro-chemical reactions)

(b) Research of new electrodic processes for chemical industries (in fact, for example, million of tons of aluminium, chlorine, soda are produced by means of electrochemical reactions)

(c) Production of new type of batteries that can store rapidly a great quantities of energy.

iv. Quantitative Analysis [2]

One of the most important application of Voltammetry is the quantitative analysis of trace of heavy metals (or, anyway, of those reducible or oxidizable chemicals) at μg/L levels or less in several different matrix like water, food, soil, air, industrial product, pharmaceutical, biology and so on.Every oxidizable or reducible substance that can be dissolved in an appropriate solvent can be analysed by Voltammetry, using appropriate conditions (supporting electrolyte, kind of electrode, scanning technique and scanning parameters) for its discharge.

Source

1. *Rubin Gulaboski and Carlos M. Pereira, Electroanalytical Techniques and Instrumentation in Food Analysis, Handbook of Food Analysis Instruments 45660_ C017.*
2. *AMEL Electrochemistry, Introduction to Modern Voltammetric and Polarographic Analysis Techniques IV Edition, 2001.*
3. *Kissinger, Peter; William R. Heineman (1996-01-23). Laboratory Techniques in Electroanalytical Chemistry, Second Edition, Revised and Expanded (2 ed.). CRC.*
4. *Zoski, Cynthia G. (2007-02-07). Handbook of Electrochemistry. Elsevier Science.*
5. *Bard, Allen J.; Larry R. Faulkner, Electrochemical Methods: Fundamentals and Applications (2 ed.). Wiley, 2001.*
6. *Gritzner, G.; J. Kuta (1984). "Recommendations on reporting electrode potentials in nonaqueous solvents". Pure Appl. Chem. 56 (4): 461–466.*
7. *Nicholson, R. S.; Irving. Shain (1964-04-01). "Theory of Stationary Electrode Polarography. Single Scan and Cyclic Methods Applied to Reversible, Irreversible, and Kinetic Systems.". Analytical Chemistry. 36 (4): 706–723.*
8. *Kounaves, Samuel. "Voltammetric Techniques". Tufts University Department of Chemistry.*
9. *Barek, J. (2003)."Eighty Years of Polarography - History and Future". Electro-analysis. 15: 467–472.*
10. *Hickling, A. (1942)."Studies in electrode polarisation. Part IV.-The automatic control of the potential of a working electrode". Transactions of the Faraday Society. 38: 27-33.*
11. *Barek, J. (2003). "Eighty Years of Polarography - History and Future".Electro-analysis. 15: 467–472.*
12. *Sanghavi, Bankim; Srivastava, Ashwini (2010)."Simultaneous voltammetric determination of aceta-minophen, aspirin and caffeine using an in situ surfactant-modified multiwalled carbon nanotube paste electrode".Electrochimica Acta.55: 8638–8648.*
13. *Heinze, J., Cyclicvoltammetry-"Electrochemical Spectroscopy, " Angew. Chem., 23, 831, 1984.*
14. *Wikipedia.*

v. Stripping Voltammetry in Food Analysis

Stripping voltammetric techniques were primarily used in the determination and study of food additives (preservatives, color agents, artificial sweeteners) and contaminants (toxic metals, pesticides, residual veterinary drugs) in food but lator on these methods had been proved to be exceedingly effective in analyzing vitamins, aromatic flavors, alkaloids and natural organic acids (Molecules of high biological significance which indicates the quality of the food product and its nutritional value). Now, stripping voltammetric technique has become widely adopted for monitoring of various food components and contaminants.

a. Stripping Voltammetric Determination of Contaminants

Chemical contaminants in food specially in milk and milk products are of great concern to the consumer, food manufacturers/Milk processors and regulatory authorities. Contamination of milk/food with chemicals may arise from environmental pollutants and industrial chemicals during food processing and packaging or from the use of agrochemicals and veterinary drugs in food production/Milch animal treatment [1].

i. Determination of Toxic Metals in Food

Stripping voltammetric techniques have been used with a great potential for the determination of trace and ultratrace concentrations of toxic metals such as mercury [2], lead [3], cadmium [4]), arsenic [5], antimony [6] and uranium [7].

(a) The determination of ultratrace levels of lead and cadmium in selected agriculture crops (over 1700 crop samples) by differential pulse anodic stripping voltammetry was reported [8].

(b) A collaborative study (13 laboratories) for the direct determination of lead in evaporated milk and apple juice by anodic stripping voltammetry was also reported [9].

(c) A collaborative study (20 laboratories) for the anodic stripping voltammetric determination of lead and cadmium in various food commodities *i.e.* beans, beef, fish, infant formula apple juice and cereal was also reported [10].

(d) A comparative study of the levels of cadmium and lead toxic metals in 36 wine samples by anodic stripping voltammetry was also reported [11].

(e) Differential Pulse anodic stripping voltammetry was used for Ultra trace determination of lead and cadmium in raw agricultural crops [12].

(f) Stripping voltammetric techniques, especially ASV method, were successfully utilized for the determination of these toxic metals content in various foodstuffs such as leafy vegetables [13], wheat and rice [14], common table salt [15], liver and fish [16], infant formulas [17] and canned soft drinks [18].

(g) Determination of Cd (II) in some foodstuffs by anodic stripping pulse voltammetry (ASPV) using glassy carbon electrode was suggested. Linear calibration curves were obtained in the concentration range

between 1.5×10^{-9} – 2×10^{-10} M. This method was successfully applied to the determination of Cd (II) in some foodstuffs (wheat and its products, vegetables) after acid digestion. [19].

(*h*) Zinc (Zn) Analysis in Milk by Microwave Oven Digestion and Differential Pulse Anodic Stripping Voltametry (DPASV) Technique. The direct determination of Zinc (Zn) heavy metal in milk samples of different brands was carried out by differential pulse anodic stripping Voltammetric technique at Hanging Mercury Drop Electrode (HMDE). Milk samples were processed by microwave oven digestion using HP/VHP Vessels and TFM Liners and nitric acid (HNO3). Determination of Zn was made in acetate buffer (pH 4.6) with a sweep rate (scan rate) of 59.5 mV/s and pulse amplitude 50mV by HMDE by standard addition method. The solution was stirred during pre-electrolysis at -1150mV (vs. Ag/AgCl) for 90 seconds and the potential was scanned from -1150V to +100V (vs. Ag/AgCl). The zinc ions were deposited by reduction at -1150 mV on HMDE. The stripping current arising from the oxidation of metal was correlated with the concentration the metal in the sample. As a result the minimum level of Zn observed in the milk sample of different brands was determined as 2.28 mgL^{-1} [19a].

ii. Determination of Pesticides Residuals in Food

Pesticides including fungicides, herbicides and insecticides are routinely used in most food production to control pests *e.g.* insect, fungi, rodents, warms and herbs that would otherwise destroy or reduce the food production.

(*a*) Stripping voltammetric technique proved to be highly effective in analyzing various pesticides molecules found as contaminants in foodstuffs, soil [20], tap water [21] and irrigation water [22].

(*b*) The adsorptive stripping voltammetric (AdSV) technique has been used to develop a method for the determination of fenthion pesticide in olive oil after the extraction of fenthion by solid–liquid extraction procedure using silica cartridge [23].

(*c*) AdSV method has also been applied successfully to the determination of the trace concentration of pyridafenthion insecticide in wine sample [24] and phosalone and carbophos insecticides in potato and tomato [25].

(*d*) Anodic stripping voltammetric (ASV) technique has been used for the ultratrace measurements of ziram (a dithiocarbamate fungicide) in rice samples using a static mercury drop electrode [26].

(*e*) Sensitive adsorptive stripping voltammetric procedures for dinoseb herbicides at nanomolar levels in spiked apple juice [27] and potatoes [28] were used.

iii. Determination of Veterinary Drugs Residuals in Food

Veterinary drugs are used to treat animals for bacterial infections, parasitic infections, to enhance growth, to control fertility and reproduction or to alter the behavior. There is some possibility that a residual of these veterinary drugs find its way to human diet via the cattle and poultry products such as milk, meat and dairy:

(a) Stripping voltammetric methods are very well known established as an analytical technique for the determination of numerous of pharmaceutical drugs in general [29-32] and some veterinary drug.

(b) Quantitative analysis of three antibiotic fluoroquinolones veterinary drugs in bird feedstuffs was carried out using linear sweep stripping voltammetrey (LSSV) method [33].

(c) Both differential pulse AdSV and square wave AdSV techniques were used for the determination of anthelmintic veterinary drug nitroxnil [34].

iv. Determination of Fertilizers Residuals in Food

As in the case of pesticides and veterinary drugs, residual fertilizers may find its way to the food products. The build up of high fertilizer concentrations has caused some health as well as environmental impacts:

(a) A sensitive anodic stripping voltammetric procedure was used for determination of trace nitrite fertilizer in meat and food products [35].

(b) Similar anodic voltammetric approach but with different chemically modified electrode has been developed for the electroanalytical determination of nitrite in food samples such as sausages and pickled vegetables [36].

b. Stripping Voltammetric Determination of Trace Essential Elements

Essential elements are any elements required by living organisms to ensure normal growth, development maintenance. The trace essential elements usually occurred in living tissues at low concentration levels and thus their requirements are much less. The most important elements are Fe, Mn, Zn, Cu, I, Co, Se, Mo, Cr and Si and each element may fulfill one or more of a variety of metabolic roles. The determination of food minerals and essential elements can be done using fast, sensitive, inexpensive while still being adequately precise stripping voltammetric electroanalytical methods:

(a) Several scientific workers [37-39] have devoted to emphasize and illustrate the wide spectrum and scope of stripping voltammetric applications and its potentialities in the analysis of metal ions.

(b) A study of the square wave AdSV determination of iron essential element in olive oils has been reported [40].

(c) Cathodic stripping voltammetry determination of trace amounts of selenium in rapeseed oils and seed [41] has been made.

(d) The applicability of AdSV technique for the determination of trace levels of cobalt in vegetable animal foodstuffs [42] and milk powder [43] was demonstrated.

(e) The usefulness of AdSV method in the determination of nickel in infant formulas has also been described [44].

(*f*) Aluminum content in samples of commercial tea leaves and coffee beans was determined by adsorptive striping voltammetry [45].

(*g*) The developed stripping voltammetric procedure was applied for the determination of tin in fruit juices [46].

(*h*) The application of cathodic stripping voltammetric approach in the determination of iodine mineral in table salt and eggs was demonstrated [47].

c. Stripping Voltammetric Determination of Food Additive Dyes

Food additives are substances added to the food during its manufacture or processing in order to improve its flavor, color, texture, stability and shelf life and these additives are usually present in minute quantities. Natural (from vegetable, animal or mineral origin) or synthetic food dyes provide food with an attractive appearance. On the other hand, use of dyes or chemical impurities in colorant agents can have undesirable side effects. These can be analyzed by electrochemical methods, most recently using stripping voltammetric techniques:

(*a*) The utilization of stripping voltammetric techniques for the determination of several synthetic dyes usually used in textile industry was reviewed [48-49].

(*b*) Stripping voltammetric method was effectively employed for the determination of various food additive dyes. It was demonstrated that 16 food synthetic coloring matters can adsorb strongly on a hanging mercury drop electrode and to be amenable to determination by differential pulse adsorptive stripping voltammetry [50].

(*c*) A quantitative chemometric approach was used for the simultaneous adsorptive voltammetric analysis of mixed colorants solution of some food dyes namely amaranth, sunset yellow, tartrazine and ponceau 4R [51] in some food samples such as drink, candy, and juice.

The summary of the applicability of stripping voltammetric techniques for the determination of several essential elements, toxic metals in food commodities, additive dyes in foodstuff products and Pesticide residues are given in Table given on next page.

d. Stripping Voltammetric Determination of other Organic Molecules

Many other organic compounds some of them are of biological significance can also be measured in foodstuffs samples using stripping voltammetric technique. Application of stripping voltammetric techniques to the determination of several vitamins [52] and food samples is being extensively explored:

(*a*) A sensitive and selective method for the determination of vitamin C (ascorbic acid) in fruit and vegetable juice using stripping voltammetry on a glassy carbon electrode has been reported [53].

(*b*) A cathodic stripping voltammetric method has been developed to quantify minute amounts of three cytokinins plant hormones (Phytohormones) in apples [54].

Applications of Stripping Voltammetry in the Determination of Toxic Metals, Essential Elements in Food Samples, Additive Dyes and Pesticide Residues in Foodstuff Products

Sl. No.	*Food Sample*	*Toxic Metals*	*Stripping Voltammetric Technique*	*Working Electrode*	*Reference*
A. Toxic Metals					
1	Cow's milk	Pb(II)	ASV	HMDE	Inam and Somer (2000)
2	Honey	Pb(II), Cd(II)	ASV	Mercury Micro electrode	Sanna *et al.* (2000)
3	Table salt	Hg(II)	ASV	Gold film electrode	Okcu *et al.* (2008)
4	Wheat, maize	Sb(III)	ASV	HMDE	Clinio and Torsi (2004)
5	Sugar	U(VI)	CSV	Glassy carbon electrode	Abo-Maali and Elhady (1999)
6	Canned juice, stewed fruit	Pb(II)	ASV	Modified silver electrode	Fomintseva *et al.* (1997)
7	Honey, apple drink	Pb(II), Cd(II)	ASV	Modified glassy carbon electrode	Buckova *et al.* (1996)
8	Edible oils	Pb(II), Cd(II)	ASV	Modified glassy carbon electrode	Wahdat *et al.* (1998)
B. Essential elements					
1	Milk	Zn(II)	ASV	Bismuth film electrode	Guo *et al.* (2005)
2	Canned food	Sn(IV)	AdSV	HMDE	Li *et al.* (2005)
3	Oil, vegetables, milk	Ni(II)	AdSV	HMDE	Tripathi *et al.* (1994)
4	Cow's milk	Se(IV)	CSV	HMDE	Inam and Somer (2000)
5	Milk	Ni(II)	AdSV	HMDE	Zhang *et al.* (1991)
6	Beet sugar	Ni(II), Co(II)	CSV	HMDE	Sancho *et al.* (2000a)
7	Beer	Cu(II)	ASV	Nafion coated mercury film	Gutierrez *et al.* (1999)
8	Canned vegetables	Ni(II)	CSV	HMDE	Meryan *et al.* (1997)
9	Rice	Se(IV)	CSV	HMDE	Panigate and Falciola (2007)
10	Apple, potato	Fe(II)	AdSV	HMDE	Gao and Siow (1996b)
11	Food supplements	Se(IV)	CSV	HMDE	Holak and Specchio (1994)
C. Additive dyes in foodstuff products					
12	Carmine, E Number E120		AdSV	Ice cream, soft drinks	Alghamdi (in press)
13	Erythrosine B, E Number-E127		AdSV	Ice cream, soft drinks	Alghamdi *et al.* (in preparation)

Contd.

Sl. No.	Food Sample	Toxic Metals	Stripping Voltammetric Technique	Working Electrode	Reference
14	Sunset yellow, E Number-E110		AdSV	Soft drinks	Nevado *et al.* (1997a)
15	Tartrazine, E Number-E100		AdSV	Soft drinks	Nevado *et al.* (1997b)
16	Brilliant blue FCF, E Number-E133		CSV	Commercial food products	Florian *et al.* (2002)
D. Pesticide residues in foodstuff products					
17	Carbaryl Carbamate		Biosensor]	Milk	Zhang,Y *et al.*,2005
18	Paraquat Bipyridinium		Au UME	Fruit juice	De Souza, *et al.*, 2005
19	Carbaryl Carbamates		Biosensor]	Honey	Del Carlo. M. *et al.*, 2004
20	Paraoxon Organophosphate		Biosensor	Milk	Zhang,Y *et al.*,2005
21	Deltamethrin Pyrethroid		HMDE	Vegetables and cereals	Samatha, K., *et al.*, 1999
22	Clorofenvinphos Organo-phosphate		DME	Cereals	Sreedhar, N.Y.et al, 1997
23	Paraoxon Organophosphate		Biosensor	Baby food	Schulze, H. *et al.*,2002
24	Desmetryne Triazines		HMDE	Fruit juice	Pedrero, M. *et al.*, 2001

(*c*) An electrochemical method based on adsorptive stripping voltammetry was used for the study and determination of kinetin (growth regulator in plant) in apple extracts [55].

(*d*) A cathodic stripping voltammetric method has been used for the determination of naringin (flavonoid compound found in some citrus fruits) in grapefruit juice [56].

(*e*) Saponin molecules (phtyo-chemical which is found in various vegetables) were also determined in sugar juices from sugar refining processes by stripping voltammetry [57].

(*f*) An adsorptive stripping voltammetric method for the determination of the flavor enhancer inosinic acid in dehydrated soup samples was reported [58].

Source

1. *Ahmad H. Alghamdi, Applications of stripping voltammetric techniques in food analysis.*
2. *D.Sancho, L.Deban, F.Barbosa, R.Pardo, M.Vega, Food Chem., 74 (2001), p. 527.*
3. *R.D. Satzger, R.W. Kuennen, F.L. Fricke, J. Assoc. Anal. Chem., 66 (1983), p. 985.*
4. *R.J.Gajan, S.G.Capar, C. A.Subjoc, M. Sanders, J. Assoc. Anal. Chem., 65 (1982), p.970.*

5. *D. Sancho, M. Vega, L. Deban, R. Pardo, G.Gonzalez, Analyst, 123 (1998), p. 743.*
6. *L.Clinio, G. Torsi, Microchem.J., 78 (2004), p.175.*
7. *N. Abo-Maali, D. A. Elhady, Electro-analysis, 7 (1999), p. 201.*
8. *R.D. Satzger, C.S. Clow, F. Bonnin, F.L. Fricke, J. Assoc. Anal. Chem., 65 (1982), p. 987.*
9. *E.W. Zink, P.H.Davis, R.M.Griffin, W.R. Matson, R. A. Moffitt, D. T. Sakai, J. Assoc. Anal.Chem., 66 (1983), p.1414.*
10. *S.C. Capar, R.J. Gajan, E. Madzsar, R.H. Albert, M. Sanders, J. Zyren, J. Assoc. Anal. Chem., 65 (1982), p. 978.*
11. *J. Golimowski, P. Valenta, M. Stoeppler, H.W. Nurnberg, Z. Lebensm. Unters. Forsch., 29 (1979), p. 439.*
12. *R.D. Satzger, F. Bonnin, F.L. Fricke, J. Assoc. Anal. Chem., 67 (1984), p.1138.*
13. *M.H.Matloob, East Mediterr. Health, 9 (2003), pp. 28–36.*
14. *B. Ogorevc, A. Krasna, V. Hudnik, Anal.Chim. Acta, 196 (1987), p. 183.*
15. *A.M. Ali, J. AOAC Int., 82 (1999), p. 1413.*
16. *S.B. Adeloju, A.M. Bond, Anal. Chim. Acta, 148 (1983), p. 59.*
17. *M.J. Esteve, R. Farre, A. Frigola, Nahrung, 38 (1994), p. 386.*
18. *S.M.Sabry, A. A.M.Wahbi, Anal.Chim. Acta, 401 (1999), p.173.*
19. *O. A.Farghaly, R.S. AbdelHameed, Abd-AlhakeemH. Abu-Nawwas, Analytical Application Using Modern Electrochemical Techniques, Int. J. Electrochem. Sci., 9 (2014) 3287 - 3318.*

19a. *Mohineesh, A. Raina, J. Raj and T. D. Dogra, 1 Department of Forensic Medicine and Toxicology, All India Institute of Medical Sciences (AIIMS), New Delhi, India, EDP Sciences, 2013.*

20. *P. Mainisankar, S. Viswanathan, A.M. Pusphalatha, C. Rani, Anal. Chim. Acta, 528 (2005), p. 157.*
21. *M.S. Ibrahim, K.M. Almagboul, M.M. Kamal, Anal.Chim. Acta, 432 (2001), p. 21.*
22. *M. Pedrero, V.Calvov, F.J.M.Devillena, J.M.Pingarron, L.M.Polo, Analyst, 118 (1993), p.1405.*
23. *T.G.Diaz, A.G. Cabanillas, M.D. Soto, J.M. Ortiz, Talanta, 76 (2008), p. 809.*
24. *M.C.Sampedro, Z.G.de Balugera, A. Goicolea, R.J. Barrio, Food Additiv. Contamin., 15 (1998), p.793.*
25. *N. A.Ulakhovich, E.V.Priimau, E.P. Medyantseva, L. A. Anisimova, M. A. Algakhri, Zh. Anal. Khim., 53 (1998), p.167.*
26. *L.Mathew, M.L.P.Reddy, T.P. Rao, C.S.P. Iyer, A.D.Damodaran, Talanta, 43 (1996), p.73.*
27. *M. Pedrero, B. Casado, F.J.M. Devillen, J.M. Pingarron, Fresenius J. Anal. Chem., 349 (1994), p.546.*

28. *M. Pedrero, F.J.M. Devillen, J.M. Pingarron, L.M. Polo, Electroanalysis, 3 (1991), p.419.*
29. *A.H. Alghamdi, J. Saudi Chem. Soc., 2 (2002), p. 185.*
30. *A.Z. Abu Zuhri, W.Voelter Fresenius, J. Anal. Chem., 360 (1998), p.1.*
31. *P. Gratteri, S. Pinzauti, G.P.E.Porta, S. Furlanetto, IL Farmaco, 47 (1992), p. 695.*
32. *J.C. Vire, M. Kauffmann, G.J. Patriache, J. Pharma. Biomed. Anal., 7 (1989), p. 1323.*
33. *Y. Ni, Y. Wang, S. Kokot, Talanta, 69 (2006), p. 216.*
34. *M.M.Ghoneim, M. Elries, A.M. Hassanein, A.M. Abd-Elaziz, J.Pharm. Biomed. Anal., 41 (2006), p. 1268.*
35. *L.Guanghan, J. Hong, S.Dandan, Food Chem., 59 (1997), p.583, 1997.*
36. *W.J.R.Santos, P.R. Lima, A. A. Tanaka, S.M.C.N. Tanaka, L. T. Kubota, Food Chem., 113 (2009), pp. 1206–1211.*
37. *A.Z. Abu Zuhri, W.Voelter Fresenius, J. Anal. Chem., 360 (1998), p.1.*
38. *P.M. Zaitsev, R.M.F. Salikhdzhanova, N.K. Zaitsev, Indus. Lab Diagno. Mater., 65 (1999), pp. 1-15.*
39. *K.H. Brainina, N. A. Malakhova, N.Yu. Stojko, Fresenius J. Anal. Chem., 368 (2000), p. 307.*
40. *D. T. Galeno, A. Guiberteau, S.M.D. Lopez, J.M. Ortiz, J. Agric. Food Chem., 18 (2003), p.3743.*
41. *M.W.Blades, J. A. Dalziel, C.M. Elson, J. Assoc. Anal.Chem., 59 (1976), pp. 1234–1239.*
42. *S. Giroussi, A. Voulgaropoulos, A.K. Ayiannidis, J. Golimowski, M. Janicki, Sci. Total. Environ., 176 (1995), p. 135.*
43. *J.Koops, H.Klomp, Neth. Milk Dairy J., 45 (1991), p.127.*
44. *R. Barbera, M.J. Esteve, R. Farre, J.C. Lopez, Food Chem., 49 (1994), p. 427.*
45. *E.Stryjewska, S.Rubel, J. Sadowska, M. Karpiuk, Chim. Anal., 38 (1993), p.175.*
46. *S.B.O. Adeloju, F. Pablo, Anal. Chem. Acta, 270 (1992), p. 143.*
47. *S.X. Yang, S.J. Fu, M.L. Wang, Anal. Chim., 63 (1991), p. 2970, Yang et al., 1991.*
48. *A.G. Fogg, Portugal. Electrochem. Acta, 16 (1998), p. 5.*
49. *J. Zima, J. Barek, J.C. Moreira, V. Meejestic, A.G. Fogg, Crit. Rev. Anal. Chem., 29 (1999), p. 125.*
50. *A.G. Fogg, A. A. Barros, J.O. Cabral, Analyst, 111 (1986), p. 381.*
51. *Y.Ni, J.Bai, L. Jin, Anal. Chim. Acta, 329 (1996), p. 65.*
52. *A.H. Alghamdi, J. Saudi Chem. Soc., 2 (2002), p. 185.*
53. *G. Lu, Y.Wang, L.Yao, S.Hu, Food Chem., 51 (1994), p.237.*
54. *M.H. Blanco, M.D. Quintana, L. Hernandez, Electro-analysis, 12 (2000), p.147.*
55. *P.Hernandez, F.Paton, L.Hernandez, Electro-analysis, 9 (1997), p. 1372.*

56. *E. Reichart, D. Obendorf, Anal. Chim. Acta, 360 (1998), p.179.*

57. *D. Sancho, L.Deban, I.Campos, R.Pardo, M. Vega, Food Chem., 71 (2000), p. 139.*

58. *M.J. Portela, R.J. Barrio, M. A. Goicolea, Z.G. Debalugera, Analyst, 119 (1994), p. 2183.*

Instrumentation

The modern electroanalytical system for voltammetric measurements are usually composed of three modules: a potentiostat, a personal computer, and an electrochemical cell. In some cases, the potentiostat and computer are packed into one part, whereas in most of the electrochemical systems the computer and the converters and microcontrollers are separate, so the potentiostat can operate autonomously.Basic apparatus/materials required for Voltammetry are as follow:

Cell

A container to hold the analyte solution. An electrochemical cell is considered to be a sample holder, in which the corresponding analyte is dissolved in an appropriate solvent, and placed there after in an ionic electrolyte, where usually three electrodes (working, reference, and counter) are situated. The cells come commercially in various designs, built mainly of glass, Teflon, or polyethylene material.

Electrode

In Classical technique, two electrodes -Reference (RE) and Working (WE) whereas in modern techniques, three electrodes-Reference (RE), Working (WE) and Auxillary or Counter electrode (CE)] are being used. The electrodes of the electrochemicals system are usually all submerged in the electrochemical cell, whereas in some systems the reference electrode is placed in a separate compartment to avoid contamination, and it is connected to the cell via an electrolyte bridge. The details of the aforesaid electrodes are already given under Principle of Voltammetry [1].

The main criteria for a reference electrode are to provide a reversible half-reaction with Nernstian behavior, to have a constant potential over time, and to be easy to assemble and maintain [2]. Among the most widespread reference electrodes used for experiments performed in aqueous solutions are the calomel electrode, whose potential is determined by the reaction Hg_2Cl_2 (s)+2e = 2Hg (l)+2Cl, and the silver/silver chloride electrode (Ag/AgCl), whose potential is defined by the reaction AgCl (s)+ e= Ag (s)+Cl. These electrodes are commercially available in a variety of sizes and shapes [1].

With respect to the counter electrodes, Pt wire, graphite, or a thin piece of gold are the common variants in most of the voltammetric techniques. The task of the counter electrode is to preserve electroneutrality in the system, which is done by the occurrence of an electrochemical reaction (usually H^+ reduction, or water oxidation) counter to that taking place at the working electrode [1].

A wide range of working electrodes are available. The metallic electrodes, such as Hg drops andPt or Au disks, are among the most frequently used [3]. Mercury is very practical because of its highly negative overpotential for hydrogen ions reduction, and because of its constantly renewed surface [1,9,4]. The main shortcoming with the mercury electrode is its low potential of oxidation (about -+0.1 to +0.6 V depending on pH), because of which many compounds cannot be studied in the oxidation mode. Mercury can be used in different measuring modes such as hanging mercury drop electrode (HMDE) [5,6], DME [7,8], static mercury drop electrode (SMDE) [9], or even as AQ3 mercury film electrodes [10,11]. Carbon electrodes are extremely lipophilic, and hence appropriate for studying the electrochemical features of lipophilic organic compounds. Carbon electrodes can also be found in a large variety of forms such as graphite composite electrodes [12], glassy carbon electrodes [13], carbon-paste electrodes [14,15], screen-printed electrodes [16], diamond electrodes [17], and bismuth-coated carbon electrodes [18], among others examples. Gold [19,20], bismuth films [21], biosensors [22], and immunosensors [23] are also part of the paraphernalia of electrode materials used with some advantage to solve analytical problems in food analysis.

Potentiostat/Voltage Supply (Variable DC/AC)

The potentiostat is considered as the "heart" of every electroanalytical instrumentation. In voltammetric techniques, the task of the potentiostat is to apply an exact potential and to observe the current changes in the system. In all modern electroanalytical instruments, the potentiostat package includes electrometer circuits, various converters and amplifiers, as well as microprocessors with internal memory. All the modern potentiostats designed after 1980s operate in a digital (incremental) fashion. In these, a 'staircase' modulated potential is generally used, with the 'steps' having constant potential increment. The major benefits of staircase-modulated potentials are seen by the significant discrimination over the capacitative (nonfaradaic) currents [24]. The most frequent waveforms in modern potentiostats are linear scan, differential pulse, and square wave. Now-a-days, various types of potentiostats are available.

Voltameter and Ammeter

These are also make the parts of the instrument.

Source

1. *Rubin Gulaboski and Carlos M. Pereira, Electroanalytical Techniques and Instrumentation in Food Analysis, Handbook of Food Analysis Instruments 45660, C017.*

2. *Kahlert, H., Reference electrodes, in Electroanalytical Methods, Scholz, F. (Ed.), Springer, Berlin/Heidelberg/NewYork, 2001, chap. III.*

3. *Komorsky-Lovric, S., Working electrodes, in Electroanalytical Methods, Scholz, F. (Ed.), Springer, Berlin/Heidelberg/New York, 2001, chap. III.*

4. *Wang, J., Analytical Electrochemistry, 3rd ed., Wiley-VCH, 2006.*

5. *Castrillejo, Y. et al., Determination of food additive azo dyes at an HMDE with adsorptive stripping voltammetry, Electroanalysis, 2, 553, 1990.*

6. *Qiong, L. et al., Determination of trace aluminum in foods by stripping voltammetry, Food Chem., 97, 176, 2006.*

7. *Inam, R. and Toprak, C., Polarographic determination of some toxic trace elements in fish muscles, Fresenius Environ. Bull., 14, 489, 2005.*

8. *Obendorf, D. and Reichart E., Determination of hesperidin by cathodic stripping voltammetry in orangejuice and helopyrin, a phytopharmaceutical preparation, Electroanalysis, 7, 1075, 1995.*

9. *Carrai, P., Nucci, L., and Pergala, F., Polarographic behavior of alachlor application to analytical determination, Anal. Lett., 25, 163, 1992.*

10. *Babkina, S.S. and Ulakhovich, N. A., Amperometric biosensor based on denatured DNA for the study of heavy metals complexing with DNA and their determination in biological, water and food samples, Bioelectrochemistry, 63, 261, 2004.*

11. *Lo Coco, F. et al., Determination of lead (II) and cadmium (II) in hard and soft wheat by derivative potentiometric stripping analysis, Anal. Chim. Acta, 409, 93, 2000.*

12. *Luque, M. et al., Supported liquid membranes for the determination of vanillin in food samples with amperometric detection, Anal. Chim. Acta, 410, 127, 2000.*

13. *Desimoni, E., Brunetti, B., and Cosio, M.S., Determination of patent blue V (E131) at a Nafion-modifiedglassy carbon electrode, Electroanalysis, 18, 231, 2006.*

14. *Jayasri, D. and Narayanan, S.S., Manganese (II) hexacyanoferrate based renewable amperometric sensor for the determination of butylated hydroxyanisole in food products, Food Chem., 101, 607, 2007.*

15. *Li, Y.H. et al., Determination of trace tin by anodic stripping voltammetry at a carbon paste electrode, Electroanalysis., 18, 976, 2006.*

16. *Miserere, S. et al., Biocompatible carbon-based screen-printed electrodes for the electrochemical detection of nitric oxide, Electrochem. Commn., 8, 238, 2006.*

17. *Bouamrane, F. et al., Electro- chemical study of diamond thin films in neutral and basic solutions of nitrate, J. Electroanal.Chem., 405, 95, 1996.*

18. *Wang, J. et al., Bismuth-coated carbon electrodes for anodic stripping voltammetry, Anal. Chem., 72, 3218, 2000.*

19. *De Souza, D., Machado, S. A.S., and Pires, R.C., Multiple square wave voltammetry for analytical determination of paraquat in natural water, food, and beverages using microelectrodes, Talanta, 69, 1200, 2006.*

20. *Radke, S. A. and Alocilja, E.C., A high density microelectrode array biosensor for detection of E-coli O157: H7, Biosens. Bioelectron., 20, 1662, 2005.*

20a. *Sánchez Arribas, A. et al., Voltammetric detection of the herbicide metamitron at a bismuth film electrode in nondeaerated solution, Electroanalysis, in press.*

21. *Schulze, H. et al., Development, validation, and application of an acetylcholinesterase-biosensor test for the direct detection of insecticide residues in infant food, Biosens. Bioelectron., 17, 1095, 2002.*

22. *Adanyi, N. et al., Development of new immunosensors for determination of contaminants in food, Curr. App. Phys., 6, 279, 2006.*

23. *Heinze, J., Cyclic voltammetry-"Electrochemical Spectroscopy," Angew. Chem., 23, 831, 1984.*

24. *Bard, A.J. and Faulkner, L., Electrochemical Methods, Wiley, New York, 2001.*

Different Available Models

The details of the some available models (Portable and Non Portable) of electrochemical instruments are given below:

1. Portable Metal Monitor (PDV6000 ultra-Modern water)

Voltammetry offers an internationally accepted alternative to laboratory analysis. Modern Water's PDV provides better on-site characterisation, with pollution hot spots and contamination sources more easily identified. The PDV6000 ultra has been upgraded to allow much improved stand-alone operation and an indication of blank and standard run performance. A larger screen allows simple voltammograms to be shown. The USB connector replaces the serial port. The upgraded product has a battery life indicator and can be run either on 4 x 1.5VAA batteries, or from the mains using the standard 8 - 12V DC transformer.

The PDV6000 ultracomes with the VAS software package, which is easy to use and is compatible with Windows XP and 7. VAS enables storage and manipulation of voltammograms, operating data and in-depth data analysis.

SV LabCell

PDV6000 ultracomes equipped with a standard analytical cell which can detect a wide range of different metals. The SV LabCell is an optional extra that allows Cathodic StrippingVoltammetry (CSV) to be used on the PDV.

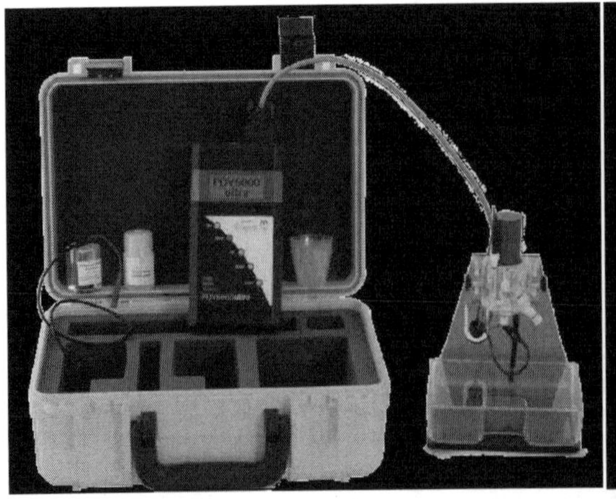

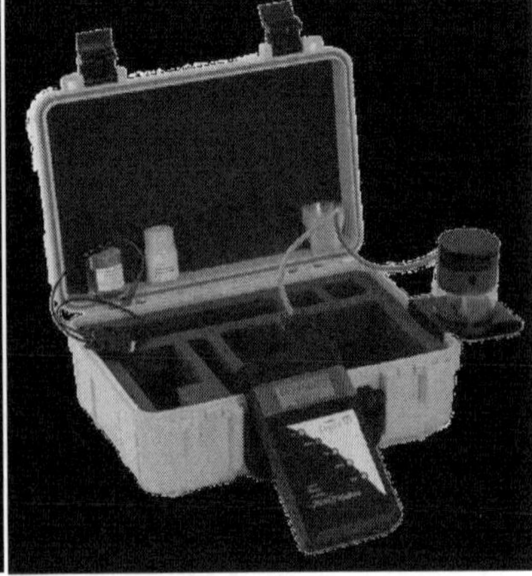

The SV LabCell extends the PDV's range of metals to include molybdenum and uranium; it also gives a better response for nickel, cobalt and chromium at low levels.

Detection

The PDV6000 ultra can detect a range of different metals (for example As, Cd, Cr, Cu, Hg, Ni, Pb, Se, Tl, Zn) and others.With the optional add-on SV Lab Cell, the PDV6000ultra can detect Co, Cr and Ni more easily, typically to 1ppb levels. The SV LabCell can also detect Mo and U.Colour or turbidity does not affect the method. Ideally suited for water analysis, dirty water samples and soil samples may require simple on-site sample preparations to remove interferences. Food samples will require digestion to prepare them for analysis. Samples can be pre-treated to remove interferences.In voltammetry metals are drawn onto the working electrode when a specific voltage is applied to the water sample under test.When a stripping voltage is applied, the metals return to the sample solution, generating a small current. Each metal has a specific voltage at which it returns to solution. So the metal is identified by its stripping voltage, and the current generated indicates the concentration of metal in the sample.

Specification and Characteristics

Sl.No.	*Characteristics*	*Values*
1	Power Supply	AC, 110 - 240V or DC 8 - 12V or 4 x AA batteries
2	Working Electrode, Std. Cell	Glassy carbon, used with a variety of films, or solid gold
3	Working Electrode, SV LabCell	Glassy carbon with bismuth film
4	Counter Electrode	Platinum
5	Reference Electrode	Ag/AgCl in KCl
6	Cell Material	Acrylic and PTFE
7	Cell Stirrer	DC magnetic motor and magnetically coupled stirrer.
8	Display	LCD graphic screen
9	Operating System	Operating Software Windows XP and 7; VAS internal firmware Communications USB. Portable laboratory instrument connected to PC or laptop Windows OS: XP and Windows 7 VAS software, making the instrument a top of the range voltammetry instrument.
10	Analysis Methods Available	Anodic stripping, Cathodic stripping
11	Waveforms Available	Linear sweep, square wave and differential pulse
12	Voltammetry Range	2.0V to +2.0V (switchable to -3.3V to +3.3V if required)
13	Sensitivity	2 Na
14	Variation (per cent CV)	5 to 10 per cent
15	Result Output	Voltammetry curves, element concentration(s), historical data
16	Others	Automatic data saving, graph optimisation, print facility for all data, reports and graphs.

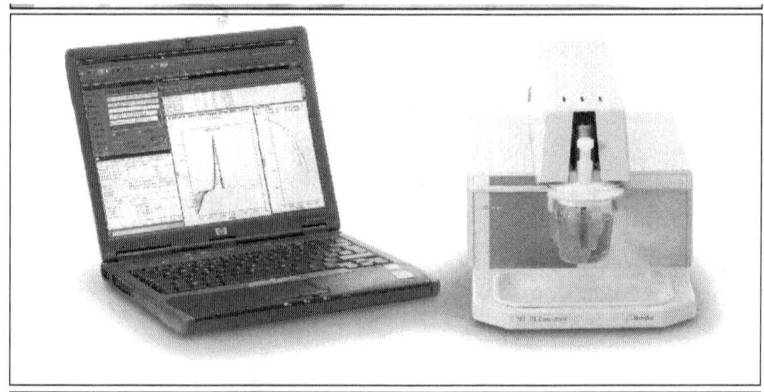

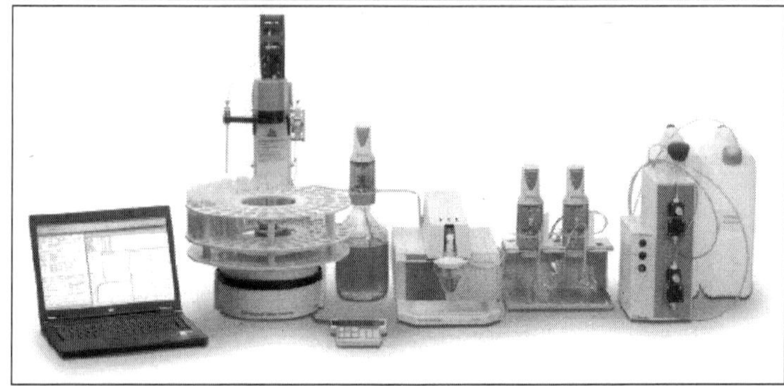

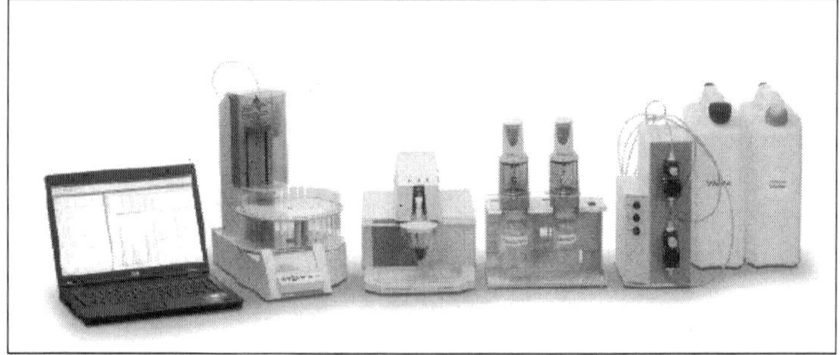

Main Assenbly and Accessories of 797 VA Computrace (Metrohm).

Source

1. *Technical Brochure of PDV6000 ultra- Modern water.*

2. Non Portable Electrochemical Instruments

Some Non portable electro-chemical istruments are available. Currently, several companies manufacture high-quality potentiostats/Voltammetric instruments capable of performing various voltammetric analyses. Among them, EG and G Princeton Applied Research, EcoChemieNetherlands, ACM Instruments, Cypress Systems, Bioanalytical Systems, USA, Chi Instruments, Electro chemical work station, France and Metrohm Electro Chemical Instruments, Switzerland and Radiometer are the leaders.

797 VA Computrace (Metrohm)

For a fraction of the purchase price of an AAS or ICP instrument it is possible to carry out metal analyses with the same or improved sensitivity. The only thing required apart from small amounts of reagents is a small amount of pure nitrogen. No expensive flammable gases, nospecially constructed fume hoods in the laboratory, nocostly metal vapor lamps.

Spectroscopic methods can only determine the total concentration of the metals. With voltammetry it is also possible to differentiate between various oxidation states of metal ions or between free and bound metal ions. This allows statements to be made about the biological availability and toxicity of heavy metals and makes this technique a valuable tool for environmental analysis. In contrast, spectroscopy requires complicated separation of the metal species before comparable statements can be made.

Detection Limits of 797 VA Computrace for Metal Analysis

The detection Limits of 797 VA Computrace for different metalsare given in following table:

Sl.No.	*Metals*	*Limits*
1	Antimony SbIII/SbV	200 ppt
2	Arsenic AsIII/AsV	100 ppt
3	Bismuth Bi	500 ppt
4	Cadmium Cd	50 ppt
5	Chromium CrIII/CrVI	25 ppt
6	Cobalt Co	50 ppt
7	Copper Cu	50 ppt
8	Iron FeII/FeIII	50 ppt
9	Lead Pb	50 ppt
10	Mercury Hg	100 ppt
11	Molybdenum Mo	50 ppt
12	Nickel Ni	50 ppt
13	Platinum Pt	0.1 ppt
14	Rhodium Rh	0.1 ppt
15	Selenium SeIV/SeVI	300 ppt
16	Thallium Tl	50 ppt
17	Tungsten W	200 ppt
18	Uranium U	25 ppt
19	Zinc Zn	50 ppt
20	Note- 1 ppt = 1 ng/kg	

Technical Specifications

797 VA Computrace Voltammetric measuring stand with built-in potentiostat and galvanostat. It has following characteristics:

Sl.No.	*Characteristics*	*Values*
1	Working electrodes	Multi-Mode Electrode pro (MME pro) 6.1246. 120, pneumatically operated as DME, HMDE or SMDE; requires mercury 99.9999 per cent and nitrogen 99.995 per cent, whose pressure must be reduced with a reduction valve to 1.0 ± 0.2 bar (100 ± 20 kPa). Rotating disk electrodes (RDE) with exchangeable electrode tips made from various materials- Ultra Trace graphite, glassy carbon, gold, silver, platinum.
2	Reference electrode	6.0728.02 or 6.0728.03 Ag/AgCl/KCl 3 mol/Lwith 6.1245.010 electrolyte vessel.
3	Auxiliary electrodes	6.0343.000 Platinum auxiliary electrode with plastic shaft 6.1241.020+6.1247.000 glassy-carbon auxiliary electrode (option).
4	Stirrer	Speed 200 to 3'000 min–1,Speed stability ± 5 per cent, Material PET.
5	Measuring cell	Working volume 10-70 mL, 50-150 mL (option: 5 mL-70mL). Measuring cell with thermostat jacket as option.
6	Potentiostat/galvanostat	Voltage range ± 5 V,Output voltage ± 12 V Current range ± 80 mA, Current measurement 7 ranges (10 nA to 10 mA),Sweep rate (CV) < 1 m V/s-3 V/s (at 1 mV resolution) < 1 mV/s-35 V/s (at 10 mV resolution).
7	Power supply	Voltage 100.240 V, Frequency 50-60 Hz,Power consumption-120 VA.
8	PC requirements	For perfect functioning we recommend a Pentium III processor with a clock rate of 1 GHz or higher.Operating system-Microsoft Windows TM 2000, XP Professional, Vista, or Windows 7 (32 bit versions only) with the following configuration- RAM -256 MB, Program files- 30 MB, Hard disk- Min. 200 MB free,Graphics adapter//screen- Min. resolution 1024 x768 pixel or higher, Printer -Any supported by Windows TM, Connection -1 free USB interface.

This instrument is being used in analysis of followings:

Food and Drinks

Ascorbic acid (Vitamin C) in foodstuffs and pharmaceuticals, Release of lead and cadmium from crockery and glassware, Cadmium, lead and copper in foodstuffs after digestion, Quinine in drinks and tablets, Cystine and Cysteine simultaneously in biological samples, Nicotinamide in fruit juices and vitamin preparations, Tocopherols (Vitamin E) in edible fats and oils

Pharmaceuticals, Biology

Chromium in water samples and biological materials, Cinchocaine in pharmaceuticals, Diazepam in body fluids, Folic acid (Vitamin B9, Vitamin Bc) in mono vitamin tablets, Nitrate in soil, plants, vegetable juices, meat, sausages,

Pyridoxine (Vitamin B6) in vitamin preparations, Riboflavin (Vitamin B2) in vitamin preparations, Selenium in various matrices, Thiamine (Vitamin B1) in vitamin preparations,

Source

1. *Technical Brochure of 797 VA Computrace (Metrohm).*

Polarography

Polarography is a special case of voltammetry referring to the current-voltage measurement acquired using dropping mercury electrode with a constant flow of mercury drop. Polarography is a subclass of voltammetry where the **working electrode is a dropping mercury electrode (DME), Hanging mercury drop electrode (HMDE) or a static mercury drop electrode (SMDE), which are useful for their wide cathodic ranges and renewable surfaces and auxillary electrode is a pool of mercury at the base of the cell.** The using of mercury drop as electrode provides us uniform area and fresh uncontaminated electrode surface for use over a range of applied potential.Czech chemist called Jaroslav-Heyrovsky first introduced Polarography in 1922, for which he received the 1959 Nobel Prize in chemistry [1-6].

Polarography is an voltammetric measurement whose response is determined by combined diffusion/convection mass transport. Polarography is a specific type of measurement that falls into the general category of linear-sweep voltammetry where the electrode potential is altered in a linear fashion from the initial potential to the final potential. As a linear sweep method controlled by convection/diffusion mass transport, the current vs. potential response of a polarographic experiment has the typical sigmoidal shape. What makes polarography different from other linear sweep voltammetry measurements is that polarography makes use of the dropping mercury electrode (DME). Close-up of the DME.

Theory of Operation [16]

Polarography is a voltammetric measurement whose response is determined by combined diffusion/convection mass transport. The simple principle of polarography is the study of solutions or of electrode processes by means of electrolysis with two electrodes, one polarizable and one unpolarizable, the former formed by mercury regularly dropping from a capillary tube. Polarography is a specific type of measurement that falls into the general category of linear-sweep voltammetry where the electrode potential is altered in a linear fashion from the initial potential to the final potential. As a linear sweep method controlled by convection/diffusion mass transport, the current vs. potential response of a polarographic experiment has the typical sigmoidal shape. What makes polarography different from other linear sweep voltammetry measurements is that polarography makes use of the dropping mercury electrode (DME) or the static mercury drop electrode.

A plot of the current vs. potential in a polarography experiment shows the current oscillations corresponding to the drops of Hg falling from the capillary. If one connected the maximum current of each drop, a sigmoidal shape would result. The limiting current (the plateau on the sigmoid), called the diffusion current

Heyrovský's Polarograph [16].

Heyrovský's Polarograph and DME [16].

because diffusion is the principal contribution to the flux of electroactive material at this point of the Hg drop life.

Limitations [16]

There are various limitations in particular for the classical polarography experiment for quantitative analytical measurements. Because the current is continuously measured during the growth of the Hg drop, there is a substantial contribution from capacitive current. As the Hg flows from the capillary end, there is initially a large increase in the surface area. As a consequence, the initial current is dominated by capacitive effects as charging of the rapidly increasing interface occurs. Toward the end of the drop life, there is little change in the surface area which diminishes the contribution of capacitance changes to the total current. At the same time, any redox process which occurs will result in faradaic current that decays approximately as the square root of time (due to the increasing dimensions of the Nernst diffusion layer). The exponential decay of the capacitive current is much more rapid than the decay of the faradaic current; hence, the faradaic current is proportionally larger at the end of the drop life. Unfortunately, this process is complicated by the continuously changing potential that is applied to the working electrode (the Hg drop) throughout the experiment. Because the potential is changing during the drop lifetime (assuming typical experimental parameters of a 2 mV/s scan rate and a 4 s drop time, the potential can change by 8 mV from the beginning to the end of the drop), the charging of the interface (capacitive current) has a continuous contribution to the total current, even at the end of the drop when the surface area is not rapidly changing. As such, the typical signal to noise of a polarographic experiment allows detection limits of only approximately 10^{-5} or 10^{-6} M.

Improvements [16]

The better discrimination against the capacitive current can be obtained using the tast and pulse polarographic techniques. These have been developed with introduction of analog and digital electronic potentiostats. A first major improvement is obtained, if the current is only measured at the end of each drop lifetime (tast polarography). An even greater enhancement has been the introduction of differential pulse polarography. Here, the current is measured before the beginning and before the end of short potential pulses. The latter are superimposed to the linear potential-time-function of the voltammetric scan. Typical amplitudes of these pulses range between 10 and 50 mV, whereas pulse duration is 20 to 50 ms. The difference between both current values is that taken as the analytical signal. This technique results in a 100 to 1000-fold improvement of the detection limit, because the capacitive component is effectively suppressed.

Qualitative Information [16]

It can also be determined from the half-wave potential of the polarogram (the current vs. potential plot in a polarographic experiment). The value of the half-wave potential is related to the standard potential for the redox reaction being studied.

This technique and especially the differential pulse anodic stripping voltammetry (DPASV) method can be used for environmental analysis, and especially for marine study for characterisation of organic matter and metals interactions [7].

Quantitative Information [16]

A plot of the current vs. potential in a polarography experiment shows the current oscillations corresponding to the drops of Hg falling from the capillary. If one connected the maximum current of each drop, a sigmoidal shape would result. The limiting current (the plateau on the sigmoid), called the diffusion current because diffusion is the principal contribution to the flux of electroactive material at this point of the Hg drop life, is related to analyte concentration by the Ilkovic equation.

The Ilkovic equation is a relation used in polarography relating the diffusion current (I_d) and the concentration of the depolarizer (c), which is the substance reduced or oxidized at the dropping mercury electrode. The **Ilkovic equation** has the form:

$$id = KnFD^{1/2}\ m^{2/3}\ t^{1/6}C$$

Where k is a constant which includes π and the density of mercury, and with the Faraday constant (F) has been evaluated at 708 for max current and 607 for average current, D is the diffusion coefficient of the depolarizer in the medium (cm^2/s), n is the number of electrons exchanged in the electrode reaction, m is the mass flow rate of Hg through the capillary (mg/sec), t is the drop lifetime in seconds, and c is depolarizer concentration in mol/cm3.

The equation is named after the scientist who derived it, the Slovak chemist, Dionýz Ilkoviè (1907-1980).

Instrumentatation [8]

The main parts of a polarographic analyser are as follows:

(i) **The polarographic cell**-The polarographic cell is a glass or teflon vessel in which are inserted a tube for the bubbling of nitrogen and three electrodes:

The working electrode -A capillary connected with a mercury container or a solid electrode, **The reference electrode** -A simply Ag/AgCl, saturated KCl electrode. **The auxiliary electrode** -A platinum wire inserted on a teflon rod. The stirring of the sample solution is allowed by a magnetic stirrer and a magnetic rod.

(ii) **The potentiostat**-The potentiostat allow to impose the desired potential scanning between working and reference electrodes while the current flowing in the circuit is recorded between working and auxiliary electrodes.

(iii) **A PC with the application software**-The PC is interfaced to measuring system and the software has the following functions:

(a) Send to the potentiostatic analyser the scanning parameters.

(b) Check if the execution of the scanning is correct or not.

(c) Handle the output data (current and potential).

Detection Limits for different Polarographic Techniques

Electrochemical detection limits for several polarographic techniques are given in following table [8]:

Sl.No.	*Techniques*	*Lower Detection Limits (mol dm^{-3})*
1	Classical polarography	5×10^{-3}
2	Sampled DC polarography	1×10^{-5}
3	Normal pulse polarography	10^{-7} -10^{-8}
4	Differential pulse polarography	10^{-8} –5×10^{-8}
5	Square-wave polarography	1×10^{-8}
6	Anodic stripping voltammetry	10^{-10} -10^{-11}

Source

1. *Reinmuth, W. H. (1961-11-01). "Theory of Stationary Electrode Polarography". Analytical Chemistry. 33 (12): 1793-1794.*

2. *Nicholson, R.S.; Irving. Shain (1964-04-01). "Theory of Stationary Electrode Polarography. Single Scan and Cyclic Methods Applied to Reversible, Irreversible, and Kinetic Systems.". Analytical Chemistry. 36 (4): 706-723.*

3. *Skoog, Douglas A.; Donald M. West; F. James Holler (1995-08-25). Fundamentals of Analytical Chemistry (7th ed.). Harcourt Brace College Publishers.*
4. *Kissinger, Peter; William R. Heineman (1996-01-23). Laboratory Techniques in Electro-analytical Chemistry, Second Edition, Revised and Expanded (2 ed.). CRC.*
5. *Bard, Allen J.; Larry R.Faulkner (2000-12-18).Electro-chemical Methods: Fundamentals and Applications (2 ed.).Wiley.*
6. *Zoski, Cynthia G. (2007-02-07). Handbook of Electro chemistry. Elsevier Science.*
7. *Louis, Yoann; Cédric Garnier; Véronique Lenoble; Dario Omanoviæ; Stéphane Mounier; Ivanka Pižeta (2009)."Characterisation and modelling of marine dissolved organic matter interactions with major.*
8. *AMEL Electro-chemistry, Introduction to Modern Voltammetric and Polarographic Analysis Techniques IV Edition.*
9. *Wind trace cations". Marine Environmental Research. 67 (2): 100–107*
10. *Quantitative chemical analysis, Daniel C.Harris, W.H. Freeman and Co.*
11. *Fundamentals of electroanalytical chemistry, Paul M.S. Monk (Wiley).*
12. *Analytical Electrochemistry, Joseph Wang.*
13. *Electrochemical Methods, Fundamentals and Applications, Allen J. Bard and Larry R. Falkner (Wiley).*
14. *Modern Techniques in Electroanalysis, Peter Vangsek.*
15. *Electroanalysis: Theory and Application in Aqueous and Non-Aqueous Media and in Automated Chemical Control (Techniques and Instrumentation in Analytical) E. A.M.F. Dahmen.*
16. *Wikipedia.*

Applications of Polarography [1]

Most metal ions are reducible at DME, and multicomponent mixtures can frequent be analysed through selecting an appropriate supporting electrolyte so in which the half-wave potentials of two ions are differed through about -0.2 V vs SCE or by using complexing agents by taking the advantage of complexing ability of the metal ions.

Based on this, polarography is used predominantly for trace metal analysis of alloys, ultra-pure metals, minerals/metallurgy, environmental analysis (air, water, and soil and sea water contaminants), foodstuffs, beverages and body-fluids, toxicology and clinical analysis. Reducible anions such as BrO^{-}, IO^{3-}, and $Cr2O2^{-}$ and NO^{2-} could also be determined using well buffered solutions.

It is also meaningful in the analysis of biological systems to determine vitamins, hormones, alkaloids, terpenoid substances and natural colouring substance, analysis of drugs and pharmaceutical preparations, determination of pesticide or herbicide residues in foods, in the structure determination of several organic compounds *etc.*

Different Samples and Element or Compound Determined

Sl.No.	Type of Sample	Parts or Compound Determined
A. Food and allied products		
1	Milk, pharmaceuticals	Riboflavin
2	Fruit and vegetables	Vitamin C
3	Sea water, gases	Oxygen
4	High-purity salts	Transition metals
5	Foodstuffs	Cu, Pb, Sn, Zn
6	Beer and soft drinks	Sn, Pb
B. Other Products		
1	Petroleum fractions	Free sulphur
2	Fuels	Antioxidants
3	Steels	Cu, Pb, Ni, Co

Source

1. *Nathanian, Application of Polarography. Amperometric Titration and Voltammetry, Unit, 9.*

Difference between Voltammetry and Polarographic Technique

Voltammetry is a category of electroanalytical methods used in analytical chemistry and various industrial processes. In voltammetry, information about an analyte is obtained by measuring the current as the potential is varied. Voltammetry, and especially polarography, were advanced by Nobel Prize winner Jaroslav Heyrovsky. It is well described under polarography.

Section C

Instrumentation in Food Safety and Microbiological Analysis

Chapter 6

Flow Cytometry

It is a laser-based, bio-physical technology employed in cell counting, cell sorting, biomarker detection and protein engineering, by suspending cells in a stream of fluid and passing them by an electronic detection apparatus. It allows simultaneous multi parametric analysis of the physical and chemical characteristics of up to thousands of particles per second [1].

Flow cytometry (FCM) is a quantitative technology for the rapid individual analysis of large numbers of cells in mixture, using light scattering and fluorescence measurements. It is defined as the simultaneous measurement of multiple physical characteristics of a single cell as the cell flows in suspension through a measuring device. It is a subset of the field of cytometry based on the use of specialized instruments [2].

The properties measured by FCM, include the followings (2a):

(i) A particle's relative size (represented by forward angle light scatter),

(ii) Relative granularity or internal complexity (represented by right-angle scatter),

(iii) Relative fluorescence intensity.

These characteristics are determined using an optical-to-electronic coupling system which records how the cell or particle scatters incident laser light and emits fluorescence [2a]. Flow cytometry is a method of studying cells and chromosomes. Thousands of these microscopic particles can be analyzed every second. This is done with detection apparatus while the cells are held in fluid.Flow cytometry, a technique that can help characterize microscopic particles such as cells and chromosomes, enables concurrent analysis of the physical and chemical characteristics of thousands of particles per second, and on various different scientific parameters. Fluorescence-activated cell sorting, which is a specialized form of flow cytometry, can also individually sort a heterogeneous mixture of biological cells into two or more streams, based on their fluorescent light scattering properties [2b]. Flow Cytometry is process for quantifying cells, measures different property of cells, able to categorize and quantify and even able to separate out subpopulations of cells [2c]. **Flow cytometry has been described as automated microscopy** that has the advantages of automation, objectivity and speed (flow cytometers can analyze many thousands of cells per second). To do this flow cytometers quantitatively measure the optical characteristics of cells (or other particles) as they are presented in single file in front of a focused light beam [2a].

Flow cytometry is routinely used in the diagnosis of health disorders, especially blood cancers, but has many other applications in basic research, clinical practice and clinical trials. A common variation is to physically sort particles based on their properties, so as to purify populations of interest.

History [2]

The details related to development of FCM are given below:

(*i*) The first impedance-based flow cytometry device, using the Coulter principle, was disclosed in U.S. Patent issued in 1953 to Wallace H. Coulter.

(*ii*) Mack Fulwyler was the inventor of the fore runner to today's flow cytometers - particularly the cell sorter. Fulwyler developed this in 1965 with his publication in Science [3].

(*iii*) The first fluorescence-based flow cytometry device (ICP11) was developed in 1968 by Wolfgang Göhde from the University of Münster, filed for patent on 18 December 1968 [4] and first commercialized in 1968/69 by German developer and manufacturer Partec through Phywe AG in Göttingen. At that time, absorption methods were still widely favored by other scientists over fluorescence methods [5].

(*iv*) Soon after, flow cytometry instruments were developed, including the Cytofluorograph (1971) from Bio/Physics Systems Inc. (later: Ortho Diagnostics), the PAS 8000 (1973) from Partec, the first FACS (Fluorescence-activated cell sorting) instrument from Becton Dickinson (1974), the ICP 22 (1975) from Partec/Phywe and the Epics from coulter (1977/78).

Name of the Technology [2]

(*i*) The original name of the fluorescence-based flow cytometry technology was "pulse cytophotometry", based on the first patent application on fluorescence-based flow cytometry.

(i*i*) At the 5th American Engineering Foundation Conference on Automated Cytology in Pensacola (Florida) in 1976 - eight years after the introduction of the first fluorescence-based flow cytometer (1968) - it was agreed to commonly use the name "flow cytometry" (FCM), a term that quickly became popular [6]. The technology that has greatly evolved together with the development of fluorescent dyes is Flow cytometry (FCM) [2].

Source

1. *Antony C.Bakke, PhD, Department of Pathology, Oregon Health Sciences University, Portland, The Principles of Flow Cytometry, CE update.*
2. *Wikipedia.*

2a. *Raghu H V., Naresh Kumar and Pradip Kumar Sharma, Flow Cytometry Based Rapid Method for Enumeration of Hygiene Indicator Organisms in Milk, Microbial Biosensors, Food Safety and Quality Assurance Lab, Compendium on Advance Techniques and Novel Approaches for Quality and Safety Evaluation of Dairy Foods, NDRI-2015.*

2b. *Disadvantages of Flowcytometers, Creative pen, Wednesday, November 27, 2013.*

2c. *Dr. Jeffrey Lyczak, http://courses.dce.harvard. edu/-bios 169c/resource main.html.*

3. *Fulwyler, M.J. (1965)."Electronic separation of biological cells by volume".Science 150 (698): 910–911.*
4. *Wolfgang Dittrich and Wolfgang Göhde, "Flow-through Chamber for Photometers to Measure and Count Particles in a Dispersion Medium".*
5. *Kamentsky in Proceedings of the 1968 Conference „Cytology Automation" (1970), edited by D. M. D. Evans.*
6. *Sack et al., Ulrich. Zelluläre Diagnostik. Karger Publishers (2006).*

Advantages, Disadvantages and Limitation of FCM

1. Advantages

(i) It allows characterizing individual cells within a cell population, and to discriminate between various VBNC and physiological states [1-8], which is a major advantage over culture-based methods.

(ii) Now-a-days, FC appears as the most promising labeling technology [9,10] and being used successfully in food microbiology.

(iii) A large choice of dyes is available to target cell components (nucleic acids, proteins, lipids), intracellular pH, membrane integrity, intracellular ions, viability markers, and membrane energization [1,3.5,7,11,12,14-16,17].

(iv) FC has also been proposed as a means to enumerate viable probiotic populations in commercial products [18].

(v) Correlations between FC enumerations and plate counts are often very good during exponential growth and when cells are not submitted to stress [7,19,]. In other situations FC counts generally outnumber the plate counts by 0.3-1 log of CFU/g.

(vi) The shorter analysis time and possible automation are among the other advantages of FC over traditional culture methods [10,15,18].

(vii) In addition, the quantification - and sometimes detection - limit is relatively high, for instance 10^3-10^4 cells/ml in milk [11,20].

(viii) If flow cytometry is used to study heterogeneous populations of cells, then it will analyze the **subpopulations** in a few minutes. Not only is it much faster than other options, the data it produces are also detailed [13].

(ix) During study of uniform cells populations, FC has the benefit of always highlighting any non-uniformity. It also takes off any debris or dead cells when providing the final data [13].

(x) The main advantage is the ability to characterize antigen expression (*i.e.* immuno phenotyping) on a cell-by-cell basis on large populations of cells. This is important in the diagnosis of hematopoietic disorders [13a].

2. Disadvantage [13,13a]

(i) In studying a uniform population of cells, the desired data will be the average receptor densities.Flow cytometry can handle this job easily, but is more expensive than alternatives such as radio immunoassay and enzyme-linked immunosorbent assay. The alternative methods can do the job just as quickly and can even produce more samples per day.

(ii) **Low cell throughput rate-**One major disadvantage with the flow cytometer is its low cell throughput rate.Even for high-speed sorters, this is still less than a few thousand cells per second [2a]. A high-speed sorter can yield up to 106 cells per hour when dealing with a subpopulation that makes up 20 per cent of the whole population. This rate is too low for many experiments.Flow cytometry sorters are very accurate and purify small or complex subpopulations [13]. Many experiments however require very large number of cells. This implies that even high-speed sorters need to run for long durations, which is not only an expensive proposition but may also pose quality issues because the cells sorted from such long runs may no longer be usable in scientific experiments. This problem may be further aggravated when the sorted cells need to be sterile. Although high speed flow cytometers can give sterile cells, but this makes the operation complex and further reduces the throughput [2a].

(iii) Requires live cells [13a].

(iv) Does not work well for cells that tend to stick together (*e.g.* carcinoma or sarcomas) [13a].

3. Limitation

There are however some severe limitations in using this scientific technique. Some are given below:

(i) Since the resulting data from flow cytometric analysis is at an aggregate level, it is therefore not easy to observe and measure individual cell behavior. Also, as flow cytometry technique requires the passing of these cells through a fluid stream, it restricts the analysis to only cell suspension solutions. For the same reason, solid tissue cells also have to be disaggregated, by treating the intact tissues with an enzyme, so as to release individual cells for further analysis [2a].

(*ii*) **Require highly trained operators** -Since flow cytometer has very sophisticated instrumentation, only skilled and highly trained operators can run it and get any acceptable levels of performance from such an apparatus [2a]. Protocol development and data analyses also require some experience [10].

(*iii*) Indeed, depending on their composition some samples need additional preparation, such as a reduction of protein amounts [16,20].

(*iv*) False-positives or data dispersion may be observed with some dyes [11]. In spite of the few aforesaid limitations, FC appears as a very promising tool for the food industry [9,10,15,].

(*v*) **Expensive**-Flow cytometers are expensive instruments to purchase and maintain. Flow cytometers with the additional sorting capability can cost almost double of their cheaper versions. Additionally, operating a high speed sort is another recurring expense that typically costs hundreds of dollars for each run [2a].

Source

1. *Sheehan A., O'Loughlin C., O'Cuinn G., Fitzgerald R. J., Wilkinson M. G. (2005). Cheddar cheese cooking temperature induces differential lactococcal cell permeabilization and autolytic responses as detected by flow cytometry: implications for intracellular enzyme accessibility. J. Appl. Microbiol. 99 1007-1018.*

2. *Lahtinen S.J., Ouwehand A.C., Reinikainen J.P., Korpela J. M., Sandholm J., Salminen S.J. (2006). Intrinsic properties of so-called dormant probiotic bacteria, determined by flow cytometric viability assays. Appl.Environ.Microbiol. 72 5132-5134.*

2a. *Disadvantages of Flowcytometers, Creative pen, Wednesday, November 27, 2013.*

3. *Papadimitriou K., Pratsinis H., Nebe-Von-Caron G., Kletsas D., Tsakalidou E. (2006). Rapid assessment of the physiological status of Strepto-coccus macedonicus by flow cytometry and fluorescence probes. Int. J.Food Microbiol. 111, 197–205.*

4. *Papadimitriou K., Pratsinis H., Nebe-Von-Caron G., Kletsas D., Tsakalidou E. (2007). Acid tolerance of Streptococcus macedonicus as assessed by flow cytometry and single-cell sorting. Appl. Environ. Microbiol. 73, 465-476.*

5. *Quiros C., Herrero M., Garcia L. A., Diaz M. (2007). Application of flow cytometry to segregated kinetic modeling based on the physiological states of microorganisms. Appl.Environ. Microbiol. 73, 3993–4000.*

6. *Sunny-Roberts E. O., Knorr D. (2008). Evaluation of the response of Lacto-bacillus rhamnosus VTT E-97800 to sucrose-induced osmotic stress. Food Microbiol. 25 , 183–189.*

7. *El Arbi A., Ghorbal S., Delacroix-Buchet A., Bouix M. (2011). Assessment of the dynamics of the physiological states of Lactococcus lactis sp. cremoris SK11 during growth by flow cytometry. J. Appl. Microbiol. 111, 1205-1211.*

8. *Zotta T., Guidone A., Tremonte P., Parente E., Ricciardi A. (2012). A comparison of fluorescent stains for the assessment of viability and metabolic activity of lactic acid bacteria. World J. Microbiol. Biotechnol. 28, 919–927.*

9. *Tracy B.P., Gaida S.M., Papoutsakis E. T. (2010). Flow cytometry for bacteria: enabling metabolic engineering, synthetic biology and the elucidation of complex phenotypes. Curr. Opin. Biotechnol. 21, 85-99.*

10. *Davey H. M. (2011). Life, death, and in-between: meanings and methods in microbiology. Appl.Environ. Microbiol. 77, 5571-5576.*

11. *Maecker H. T., Frey T., Nomura L. E., Trotter J. (2004). Selecting fluorochrome conjugates for maximum sensitivity. Cytometry A 62, 169-173.*

12. *Berney M., Hammes F., Bosshard F., Weilenmann H. U., Egli T. (2007). Assessment and interpretation of bacterial viability by using the Live/Dead BacLight Kit in combination with flow cytometry. Appl. Environ. Microbiol. 73, 3283–3290.*

13. *Martin Cole; Advantages and Disadvantages of Flow Cytometry, Updated April 25, 2017.*

13a. *Nathan Cuka, Hematopathology fellow and slide wrangler.*

14. *Comas-Riu J., Rius N. (2009). Flow cytometry applications in the food industry. J. Ind. Microbiol. Biotechnol. 36, 999-1011.*

15. *Díaz M., Herrero M., García L. A., Quirós C. (2010). Application of flow cytometry to industrial microbial bioprocesses. Biochem. Eng. J. 48, 385-407.*

16. *Doherty S.B., Wang L., Ross R.P., Stanton C., Fitzgerald G.F., Brodkorb A. (2010). Use of viability staining in combination with flow cytometry for rapid viability assessment of Lactobacillus rhamnosus GG in complex protein matrices. J. Microbiol. Methods 82, 301-310.*

17. *Zotta T., Guidone A., Tremonte P., Parente E., Ricciardi A. (2012). A comparison of fluorescent stains for the assessment of viability and metabolic activity of lactic acid bacteria. World J.Microbiol. Biotechnol. 28, 919-927.*

18. *Maukonen J., Alakomi H. L., Nohynek L., Hallamaa K., Leppämäki S., Mättö J., et al. (2006). Suitability of the fluorescent techniques for the enumeration of probiotic bacteria in commercial non-dairy drinks and in pharmaceutical products. Food Res. Int. 39, 22–32.*

19. *Bunthof C.J., Abee T. (2002). Development of a flow cytometric method to analyze subpopulations of bacteria in probiotic products and dairy starters. Appl. Environ. Microbiol. 68, 2934-2942.*

20. *Gunasekera T. S., Attfield P. V., Veal D. A. (2000). A flow cytometry method for rapid detection and enumeration of total bacteria in milk. Appl. Environ. Microbiol. 66, 1228-1232.*

Working Principle of Flow Cytometer [1]

Flow cytometry uses the principles of light scattering, light excitation and the emission of fluorochrome molecules to generate specific multi-parameter data from particles and cells within the size range of 0.5um to 40 um diameter. The Working Principle can be easily understood by the Figure given on next page.

A source of monochromatic light, usually a laser diode (A) sends a beam that is collimated (L1) and focused (L2) by lenses on the laminar flow chamber (C), after

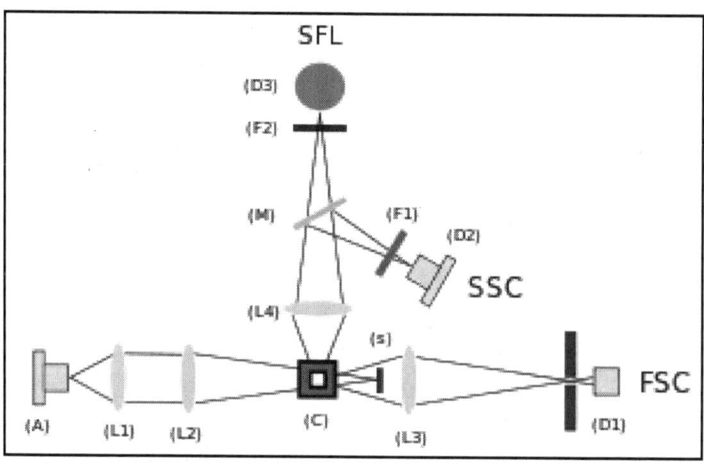

Diagram of the Optics of a Flow Cytometer.

passing through the camera, the direct light beam is arrested by a screen (s), while the scattered light is focused by another lens (L3) on a photodiode (D1), this constitutes the Forward SCatter detector FSC. The side scattered light and the fluorescence are focused by a lens onto a dichroic mirror (M) that reflects most of the light of wavelength equal to that produced by the source (A), goes through a filter (F1) and impinges on a photodiode detector. This constitutes the Side SCatter detector (SSC). The light that is not reflected by the dichroic mirror and which has a wavelength different from that emitted and impinges on it goes through an interference filter (F2) which can be adjusted to a specific wavelength, thus discriminating between different fluorophores, to ultimately impinge on a photomultiplier tube (D3), this is the Side FLuorescence detector (SFL).

A beam of light (usually laser light) of a single wavelength is directed onto a hydrodynamically focused stream of liquid. A number of detectors are aimed at the point where the stream passes through the light beam: one in line with the light beam (Forward Scatter or FSC) and several perpendicular to it (Side Scatter or SSC) and one or more fluorescence detectors. Each suspended particle from 0.2 to 150 micrometers passing through the beam scatters the ray, and fluorescent chemicals found in the particle or attached to the particle may be excited into emitting light at a longer wavelength than the light source. This combination of scattered and fluorescent light is picked up by the detectors, and, by analysing fluctuations in brightness at each detector (one for eachfluorescent emission peak), it is then possible to derive various types of information about the physical and chemical structure of each individual particle.

Light scattering occurs when a particle or a cell deflects incident laser light. The extent to which this occurs depends on the physical properties of a cell, namely its size and internal complexity. Factors that affect light scattering are as follows [2a]:

(*i*) The cell membrane.

(ii) The nucleus.

(iii) Any granular material inside the cell.

(iv) Cell Shape and surface topography also contribute to the total light scatter.

FSC correlates with the cell volume and SSC depends on the inner complexity of the particle (*i.e.*, shape of the nucleus, the amount and type of cytoplasmic granules or the membrane roughness). This is because the light is scattered off of the internal components of the cell. Some flow cytometers have eliminated the need for fluorescence and use only light scatter for measurement. Other flow cytometers form images of each cell's fluorescence, scattered light, and transmitted light.

Due to differences between the refractive indices of cells and their surrounding media, light impinging upon the cells is scattered. The forward scatter light (FSC, light scattered at low angles) provides information on cell size, although there is no direct correlation between size and FSC [1b]. Light scattered in an orthogonal or near-orthogonal direction can also be collected by a different detector (a side scatter or SSC detector), which provides information about granularity and cell morphology. Fluorescence produced by intracellular compounds with specific fluorescence [like nicotinamide adenosine dinucleotide phosphate (NADPH)] or specific compounds known as fluorochromes is detected by photomultiplier tubes, allowing certain cell components to be selectively assayed. Several compounds with high quantum efficiency are used in microbiological applications of flow cytometry, where low signals are expected due to the small size of the analysed cells. The scatter and Fluorescence signals detected can be combined in various ways to allow the detection of subpopulations. The use of autofluorescence or fluorescent signals associated with cellular compounds or specific labelling can be used to distinguish cells in media in which nonbiological particles are also present [1c].

The most common type of quantitative analysis using flow cytometry data involves creating a histogram of fluorescence events to count the number of cells with the attached probe. This effectively creates a set of data which gives a ratio of the cells in a population with a particular structural parameter or else with a specific functional property [2a]

Instrumentation-Flow Cytometers [1]

Modern flow cytometers are able to analyze several thousand particles every second, in "real time," and can actively separate and isolate particles having specified properties. A flow cytometer is similar to a microscope, except that, instead of producing an image of the cell, flow cytometry offers "high-throughput" (for a large number of cells) automated quantification of set parameters. To analyze solid tissues, a single-cell suspension must first be prepared.

Flow cytometers may be compared with fluorescence microscopes. With a microscope, the light source is a lamp, the objective lens focuses the image, and the eye is the detector. With a flow cytometer, the light source is a laser, an objective lens focuses the light, and the detectoris a photomultiplier tube.With a fluorescence microscope, the observer views cells and recognizes their different sizes. If a fluorescent probe is bound to the cell, a specific-colored fluorescence isemitted [1a].

Some models of Flow cytometers are given below:

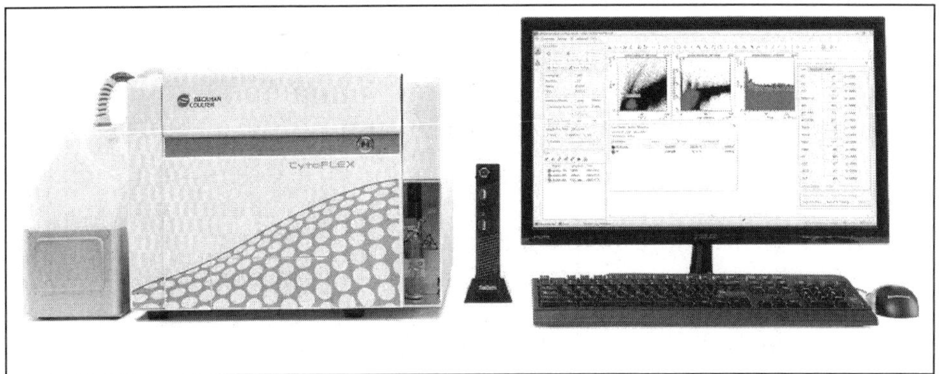

Cyto FLEX Flow Cytometer Backman Coulter Life sciences.

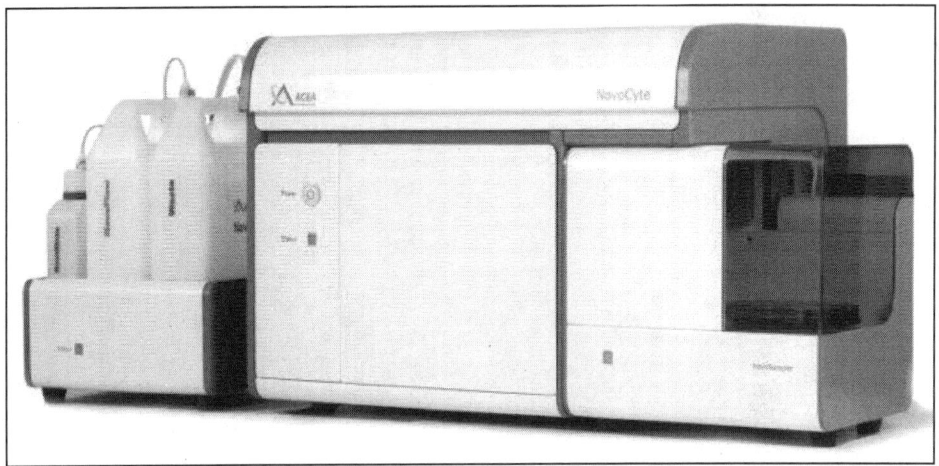

NovoCyte Flow Cytometer, ACEA Biosciences. Inc.

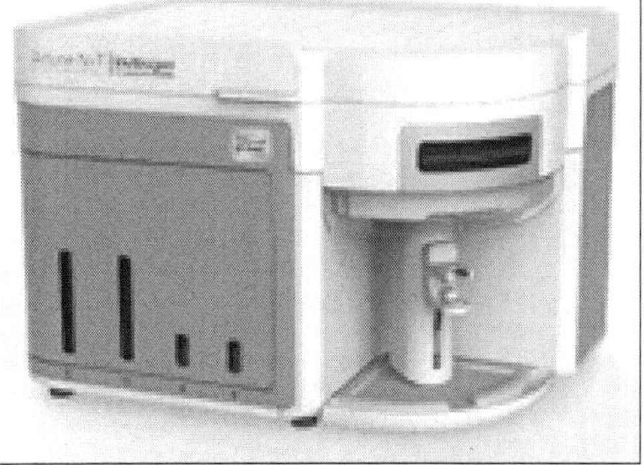

Better Flow Cytometry ThermoFisher. **Attune NxT Flow Cytometer Thermo Fisher.**

Desktop Flow Cytometer-Becton-Dickinson Fluorescence Activated Cell Sorter (FACS).

Computerised Flow Cytometer.

Source

1. *Wikipedia, ACEA Biosciences. Inc., Thermo Fisher Scientific, Backman Coulter Life sciences.*

Components of Flow Cytometer [1]

A common flow cytometer is formed by following basic units [2a,2b]:

(*i*) A Light Source (lasers are most often used).

(ii) A flow cell.

(iii) Hydraulic fluidic system.

(iv) Several optical filters to select specific wavelengths.

(v) A group of photodiodes or photomultiplier tubes to detect the signals of interest

(vi) A data processing unit.

i. A Flow Cell (Fluidics)

liquid stream (sheath fluid), which carries and aligns the cells so that they pass single file through the light beam for sensing [1]. The flow cytometer has 2 sources of fluid passing through it -i- a sheath fluid that is usually a buffered saline and ii-the specimen containing the cells to be analyzed. The specimen forms a core, which is injected into the middle of the sheath fluid, and both pass through the flow chamber. The outer sheath fluid causes hydrodynamic focusing, which compresses the specimen core into a small area. Cells are forced to pass singly through the focalpoint of the laser, which is crucial for collecting accurate information about each cell in the specimen [1a].

ii. A Measuring System

Commonly used are measurement of impedance (or conductivity) and optical systems (Optics) - lamps (mercury, xenon); high-power water-cooled lasers (argon, krypton, dye laser); low-power air-cooled lasers (argon (488 nm), red-HeNe (633 nm), green-HeNe, HeCd (UV)); diode lasers (blue, green, red, violet) resulting in light signals. A laser beam is used in most flowcytometers as the light source to excite a fluorescent tag (*e.g.*, a fluorescent antibody)bound to the cells in the specimen. An argon laser, which produces bluelight, is the most commonly used, although some instruments add a red helium-neon laser, and occasionally a mercury arc lamp is substituted [1a].

iii. A Detector and Analogue-to-Digital Conversion (ADC) System

Which generates FSC and SSC as well as fluorescence signals from light into electrical signals that can be processed by a computer [1]. The voltage signal is an analog signal and cannot be recognized by the computer. It is converted by an analog-to digital converter into a digital signal, called a channel, which can range from 0 to 1,023 channels. The channel is a counter, and all signals of a specific voltage are counted in a channel corresponding to the strength of the signal [1a].

iv. An Amplification System

The amplification of the initial signal can be either **linear or logarithmic**.

(a) **Linear amplification** provides a direct visual relationship and is useful for scatter signals and fluorescent measurements of DNA. Linear amplification allows a dynamic range of approximately 200- to1,000-fold.

(b) **Logarithmic amplification** isused for most other biologic signals, including immunofluorescence, owing to the extreme dynamic range of these signals.

Logarithmic amplification gives a dynamic range of 2,000- to 10,000-fold [1a].

v. A Computer for Analysis of the Signals

The process of collecting data from samples using the flow cytometer is termed 'acquisition'. Acquisition is mediated by a computer physically connected to the flow cytometer, and the software which handles the digital interface with the cytometer. The software is capable of adjusting parameters (*i.e.* voltage, compensation, *etc.*) for the sample being tested, and also assists in displaying initial sample information while acquiring sample data to insure that parameters are set correctly. Early flow cytometers were, in general, experimental devices, but technological advances have enabled widespread applications for use in a variety of both clinical and research purposes. Due to these developments, a considerable market for instrumentation, analysis software, as well as the reagents used in acquisition such as fluorescently labeled antibodies has developed.

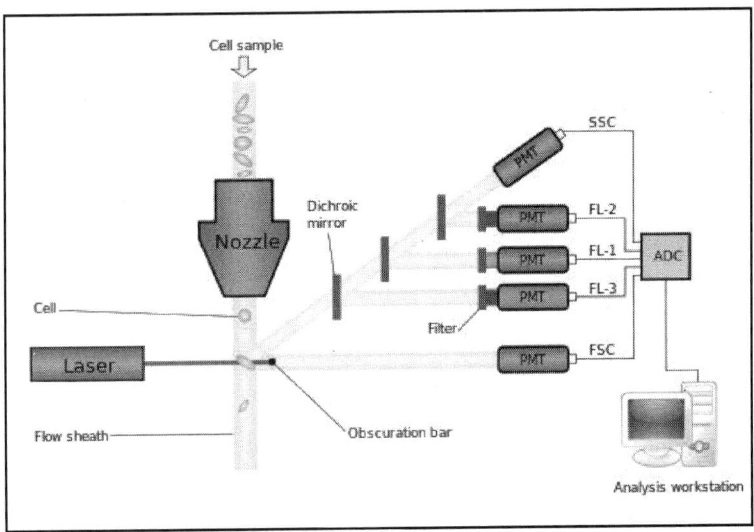

Schematic Diagram of a Flow Cytometer.

The above diagram is showing focusing of the fluid sheath, laser, optics (in simplified form, omitting focusing), photomultiplier tubes (PMTs), analogue-to-digital converter, and analysis work station.

Modern instruments usually have multiple lasers and fluorescence detectors. The current record for a commercial instrument is five lasers [2] and 18 fluorescence detectors. Increasing the number of lasers and detectors allows for multiple antibody labeling, and can more precisely identify a target population by their phenotypic markers. Certain instruments can even take digital images of individual cells, allowing for the analysis of fluorescent signal location within or on the surface of cells.

Data Analysis

The following techniques are being used to analyses the data obtained during measurement:

1. Traditional Gating Tehnique

The data generated by flow-cytometers can be plotted in a single dimension, to produce a histogram, or in two-dimensional dot plots or even in three dimensions. The regions on these plots can be sequentially separated, based on fluorescence intensity, by creating a series of subset extractions, termed "**gates.**"

The plots are often made on logarithmic scales. Because different fluorescent dyes' emission spectra overlap [3], signals at the detectors have to be compensated electronically as well as computationally.Data accumulated using the flow cytometer can be analyzed using software, *e.g.*, WinMDI [4], Flowing Software [5] and web-based Cytobank [6], FCS Express, Flowjo, FACSDiva, CytoPaint (akaPaint-A-Gate) [7], VenturiOne, CellQuestPro, or Cytospec [8]. Once the data are collected, there is no need to stay connected to the flow cytometer. For this reason, analysis is most often performed on a separate computer. This is especially necessary in core facilities where usage of these machines is in high demand.

2. Computational Analysis

The automated population identification using computational methods has offered an alternative to traditional gating strategies. The rare and hidden populations could be determined with the potentially help of automated identification systems. Some of the representative automated methods are:

(i) FLOCK [9] in Immunology Database and Analysis Portal (ImmPort) [10].

(ii) FLAME [11] in GenePattern and flowClust [12,13,14].

(iii) in Bioconductor. Collaborative efforts have resulted in an open project called Flow Cytometry: Critical Assessment of Population Identification Methods (FlowCAP) [15], to provide an objective way to compare and evaluate the flow cytometry data clustering methods, and also to establish guidance about appropriate use and application of these methods.

3. Fluorescence-Activated Cell Sorting (FACS)

Fluorescence-activated cell sorting (FACS) is a specialized type of flow cytometry. It provides a method for sorting a heterogeneous mixture of biological cells into two or more containers, one cell at a time, based upon the specific light scattering and fluorescent characteristics of each cell. It is a useful scientific instrument, as it provides fast, objective and quantitative recording of fluorescent signals from individual cells as well as physical separation of cells of particular interest. These powerful and rapid cell-sorting techniques could reduce the time needed to determine probiotic strain abundance, size, and metabolic activity. Fluorescent DNA stains, nucleic acid probes, and immuno- fluorescence probes directed at cell proteins, extend the capabilities of the technique, enabling cells to be discriminated based on amount and type of nucleic acids, amount of respiratory enzymes, or membrane integrity [15a]. The acronym FACS is trademarked and owned by Becton, Dickinson and Company [16].

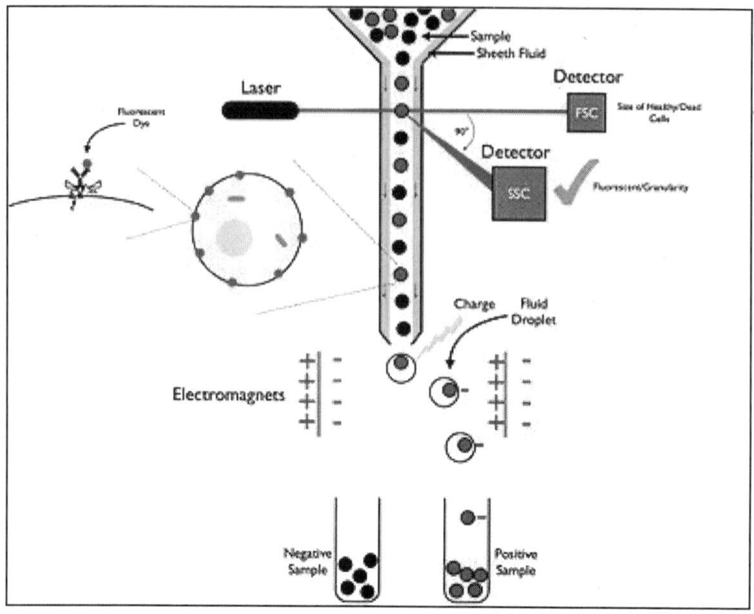

Fluorescence Assisted Cell Sorting (FACS).

Among the large majority of researchers who use this technology for sorting or analysis, this term has become generic in common usage, much like xerox or kleenex. The first cell sorter was invented by Mack Fulwyler in 1965, using the Coulter principle, a relatively difficult technique and one no longer used in modern instruments. The technique was expanded by Len Herzenberg, who was responsible for coining the term FACS [17]. Herzenberg won the Kyoto Prize in 2006 for his seminal work in flow cytometry.

The cell suspension is entrained in the center of a narrow, rapidly flowing stream of liquid. The flow is arranged so that there is a large separation between cells relative to their diameter. A vibrating mechanism causes the stream of cells to break into individual droplets. The system is adjusted so that there is a low probability of more than one cell per droplet. Just before the stream breaks into droplets, the flow passes through a fluorescence measuring station where the fluorescent character of interest of each cell is measured. An electrical charging ring is placed just at the point where the stream breaks into droplets. A charge is placed on the ring based on the immediately prior fluorescence intensity measurement, and the opposite charge is trapped on the droplet as it breaks from the stream. The charged droplets then fall through an electrostatic deflection system that diverts droplets into containers based upon their charge. In some systems, the charge is applied directly to the stream, and the droplet breaking off retains charge of the same sign as the stream. The stream is then returned to neutral after the droplet breaks off.

Labels [1]

The following types of labels are used in determination/identification of cells by Flowcytometer:

1. Traditional Fluorescent Labels

A wide range of fluorophores can be used as labels in flow cytometry. Fluorophores, or simply "fluors", are typically attached to an antibody that recognises a target feature on or in the cell; they may also be attached to a chemical entity with affinity for the cell membrane or another cellular structure.Each fluorophore has a characteristic peak excitation and emission wavelength and the emission spectra often overlap. Consequently, the combination of labels which can be used depends on the wavelength of the lamp (s) or laser (s) used to excite the fluorochromes and on the detectors available [18]. The maximum number of distinguishable fluorescent labels is thought to be 17 or 18, and this level of complexity necessitates laborious optimization to limit artifacts, as well as complex deconvolution algorithms to separate overlapping spectra [19].

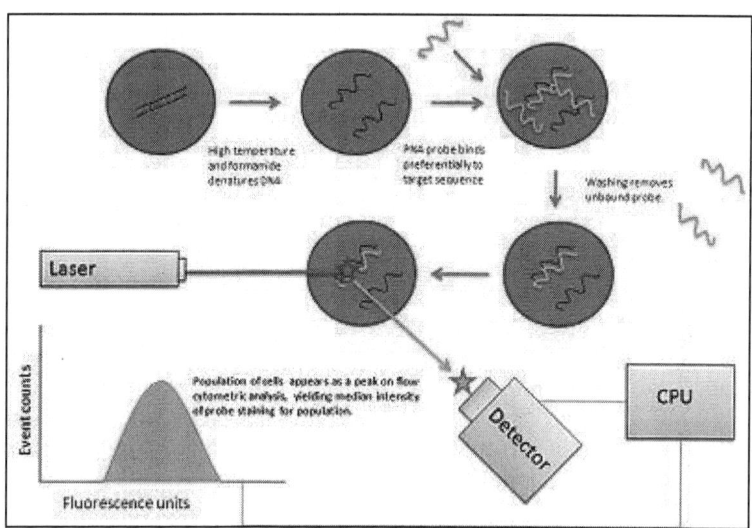

Use of Flow Cytometry to Measure Copy Number Variation of a Specific DNA Sequence (Flow-FISH).

2. Quantum Dots

Quantum dots are sometimes used in place of traditional fluorophores because of their narrower emission peaks.

3. Isotope Labeling

To overcoming the fluorescent labeling limit, lanthanide isotopes are attached to antibodies. This method could theoretically allow the use of 40 to 60 distinguishable labels and has been demonstrated for 30 labels [20]. Cells are introduced into a plasma, ionizing them and allowing time-of-flight mass spectrometry to identify the associated isotopes. Although this method permits the use of a large number of labels, it currently has lower throughput capacity than traditional flow cytometry. It also destroys the analysed cells, precluding their recovery by sorting [22].

Measurable Parameters [1]

A large numbers of parameters are being measured and there is constant increase in measuring parameters. Some examples of the measuring parameters are given below:

(i) Used for confirming diagnosis of chronic lymphocytic leukemia.

(ii) Volume and morphological complexity of cells.

(iii) Cell pigments such as chlorophyll or phycoerythrin.

(iv) Total DNA content (cellcycleanalysis).

(v) Cell kinetics, proliferation, ploidy, aneuploidy, endoreduplication, etc.).

(vi) Total RNA content.

(vii) DNA copy number variation (by Flow-FISH or BACs-on-Beads technology).

(viii) Chromosome analysis and sorting (library construction, chromosome paint).

(ix) Protein expression and localization.

(x) Protein modifications, phosphoproteins.

(xi) Transgenic products *in vivo*, particularly the Green fluorescent protein or related Fluorescent Proteins.

(xii) Cell surface antigens (Cluster of differentiation (CD) markers).

(xiii) Intracellular antigens (various cytokines, secondary mediators, *etc.*).

(xiv) Nuclear antigens.

(xv) Enzymatic activity.

(xvi) Intra-cellular ionized pH, calcium, magnesium, membrane potential.

(xvii) Membrane fluidity.

(xviii) Apoptosis (quantification, measurement of DNA degradation, mitochondrial membrane potential, permeability changes, caspase activity).

(xix) Cell viability.

(xx) Monitoring electro-permeabilization of cells.

(xxi) Oxidative burst characterising multidrug resistance (MDR) in cancer cells.

(xxii) Glutathione.

(xxiii) Various combinations (DNA/surface antigens, *etc.*).

(xxiv) Cell adherence (for instance pathogen-host cell adherence).

Source

1. *Wikipedia.*

1a. *Antony C. Bakke, PhD, Department of Pathology, Oregon Health Sciences University, Portland, The Principles of Flow Cytometry, CE update.*

1b. *Julià O, Comas J, Viver-Rego J (2000) Second-order functions are the simplest correlations between flow cytometric light scatter and bacterial diameter. J Microbiol Meth 40: 57-61.*

1c. *Jaume Comas-Riu · Núria Rius Flow cytometry applications in the food industry, J Ind Microbiol Biotechnol (2009) 36: 999-1011.*

2. *UniCelDxH800.*

2a. *Raghu H V., Naresh Kumar and Pradip Kumar Sharma, Flow Cytometry Based Rapid Method for Enumeration of Hygiene Indicator Organisms in Milk, Microbial Biosensors, Food Safety and Quality Assurance Lab, Compendium on Advance Techniques and Novel Approaches for Quality and Safety Evaluation of Dairy Foods, NDRI-2015.*

2b. *Diaz M., Herrer, M., Garcia L. A. and Quiros C. (2010). Application of flow cytometry to industrial microbial bioprocesses. Biochemical Engineering Journal 48 (3): 385-407.*

3. *http://pingu.salk. edu/flow/fluo.html.*

4. *"TSRI Cytometry Software Page". Retrieved 2009-09-03.*

5. *"Flowing Software Web Page". Retrieved 2013-02-22.*

6. *"Cytoank Main Page". Retrieved 2013-05-28.*

7. *http://leukobyte.com.*

8. *"PUCL Cytometry Software Page". Retrieved 2011-07-07.*

9. *Qian, Yu; Wei, Chungwen; Eun-Hyung Lee, F.; Campbell, John; Halliley, Jessica; Lee, Jamie A.; Cai, Jennifer; Kong, Y. Megan et al. (2010). "Elucidation of seventeen human peripheral blood B-cell subsets and quantification of the tetanus response using a density-based method for the automated identification of cell populations in multidimensional flow cytometry data".Cytometry Part B: Clinical Cytometry 78B: S69.*

10. *"Immunology Database and Analysis Portal". Retrieved 2009-09-03.*

11. *"Flow analysis with Automated Multivariate Estimation (FLAME)".Retrieved 2009-09-03.*

12. *"FlowClust". Retrieved 2009-09-03.*

13. *http://www3.interscience.wiley.com/journal/117925662/abstract?CRETRY=1 and SRETRY=0.*

14. *http://www. biomed central.com/1471-2105/10/145.*

15. *"FlowCAP-FlowCytometry: Critical Assessment of Population Identification Methods". Retrieved 2009-09-03.*

15a. *Raghu H V., Naresh Kumar and Pradip Kumar Sharma, Flow Cytometry Based Rapid Method for Enumeration of Hygiene Indicator Organisms in Milk, Microbial Biosensors, Food Safety and Quality Assurance Lab, Compendium on Advance Techniques and Novel Approaches for Quality and Safety Evaluation of Dairy Foods, NDRI-2015.*

16. *"FACS MultiSET System" (PDF). Becton Dickinson. Retrieved 2007-02-09.*

17. *Herzenberg, LA; Julius MH, Masuda T (1972). "Demonstration that antigen-binding cells are precursors of antibody-producing cells after purification with a fluorescence-activated cell sorter." PNAS 69 (7): 1934-8.*

18. *Loken MR (1990).Immunofluorescence techniques in Flow Cytometry and Sorting (2nd ed.).Wiley. pp.341-53.*

19. *Ornatsky, O., Bandura, D.; Baranov, V., Nitz, M.; Winnik, M. A., Tanner, S. (2010)." Highly multiparametric analysis by mass cytometry". Journal of Immunological Methods 361 (1–2): 1-20.*

Quality Control Steps Required to get Accurate Test Results from Cytometer [1]

There are following kinds of quality control measures:

i. First QC Measure

The flow cytometer itself must be evaluated for proper function. This is usually accomplished with standardized fluorescent beads. These give very precise, reproducible patterns, which quickly assess instrument function. In addition, the beads are fairly stable and can be used daily with little change in values. The bead values for instrument settings and signal intensity are used to generate Levy-Jennings plots and determine instrument errors.

ii. Second QC Measure

The quality control material is used to set up the appropriate instrument settings for the type of staining used. These can be beads or antibody-stained cells. One of the primary functions of this second control is to establish the compensation settings for the instrument; therefore, either beads or cells should mimic the staining that will be used on actual specimens.

iii. Third QC Measure

Third level of quality control is a control substance that mimics actual specimens. These controls are available commercially and usually consist of stabilized blood, sometimes with added tissue-culture cells that mimic a specific cancer cell. These controls can be used to assess the sensitivity of the instrument.

Source

1. *Antony C. Bakke, PhD, Department of Pathology, Oregon Health Sciences University, Portland, The Principles of Flow Cytometry, CE update.*

Applications of FCM in Food/Dairy Industry

The review [1] describes the applications to industrial microbial bioprocesses including dairy industry applications. The technology has applications in a number of fields, including molecular biology, pathology, immunology, plant biology and marine biology. It has broad application in medicine (especially in transplantation, hematology, tumor immunology and chemotherapy, prenatal diagnosis, genetics and sperm sorting for sex preselection). In marine biology, the autofluorescent properties of photosynthetic plankton can be exploited by flow cytometry in order to characterise abundance and community structure. In protein engineering, flow cytometry is used in conjunction with yeast display and bacterial display to identify cell surface-displayed protein variants with desired properties.Beyond the simple enumeration of cells, FC can provide higher knowledge about microbial fitness and metabolic activities during bioprocesses. Therefore, this will improve optimization of technological processes involving dairy bacteria, prediction of microbial performances along the whole process and the presence/absence of activity during storage. Such application would, for instance, benefit the quality control of probiotic products during their shelf life, and inclusion of FC as a validated technology for this purpose is supported [2]. Today, existing kits and automated systems are rather dedicated to the detection of industrial and environmental contaminants [1]. But **FC is currently used for milk quality control, and an ISO standardized method is being set-up under coordination of the International Dairy Federation, for enumeration of LAB in starter cultures and their applications [3].** An automated system is being commercialized by AES-Chemunex to enumerate viable bacteria in industrial products [4]. An important step is still to overcome for FC routine analysis, and developed applications require thorough calibration and validation of their performances and limitations. Promising FC methods were used for analyses of mixed cultures in milk, yogurt, and other fermented milk products. These methods could be applied for quality controls during processes or during product shelf life. Nevertheless, they would require careful optimization to prevent artifacts due to the presence of milk proteins. One must also keep in mind that the quantification limits remain high and that current probes are not specific for a species or genus but for a target cell component. Therefore FC would rather be used to characterize physiological states than for specific counts [5].

FCM can also be used for the assessment of physiologically active bacteria based on measurements of intracellular esterase activity. A variety of cell permeant, esterified fluorogenic substrates are available that are colorless until cleaved by active intracellular enzymes and the fluorescent product is accumulated within the cell. Some of these are given below [6]:

(i) Chem Chrome B [7],

(ii) Fluorescein diacetate (FDA),

(iii) Calcein acetoxymethyl ester (Calcein-AM) [8],

(iv) Carboxyfluorescein diacetate (CFDA), a derivative of FDA with enhanced retention characteristics [9], has also been applied for the rapid enumeration of esterase active bacteria in water used in the manufacturing

process of pharmaceutical products [10] and in fresh water environments [11].

Source

1. *Díaz M., Herrero M., García L. A., Quirós C. (2010). Application of flow cytometry to industrial microbial bioprocesses. Biochem. Eng. J. 48, 385–407.*
2. *Lahtinen S.J., Ouwehand A. C., Reinikainen J. P., Korpela J. M., Sandholm J., Salminen S. J. (2006). Intrinsic properties of so-called dormant probiotic bacteria, determined by flow cytometric viability assays. Appl. Environ. Microbiol. 72, 5132-5134.*
3. *International Dairy Federation, 2012.*
4. *http://www.aeschemunex. com.*
5. *Danièle Sohier, Sonia Pavan, Armelle Riou, Jérôme Combrisson, and Florence Postollec, Evolution of microbiological analytical methods for dairy industry needs, doi: 10.3389/fmicb.2014.00016.*
6. *Raghu H V., Naresh Kumar and Pradip Kumar Sharma, Flow Cytometry Based Rapid Method for Enumeration of Hygiene Indicator Organisms in Milk, Microbial Biosensors, Food Safety and Quality Assurance Lab, Compendium on Advance Techniques and Novel Approaches for Quality and Safety Evaluation of Dairy Foods, NDRI-2015.*
7. *Reynolds D. T., Fricker E.J., Purdy D. and Fricker C. R. (1997). Development of a rapid method for the enumeration of bacteria in potable water. Water Sci. Technol. 35: 433–436.*
8. *Diaper J. P. and Edwards C. (1994). The use of fluorogenic esters to detect viable bacteria by flow cytometry. J. Appl.Bacteriol. 77: 221–228.*
9. *Haugland R. P. (1999). Molecular Probes Handbook of Fluorescent Probes and Research Chemicals.*
10. *Kawai M., Yamaguchi N. and Nasu M. (1999). Rapid enumeration of physiologically active bacteria in purified water used in the pharmaceutical manufacturing process. J. Appl. Microbiol. 86: 496–504.*
11. *Porter J., Diaper J., Edwards C., Pickup R. (1995). Direct measurements of natural planktonic bacterial community viability by flow cytometry. Appl. Environ. Microbiol. 61: 2783-2786.*

A. Applications of Flow Cytometry in Dairy Industry

1. Potential for Broad Applications of Flow Cytometry and Fluorescence Techniques in Microbiological and Somatic Cell Analyses of Milk

Monitoring the quality and safety of milk requires careful analysis of microbial and somatic cell loading. The aim was to demonstrate proof of the principle that flow cytometry (FCM), coupled with fluorescence techniques for distinguishing between cell types, could potentially be employed in a wide variety of biological assays relevant to the dairy industry. To this end, researchers studied raw milk samples and ultraheat-treated milk, into which known numbers of bacteria or mouse cells were inoculated. For bacterial analyses, protein and lipids were removed, whereas only centrifugal lipid clearing was needed for somatic cell analyses. Cleared samples were stained with fluorescent dyes or with bacterial-specific fluorescent-labeled oligonucleotides and analyzed by FCM. A fluoresceinated peptide nucleic acid probe enabled efficient enumeration of bacteria in milk. Dual staining of samples with fluorescent dyes that indicate live (5-cyanol-2,3-ditolyl tetrazolium chloride, CTC or SYTO 9) or damaged cells (oxonol or propidium iodide, PI) enabled determination of viable bacteria in milk. Gram-positive and -negative bacteria were distinguished using hexidium iodide and SYTO 13 in dual staining of cleared milk samples. An FCM-based method gave a good correlation (r=0.88) with total microscopic counts of somatic cells in raw milk. The FCM method also correlated strongly (r=0.98) with the standard Fossomatic method for somatic cell detection. It is concluded that FCM, coupled with fluorescence staining techniques, offers potentially diverse and rapid approaches to biological safety and quality testing in the dairy industry. Potential application of flow cytometers to a broad range of assays for milk biological quality should make this instrumentation more attractive and cost effective to the dairy industry and indeed the broader food industry.

Source

1. *Gunasekera TS, Veal DA, Attfield PV., Centre for Fluorimetric Applica-tions in Biotechnology, Department of Biological Sciences, Macquarie University, Sydney, New South Wales 2109, Australia.*

2. Rapid Detection of Food Borne Pathogens [1]

i. By Flow Cytometry

This may be considered as the form of automated fluorescence microscopy in which instead of sample being fixed to a slide, it is injected into a fluid (dye), which passes through a sensing medium of flow cell. In flow cytometer the cells are carried by laminar flow of water through a focus of light the wavelength of which matches the absorption spectrum of the dye with which the cells have been stained. On passing through the focus each cell emits a pulse of fluorescence and the scattered light is collected by lenses and directed on to selective detectors (photomultiplier tubes). These detectors transform the light pulses into an equivalent electrical signal. The light scattering of the cells gives information on their size, shape and structure. This system is highly effective means for rapid analysis of individual cells at the rate of thousand cells per second.

ii. By Solid Phase Cytometry (SPC)

SPC is a novel technique that allows rapid detection of bacteria at single cell level, without the need for growth phase [2]. The short time detection inherent in this approach is of considerable advantage over conventional plating techniques especially for slow growing bacteria.

SPC combines aspects of flow cytometry and Epifluorescene microscopy. The microbes are isolated from their matrix from membrane filter fluoresecently labeled with argon laser excitable dye and automatically counted by laser scanning device. During 3 min scanning process the entire membrane filter surface is scanned yielding a theoretical detection limit of one cell per membrane filler. During scanning two photo multiplier tubes with wavelength 500 to 530 nm (green) and 540-585 nm (amber) detects the fluorescent light emitted by labeled cells. The signals are processed with software's which differentiate between viable signals (target cells) and back ground noises (electronic noise and fluorescent panicles). Scanned results are displayed as primary and secondary maps. Actual nature of each fluorescent spot can be further examined by epifluorescent microscope.

Source

1. *P.K. Mandal, A.K. Biswas, K. Choi and U.K. Pal, Methods for Rapid Detection of Foodborne Pathogens: An Overview, Journal of Food Technology, 6: 87-102*
2. *Haese, E.D. and H.J. Nelis, 2002. Rapid detection of single cell bacteria as a novel approach in food microbiology. J. AOAC. Int., 85: 979-983.*

3. Flow Cytometric Assessment of Viability of Lactic Acid Bacteria [1]

Lactic acid bacteria (LAB) are applied in food production for their useful metabolic properties. They are used as starters and as probiotics. However, these applications imply that the LAB are exposed to various stress conditions that may affect the physiological status of the microbes. LAB are employed as starter cultures in the production of fermented foods, such as cheese, yogurts, wines, and fermented meats. The starter cultures are often stored in freeze-dried form, which decreases the number of CFU significantly [2]. Cell proliferation and metabolic activity are crucial for success of fermentation processes, such as in cheese production. The starter bacteria multiply after being added to the curd, convert lactose to lactic acid, and degrade casein to peptides and amino acids. These are essential functions for the development of texture and flavor [3]. At the same time the conditions of the fermentation process, in particular the decline of the pH, the temperature, and the high salt concentration, affect the physiological status of the bacteria.Besides being used in dairy fermentations, several LAB species are employed as probiotics. Probiotics are living microorganisms which upon ingestion in certain numbers should exert health effects beyond inherent basic nutrition [4].

The usefulness of flow cytometry (FCM) for viability assessment of lactic acid bacteria was established. The esterase substrate carboxyfluorescein diacetate (cFDA) and the dye exclusion DNA binding probes propidium iodide (PI) and TOTO-1were tested for live/dead discrimination using a *Lactococcus,* a *Strepto-coccus,* three *Lactobacillus,* two *Leuconostoc,* an *Enterococcus,* and a *Pedio-coccus* species. Plate count experiments were performed to validate the results of the FCM assays. The results showed that cFDA was an accurate stain for live cells; in exponential-phase cultures almost all cells were labeled, while 70°C heat-killed cultures were left unstained. PI did not give clear live/dead discrimination for some of the species. TOTO-1, on the other hand, gave clear discrimination between live and dead cells. The combination of cFDA and TOTO-1 gave the best results. Well-separated subpopulations of live and dead cells could be detected with FCM. Cell sorting of the subpopulations and subsequent plating on agar medium provided direct evidence that cFDA labels the culturable subpopulation and that TOTO-1 labels the nonculturable subpopulation. Applied to cultures exposed to deconjugated bile salts or to acid, cFDA and TOTO-1 proved to be accurate indicators of culturability. Our experiments with lactic acid bacteria demonstrated that the combination of cFDA and TOTO-1 makes an excellent live/dead assay with versatile applications.

FCM is a rapid technique for cell-by-cell multiparameter analysis that is often used in combination with fluorescent labeling [5]. FCM in combination with fluorescent labeling is increasingly applied in microbiology. It is used in counting the total number of bacteria and in detecting specific strains by 16S rRNA sequence or by antigen expression. It is also used for characterizing and quantifying cellular physiological parameters such as DNA content, enzyme activity, respiration, membrane potential, intracellular pH, and membrane integrity [6,7,8].

Two groups of dye exclusion probes are of importance: phenanthridium nucleic acid dyes and cyanine nucleic acid dyes. The phenanthridium nucleic acid dyes include ethidium bromide, propidium iodide (PI), and ethidium homodimer-1. These probes have been used almost exclusively to evaluate cell membrane integrity of bacterial as well as eucaryotic cells [9,10,11]. Cyanine nucleic acid dyes are compounds that also have the chemical characteristics necessary for a viability assay based on dye exclusion. The group comprises the compounds of the monomeric TO-PRO series, the dimeric TOTO series, and the SYTOX series (Molecular Probes Inc., Eugene, Oreg.). These probes bind to DNA with little specificity, have very high fluorescence enhancement factors, and have a range of spectra covering the entire visible spectrum [10,12].

The rapid FCM analysis of LAB, in particular, the assessment of survival when LAB are exposed to bile salts or to acid, was studied with aim for an FCM assay that accurately indicates culturability, with proven validity for the given stress conditions. A selection of nine LAB species was tested, including species from the different genera of dairy LAB as well as species used in various dairy products and probiotics [13,14]. The carboxyfluorescein diacetate (cFDA) was evaluated as a live stain using the labeling protocol for *Lactococcus lactis* analysis by fluorescence microscopy developed in an earlier study [9]. Further more, the impermeant nucleic acid stains PI and TOTO-1 were evaluated for their capacity to stain dead LAB cells using FCM.

Materials, Treatment and Methods for Estimation

The information related to the different steps required for estimation and method of estimation are given below:

A. Materials-Bacterial Strains and Culture Conditions

Leuconostoc lactis L60, *Lactobacillus helveticus* T97, *Lactobacillus casei* R, and *Lactobacillus delbrueckii* subsp.*bulgaricus* 2 (NIZO Food Research, Ede, Netherlands) were used. The strains were maintained as freezer stocks at –80°C in 40 per cent glycerol. After inoculation from the freezer stocks, the cultures were grown to stationary phase (16 to 30 h).

The details of Bacterial cultures and their growing conditions are given below:

Sl.No.	Bacterial Cultures	Culture Conditions/Growing Conditions
1	*Lactococcus lactis, Enterococcus faecium,* and *Pediococcus acidilacti*	At 30°C in M17 broth supplemented with 0.5 per cent (wt/vol) lactose (LM17).
2	*Streptococcus salivarius* subsp. *Thermophilus.*	At 42°C in LM17.
3	*Leuconostoc mesenteroides,Leuco-nostoc lactis.*	At 30°C in MRS broth.
4	*Lactobacillus casei.*	At 37°C in MRS broth.
5	*Lactobacillus delbrueckii* subsp. *Bul-garicus* and *Lactobacillus helveticus.*	At 42°C in MRS broth.

The cultures were then diluted (1: 9) in fresh medium and grown to mid-exponential phase. The cultures were harvested at an optical density at 620 nm (OD_{620}) of approximately 0.7 by centrifugation at 4,000 × *g* for 10 min at 10°C. Harvested cells were washed twice, concentrated to an OD_{620} of 20, and kept on ice until use.

B. Treatments

Portions of 200 µl of concentrated cell suspension (OD_{620} of 20) were exposed to heat, acid, or bile salts. Cells were heat killed by exposure to 70°C for 10 min. As a control, cells were incubated at pH 7.0 and 30°C for 60 min. After the incubations, the cells were spun down, resuspended in buffer, and put on ice until use.

C. Measurement of Culturability

Tenfold serial dilutions of control and treated samples were made in buffer, and triplicate aliquots of 100 µl of the appropriate dilution were spread out on agar plates.*Lactococcus lactis, E. faecium,* and *P. acidilacti* were plated on LM17 medium. The other species were plated on MRS medium. After 3 days of aerobic incubation at 30°C, the colonies were counted.

D. Labelling and its Measurement

a. Fluorescence Labeling

cFDA, PI, and TOTO-1 (Molecular Probes, Inc.) were used.Stock solutions of 100 µM cFDA in 50 mM KP_ibuffer (pH 7.0), 1.5 mM PI in distilled water, and 100 µM TOTO-1 in dimethyl sulfoxide were prepared.

(i) **TOTO-1**-1,1 (4,4,7,7-tetramethyl-4,7-diazaundecamethylene) -bis - 4- [3-methyl -2,3 dihydro (benzo-1,3-oxazole) -2-methylidene] -1-(3-tri-methyl ammonium propyl) -pyridinium tetraiodide. TOTO-1 has a molecular mass of 1,303 g per mol and a very high fluorescence enhancement of 1,400-fold.

(ii) **cFDA (carboxyfluorescein diacetate)** is an esterase substrate yielding the fluorescent carboxyfluorescein (cF) upon hydrolysis. PI and TOTO-1 bind to DNA.

(iii) **PI** has a molecular mass of 668 g per mol and a fluorescence enhancement of 20- to 30-fold upon binding.

For single-probe labeling, concentrated cell suspensions (OD_{620} of 10) were incubated with 50 µM cFDA, 30 µM PI, or 1 µM TOTO-1 at 30°C for 10 min. After incubation with cFDA, the cells were washed once. For double labeling, concentrated cell suspensions (OD_{620} of 10) were incubated with 50 µM cFDA and 1 µM TOTO-1 simultaneously at 30°C for 10 min, after which the cells were washed once. All labeled cell suspensions were kept on ice until use.

b. Fluorescence Microscopy

Labeled cell suspensions were diluted to approximately 10^9 cells per ml and microscopically analyzed with an Axioskop epifluorescence microscope. Photomicrographs were made with simultaneous light and epifluorescence microscopy, a low transmitted-light intensity, and an exposure time of 15 s on Kodak 400 ASA color films. In these photomicrographs both the labeled cells and the nonlabeled cells were visible.

c. Spectrofluorimetry

To measure the cF labeling capacity, *i.e.*, the amount of cF in the cells per milligram of protein, labeled cells were lysed by incubation at 70°C for 15 min and the debris was removed by centrifugation. The fluorescence of the supernatant was measured with a Perkin-Elmer LS 50B luminescence spectrometer. The cF concentration was calculated from a calibration curve with a cF concentration range. The cell protein concentrations were analyzed by the Lowry method.

E. FCM

FCM analyses were performed on a FACSCalibur flow cytometer equipped with a air-cooled argon ion laser and a cell-sorting catcher tube. Cell samples were diluted to approximately 10^6 cells per ml and delivered at the low flow rate, corresponding to 150 to 500 cells per s. FSC, SSC, and three fluorescence signals were measured. A band pass filter of 530 nm was used to collect the green fluorescence (FL1), a band pass filter of 585 nm was used to collect the yellow-orange fluorescence (FL2), and a long-pass filter of 670 nm was used to collect the red fluorescence (FL3).FSC was collected with a diode detector. SSC and the three fluorescence signals were collected with photomultiplier tubes. All signals were collected by using logarithmic amplifications. A combination of FSC and SSC was used to discriminate bacteria from background. Data were analyzed with the CELL Quest program and the WinMDI program.

F. Sorting

Sorting experiments were performed with *Lactococcus lactis*. Exponential-phase cell suspensions that were left unstained or incubated with cFDA, as well as 1: 1 mixtures of exponential-phase cells and 70°C heat-killed cells labeled with cFDA or TOTO-1, were used. Further more, cell suspensions exposed to 0.10 per cent bile salts labeled with cFDA or TOTO-1 were used. All sample handling was done aseptically. In the dot plot of FL1 and FL2, regions of nonlabeled, cF-labeled, and TOTO-1-labeled cells were defined to use as sort gates. One region was sorted at a time. Culturability was tested by plating 100-µl samples of the sorted subpopulations directly out of the sort collection tubes.

G. Results

The results observed on staining, Viability assessment after treatment with bile salts or acid and cell sortingare given below:

i. Staining with cFDA, PI, or TOTO-1

(*a*) **cFDA** gave good results. Incubation of exponential-phase cells of the nine selected LAB species with cFDA resulted in a high fraction of cF-labeled cells, in most cases above 90 per cent, whereas no cells of 70°C heat-killed suspensions were labeled. In the FCM analysis of all species the labeled population gave a peak in the green fluorescence histogram, which was resolved from the signal of nonlabeled cells.

(*b*) **PI** gave satisfactory results for only some of the tested species. Good results were obtained for the species *S.salivarius*subsp. *thermophilus, Lactobacillus delbrueckii subsp.Bulgaricus, Lactobacillus casei,* and *Lactobacillus mesenteroides.* The FCM results agreed with microscopic observations. However, for the other species PI labeling did not give clear FCM results.

(*c*) **TOTO-1** labeling enabled easy identification of the live and dead cells by FCM for most of the LAB species.Dead cells were labeled with high-fluorescence intensity, while live cells were left unstained. For *Lactobacillus delbrueckii* subsp. *bulgaricus* and *Leuconostoc lactis* there was some overlap between the FCM fluorescence histogram peaks of labeled and nonlabeled populations. In general, the TOTO-1 labeling gave clearer results than the PI labeling, and peaks of live and dead cells in the FCM fluorescence histograms were better resolved.

Because cFDA and TOTO-1 gave the best results and appeared to be useful for the different LAB species, these probes were chosen for further testing as stains for live and dead cells.

ii. Viability Assessment after Treatment with Bile Salts or Acid

FCM was applied for viability assessment of *Lactococcus lactis, Lactobacillus helveticus* and *Leuconostoc mesenteroides* using cFDA and TOTO-1 after exposure to deconjugated bile salts or acid. The results of the FCM were compared with the survival tested by plate counts. For visual impressions of labeling after stress, the cell suspensions were also analyzed by fluorescence microscopy. The microscopic observations agreed with the FCM results.

iii. Cell Sorting

To establish a direct relationship between labeling and culturability, the labeled and nonlabeled populations were sorted and plated. *Lactococcus lactis* exponential-phase nonlabeled cell suspensions gave a number of colonies corresponding to approximately 80 per cent of the number of cells actually sorted. Similarly, cFDA-stained nontreated cell suspensions sorted on cF labeling gave a number of colonies corresponding to approximately 80 per cent of the number of cells recovered in the sorting tube.

H. Conclusion

(I) The usefulness of FCM for viability assessment of LAB was established. Three fluorescent probes were tested for their usefulness for live/dead discrimination: cFDA as a stain for live cells and PI and TOTO-1 as stains for dead cells. The results showed that cFDA was successful in labeling the live cells and leaving the dead cells unstained. TOTO-1 appeared to be better than PI as a stain for dead cells.

(ii) Generally, the viability is not affected when LAB are incubated with hydrochloric acid at pH 3.5 or higher, while at lower pH the survival decreases to less than 1 per cent, at pH values that are dependent on species and strain [15,16,17].

(iii) Resistance to bile is recognized as an important feature for LAB used as probiotics [18,19,20]. In the human intestinal tract the concentration is variable, with a maximum of 2 per cent [21]. The results on survival in buffer show that the concentration of 0.10 per cent deconjugated bile salts falls within the critical range, but different survival fractions were found for the three species. At 0.25 per cent almost no surviving cells were detected.

(iv) Live/dead assays based on two probes have advantages over assays with one probe that labels either live or dead cells. In FCM, the identification of cells is facilitated when all cells are fluorescently labeled, especially when the background is high. Evaluation of FCM in combination with cFDA, PI, and TOTO-1 for identifying live and dead cells are tabulated below:

Sl.No.	*Micro-organism*	*Result with probe*		
		cFDA	*PI*	*TOTO-1*
1	*Lactococcus lactis* NCDO 712	+	-	+
2	*Streptococcus salivarius* subsp. *thermophilus*	+	+	+
3	*Lactobacillus delbrueckii* subsp. *bulgaricus 2*	+	+	+/-
4	*Lactobacillus casei* R	+	+	+
5	*Lactobacillus helveticus* T97	+	-	+
6	*Leuconostoc mesenteroides* DSM 20343	+	+	+
7	*Leuconostoc lactis* L60	+	-	+/-
8	*Enterococcus faecium* 93-828-3	+	-	+
9	*Pediococcus acidilacti*	+	+/-	+

Source

1. *C. J. Bunthof, K. Bloemen, P. Breeuwer, F. M. Rombouts, and T. Abee, Appl. Environ. Microbiol. 67: 2326-2335, 2001.*

2. *Champagne C P, Mondou F, Raymond Y, Roy D. Effect of polymers and storage temperature on the stability of freeze-dried lactic acid bacteria. Food Res Int. 1996; 29: 555–562.*
3. *Crow V L, Coolbear T, Holland R, Pritchard G G, Martley F G. Starters as finishers: starter properties relevant to cheese ripening.Int Dairy J. 1993; 3: 423–460.*
4. *Guarner F, Schaafsma G J. Probiotics. Int J Food Microbiol.1998; 39: 237–238.*
5. *Mortimer F C, Mason D J, Gant V A. Flow cytometric monitoring of antibiotic-induced injury in Escherichia coli using cell-impermeant fluorescent probes. Antimicrob Agents Chemother.2000; 44: 676–681.*
6. *Davey H M, Kell D B. Flow cytometry and cell sorting of heterogeneous microbial populations: the importance of single-cell analyses.Microbiol Rev. 1996; 60: 641–696.*
7. *Lloyd D, editor. Flow cytometry in microbiology. London, United Kingdom: Springer-Verlag; 1993.*
8. *Porter J, Deere D, Pickup R, Edwards C. Fluorescent probes and flow cytometry: new insights into environmental bacteriology. Cytometry. 1996; 23: 91-96.*
9. *Bunthof C J, van den Braak S, Breeuwer P, Rombouts F M, Abee T. Rapid fluorescence assessment of the viability of stressed Lactococcus lactis. Appl Environ Microbiol. 1999; 65: 3681-3689.*
10. *Haugland P. Molecular probes-handbook of fluorescent probes and research chemicals. 7th ed. 1999. Molecular Probes, Inc., Eugene, Oreg. CD-ROM.*
11. *Haugland P. Molecular probes-handbook of fluorescent probes and research chemicals. 7th ed. 1999. Molecular Probes, Inc., Eugene, Oreg. CD-ROM.*
12. *Hirons G T, Fawcett J J, Crissman H A. TOTO and YOYO: new very bright fluorochromes for DNA content analyses by flow cytometry. Cytometry. 1994; 15: 129-140.*
13. *Cogan T M. History and taxonomy of starter cultures. In: Cogan T M, Accolas J-P, editors; Cogan T M, Accolas J-P, editors.Dairy starter cultures. New York, N.Y: VCH Publishers, Inc.; 1996. pp. 1–24.*
14. *Limsowtin G K Y, Powell I B. Types of starters. In: Cogan T M, Accolas J-P, editors; Cogan T M, Accolas J-P, editors. Dairy starter cultures. New York, N.Y: VCH Publishers, Inc.; 1996. pp.101-130.*
15. *Bankes P, Rowe D, Betts RP (1991) The rapid detection of yeastspoilage using the ChemXow system (Technical Memorandum).*
16. *Connil N, Dousset X, Onno B, Pilet MF, Breuil MF, Montel MC (1998) Enumeration of Carnobacterium divergensV41, Carnobacterium piscicola V1 and Lactobacillus brevis LB62 byin situ hybridization-Xow cytometry. Lett Appl Microbiol 27: 302-306.*
17. *Dowhanick TM, Russel I (1993) Advances in detection and identification methods applicable to the brewing industry. In: GumpBH (ed) Beer and wine production. Analysis, characterization, and technological advances. American Chemical Society, Washington, DC, pp 13–32.*

18. *Boyd AR, Gunasekera TS, AttWeld PV, Simic K, Vincent SF, VealDA (2003) A Xow cytometric method for determination of yeast viability and cell number in a brewery. FEMS Yeast Res 3: 11–16.*

19. *Buckenhüskes HJ (1993) Selection criteria for lactic acid bacteriato be used as starter cultures for various food commodities. FEMSMicrobiol Lett 12: 253–272.*

20. *Fleet GH (1999) Microorganisms in food ecosystems. Int J FoodMicrobiol 50: 101–117.*

21. *Bianchi MA, Del Río D, Pellegrini N, Sansebastiano G, Neviani E, Brighenti F (2004) A Xuorescence-based method for the detectionof adhesive properties of lactic acid bacteria to Caco-2 cells.Lett Appl Microbiol 39: 301–305.*

4. Principles and Applications of Flow Cytometry (FCM) in Dairy Cattle Sex Determining

More dairy cattle offspring have been produced using AI (artificial insemination) and other means using spermatozoa separated by flow cytometer. Flow cytometric sperm sorting based on differences in their DNA content is the best method for separation of X-and Y-chromosome bearing spermatozoa. Recently, high-speed sperm sorting has resulted in production of 90 per cent pure X-and Y-sperm at rate of 1.5~2×108 sperm per hour. Application of this new technique will enable conduct of more conventional technologies for both AI and cryopreservation in dairy cattle and in other farm animals using X or Y sperm.

Source

Li Nan Wang Xiuli, Cui Yudong, Gan Wenping and Zhang Hongtao, Principles and Applications of Flow Cytometry (FCM) in Dairy Cattle Sex Determining, Institute of Animal Science and Veterinary Medicine, Heilongjiang Academy of Land-reclamation Sciences, Harbin 150038).

5. Flow Cytometry Method for Rapid Detection and Enumeration of Total Bacteria in Milk

Application of flow cytometry (FCM) to microbial analysis of milk is hampered by the presence of milk proteins and lipid particles. The development of a rapid (≤1-h) FCM assay based on enzymatic clearing of milk to determine total bacteria in milk was reported. When bacteria were added to ultra-heat-treated milk, a good correlation ($r \geq 0.98$) between the FCM assay and the more conventional methods of plating and direct microscopic counting was achieved. Raw milk data showed a significant correlation ($P < 0.01$) and a good agreement ($r = 0.91$) between FCM and standard plate count methods.

There are several methods available for detection and enumeration of microorganisms in raw and processed milks. To alleviate problems associated with culture-based detection systems and direct microscopic methods, various other methods have been developed [1,2,3]. Although biochemically and physiologically based assays are fast, they lack absolute specificity.

BactoScan was developed as an automated instrument for routine testing of the bacteriological quality of raw milk [4,5]. However, this instrumentis limited to measuring either total microbial or somatic cell counts and not able to microbial differentiation. The flow cytometers may suit the broad and specific needs of microbial analysis of milk and dairy products within one type of instrument [6,7]. Use of fluorescent stains or fluorogenic substrates in combination with FCM allows the detection and discrimination of viable culturable, viable nonculturable, and nonviable organisms [8,9,10]. A rapid method for detecting total bacteria by FCM was developed which includes the following steps:

A. Milk-Clearing Treatments

(i) *Escherichia coli* and *Staphylococcus aureus* were chosen to represent gram-negative rods and gram-positive cocci that are potential contaminants of milk. Bacteria were grown overnight in Trypticase soy broth at 28°C on a rotary shaker (180 rpm) for 16 h.

(ii) Pure populations of *E. coli* and *S. aureus* were easily detected by FCM when they were suspended in phosphate-buffered saline (PBS) but when they were inoculated into ultra-heat-treated (UHT) milk, no distinct separation appeared. This is due to the presence of proteins and lipid globules that can bind nonspecifically to fluorescent stains and interfere with staining and detection of bacteria.

(iii) Treatment of milk by centrifugation to remove lipids without also treating samples with proteases was insufficient to allow definition of bacteria. Thus, the most critical barrier to FCM analysis of milk is the presence of protein globules.

(iv) Enzymatic treatment to remove or modify proteins is applied and thereby enable distinction of bacteria by flow cytometry. 0.05 mg of proteinase K or 10 µl of savinase were used to treat 100 µl of UHT milk samples and 50 µl of savinase plus 50 µl of 0.1 per cent Triton X-100 to treat 100 µl of raw milk samples.

(v) Treated milk samples were incubated at 37°C for 30 to 45 min, after which 900 µl of 150 mM NaCl was added and mixed by inversion of tubes. Samples were then centrifuged at 14,000 × g for 10 min and lipids (top layer) and digested proteins of the milk were drawn off with a micropipette without disturbing the pelleted material, which contained bacteria. The pellet was resuspended in 100 µl of 150 mM NaCl.

B. Detection of Bacteria in Milk by using FCM

The above treated samples were then stained and analyzed using FCM.

(i) **Staining-**Fluorescent staining and FMC.Bacterial cells in treated milk samples were stained with SYTO BC. This stain is a high-affinity nucleic acid stain that penetrates both gram-positive and gram-negative bacteria, giving a bright green fluorescent signal. SYTO BC, diluted 1: 20 in dimethylsulfoxide, was mixed 1: 50 with resuspended bacteria and incubated in the dark at 37°C for 5 to 10 min. The viabilities of bacterial cultures were determined by dual staining of the subsamples.

(*ii*) **FCM Analysis**-Stained samples were analyzed using FACScan (analysis only) or FACSCalibur (cell sorting) flow cytometers. The detection threshold was adjusted for FL1 to eliminate particles emitting green fluorescence at a level significantly below that of bacteria suspended in PBS and stained with SYTO BC. Compensation was set so as to remove FL2 fluorescence from the FL1 channel and FL1 from FL2. Bacterial counts by FCM were obtained by normalizing the numbers of events occurring in regions on dot plots that defined bacterial populations to the volume of sample analyzed. Data acquired from FCM were converted from Hewlett Packard to PC format by using the computer program HP-Reader and analyzed with the computer program Windows Multiple Document Interface FCM application. For cell sorting, a FACSCalibur flow cytometer was used. Sort regions were defined on bivariate dot plots that delineated distinct populations. Droplets were collected onto 0.02-µm-pore-size filters through a Sort Stage and examined using epifluorescence microscopy. Sorted samples were also diluted, and 100 µl was spread onto differential agar media to confirm identity and viability.

C. Microscopic and Plate Count Methods

(*i*) **Microscopic Method**-A Carl Zeiss, Axioskop 2 epifluorescence microscope fitted with 10× eyepieces and 40× and 100× (oil immersion) objectives was used to confirm and count cells. Excitation of SYTO BC- or PI-stained cells was by a 100-W Hg vapor arc lamp with an appropriate filter block giving excitation at 450 to 490 nm and examination at 520 nm. Direct microscopic counts of bacterial suspensions were carried out using bright-field microscopy and Thoma counting chamber procedures [11].

(*ii*) **Plate count method**-Plate count methods were used to determine viable cell numbers. Suspensions of *E. coli* or *S. aureus* in milk were serially diluted in PBS, and 100-µl volumes of dilutions were spread plated in triplicate. To differentiate *E. coli*, Chromocult coliform agar was used, whereas Baird-Parker agar supplemented with egg yolk tellurite emulsion was used for *S. aureus*. Colonies were counted after incubation for 24 to 48 h at 37°C. The total viable microbial counts of raw milk were determined by standard plate count using the pour plate method [12]. Analysis of variance and Student's *t* test were used to detect significant differences between different methods. Correlations between flow cytometric method and total microscopic counts and plate count methods were calculated with statistical software.

D. Comparison between FCM, Total Microscopic Count, and Plate Count Methods

Dual staining of pure bacterial cultures with PI and SYTO BC showed that inocula (16-h bacteria cultures) were >99 per cent viable. UHT milk was inoculated with *E. coli* and *S. aureus*cells at different cell concentrations between 10^3 and 10^8 ml^{-1} and analyzed using FCM and traditional techniques such as plate count and direct microscopy. Strong correlations (r = ≥0.98) between FCM and plate counts or direct microscopic counts were obtained for both *E. coli* and *S. aureus*.

E. *Potential of FCM in Milk Microbiology*

The aforesaid study demonstrates the ability of FCM to determine total bacterial numbers after clearing of milk and staining of bacteria with a readily available fluorescent stain. The sensitivity of the FCM procedure was $\leq 10^4$ total bacteria ml of milk^{-1}. A limit of $\leq 10^4$ total bacteria ml of raw milk^{-1} is below the level of detection required to satisfy legislation in many countries and states [14]. The assay takes 45 to 60 min depending on whether processed or raw milk is being analyzed. The total FCM analysis time is between 30 s and 2 min, depending on numbers of bacteria in milk. This time frame compares favorably with current culturing methods, which take 72 h [12].

In additional studies, it was found that FCM is suitable for monitoring the growth of bacteria in refrigerated raw, pasteurized, homogenized milk and flavored milk drinks and enumerating viable bacteria [8]. These combined works indicate the potential breadth of FCM for applications in dairy microbiology. These applications include testing of raw milk for conformation with standards, specifications and regulatory compliance, monitoring efficiency of manufacturing, cleaning, and sanitation practices, and predicting the shelf life of milk products. The ability to use a single instrument for numerous rapid microbiological assay procedures has obvious advantages for the dairy industry over the current situation, where culturing, microscopy, or several dedicated instruments are needed.

Source

1. *Fung D.Y.C. (1994) Rapid methods and automation in food microbiology: a review. Food Rev. Int. 10: 357–375.*
2. *McClelland R.G., Pinder A. C. (1994) Detection of low levels of specific Salmonella species by fluorescent antibodies and flow cytometry.J. Appl. Bacteriol.77: 440–447.*
3. *Vasavada P. C., White C.H. (1993) Symposium: developing methodology for microbial evaluation of milk and dairy products. J. Dairy Sci. 76: 3099–3100.*
4. *Cunningham J.D., Saunders C.L. (1988)Collaborative study of the Bactoscan-an automated method for determinations of total bacteria in raw milk. Can. Inst. Food Sci. Technol. J. 21: 464–466.*
5. *Vasavada P.C. (1993) Rapid methods and automation in dairy microbiology. J.Dairy Sci. 76: 3101–3113.*
6. *McClelland R. G., Pinder A. C. (1994) Detection of low levels of specific Salmonella species by fluorescent antibodies and flow cytometry. J. Appl. Bacteriol.77: 440-447.*
7. *Sorhaug T., Stepaniak L. (1997) Psychrotrophs and their enzymes in milk and dairy products: quality aspects. Trends Food Sci. Technol. 8: 35-41.*
8. *Attfield P., Gunasekera T., Boyd A., Deere D., Veal D. (1999)Application of flow cytometry to microbiology of food and beverage industries. Australas.Biotechno l.9: 159-166.*

9. *Davey H. M., Kell D. B. (1996) Flow cytometry and cell sorting of heterogeneous microbial populations: the importance of single-cell analyses. Microbiol. Rev. 60: 641-696.*

10. *Wallner G., SteinmetzI., Bitter-Suermann D., Amann R. (1996) combination of rRNA-targeted hybridisation probes and immuno-probes for the identification of bacteria by flow cytometry. Syst. Appl. Microbiol.19: 569–576.*

11. *Cruickshank R., DuguidJ.P., Marmion B. P., Swain R.H. A.Medical microbiology 12th ed. 21975: the practice of medical microbiology. Churchill Livingstone, Edinburgh, United Kingdom.*

12. *Desmasures N., Gueguen M. (1997) Monitoring the microbiology of high quality milk by monthly sampling over 2 years. J. Dairy Res.64: 271-280.*

6. Development of a Flow Cytometric Method to Analyze Subpopulations of Bacteria in Probiotic Products and Dairy Starters

Flow cytometry (FCM) is a rapid and sensitive technique that can determine cell numbers and measure various physiological characteristics of individual cells by using appropriate fluorescent probes. After development of an FCM assay with the viability probes cFDA and TOTO-1 for (stressed) lactic acid bacteria, now this assay is used to study the viability of bacterial suspensions in milk, dairy fermentation starters, and probiotic products. To facilitate FCM analysis of bacteria in milk, a commercially available milk-clearing solution was used. The procedure was optimized to increase the signal-to-noise ratio. FCM enumerations were accurate down to a concentration of 10^5cells ml^{-1}. The level of retrieval of *Lactobacillus plantarum* WCFS 1 suspended in milk was high, and viability was not affected by the procedure. The plate counts for cleared samples of untreated cell suspensions were nearly as high as the total FCM counts, and the correlation was strong ($r > 0.99$). In dairy fermentation starters and in probiotic products the FCM total cell counts were substantially higher than the numbers of CFU. The following three functional populations could be distinguished:

(i) Culturable cells.

(ii) Cells that are intact and metabolically active but not culturable

(iii) Permeabilized cells.

The proportions of the populations differed in the products tested. This FCM method provides tools to assess the functionality of different populations in fermentation starters and probiotic products.

Source

1. *Christine J. Bunthof and Tjakko Abee, Laboratory of Food Microbiology, Department of Agrotechnology and Food Sciences, Wageningen University, 6700 EV Wageningen, The Netherlands.*

7. Detection of *Salmonella typhimurium* in Dairy Products with Flow Cytometry and Monoclonal Antibodies

Flow cytometry, combined with fluorescently labelled monoclonal antibodies, offers advantages of speed and sensitivity for the detection of specific pathogenic bacteria in foods. The detection of Salmonella typhimurium in eggs and milk was investigated. Using a sample clearing procedure, it was determined that the detection limit was on the order of 10 (3) cells per ml after a total analysis time of 40 min. After 6 h of nonselective enrichment, the detection limits were 10 cells per ml for milk and 1 cell per ml for eggs, even in the presence of a 10,000-fold excess of Escherichia coli cells.

Source

1. *R G McClelland and A C Pinder, Food Biophysics Department, Institute of Food Research Norwich Laboratory, Colney, United Kingdom. Appl Environ Microbiol. 1994 December; 60 (12): 4255–4262.*

8. Flow Cytometric Technique for Quantification and differentiation of Bacteria in Bulk Tank Milk

A total of 75 bulk tank milk samples exceeding the grading level of $3x10^{-1}$ CFU ml^{-1} were examined by both flow cytometry and traditional microbiological analyses. The correlation coefficient (r) between the two methods was 0.71. or the differential analyses of the dominant bacterial populations four different parameters were used to give a species-characteristic pattern. The four parameters were as follows:

(i) Staining with Oregon Green conjugated wheat germ agglutinin that binds to the cell wall of bacteria.

(ii) Staining with hexidium iodide that binds to all bacterial DNA.

(iii) The flow cytometric forward scatter.

(iv) The flow cytometric side scatter.

The flow cytometric plots were observed. Three regions in the flow cytometric plot were dened as:

(i) **Region 1-** includes bacteria mainly associated with poor hygiene.

(ii) **Region 2-** includes psychrotrophic hygiene bacteria.

(iii) **Region-3-** includes bacteria mainly related to mastitis.

The ability of the flow cytometric technique to predict the main cause of elevated bacterial counts on routine samples was examined. Comparing these results with results obtained by traditional microbiological analyses for identication showed that for 81 per cent of the samples the two techniques agreed on the main cause of an elevated bacterial count. This study described the first step in development of a technique suitable for routine analyses of bulk tank milk samples. A technique indicating the main cause of an elevated count will enable the farmer to eliminate the contamination source within a short time limit.

Source

1. *C. Holm, T. Mathiasen, and L. Jespersen, Journal of Applied Microbiology, 2004, 97, 935–941.*

9. Rapid Method for Enumeration of Hygiene Indicator Organisms in Milk [1]

The primary hazard to safety, quality, and wholesomeness of milk and dairy foods is because of the undesirable microorganisms. Due to unhygienic condition/ lapses in GMP prevailed during milk production, storage, transportation and distribution may add large number of microorganisms into milk. Several methods for rapid detection, isolation, enumeration, and characterization of microorganisms are being adapted by the dairy industry but now more emphasis is given to rapid D-count Flow cytometry based detection of hygiene indicator organisms in milk. D-count is a closed FCM system which involves fluorescent cell labelling, automated flow cytometry detection and counting of viable micro-organisms on a cell to cell basis after labelling of targeted microorganisms with a specific fluorescent dye. It gives qualitative and quantitative detection of bacteria, Gram negative bacteria, yeasts and molds in products including raw milk, UHT milk, yoghurts, deserts, fermented milk products and soya milk. It has Real time detection within 30 sec. to 2 min (after sample injection). System has 2 robotic arms; one dedicatedly working for preparation of samples and another for presenting the sample to the analyzer. It also has automatic cleaning during samples processing with cleaning solution.

Method

(i) D-count FC detect of total viable counts in raw, pasteurized and UHT milk, Enterobacteriaceae family, yeast in fermented milk products after an enrichment phase.

(ii) Samples are treated with reagents that label viable microorganisms when potentially present.

(iii) Only viable microorganisms are able to enzymatically cleave the non-fluorescent substrate inside the cell to liberate and to accumulate the resulting fluorochrome.

(iv) Then, the sample is injected into the quartz flow cell of the D-CountR, forming a narrow and laminar flow stream that ensures microorganisms pass the laser excitation beam one by one.

(v) Sensitive detectors collect the fluorescent signals from each cell.

Source

1. *Raghu H V., Naresh Kumar and Pradip Kumar Sharma, Flow Cytometry Based Rapid Method for Enumeration of Hygiene Indicator Organisms in Milk, Microbial Biosensors, Food Safety and Quality Assurance Lab, Compendium on Advance Techniques and Novel Approaches for Quality and Safety Evaluation of Dairy Foods, NDRI-2015.*

Instruments based on FCM

The principle behind flow cytometry is quite simple. In this, a suspension of cells is stained and forced through a capillary tube, which is illuminated in front of microscope objective. Every passing cell is registered by photo electronics attached to the microscope. Most microorganisms are optically too similar to resolve from each other or from debris, therefore, labeling with fluorescent dyes can be used to probe the viability (SYTO-9 in combination with propidium iodide) and metabolic state (*e.g.*, ChemChrome (AES Chemunex) of microorganisms. By using nonspecific and specific fluorochromes, as well as different wavelengths, and by measuring at different angles, it is possible to discriminate between bacteria in mixed populations. For routine analyses of milk, water, beverage, or dairy industries flow cytometry is used. The examples of this method for practical use in the field is BactoScan (Foss Electric, Denmark), ChemFlow (Chemunex, France), Argus Flow Cytometer (Skatron, Norway).

Source

1. *Technews, Technological Advancement in Methods of Microbiology in Milk Analysis-Part 1 (Direct methods), NDDB, Issue no.91, October-December 2016.*

1. BactoScan

BactoScan (M/S FOSS) measures the hygienic quality of milk by analysing bacteria in raw milk. Delivering results in just minutes, BactoScan allows farmers or milk testing laboratories to take action fast to preserve and enhance milk quality. The principle of Bactoscan 8000 is direct microscopic counting. It is a fully automated instrument in which the bacteria are centrifuged and separated from the milk, stained with fluorescent dye (Acriodenic Orange) and counted electronically as light impulses in a continuous flow fluorescent microscope. The sample is treated with a lysing solution (Suhran *et al.*, 1999) and so bacterial clumps are dissolved, a fact that leads to more accurate counts. The main advantage is speed (approximately 80 samples/hour). Lachowsky *et al.* (1997) stated that automated Bactoscan 8000 can provide a reasonable alternative to the classical culture methods for enumeration of bacteria in bulk tank raw milk when bacterial countlie within the instrument"s specified applicable rangeof 40000-80000 cfu ml^{-1}.

BactoScan FC and BactoScan FC+

The newer development for the counting of individual cell, is the introduction of BactoScanFC. **BactoScan FC+ is based on flowcytometry technology** that enables precise and instant milk bacteria analysis. Flowcytometry is a technique that counts and characterises bacteria cells. In this the DNA/RNA of the bacteria is stained with fluorescent dye (Ethidium bromide).Certain buffers and enzymes are added during sample preparation to reduce the influence of other milk constituents and bacterial clusters are also separated into single bacteria.

Fluorescence is induced by the light emitted is detected when stained particles pass as a hydrodynamically focused stream through a fluorescence detector. The new BactoScanFC+ is 30 per cent faster than the previous version of BactoScan FC and the maximum number of samples that the instrument can handle has been increased from 150 to 200 samples per hour.BactoScan FC+ can now analyse up to 200 milk samples per hour with upgrades available to meet the requirements of all milk testing laboratories. This high performance capacity enables a high throughput of daily samples and frees up testing staff for other purposes. Bactoscan FC+ is

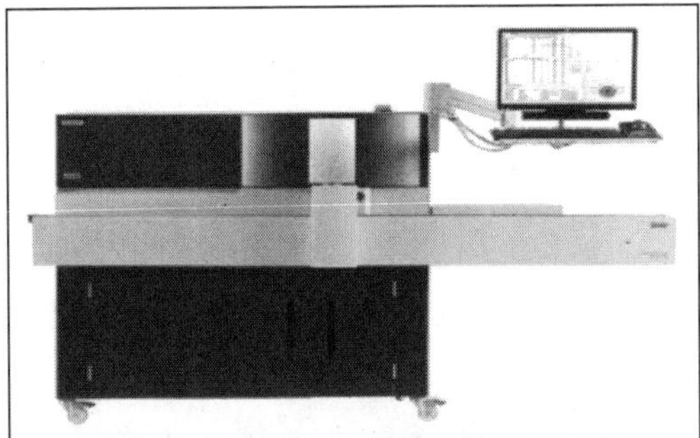

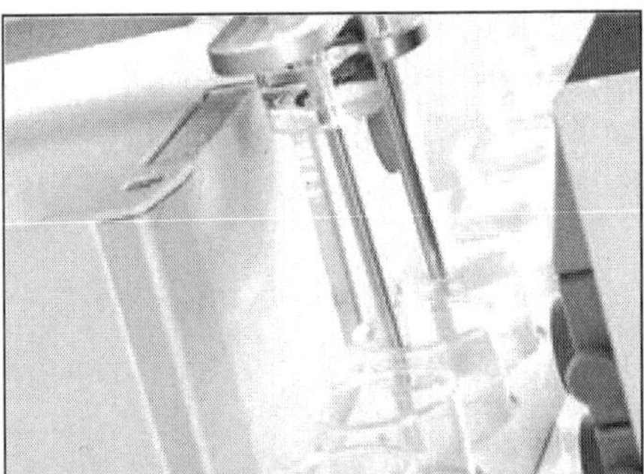

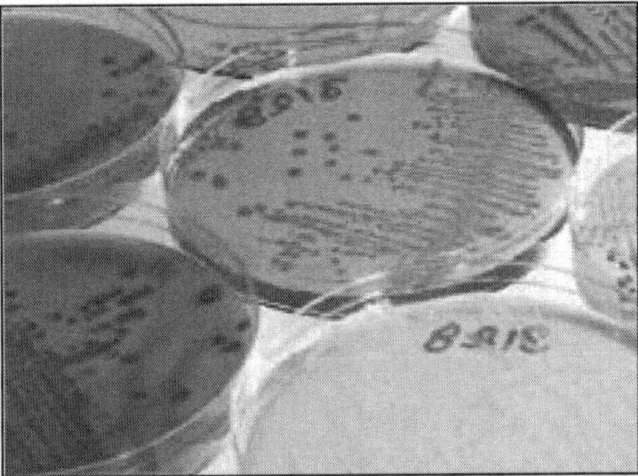

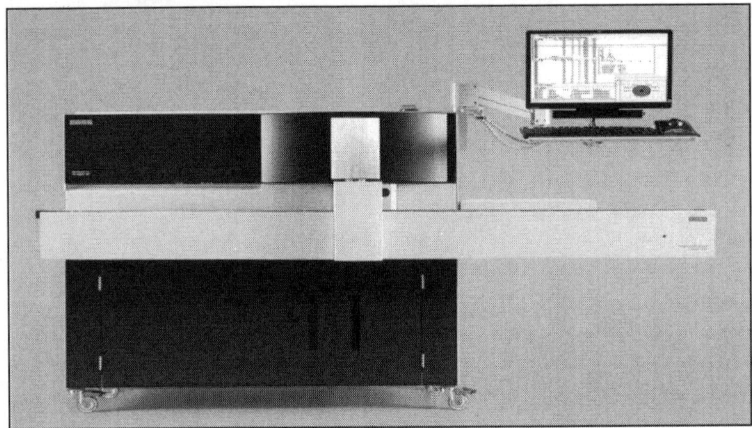

available in three models- 65L, 65H: 65 samples per hour, 130L, 130H: 130 samples per hour and 200L, 200H: 200 samples per hour.

BactoScan FC+ Semi-automatic

The BactoScan FC+ semi-automatic employs the proven BactoScan analysis method but is scaled down to fit laboratories with limited sample throughput by removing the sample conveyer. It can analyse 65 samples per hour with same high performance as the fully automated version.Because the BactoScan FC+ counts all bacteria as single cells and not clusters it gives a uniform high performance with accuracy, repeatability and reproducibility of results. Unlike methods such as the plate counting method, BactoScan FC+ cannot be influenced by the operator and it can analyse raw milk samples directly without prior heating or dilution, which can affect precision.

Superior Accuracy through Individual Bacteria Count

Accuracy

Typical Sy, x < 0.25 log units in the entire measuring range.

Reference method: Standard Plate Count (IDFStandard 100B: 1991).

BactoScan™ FC+ uses the same efficient conveyor system as other FOSS instruments such as Fossomatic, CombiFoss FT+ and MilkoScan FT+. For laboratory personnel this means they only have to learn how to use one user interface and can enjoy the benefits of easy integration of data.Integrated software platform.

BactoScan FC+, Fossomatic, MilkoScan FT+ and CombiFoss FT+ all share the same powerful Foss Integrator software platform offering advanced data handling and storage facilities for traceability while a common user interface promotes more efficient laboratory operations.

Technical Specification of Bactoscan

Sl.No.	*Parameters*		*Models*	
			SR (log-units)	*Typical SR (log units)*
1	Repeatability*	*Range (IBC/µl)*		
		10 – 50	0.07	0.06
		51-200	0.05	0.04
		> 200	0.04	0.02
2	Reproducibility* (between instruments):	10 – 50	0.11	0.08
		51 - 200	0.07	0.06
		> 200	0.06	0.04
3	Carry-over effect	-	< 0.5 per cent (uncompensated)	
4	Working factor	-	Standard 300, (optionally: 95, 600 and 1200)	
5	Accuracy	-	Typical Sy.x < 0.25 log units in the entire measuring range	
6	Reference or anchor method	-	Standard Plate Count (SPC) (IDF Standard 100B:1991)	

Bactoscan-BactoScan FC

Models 65H, 130H, 200H = 65, 130, 200 samples per hour BactoScan™ FC+ Semiautomatic is only available at a capacity of 65 samples per hour. This version does not include Conveyor and Stirrer system.

Sl.No.	Parameters	Range
1	Analysis time	9 minutes
2	Sample intake	Approx. 4.5 ml
3	Sample temperature	2 – 42°C (35.6-107.6 F)
4	Sample quality	Raw milk of normal composition and good quality. Unpreserved or preserved with azidiol

Standards and Approvals

BactoScan™ FC+ is CE-labelled and complies with the following directives and regulations:

- *(i)* EMC (ElectroMagnetic Compatibility) Directive 2004/108/EC.
- *(ii)* LVD (Low Voltage) Directive 2006/95/EC.
- *(iii)* Machinery Safety Directive 2006/42/EC.
- *(iv)* Regulation (EC) 1272/2008 on classification, labelling and packaging of substances and mixture, CLP (EC).
- *(v)* WEEE Directive 2002/96/EC.
- *(vi)* Packaging and packaging waste Directive 94/62/EC.
- *(vii)* REACH 1907/2006/EC.

BactoScan™ FC+ technology complies with

- *(i)* EURL/Microval approved
- *(ii)* FDA/NCIMS approved
- *(iii)* SO/IDF and AOAC standards
- *(iv)* Several national approvals *e.g.*: Germany: Bundesanstalt für Milchforshung, France: Ceca La

Source

1. *Product literature, FOSS.*

2. Bentley Bacto Count IBC

Bacteria Enumeration (50, 100 or 150 Samples Per Hour)-The **BactoCount IBC is a fully automated instrument that uses flow cytometry** for the rapid enumeration of individual bacteria in raw milk. The high processing speed of this model makes it the ideal solution for mid to large size laboratories that need an easy to maintain, exceptionally fast bacteria counting system.

This model is available in three different configurations (BactoCount IBC 50, BactoCount IBC 100, and BactoCount IBC 150), making it easy to select an instrument that meets the needs of your laboratory or dairy.

- *(i)* Capable of analyzing 50 – 150 samples per hour.
- *(ii)* Use of a standard computer offers flexible data output options.

(iii) Fully automated Autosampler provides a steady flow of samples and

(iv) Low maintenance design.

Evaluated by the CECALAIT and IAI Reference Laboratories

The IBC can be used as an alternative method to the IDF 100B: 1991 and AOAC 986.33 reference methods for the rapid and accurate determination of the hygienic quality of cow, sheep, and buffalo raw milk for payment purposes.

Technical Overview and Principle of Operation

The IBC is a fully automated instrument that uses a proprietary **process based on flow cytometry** for the rapid enumeration of individual bacteria in raw milk. The practical application follows a simple general principle: the cells are stained with a fluorescent dye that inserts into the intracellular DNA followed by detection and enumeration by a flow cytometry based cell counter.

Modules of IBC

The IBC consists of following five modules:

1. The Counter

The counting assembly includes a powerful solid state laser, a flow cell, set of optics, a narrow band filter, and a highly sensitive photo multiplier. The bacteria are aligned in the flow cytometer and exposed to an intense laser beam to excite the fluorescent marker intercalated into the bacterial DNA. The fluorescence pulses are collected with the optics, filtered with a narrow band filter, and detected with the photo multiplier. The pulses are sorted based on duration and intensity and then translated into individual bacteria counts (IBC's). Using parallel results of IBC's and traditional plate counts the instrument can be calibrated to output results in CFU/mL. The counting assembly is compact, enclosed and regulated at 30°C (86°F) to provide a highly stable environment.

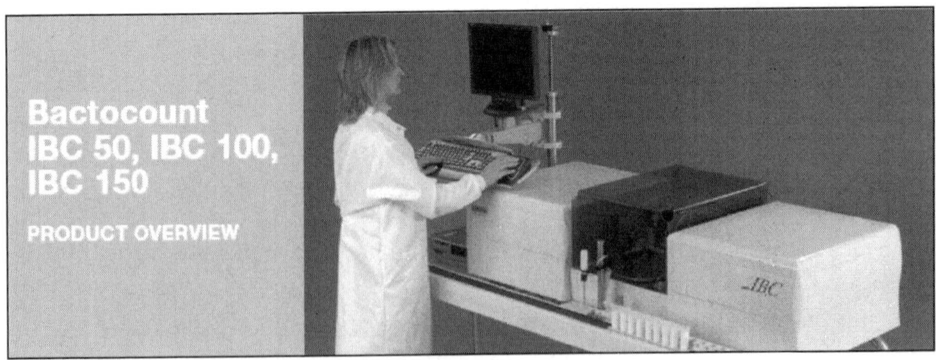

Bactocount.

2. The Carousel

The carousel assembly consists of the carousel, a sonic station, a wash station for the carousel, and a pipetting station. The carousel has 33 sample wells and is temperature regulated at 50°C (122°F). The milk and incubation reagent is dispensed simultaneously into the wells and subjected to a chemical treatment, heat treatment and two sonications in order to stain the bacteria, separate bacterial clumps and chains, and break down all interfering components. The wells and lines are automatically rinsed after each analysis.

3. The Pumping and Filtration

The pumping/filtering of the working solution and the waste removal is done primarily from the diaphragm pumps. The filtration station is located in the same area as the diaphragm pumps and the working solution reservoir. The filtration station consists of three 0.2 μm filters. The working solution reservoirs are equipped with level sensors to monitor the liquid levels and provide warnings if necessary.

4. The Autosampler

The IBC uses a belt driven conveyor that is typically setup with a 20-vial rack. The conveyor's pipette is used as both a stirrer and an empty bottle probe to detect fluid levels. The conveyor's wash station washes the outer part of the pipette while the internal part of the pipette is back flushed with a cleaning solution after each sample.

5. The Computers

The IBC is equipped with a powerful internal industrial computer. The internal computer is used to operate the instrument and send all the data and diagnostics to

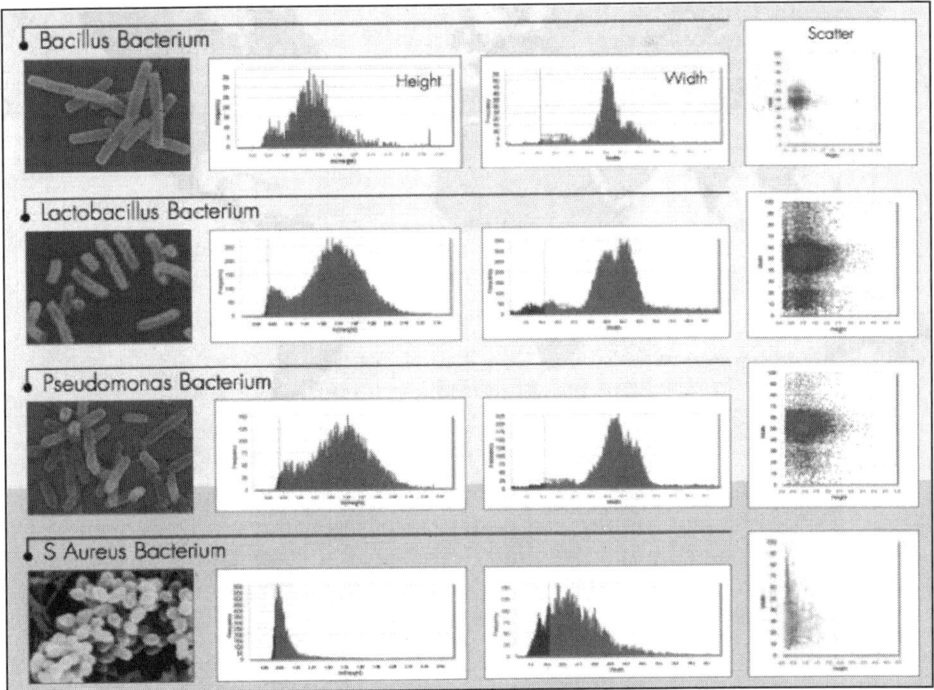

the external computer. The external computer is a standard personal computer or workstation running a version of Microsoft Windows (currently the system has been tested using Microsoft Windows 95, Windows 98, Windows 2000 and Windows XP).

- ☆ **Analysis performance:** The graphs given on back page display typical peak, width and scatter distributions obtained with the BactoCount method of the predominant bacteria species used for the assessment of the hygienic quality of raw milk.

Different Models: Two IBC models are available, making it easy to select an instrument that fits the needs of your workflow.

- ☆ **Bacto Count IBCM:** A semi-automated instrument designed for small to medium analysis. Note: the IBCm can perform somatic cell counting as well.
- ☆ **Bacto Count IBC:** A fully automated instrument designed for medium to large laboratories and dairies (Available in three sizes- will process 50, 100 or 150 samples per hour).

Bacto Count IBCM

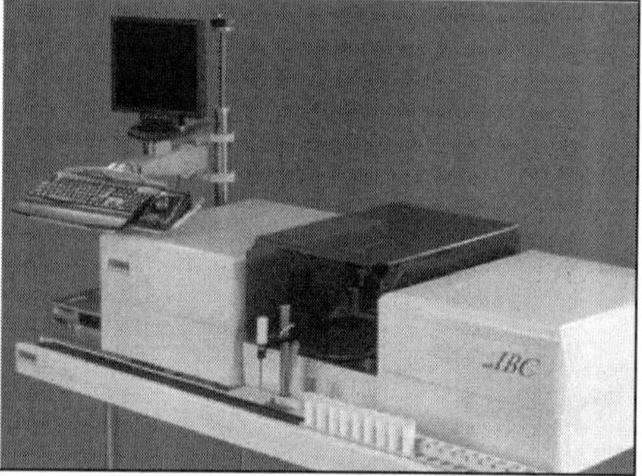

Bacto Count IBC.

Bentley Somatic Cell Counting- Early Mastitis Detection

Bentley Instruments is an industry leader in reliable and precise determination of somatic cell counts in raw milk. Four different models are available, which are making it easy to select the right instrument for your operation:

- ☆ **Soma Count FCM**-The next big innovation from Bentley Instruments includes one of the fastest somatic cell counters available. **It uses the principles of flow cytometry** to give accurate results in 7 seconds or less, making it a perfect choice for central laboratory testing.
- ☆ **Soma Count150-**The best compact cell counter in the world. Robust and reliable, it is used in both regulatory and high performing dairy laboratories around the globe. Capable of analyzing approximately 150 samples per hour. Flow cytometry at its best. For any laboratory where precision and performance is essential.

☆ **Soma Count300**-Designed for analyzing 300 samples per hour, this model uses the same technology as the Somacount 150.

☆ **Soma Count500**-Designed for analyzing 500 samples per hour, this model uses the same technology as the Somacount 150.

Source

1. *Product literatures, Bentley.*

B. Applications in the Food Industry [1]

Flow cytometry (FCM) has become a valuable tool in food microbiology. By analysing large numbers of cells individually using light-scattering and fluorescence measurements, this technique reveals both cellular characteristics and the levels of cellular components. Flow cytometry has been developed to rapidly enumerate microorganisms; to distinguish between viable, metabolically active and dead cells, which is of great importance in food development and food spoilage; and to detect specific pathogenic microorganismsby conjugating antibodies with fluorochromes, which is of great use in the food industry. In addition, high-speed multiparametric data acquisition, analysis and cell sorting, which allow other characteristics of individual cells to be studied, have increased the interest of food microbiologists in this technique.

FCM has been widely used in research since early 1975 for yeast and bacterial cell analysis, but the routine application of FCM in industrial microbiology was nonexistent before 1988. FCM is broadly being used for determination of the following properties of cells:

(i) Membrane integrity and enzyme activities.

(ii) Membrane potential.

(iii) Pathogen detection.

(iv) Respiratory activity.

(v) Intracellular pH.

Specifically FCM is successfully used for determining the microbiological properties of food. The advantage of FCM is its ability to detect micro-organisms at relatively low concentrations in a short time.Multiple labelling allows the detection of different organisms or different stages in the same sample, such as foodmatrices. The details of some food microbiological applications of FCM are given below:

(i) The application of Flow cytometry directly to food samples using a commercial Xow cytometer for the multiparameter analysis of bacteria was first described by Patchett *et al.* [2]. Use of the Skatron Argus 100 Flow cytometer enabled the counting of pure cultures of bacteria to$<10^3$cfu/mL within a few minutes. Application of the technique to meat samples gave a good correlation (r= 0.95) with plate counts and enabled enumeration down to approximately $1x10^4$ cells/mL of supernatant (equivalent to$1x10^5$cfu/g meat) within minutes.

(ii) The application of Flow cytometry to foods is restricted by the presence of high levels of suspended material. The correctly setting the discriminator gates to select bacterial counts from background signals is a critical task and could be asignificant problem with food samples containing an unknown or variable microbiota. Gunasekera *et al.* [3] developed a rapid (≤1 h) FCM assay based on enzymatic clearing of milk to determine total bacteria in milk.

(iii) The ability of FCM to determine total and viable bacterial numbers after dual staining of bacteria with SYTO BC and propidium iodide (PI) was demonstrated.

(iv) The suitability of FCM for the **brewing** [4,5] and **oenology industries** [6,7,8] has been established and demonstrated. The detection limit for yeasts was on the order of 10^3 cells/mL. Viable, dead and injured Saccharomyces cerevisiae cells were enumerated in high-sugar Chardonnay fermentations using thiazole orange (TO) and propidium iodide (PI) [8]. A combination of gating on Fluorescence and scatteral lows yeast cells to be distinguished from debris. Flowcytometry was also applied to simultaneously detect malolactic bacteria and yeasts inoculated in wine. The suitability of FCM for detecting lactic acid and acetic acid bacteria in yeast starter cultures and sparkling wines after dual microbial staining with SYTO13 (a permeant-DNA stain) and PI was also studied.

(v) **Spoilage and pathogen microorganism contamination-**Contamination of food products by bacteria, yeasts and moulds is a major concern of industries involved in food transformation. The presence of these microorganisms in products, even at low concentrations, can severely affect their quality. The best-known approach that utilises Flow cytometry for the specific detection of microorganisms is the use of mono-clonal or polyclonal antibodies conjugated to Fluorochromes such as Fluorescein isothiocyanate (FITC) or phycoerythrin [9,10]. Chemunex S. A. developed and marketed an instrument (which, together with the reagents, is called the **Chem Flow System**) to detect contaminating yeast in dairy and fruit products [11]. Detection of specific food contaminants can be achieved by FISH coupled with FCM (FLOW–FISH). For example, rapid and sensitive FLOW–FISH methods have been used for detecting and enumerating carnobacteria and Lactobacillus brevis in seafood products [12] and lactic acid bacteria used as starter cultures [13].

(vi) **Starters-** Starter culture strains are selected for their fermentation capacities and their Flavour formation characteristics. Other important criteria are the robustness of the cells during processing and the maintenance of different strains in a mixed starter [14]. In the food industry, lactic acid bacteria (LAB) and yeasts are used for the production of fermented foods, such as cheese, yoghurt, wine, and fermented meat product.

   (a) **In cheese production-** LAB are responsible for the rapid acidification of the milk. During a late stage of the process, intracellular enzymes of the starter bacteria are released through permeabilization or

complete autolysis of the cell into the cheese, playing a crucial role in the generation of flavour components. By selecting rapidly lysing strains and process conditions that favour lysis, flavour development may be enhanced during ripening [15]. FCM has been proven useful for measuring the lysis of cheese bacteria. Using the LIVE/DEAD BacLight bacterial viability kit, the lysis of Lactococcus lactis was monitored by counting the number of intact and permeable cells at different time points [16,17].

(b) **Another complete FCM assay**, with CFDA and TOTO-1, was developed for measuring the viability of dairy fermentation starters [18, 19].

(c) **Detection of infection in starter culture**-The infection of starter cultures by virulent bacteriophages is a serious problem for the dairy industry, leading to either complete loss of the fermentation batch or, if only one out of several strains is affected, an altered flavour. Michelsen *et al.* [20] have recently used FCM for fast and early detection of phage-infected Lactococcus lactis. Infected cells at the lysis phase were clearly separated from the uninfected cells by Flow cytometry.

(vi) **Probiotics-**Oral probiotics are living microorganisms that, upon ingestion in certain numbers, exert health effects beyond inherent basic nutrition. Probiotics may be consumed as either a food component or a non-food preparation [21]. In the development of probiotic foods intended for human consumption, strains of lactic acid bacteria such as Lactobacillus and Bifidobacterium have been used most commonly. The use of FCM to enumerate LAB was already addressed earlier [18,22,23]. Bunthof and Abee [18] used FCM with CFDA and TOTO-1 to quantify intact cells and cells with damaged membranes in three probiotic products - a powder, a drink and a yoghurt.FCM with cell sorting may provide an ideal tool to analyse the activity and stability of probiotics under stress conditions [24].

(vii) **Food preservation**-Although heat is still the treatment most commonly used to preserve foodstuffs, thermal processing has several undesirable effects on the nutritional and sensory attributes offoods. Alternative methods to conventional heat treatments for preservation purposes have received increasing attention in the food industry, *e.g.* **high hydrostatic pressure, high-intensity ultrasound and pulsed electric field treatments**.

The resistance of microorganisms to high pressure is variable, and the reduction in the microbial population is directly related to the amount of **hydrostatic pressure applied [25].** Cell lysis and the viability of different populations of saprophytic, pathogenic and phytopathogenic microorganisms following pressurization treatments were measured by FCM.Flow cytometry has been proven to be a useful tool for establishing the efficacy of pressurisation by determining the percentages of dead, living and metabolically inactive cells [26]. The effectiveness of Flow cytometric multiparameter analysis in characterising the physiological state of cells after pressurisationas validated by other analytical methods. Flow cytometry with CFDA and PI staining was also used to monitor pressure-induced changes in Lactobacillus rhamnosus [27]. Their results correlate well with those found in *L. monocytogenes*-pressurised cells [28]. Ananta *et al.* [29] studied the effect of high-intensity ultrasound (**thermosonication**) on *L. rhamnosus* and *E. coli*. The Gram-positive species was more resistant to the lethal effect of ultrasound than the Gram-negative species.CFDA and PI staining in combination with Flow cytometry demonstrated the rupture of the lipopolysaccharide layer of the outer membranes of Gram-negative bacteria by ultrasound.**Pulsed electric field (PEF)** treatment is the application of pulses with very high field strength for a short time (micro-seconds) to foods placed between two electrodes.Membrane permeabilization due to PEF treatment of Lactobacillus species was investigated by Wouters *et al.* [30] using PI uptake and single-cell analysis with Flow-cytometry.

Source

1. *Comas-Riu, J. and Rius, N. J Ind Microbiol Biotechnol (2009) 36: 999.*
2. *Patchett RA, Back JP, Pinder AC, Kroll RG (1991) Enumeration of bacteria in pure cultures and in foods using a commercial Flowcytometer. Food Microbiol 8: 119–125.*
3. *Gunasekera TS, Affield PV, Veal DA (2000) A FLow cytometry method for rapid detection and enumeration of total bacteria inmilk. Appl Environ Microbiol 66: 1228–1232.*
4. *Boyd AR, Gunasekera TS, Attfield PV, Simic K, Vincent SF, VealDA (2003) A Flow cytometric method for determination of yeastviability and cell number in a brewery. FEMS Yeast Res 3: 11–16.*
5. *Jespersen L, Lassen S, Jakobsen M (1993) Flow cytometric detection of wild yeast in lager breweries. Int J Food Microbiol 17: 321–328.*
6. *Attfield PV, Kletas S, Veal SA, van Rooijen R, Bell PJL (2000)Use of Flow cytometry to monitor cell damage and predict fermentation activity of dried yeasts. J Appl Microbiol 89: 207–214.*
7. *Bruetschy A, Laurent M, Jacquet R (1994) Use of Flow cytometryin oenology to analyse yeasts. Lett Appl Microbiol 18: 343–34.*
8. *Farthing JB, Rodriguez SB, Thornton RJ (2007) Flow cytometric analysis of Saccharomyces cerevisiae populations in high-sugarChardonnay fermentations.J Sci Food Agric 87: 527–533.*
9. *McClelland RG, Pinder AC (1994) Detection of Salmonella typhimurium in dairy products with flow cytometry and monoclonal antibodies. Appl Environ Microbiol 60: 4255–4262.*
10. *Donnelly CW, Baigent GT (1986) Method for Flow cytometric detection of Listeria monocytogenes in milk. Appl Environ Microbiol 52: 689–695.*
11. *Bankes P, Rowe D, Betts RP (1991) The rapid detection of yeastspoilage using the Chemflow system (Technical Memorandum 621).Campden Food and Drink Research Association, ChippingCampden.*

12. *Connil N, Dousset X, Onno B, Pilet MF, Breuil MF, Montel MC (1998) Enumeration of Carnobacterium divergens V41, Carnobacterium piscicola V1 and Lactobacillus brevis LB62 byin situ hybridization-Flow cytometry. Lett Appl Microbiol 27: 302-306.*

13. *Friedrich U, Lenke J (2006) Improved enumeration of lactic acid bacteria in mesophilic dairy starter cultures by using multiplexquantitative real-time PCR and Flow cytometry-Fluorescence insitu hybrid-ization. Appl Environ Microbiol 72: 4163–4171.*

14. *Buckenhüskes HJ (1993) Selection criteria for lactic acid bacteria to be used as starter cultures for various food commodities. FEMS Microbiol Lett 12: 253–272.*

15. *Crow VL, Coolbear T, Holland R, Pritchard GG, Martley FG (1993) Starters and Wnishers: starters properties relevant to cheeseripening. Int Dairy J 3: 423–460.*

16. *Bunthof CJ, van Schalkwijk S, Meijer W, Abee T, Hugenholtz J (2001) Fluorescent method for monitoring cheese starter permeabilization and lysis. Appl Environ Microbiol 67: 4264–4271.*

17. *Sheehan A, O'Loughlin C, O'Cuinn G, Fitzgerald RJ, Wilkinson MG (2005) Cheddar cheese cooking temperature induces differential lactococcal cell permeabilization and autolytic responses as detected by Flow cytometry: implications for intracellular enzymeaccessibility.J Appl Microbiol 99: 1007-1018.*

18. *Bunthof CJ, Abee T (2002) Development of a flow cytometric method to analyse subpopulations of bacteria in probiotic products and dairy starters. Appl Environ Microbiol 68: 2934-2942.*

19. *Bunthof CJ, Bloemen K, Breeuwer P, Rombouts FM, Abee T (2001) Flow cytometric assessment of the viability of lactic acidbacteria. Appl Environ Microbiol 67: 2326–2335.*

20. *Michelsen O, Cuesta-Dominguez A, Albrechtsen B, Jensen PR (2007) Detection of bacteriophage-infected cells of Lactococcuslactis by using flow cytometry. Appl Environ Microbiol 73: 7575-7581.*

21. *Guarner F, Schaafsma GJ (1998) Probiotics.Int J Food Microbiol39: 237-238.*

22. *Ben Amor K, Breeuwer P, Verbaarschot P, Rombouts FM, Akkermans AD, de Vos WM, Abee T (2002) Multiparametric Flow-cytometry and cell sorting for the assessment of viable, injured, and dead Bifidobacterium cells during bile salt stress. Appl Environ Microbiol 68: 5209-5216.*

23. *Lahtinen SJ, Ouwehand AC, Reinikainen JP, Korpela JM, Sandholm J, Salminen SJ (2006) Intrinsic properties of so-called dormant probiotic bacteria, determined by flow cytometric viability assays. Appl Environ Microbiol 72: 5132-5134.*

24. *Nebe von Caron G, Stephens P, Badley RA (1998) Assessment of bacterial viability status by Flow cytometry and single cell sorting.J Appl Microbiol 84: 988–99.*

25. *Arroyo G, Sanz PD, Préstamo G (1997) Effect of high pressure onthe reduction of microbial populations in vegetables. J ApplMicrobiol 82: 735–742.*

26. *Arroyo G, Sanz PD, Préstamo G (1999) Response to high-pressure, low-temperature treatment in vegetables: determination of survival rates of microbial propulations using Flow cytometry and detection of peroxidase activity using confocal microscopy. J ApplMicrobiol 86: 544–556.*

27. Ananta E, Heinz V, Knorr D (2004) Assessment of high pressure induced damage on Lactobacillus rhamnosus GG by Flow cytometry. Food Microbiol 21: 567–577.

28. Ritz M, Tholozan JL, Federighi M, Pilet MF (2001) Morphological and physiological characterization of Listeria monocytogenes subjected to high hydrostatic pressure. Appl Environ Microbiol67: 2240–2247.

29. Ananta E, Voight D, Zenker M, Heinz V, Knorr D (2005) Cellular injuries upon exposure of Escherichia coli and Lactobacillus rhamnosus to high-intensity ultrasound. J Appl Microbiol 99: 271-278.

30. Wouters PC, Bos AP, Ueckert J (2001) Membrane permeabilization in relation to inactivation kinetics of Lactobacillus species due to pulsed electric Welds. Appl Environ Microbiol 67: 3092–3101.

Chapter 7

Polymerase Chain Reaction and Real Time PCR

Advances in biotechnology have led to the development of a diverse array of assay for detection of food pathogens. Rapid analysis that used nucleic acid hybridization and nucleic acid amplification techniques offer more sensitivity and specificity than culture based methods as well as dramatic reduction in the time to get results. Many methods have also achieved the high level automation, facilitating their application as routine sample screening assays [1,2].

Although molecular techniques have improved food microbiology to a great extent. Certain techniques and methods look good and work well if used in research laboratories by skillful technician, but are not useful for routine testing of food pathogens [3]. The essential principle of nucleic acid based assays is the specific formation of double stranded nucleic acid molecules from two complementary single stranded molecules under defined physical and chemical conditions.

There are **many nucleic acid based assays but only DNA probe and PCR** has been developed commercially for detecting food pathogens [4]. **Nucleic acid amplification strategies (PCR) and advanced amplicon detection methodoloies (Real Time PCR)** have been key factors in the progress of molecular microbiology and have been used widely for the detection of food borne pathogens like Shiga toxin producing Escherichia coli (E. coli O157: H7), Listeria monocytogene, Salmonella sp., Vibrio cholerae, Vibrio parahaemolyticus, Campylobacter jejuni, Enterobacter sakazakii and Staphyloccus aureus [4a]. Recently a number of DNA based molecular typing methods, including Pulse Field Gel Electrophoresis (PFGE), Restriction Fragment Length Polymorphism (RFLP) and ribotyping have also been developed [5].

The **PCR technique** is a three step cyclic in vitro procedure based on the ability of the DNA polymerase to copy a strand of DNA. **Real-time PCR** is a rapid tool for screening of samples, still in case of positive PCR results, it should be tempted to confirm the positive result by means of the culture-based method [5a].

1. DNA Hybridization

The identification of bacteria by DNA probe hybridization is based on the presence or absence of particular genes. A gene probe is composed of nucleic acid molecules, most often double stranded DNA. It consists of either an entire gene or a fragment of a gene with a known function. Alternatively, short pieces of single stranded DNA can be synthesized, based on the nucleotide sequence of the known gene [6]. Double stranded DNA probes must be denatured before hybridization reaction, whereas, oligonucleotide and RNA probes, which are single stranded need not be denatured. Gene probes can be labeled with following two ways:

(i) **Labeling by radioactive substances**- It is done by two methods. Nick translation and random priming technique. Oligonucleotide probes are usually labeled at 5 with 32p, using bacteriophage T4 polynucleotide kinase and gamma AT 32p. Although, radioactive probes seem to have the greatest sensibility in hybridization process, they are potential hazards and disposal of radioactive wastes can be expensive.

(ii) **Labeling by non radioactive substances-**Currently, labeling of probes with non-radioactive substances such as alkaline phosphatase have been used without effecting the kinetics or specificity of the hybridization [7].

Target **nucleic acids** are denatured by **high temperature** (above 95°C) or high pH (above 12) and then the labeled gene probe is added. If the target nucleic acid in the sample contains the same nucleotide sequence as that of the gene probe, the probe will form hydrogen bond with the target. The unreacted, labeled probe is removed by washing the solid support and presence of probe target complexes is signaled by the bound label and detected by autoradiography [6].

2. DNA Microarrays (Gene Chip Technology)

This technology used photo-lithography, which was developed by computer chip makers. Microarrays or gene chips provide a miniaturized system for the simultaneous analysis of hybridization of fluorescent labelled single-strand nucleotide chains to an array of oligonucleotide probes immobilized on a support such as glass or a synthetic membrane. PCR amplification is often used prior to hybridization to increase the sensitivity of detection. A gene chip can be made of glass or nylon membrane and there are following two basic variants:

(i) In one format, target DNA is amplified by PCR and spread on to a membrane, which is then probed either singly or simultaneously by hundreds of labeled probes to determine specific hybridization.

ii In other format, an array of oligonucleotides are synthesized directly on a glass chip and then exposed to labeled target DNA.

Applications

(i) DNA chip technology also makes it possible to detect diverse individual sequence in complex DNA samples [7]. Development of this approach is continuing at a rapid pace and for microbiologists, this technology will be one of the major tools for the future.

(ii) DNA microarrays may be very useful for detecting multiple bacteria simultaneously on a single glass slide. The complexity of the food matrix is a major drawback of microarrays to be used as a detection method. Therefore, microarrays could be described as a tool for identification, genotyping, and pathotyping (detection of appropriate virulence factors) or characterization of bacterial isolates [5a].

(iii) Since DNA arrays allow simultaneous measurements of thousands of interactions between mRNA-derived target molecules and genome-derived probes, they are rapidly producing enormous amounts of raw data never before encountered by biologists. The analysis of data on this scale is a major current challenge and therefore needs permanent attention to the issue of sample preparation and genetic material purification for successful and reliable analysis [5a].

Source

1. *P.K.Mandal, A.K. Biswas, K. Choi and U.K. Pal, 2011. Methods for Rapid Detection of Foodborne Pathogens-An Overview. American Journal of Food Technology, 6: 87-102.*

2. *Wang, R.F., W.W. Cao and C.E. Cerniglia, 1997. A universal protocol for PCR detection of 13 species of food borne pathogens in food. J. Appl. Microbiol., 83: 727-736.*

3. *Rijpens, N.P. and L.M. Herman, 2002. Molecular methods for identification and detection of bacterial food pathogens. J. AOAC. Int., 85: 984-995.*

4. *Wang, H., 2002. Rapid method for detection and enumeration of Campylobacter spp. in Food. J. AOAC. Int., 85: 996-999.*

4a. *Rashmi H. M., Jitender Singh, Sunita Grover and V. K. Batish, PCR Based Techniques for Detection of Pathogens in Food, Molecular Biology Unit, Dairy Microbiology Division, NDRI, Karnal.*

5. *Ritcher, E.R., 1993. Biosensors: Application for dairy food industry. J. Dairy Sci., 76: 3114-3117.*

5a. *Technews, NDDB, Technological Advancement in Methods of Microbiology in Milk Analysis- Part 1 (Direct method, Issue no.91, October-December 2016.*

6. *Laizrd, P.W., A. Zijderveld, K. Linders, M. A. Rudnicki and A. Berns, 1991. Simplified marnmalian DNA isolation procedure. Nucleic Acid Res., 19: 4293-4293.*

7. *Pitcher, D.G., N. A. Saunders and R.J. Owen, 1989. Rapid extraction of bacterial genomic DNA with guanidium thiocyanate. Lett. Applied Microbiol., 8: 151-156.*

Fluorescent *In situ* Hybridization (FISH) [1]

FISH is a cytogenetic technique used to detect the presence/absence of specific rRNA sequences. It uses fluorescent probes that bind to only to target sequence with which they show a high degree of sequence similarity. Fluorescence microscopy can be used to find out where the fluorescent probe bound to the chromosomes. The choice to target RNA instead of DNA results in a more sensitive technique (higher copy numbers available) and the link to viability. The enumeration involves the following steps:

(i) In the elaboration of FISH, microbial cells are treated with appropriate chemical fixatives and then hybridized under stringent conditions on a glass slide or in solution with oligonucleotide probes.

(ii) Generally, these probes are 15-25 nucleotides in length and are labelled covalently at the 50 end with a fluorescent dye.

(iii) After stringent washing to remove unbound probe, specifically stained cells are detected via epifluorescence microscopy.

(iv) The limit of detection is approximately 104 cfu/ml. After prior enrichment (usually overnight) to attain these levels of detection, results are available in 3 h, while the hands-on time required per analyse is only a few minutes.

FISH is commercially exploited by, for example, Vermicon, which has a detection kit for food pathogens like Campylobacter, E coli, Enterobacter, Listeria, Salmonella, Staphylococcus and Yersinia.

Source

1. *Technews, NDDB, Technological Advancement in Methods of Microbiology in Milk Analysis- Part 1 (Direct method, Issue no.91, October-December 2016.*

Polymerase Chain Reaction (PCR)

It is a biochemical technology in molecular biology to amplify a single or a few copies of a piece of DNA across several orders of magnitude, generating thousands to millions of copies of a particular DNA sequence.Sometimes called **"molecular photocopying,"** the polymerase chain reaction (PCR) is a fast and inexpensive technique used to "amplify"- copy-small segments of DNA. Because significant amounts of a sample of DNA are necessary for molecular and genetic analyses, studies of isolated pieces of DNA are nearly impossible without PCR amplification.

The PCR based methods are not only fast but also highly specific as they are based on the genetic makeup of the organisms and are frequently used for detection and serotyping of food pathogens. The development of PCR-based assays typically includes steps of identifying the specific genomic markers, designing PCR primers, validating the specificity of the assay using reference bacterial strains, and testing and optimizing the PCR assays for detection in broth cultures and food samples.

The details related to the history or developmental journey of PCR are tabulated below [9]:

Sl.No.	Step-wise Details for PCR Development
1	A 1971 paper in the Journal of Molecular Biology by Kleppe and co-workers first described a method using an enzymatic assay to replicate a short DNA template with primers *in vitro* [1]
2	The discovery in 1976 of Taq polymerase - a DNA polymerase purified from the thermophilic bacterium, Thermus aquaticus, which naturally lives in hot (50 to 80 °C (122 to 176 °F) environments [6] such as hot springs - paved the way for dramatic improvements of the PCR method.
3	However, this early manifestation of the basic PCR principle did not receive much attention, and the invention of the polymerase chain reaction in 1983 is generally credited to Kary Mullis [2] who was working in, California for Cetus Corporation(one of the first biotechnology companies).
4	In Scientific American, Mullis summarized the procedure: "Beginning with a single molecule of the genetic material DNA, the PCR can generate 100 billion similar molecules in an afternoon. The reaction is easy to execute. It requires no more than a test tube, a few simple reagents, and a source of heat [3]. He was awarded the Nobel Prize in Chemistry in 1993 for his invention [4].
5	At the core of the PCR method is the use of a suitable DNA polymerase able to withstand the high temperatures of >90 °C (194 °F) required for separation of the two DNA strands in the DNA double helix after each replication cycle.The DNA Polymerases initially employed for in vitro experiments presaging PCR were unable to withstand these high temperatures [5]. So the early procedures for DNA replication were very inefficient and time consuming, and required large amounts of DNA polymerase and continuous handling throughout the process.
6	The DNA polymerase isolated from *T. aquaticus* is stable at high temperatures remaining active even after DNA denaturation [7], thus obviating the need to add new DNA polymerase after each cycle [8].[4] This allowed an automated thermocycler-based process for DNA amplification.

"Baby Blue", a 1986 Prototype Machine for doing PCR.

Source

1. *Kleppe K, Ohtsuka E, Kleppe R, Molineux I, Khorana HG (1971). "Studies on polynucleotides. XCVI. Repair replications of short synthetic DNA's as catalyzed by DNA polymerases".J. Mol. Biol.56 (2): 341–361.*

2. *Rabinow, Paul (1996). Making PCR: A Story of Biotechnology. Chicago: University of Chicago Press. ISBN 0-226-70146-8.*

3. *Mullis, Kary (1990). "The unusual origin of the polymerase chain reaction". Scientific American262 (4): 56–61, 64–5.*
4. *Kary B. Mullis – Nobel Lecture: The Polymerase Chain Reaction.*
5. *Saiki, R.; Scharf, S.; Faloona, F.; Mullis, K.; Horn, G.; Erlich, H.; Arnheim, N. (1985). "Enzymatic amplification of beta-globin genomic sequences and restriction site analysis for diagnosis of sickle cell anemia". Science. 230 (4732): 1350–1354.*
6. *Chien A, Edgar DB, Trela JM (1976). "Deoxyribonucleic acid polymerase from the extreme thermophile Thermus aquaticus". J. Bacteriol. 127 (3): 1550–1557.*
7. *Lawyer, F.; Stoffel, S.; Saiki, R.; Chang, S.; Landre, P.; Abramson, R.; Gelfand, D. (1993). "High-level expression, purification, and enzymatic characterization of full-length Thermus aquaticus DNA polymerase and a truncated form deficient in 5′ to 3′ exonuclease activity". PCR methods and applications. 2 (4): 275–287.*
8. *Saiki, R.; Gelfand, D.; Stoffel, S.; Scharf, S.; Higuchi, R.; Horn, G.; Mullis, K.; Erlich, H. (1988). "Primer-directed enzymatic amplification of DNA with a thermostable DNA polymerase". Science. 239 (4839): 487–491.*
9. *Wikipedia.*

Present Status of PCR

After development in 1983 by Kary Mullis [1,2], PCR is now a common, revolutionary and often indispensable technique used in medical and biological research labs for a variety of applications [3,4,6]. These include DNA cloning for sequencing, DNA-based phylogeny, or functional analysis of genes; the diagnosis of hereditary diseases; the identification of genetic fingerprints (used in forensic sciences and paternity testing); and the detection and diagnosis of infectious diseases. Often heralded as one of the most important scientific advances in molecular biology, PCR revolutionized the study of DNA to such an extent that its creator, Kary B. Mullis, was awarded the Nobel Prize for Chemistry with Michael Smith in 1993 [5].

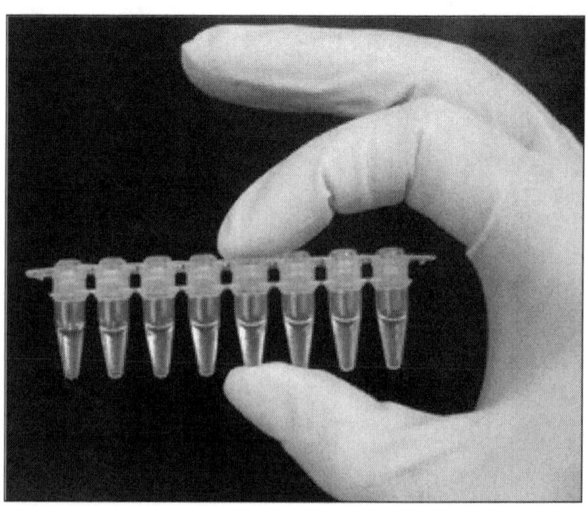

A Strip of Eight PCR Tubes, each Containing a 100 µl Reaction Mixture.

The method relies on thermal cycling, consisting of cycles of repeated heating and cooling of the reaction for DNA melting and enzymatic replication of the DNA. Primers (short DNA fragments) containing sequences complementary to the target region along with a DNA polymerase (after which the method is named) are key components to enable selective and repeated amplification. As PCR progresses, the DNA generated is itself used as a template for replication, setting in motion a chain reaction in which the DNA template is exponentially amplified. PCR can be extensively modified to perform a wide array of genetic manipulations.

PCR is based on using the ability of DNA polymerase to synthesize new strand of DNA complementary to the offered template strand. Because DNA polymerase can add a nucleotide only onto a preexisting 3′-OH group, it needs a primer to which it can add the first nucleotide. This requirement makes it possible to delineate a specific region of template sequence that the researcher wants to amplify. At the end of the PCR reaction, the specific sequence will be accumulated in billions of copies (amplicons).

Almost all PCR applications employ a heat-stable DNA polymerase, such as Taq polymerase (an enzyme originally isolated from the bacterium Thermus aquaticus). This DNA polymerase enzymatically assembles a new DNA strand from DNA building-blocks, the nucleotides, by using single-stranded DNA as a template and DNA oligonucleotides (also called DNA primers), which are required for initiation of DNA synthesis.

The vast majority of PCR methods use thermal cycling, *i.e.*, alternately heating and cooling the PCR sample through a defined series of temperature steps.

(*i*) **In the first step**, the two strands of the DNA double helix are physically separated at a high temperature in a process called DNA melting.

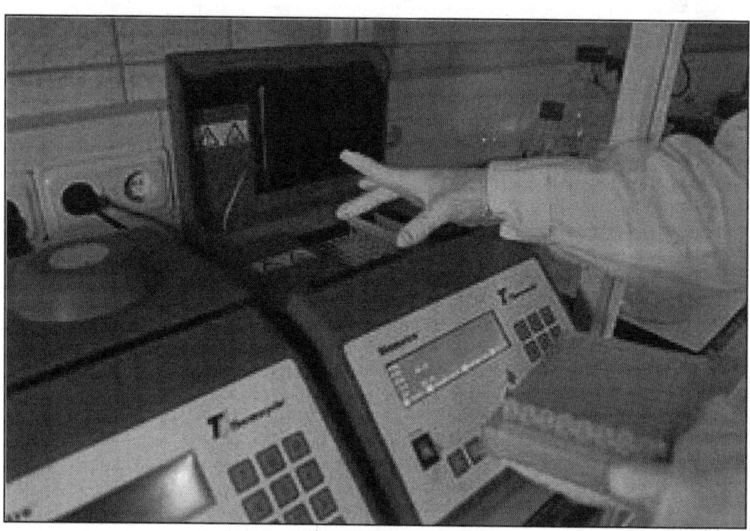

Placing a Strip of Eight PCR Tubes, each Containing a 100 µl Reaction Mixture, into the Thermal Cycler.

(ii) **In the second step,** the temperature is lowered and the two DNA strands become templates for DNA polymerase to selectively amplify the target DNA. The selectivity of PCR results from the use of primers that are complementary to the DNA region targeted for amplification under specific thermal cycling conditions.

PCR follows several steps and have some important characteristics. These are summerised below [7]:

Sl.No.	*Steps and Characteristics*
1	The region to be amplified is specified by the choice of primers. Primers are short oligonucleotides, usually 20-30 nucleotides in length, whose sequence matches the end of the region of interest. Amplification takes place over a number of cycles.
2	During each cycle, the double-stranded DNA template is denatured by heating to produce single strands. The reaction mixture is then cooled, allowing the primers to bind to the single strands.
3	This provides an active site for thermo-stable DNA polymerase, which synthesizes the complementary strand, producing again double-stranded DNA.
4	In subsequent cycles, primers will bind to both the original DNA and the newly synthesized DNA, resulting in an exponential increase in the number of copies.
5	The presence of even one copy of the template within the reaction mixture can be detected within a couple of hours as about a million fold of copies are created.
6	The results of PCR are traditionally (conventional PCR) detected by agarose gel electrophoresis and staining.
7	This enables the amplified DNA to be visualized as bands differing in size. Specificity of the bands may be further identified by sequencing. PCR as such is taking only 30 to 90 min.
8	Indeed, PCR methods for detection of pathogens in foods recommended a 6 to 8 hrs up to 24 hrs prior enrichment step before execution of PCR, as only a small volume (1ml) is processed to extract DNA.
9	The inclusion of an internal control is recommended to highlight inhibition of the PCR reaction.

Source

1. *Bartlett, J. M. S.; Stirling, D. (2003). "A Short History of the Polymerase Chain Reaction". PCR Protocols 226. pp. 3–6.*
2. *US4683195.*
3. *Saiki, R., Scharf, S., Faloona, F., Mullis, K., Horn, G.; Erlich, H., Arnheim, N. (1985). "Enzymatic amplification of beta-globin genomic sequences and restriction site analysis for diagnosis of sickle cell anemia". Science 230 (4732): 1350–1354.*
4. *Saiki, R.; Gelfand, D., Stoffel, S, Scharf, S., Higuchi, R., Horn, G., Mullis, K., Erlich, H. (1988). "Primer-directed enzymatic amplification of DNA with a thermostable DNA polymerase". Science 239 (4839): 487–491.*
5. *Kary Mullis Nobel Lecture, December 8, 1993.*
6. *Wikipedia.*
7. *Technews, NDDB, Technological Advancement in Methods of Microbiology in Milk Analysis- Part 1 (Direct method, Issue no.91, October-December 2016.*

Components and Reagents for PCR [2]

A basic PCR set up requires following components and reagents [1]:

(i) **DNA template-**It contains the DNA region (target) to be amplified (the sample DNA that contains the target sequence. At the beginning of the reaction, high temperature is applied to the original double-stranded DNA molecule to separate the strands from each other).

(ii) **Two primers-**They are complementary to the 3′ (three prime) ends of each of the sense and anti-sense strand of the DNA target. These are short pieces of single-stranded DNA that are complementary to the target sequence. The polymerase begins synthesizing new DNA from the end of the primer.

(iii) **Taq polymerase or another DNA polymerase** with a temperature optimum at around 70 °C. It is a type of enzyme that synthesizes new strands of DNA complementary to the target sequence. The first and most commonly used of these enzymes is *Taq* DNA polymerase (from *Thermis aquaticus*), whereas *Pfu* DNA polymerase (from *Pyrococcus furiosus*) is used widely because of its higher fidelity when copying DNA. Although these enzymes are subtly different, they both have two capabilities that make them suitable for PCR-a- they can generate new strands of DNA using a DNA template and primers, and b-they are heat resistant.

(iv) **Nucleotides** [Deoxynucleoside triphosphates (dNTPs), sometimes called "deoxynucleotide triphosphates"-Nucleotides containing triphosphate groups] -, single units of the bases A, T, G, and C, which are essentially "building blocks" for new DNA strands.

(v) **Buffer solution,** providing a suitable chemical environment for optimum activity and stability of the DNA polymerase.

A Thermal Cycler for PCR.

An Older Model Three-Temperature Thermal Cycler for PCR.

Different Models of Thermal Cycler.

(vi) **Divalent cations, magnesium or manganese ions**; generally Mg^{2+} is used, but Mn^{2+} can be utilized for PCR-mediated DNA mutagenesis, as higher Mn^{2+} concentration increases the error rate during DNA synthesis [9].

(vii) **Monovalent cation potassium ions.**

The PCR is commonly carried out in a reaction volume of 10–200 µl in small reaction tubes (0.2–0.5 ml volumes) in a thermal cycler. The thermal cycler heats and cools the reaction tubes to achieve the temperatures required at each step of the reaction. Many modern thermal cyclers make use of the Peltier effect, which permits both heating and cooling of the block holding the PCR tubes simply by reversing the electric current. Thin-walled reaction tubes permit favorable thermal conductivity to allow for rapid thermal equilibration. Most thermal cyclers have heated lids to prevent condensation at the top of the reaction tube. Older thermocyclers lacking a heated lid require a layer of oil on top of the reaction mixture or a ball of wax inside the tube.

Source

1. *Joseph Sambrook and David W. Russel (2001). Molecular Cloning: A Laboratory Manual (3rd ed.). Cold Spring Harbor, N.Y.: Cold Spring Harbor Laboratory Press. ISBN 0-879-69576-5. Chapter 8: In vitro Amplification of DNA by the Polymerase Chain Reaction.*

2. *Wikipedia.*

PCR Principles and Procedure [15]

1. Princple

PCR is used to amplify a specific region of a DNA strand (the DNA target). Most PCR methods typically amplify DNA fragments of between 0.1 and 10 kilo base pairs (kb), although some techniques allow for amplification of fragments up to 40 kb in size [1]. The amount of amplified product is determined by the available substrates in the reaction, which become limiting as the reaction progresses [2]. The PCR is an in-vitro method used to increase number of specific DNA sequence in a sample. PCR is used increasingly in research in food microbiology because of its high sensibility or specificity. By this method, a specific DNA fragment is amplified during a following cyclic 3-step process [3]:

(i) The target DNA is denatured at **high temperature**.

(ii) Two synthetic oligonucleotides (primers) one annealed to opposite strands at a temperature that only allows hybridization to correct target.

(iii) Polymerization is performed with the oligonucleotide as primers for the enzymes and the target DNA as template.

When this is performed over and over with namely synthesized DNA as template in addition to original target DNA, an exponential amplification of the DNA fragment between 2 primers is obtained [4]. Theoretically, PCR can amplify a single copy of DNA a million fold in less than 2 h, hence, it has the potential to

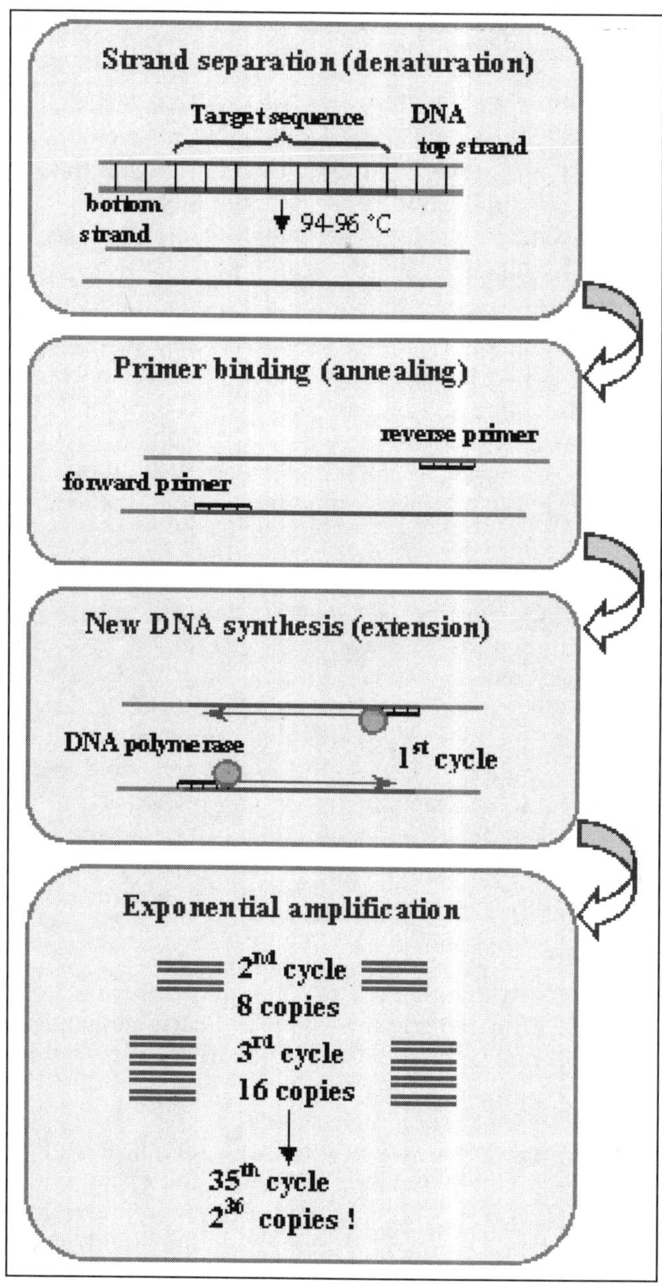

disseminate or greatly reduce the dependence on cultural enrichment. In a PCR system, assuming a sensitivity of 1 cell/reaction tube, approximately 10^3 bacteria mL^{-1} sample required to ensure a reliable and repeatable amplification [5].

The working of PCR involves the following steps:

(i) To amplify a segment of DNA using PCR, the sample is first heated so the DNA denatures, or separates into two pieces of single-stranded DNA.

(ii) Next, an enzyme called "Taq polymerase" synthesizes - builds - two new strands of DNA, using the original strands as templates. This process results in the duplication of the original DNA, with each of the new molecules containing one old and one new strand of DNA.

(iii) Then each of these strands can be used to create two new copies, and so on, and so on.

(iv) The cycle of denaturing and synthesizing new DNA is repeated as many as 30 or 40 times, leading to more than one billion exact copies of the original DNA segment. The entire cycling process of PCR is automated and can be completed in just a few hours. It is directed by a machine called a thermocycler, which is programmed to alter the temperature of the reaction every few minutes to allow DNA denaturing and synthesis.

2. Procedure [15]

The purpose of a PCR (Polymerase Chain Reaction) is to make a huge number of copies of a gene. This is necessary to have enough starting template for sequencing. Typically, PCR consists of a series of 20-40 repeated temperature changes, called cycles, with each cycle commonly consisting of 2-3 discrete temperature steps, usually three. The cycling is often preceded by a single temperature step (called hold) at a high temperature (>90°C), and followed by one hold at the end for final product extension or brief storage. The temperatures used and the length of time they are applied in each cycle depend on a variety of parameters. These include the enzyme used for DNA synthesis, the concentration of divalent ions and dNTPs in the reaction, and the melting temperature (Tm) of the primers [6].

i. Initialization Step

This step consists of heating the reaction to a temperature of 94–96 °C (or 98 °C if extremely thermostable polymerases are used), which is held for 1–9 minutes. It is only required for DNA polymerases that require heat activation by hot-start PCR [7].

ii. Denaturation Step

This step is the first regular cycling event and consists of heating the reaction to 94–98 °C for 20–30 seconds. It causes DNA melting of the DNA template by disrupting the hydrogen bonds between complementary bases, yielding single-stranded DNA molecules (During the denaturation, the double strand melts open to single stranded DNA, all enzymatic reactions stop -for example: the extension from a previous cycle).

iii. Annealing Step

The reaction temperature is lowered to 50–65 °C (54°C) for 20-40 seconds allowing annealing of the primers to the single-stranded DNA template.

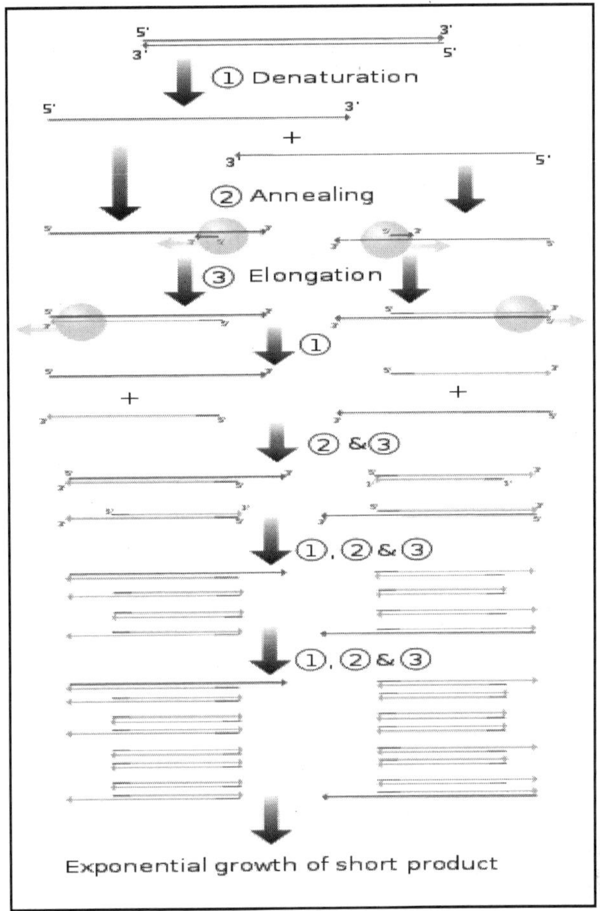

Schematic Drawing of the PCR Cycle (Four cycles are shown here)

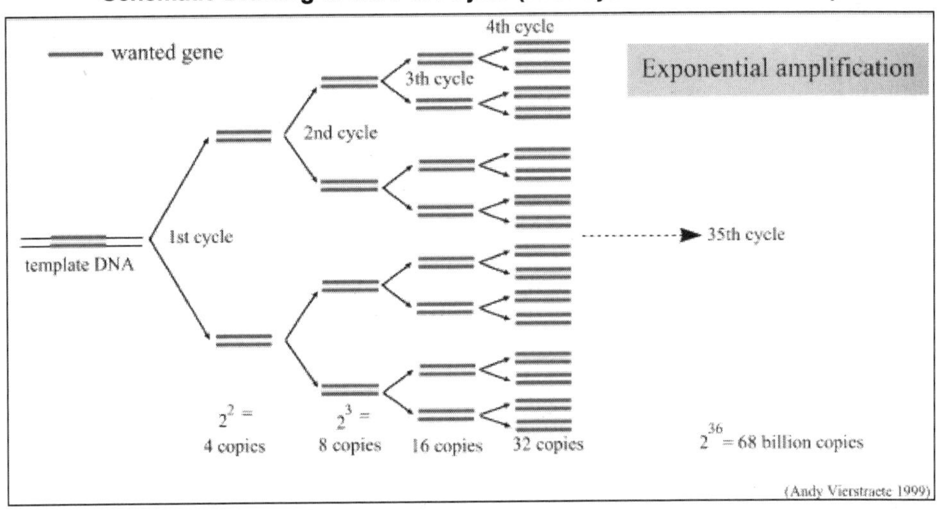

The Exponential Amplification of the Gene in PCR.

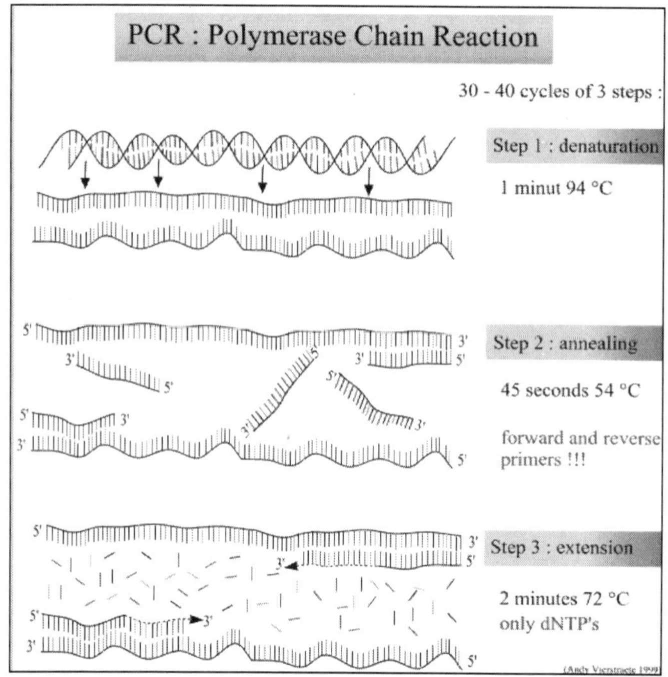

Different Steps of PCR.

Typically the annealing temperature is about 3-5 degrees Celsius below the Tm of the primers used. Stable DNA-DNA hydrogen bonds are only formed when the primer sequence very closely matches the template sequence. The polymerase binds to the primer-template hybrid and begins DNA formation. The primers are jiggling around, caused by the Brownian motion. Ionic bonds are constantly formed and broken between the single stranded primer and the single stranded template. The more stable bonds last a little bit longer (primers that fit exactly) and on that little piece of double stranded DNA (template and primer), the polymerase can attach and starts copying the template. Once there are a few bases built in, the ionic bond is so strong between the template and the primer, that it does not break anymore.

iv. Extension/Elongation Step

The temperature at this step depends on the DNA polymerase used; Taq polymerase has its optimum activity temperature at 75-80 °C [8,9], and commonly a temperature of 72 °C is used with this enzyme. At this step the DNA polymerase synthesizes a new DNA strand complementary to the DNA template strand by adding dNTPs that are complementary to the template in 5′ to 3′ direction, condensing the 5′-phosphate group of the dNTPs with the 3′-hydroxyl group at the end of the nascent (extending) DNA strand. The extension time depends both on the DNA polymerase used and on the length of the DNA fragment to be amplified. As a rule-of-thumb, at its optimum temperature, the DNA polymerase will polymerize a thousand bases per minute.Under optimum conditions, *i.e.*, if there are no limitations due to limiting substrates or reagents, at each extension step, the amount of DNA target is doubled, leading to exponential (geometric) amplification of the specific DNA fragment.

v. Final Elongation

This single step is occasionally performed at a temperature of 70–74 °C for 5-15 minutes after the last PCR cycle to ensure that any remaining single-stranded DNA is fully extended.

vi. Final Hold

This step at 4–15 °C for an indefinite time may be employed for short-term storage of the reaction.

Checking of DNA Fragmentation

To check whether the PCR generated the anticipated DNA fragment (also sometimes referred to as the amplimer or amplicon), agarose gel electrophoresis is employed for size separation of the PCR products. The size (s) of PCR products is determined by comparison with a DNA ladder (a molecular weight marker), which contains DNA fragments of known size, run on the gel alongside the PCR products.

Following figure shows ethidium bromide-stained PCR products after gel electrophoresis. Two sets of primers were used to amplify a target sequence from three different tissue samples.No amplification is present in sample 1; DNA bands in sample 2 and 3 indicate successful amplification of the target sequence. The gel also shows a positive control, and a DNA ladder containing DNA fragments of defined length for sizing the bands in the experimental PCRs.

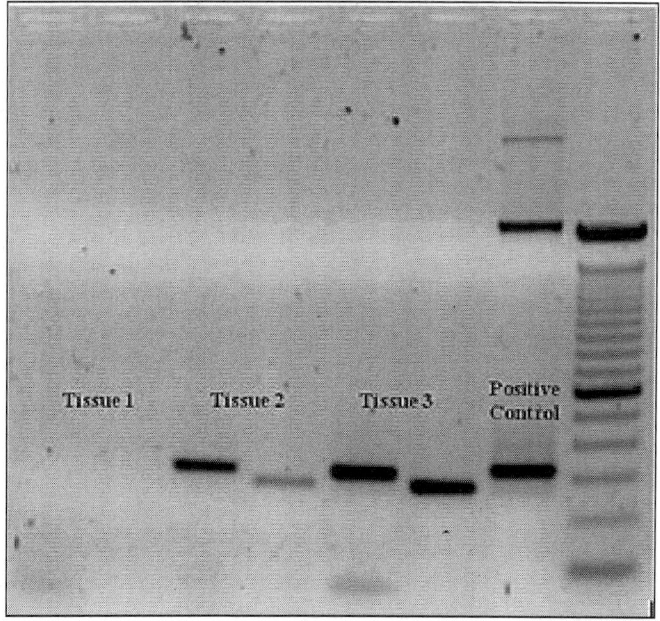

Ethidium Bromide-Stained PCR Products after Gel Electrophoresis.

Following figure shows the ladder is a mixture of fragments with known size to compare with the PCR fragments. Notice that the distance between the different fragments of the ladder is logarithmic. Lane 1: PCR fragment is approximately 1850 bases long. Lane 2 and 4: the fragments are approximately 800 bases long. Lane 3: no product is formed, so the PCR failed. Lane 5: multiple bands are formed because one of the primers fits on different places.

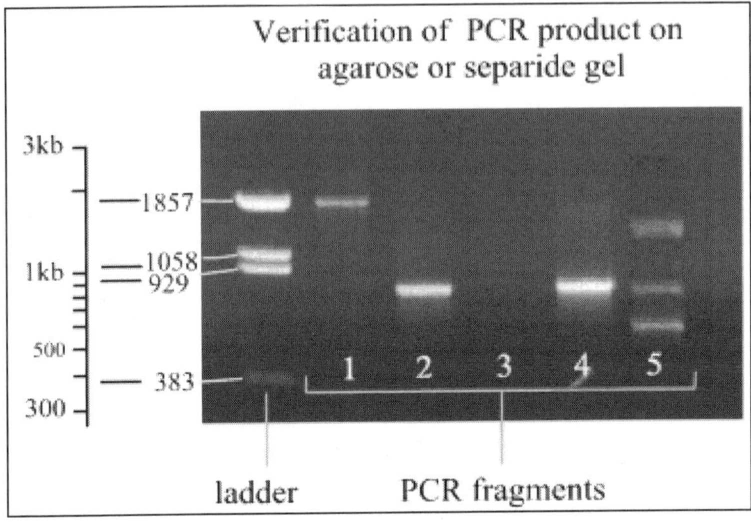

Verification of PCR Products on Gel.

PCR Stages

The PCR process can be divided into three stages:

(i) **Exponential amplification-**At every cycle, the amount of product is doubled (assuming 100 per cent reaction efficiency). The reaction is very sensitive: only minute quantities of DNA need to be present [10]. Because both strands are copied during PCR, there is an **exponential** increase of the number of copies of the gene. Suppose there is only one copy of the wanted gene before the cycling starts, after one cycle, there will be 2 copies, after two cycles, there will be 4 copies, three cycles will result in 8 copies and so on.

(ii) **Leveling off stage-**The reaction slows as the DNA polymerase loses activity and as consumption of reagents such as dNTPs and primers causes them to become limiting.

(iii) **Plateau-**No more product accumulates due to exhaustion of reagents and enzyme.

PCR Optimization [15]

In practice, PCR can fail for various reasons, in part due to its sensitivity to contamination causing amplification of spurious DNA products. Because of this, a number of techniques and procedures have been developed for optimizing PCR conditions [11,12]. Contamination with extraneous DNA is addressed with lab protocols and procedures that separate pre-PCR mixtures from potential DNA contaminants [13]. This usually involves spatial separation of PCR-setup areas from areas for analysis or purification of PCR products, use of disposable plasticware and thoroughly cleaning the work surface between reaction setups.Primer-design techniques are important in improving PCR product yield and in avoiding the formation of spurious products, and the usage of alternate buffer components or polymerase enzymes can help with amplification of long or otherwise problematic regions of DNA. Addition of reagents, such as formamide, in buffer systems may increase the specificity and yield of PCR [14]. Computer simulations of theoretical PCR results (Electronic PCR) may be performed to assist in primer design [15].

Source

1. *Cheng, S.; Fockler, C.; Barnes, W. M.; Higuchi, R. (1994). "Effective Amplification of Long Targets from Cloned Inserts and Human Genomic DNA". Proceedings of the National Academy of Sciences91 (12): 5695–5699.*

2. *Carr AC, Moore SD (2012). "Robust quantification of polymerase chain reactions using global fitting". In Lucia, Alejandro. PloS ONE7 (5): e37640.*

3. *Olsen, J.E., S. Aabo, W. Hill, S. Notermans and K. Wernars et al., 1995. Probes and PCR for detection of food-borne bacterial pathogens. Int. J. Food Microbiol., 28: 1-78.*

4. *Biswas, A.K., N. Kondaiah, K.N. Bheilegaonkar, A.S.R. Anjaneyulu and S.K. Mendiratta et al., 2008. Microbial profiles of frozen trimmings and silver sides prepared at Indian buffalo meat packing plants. Meat Sci., 80: 418-422.*

5. *Wang, R.F., W.W. Cao and C.E.Cerniglia, 1997. A universal protocol for PCR detection of 13 species of food borne pathogens in food. J. Appl. Microbiol., 83: 727-736.*

6. *Rychlik W, Spencer WJ, Rhoads RE (1990). "Optimization of the annealing temperature for DNA amplification in vitro". Nucl Acids Res18 (21): 6409-6412.*

7. *Sharkey, D. J.; Scalice, E. R.; Christy, K. G.; Atwood, S. M.; Daiss, J.L. (1994)."Antibodies as Thermolabile Switches: High Temperature Triggering for the Polymerase Chain Reaction". Bio/Technology12 (5): 506–509.*

8. *Chien A, Edgar DB, Trela JM (1976)."Deoxyribonucleic acid polymerase from the extreme thermophile Thermus aquaticus". J. Bacteriol127 (3): 1550–1557.*

9. *Lawyer, F.; Stoffel, S.; Saiki, R.; Chang, S.; Landre, P.; Abramson, R.; Gelfand, D. (1993). "High-level expression, purification, and enzymatic characterization of full-length Thermus aquaticus DNA polymerase and a truncated form deficient in 5′ to 3′ exonuclease activity".PCR methods and applications 2(4): 275-287.*

10. *Campbell Biology, 7th edition, 15-PCR from problematic templates. Focus 22: 1 p.10 (2000).*

11. *Helpful tips for PCR. Focus 22: 1 p.12 (2000).*

12. *Sarkar, G.; Kapelner, S.; Sommer, S. (1990)."Formamide can dramatically improve the specificity of PCR".Nucleic Acids Research 18 (24): 7465.*

13. *Joseph Sambrook and David W. Russel (2001).Molecular Cloning: A Laboratory*

Manual (3rd ed.). Cold Spring Harbor, N.Y.: Cold Spring Harbor Laboratory Press. ISBN 0-879-69576-5. Chapter 8: In vitro Amplification of DNA by the Polymerase Chain Reaction.

14. *"Electronic PCR". NCBI - National Center for Biotechnology Information. Retrieved 13 March 2012.*

15. *Wikipedia.*

Limitations of PCR [1]

Although PCR is a powerful technology, the reactions can be dramatically affected by the presence of inhibitory compounds in foods and selective microbiological media like bile salts and acriflavin. A problem to routine use of PCR in food testing lab is that the procedures are rather complicated and very clean environment is needed to perform the tests. Further, PCR can not distinguish between live and dead cells and hence providing more false negative results [2].

Source

1. *P.K. Mandal, A.K. Biswas, K. Choi and U.K. Pal, 2011. Methods for Rapid Detection of Foodborne Pathogens: An Overview. American Journal of Food Technology, 6: 87-102.*

2. *Biswas, A.K., N. Kondaiah, K.N. Bheilegaonkar, A.S.R. Anjaneyulu and S.K. Mendiratta et al., 2008. Microbial profiles of frozen trimmings and silver sides prepared at Indian buffalo meatpacking plants. Meat Sci., 80: 418-422.*

International Standards for Methods for PCR

The following are some internatonal standards for PCR Methods for microbiological analysis:

1. ISO Methods for PCR

(i) ISO. (2005a). Microbiology of Food and Animal Feeding Stuffs - Polymerase Chain Reaction (PCR) for the Detection of Food-Borne Pathogens - Performance Testing for Thermal Cyclers.

(ii) ISO. (2005b). Microbiology of Food and Animal Feeding Stuffs. Polymerase Chain Reaction (PCR) for the Detection of Food-Borne Pathogens. General Requirements and Definitions.

(iii) ISO. (2006a). Microbiology of Food and Animal Feeding Stuffs - Polymerase Chain Reaction (PCR) for the Detection of Food-Borne Pathogens - Requirements for Amplification and Detection for Qualitative Methods.

(iv) ISO. (2006b). Microbiology of Food and Animal Feeding Stuffs - Polymerase Chain Reaction (PCR) for the Detection of Food-Borne Pathogens - Requirements for Sample Preparation for Qualitative Detection.

(v) ISO. (2006c). Milk, Milk Products and Mesophilic Starter Cultures -Enumeration of Citrate-Fermenting Lactic Acid Bacteria -Colony-Count Technique at 25 Degrees.

(vi) ISO. (2007). Microbiology of Food and Animal Feeding Stuffs -General Requirements and Guidance for Microbiological Examinations.

(vii) ISO. (2010a). Fermented Milk Products - Bacterial Starter Cultures - Standard of Identity.

(viii) ISO. (2010b). Milk Products - Enumeration of Presumptive Bifidobacteria - Colony Count Technique at 37 Degrees.

(ix) ISO. (2011a). Microbiology of Food and Animal Feeding Stuffs - Polymerase Chain Reaction (PCR) for the Detection and Quantification of Food-Borne Pathogens – Performance Characteristics.

(x) ISO. (2011b). Microbiology of Food and Animal Feeding Stuffs - Real-Time Polymerase Chain Reaction (PCR) for the Detection of Food-Borne Pathogens - General Requirements and Definitions.

(xi) ISO. (2012). Microbiology of Food and Animal Feed - Real-Time Polymerase Chain Reaction (PCR)-Based Method for the Detection of Food-Borne Pathogens -Horizontal Method for the Detection of Shiga Toxin-Producing Escherichia coli (stec) and the Determination of o157, o111, o26, o103 and o145 Serogroups.

(xii) ISO. (2013). Microbiology of Food and Animal Feed - Horizontal Method for Determination of Hepatitis A Virus and Norovirus in Food using Real-Time RT-PCR – Part 1: Method for Quantification.

International Patent for PCR

The PCR technique was patented by **Kary Mullis** and assigned to Cetus Corporation, where Mullis worked when he invented the technique in 1983. The *Taq* polymerase enzyme was also covered by patents. There have been several high-profile lawsuits related to the technique, including an unsuccessful lawsuit brought by DuPont. The pharmaceutical company Hoffmann-La Roche purchased the rights to the patents in 1992 and currently holds those that are still protected.

A related patent battle over the Taq polymerase enzyme is still ongoing in several jurisdictions around the world between Roche and Promega. The legal arguments have extended beyond the lives of the original PCR and Taq polymerase patents, which expired on March 28, 2005 [1,2].

Source

1. *Isenbarger TA, Finney M, Ríos-Velázquez C, Handelsman J, Ruvkun G (2008). "Miniprimer PCR, a New Lens for Viewing the Microbial World". Applied and Environmental Microbiology. 74 (3): 840–9.*
2. *Wikipedia.*

Real Time PCR [1,2]

The most used arsenal for the rapid detection of food borne pathogens is Real Time PCR. This is also known as **quantitative PCR or q-PCR**. It incorporates fluorescence creating the ability to directly measure and quantify the amplification reaction. Reactions are characterized by the point in time during cycling when amplification of a PCR product is first detected rather than the amount of PCR product accumulated after a fixed number of cycles. The specific primers and probes designed, bind to the target DNA of interest and produce fluorescence. Even a single copy of target DNA can be detected due to high dynamic range.

Different Chemistries Involved in the Real Time Identification

There are around six different chemistries involved in the real time identification. The details are given below:

Sl.No.	*Different Chemistries*	*Details*
1	Intercalating Dyes	These dyes non-specifically bind to the double stranded DNA and fluoresce on excitation by appropriate source. SYBR Green is commonly used. They need melt curve analysis for the detection of specific target sequence. These SYBR Green PCR assays dyes have been extensively developed, used and validated for the detection of wide range of pathogens [3,4,5].
2	5' Nuclease Assay	This chemistry is based upon the 5' exonuclease activity of DNA polymerase and has been extensively used by different investigators for the detection of various pathogens, e.g *Salmonella spp.* detection [6].
3	Molecular Beacons	These are DNA hybridization probes that form a stem-loop structure. When a molecular beacon hybridizes to the specific target template, the hairpin structure is broken and reporter dye is no longer quenched. The fluorescence thus produced is proportional to target concentration. Use of A Molecular Beacon-Based Duplex Real-Time Polymerase Chain Reaction Assay for Simultaneous Detection of *Escherichia coli* O157:H7 and *Listeria monocytogenes* in Milk and Milk Products have reported [7].
4	Hybridization probes -FRET (Fluorescence Resonance Energy Transfer) probe	Hybridization probe or system involves two separate sequence specific oligos, which are fluorescently labeled. The upstream probe has a donor at 3'end and the downstream probe has an acceptor molecule at 5' end. Hybridization probes have also been used for detection of high-risk food pathogens.
5	Scorpion probes	These involve a hairpin loop which is linked to the 5' end of a primer via PCR stopper.After extension of primer during PCR amplification, the specific probe sequence is able to bind to its complementary sequence within the same strand of DNA. This hybridization event opens the hairpin loop so that fluorescence is no longer quenched and increase in signal is observed [8].
6	Amplifluor Technology	This technology deploys Amplifluor primer consisting of 5' intra complementary sequence labeled with reporter and quencher. The sequences 3' to the hairpin loop is the target specific primer region. PCR Based Techniques for Detection of Pathogens in Food.

Difference between PCR and Real Time PCR

(*i*) Real-time PCR or quantitative PCR is different from simple PCR whereby it does not require agarose gel electrophoresis for the detection of PCR products [2].

(ii) Real Time PCR modifies the technique in a way that reduces the chance of false positives observed in traditional PCR by 99.9 per cent.

Source

1. *Rashmi H. M., Jitender Singh, Sunita Grover and V. K. Batish, PCR Based Techniques for Detection of Pathogens in Food, Compendium on advanced Techniques and Novel Approaches for Quality and Safety Evaluation of Dairy Foods-2015, NDRI, Karnal.*
2. *Zhao X., Lin C. W., Wang J. and Oh D. H. (2014). Advances in rapid detection methods for food borne pathogens.Journal of Microbiology and Biotechnology. 24: 297–312.*
3. *Cady, N. C., Stelick, S., Kunnavakkam, V. and Batt, C. A. (2005). Real-Time PCR detection of Listeria monocytogenes using an integrated microfluidics platform. Sensors and Actuators B. 107: 332 – 341.*
4. *Nam H. M., Srinivasan V., Gillespie B. E., Murinda S. E. and Oliver S. P. (2005) Application of SYBR green real-time PCR assay for specific detection of Salmonella spp. in dairy farm environmental samples. International Journal of Food Microbiology 102: 161-171.*
5. *Brusa V., Galli L., Linares L. H., Ortega E.E., Liron JP, et al. (2015) Development and validation of two SYBR green PCR assays and a multiplex real-time PCR for the detection of Shiga toxin-producing Escherichia coli in meat. Journal of Microbiological Methods 119: 10-17.*
6. *Singh J, Batish VK, Grover S. (2012). Simultaneous detection of Listeria monocytogenes and Salmonella spp. in dairy products using real time PCR-melt curve analysis. Journal of food science and technology 49: 234-239.*
7. *Jitender Singh, Virender K. Batish, and Sunita Grover. (2009b). A Molecular Beacon based duplex RTi-PCR assay for simultaneous detection of E. coli. O157: H7 and Listeria monocytogenes in milk and milk products. Food borne pathogens and disease. 6 (10): 1195-1201.*
8. *Jitender Singh, Virender K. Batish, and Sunita Grover. (2009a). A Scorpion Probe–Based Real-Time PCR Assay for Detection of E. coli O157: H7 in Dairy Products. Food borne pathogens and disease. 6 (3): 395-400.*

Variations on the Basic PCR Technique [30]

The following types of PCR are being used:

i. Allele-specific PCR

A diagnostic or cloning technique based on single-nucleotide variations (SNVs not to be confused with SNPs) (single-base differences in a patient). It requires prior knowledge of a DNA sequence, including differences between alleles, and uses primers whose 3′ ends encompass the SNV (base pair buffer around SNV usually incorporated). PCR amplification under stringent conditions is much less efficient in the presence of a mismatch between template and primer, so successful amplification with an SNP-specific primer signals presence of the specific SNP in a sequence [1].

ii. Assembly PCR or Polymerase Cycling Assembly (PCA)

Artificial synthesis of long DNA sequences by performing PCR on a pool of long oligonucleotides with short overlapping segments. The oligonucleotides alternate between sense and antisense directions, and the overlapping segments determine the order of the PCR fragments, thereby selectively producing the final long DNA product [2].

iii. Asymmetric PCR

Preferentially amplifies one DNA strand in a double-stranded DNA template. It is used in sequencing and hybridization probing where amplification of only one of the two complementary strands is required. PCR is carried out as usual, but with a great excess of the primer for the strand targeted for amplification. Because of the slow (arithmetic) amplification later in the reaction after the limiting primer has been used up, extra cycles of PCR are required [3].

iv. AFLP PCR [29]

v. Colony PCR [29]

vi. Dial-out PCR

A highly parallel method for retrieving accurate DNA molecules for gene synthesis. A complex library of DNA molecules is modified with unique flanking tags before massively parallel sequencing. Tag-directed primers then enable the retrieval of molecules with desired sequences by PCR [5].

vii. Digital PCR (dPCR)

Used to measure the quantity of a target DNA sequence in a DNA sample. The DNA sample is highly diluted so that after running many PCRs in parallel, some of them will not receive a single molecule of the target DNA. The target DNA concentration is calculated using the proportion of negative outcomes. Hence the name 'digital PCR'.

viii. Gradient PCR [29]

ix. Helicase-Dependent Amplification

Similar to traditional PCR, but uses a constant temperature rather than cycling through denaturation and annealing/extension cycles. DNA helicase, an enzyme that unwinds DNA, is used in place of thermal denaturation [6].

x. Hot Start PCR

A technique that reduces non-specific amplification during the initial set up stages of the PCR. It may be performed manually by heating the reaction components to the denaturation temperature (*e.g.*, 95°C) before adding the polymerase [7]. Specialized enzyme systems have been developed that inhibit the polymerase's activity at ambient temperature, either by the binding of an antibody [8,8a] or by the presence of covalently bound inhibitors that dissociate only after a high-temperature activation step. Hot-start/cold-finish PCR is achieved with new hybrid polymerases that are inactive at ambient temperature and are instantly activated at elongation temperature.

xi. In Silico PCR (digital PCR, virtual PCR, electronic PCR, e-PCR)

Refers to computational tools used to calculate theoretical polymerase chain reaction results using a given set of primers (probes) to amplify DNA sequences from a sequenced genome or transcriptome. In silico PCR was proposed as an educational tool for molecular biology [9].

xii. Intersequence-Specific PCR (ISSR)

A PCR method for DNA fingerprinting that amplifies regions between simple sequence repeats to produce a unique fingerprint of amplified fragment lengths [10].

xiii. Inverse PCR

It is commonly used to identify the flanking sequences around genomic inserts. It involves a series of DNA digestions and self ligation, resulting in known sequences at either end of the unknown sequence [1].

xiv. Insetu PCR [29]

xv. Ligation-Mediated PCR

Uses small DNA linkers ligated to the DNA of interest and multiple primers annealing to the DNA linkers; it has been used for DNA sequencing, genome walking, and DNA footprinting [12].

Late PCR [29]

A recent modification on this process, known as Linear-After-The-Exponential-PCR (LATE-PCR), uses a limiting primer with a higher melting temperature (Tm) than the excess primer to maintain reaction efficiency as the limiting primer concentration decreases mid-reaction [4].

xvi. Long PCR [29].

xvii. Methylation-Specific PCR (MSP)

Developed by Stephen Baylin and Jim Herman at the Johns Hopkins School of Medicine [13], and is used to detect methylation of CpG islands in genomic DNA. DNA is first treated with sodium bisulfite, which converts unmethylated cytosine bases to uracil, which is recognized by PCR primers as thymine. Two PCRs are then carried out on the modified DNA, using primer sets identical except at any CpG islands within the primer sequences. At these points, one primer set recognizes DNA with cytosines to amplify methylated DNA, and one set recognizes DNA with uracil or thymine to amplify unmethylated DNA. MSP using qPCR can also be performed to obtain quantitative rather than qualitative information about methylation.

xviii. Miniprimer PCR

Uses a thermostable polymerase (S-Tbr) that can extend from short primers ("smalligos") as short as 9 or 10 nucleotides. This method permits PCR targeting to smaller primer binding regions, and is used to amplify conserved DNA sequences, such as the 16S (or eukaryotic 18S) rRNA gene [14].

xix. Multiplex Ligation-Dependent Probe Amplification (MLPA)

Permits multiple targets to be amplified with only a single primer pair, thus avoiding the resolution limitations of multiplex PCR.

xx. Multiplex-PCR

Consists of multiple primer sets within a single PCR mixture to produce amplicons of varying sizes that are specific to different DNA sequences. By targeting multiple genes at once, additional information may be gained from a single test-run that otherwise would require several times the reagents and more time to perform. Annealing temperatures for each of the primer sets must be optimized to work correctly within a single reaction, and amplicon sizes. That is, their base pair length should be different enough to form distinct bands when visualized by gel electrophoresis.

xxi. Nanoparticle-Assisted PCR (NanoPCR)

In recent years, it has been reported that some nanoparticles (NPs) can enhance the efficiency of PCR (thus being called nanoPCR), and some even perform better than the original PCR enhancers. It was also found that quantum dots (QDs) can improve PCR specificity and efficiency. Single-walled carbon nanotubes (SWCNTs) and multi-walled carbon nanotubes (MWCNTs) are efficient in enhancing the amplification of long PCR. Carbon nanopowder (CNP) was reported be able to improve the efficiency of repeated PCR and long PCR. ZnO, TiO2, and Ag NPs were also found to increase PCR yield. Importantly, already known data has indicated that non-metallic NPs retained acceptable amplification fidelity. Given that many NPs are capable of enhancing PCR efficiency, it is clear that there is likely to be great potential for nanoPCR technology improvements and product development [15,16].

xxii. Nested PCR

Increases the specificity of DNA amplification, by reducing background due to non-specific amplification of DNA. Two sets of primers are used in two successive PCRs. In the first reaction, one pair of primers is used to generate DNA products, which besides the intended target, may still consist of non-specifically amplified DNA fragments. The product (s) are then used in a second PCR with a set of primers whose binding sites are completely or partially different from and located 3′ of each of the primers used in the first reaction. Nested PCR is often more successful in specifically amplifying long DNA fragments than conventional PCR, but it requires more detailed knowledge of the target sequences.

xxiii. Overlap-Extension PCR or Splicing by Overlap Extension (SOEing)

A genetic engineering technique that is used to splice together two or more DNA fragments that contain complementary sequences. It is used to join DNA pieces containing genes, regulatory sequences, or mutations; the technique enables creation of specific and long DNA constructs. It can also introduce deletions, insertions or point mutations into a DNA sequence [17,18].

xxiv. PAN-AC

Uses isothermal conditions for amplification, and may be used in living cells [19,20].

xxv. Quantitative PCR (qPCR)

Used to measure the quantity of a target sequence (commonly in real-time). It quantitatively measures starting amounts of DNA, cDNA, or RNA. quantitative PCR is commonly used to determine whether a DNA sequence is present in a sample and the number of its copies in the sample. Quantitative PCR has a very high degree of precision. Quantitative PCR methods use fluorescent dyes, such as Sybr Green, EvaGreen or fluorophore-containing DNA probes, such as TaqMan, to measure the amount of amplified product in real time. It is also sometimes abbreviated to **RT-PCR (real-time PCR)** but this abbreviation should be used only for reverse transcription PCR. qPCR is the appropriate contractions for quantitative PCR (real-time PCR).

xxvi. Reverse Transcription PCR (RT-PCR)

For amplifying DNA from RNA. Reverse transcriptase reverse transcribes RNA into cDNA, which is then amplified by PCR. RT-PCR is widely used in expression profiling, to determine the expression of a gene or to identify the sequence of an RNA transcript, including transcription start and termination sites. If the genomic DNA sequence of a gene is known, RT-PCR can be used to map the location of exons and introns in the gene. The 5′ end of a gene (corresponding to the transcription start site) is typically identified by RACE-PCR (Rapid Amplification of cDNA Ends).

xxvii. Solid Phase PCR

Encompasses multiple meanings, including Polony Amplification (where PCR colonies are derived in a gel matrix, for example), Bridge PCR [21] (primers are covalently linked to a solid-support surface), conventional Solid Phase PCR (where Asymmetric PCR is applied in the presence of solid support bearing primer with sequence matching one of the aqueous primers) and Enhanced Solid Phase PCR [22] (where conventional Solid Phase PCR can be improved by employing high Tm and nested solid support primer with optional application of a thermal 'step' to favour solid support priming).

xxiii. Single Cell PCR [29]

xxix. Standard PCR [29]

xxx. Thermal Asymmetric Interlaced PCR (TAIL-PCR)

For isolation of an unknown sequence flanking a known sequence. Within the known sequence, TAIL-PCR uses a nested pair of primers with differing annealing temperatures; a degenerate primer is used to amplify in the other direction from the unknown sequence [23].

xxxi. Touchdown PCR (Step-down PCR)

A variant of PCR that aims to reduce nonspecific background by gradually lowering the annealing temperature as PCR cycling progresses. The annealing temperature at the initial cycles is usually a few degrees (3-5°C) above the T_m of the primers used, while at the later cycles, it is a few degrees (3-5°C) below the primer T_m. The higher temperatures give greater specificity for primer binding, and the lower temperatures permit more efficient amplification from the specific products formed during the initial cycles [24].

xxxii. Universal Fast Walking

For genome walking and genetic fingerprinting using a more specific 'two-sided' PCR than conventional 'one-sided' approaches (using only one gene-specific primer and one general primer — which can lead to artefactual 'noise') [25] by virtue of a mechanism involving lariat structure formation. Streamlined derivatives of UFW are LaNe RAGE (lariat-dependent nested PCR for rapid amplification of genomic DNA ends) [26], 5'RACE LaNe [27] and 3'RACE LaNe [28].

Source

1. *Newton CR, Graham A, Heptinstall LE, Powell SJ, Summers C, Kalsheker N, Smith JC, and Markham AF (1989). "Analysis of any point mutation in DNA. The amplification refractory mutation system (ARMS)".Nucleic Acids Research17 (7): 2503–2516.*
2. *Stemmer WP, Crameri A, Ha KD, Brennan TM, Heyneker HL (1995). "Single-step assembly of a gene and entire plasmid from large numbers of oligodeo-xyribonucleotides". Gene164 (1): 49–53.*
3. *Innis MA, Myambo KB, Gelfand DH, Brow MA. (1988). "DNA sequencing with Thermus aquaticus DNA polymerase and direct sequencing of polymerase chain reaction-amplified DNA".Proc Natl Acad Sci USA85 (24): 9436–4940.*
4. *Pierce KE and Wangh LJ (2007). "Linear-after-the-exponential polymerase chain reaction and allied technologies Real-time detection strategies for rapid, reliable diagnosis from single cells". Methods Mol Med. Methods in Molecular Medicine™ 132: 65–85.*
5. *Schwartz JJ, Lee C, Shendure J. (2012). "Accurate gene synthesis with tag-directed retrieval of sequence-verified DNA molecules". Nature Methods 9 (9): 913–915.*
6. *Vincent, Myriam, Xu, Yan, and Kong, Huimin (2004). "Helicase-dependent isothermal DNA amplification". EMBO Reports 5 (8): 795–800.*
7. *Q. Chou, M. Russell, D.E. Birch, J. Raymond and W. Bloch (1992). "Prevention of pre-PCR mis-priming and primer dimerization improves low-copy-number amplifications". Nucleic Acids Research20 (7): 1717–1723.*
8. *Kellogg D E, Rybalkin I, Chen S, Mukhamedova N, Vlasik T, Siebert P D, Chenchik A. 1994. TaqStart Antibody: hot start PCR facilitated by a neutralizing monoclonal antibody directed against Taq DNA polymerase. BioTechniques 16 (6): 1134-1137.*

8a. *Sharkey, D. J.; Scalice, E. R.; Christy, K. G.; Atwood, S. M.; Daiss, J.L. (1994)."Antibodies as Thermolabile Switches: High Temperature Triggering for the Polymerase Chain Reaction".Bio/Technology12 (5): 506–509.*

9. *San Millan RM, Martinez-Ballesteros I, Rementeria A, Garaizar J, Bikandi J (2013). "Online exercise for the design and simulation of PCR and PCR-RFLP experiments". BMC Research Notes6: 513.*

10. *E.Zietkiewicz, A.Rafalskiand D.Labuda (1994). "Genome fingerprinting by simple sequence repeat (SSR)-anchored polymerase chain reaction amplification". Genomics20 (2): 176–83.*

11. *Ochman H, Gerber AS, Hartl DL (1988). "Genetic Applications of an Inverse Polymerase Chain Reaction". Genetics120 (3): 621-623.*

12. *Mueller PR, Wold B (1988). "In vivo footprinting of a muscle specific enhancer by ligation mediated PCR". Science246 (4931): 780–786.*

13. *Herman JG, Graff JR, Myöhänen S, Nelkin BD, Baylin SB (1996). "Methylation-specific PCR: a novel PCR assay for methylation status of CpG islands".Proc Natl Acad Sci USA93 (13): 9821–9826.*

14. *Isenbarger TA, Finney M, Ríos-Velázquez C, Handelsman J, Ruvkun G (2008). "Miniprimer PCR, a New Lens for Viewing the Microbial World". Applied and Environmental Microbiology74 (3): 840–9.*

15. *Cenchao Shen, Wenjuan Yang, Qiaoli Ji, Hisaji Maki, Anjie Dong, Zhizhou Zhang (2009). "NanoPCR observation: different levels of DNA replication fidelity in nanoparticle-enhanced polymerase chain reactions". Nanotechnology20: 455103.*

16. *Shen, Cenchao (2013). An Overview of NanoparticleAssisted Polymerase Chain Reaction Technology.US: Wiley-Blackwell Publishing Ltd. pp. 97–106.*

17. *Horton RM, Hunt HD, Ho SN, Pullen JK, Pease LR (1989). "Engineering hybrid genes without the use of restriction enzymes: gene splicing by overlap extension". Gene77 (1): 61–68.*

18. *Moller, Simon (2006). PCR (THE BASICS). US: Taylor and Francis Group. p. 144.*

19. *David, F. and Turlotte, E. (1998). "An Isothermal Amplification Method". C.R. Acad. Sci Paris, Life Science321 (1): 909-914.*

20. *Fabrice David (September–October 2002)."Utiliser les propriétés topologiques de l'ADN: une nouvelle arme contre les agents pathogènes" (PDF) 92. Fusion. (in French).*

21. *Bing, D. H., C. Boles, F. N. Rehman, M. Audeh, M. Belmarsh, B. Kelley, and C. P. Adams. (1996). "Bridge amplification: a solid phase PCR system for the amplification and detection of allelic differences in single copy genes".Genetic Identity Conference Proceedings, Seventh International Symposium on Human Identification.*

22. *Khan Z, Poetter K, Park DJ (2008). "Enhanced solid phase PCR: mechanisms to increase priming by solid support primers". Analytical Biochemistry375 (2): 391–393.*

23. *Y.G. Liu and R. F. Whittier (1995). "Thermal asymmetric interlaced PCR: automatable amplification and sequencing of insert end fragments from P1 and YAC clones for chromosome walking". Genomics 25 (3): 674–81.*

24. *Don RH, Cox PT, Wainwright BJ, Baker K, Mattick JS (1991). "Touchdown' PCR to circumvent spurious priming during gene amplification".Nucl Acids Res19 (14): 4008.*

25. *Myrick KV, Gelbart WM (2002). "Universal Fast Walking for direct and versatile determination of flanking sequence".Gene284 (1–2): 125–131.*

26. *Park DJ Electronic Journal of Biotechnology (online). 15 August 2005, vol. 8, no. 2.*

27. *Park DJ (2005). "A new 5′ terminal murine GAPDH exon identified using 5′RACE LaNe". Molecular Biotechnology29 (1): 39–46.*

28. *Park DJ (2004). "3′RACE LaNe: a simple and rapid fully nested PCR method to determine 3′-terminal cDNA sequence".Biotechniques36 (4): 586–588, 590.*

29. *Kavya SR, POCR Technique with its application, Review-Journal of Microbiology and Biotechnology, March 2015.*

30. *Wikipedia.*

Application of PCR [1]

The PCR based methods are not only fast but also highly specific as they are based on the genetic makeup of the organisms and are frequently used for detection and serotyping of food pathogens. The development of PCR-based assays typically includes steps of identifying the specific genomic markers, designing PCR primers, validating the specificity of the assay using reference bacterial strains, and testing and optimizing the PCR assays for detection in broth cultures and food samples. The PCR is generally used for the following purposes:

1. In Microbiology and Atomic Science

PCR is utilized as a part of exploration labs in the following fields:

(i) DNA cloning techniques.

(ii) Southern smudging.

(iii) DNA sequencing.

(iv) Recombinant DNA innovation.

a. Selective DNA Isolation

PCR allows isolation of DNA fragments from genomic DNA by selective amplification of a specific region of DNA. This use of PCR augments many methods, such as generating hybridization probes for Southern or northern hybridization and DNA cloning, which require larger amounts of DNA, representing a specific DNA region. PCR supplies these techniques with high amounts of pure DNA, enabling analysis of DNA samples even from very small amounts of starting material.

Other applications of PCR include DNA sequencing to determine unknown PCR-amplified sequences in which one of the amplification primers may be used in Sanger sequencing, isolation of a DNA sequence to expedite recombinant DNA technologies involving the insertion of a DNA sequence into a plasmid or the genetic material of another organism. Bacterial colonies (E. coli) can be rapidly screened by PCR for correct DNA vector constructs [2].

Source

1. *Wikipedia*

2. *Pavlov AR, Pavlova NV, Kozyavkin SA, Slesarev AI (2006). "Thermostable DNA Polymerases for a Wide Spectrum of Applications: Comparison of a Robust Hybrid TopoTaq to other enzymes". In Kieleczawa J. DNA Sequencing II: Optimizing Preparation and Cleanup. Jones and Bartlett. pp. 241–257.*

b. Amplification and Quantification of DNA

Because PCR amplifies the regions of DNA that it targets, PCR can be used to analyze extremely small amounts of sample. This is often critical for forensic analysis, when only a trace amount of DNA is available as evidence. PCR may also be used in the analysis of ancient DNA that is tens of thousands of years old. These PCR-based techniques have been successfully used on animals, such as a forty-thousand-year-old mammoth, and also on human DNA, in applications ranging from the analysis of Egyptian mummies to the identification of a Russiantsar and the body of British king Richard III [1,2].

Quantitative PCR methods allow the estimation of the amount of a given sequence present in a sample-a technique often applied to quantitatively determine levels of gene expression. Quantitative PCR is an established tool for DNA quantification that measures the accumulation of DNA product after each round of PCR amplification.

Source

1. *Wikipedia.*
2. *"Chemical Synthesis, Sequencing, and Amplification of DNA (class notes on MBB/ BIO 343)". Arizona State University. Retrieved 2007-10-29.*

2. Medical Applications [1]

PCR innovation encourages the location of DNA or RNA of pathogenic living beings and helps in clinical analytic tests for a scope of irresistible specialists like infections, microbes, protozoa [6] and so on. These PCR-based tests have various points of interest over routine counter acting agent based indicative systems that focus the body's resistant reaction to a pathogen [7]. PCR-based tests have been produced to specify the measure of infection in a man's blood ('viral burden') in this manner permitting doctors to check their patients' sickness movement and reaction to treatment. This has fantastic potential for enhancing the clinical administration of ailments created by popular disease, including AIDS [8] and hepatitis [9], appraisal of viral load all through and after treatment [10]. In clinical microbiology labs PCR is priceless for the analysis of microbial diseases and **epidemiological studies** [2]. PCR has been applied to a large number of medical procedures.

a. PCR in Diagnosis of Diseases

(i) PCR permits early diagnosis of malignant diseases such as leukemia and lymphomas, which is currently the highest-developed in cancer research and is already being used routinely. PCR assays can be performed directly on genomic DNA samples to detect translocation-specific malignant cells at a sensitivity that is at least 10,000 fold higher than that of other methods.

(ii) PCR also permits identification of non-cultivatable or slow-growing microorganisms such as mycobacteria, anaerobic bacteria, or viruses from tissue culture assays and animal models. The basis for PCR diagnostic applications in microbiology is the detection of infectious agents and the discrimination of non-pathogenic from pathogenic strains by virtue of specific genes [1,3].

(iii) Viral DNA can likewise be detected by PCR. The primers used need to be specific to the targeted sequences in the DNA of a virus, and the PCR can be used for diagnostic analyses or DNA sequencing of the viral genome. The high sensitivity of PCR permits virus detection soon after infection and even before the onset of disease. Such early detection may give physicians a significant lead time in treatment. The amount of virus ("viral load") in a patient can also be quantified by PCR-based DNA quantitation techniques.

(iv) **Infectious disease applications-**The Characterization and detection of infectious disease organisms have been revolutionized by PCR. PCR-based diagnostics tests are accessible for identifying and/or evaluating various pathogens, including [2]:

(a) **HIV [1]** -The Human Immunodeficiency Virus (or HIV), of antibodies to the virus circulating in the blood stream. However, antibodies don't appear until many weeks after infection, maternal antibodies mask the infection of a newborn, and therapeutic agents to fight the infection don't affect the antibodies.PCR can be utilized for HIV test (vicinity of the HIV infection that causes AIDS can be resolved utilizing PCR on platelets). PCR tests have been created with the goal that it can recognize one viral genome among the DNA of more than 50,000 host cells [4,5]. Infections can be detected earlier, donated blood can be screened directly for the virus, newborns can be immediately tested for infection, and the effects of antiviral treatments can be quantified.

(b) **Hepatitis B and C** infections, may prompt liver malignancy.

(c) **Human Papillomavirus**, may bring about cervical growth.

(d) **Chlamydia trachomatis**, may prompt fruitlessness in ladies.

(e) **Neisseria gonorrhoeae**, may prompt pelvic provocative illness in ladies

(f) **Cytomegalovirus**, may bring about existence debilitating illness in transplant patients and other **immunocompromised** individuals, including HIV-1/AIDS patients.

(g) **Tuberculosis** *etc.*-Mycobacterium tuberculosis, which in its dynamic state causes tuberculosis and can prompt tissue harm of tainted organs. Some disease organisms, such as that for Tuberculosis, are difficult to sample from patients and slow to be grown in the laboratory. PCR-based tests have allowed detection of small numbers of disease organisms (both live or dead), in convenient samples. Detailed genetic analysis can also be used to detect antibiotic resistance, allowing

immediate and effective therapy. The effects of therapy can also be immediately evaluated.

The spread of a disease **organism** through populations of domestic or wild animals can be monitored by PCR testing. In many cases, the appearance of new virulent sub-types can be detected and monitored. The sub-types of an organism that were responsible for earlier epidemics can also be determined by PCR analysis.

Source

1. *Wikipedia.*
2. *Kavya SR, POCR Technique with its application, Review-Journal of Microbiology and Biotechnology, March 2015.*
3. *Salis AD (2009)."Applications in Clinical Microbiology".Real-Time PCR: Current Technology and Applications. Caister Academic Press. ISBN 978-1-904455-39-4.*
4. *Kwok S et al., "Identification of HIV sequences by using in vitro enzymatic amplification and oligomer cleavage detection." J. Virol. vol. 61 (5) pp. 1690-4 (1987).*
5. *Barnor JS et al., Establishment of In-House Quantitative Real-Time RT-PCR Assay for HIV-1 Viral Load Measurement: Application to Evaluate Efficacy of ART in Ghanaian Patients in an Urban Setting. J AIDS Clin Res. 2014; 5: 305.*
6. *Park SW and Wei LN. Can GPCR, such as Kappa Opioid Receptor, be a Viable Therapeutic Target for Reducing or Preventing Neuroinflammation? ClinExpPharmacol. 2014; 4: e127.*
7. *Iida T et al., Calcified Granulomatous Lung Lesions Contain Abundant Mycobacterium tuberculosis Components. J Mycobac Dis. 2014; 4: 142.*
8. *Huws SA et al., Differential Colonization of Plant Parts by the Rumen Microbiota is likely to be due to Different Forage Chemistries. J MicrobBiochem Technol. 2014; 6: 080-086.*
9. *Olusegun AF et al., Co- Endemicity of Plasmodium falciparum and HIV-Infections in Treated Patients is Uncorrelated in Benin City, Nigeria. J Bacteriol Parasitol.2013; 4: 176.*
10. *Shakal MA et al., Surveillance on Avian Influenza H5N1 and H9N2 Subtypes In Egypt 2012-2013. Poult Fish Wildl Sci. 2013; 2: 111.*

b. Genetic Analysis [3]

The Polymerase Chain Reaction (PCR) has found widespread application in many areas of genetic analysis. For most, if not all of these applications, alternative methods of DNA (or RNA) amplification, can be substituted. For example Helicase-dependent amplification (HDA) and Recombinase Polymerase Amplification (RPA) are isothermal alternatives to PCR that can be used for the same purposes.

(i) The first application of PCR [1] was for genetic testing, where a sample of DNA is analyzed for the presence of genetic disease mutations. Prospective parents can be tested for being genetic carriers, or their children might be tested for actually being affected by a disease. DNA samples for Prenatal

testing can be obtained by amniocentesis, chorionic villus sampling, or even by the analysis of rare fetal cells circulating in the mother's bloodstream. PCR analysis is also essential to Preimplantation genetic diagnosis, where individual cells of a developing embryo are tested for mutations.

(ii) PCR can also be used as part of a sensitive test for tissue typing, vital to organ transplantation. As of 2008, there is even a proposal to replace the traditional antibody-based tests for blood type with PCR-based tests [2].

(iii) Many forms of cancer involve alterations to oncogenes. By using PCR-based tests to study these mutations, therapy regimens can sometimes be individually customized to a patient.

(iv) PCR can be utilized for hereditary testing, where an example of DNA is broke down for the vicinity of hereditary illness transformations [4]. PCR can be utilized as a major aspect of a touchy test for tissue writing, key to organ transplantation [5].

Source

1. *Saiki RK et al., "Enzymatic Amplification of β-globin Genomic Sequences and Restriction Site Analysis for Diagnosis of Sickle Cell Anemia" Science vol. 230 pp. 1350-54 (1985).*

2. *Quill E "Blood-Matching Goes Genetic" Science Magazine (14 March 2008) pp. 1478-1479.*

3. *Wikipedia.*

4. *Hartshorne T et al., A High-throughput Real-time PCR Approach to Pharmacogenomics Studies. J Pharmacogenomics Pharmacoproteomics. 2014; 5: 133.*

5. *Valenzuela-Muñoz V et al., Molecular Characterization and Transcription Analysis of P-Glycoprotein Gene from the Salmon Louse Caligusrogercresseyi. J Aquac Res Development. 2014; 5: 236.*

3. Forensic Applications [1]

PCR is additionally utilized as a part of crime scene investigation labs and is particularly valuable on the grounds that just a little measure of unique DNA is needed, for instance, adequate DNA can be gotten from a bead of blood or a solitary hair [2,3]. The development of PCR-based genetic (or DNA) **fingerprinting protocols** has seen widespread application in forensics. PCR can be utilized for **DNA fingerprinting** that can help in Parental testing (DNA sequencing) [4]. PCR may also be used for **genetic fingerprinting;** a forensic technique used to identify a person or organism by comparing experimental DNAs through different PCR-based methods. Some PCR 'fingerprints' methods have high discriminative power and can be used to identify genetic relationships between individuals, such as parent-child or between siblings, and are used in paternity testing.

(i) In its most discriminating form, Genetic fingerprinting can uniquely discriminate any one person from the entire population of the world.

Minute samples of DNA can be isolated from a crime scene, and compared to that from suspects, or from a DNA database of earlier evidence or convicts. Simpler versions of these tests are often used to rapidly rule out suspects during a criminal investigation. Evidence from decades-old crimes can be tested, confirming or exonerating the people originally convicted.

(ii) Less discriminating forms of DNA fingerprinting can help in Parental testing, where an individual is matched with their close relatives. DNA from unidentified human remains can be tested, and compared with that from possible parents, siblings, or children. Similar testing can be used to confirm the biological parents of an adopted (or kidnapped) child. The actual biological father of a newborn can also be confirmed (or ruled out).

Source

1. *Wikipedia.*
2. *Hewedy OA et al., Phenotypic and Genotypic Diversity of Rhizobia NodulatingFaba Bean from Various Egyptian Locations. J Bioprocess Biotech.2014; 4: 170.*
3. *Kavya SR, POCR Technique with its application, Review-Journal of Microbiology and Biotechnology, March 2015.*
4. *Poh JJ and Gan SKE. The Determination of Factors Involved in Column-Based Nucleic Acid Extraction and Purification. J Bioprocess Biotech. 2014; 4: 157.*

4. Research Applications [1]

PCR has been applied to many areas of research in molecular genetics.

(i) PCR allows rapid production of short pieces of DNA, even when nothing more than the sequence of the two primers is known. This ability of PCR augments many methods, such as generating hybridization probes for Southern or northern blot hybridization.PCR supplies these techniques with large amounts of pure DNA, sometimes as a single strand, enabling analysis even from very small amounts of starting material. The task of DNA sequencing can also be assisted by PCR. Known segments of DNA can easily be produced from a patient with a genetic disease mutation. Modifications to the amplification technique can extract segments from a completely unknown genome, or can generate just a single strand of an area of interest.

(ii) **DNA cloning**-PCR can be utilized for DNA cloning. It can concentrate fragments for insertion into a vector from a bigger genome, which may be accessible in little amounts [5]. PCR has numerous applications to the more traditional process of DNA cloning. It can extract segments for insertion into a vector from a larger genome, which may be only available in small quantities. Using a single set of 'vector primers', it can also analyze or extract fragments that have already been inserted into vectors. Some alterations to the PCR protocol can **generate mutations** (general or site-directed) of an inserted fragment.Sequence-tagged sites is a process

where PCR is used as an indicator that a particular segment of a genome is present in a particular clone. The Human Genome Project found this application vital to mapping the cosmid clones they were sequencing, and to coordinating the results from different laboratories.

(iii) PCR can be utilized for **Genetic fingerprinting** (scientific science) can remarkably separate one individual from the whole populace of the world [2]. Moment tests of DNA can be detached from a wrong doing scene and contrasted with that from suspects or from a DNA database of prior confirmation or convicts [3].

(iv) **Phylogenic analysis of DNA**-An exciting application of PCR is the phylogenic analysis of DNA from ancient sources, such as that found in the recovered bones of Neanderthals, or from frozen tissues of Mammoths. In some cases the highly degraded DNA from these sources might be reassembled during the early stages of amplification.

(v) **Patterns of gene expression**- A common application of PCR is the study of patterns of gene expression. Tissues (or even individual cells) can be analyzed at different stages to see which genes have become active, or which have been switched off. This application can also use quantitative PCR to quantitate the actual levels of expression.

(vi) **Genetic mapping**-The ability of PCR to simultaneously amplify several loci from individual sperm [4] has greatly enhanced the more traditional task of genetic mapping by studying chromosomal crossovers after meiosis. Rare crossover events between very close loci have been directly observed by analyzing thousands of individual sperms. Similarly, unusual deletions, insertions, translocations, or inversions can be analyzed, all without having to wait (or pay for) the long and laborious processes of fertilization, embryogenesis, *etc.*

Source

1. *Wikipedia.*

2. *Poh JJ and Gan SKE. The Determination of Factors Involved in Column-Based Nucleic Acid Extraction and Purification. J Bioprocess Biotech. 2014; 4: 157.*

3. *Ebrahimzadeh A et al., Nested Polymerase Chain Reaction (PCR) on Fixed Stained Slides in Comparison to Whole Blood as a Source of DNA in Southeast of Iran. J Trop Dis. 2014; 2: 136.*

4. *Boehnke M et al., "Fine-structure genetic mapping of human chromosomes using the polymerase chain reaction on single sperm." Am J Hum Genet vol. 45 (1) pp. 21-32 (1989).*

5. *Claudia V etal. Targeting Cancer Related Genes by Multiplex PCR Followed by High Throughput Parallel Sequencing. Int J Genomic Med. 2014; 2: 115.*

5. In Nourishment Science [1]

PCR has turn out to be progressively essential to the horticultural and sustenance businesses as an important distinct option for customary identification strategies [2].

Source

1. *Kavya SR, POCR Technique with its application, Review-Journal of Microbiology and Biotechnology, March 2015.*
2. *Ahmed OB et al., Detection of Salmonella in Food Samples by Culture and Polymerase Chain Reaction Methods. J BacteriolParasitol. 2014; 5: 187.*

6. Application of PCR in Dairying

The some of the applications of PCR in dairying are detailed below:

A. Microbiological Quality Indicators in Waters of Dairy Farms-Detection of Pathogens by PCR in Real Time

When contaminated water is used to wash the udders of dairy cattle and milking utensils, raw milk may become contaminated with pathogens. Washing with high quality water is essential to reduce the microbial contamination of milk. Further more, the wastewater generated in dairy herds also contains high populations of pathogens, antibiotics and nutrients that more often are thrown into the water bodies without any treatment. Both supply water and wastewater from 20 dairy farms from Antioquia, Colombia was monitored for 10 months to determine the presence of pathogenic microorganisms. The Polymerase Chain Reaction (PCR) technique in real time was successfully used to determine both Cryptosporidium and Fasciola. The results showed that the supply water used for drinking and activities involving the herd, has high populations of Fasciola hepatica and Cryptosporidiumparvum, with percentages of about 53.7 per cent and 64.75 per cent respectively. Additionally high populations of Pseudomonas aeruginosa, Shigella, Salmonella, total coliforms and Escherichia coli were found in both types of water, with values around 9.4×10^7, 2.1×10^7, 1.8×10^7, 1.9×10^{10} and 1.5×10^{10} UFC/100 ml respectively for the waste water and 3.1×10^4, 1.9×10^4, 7.3×10^3, 1.2×10^5 and 6.2×10^3 UFC/100 ml for the supply water.

Source

1. *Rodríguez DC, Pino N, Peñuela G., Sci Total Environ. 2012 Jun 15; 427-428: 314-8.*

B. Detection and Quantification of Lactose-Fermenting Enterobacteriaceae in the Dairy and other Food Industries by a Novel Real-Time Polymerase Chain Reaction-based Method

The presence of lactose-fermenting Enterobacteriaceae and coliforms is routinely assessed to determine the hygienic quality of water and foods, particularly dairy products. In this study, the use of **lacZ-specific primers** in an **SYBR green I-based real-time PCR method** was reported for the easy and rapid detection of coliforms in dairy products. The sensitivity of the method was assessed using artificially contaminated cheeses. The limit of detection was 1 coliform cell in cheese samples enriched for 8h in a culture medium. The entire procedure, including sample processing, enrichment, DNA extraction, and real-time PCR amplification, can be completed within 10 to 12h, making it a single-day assay.

Source

1. *Martín MC, Martínez N, del Rio B, Ladero V, Fernández M, Alvarez MA., J.Dairy Sci. 2010 Mar; 93 (3): 860-7.*

C. Confirmation of Presence of Selected Pathogenic Bacteria along Milk Value Chain in Tanga Region by the Use of Polymerase Chain Reaction (PCR)

Upto 90 per cent of all dairy-related diseases are due to pathogenic bacteria found in milk. Emergence of new milk-borne bacterial pathogens with more serious challenges such as **E.coli O157: H7** has been reported. This study has both main and specific objectives. The details are given below:

(a) The main objective of the study was to determine milk-borne zoonotic risks associated with milk handling and storage along the dairy value chain.

(b) Specific objectives included:

  (i) To assess possible sources of microbial contamination of milk from farm to consumer.

  (ii) To identify practices that lead to microbial contamination along the milk value chain.

  (iii) To enumerate total number of bacteria and coliforms present in milk.

  (iv) To assess presence of *E. coli* and *B. abortus* along the milk value chain.

Methods

Milk samples and questionnaires from all the conbsumers along the dairy value chains in Tanga region (Handeni and Lushoto districts) were collected. The questionnaires were analyzed while standard microbiological cultures and polymerase chain reaction were used to identify and confirm microbial contamination in milk.

Results

A total of 184 [from different segments including farmers118, collection centres (4), restaurants (16), vendors (35), consumers (11)] milk samples were collected and a total of 81 were analysed. From the sociological data, different practices that could lead to microbial contamination of milk during milking, storage and delivery were identified. Preliminary analysis showed that only 43 per cent of farmers cleaned the animal shed before milking, only 71 per cent washed hands before milking, and, only 71 per cent of farmers washed the cows' teats before milking.

Source

1. *Shija, F., Misinzo, G., Nonga, H., Kurwijila, L. R., Roesel, K., Grace, D., International Livestock Research Institute.International conference of the Association of Institutions for Tropical Veterinary Medicine (AITVM), Johannesburg, South Africa, 25-29 August 2013.*

D. Development and Validation of PCR Primers to Assess the Diversity of *Clostridium* spp. in Cheese by Temporal Temperature Gradient Gel Electrophoresis [25]

The growth of spore-forming bacteria of the genus *Clostridium* causes severe defects in cheeses with long ripening times, such as Beaufort, Emmental, and Comté, as a consequence of late blowing or butyric blowing. The biochemical basis of this defect is the fermentative transformation of lactic acid to butyric acid (causing rancid flavor), acetic acid, carbon dioxide, and hydrogen. Almost all the commercial value of the contaminated cheese is lost, which represents a high economic impact in cheese production. The *Clostridium* species present in cheese originate from milk which is contaminated by these spores during milking.the species *C. tyrobutyricum* and *C. sporogenes* were the main organisms associated with late gas blowing in the cheese process because of their ability to ferment lactic acid [1,2,3,4,5]. Other species of clostridia producing butyric acid, like *C. butyricum, C. beijerinckii, C. pasteurianum, C. tertium, C. perfringens,* and *C. tetanomorphum,* have also been detected in both milk and cheese [2,3,6].

Conventional methods for detecting clostridia in foods involve heat shock of the samples to destroy vegetative cells.**Molecular methods** would allow daitymen to overcome the inherent limitations of these cultivation-based approaches and to detect directly the vegetative cells that lead to cheese defects.PCR or nested-PCR approaches with species-specific primers were used earlier, in order to specifically identify *C. tyrobutyricum, C. beijerinckii, C. sporogenes,* and *C. butyricum* [7,8]. Denaturing gradient gel electrophoresis (DGGE) [9], temperature gradient gel electrophoresis [10], or temporal temperature gradient gel electrophoresis (TTGE) [11] is suitable for the analysis of mixed microbial communities [12,13] and can also be focused on the numerically subdominant bacteria by specific amplification of 16S rRNA genes [14].

Materials

(*i*) The clostridial and nonclostridial strains were used. All these strains may be found in milk ecosystems or from bovine dung.

(*ii*) Clostridium strains were grown anaerobically in Trypticase-yeast extract-glucose broth at 37°C. (Anaerobic conditions were created by boiling the medium and flushing with it with nitrogen gas during the cooling of broth).

(*iii*) *Lactobacillus* strains were cultivated in MRS broth at 37°C, *Streptococcus*spp were grown in M17 broth at 37°C.

(iv) *Propioni bacterium freudenreichii* was grown in reinforced clostridial medium.

(v) *Bacillus cereus* and *Escherichia coli* were grown in nutrient broth (Difco) at 37°C.

Methods

A nested-PCR temporal temperature gradient gel electrophoresis (TTGE) approach was developed for the detection of bacteria belonging to phylogenetic cluster I of the genus *Clostridium* in cheese suspected of late blowing. TTGE profiles of the PCR products, comprising the V5-V6 region of the 16S rRNA gene, allowed to distinguish the majority of cluster I species. PCR-TTGE was applied to analyze commercial cheeses with defects. All cheeses gave a signal after nested PCR, and on the basis of band comigration with TTGE profiles of reference strains, all the bands could be assigned to a clostridial species. The method involves the following steps:

i. Collection and Nucleic Acid Extraction from Cheese Samples

Samples of cheese from commercial sources (20 with and 1 without the presence of anomalous fermentation) were treated or stored as 250-mg aliquots in 2-ml screw-cap tubes at –80°C until processing. Total DNA was extracted as previously described [15]. The concentration and integrity of the nucleic acids were determined visually by electrophoresis on a 1 per cent agarose gel containing ethidium bromide.

ii. Primer Design

Primers were designed based on the 16S rRNA gene sequence, and the specificity was confirmed in PCRs performed with DNAs from cluster I and non-cluster I species as the templates. 16S rRNA sequences of the target group were obtained from the GenBank database [16] and were aligned using the program Clustal W [17]. After alignment, the primer sequences were visually searched. Potential target sites were further analyzed by BLAST to search for homologous nucleotide sequences in the database. Primers were synthesized commercially. The appropriate annealing temperatures and primer specificities were determined using the genomic DNA of pure-culture control strains.

iii. PCR Amplification

The PCR was carried out by respectively amplifying the V5-V8 region of the 16S rRNA gene of phylogenetic cluster I. PCR amplifications were performed in a final volume of 25 µl which included 12 µl of premix (0.5 µM of forward and reverse primers, 4 per cent final concentration of dimethyl sulfoxide), 12 µl of HotStar *Taq*Master Mix, and 5 ng of genomic DNA. The samples were subjected to amplification in a PTC 100 Thermal Cycler. The PCR program consisted of 15 min at 95°C to allow HotStar *Taq*activation and 30 or 40 cycles of 94°C for 20 s, 58°C for 30 s, and 72°C for 30 s. A 5-min final extension at 72°C was performed at the end of cycling steps, and reaction mixtures were stored at 4°C. The overall time used for this cycle was 120 min. Five microliters of the PCR products was analyzed on 2 per cent agarose gels containing ethidium bromide. Negative controls (without DNA) were run in all amplifications.

iv. Nested-PCR Amplification

Serial dilutions of PCR products were tested to determine the optimal DNA concentrations for nested PCR which amplified 290 bp of the V5-V6 region of the 16S rRNA gene. From these dilutions 1 µl was used as template. The PCR was performed with primers S-D-Bact-1060-a-A-21 (P983) and S-D-Bact-0786-a-S-20 (P984) but with 25 cycles run and with the annealing temperature set at 64°C. The GC clamp at the 52 end of primer P984 was coamplified and prevented complete dissociation of the two DNA strands during TTGE [18]. The overall time used for this cycle was 70 min. The dilution producing the cleanest PCR band was used for TTGE analysis.

v. Detection Limit of PCR

In order to estimate the detection limit of PCR, a clostridium-free cheese sample of 250 mg was artificially contaminated with 250 µl of 10-fold dilutions of pure culture of *C. tyrobutyricum* at 10^9 cells · ml^{-1}. The sample from each tube of the different dilutions was subjected to DNA extraction.

vi. Sensitivity of the TTGE Technique

The sensitivity of TTGE for the detection of subdominant bacterial species in cheese was tested. Overnight liquid cultures of two strains (*C. tyrobutyricum*and *C. beijerinckii, C. tyrobutyricum* and *C. Sporogenes* and *C. beijerinckii* and *C. sporogenes*) were mixed in different ratios (1/1, 1/10, and 1/100) and added to clostridium-free cheese samples. The mixtures were then subjected to DNA extraction and PCR-TTGE.

vi. Analysis of PCR Products by TTGE

TTGE analysis of PCR amplicons was performed using a DCode Universal Mutation Detection System according to the earlier described procedures [19] with the following modifications:

(*a*) The electrophoresis was performed in an 8 per cent (wt/vol) polyacrylamide-bisacrylamide gel and 7 M urea with 1.25× TAE buffer (Tris-acetic acid-EDTA; Bio-Rad Laboratories) at a constant voltage of 70 V for 16 h.

(*b*) A temperature gradient from 63 to 70°C and a temperature ramp rate of 0.4°C/h were applied.

(*c*) Five-microliter samples of PCR products were loaded in wells.

(*d*) After electrophoresis, the gels were stained for 30 min with SYBR Green I nucleic acid gel stain.

(*e*) Their images were digitized using a Storm system.

vii. Clostridium Identification Ladder Cluster I

TTGE gels were standardized by including an identification ladder obtained by mixing PCR products of the V5-V6 region of 16S rRNA from pure cultures of seven *Clostridium* reference strains (*C. perfringens, C. butyricum, C. pasteurianum, C. subterminale, C. beijerinckii, C. tyrobutyricum,* and *C. sporogenes*).

viii. Gel Analysis

For direct identification of bands in given TTGE profiles, a database was created by using the GelCompar II software. This database contained the V5-V6 bands of *Clostridium* species of cluster I. The inclusion of a reference pattern consisting of amplicons from seven *Clostridium* reference strains every eight lanes on each TTGE gel allowed a digital normalization of patterns in the GelCompar database. After normalization, tentative identification of *Clostridium* strains was performed by comparing positions of bands in sample profiles with those of reference strains present in the database.

ix. Sequencing of Excised TTGE Bands

The selected DNA bands were visualized on a UV-Transluminator and excised directly from the gels with a sterile pipette tip. They were allowed to diffuse overnight at 4°C in 100 µl of sterile water. Three microliters of eluate from individual bands was used to reamplify the PCR products. Before sequencing, the purity of each PCR-reamplified TTGE band was checked by running on a denaturing gradient gel in parallel to the original cheese sample. The PCR products were sent to a sequencing facility. Recovered sequences were aligned with 16S rRNA gene fragments. The BLAST program was used to determine the closest known relatives.

x. Volatile Fatty Acid Analysis

Butyric acid fermentation in cheeses was detected by gas-liquid chromatography analysis of butyric acid. Twenty grams of cheese was mixed with 50 ml of sulfuric acid and homogenized with a stomacher for 5 min. The liquid phase was then neutralized by a 1 per cent NaOH solution containing phenolphthalein. The soaps obtained were dried by vacuum [20]. Samples were stored at 4°C before gas-liquid chromatography analysis. Five microliters of each of the supernatants was injected in a gas chromatograph equipped with a flame ionization detector and connected to an integrator. A glass column (2 m by 4 mm) packed with 10 per cent SP 1000 plus 1 per cent H_3PO_4 on Chromosorb W AW (100/120 mesh) was used. The instrument was operated at 125°C for 6 min and then programmed at 20°C min^{-1} to 145°C. The operating conditions were as follows: injector temperature, 200°C, detector temperature, 250°C, carrier gas (nitrogen) flow rate, 30 ml min^{-1} [21].

Observations/Result and Conclusion

(*i*) **Application of PCR-TTGE to detection of *Clostridium* in cheese samples:** Experimental testing of the protocol on cheese samples revealed that the number of cycles of PCR needed to be increased to 40 cycles. The detection limits of *Clostridium* in cheese were investigated at different concentrations. After nested PCR the detection sensitivity was increased 10 times compared to the first PCR. It was possible to detect 100 CFU/g of cheese. In all cheeses with gas production (holes in the cheese matrix) and elevated concentrations of butyric acid, nested PCR resulted in a specific signal. The TTGE analysis of 16S rRNA genes of *Clostridium* from cluster I demonstrated the presence of one or two detectable species in each cheese sample. After normalization, band positions were compared with the GelCompar database of reference strains. Nine samples showed the presence of *C. tyrobutyricum* and *C. beijerinckii*, one showed the presence of *C. tyrobutyricum* and *C. sporogenes*, and one showed the presence of *C. beijerinckii* and *C. butyricum*. For all other samples, only one species was detected.Five samples showed the presence of *C. beijerinckii*, and four showed the presence of *C. tyrobutyricum*.

(ii) A16S rRNA gene-based PCR protocol for the detection and identification of *Clostridium* of cluster I in cheese samples by using TTGE was developed. A nested-PCR approach was applied, combining a first cluster-specific PCR step with a second bacterial PCR step. The PCR primers were subjected to theoretical testing with database information and experimental testing with pure cultures and DNA from cheese samples. The PCR system combining P930 and P932 was highly specific for *Clostridium*. In a nested setup, the DNA amplification system as designed in 1987 by Woese [22] was used in order to generate a clamped PCR fragment for TTGE separation. Using our primers, a fragment of 290 bp, suitable for TTGE analysis, was amplified.

The potential of TTGE fingerprints, based upon 16S rRNA gene fragments amplified via nested PCR from community DNA, to characterize the clostridial population in cheese, was demonstrated. The applicability of the technique was demonstrated by the analysis of 21 cheese samples. The results have shown that DNA extracted from cheese according to the protocol developed by Godon *et al.*, can always be successfully used as template DNA in PCR [23]. Further more, the extraction protocol used, including bead beating, provided sufficient DNA for amplification of *Clostridium*, as even spores were sufficiently lysed [24]. All cheeses with splitting and/or rancid flavor gave a signal after nested PCR. After normalization, comparison of the amplicon migration patterns in TTGE gels with those of reference strains allowed us to identify the bacteria. All the bands present on the TTGE gel could be linked to a given clostridial species.

The direct identification of *Clostridium* spp. was confirmed by sequencing of excised bands. In all the 20 cheeses with defects, only four different species were detected: *C. tyrobutyricum, C. beijerinckii, C. butyricum,* and*C. sporogenes*. The most important species detected with PCR-TTGE were *C.beijerinckii* and *C. tyrobutyricum*, which contaminated 15 and 14 samples of the 20 tested, respectively and *C.butyricum* and *C. sporogenes* were detected in one cheese sample. These results demonstrated the efficacy of the PCR-TTGE method to identify *Clostridium*in cheeses. The sensitivity of the method was estimated to be 100 CFU/g.

This is allowing the detection and differentiation of all clostridial species present in cheese with a single test. It is noteworthy that clostridia such as *C. butyricum* and *C.beijerinckii* cannot easily be identified by normal culture and biochemical techniques but can readily be differentiated by PCR-TTGE with P930-P932 primers.

Moreover, this technique is rapid since no cloning and sequencing of the TTGE bands are required, provided that an identification match is obtained with the database. The developedprocedure can be implemented as a routine detection method in the industry in order to screen cheese during production and to have a complete study of the diversity of *Clostridium* species in defective cheeses.

Source

1. *Bergère, J.L., P.Gouet, J.Hermier, and G. Mocquot.1968. Les Clostridium du groupe butyrique dans les produits laitiers. Ann. Inst. Pasteur Lille19: 41-54.*
2. *Hood, E.G., and J. F. Bowen.1950. A new type of bacterial spoilage in Canadian process cheese.Sci. Agr.30: 38-42.*
3. *Ingham, S. C., J. R. Hassler, Y. W. Tsai, and B.H.Ingham.1998.Differentiation of lactate-fermenting, gas-producing Clostridium spp.isolated from milk. Int. J. Food Microbiol.43: 173-183.*
4. *Matteuzzi, D., S. Annibaldi and P. Sabatini.1972. Importanza di Clostridium sporogenes come agente di gonfiore del formaggio Grana. Ann. Microbiol. Enzimol.22: 145-154.*
5. *Toyoda, S., Y. Kobayashi, and K. Ahiko.1990. Isolation and characterization of Clostridia from Gouda cheese with late-blowing. Jpn. J. Zootech. Sci.61: 591-598.*
6. *Goudkov, A. V., and M. E. Sharpe.1965. Clostridia in dairying.J. Appl. Bacteriol. 28: 63-73.*
7. *Herman, L., J. De Block, and R. Van Renterghem.1997. Isolation and detection of Clostridium tyrobutyricum cells in semi-soft and hard cheeses using the polymerase chain reaction.J. Dairy Res.64: 311-314.*
8. *Klijn, N., F.F. Nieuwenhof, J. D. Hoolwerf, C. B. van der Waals, and A. H. Weerkamp.1995. Identification of Clostridium tyrobutyricum as the causative agent of late blowing in cheese by species-specific PCR amplification. Appl.Environ. Microbiol.61: 2919-2924.*
9. *Fischer, S.G., and L.S. Lerman.1983. DNA fragments differing by single base-pair substitutions are separated in denaturing gradient gels: correspondence with melting theory. Proc. Natl. Acad.Sci.USA80: 1579-1583.*
10. *Rosenbaum, V., and D.Riesner.1987. Temperature-gradient gel electrophoresis. Thermodynamic analysis of nucleic acids and proteins in purified form and in cellular extracts. Biophys. Chem.26: 235-246.*
11. *Yoshino, K., K. Nishigaki, and Y. Husimi.1991. Temperature sweep gel electrophoresis: a simple method to detect point mutations. Nucleic Acids Res.19: 3153.*
12. *Cocolin, L., N. Innocente, M. Biasutti, and G. Comi.2004. The late blowing in cheese: a new molecular approach based on PCR and DGGE to study the microbial ecology of the alteration process. Int. J. Food Microbiol.90: 83-91.*
13. *Muyzer, G., and K. Smalla.1998. Application of denaturing gradient gel electrophoresis (DGGE) and temperature gradient gel electrophoresis (TGGE) in microbial ecology. Antonie Leeuwenhoek 73: 127-141.*

14. *Heilig, H. G., E. G. Zoetendal, E. E. Vaughan, P. Marteau, A. D. Akkermans, and W. M. de Vos.2002. Molecular diversity of Lactobacillus spp. and other lactic acid bacteria in the human intestine as determined by specific amplification of 16S ribosomal DNA. Appl. Environ. Microbiol.68: 114-123.*

15. *Godon, J., E. Zumstein, P. Dabert, F. Habouzit, and R. Moletta.1997.Molecular microbial diversity of an anaerobic digestor as determined by small-subunit rDNA sequence analysis. Appl. Environ. Microbiol.63: 2802-2813.*

16. *Benson, D. A., I. Karsch-Mizrachi, D. J. Lipman, J. Ostell, B. A.Rapp, and D. L.Wheeler.2000. GenBank.Nucleic Acids Res.28: 15-1.*

17. *Thompson, J. D., D. G. Higgins, and T. J. Gibson.1994. CLUSTAL W: improving the sensitivity of progressive multiple sequence alignment through sequence weighting, position-specific gap penalties and weight matrix choice.Nucleic Acids Res.22: 4673-4680.*

18. *Muyzer, G., E. C. de Waal, and A. G. Uitterlinden.1993. Profiling of complex microbial populations by denaturing gradient gel electrophoresis analysis of polymerase chain reaction-amplified genes coding for 16S rRNA. Appl. Environ. Microbiol.59: 695-700.*

19. *Seksik, P., L. Rigottier-Gois, G. Gramet, M. Sutren, P. Pochart, P. Marteau, R. Jian, and J. Doré.2003. Alterations of the dominant faecal bacterial groups in patients with Crohn's disease of the colon.Gut52: 237-242.*

20. *Berdague, J.L. 1986.Methode d'extraction rapide des acides gras volatils libres des fromages. Lait66: 233-246.*

21. *Carlier, J.1985.La chromate-graphie en phase gazeuse des produits de fermentation: son application dans le diagnostic des bactéries anaérobies. Bull. Inst. Pasteur83: 57-69.*

22. *Woese, C. R.1987. Bacterial evolution. Microbiol. Rev.51: 221-271.*

23. *Godon, J., E.Zumstein, P. Dabert, F. Habouzit, and R. Moletta.1997.Molecular microbial diversity of an anaerobic digestor as determined by small-subunit rDNA sequence analysis. Appl. Environ. Microbiol.63: 2802-2813.*

24. *Reif, T. C., M. Johns, S. D. Pillai, and M. Carl.1994. Identification of capsule-forming Bacillus anthracis spores with the PCR and a novel dual-probe hybridization format. Appl. Environ. Microbiol.60: 1622-1625.*

25. *Anne-Gaëlle Le Bourhis, Katiana Saunier, Joël Doré, Jean-Philippe Carlier, Jean-François Chamba, Michel-Robert Popoff and Jean-Luc Tholozan, Development and Validation of PCR Primers to Assess the Diversity of Clostridium spp. in Cheese by Temporal Temperature Gradient Gel Electrophoresis.*

E. Detection of Mastitis with PCR and Real Time PCR Technology

The technical definition of mastitis is "inflammation of the mammary gland" but on a practical basis, almost all bovine mastitis is caused by bacteria [1]. Appropriate mastitis control is based on knowledge of the etiology, thus identification of pathogens is a fundamental aspect of mastitis control programs. Mastitis occurs after an infective dose of a pathogenic organism passes through the streak canal and progresses to a subclinical or clinical state but many intramammary infections are spontaneously cured as a result of the cows immune response [2]. ***There are following two types of Mastitis:***

Subclinical Mastitis

By definition, milk obtained from mammary gland quarters of cows experiencing *subclinical mastitis* appears visually but the milk contains an excessive number of somatic cells, (with or without the detectable presence of pathogenic organisms) [3]. The SCC of healthy quarters is usually well below 100,000 cells/mL and a threshold of < 200,000 cells/mL is usually considered to be the most practical value to use to define a mammary quarter as healthy [3-6].

Clinical Mastitis

Inflammation that results in visible abnormalities of milk is defined as clinical mastitis, regardless of SCC level. Most symptoms of clinical mastitis are quite mild and cannot be detected unless foremilk is observed before attaching the milking cluster.

Economic losses of bovine mastitis caused by Strep. agalactiae were contributed to production losses, milk quality penalties and decreases in milk quality [8,9]. The mammary gland infection referred to as infectious mastitis causes dairy producers to lose billions of dollars every year world widely.Strep. agalactiae are mainly responsible for the mastitis situation in many farms [10]. It is one of the major pathogens caused subclinical mastitis which can be diagnosed only in the laboratory [11]. It is important to identify Strep. agalactiae as early as possible for dairy producer to prevent serous infection in the herd.

Source

1. *MC Laboratory Handbook on Bovine Mastitis, Revised Ed., N.M. Council, Editor.1999: Madison, WI.*
2. *Oviedo-Boyso, J., J. J. Valdez-Alarcon, M. Cajero-Juarez, et al., Innate immune response of bovine mammary gland to pathogenic bacteria responsible for mastitis. Journal of Infection, 2007. 54 (4): p. 399-409.*
3. *Dohoo, I.R. and K.E.Leslie, Evaluation of Changes in Somatic-Cell Counts as Indicators of New Intramammary Infections. Preventive Veterinary Medicine, 1991. 10 (3): p. 225-237.*
4. *Schepers, A.J., T.J.G.M. Lam, Y.H. Schukken, et al., Estimation of variance components for somatic cell counts to determine thresholds for uninfected quarters. Journal of Dairy Science, 1997. 80 (8): p. 1833-1840.*
5. *Djabri, B., N. Bareille, F. Beaudeau, et al., Quarter milk somatic cell count in infected dairy cows: a meta-analysis. Veterinary Research, 2002. 33 (4): p. 335-357.*
6. *Pantoja, J.C.F., C. Hulland, and P.L.Ruegg, Dynamics of somatic cell counts and intramammary infections across the dry period. Preventive Veterinary Medicine, 2009. 90 (1-2): p. 43-54.*
7. *Pamela L. Ruegg, DVM, MPVM, University of WI, Dept. of Dairy Science, Madison WI USA.*

8. *Keefe, G. P. 1997. Strep. agalactiae mastitis: a review. Can. Vet. J.38: 429-437.*

9. *Kalorey, D. R., N. V. Kurkure, N. K. Nigot, M. P. Patil and V. P.Pathak. 2001. Effect of subclinical mastitis on milk of crossbred Sahiwas×Jersey cows: a biochemical study. Asian-Aust. J. Anim. Sci. 14: 382-383.*

10. *Merl, K., A. Abdulmawjood, C. Lammler and M. Zschock. 2003.Determination of epidemiological relationships of Strep. agalactiae isolated from bovine mastitis. FEMS Microbiol.Lett. 226: 87-92.*

11. *Meiri, B., I. E. Lipkin, A. Friedmann, G. Leitner, A. Saran, S.Friedman and Y. Kashi. 2002. A PCR-based method for the detection of Strep. agalactiae in milk. J. Dairy Sci. 85: 1717-1723.*

1. Detection of Mastitis with PCR [1]

The use of PCR is very basic to many molecular procedures. A PCR is based on the concept that the nucleus of bacterial cells contains DNA composed of nucleotides with unique sequences. When PCR is used for identification of mastitis pathogens, bacterial DNA is extracted from milk samples and then mixed with "primers" which are basically templates of known nucleotide sequences from particular bacterial species (not strains). The primers duplicate matching sequences until enough copies are made so that they can be matched to a bacterial species in the library of known bacterial sequences. A key concept for interpretation of results of these tests is to understand that PCR tests can only identify organisms for which a specific primer is included in the "test mix."

One real-time PCR method for diagnosis of bovine mastitis pathogens is widely commercially available (Pathoproof®; Thermo Fisher Scientific). The Pathoproof™ kit contain of primers that can bind with and multiply DNA from a number of organisms. Some kits include primers that can identify bacterial DNA from just a few organisms (M. bovis, Staph aureus and Strep agalactiae) and other kits include primers for DNA of up to 16 organisms. Intramammary infection is not the only source of bacterial DNA found in milk samples, as teat skin, the streak canal and sampling methods can contaminate milk with bacterial DNA. To ensure a useful result, milk samples used for PCR testing must be collected using aseptic technique.

PCR testing is useful to detect the presence of obligate udder pathogens (such as *S. agalactiae*) in bulk milk, especially when there is a low prevalence of infected cows within the herd [2]. However interpretation of PCR test results for other bacteria found in bulk milk is difficult. PCR test results are interpreted relative to the cycling threshold values (Ct) which indicate the number of PCR cycles that are required to make enough copies to reach the signaling threshold. In general, the lower the Ct value the greater the amount of the specified DNA in the sample.

DNA Fingerprinting

When determination of individual strains of a bacterial species is desired, then a process of DNA fingerprinting is used. There are several different methods for comparing the DNA of a bacterial species, but they are all based on extraction of bacterial DNA followed by separation of the bacterial DNA into columns based on size of specific fragments. Some of the methods include Pulsed Field Gel Electrophoresis (PFGE) or Random Fragment Length Polymorphism (RFLP) [3]. These methods are performed by adding restriction enzymes that cut the bacterial DNA in specific areas. In PFGE, those cut pieces are then put in a gel column and separated by use of an electrical field to resemble a bar code. The resulting bands of DNA fragments are then compared to each other and form the basis for deciding if strains are identical, similar or different. Bands that are identical or very similar are considered to be the same strain or a slightly different substrain.

Source

1. *Pamela L. Ruegg, DVM, MPVM, University of WI, Dept. of Dairy Science, Madison WI USA.*
2. *Katholm, J., T.W. Bennedsgaard, M. T. Koskinen, et al., Quality of bulk tank milk samples from Danish dairy herds based on real-time polymerase chain reaction identification of mastitis pathogens. Journal of Dairy Science, 2012. 95 (10): p. 5702-5708.*
3. *Zadoks, R.N. and Y.H. Schukken, Use of molecular epidemiology in veterinary practice. Veterinary Clinics of North America-Food Animal Practice, 2006. 22 (1): p. 229.*

2. Detection of Mastitis with Real Time PCR Technology [1]

In dairy herds, mastitis continues to be the most common disease, costing the industry hundreds of millions annually in lost production and decreased milk quality.

Identifying the mastitis pathogens is traditionally undertaken through bacteriology, where the sample is grown in the laboratory. Bacteriology is still an important tool, especially in looking for the potential mastitis pathogens in a bulk tank milk sample.

A recent drive to improving mastitis management has highlighted the need for a fast, accurate and convenient method of mastitis detection. Recently a new system

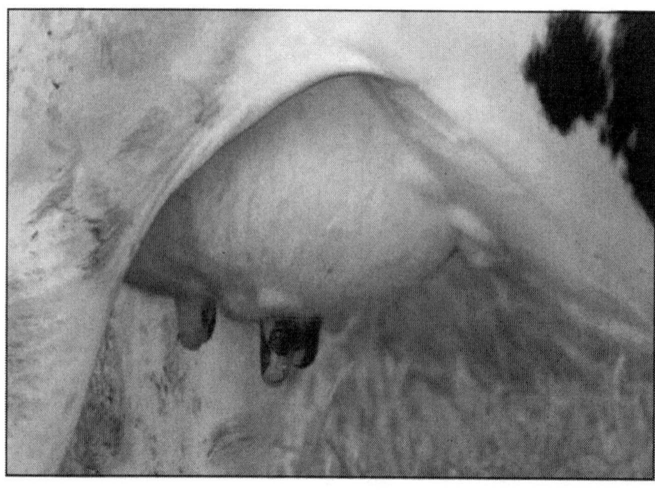

of identifying the bacteria present in milk has become available. This is called PCR (Polymerase Chain Reaction) and uses the latest diagnostic equipment to detect bacterial DNA in the milk sample. With these advances, dairy producers now have the latest technology available to them when addressing milk hygiene problems.

Source

1. *WestPoint Veterinary Group Ltd. Dawes Farm, Bognor Road, Warnham, West Sussex, RH12 3SH.*

a. Multiplex Polymerase Chain Reaction as a Mastitis Screening Test for *Staphylococcus aureus, Streptococcus agalactiae, Streptococcus dysgalactiae* and *Streptococcus uberis* in Bulk Milk Samples [1]

Regular monitoring of bulk tank milk is a vital part of dairy herd health programmes to control mastitis.PCR is a rapid and sensitive method to identify bacteria and a multiplex polymerase chain reaction (PCR) assay for simultaneous detection of Staphylococcus aureus, Streptococcus agalactiae, Streptococcus dysgalactiae and Streptococcus uberis was used to assess its applicability as an alternative method for monitoring mastitis caused by these organisms at the herd level. PCR was used to detect the presence of these organisms in bulk milk samples. Bulk milk tests that primarily relate to mastitis are **bulk milk somatic cell counts (BMCC) and bulk milk cultures.** A persistently high BMCC indicates a mastitis problem in a herd. It has been estimated that for each 100 000 cells/ml, approximately 10 per cent of cows in a herd have mastitis [2]. Bulk milk culture has been used as a screening test for the presence of mastitis pathogens within a herd. This method reduces the cost and overcomes the problems of collecting and processing large numbers of samples. There are a number of limitations associated with bulkmilk culture. No standardized methods have been established for the collection and culture of bulk milk samples [3]. Bulk milk culture is also reported to have a low sensitivity for determining the presence of major bacterial mastitis pathogens [3].

Materials and Methods

(a) Bulk milk samples were provided by a large milk processing co-operative operating in south-eastern Australia, drawing milk from farms using pasture-based dairying systems. Samples provided were those collected from supplier herds for routine milk quality analysis of butterfat and protein content, cell count and bacteria count. Herds were selected to represent those with low, medium or high rolling average BMCC. A total of 176 bulk milk samples from 42 herds were forwarded by the factory laboratory immediately after the completion of routine processing.

(b) The 42 herds were divided into following three groups:

- *(i)* Group 1 consisted of 14 herds that had low BMCCgm (<200 000 cells/ml).
- *(ii)* Thirteen herds with medium BMCCgm (200 000–400 000 cells/ml) were allocated to group 2.

(iii) 15 herds with high BMCCgm (>400 000 cells/ml) were allocated to group 3.

Data obtained from the milk factory Bulk milk somatic cell counts were determined using a Fossomatic cell counter (FOSS A/S, Hillerød, Denmark) and routinely recorded by the milk factory. Bacteria counts, as determined using Bactoscan (FOSS A/S, Hillerød, Denmark), and thermoduric bacteria counts were also provided by the milk factory.

c. Sample Preparation for PCR

DNA was extracted from milk samples using the Rneasyspin column method. One millilitre of milk was inoculated into 9 ml Streptosel and the culture incubated for 8–12 h prior to extraction. One hundred microlitres of broth culture was taken and mixed with 350 ml RLT buffer containing 1 per cent 2-mercaptoethanol. After mixing vigorously, 300 ml of 100 per cent ethanol was added, and 15 ml Qiaex II matrix was then added. The suspension was loaded onto a spin column. Columns were then centrifuged at 9000 g for 30 s and the flow-through discarded. Columns were washed once with 600 ml RLT buffer and then washed twice with 500 ml RPE buffer. For each wash, columns were centrifuged at 9000 g for 30 s. After the final wash, columns were centrifuged at 16 000 g for an additional 90 s to remove any excess ethanol. Thirty microlitres of Rnase-free water was placed directly on the matrix.DNA was then eluted from the matrix by centrifugation at 9000 g for 1 min and 5 ml of the extract was subjected to PCR.

d. Oligonucleotide Primers

Oligonucleotide primers were derived from published sequences [4] with primer sequences for Str. agalactiae, Str. dysgalactiae and Str. uberis adjusted from the published sequences to have similar Tm values [5].

e. PCR Amplification

Multiplex PCR was performed with 2.5 U Taq DNA polymerase in 50-ml reactions containing 500 ng of extracted DNA in 10 mM-Tris-HCl (pH 8.3), 2.5 mM-MgC12, 50mM-KCl,200 mM each of dATP, dTTP, dCTP and dGTP, 0.4 mM each of the Str. agalactiae and Str. dysgalactiae primers,0.1 mM of the Str. uberis primers and 0.3 mM each of the Staph. aureus primers. Reactions were incubated at 95.8Cfor 5 min, then 36 cycles of 95.8C for 1 min, 59. 8C for 30 s and 72. 8C for 30s and finally incubated at 72 8C for7 min. A tube containing the PCR mixture without template DNA was routinely included as a negative control.

f. Analysis of PCR Products

Amplification products were analysed by electrophoresis through 2 per cent agarose gels or 20 per cent polyacrylamide gels. Twelve microlitres of amplified sample was electrophoresed in a 2 per cent agarose gel containing ethidium bromide at 0.1 mg/ml. Gels were run at 6 V/cm for 45 min and PCR products were visualized by u.v. transillumination. Seven microlitres of amplified product was used for analysis by polyacrylamide gel electrophoresis. Gels were run at 6 V/cm for 2 h and then stained with silver. Fragments from a HaeIII restriction endonuclease digest of plasmid pUC18 were used as molecular weight markers.

Statistical Analysis

Log-transformations of the BMCC, total bacteria counts (TBC) and thermoduric bacteria counts were used because the raw data were not normally distributed. The geometric means (gm) of BMCC, TBC and thermoduric bacteria counts of sequential samples collected from each herd were determined. One-way analysis of variance was used for comparing mean TBC and mean thermoduric bacteria counts between groups of herds. Fisher's exact test was used to examine the relationship between the presence of specific organisms in the first sample for each herd and either the BMCC, TBC or thermoduric bacteria count for that sample. Statistical significance was defined as P<0.05. Agreement between the detection of specific organisms in multiple samples from the same herd was analysed using the kappa statistic [6]. A kappa value >0.8 is regarded as very good agreement,0.61–0.8 good, 0.41–0.6 moderate, 0.21–0.4 fair, and <0.2 poor [7].

PCR Results

Of the 176 bulk milk samples,43 samples were negative for all four bacterial species, 72 samples were positive for one species and 61 samples were positive for more than one species. Of the samples negative by PCR, 60 per cent (26/43) were from herds with lowBMCC; 26 per cent (11/43) of such samples were from herds with medium BMCC and 13 per cent (6/43) were from herds with high BMCC. Of the samples positive for more than one species, 82 per cent (50/61) were from herds with medium and high BMCC. The species most frequently present in bulk milk was Str. uberis, being found in 78 (44 per cent) samples. Str.dysgalactiae, Str. agalactiae and Staph. aureus were present in 49 (28 per cent), 38 (22 per cent) and 28 (16 per cent) samples respectively.Str. uberis was found in at least one sample from 35 herds. Str. dysgalactiae was detected in 23 herds.Staph. aureus was detected in 14 herds and Str. Agalactiae was detected in 11 herds.

BMCC, TBC and Thermoduric Bacteria Counts

BMCC of samples from each herd were relatively stable during the study period. Relationships between BMCC and TBC and thermoduric bacteria counts were also investigated.Mean TBCgm in samples from group 3 was significantly higher than mean TBCgm in samples from group 1and group 2 (P<0.01). Similarly, mean thermoduric bacteria counts were significantly higher in group 3 herds than in either group 1 or group 2 herds, P<0.05).

Association between the Presence of Pathogens in Bulkmilk Samples and BMCC/Total Bacterial Count

There were no obvious relationships between the presence or absence of Staph. aureus, Str. dysgalactiae or Str. uberis and BMCC in herds. However, a significant relationship was found between prevalence of Str. agalactiae and BMCC (P<0.01). There were no significant differences between the prevalence of Staph. aureus, Str. dysgalactiae or Str. uberis.PCR results on five consecutive bulk milk samples and geometric means of BMCC, total bacteria counts and thermoduric bacteria counts of herds classified as having high BMCC.

A relationship between prevalence of Str. agalactiae and TBC was only found in samples from herds that were consistently positive for Str. agalactiae. The three herds that were inconsistently positive for Str. agalactiae had an averageTBC <50 000 cells/ml. No relationship was found between the presence of Staph.aureus, Str. agalactiae, Str. dysgalactiae or Str. uberis and numbers of thermoduric bacteria in bulk milk samples.

The study revealed that the multiplex PCR assay used was an easy and rapid method to detect the four major mastitis pathogens, Staph. aureus, Str. agalactiae, Str. dysgalactiae.PCR required little time to process and results were easier to interpret than bulk milk cultures, which are subject to contamination with bacteria from the environment.Str. uberis andStr. dysgalactiae were the most prevalent species in bulk milk samples. As found in previous studies [8,9], their presence in bulk milk was not significantly associated with higher BMCC or higher numbers of bacteria in bulk milk. As these species are ubiquitous in the cow's environment, it is more likely that these organisms enter the bulk milk from non-specific contamination rather than in milk from infected cows. The inconsistency of the detection of these organisms in sequential samples also suggests sporadic contamination of the bulk milk samples by bacteria from the environment. Nevertheless, the consistent detection of Str. dysgalactiae in a herd with high BMCC might suggest the presence of Str. dysgalactiae mastitis in the herd. Clinical mastitis due to Str. dysgalactiae has been found to be more common in herds with high BMCC [10].

A considerably higher prevalence and incidence of clinical mastitis due to Staph. aureus has been observed by analysis of quarter milk samples in herds with high BMCC [10,11]. There was no significant association, however, between presence of Staph. aureus in bulk samples and BMCC in the present study. Although it has been proposed that presence of Staph. aureus in bulk milk is a reliable indicator of infection of the herd due to this organism [3,12], interpretation of the detection of Staph. aureus in bulk milk is not as reliable as interpretation of the detection of Str. agalactiae. Staph. aureus may enter bulk milk from non-milk sources as there is evidence of contamination of both composite and bulk milk by Staph. aureus strains from milkers [13]. There was a close correlation between detection of Str.agalactiae and either high BMCC or total bacteria counts, which concurs with previous reports [9,14,15,16]. It was found that most herds consistently positive for Str. agalactiae had high total bacteria counts, suggesting either a relatively high prevalence of Str. agalactiae infections in these herds or that the infections in these herds were in a phase of high shedding. However, none of the herds from which Str. agalactiae was inconsistently detected had average total bacterial counts of more than50 000 cells/ml. Intermittent recovery of Str. agalactiae in these herds may suggest that prevalence of Str. agalactiae infections in these herds was low.

Str. agalactiae is an obligate parasite of the udder and its presence in bulk milk is a conclusive demonstration of shedding from infected quarters [17]. **Hence, identification of Str. agalactiae in bulk milk is indicative of the presence of Str. agalactiae mastitis in the herd**. Further investigation is necessary to determine the sensitivity of the PCR assay for detection of the presence of Str.agalactiae infection in the herd. Sensitivity of bulk milk cultures for detecting Str. agalactiae varies from 20 per cent to 84 per cent [12,14,17,18]. The variation is thought to be due to differences in methods used as well as to prevalence of infected cows within the herd and rate of bacterial shedding, which is related to the stage of infection [14].

It has been reported that at least 5 per cent of the quarters in a herd must be infected with Str. agalactiae before the organism can be detected in bulk milk cultures [18]. The use of enrichment before DNA extraction and the high sensitivity of PCR itself might be expected to yield a higher sensitivity for detection of Str. agalactiae.

The results indicated that most herds with low BMCC were free of Str. agalactiae infections, as Str. agalactiae was found in only a single herd with a low BMCC and then in only the first two samples. Conversely, prevalence of Str. agalactiae in samples from high BMCC herds was high, accounting for 60 per cent of the high BMCC herds evaluated. Although Str. agalactiae can be readily eliminated [14], the results indicate that this organism remains a significant cause of mastitis in many herds in Australia.

Regular analysis of bulk milk by PCR may be a useful tool for determining the herd status with regard to Str.agalactiae, but appears to be of less value for monitoring the occurrence of mastitis caused by Staph. aureus, Str.dysgalactiae and Str. uberis. Use of PCR in conjunction with BMCC could be a rapid and convenient means to determine if Str. agalactiae has been eradicated from infected herds after treatment and whether appropriate control programmes have been implemented. Alternatively, PCR could be used to detect if infected animals have been introduced into herds that were free of this organism. It may also be used to evaluate the efficacy of control practices adopted for herds. Relationship between positive PCR results and prevalence of infected cows in the herd requires further clarification.

Source

1. *Patchara Phuektes, Glenn F Browning, Garry Anderson and Peter D Mansell.Faculty of Veterinary Science, University of Melbourne, Parkville 3010, Victoria, Australia.*
2. *Ryan D 1991 Investigation of mastitis problem farms. In Dairy Medicine and Production.Sydney: University of Sydney Postgraduate Foundation in Veterinary Science.*
3. *Godkin MA and Leslie KE 1993 Culture of bulk tank milk as a mastitis screening test: a brief review. Canadian Veterinary Journal 34601-605.*
4. *Forsman RJ, Tilsalatimisjarvi A and Alatossava T 1997 Identification of staphylococcal and streptococcal causes of bovine mastitis using16S-23S rRNA spacer regions. Microbiology 143 3491–3500.*
5. *Phuektes P, Mansell PD and Browning GF 2001 Multiplex polymerase chain reaction assay for simultaneous detection of Staphylococcus aureus and Streptococcal causes of bovine mastitis. Journal of Dairy Science 84 1140–1148.*
6. *Fleiss JL 1981 Statistical methods for rates and proportions.2nd edition, John Wiley and Sons, New York.*

7. Altman DG 1991 Practical Statistics for Medical Research. London: Chapman and Hall.

8. Bramley AJ, McKinnon CH, Staker RT and Simpkin DL 1984 The effect of udder infection on the bacterial flora of the bulk milk of ten dairyherds. Journal of Applied Bacteriology 57 317–323.

9. Hogan JS, Pankey JW, Murdough P and Howard DB 1986 Survey of bulktank milk using blood-esculin agar counts. Journal of Food Protection49 990–993.

10. Barkema HW, Schukken YH, Lam T, Beiboer ML, Wilmink H, BenedictusG and Brand A 1998 Incidence of clinical mastitis in dairy herds grouped in three categories by bulk milk somatic cell counts. Journal of Dairy Science 81 411–419.

11. Erskine RJ, Eberhart RJ, Hutchinson LJ, Spencer SB and Campbell MA 1988, Incidence and types of clinical mastitis in dairy herds with high and low somatic cell counts. Journal of the American Veterinary MedicalAssociation 192 761-766.

12. Gonzalez RN, Jasper DE, Bushnell RB and Farver TB 1986 Relationship between mastitis pathogen numbers in bulk milk and bovine infectionsin California dairy herds. Journal of the American Veterinary MedicalAssociation 189 442–445.

13. Adesiyun AA, Webb LA and Romain HT 1997 Relatedness of Staphylo-coccusaureus strains isolated from milk and human handlers in dairyfarms in Trinidad. Journal of Veterinary Medicine (Series B) 44 551–560.

14. Keefe GP 1997 Streptococcus agalactiae: A review. Canadian Veterinary Journal 38 429–437.

15. Erskine RJ, Eberhart RJ, Hutchinson LJ and Spencer SB 1987 Herd management and prevalence of mastitis in dairy herds with high andlow somatic cell counts. Journal of the American Veterinary MedicalAssociation 190 1411–1416.

16. Ward GE and Schultz LH 1972 Relationship of somatic cells in quarter milk to type of bacteria and production. Journal of Dairy Science 55, 1428-1431.

17. Bartlett PC, Miller GY, Lance SE and Heider LE 1991 Use of bulk tank and milk filter cultures in screening for Streptococcus agalactiae and coagulase-positive staphylococci. Journal of Food Protection 54 848–851.

18. Postle DS 1968 Evaluation of a selective medium for screening bulk milk samples for Streptococcus agalactiae. American Journal of Veterinary Research 29 669–678.

3. Rapid PCR Test for Identification of *Streptococcus agalactiae* in Milk Samples Collected on Filter Paper Disks [1]

There are several methods for the collection of liquid biological sample, *e.g.* filter paper disks [2,3] and the conventional tube-like container samples. The filter paper disks and strips are inexpensive, bio-safe, reliable and applicable method in field tests with limited technical facilities to collect, transport and store samples.

Materials and Method

i. Filter Paper Disks Preparation

Xinhua grade 3 chr paper was used. It is 0.92 mm thick and highly absorbent pure cellulose chromatography paper with a flow rate of190 mm/30 min. Filter paper disks for analysis were produced by a standard punch device, which punched out0.50 cm disks, and handled using disposable gloves. The filter paper disks were divided into five groups. The first to 5th groups were treated with:

(a) Double distilled water.

(b) Tris-HCl buffer (20 mM, pH 8.0).

(c) Tris-HCl buffer (20mM, pH 8.0) with 5 mM EDTA.

(d) Tris-HCl buffer (20mM, pH 8.0) with 10 mM EDTA.

(e) Tris-HCl buffer (20mM, pH 8.0) with 20 mM EDTA overnight respectively.

Then, the paper disks were sterilized at 121°C for 15minutes and allowed to dry at 60°C overnight. One hundred blank filter paper disks were used to analyze the volume of milk saturating the disks by an analytical balance.

ii. Sample Collection and Field Test

826 CMT-positive quarter milk samples were collected from 620 Holstein cows in the 2nd, 3rd, 4th lactations during the middle stage of the lactation cycle. Udder and teats were washed with water, and 70 per cent ethyl alcohol was applied after drying. The first several streams were discarded and a CMT was conducted before 5ml of milk were collected in sterile tubes. Meanwhile each milk sample was dropped on filter paper disks mounted on pen-like samplers (our laboratory invented, a container which could accelerate the dry and protect the sample) and threw away the redundant milk immediately. Milk samples were transported to the laboratory on ice. Filter paper disks were stored at room temperature for two weeks. After that, the filter paper disks were used to extract DNA for field tests. The liquid milk samples were cultured in blood agar within 6 h to isolate the Strep. agalactiae.

iii. Bacterial Strains and Growth Conditions

The organisms used include Staphylococcus aureus Nu305 and two clinical isolates from cases of bovine mastitis, Strep. agalactiae 2002-007, and Strep. dysgalactiae 2002-004, which were identified biochemically by the API Strep systems. All organisms were cultured in trypticase soy broth (TSB) at 37°C with a horizontal rotator for about 18 h before DNA extraction.

iv. The Recovery of Strep. agalactiae Artificially Contaminated on Filter Paper Disks

The culture of Strep. agalactiae in TSB was diluted to 2×104 CFU/ml. Twenty µl of the diluted solution was used to artificially contaminate pretreated filter paper disks and dried by chemical desiccanton sealed containers. After that they were kept on 25°C and 37°C for one to four weeks in the original container. The filter paper disks were put into a 1.5 ml sterilized eppendorf tube, one hundred µl of incubation buffer (20 mM Tris-HCl, pH 8.0, sterilized by filtration) was added and mixed in a vortex mixer for 20 min. The mixtures were cultured at TSB for 48 h. The change of TSB was recorded.

iv. Preparation of Bacterial DNA for PCR

About 1.5 ml of the bacterial culture or 1 ml of liquid milk sample was transferred into an eppendorf tube and centrifuged at 5,000 rpm for 10 min. The pellets were resuspended in 500 µl of enzyme incubation buffer (20 MmTris-HCl, pH 8.0, 1.2 per cent Triton x-100, 20 mg/ml of lysozyme, 2 mM EDTA) followed 30 min, incubated for37°C, followed by the addition of 25 µl of proteinase K (20mg/ ml) and incubation at 56°C for 2 h. After that, the DNA was extracted with phenol-chloroform-isoamylol (25: 24: 1)and precipitated by ethanol. The DNA was dissolved in 500µl TE (10 mM Tris-HCl, 1 mM EDTA, pH 8.0).

v. Preparation of Bacterial DNA from Filter Paper Disk

The filter paper disks were put into a 1.5 ml eppendorf tube, Five hundred µl of incubation buffer (20 mM Tris-HCl, pH 8.0, 1.2 per cent Triton x-100, 2 mM EDTA, sterilization by filtration) was added; mix in a vortex mixer for 20 min. Then add 1 mg of lysozyme and followed by an incubation period of 30 min at 37°C, followed by the addition of 25 µl of proteinase K and incubation at 56°C for 2 h. After that, the DNA was extracted with phenol-chloroform-isoamylol (25: 24: 1) and precipitated by ethanol. The DNA was dissolved by 50 µl of TE (10 mM Tris-HCl, 1 mM EDTA, pH 8.0), then stored at 4°C for later usage.

vi. Amplification and Detection of Strep. agalactiae by PCR

The oligonucleotide primers were selected from the DNA coding for the 16S to 23S rRNA ISR based on previously published paper (Phuektes *et al.*, 2001). The following primers were used to amplify a 270 bp DNA sequence:

(*a*) **Sense primer** (AAGGAAACCTGCCATTTG).

(*b*) **Antisense primer** (TTAACCTAGTTTCTTTAAAACTAGAA).

Reaction mixtures (50 µl each) were prepared by using 2.5 U of Taq DNApolymerase (Biobasic Inc, USA). 1× Taq polymerase buffer minus Mg, 6 µl of 25 mM MgCl2, and 0.2mM of each dNTPs, 0.5 µm each primer, and 5 µl template DNA. A pre-PCR step at 94°C for 3 min was applied. The reaction mixture was cycled 40 times in a GeneAmp PCR System 2400 denaturation at 94°C for 30 s, annealing at 58°C for 45 s, and extending at 72°C for 1 min. After the final cycle, the reaction mixture was kept at 72°C for 7 min to complete the reaction. The resulting products were resolved by electrophoresis on 1.5 per cent agarose (1× Tris-borate-EDTA)Gels were visualized under UV illumination and photographed.

vii. Statistics Analysis

The specificity and sensitivity and its 95 per cent confidence intervals of the PCR method was calculated in comparisonwith *in vitro* culture method. A Kappa test was performed to test the fitness of PCR to bacterial culture. A Chi-square test was carried to calculate the impact of preservation time and temperature, and pretreatment of media on the reviving rate of bacterium artificially contaminated on filter paper disks.

viii. Sensitivity of PCR Assay on Filter Paper Disk Collected Sample

Agarose gel electrophoresis demonstrated that all PCR results of 2×10^5 CFU/ml to 2×10^2 CFU/ml dilution have unique 270 bp amplification product. With the dilution increasing, the lightness of electrophoresis bands decreased and completely disappeared at 2 CFU/ml dilution.

ix. Observations-Field Test

Strep. agalactiae was isolated from 182 samples from the 826 milk samples of field tests. The following table revealed that there are 184 samples showing positive in a PCR test among the 826 milk samples preserved on filter paper disks. 175 PCR positives amples show the same results as Strep. agalactiae *in vitro* culture. Other 9 PCR positive samples show no Strep. agalactiae growth. Seven Strep. agalactiae culture positives amples presented negative in PCR tests:

Comparison of Bacterial Culturing and PCR Results in Field Tests

Sl.No.	*No. of Samples*	*Number of Strep. agalactiae Culture Positive*	*Number of Strep. agalactiae Culture Negative*	*Total*
1	Number of PCR positive	175	8	184
2	Number of PCR negative	7	635	642
3	Total	182	644	826

The diagnosis sensitivity is 96.15 per cent (95 per cent confidence intervals is 91.92-98.30 per cent) and the specificity is 98.60 per cent (95 per cent confidence intervals is 97.27-99.32 per cent).

a. Merits

PCR based on milk samples collected on filter paper disks is less labor-intensive, sensitive, more rapid and portable methods than conventional bacterial identification. The present PCR protocol based on milk samples collected on dried filter paper disks can get the results within 6 hours. PCR based on milk samples collected on filter paper disks presented no cross-reaction with other major. It enables to detect one bacterium in 1 ml of raw milk and can be easily incorporated as a part of routine screening protocol in dairy industry pathogens. The preservation test demonstrated that the sensitivity and specificity of PCR for the milk collected on filter paper disk could keep stable at room temperature for up to four weeks. The suitability of a protocol for routine diagnosis depends on several factors such as specificity, sensitivity, time required and its applicability to large numbers of samples.

It was suggested that the sensitivity and specificity of PCR based on the milk samples collected on filter paper disks were quite good in comparison with bacterial culturing. It is hoped that the PCR diagnosis protocol for the milk samples kept on filter paper disks become a common tool used to identify Strep. agalactiae in bovine mastitis.

b. Demerits

There are several factors that can cause the sensitivity of PCR to decrease. The physical and chemical features of the filter paper can cause following problem:

(i) Firstly, the DNA molecule can adhere to the fiber matrix of paper which may decrease the DNA concentration of extract (PCR template).

(ii) Secondly, the chemical substance in the filter paper can also change the pH value and DNA configuration, thus it causes the sensitivity to reduce.

Source

1. *Wu Jiusheng, Liu Yuehuan, Hu Songhua and Zhou Jiyong, Development of a Rapid PCR Test for Identification of Streptococcus agalactiae in Milk Samples Collected on Filter Paper Disks, Asian-Aust. J. Anim. Sci.Vol. 21, No. 1: 124-130.*

2. *Abdul Razzaq, Y. M. and A. Ibrahim. 2001. Determination of aminoacids by ion-exchange chromatography on filter paper spotted blood samples stored at different temperatures and for different periods: comparison with capillary and venous blood. Clin.Biochem. 34: 399-406.*

3. *Wollants, E., P. Maes, I. Thoelen, F. Vanneste, M. Rahman and M.Van Ranst. 2004. Evaluation of a norovirus sampling method using sodium dodecyl sulfate/EDTA-pretreated chromatography paper strips. J. Virol. Methods. 122: 45-48.*

F. Application of PCR-Based Methods to Dairy Products and to Non-Dairy Probiotic Products

Microorganisms are very important components of fermented dairy products, including probiotic food, as well as of probiotic food supplements and pharmaceutical preparations.PCR-based methods have become indispensable in the microbiological analysis of these groups of products. In the field of fermented dairy products, several applications based on PCR have been developed with the aim to detect, identify and quantify either unwanted bacteria, which may negatively influence the sensory properties of food or may be pathogenic, or beneficial microorganisms which are added as starter cultures or probiotic cultures. Beside PCR analysis of DNA, reverse transcription real-time PCR analysis of mRNA transcript is particularly useful, especially in studies of the physiology and functionality of bacteria in the food environment. In the probiotic field, PCR is expected to be increasingly applied in quality control in terms of detection and quantification of labeled probiotic bacteria in probiotic food supplements or pharmaceutical preparations, and inviability analysis of probiotics in the products. In addition to the already well-established methods, ever easier access to the next generation sequencing may replace some PCR approaches as molecular fingerprint, metagenomic and metatranscriptomic analyses. The access to increasing number of complete bacterial genomes may also facilitate the strain-specific analysis of probiotics or other bacteria through identification of strain-specific sequences.

Source

1. *Christophe Monnet and Bojana Bogovic Matijasic (2012). Application of PCR-Based Methods to Dairy Products and to Non-Dairy Probiotic Products, Polymerase Chain Reaction, Dr Patricia Hernandez-Rodriguez (Ed.), ISBN: 978-953-51-0612-8,*

G. Detection of *Staphylococcus aureus* in Milk by PCR

A comparative study for detection of *Staphylococcus aureus* (*S. aureus*) in raw milk samples by polymerase chain reaction (PCR) with detection by cultural methods was conducted. A total of 100 milk samples were collected from individual cattle in industrial dairy herds. DNA was extracted from all samples directly. PCR amplifications were performed with a pair of primers specific for the *coa* gene of *S. aureus*. In order to isolate *S. aureus* by cultural methods, bacteriological examinations were done on all samples. Of the 100 milk samples tested, 14 per cent were positive on the basis of the cultural results while 21 per cent of the samples were positive using amplification of the *coa* gene. There were no culture positive samples that were negative using PCR. PCR amplification of the *coa* gene could be used as a powerful tool for the identification of *S. aureus* in milk samples.

Source

1. *Ahmadi, M., Rohani, S.M.R. and Ayremlou, N. Comp Clin Pathol (2010) 19: 91.doi: 10.1007/s00580-009-0901-0.*

H. Polymerase Chain Reaction to Detect Adulteration of Sheep's Milk with Goats' Milk, Goats' Milk Adulteration by Cow Milk and Adulteration of Buffalo Milk with Cow Milk

i. Detection of Adulteration of Sheep's Milk with Goats' Milk

The polymerase chain reaction has been applied for the specific detection of goats' milk in sheep's milk using primers targeting the mitochondrial 12S ribosomal RNA gene. The use of goat-specific primers yielded a 122-bp fragment from goats' milk DNA, where as no amplification signal was obtained in sheep's, cows', and water buffaloes' milk DNA. Polymerase chain reaction analysis of raw and heat-treated milk binary mixtures of sheep/goat enabled the specific detection of goats' milk with a sensitivity threshold of 0.1 per cent. This study demonstrates the usefulness of the proposed polymerase chain reaction assay for authentication of milk products in routine analysis.

Source

1. *López-Calleja I, González I, Fajardo V, Martín I, Hernández PE, García T, Martín R, Application of polymerase chain reaction to detect adulteration of sheep's milk with goats' milk, J Dairy Sci. 2005 Sep; 88 (9): 3115-20.*

ii. Detection of Goats' Milk Adulteration by Cow Milk

Numerous methods based on DNA analysis have been employed in the food industry to monitor adulterations of food products of animal origin. Among them the most frequently used are: polymerase chain reaction (PCR) amplification of a marker gene fragment (s) with universal primers, or amplification of DNA with species-specific primers. PCR-products of different origin can be discriminated by size, restriction fragment length polymorphism (RFLP) or single stranded conformational polymorphism (SSCP) analysis.

These methods have been used for identification, and differentiation between, the animal origins of raw or heat-treated meat and meat products (Chikuni *et al.*, 1994; Meyer *et al.*, 1994,1995; Zehner *et al.*1998; Behrens *et al.*, 1999; Guoli *et al.*1999; Hopwood *et al.*, 1999; Matsunaga *et al.*, 1999; Wolf *et al.*, 1999). These approaches are also applicable to the analysis of dairy products. However, adulterations of goats' milk and its products are traditionally tested by immunological and/or electrophoretic methods (Amigo *et al.*, 1992; Levieux and Venien, 1994; Mimmo and Pagani, 1998). So far, only a few DNA-based techniques designed to detect the presence of bovine DNA in goats' milk have been described (Plath *et al.*, 1997; Branciari *et al.*, 2000). This study presents a one-step PCR procedure for detection of adulteration of goats' milk with cows' milk. The method, employing bovine-specific primers for amplification of a 274 bp fragment of cytochrome b DNA, seems to be simple, fast, specific and sensitive.

Source

1. *JacekBania, Maciej Ugorski, Antoni Polanowski and Eryk Adamczyk, DOI: https://doi.org/10.1017/S0022029901004708.*

iii. Detection of Cow Milk Adulteration in Buffalo Milk [1,2]

It is very common in India that with low priced cow milk adulterated high priced buffalo milk. Currently a PCR based method was developed to identify cow milk in raw buffalo milk. DNA was isolated from milk by DNeasy Mericon food kit which resulted in DNA of requisite quality for downstream applications. A bovine specific primer targeting D-loop (displacement) of mt- DNA (mitochondrial) was selected and standardized to amplify cow DNA. Specificity of primer was tested across the species in the genomic DNA isolated from both milk and blood. The protocol can be sensitive to detect upto 5 per cent level of cow milk in the buffalo milk. Designed protocol was efficient, robust and sensitive and could be used as a platform test in routine quality analysis laboratory. Hence, it could be a great alternative for other protein based methods to identify cow milk in buffalo milk.

Source

1. *Tanmay Hazra, Vivek Sharma, Rekha Sharma and Sumit Arora, PCR based assay for the detection of cow milk adulteration in buffalo milk, Indian Journal of Animal research, DOI: 10.18805/ijar.11331.*
2. *Calleja, I.L., Alonso, G.I., Fajardo, V., Rodriguez, M. A., HernAndez, P. A., Garcia, T. and Martýin, R. (2005) PCR detection of cows' milk in water buffalo milk and mozzarella cheese. Int. Dairy J., 15: 1122-1129.*

I. Rapid PCR Test for Detection of Yeast and Mould in Finished Dairy Products [1]

A new PCR yeast and mold test developed by Germany-based Biotecon Diagnostics provides results in just four hours is available for finished dairy products. This foolproof yeast and mold quantification test has demonstrated 100 per cent specificity with 290 strains, representing 260 species with all phylogenetic groups included [2].

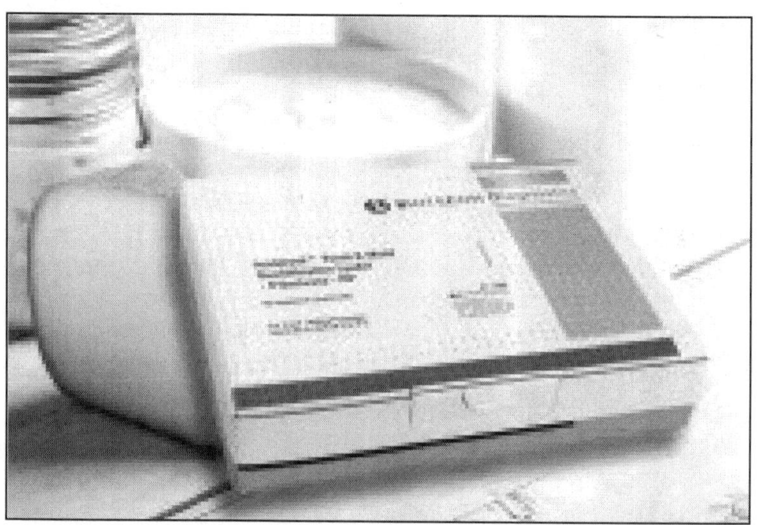

It also demonstrated 100 per cent exclusivity with 60 strains of non-target microorganisms typically found in similar ecological niches (Sensitivity is 101-102 cells/g depending on sample type). This test is based on well-established real-time PCR technology [3]. There is a high degree of correlation between conventional methods and PCR. Results are displayed in CFU per gram and only viable organisms are detected. The real cost benefits will be achieved by how much more quickly can product be shipped and how many recalls avoided.

Source

1. *Food SafetyTech.*
2. *A Release from Weber Scientific.*
3. *Phil Coombs, product specialist at Weber Scientific.*

j. Detection of Bacteriophages in Dairy Industry [1,2]

Bacteriophages that infect LAB/*Streptococcus thermophilus* have been claimed as one of the principal sources of fermentation failure (spoilage or delay) on the manufacture of many dairy products [3,4,5]. Some estimates assume that virulent phages are the primary direct responsible of the largest-economic loss of dairy factories, since they affect negatively up to the 10 per cent of all milk fermentations [6].

Starter cultures consisting in selected bacterial strains are added to the fermentation vats to enhance the fermentative process and also to improve or influence the flavor and texture ofthe cultured products. The starter culture population grows through the fermentation process and reaches high levels inside the industrial vat. This is the perfect environment where bacteriophages can infect sensitive bacteria. The lysis of the infected host-bacteria can decrease several folds the total number of starter cells with consequences ranging from the delay of the acidification with quality changes of the final product, to the total failure of the fermentation. Even with frequent cleaning, disinfestations and sterilization of all the facilities, the total absence of phages in the dairy plants is not possible.

The fermentation of milk by *Streptococcus thermophilus* is a widespread industrial process that is susceptible to bacteriophage attack. *Streptococcus thermophilus* is a gram-positive thermophilic lactic acid bacterium used, along with *Lactobacillus*-spp., as a starter culture for the manufacture of important fermented dairy foods, (yogurt and Swiss- or Italian-type hard cooked cheeses *etc.*) [7]. These bacteria are susceptible to infection by bacteriophages during the fermentation process, a phenomenon that ultimately results in fermentation failure. The common features of *S. thermophilus* phages include double-stranded DNA genomes and they are currently divided into two groups (*cos* and *pac* types) based on the genome encapsidation machinery [8].

Since dairy bacteriophages are one of the major dairy threats, great research efforts have been made to reduce its load on dairy plants and to design new monitoring methods for their early detection.

Detection of bacteriophages in milk is normally carried out using standard microbiological methods (plaque assays, activity tests, *etc.*) [9], but these methods are time-consuming. Due to all these disadvantages, classical microbiological methods have being ousted by more sensitive, faster, and accurate genetic tools as the polymerase chain reaction (PCR). This is a specific and sensitive technology used to detect and identify (traditional PCR), and also to quantify (quantitative PCR) minimal amounts of bacteriophage DNA from samples of different dairy sources. Therefore, the PCR is a fast and reliable approach to screen a great number of samples for phage DNA. In most cases, low quantities of the sample (milk, whey, yogurt, *etc.*) can be directly checked for bacteriophage presence by PCR. To speed up the analysis, PCR techniques have been used to detect phages in different kinds of dairy samples [10,11,12]. To meet the increasing demand for quantitative, more sensitive, and quicker procedures, prompts the development of real-time quantitative PCR (qPCR) methods and hence a fast multiplex qPCR method that allows quantitative detection and identification of *cos*- and *pac*-type *S. thermophilus* bacteriophages in milk samples was developed.

One of the advantages of the PCR based methods is that they can be applied in the raw milk received in the fermentation plant and within hours, each milk batch can be classified for its phage type content and it could be rapidly distributed. For example, if Lb. delbrueckii phages are detected in the raw milk, the batch cannot be used for the elaboration of yoghurt but the milk could be redirected to elaborate cheese, buttermilk, bottled to drink milk or dried to skim milk, thus allowing foran efficient distribution of the raw material. PCR-based approaches have been also applied for routine checking of lysogenic strains in starter cultures, which contain hidden temperate bacteriophages that have the potential to become lytic and compromise the fermentation process [13].

Source

1. *Beatriz del Río, María Cruz Martín, Víctor Ladero, Noelia Martínez, Daniel M. Linares, María Fernández and Miguel A. Alvarez, Bacteriophages in Dairy Industry: PCR Methods as Valuable Tools, Instituto de Productos Lácteos de Asturias, IPLA-CSIC, Spain.*
2. *Beatriz del Rio, María Cruz Martín, Noelia Martínez, Alfonso H. Magadán and Miguel A. Alvarez, Multiplex Fast Real-Time PCR for Quantitative Detection and Identification of cos- and pac-Type Streptococcus thermophilus Bacterio-phages, Appl. Environ. Microbiol. August 2008 vol. 74 no. 15 4779-4781.*
3. *Brüssow, H., Bruttin, A.; Desiere, F., Lucchini, S. and Foley S. (1998). Molecular ecology and evolution of Streptococcus thermophilus bacteriophages-a review. Virus Genes.Vol.16, No.1, pp. 95-109, ISSN 0920-8569.*
4. *Josephsen, J. and Neve, H. (1998). Bacteriophages and lactic acid bacteria. In: Lactic Acid Bacteria. Microbiology and Functional Aspects. Eds. Salminen, S., and S. von Wright, Marcel Dekker Inc, pp. 385-436, ISBN 082470133X, New York.*
5. *Garneau, J.E. and Moineau, S. (2011). Bacteriophages of lactic acid bacteria and their impact on milk fermentations. Microbial Cell Factories, 10 (Suppl 1): S20, (August 2011), ISSN 1475-2859.*
6. *Moineau, S. and Levesque, C. (2005). Control of bacteriophages in industrial fermentations. In: Bacteriophages: Biology and Applications, E. Kutter and A. Sulakvelidze, (Eds.), pp. 285-296, CRC Press, ISBN 0-8493-1336-8, Boca Raton, USA.*
7. *Delcour, J., T. Ferain, and P. Hols.2000. Advances in the genetics of thermophilic lactic acid bacteria. Curr. Opin. Biotechnol.11: 497-504.*
8. *Le Marrec, C., D. van Sinderen, L. Walsh, E. Stanley, E. Vlegels, S. Moineau, P. Heinze, G. Fitzgerald, and B. Fayard. 1997. Streptococcus thermophilus bacteriophages can be divided into two distinct groups based on mode of packaging and structural protein composition. Appl. Environ. Microbiol. 63: 3246-3253.*
9. *Everson, T. C. 1991. Control of phage in the dairy plant. Bull. Int. Dairy Fed. 263: 24-28.*
10. *Binetti, A. G., B. del Rio, M. C. Martín, and M. A. Alvarez. 2005. Detection and characterization of Strepto-coccus thermophilus bacteriophages by use of the antireceptor gene sequence. Appl. Environ. Microbiol. 71: 6096-6103.*
11. *Brüssow, H., M. Fremont, A. Bruttin, J. Sidoti, A. J. Constable, and V. Fryder. 1994. Detection and classification of Streptococcus thermophilus bacteriophages isolated from industrial milk fermentation. Appl. Environ. Microbiol. 60: 4537-4543.*
12. *Del Rio, B., A. G. Binetti, M. C. Martín, M. Fernández, A. H. Magadán, and M. A. Alvarez. 2007. Multiplex PCR for the detection and identification of dairy bacteriophages in milk. Food Microbiol. 24: 75-81.*

13. *Martín, M.C.; Ladero, V. and Alvarez, M. A. (2006). PCR Identification of lysogenic Lactococcus lactis strains. Journal of Consumer Protection and Food Safety, Vol.1, No.2, pp. 121-124, ISSN 1661-5751.*

I. *Multiplex Fast Real-Time PCR for Quantitative Detection and Identification of cos- and pac-Type Streptococcus thermophilus Bacteriophages [1]*

Multiplex Fast Real-Time PCR for Quantitative Detection and Identification of *cos-* and *pac-*Type *Streptococcus thermophilus* Bacteriophages includes following steps:

1. *Primer and Probe Design*

The databases were screened to select the most conserved genes of *cos-* and *pac-*type *S. thermophilus* phages. The following genes were selected and aligned, using the CLUSTAL W algorithm [8], with the sequences of the orthologous genes available in the GenBank database:

(i) ***orf1510*** encoding the putative minor tail protein of the Sfi11 bacteriophage (*pac*type)

(ii) ***orf18*** encoding the antireceptor protein of the Sfi21 bacteriophage (*cos* type).

Highly similar sequences were selected to design primers qPac1, qPac2, qCos1, and qCos2 and probes mgbPac2 and mgbCos using Primer Express software. The species specificity of the primers was assessed by using BLAST 2.2.15 (Basic Local Alignment Search Tool) to ensure that they amplify only the corresponding *S. thermophilus* bacteriophage sequences. Both the mgbPac2 and mgbCos probes were synthesized with a minor groove binder (MGB) nonfluorescent quencher attached to the 32 end and with a different reporter dye attached to the 52 end (VIC and 6-carboxyfluorescein [FAM], respectively) in order to combine them in the same sample.PCR primers and TaqMan MGB probes were used.

2. *Quantification Range and Sensitivity*

(i) **pEM212**, the plasmid used as a standard and a positive control in the qPCR for *pac*-type bacteriophages, was constructed by cloning a 1,196-bp fragment of the *orf1510* gene from the Sfi11 bacteriophage into the pCR-2.1 TOPO vector.

(ii) **pEM213**, the plasmid used as a standard and a positive control in the qPCR for *cos*-type bacteriophages, was constructed by cloning a 147-bp fragment of the *orf18* gene from the Sfi21 bacteriophage into the same vector.

The results showed that the detection limit was one plasmid molecule in 33.22 cycles with a standard deviation (SD) of ±0.7 for plasmid pEM212 and one plasmid molecule in 33.23 cycles with an SD of ±1.0 for plasmid pEM213.

3. Reproducibility and Specificity of Primers and Probes

a. Reproducibility

The reproducibility of the proposed method was determined by performing quadruplicate reactions with two independent φP13.2 and φipla124 suspensions and tenfold milk dilutions containing from 10^3 to 10^9 PFU ml^{-1} were used as templates. The SD of the C_T values obtained were calculated and ranged from a minimum of ±0.07 to a maximum of ±0.6 for φP13.2 (*pac*type) and from a minimum of ±0.04 to a maximum of ±0.92 for φipla124 (costype).

b. Specificity

The specificity was assessed by testing the following *S. thermophilus* bacteriophages:

(i) 27 different *S. thermophilus* bacteriophages previously isolated in Europe from failed industrial fermentations and characterized in laboratory.

(ii) 15 different *S. thermophilus* phages isolated in America [9,10].

(iii) The type phages Sfi11 and Sfi21.

(iv) Four different bacteriophages infecting *Lactobacillus delbrueckii* and nine bacteriophages infecting *Lactococcus lactis* were also tested.

The qPCR method designed in this work was extremely specific since only the *S. thermophilus*phages were detected. Moreover, using the function of the fluorescent dye detected, VIC or FAM, it was possible to identify the type of phage (*pac* and *cos*, respectively).

4. Observations

The proposed qPCR procedure is easy, sensitive, and specific and allows detection, quantification, and identification of the type of *S. thermophilus* phages in a short period of time and thus is suitable for routine use in factory-associated laboratories. Correct and rapid identification of bacteriophages potentially able to attack starter cultures allows speedy decisions concerning the destination of contaminated milk. Such milk might be earmarked for use in processes in which phages are deactivated, processes that do not require starters, or processes that employ starter bacteria not sensitive to the detected phage. In addition, qPCR would also be useful for detection and characterization of phages at all stages of milk product manufacture and in all the niches of the dairy industries.

Source

1. *Beatriz del Rio, María Cruz Martín, Noelia Martínez, Alfonso H. Magadán and Miguel A. Alvarez, Multiplex Fast Real-Time PCR for Quantitative Detection and Identification of cos- and pac-Type Streptococcus thermophilus Bacterio-phages, Appl. Environ. Microbiol. August 2008 vol. 74 no. 15 4779-4781.*

2. *Thompson, J. D., D. G. Higgins, and T. J. Gibson. 1994. CLUSTAL W: improving the sensitivity of progressive multiple sequence alignment through sequence weighting,*

position-specific gap penalties and weight matrix choice. Nucleic Acids Res. 22: 4673-4680.

3. *Binetti, A. G., B. del Rio, M. C. Martín, and M. A. Alvarez. 2005.Detection and characterization of Streptococcus thermophilus bacteriophages by use of the antireceptor gene sequence. Appl. Environ. Microbiol. 71: 6096-6103.*

4. *Suárez, V. B., A. Quiberoni, A. G. Binetti, and J. A. Reinheimer. 2002. Thermophilic lactic acid bacteria phages isolated from Argentinean dairy industries. J. Food Prot. 65: 1597-1604.*

II. Detection of Bacteriophages that infect the main dairy LAB species [1]

Since the traditional PCR generates DNA fragments of specific size, different protocols have been designed not only to detect but to identify, by fragment size discrimination, bacteriophages that infect the main dairy LAB species used in the dairy industry such as Lactobacillus casei/paracasei, Lactobacillus delbrueckii, Lactobacillus helveticus, Streptococcus thermophilus, and Lactococcus lactis. The detection by PCR-based methods might be easily incorporated into dairy industry routines to monitor the presence of phages and be included as part of the prevention strategy for controlling phage contamination. This methodology allows the detection of bacteriophages directly in a small input volume of industrial samples. It also reduces the screening time. Traditional PCR, Multiplex PCR and qPCR are being used to detect bacteriophages that infect the main dairy LAB species. The continuous efforts are being made in order to improve the PCR-based protocols for the detection and quantification of bacteriophages from samples of any origin within the dairyplant. PCR-based methods could be included in the **Hazard Analysis and Critical Control Points (HACCP) protocol** to prevent phage accumulation niches and reduce their impact indairy fermentations.

Traditional PCR

The PCR is a broadly used molecular biology technique, in which a single or a few copies of a specific DNA fragment present in a sample are amplified by a DNA polymerase in a thermal cycler to produce up to 2^{35} copies in few hours. At the end of the amplification reaction, the product is analyzed on an agarose gel and the DNA fragments are separated by size. To ensure the specificity of the reaction, both the target DNA sequence and the flanking primers are chosen in such a way that only samples containing the specific DNA are PCR-positives and yield DNA fragments of known size. The Figure on next page shows a graphic representation of the process to detect and identify bacteriophages that infect different LAB species from dairy samples by PCR amplification and subsequent size discrimination on agarose gel.

The PCR-based methodology has been applied in several stages of dairy product manufacturing to rapid detect, identify and characterize different bacteriophages that infect LAB strains (*e.g.* S. thermophilus, Lactobacillus sp. and Lb.casei/paracasei). S. thermophilus bacteriophages have been identified in industry milk fermentation samples as yogurt and cheese whey, after phage DNA extraction

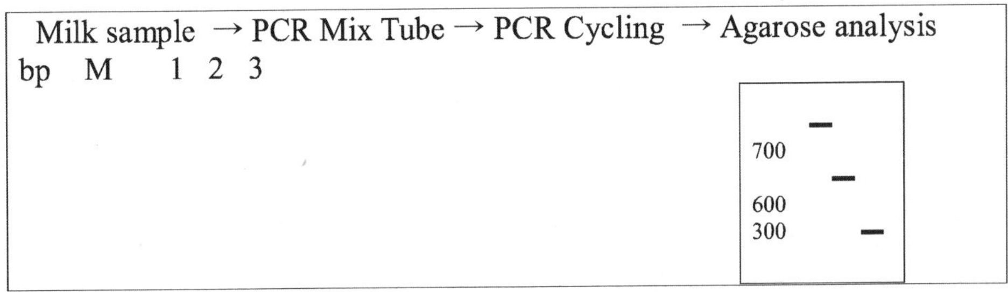

Detection of bacteriophages in dairy samples by PCR. A variety of dairy samples canbe checked for bacteriophage presence. The sample is loaded in a tube with the PCR mixand subjected to amplification in a thermocycler. At the end of the process, the amplifiedDNA is analyzed in an agarose gel and the bacteriophage is identified by the fragment size.Lane 1-*S. thermophilus* phage (750 bp), Lane 2-*Lb. delbrueckii* phage (650 bp), Lane 3-*Lc. lactis*phage (300 bp).

and subsequent PCR [2] or by direct PCR of milk [3,4]. A PCR protocol to detect Lb. casei/paracasei bacteriophages from milk and other commercial samples as fermented milk and cheese whey was also designed [5]. In addition, Lb. delbrueckii bacteriophages have been detected in milk and yogurt samples [6].

Multiplex PCR

The production of different culture milk products needs addition of starter culture. The composition of modern dairy starter cultures (made of a variety of strains) is increasing on complexity. This variety of strains raises the total number of different bacteriophage species that might affect the starter culture and therefore the quality of the final product. Usually, one PCR protocol is designed to specifically amplify and detect only one species of bacteriophage, so more than one PCR would be needed to detect all the phage types.

To solve this problem, multiplex PCR assays were developed to check for more than one phage species in a single reaction. There are different multiplex assays towards multi-type bacteriophage detection [6,7,8]. A multiplex PCR method was designed to detect in a single reaction the presence of the three Lc. Lactis phages species (problematic in dairyindustries) namely 936, c2, and P335 [8]. The PCR for the detection of two additional phage groups: bacteriophages infecting S. thermophilus and Lb. delbrueckii, starters of common use for the elaboration of yoghurt was designed [9]. This simple and rapid multiplex PCR method detects the presence of bacteriophages infecting LAB species most commonly used as starters in dairy fermentations: the three genetically distinct groups of Lc. lactisphages species (P335, 936 andc2) plus phages infecting S. thermophilus and Lb. delbrueckii.

i. Detection of Lysogenic Strains in Starter Cultures by PCR

Lysogenic bacteria mean bacteria that carry one or several inducible prophages integrated into their genomes. Numerous LAB belonging to different genera are lysogenic bacteria. Functional prophages has been identified in species usually used in dairy fermentations such as Lc. lactis subsp. lactis, Lc. lactis subsp. cremoris, S.

thermophilus, Lb. delbrueckii, Lb. casei, Lactobacillus rhamnosus or Bifidobacterium longum.PCR can be used as a tool for the rapid screening towards the identification of lysogenic strains in large culture collections [10].

The cells can be tested by a direct phage-specific PCR assay. If no amplification is obtained, the problem strains can be tagged as phage-free. However, if the strain gives positive PCRresult, it can be further assayed for prophage induction by using a known induction agent asmitomycin C (MitC).

ii. Bacteriophages Typification by PCR and Amplicon Sequencing

The rotation of starter cultures that share similar technological properties but with a different pattern of phage susceptibility is the usual and most successful strategy to reduce the impact of phage attacks in dairy industries.In some detection methods, once a sample is confirmed for bacteriophage presence by PCR amplification, it is possible to get additional information from the amplicon nucleotide sequence. In this sense it has been described a PCR assay that correlates the host range of a bacteriophage and the nucleotide sequence of the amplified fragment [11]. This test is based on the amplification of the VR2 variable region of *orf18*, the antireceptorgene of S. thermophilusphages that was claimed as responsible for host specificity [12]. It is possible to know which strains are susceptible to be infected by the detected phages. This is a very useful tool for the dairy industry, since it allows preventing phage attack by designing a rational starter rotation system based on the phage types detected.

Quantitative PCR

PCR-based technology is a fast and sensitive approach to detect bacteriophages but traditional PCR has an important disadvantage since it is a qualitative but not a quantitative method for bacteriophage screening. The traditional PCR methodology is being changed towards a faster, more sensitive and useful technique, the quantitative PCR (qPCR). This is a new technology based on the traditional PCR that not only adds specificity, accuracy and speed to former PCR screening methods, but also allows the quantification of the number of copies of the target DNA, in this case bacteriophage particles in a dairy sample. The qPCR collects the data in real time throughout the reaction and not just at the end point as the conventional PCR. This feature could be of great value for the dairy industry since it saves time and speed up the decision towards theuse of the analyzed material, in case it is contaminated with bacteriophages.

Several qPCR protocols based on DNA fluorescent probes have been described to detect, quantify and identify bacteriophages present in dairy samples. A qPCR method was developed to quantify Lb. delbrueckii bacteriophages present in milksamples [13].

Source

1. *Beatriz del Río, María Cruz Martín, Víctor Ladero, Noelia Martínez, Daniel M. Linares, María Fernández and Miguel A. Alvarez, Bacteriophages in Dairy Industry: PCR Methods as Valuable Tools, Instituto de Productos Lácteos de Asturias, IPLA-CSIC, Spain.*

2. *Brüssow, H.; Freimont, M.; Bruttin, A.; Sidoti, J.; Constable, A. and Fryder, V. (1994). Detection and classification of Streptococcus thermophilus bacteriophages isolated from industrial milk fermentation. Applied and Environmental Microbiology, Vol.60, No.12, pp. 4537-4543, ISSN 0099-2240.*

3. *Binetti, A.G.; del Rio, B.; Martin, M.C. and Alvarez, M. A. (2005). Detection and characterization of Streptococcus thermophilusphages based on the antireceptor genesequence. Applied and Environmental Microbiology, Vol.71, No.10, (October 2005), pp.6096-103, ISSN 0099-2240.*

4. *Dupont, K.; Vogensen, F.K.; Neve, H.; Bresciani, J. and Josephsen, J. (2004). Identification of the receptor-binding protein in 936-species lactococcal bacteriophages. Applied and Environmental Microbiology, Vo.70, No.10, (October 2004), pp. 5818–5824, ISSN 0099-2240.*

5. *Binetti, A.G.; Capra, M.L.; Alvarez, M. A. and Reinheimer, J. A. (2008). PCR method for detection and identification of Lactobacillus casei/paracaseibacteriophages in dairyproducts. International Journal of Food Microbiology, Vol.124, No.2, pp.147-153, ISSN 0168-1605.*

6. *del Río, B.; Martín, M. C.; Martínez, N.; Magadán, A.H. and Alvarez, M. A. (2008). Multiplexfast real-time polymerase chain reaction for quantitative detection and identification of cos and pac Streptococcus thermophilusbacteriophages. Applied and Environmental Microbiology, Vol.74, No.15, pp. 4779-4781, ISSN 0099-2240.*

7. *Labrie, S. and Moineau, S. (2000). Multiplex PCR for detection and identification of lactococcal bacteriophages. Applied and Environmental Microbiology, Vol.66, No.3, pp. 987-994, ISSN 0099-2240.*

8. *Quiberoni, A.; Tremblay, D.; Ackermann, H.W.; Moineau, S. and Reinheimer, J. A. (2006). Diversity of Streptococcus thermophilus phages in a large-production cheese factory in Argentina. Journal of Dairy Science, Vol.89, No.10, pp. 3791-3799, ISSN 0022-0302.*

9. *del Río, B.; Binetti, A.G.; Martín, M.C.; Fernández, M.; Magadán, A.H. and Alvarez M. A. (2007). Multiplex PCR for the detection and identification of dairy bacteriophages in milk. Food Microbiology, Vol.24, No.1, pp. 75-81, ISSN 0740-0020.*

10. *Martín, M.C.; Ladero, V. and Alvarez, M. A. (2006). PCR Identification of lysogenic Lactococcus lactis strains. Journal of Consumer Protection and Food Safety, Vol.1, No.2, pp. 121-124, ISSN 1661-5751.*

11. *Binetti, A.G., del Rio, B., Martin, M.C. and Alvarez, M. A. (2005). Detection and characterization of Streptococcus thermophilusphages based on the antireceptor gene sequence. Applied and Environmental Microbiology, Vol.71, No.10, pp.6096-103, ISSN 0099-2240.*

12. *Duplessis, M. and Moineau, S. (2001). Identification of a genetic determinant responsible forhost specificity in Streptococcus thermophilus bacteriophages. Molecular Microbiology, Vol.41, No.2, pp. 325-336, ISSN 0950-382X.*

13. *Martín, M.C.; del Río, B.; Martínez, N.; Magadán, A.H. and Alvarez, M. A. (2008). Fast real time polymerase chain reaction for quantitative detection of Lactobacillus delbrueckii bacteriophages in milk. Food Microbiology, Vol. 25, No.8, pp. 978-982, ISSN 0740-0020.*

K. Detection of Bifidobacteria in Raw Milk and Raw Milk Cheese [1]

Bifidobacteria are Gram-positive, non-motile and non-spore-forming bacteria. They had been considered as anaerobic and are well known for their beneficial effects on health and are used as probiotics in food. Bifidobacteria species isolated in humans are different from those isolated in animals. It should therefore be possible to determine contamination origin (human or animal).

Testing Methods Available

Numerous culture-based methods and several molecular methods for bifidobacteria detection have been used. The details are given in following table:

Culture-based Methods	*Molecular Methods*
(i) The culture-based method using propionic acid [2] and paromomycin as selective agents [3] to detect bifidobacteria in meat products and in raw milk samples is not sufficiently efficient to eliminate contaminating flora such as lactobacilli in raw milk or Clostridia in meat samples. *(ii)* Using the culture-based detection method requires knowledge of the contaminating flora and the researched Bifidobacterium species in the samples.	**A. Bifidobacterium species in humans** *(i)* PCR-Elisa method based on the 16S rRNA to detect the most common Bifidobacterium species in humans [4]. *(ii)* Pulsed-field gel electrophoresis (PFGE) and PCR targeting the 16S rRNA [5,6] and PCR in denaturing gradient gel electrophoresis (DGGE) targeting the transaldolase gene for identification, detection and enumeration of human Bifidobacterium species [7]. **B. Bifidobacterium species in animal and human origins** *(i)* PCR-RFLP method based on the 16SrRNA to detect the most common species from animal and human origins [8,9]. *(ii)* Real-time quantitative PCR from the 16S or the trans aldolase gene [10]. **C. Bifidobacterium species in Others** *(i)* They have also been used in the detection of human Bifidobacterium species from feces [11,10,12,13] of bifidobacteria as probiotics [14] or as fecal indicators in waters [15].

Source

1. *V.Delcenseriea, N. Bechouxa, B. Chinaa, G.Daubea, F. Gavinib, A PCR method for detection of bifidobacteria in raw milk and raw milk cheese: comparison with culture-based methods, Journal of Microbiological Methods.*

2. *Beerens, H., 1990. An elective and selective isolation medium for Bifidobacterium spp. Lett. Appl. Microbiol. 11, 155-157.*

3. *Beerens, H., 1998. Bifidobacteria as indicators of faecal contamination in meat and meat products. Detection, determination of origin and comparison with Escherichia coli. Int. J. Food Microbiol. 40, 203-207.*

4. *Malinen, E., Matto, J., Salmitie, M., Alander, M., Saarela, M., Palva, A., 2002. PCR-ELISA II: analysis of Bifidobacterium populations inhuman faecal samples from a consumption trial with Bifidobacterium lactisBb-12 and a galacto-oligosaccharide preparation.Syst. Appl.Microbiol.25, 249-258.*

5. *Roy, D., Ward, P., Champagne, G., 1996. Differentiation of bifidobacteria by use of pulsed-field gel electrophoresis and polymerase chain reaction. Int. J. Food Microbiol. 29, 11-29.*

6. *Bonjoch, X., Balleste, E., Blanch, A., 2004. Multiplex PCR with 16S rRNA gene-targeted primers of Bifidobacterium spp. to identify sources of fecal pollution. Appl. Environ. Microbiol. 70, 3171-3175.*

7. *Requena, T, Burton, J., Matsuki, T, Munro, K., Simon, M. A., Tanaka, R., Watanabe, K., Tannock, G., 2002. Identification, detection and enumeration of human Bifidobacterium species by PCR targeting the trans aldolase gene. Appl. Environ. Microbiol. 68, 2420-2427.*

8. *Delcenserie, V, Bechoux, N., Leonard, T., China, B., Daube, G., 2004. Discrimination between Bifidobacterium species from human and animal origin by PCR-Restriction fragment length polymorphism. J. Food Prot. 67, 1284-1288.*

9. *Roy, D., Sirois, S., 2000.Molecular differentiation of Bifidobacterium species with amplified ribosomal DNA restriction analysis and alignment of short regions of the Idhgene. FEMS Microbiol. Lett. 191, 17-24.*

10. *Requena, T, Burton, J., Matsuki, T, Munro, K., Simon, M. A., Tanaka, R., Watanabe, K., Tannock, G., 2002. Identification, detection and enumeration of human Bifidobacteriumspecies by PCR targeting the transaldolase gene. Appl. Environ. Microbiol. 68, 2420-2427.*

11. *Matsuki, T., Watanabe, K., Fujimoto, J., Miyamoto, Y., Takada, T., Matsumoto, K., Oyaizu, H., Tanaka, R., 2002. Development of 16SrRNA-gene-targeted group-specific primers for the detection and identification of predominant bacteria in human feces. Appl. Environ.Microbiol. 68, 5445-5451.*

12. *Mullié, C, Odou, M., Singer, E., Romond, M., Izard, D., 2003. Multiplex PCR using 16S rRNA gene-targeted primers for the identification of bifidobacteria from human origin. FEMS Microbiol. Lett. 222, 129-136.*

13. *Venema, K., Maathuis, A., 2003. A PCR-based method for identification of bifidobacteria from the human alimentary tract at the specieslevel. FEMS Microbiol. Lett. 224, 143-149.*

14. *Brigidi, P., Swennen, E., Vitali, B., Rossi, M., Matteuzzi, D., 2003. PCR detection of Bifidobacterium strains and Streptococcus thermophilus in feces of human subjects after oral bacterio-therapy and yogurt consumption. Int. J. Food Microbiol. 81, 203-209.*

15. *Bernhard, A., Field, K., 2000. Identification of nonpoint sources of fecal pollution in coastal waters by using host-specific 16S ribosomalDNA genetic markers from fecal anaerobes. Appl. Environ. Microbiol. 66, 1587-1594.*

Materials [1]

Samples

a. Raw milk samples

Samples were diluted until 10^{-4} and presence or absence of bifidobacteria at each dilution was compared statistically by the different detection methods as follows:

(i) 39 samples (195 dilutions) have been analyzed and results compared using three combinations of culture-based methods using mupirocin or not.

(ii) 12 samples (60 dilutions) have been analyzed and results compared by PCR from two enrichment broth using or not mupirocin.

(iii) 148 samples (740 dilutions) have been analyzed and results compared by the different PCR and culture-based methods.

b. Raw Milk Cheese Samples

Milk was collected on farms and stored in tanks at the factory at 4°C. Milk is prepared for maturation by addition of cream, ferment and surface flora. Animal rennet is added.On day 1, the following steps are successively performed-molding, a first manual turnover, amanual salting and a second turnover. On day 2, cheeses are removed from the molds and a new manual or mechanical salting is performed. Ripening is carried out for 28 days.

(i) Twenty-five raw milk cheeses at four different steps of the production chain from raw milk to the end product.

(ii) 100 samples were analyzed by the best culture-based method chosen among the three tested and by the best PCR method. The following production steps were analyzed- raw milk (Step A), after addition of rennet (StepC), after removal from the mold (Step E), during ripening (Step G). Samples were diluted until 10^{-6} to perform semiquantitative detection of bifidobacteria.

Methods

1. **Culture-based method for E. coli detection-** E. coli was numerated on the Coli ID medium [1].
2. **Culture-based methods for bifidobacteria detection-**The methods were performed in two steps, an enrichment and an isolation step. Components of enrichment and isolation media before adding mupirocin have been described by Beerens [2]. The details of the enrichment, isolation step *etc.* are given below:

   (i) **Enrichment step medium-**The propionic acid, 5 ml/l, Fe-citrate, 0.5 g/l, cysteinchlorhydrate, 0.5 g/l, yeast extract, 5 g/l, agar, 2 g/l added to the medium Brain Heart Infusion37 g/l.Mupirocin (GSK Laboratories, England) was added (BHMup) or not (BH) at the concentration equal to 80 mg/l (to be added when the medium

must be used). The final pH- 5.0, was obtained with the addition of a NaOHsolution. The medium (without mupirocin) was dispensed in 9 ml amounts. Sterilization was not needed because of the medium's low pH.

(*ii*) **Isolation step medium-**Columbia blood agar medium (France) was used with addition of Fe-citrate, 0.5 g/l; glucose, 5 g/l; cystein chlorhydrate, 0.5 g/l. Mupirocin was added (CMup) or not (C) at the concentration equal to 50 mg/l (to be added when the medium must be used). The medium (without mupirocin) was dispensed in 100 ml bottles and autoclaved at 120°C.

(*iii*) **Protocol-**The selective enrichment medium, with mupirocin added or not, was held in boiling water for 20 min to expel oxygen and cooled to 30-40°C.

3. **PCR method for bifidobacteria detection-**A method of detecting the Bifidobacterium genus was developed by PCR targeting the hsp60 gene. The genus Bifidobacterium was identified by PCR amplification of a 217-bp hsp60 gene fragment. The degenerated primer pair specific to the Bifidobacterium genus used was tested for it specificity on 127 strains. Sensitivity was measured on artificially contaminated samples. This involves the following steps:

(*i*) Target DNA preparation: **Pure strains-**127 reference strains belonging to 14 Bifidobacterium species and 37 non-Bifidobacterium strains belonging to species or genera often food-contaminating (5 Enterococcusspp., 5 Pseudomonas sp., 5 Staphylococcus aureus, 6 Lactobacillus,4 Clostridium perfringens, 6 Bacillus cereus, 5 E. coli and 1 Salmonella typhimurium) were tested for primers validation. Before testing, the Bifidobacterium strains were withdrawn from frozen storage on Rosenow medium and subcultured on Brain Heart Infusion at 37 °C for 48 to 72 h under anaerobic conditions.

One ml of bacterial cultures in BHI broth was centrifuged at 12,000 X g for 2 min using a bench top centrifuge. The pellets were transferred in sterile, demineralized water, and the DNA was extracted using WizardGenomic DNA purification kit with addition of lysozyme (10 mg/ml), as recommended for Gram-positive bacteria. DNA concentrations were spectrophotometrically estimated. DNA samples were diluted with distilled water to obtain a concentration between 25 and 50 μg/ml.

Artificially contaminated samples-These samples were prepared as per the following way:

40 ml of UHTmilk was added to 360 ml of BHI broth supplemented with mupirocin (80 mg/ml) in a sterile bottle. After homogenizing, the resulting mixture was distributed in aliquots of 10 ml. They were inoculated with 100 μl of 10-fold serial dilutions of a 48 h culture of B. pseudolongum subsp. globosum (RU224) and B. thermophilum

(DSM 20210T) in peptone sodium solution, and including a negative control without inoculation.

One ml of each aliquot was distributed on MRS medium (Oxoid) plates supplemented with mupirocin (50mg/ml) for bifidobacteria counting after 72 h anaerobic incubation at 37°C. The other part of aliquot (about 9ml) was incubated during 24h in anaerobic conditions at 37°C. The same procedure was repeated with a 48h incubation time. After this, 1 ml of each incubation broth was transferred into a microcentrifuge tube and centrifuged at 12,000 X g for 2 min using a bench-top centrifuge. The pellets were transferred in sterile, demineralized water and the DNA extracted using Wizard Genomic DNA purification kit.

Raw milk and raw milk cheese samples-DNA was extracted from cultures obtained after the enrichment step of the cultural-based method (from pure until 10^{-4} dilution for milk and until 10^{-6} dilution for raw milk cheese samples). One ml of each homogenized content was transferred in a microcentrifuge tube and centrifuged at 12,000 X g for 2 min using a bench-top centrifuge. The pellets were transferred in sterile, demineralized water, and the DNA extracted using Wizard Genomic DNA purification kit as previously described for puretrains. In case of PCR inhibition, the DNA samples were diluted 10 fold.

(ii) **Selection of primers-**The sequences of the hsp60 gene are available on Genbank for several representative Bifidobacterium species in human and animal feces (B. adolescentis, B. animalis, B. cuniculi, B. choerinum, B. pseudolongum subsp. globosum, B. pseudolongum subsp. pseudolongum, B. merycicum, B.pseudo-catenulatum, B. ruminantium, B. thermophilum). These sequences were aligned. From these sequence alignments, Bifidobacterium genus-specific degenerated primers were selected using Oligo software (Medprobe). Specificity of the primers for the Bifidobacterium-genus was checked realizing a Meg-ablast. Only Bifidobacterium DNA was fully complementary to the primers sequences.

The genus-specific amplification of a 217 bp fragment of the hsp60gene is generated using primers: B11 up: 5′-GTS CAY GAR GGY CTS AAG AA-3′, B12 down: 5′-CCR TCC TGG CCR ACC TTG T-3′.

(iii) **Controls-**The following amplification controls were run with each series:

a: positive- reaction mix containing DNA extract from a positive strain of *B.* pseudolongum.

b: two reagent controls - mix containing all reagents without sample DNA and extraction control, *i.e.* 1000 µl of distilled water processed in the same manner as the samples.

As food contains PCR inhibitors, it can be a difficult matrix for PCR testing and hence construction of an internal PCR control is needed to sure that a negative result is indeed due to absence of the target rather than to an inhibition of PCR reaction.

An artificially created DNA fragment was used as an **internal positive control** in every reaction mixture, except for the other controls. The control DNA consisted of a fragment of 315 bp of the pGEMT vector, flanked by the target for the Bifidobacterium-genus PCR primers. This product was created by a two-step PCR as follows.

Chimerical PCR primers flanked with the Bifidobacteriumgenus-specific primers were chosen: CI up: 5′-GTSCAY GAR GGY CTS AAG AAG CAG GAA AGA ACA TGT GAG CA-3′ and CI down: 5′-CCR TCC TGGCCR ACC TTG TAC GAC CTA CAC CGA ACT GAG A-3′. The first step comprised amplification of DNA from the pGEMT vector using the chimerical primers by 45 cycles at the following PCR conditions. A 5 µlpGEMT DNA was introduced in a mix containing 0.2 mol 1-1 dNTPs, 400 pmol1-1 of each chimerical PCR primers, 0.8 U of Dap Goldstar polymerase, 1 x buffer: 20 mM Tris-HCl, pH 8.0, 100 mM KC1, 0.1 mM EDTA, 1 mM DTT, 50 per cent glycerol, 0.5 per cent Nonidet P-40 and 0.5 per cent Tween-20.

The samples were subjected to an initial step of denaturation at 95 °C for 5 min, followed by 15 denaturation cycles at 95 °C for 30 s, annealing at 50°C for 30 s and extension at 72 °C for 30 s and 30 cycles of denaturation at 95 °C for 30 s, annealing at 60 °C for 30 s and extension at 72 °C for 30 s. In the second step, the amplicon of the first amplification was purified, diluted 1/1000 in distilled water and used as a template to perform a second amplification using the Bifido-bacteriumgenus-specific primers (in PCR conditions). The final amplicon was purified, diluted and used as internal control. As measured by optical density using a GeneQuant pro spectrophotometer UV, the DNA concentration was 185 µg/ml. The final dilution in distilled water of the internal control target was established empirically to reduce competition with target DNA and corresponded to 1.1 µg/µl of DNA. The **control DNA was used as a positive amplification control** in all samples.

(iv) **PCR conditions***:*

a. PCR mix-It was composed of 0.2 mol 1-1 dNTPs, 400 pmol 1-1 of each primer, 1 U of FastStart Taq Polymerase, 1 x buffer: 500 mM Tris-HCl, 100 mM KC1, 50 mM $(NH4)_2SO_4$, pH 8.3/25°C, 4 µl DNA (50-100 ng), 1 µl internal control and H_2O in a total volume of 20 µl.

b. Cycling Conditions-PCR was run using the following cycling conditions:

1 x 5 min at 95°C, 40 x 95°C for 30s, 60°C for 30s and 72°C for 30s and a final extension (5 min at 72°C). Samples were kept at 4°C or stored at -20°C before analyzing. A sample was considered as positive when the 217 bp amplicon was visible on 2 per cent agarose gels after electrophoresis and ethidium bromide staining.

Two different protocols were tested- PCR from BH (PCR 1) and PCR from BHMup enrichment broth (PCR 2).

Statistical Analysis

The Mc Nemar test [2] was choosen to statistically evaluate the different methods. Dilutions were tested as separate values. To compare results obtained at different steps of the raw milk cheese production, an ANOVA test [3] was performed.

3. Observations

A. Culture-based Methods

Thirty-nine raw milk samples (195 dilutions) were analyzed by three culture-based methods using three combinations of enrichment and isolation media- Cultural1 (BH/CMup),2 (BHMup80/C) and 3 (BHMup/CMup). The highest percentage of positives (95 per cent) was detected for pure and for 10^{-1} dilutions with Cultural 3. The comparison of the three methods by the Mc Nemar test (based on Chi-square table), which was calculated on the basis of the number of different results obtained on the 195 dilutions, with methods compared 2 by 2, was studied. In the McNemar test, Cultural 3 was confirmed to be the best method. A statistical difference was observed between Cultural 2 and Cultural 3 in favor of Cultural 3 (χ^2=5.56; P<0.025). A trend was noticed between Cultural 1 and Cultural 3 in favor of using mupirocin at both steps, enrichment and isolation (χ^2 =2.91; P<0.1). No difference was observed between methods using mupirocin only, either in the isolation medium or in the enrichment medium.

Many different selective agents were used for detection of bifidobacteria: lithium chloride, sodium propionate, nalidixic acid, neomycin sulphate, paromomycin sulphate, polymixin B sulphate [4]. In the case of raw milk samples, an enrichment step was necessary because of the possible relatively low levels of bifidobacteria (10 to 106 ml -1) compared to those in human or animal feces (10^7 to 10 g^{-1}). Beerens [5] recommended using at the enrichment step the BHI medium with addition of propionic acid, yeast extract, ironcitrate, and at the isolation step, paromomycin as selective agent.

In raw milk samples, addition of mupirocin at the enrichment step can eliminate most of the lactobacilli strains present that could hide bifidobacteria in raw milk. If some lactobacilli strains were still present after the enrichment step, one might suppose that they would be eliminated during the isolation step by mupirocin, when present.

The culture-based method presented in this study provides semiquantitative results. As none of culture-based methods are sufficiently selective to detect only bifidobacteria, the F6PPK test must be performed to confirm that isolated strains indeed belong to the genus Bifidobacterium. An alternative was to carry out the F6PPK test on the whole culture at the isolation step in order to more rapidly ascertain the contamination level of studied samples.

B. PCR Methods

(a) **Validation of the primers on pure strains-** Specificity of the primers was confirmed by PCR using chromosomal DNA extracted from 37 non-Bifidobacterium strains and from 127 Bifidobacterium strains. The primers were able to detect an expected 217bp DNA fragment from all Bifidobacterium strains using the PCR described conditions. No amplification was obtained for strains of any of the other tested species. This validation was realized intriplicate.

(b) **Detection limit of the PCR method on artificially contaminated samples-** PCR results obtained from enrichment media incubated for 24 and 48 h have been compared in relation with bifidobacteria counts on MRS plates (Oxoid) after 72 h anaerobic incubation at 37 °C. Depending on the incubation time of the enrichment media, the PCR method could detect DNA from 102 to 103 cfu ml^{-1} present in the sample when the incubation time of the enrichment medium was 24 h, and around 1^{-10} cfu ml^{-1} when it was 48 h.

(c) **Comparison of PCR methods (from BH and from BHMup80 enrichment media)-**Bifidobacteria were detected in 12 samples (60 dilutions) by PCR from BH enrichment broth (PCR 1) and from BHMup broth (PCR 2) obtained after the enrichment step of the culture-based methods. The following four possibilities of results were observed:

(i) Only the target was positive (217 bp fragment).

(ii) The target (217 bp) and the internal control (315 bp fragment) were positive. In these two cases, PCR was considered as positive for bifidobacteria detection.

(iii) Only the internal control was positive. In this case, the PCR was indeed negative for bifidobacteria detection.

(iv) Finally, if the target and the internal control were negative, it signed PCR inhibition. In this case, it was necessary to do the PCR again on a diluted sample.

C. Comparison of Culture-Based and PCR Methods

PCR 1 and 2 have been compared to the culture-based methods 1, 2 and 3. Each comparison showed a greater number of positive results by PCR. However, the difference was only significant when PCR 1 was compared to Cultural 1 (χ^2 = 9.4; $P<0.005$). No significant difference was observed when PCR 2 was compared to Cultural 2 and 3. For each comparison test, a percentage of false negatives by PCR are present, respectively equal to 6 per cent, 7 per cent and 2 per cent.

A trend was observed in favor of PCR, suggesting that the PCR method is more sensitive than culture-based methods. The method using mupirocin at both enrichment and isolation steps and the PCR method performed from the culture in BHMup enrichment medium were shown to be the most efficient. No significant difference was observed in raw milk samples between PCR from BHMup and the culture-based method BHMup/CMup, while a significant difference was noticed between the same methods in raw milk cheese samples, which would favor using PCR.

The results suggested that PCR on the hsp60 gene was convenient for a rapid detection of bifidobacteria in raw milk and raw milk cheese samples and that bifidobacteria always present throughout raw milk cheese production could be efficiently used as fecal indicators. The culture-based method BHMup/CMup was efficient since it showed that 95 per cent of raw milk and more than 95 per cent of raw milk cheese samples contained bifidobacteria. However, like all culture-based methods, this method is time-consuming and could not be easily applied to food industry controls. The PCR method performed from BHM up enrichment medium does not present this disadvantage and can be used effectively to detect bifidobacteria as fecal indicators in raw milk cheese industries, instead of or with E. coli.

Source

1 *V.Delcenseriea, N. Bechouxa, B. Chinaa, G.Daubea, F. Gavinib, A PCR method for detection of bifidobacteria in raw milk and raw milk cheese: comparison with culture-based methods, Journal of Microbiological Methods.*

2. *Leroy, P., Farnir, F., 2000.Le Chi-Carredans les tests de conformite. In: Les Editions de l'Université de Liège (Ed.) Methodes Statistiquesen Medecine Veterinaire, vol. 1, pp. 120-124.*

3. *Dagnelie, P., 1975. Analyse de la variance à un critère. In: Les Presses Agronomiques de Gembloux (Ed.) Théorie et Méthodes Statistiques, vol. 2, pp. 126-138.*

4. *Payne, J., Morris, A., Beers, P., 1999. Note: evaluation of selective media for the enumeration of Bifidobacterium sp. in milk. J. Appl.Microbiol. 86, 353-358.*

5. *Beerens, H., 1998. Bifidobacteria as indicators of faecal contamination in meat and meat products. Detection, determination of origin and comparison with Escherichia coli.Int. J. Food Microbiol. 40, 203-207.*

7. Application of PCR in Meat Industry [1]

Detection of Meat Adulteration

This is required beause of the following two major causes:

(*i*) **Health conscious and quality food-**Consumers have become health conscious and thus demand quality food. The adulterated food however, often enters the supply chain and jeopardizes the health of the people as well as sentiments.

(*ii*) **Religious considerations towards Meat Consumption-**Substantial proportion of population has got religious considerations towards the consumption of meat of a particular animal species and any deviation may result into detraction from its consumption.

The meat adulteration has got social, religious, economic and public health concerns. The legal enforcement also restricts consumption of meat of wild animal species.Hence there is an urgent need to have reliable techniques which may aid in the authentic identification of meat species.Numerous analytical methods *e.g.* histological, microscopic, traditional methods (organoleptic, anatomical, chemical, electrophoretic, chromatographic and immunological techniques) have been employed but none of these are considered to be complete for the purpose of species identification because of some limitations. For instance, immunoassays fail to differentiate the closely related species as the protein markers may possess similar isotypes in different species of animals and different components of the same animal species could behave differently making the judgment difficult.

Recently, DNA based molecular techniques have become popular and could provide satisfactory solution to the problem of identification of meats of different animal species. Targeting DNA for the animal speciation offers many advantages over the protein-based techniques. The DNA molecule is more stable and the composition of DNA is same in any cell of the individual. Hence, following DNA based techniques have been employed for the identification of animal species in the recent past:

(i) Polymerase Chain Reaction (PCR) and its variants.

(ii) Restriction Fragment Length Polymorphism (RFLP).

(iii) Random Amplified Polymorphic DNA (RAPD) finger printing.

(iv) DNA Hybridization, PCR-Sequencing.

(v) Arbitrarily Primed-PCR (AP-PCR).

(vi) PCR-Single Strand Conformation Poly-morphisms (SSCP).

Further, Real Time PCR provides a major breakthrough in the quantitative analysis of meats of different animal species. The PCR-based techniques provide high level of specificity, sensitivity, accuracy and precision than the other techniques being used for the purpose of animal species identification.

Methods Used for Speciation of Meat

The following methods can be used in the laboratories for speciation of meat after standardization in the laboratories.

Meat Speciation Using DNA-Based Molecular Techniques

The reasons why DNA is the molecule of choice for the purpose of species identification include,

(i) Its stability allows the species identification in heated and processed products.

(ii) DNA structure is conserved in all the tissues of an individual.

(iii) Higher discrimination of closely related species is possible as vast information is present in the nucleotide algorithm consisting of A, T, G and C.

The DNA isolated from any cell/tissue of an individual is identical irrespective of the organs or tissues. Further, the information available in the DNA is enormous as compared to the proteins due to the degeneracy of the genetic code. The DNA based animal speciation techniques are the most preferred techniques, since the DNA carries an organisms' total genetic information and is stably functioning as long as the animal is alive. The following different DNA-based techniques are used for animal species identification:

(i) DNA hybridization.

(ii) PCR (Polymerase Chain Reaction) and its variants.

(iii) PCR-RFLP (Polymerase Chain Reaction-Restriction Fragment Lenth Poly-morphism).

(iv) RAPD-PCR (Randomly Amplified Polymorphic DNA- Polymerase Chain Reaction).

(v) PCR-SSCP (Polymerase Chain Reaction- Single Strand Conformational Polymorphism)

(vi) PCR-sequencing.

The most widely used methods for meat species identification are:

(i) PCR (Species-specific and Multiplex)

(ii) RFLP fingerprinting

(iii) FINS (Forensically Informative Nucleotide Sequencing).

All these techniques require extraction of DNA from tissues.

1. Isolation of DNA from Tissue/Meat Samples

DNA Extraction

Any tissue like muscle, organ, skin, blood, excretaion and secretaion can act as a source of DNA.

Principle

Tissue sample is trichurated and homogenized to get cells. The cells are first ruptured to release cellular contents including DNA. Interfering substances such as proteins and RNA are lysed and then DNA is separated from by precipitation. Extracted DNA is checked for quality, puririty and concentration.

Reagents

(i) Tris saturated Phenol (equilibrated with 0.1M Tris –Cl, pH 8.0).

(ii) Chloroform.

(iii) DNA lysis buffer (pH 8).

(iv) Ribonuclease-A.

(v) Proteinase-K solution (20 mg/ml).

(vi) Isoamylalcohol.

(vii) 10 M ammonium acetate.

(viii) Absolute ethanol.

(ix) 70 per cent alcohol.

(x) 1X TE (Tris-EDTA) buffer.

Apparatus

(i) Dry bath.

(ii) Water bath.

(iii) Refrigerated Centrifuge.

(v) Scissors

(vi) Forceps.

Procedure

Phenol: Chloroform method can be used for DNA extraction as per the protocol [2]. The tissue samples (0.2-1 g) are cut into very small (<1 mm) pieces or pulverized in liquid nitrogen and 10 volumes (w/v) of DNA lysis buffer (pH 8) containing Ribonuclease-A@ 100µg/ml (20µg/ml) is added and incubated at 37°C for1 h. Then, proteinase-K solution (20 mg/ml) was added@ 200µg/ml and again incubated at 50°C for not less than 3h or overnight. During incubation, regular swirling of the tubes is undertaken gently from time to time. Equal vol. of Tris-saturated phenol (equilibrated with 0.1M Tris -Cl, pH 8.0) is added and the contents of the tubes are subjected to mixing end to end for 10 min. and the contents are then centrifuged at 6,500 RPM for 15 min. The upper aqueous phase is collected into a fresh tube and equal volume of phenol: chloroform: isoamylalcohol (25: 24: 1) mixture is added and centrifuged. The upper layer is collected and again equal volume of phenol: chloroform: isoamylalcohol (25: 24: 1) is added and centrifuged. Finally, the upper aqueous phase is collected in a fresh tube and equal volume of chloroform was added and then centrifuged. The upper phase is again collected in to a fresh tube containing 0.2 volumes of 10 M ammonium acetate and 2 volumes of absolute ethanol and is mixed well for precipitation of DNA. The mixture containing visible DNA threads were centrifuged at (10,000 RPM for 10 min). The DNA pellet is washed twice with 70 per cent alcohol by centrifugation (10,000 RPM for5 min each), dried over a dry bath at 60°C and then dissolved in 1X TE (Tris -EDTA) buffer (50- 100µl) or nuclease free water either. These DNA samples are used for PCR or stored at -20°C until further use.

Source

1. *Manual of methods of analysis of Foods (Meat and Meat Produts/Fish and Fish Products), LabManual-6. FSSAI, New Delhi.*

2. *Sambrook, J. and Russell, D. W. 2001. Molecular cloning: A laboratory manual. 3rd edition. Cold Spring Harbour Laboratory Press, Cold Spring Harbour, New York.*

2. Checking the Quality, Purity and Concentration of DNA [1]

For further molecular analysis, isolated DNA has to undergo quality checks.

A. Determination of Quality of DNA

A good quality DNA is a pre-requisite for its further molecular analysis hence lts quality should be determined.

Principle

The isolated DNA is checked for quality to ensure the absence from contaminants such as protein and RNA. The DNA must be intact and free from shearing.

Reagents

(i) Agarose gel (0.6-0.8 per cent, w/v).

(ii) TE buffer.

(iii) 6X gel loading dye (Xylene Cyanol and Bromophenol Blue).

(iv) 1 kb molecular weight marker.

(v) Ethidium bromide.

Apparatus

(i) Electrophoresis unit.

(ii) UV trans-illuminator.

(iii) Gel documenttation.

Procedure

The quality of isolated DNA is checked by agarose gel (0.6-0.8 per cent, w/v) electrophoresis. About 5 µl of dissolved DNA (in TE buffer or NFW) is mixed with 1 µl of 6X gel loading dye (Xylene Cyanol and Bromophenol Blue) and loaded on to a well along with a1 kb molecular weight marker. Electrophoresis is undertaken@ 5-8 V/cm and ethidium bromide (added@ 0.5µg/ml in the agarose) stained DNA bands are visualized under UV trans -illuminator and documented over a gel documentation system to assess the quality of isolated DNA.

B. Determination of Purity of DNA

Based on wavelength maximum of DNA and protein, purity is determined.

Reagents

(i) Nuclease free water.

Apparatus

(i) UV/VIS spectrophotometer.

Procedure

The purity of the isolated DNA is checked by using UV/VIS spectrophotometer.

The DNA samples with OD260: OD280 ratios between 1.7 and1.9 are used for PCR amplification.

Calculation

The concentration of the DNA can be estimated by using following formula:

$$\text{DNA concentration } (\mu g/ml) = \frac{\text{OD260 x Dilution Factor x 50}}{1000}$$

(1 OD value at 260 nm is equivalent to ~50µg of dsDNA/ml)

Source

1. *Manual of methods of analysis of Foods (Meat and Meat Produts/Fish and Fish Products), LabManual-6. FSSAI, New Delhi.*

Comprehensive DNA-based Methods of Meat Speciation

Of the several DNA based molecular tools, the following three methods (PCR, RFLP and FINS) are most widely used for meat speciation [6].

1. Polymerase Chain Reaction (PCR-Method-1)

PCR provides a wide range of options to the analysts by virtue of its versatility and hence it is considered as state-of-the-art technique in animal species identification. It is a rapid means of making multiple copies of specific piece of DNA sequence in vitro and possesses high level of selectivity as well as sensitivity and hence it is an ideal diagnostic analytical tool.

Types of PCR being Used for Meat Speciation/Identification of Animal Species

There are several types of PCR, but most widely used in meat speciation are species-specific PCR and multiplex PCR.

a. Species-Specific PCR [1]

Species specific DNA in femto grams (fg) and pico grams (pg) could be detected in both processed and unprocessed samples using targeted amplification of rRNA genes (12S, 16S, 18S), actin-multigene families (which are highly conserved in all the eukaryotes), satellite DNA, cytochrome-b gene, cytochromeoxidase II, growth hormone gene, melanocortnin gene, mt D-loop, myofibrillar components and SatelliteI DNA using PCR. Some highly conserved regions on the mitochondrial genes such as cytochromeb have been most widely used compared to others for the species identification where pair of specific primers amplifies a specific region in PCR. Although both genomic and mt targets are used, mt sequences are preferred since it possess high rate of mutation allowing species identification and also help in the differentiation of most closely related species.

Principle

It is a process of in vitro amplification of target DNA, a polymerase enzyme sythenises new DNA which is specific to the target specific primers and other ingradients required for DNA synthesis. PCR produces millions of copies of target

Different Steps	*Species-Specific PCR [2,3,4,5,6]*	*Multiplex PCR [2,3]*
Reagents	*(i)* PCR reaction mixture *i.e.* 2.5 µl of 10X assay buffer [160 mM (NH4)2SO4, 670 mM Tris-HCl, pH 8.8, 0.1 per cent tween-20, 25 mM MgCl2]. *(ii)* dNTP mix [Sodium salts of dATP, dCTP, dGTP and dTTP 10 mM each in water *i.e.*, 40 mM total pH 7.5]. *(iii)* Forward and reverse primers (Species -specific primers as shown in table) 20 Pico moles each. *(iv)* 1U Taq DNA polymerase. *(v)* 50 ng of purified DNA. *(vi)* Nuclease Free Water vii Agarose gel (2 per cent). *(viii)* 1X TBE buffer. *(ix)* 6X gel loading dye (1 µl) *(x)* 100 bp DNA ladder.	*(i)* PCR reaction mixture. *(ii)* 10X buffer (5 µl). *(iii)* dNTP mix (2.5 µl of 10 mM dNTP solution). *(iv)* MgCl2 (3mM, 6 µl of 25 mM MgCl2 solution). *(v)* Common forward primer (20 pmoles). *(vi)* Species-specific reverse primers (20 pmoles each). *(vii)* Taq DNA polymerase (3U). *(viii)* Purified DNA (50 ng each). *(ix)* Autoclaved milli Q water
Apparatus	*(i)* Thermal cycler. *(ii)* Pipettes 1 ml, 01 ml. *(iii)* Laminar air flow cabinet. *(iv)* Electrophoresis unit. *(v)* Gel documentation system.	*(i)* Thermocycler *(ii)* Pipettes-0.1 and 1ml
Procedure	The PCR is set up in a 25 µl volume reaction mixture consisting of 2.5 µl of 10X assay buffer [160 mM (NH4)2SO4, 670 mM Tris-HCl, pH 8.8, 0.1 per cent tween-20,25 mM MgCl2], 0.5 µl (200 µM each) of dNTP mix [Sodium salts of dATP, dCTP, dGTP and dTTP 10 mM each in water *i.e.*, 40 mM total pH 7.5], 0.5 µl (20 Pico moles) each of forward and reverse primers (Species -specific primers as shown in following Table), 1U Taq DNA polymerase, 50 ng of purified DNA and NFW to make the volume.The tubes are flash spun and the PCR was performed in a Thermal cycler.	The multiplex PCR technique as described by Matsunaga *et al.,* [3] is wodely followed for meat spectation; PCR is set up in 50 µl reaction volume. The reaction mixture is optimized by using 10X buffer (5 µl), dNTP mix (2.5 µl of 10 mM dNTP solution), MgCl2 (3mM, 6 µl of 25 mM Mgcl2 solution), common forward primer (20 pmoles) and species -specific reverse primers (20 pmoles each),Taq DNA polymerase (3U), purified DNA (50 ng each) and autoclaved milli Q water to make up the reaction volume. The multiplex reaction mixture is optimized by using species-specific reverse primers at different ratios *viz.* goat:chicken: beef:sheep:pig:horse: :0.3:2.5:0.5:2.5: 0.5:1.8. The PCR tubes are splash spun in a centrifuge to get the reactants at the bottom. The reactions are performed in a PCR System.

Contd.

Different Steps	*Species-Specific PCR [2,3,4,5,6]*	*Multiplex PCR [2,3]*
PCR Programme	The cycling conditions include an initial denaturation (95°C for 5 min) followed by 30 cycles of denaturation (95°C, 30s), primer annealing (55°C, 30s) and extension (72°C, 30s). After the final extension (95°C, 5 min), the PCR products are held at 4°C until electrophoresis.	The cycling conditions are optimized with 1 cycle of initial denaturation(at 94ºC for 5 minutes), 33 cycles each of denaturation (940 for 45 seconds), annealing (61ºC for 45 seconds), extension (72ºC for 45 seconds) and a final extension(at 72ºCfor10minutes).After completing the reaction, tubes with PCR products are held at 4ºC until further analysis.
Analysis of PCR Produts	Agarose gel (2 per cent) was prepared in 1X TBE buffer and the PCR products (5µl) stained with 6X gel loading dye (1 µl) are electrophoresed at 40-60 V for 1-3 hrs. The amplified products are visualized and confirmed over a gel documentation system. The relative molecular weight of the amplicons is calculated against a 100 bp ladder (Fermentas).	After PCR amplification, the products require to be resolved for detection of species. Reagents-i- 3 per cent (w/v) agarose gel,ii- 5X TBE buffer (54 gram Tris base, 27.5 gram boric acid and 20ml of 0.5M EDTA (pH 8.0),iii-Ethidium bromide (1:10 diluted 20 mg/ml),iv-1.0X TBE buffer,v-6X Orange loading dye and vi-100 bp DNA ladder. Apparatus-i-Horizontal submarine gel electrophoresis unit,ii-UV trans-illuminator and iii-Gel documentation system. Procedure-The amplified PCR products are analyzed using horizontal submarine gel electrophoresis.The casting tray is prepared and the acrylic gel comb was placed over the tray in such a way that a gap of 0.5 mm space is left between the tips of the comb teeth and the floor of the casting tray. A 3 per cent (w/v) agarose suspension [1.2 gram agarose + 4 ml 5X TBE buffer (54 gram Tris base, 27.5 gram boric acid and 20ml of 0.5M EDTA (pH 8.0) and volume was made upto 1 litre) + 36 ml distilled water] in 0.5X TBE buffer is prepared and heated on an electric heater until the suspension become transparent. The solution is then cooled to 60ºC and 10 µl of ethidium bromide (1:10 diluted 20 mg/ml) is added and mixed thoroughly by gentle agitation. The agarose solution is then poured into a leveled casting tray in such a way that a 4 mm thickness gel was formed. Subsequently, the agarose is allowed to set and the comb is gently removed. The casted gel along with the running tray is submerged in the electrophoresis tank containing 1.0X TBE buffer.A volume of 6 µl of PCR product is mixed with 2 µl of 6X Orange loading dye and then loaded into the well. A separate well charged with 100 bp DNA ladder is also allowed to run along with the PCR products. Electrophoresis is performed at 7 volts/cm (60-70 volts) for 2 hours and the gel is observed under UV trans-illuminator and analyzed through gel documentation system. Interpretation-The size of the amplicons obtained are 157, 227, 274,331, 398 and 439 bp in goat, chicken, cattle, sheep, pig and horse, respectively.

DNA in a span of 2-3 hours thereby producing DNA that could be documented thereby allowing meat speciation.

b. Multiplex PCR [1]

Multiplex PCR is used to detect several meat species simultaneously.

Principle

Multiplex PCR offers a short cut to many individual PCRs as many targets are simultaneously amplified.Keeping a primer common, species specific forward/ reverse primers could be used for the simultaneous detection of many species. Multiplex PCR saves time, labor and it is a good tool for the mass screening of samples that might be confirmed later by species -specific PCR.

The details related to the Reagents, Apparatus, Procedure, PCR Programme and analysis of PCR products by electrophoresis for both Species-Specific PCR and MultiplexPCR are given in Table (given on previous page).

Primers for Species-Specific PCR and Multiplex PCR

The different primers for both types of PCR are detailed in Tables giving on next page.

a. Species-Specific PCR Primers

The species-specific primers found suitable for meat speciation are given in Table giving on next page.

b. Multiplex PCR Primers

The primers for mitochondrial cytochromeb gene used in multiplex PCR are custom synthesized. The common forward primer as well as species -specific reverse primers as reported by Matsunaga *et al.* [3] are used in multiplex PCR technique. The details of the primers used are given in following table:

Common Forward Primer [3]

Sl.No.	Sequence	Length
1.	5'-GACCTCCCAGCTCCATCAAACATCTCATCTTGATGAAA-3'	38

Species-Specific Reverse Primers [3]

Sl.No.	Species	Sequence	Length
1.	Goat	5'-CTCGACAAATGTGAGTTACAGAGGGA-3'	26
2.	Chicken	5'-AAGATACAGATGAAGAAGAATGAGGCG-3'	27
3.	Cattle	5'-CTAGAAAAGTGTAAGACCCGTAATATAAG-3'	29
4.	Sheep	5'-CTATGAATGCTGTGGCTATTGTCGCA-3'	26
5.	Pig	5'-GCTGATAGTAGATTTGTGATGACCGTA-3'	27
6.	Horse	5'-CTCAGATTCACTCGACGAGGGTAGTA-3'	26

Some Species-Specific Primers found Suitable for Meat Speciation*

Species	*Target*	*Primer*	*Direction*	*Sequence*	*Primer Size*	*Temp. °C*	*Spp. Specific Amplicon*
Cattle	mtD-loop	VPH-CF	Forward	5‘-TAT CAA AAA TCC CAA TAA CTC AACACA-3‘	27bp	52.0	381 bp
		VPH-CR	Reverse	5‘-GGG CCCGGA GCG AGAAG-3‘	17bp	61.6	-
Buffalo	mtD-loop	VPH-BF	Forward	5‘-TAG AAA TAA CTG CAA CCA TCAACAC-3‘	25bp	53.0	534 bp
		VPH-BR	Reverse	5‘-GTC CAAGCA TCCCCCAAAAT-3‘	20bp	52.0	-
Sheep	mtD-loop	VPH-CF	Forward	5‘-CCA CCCACGGACACGAG-3‘	17bp	58.5	329 bp and
		VPH-CR	Reverse	5‘-AGT TCA ATG CCC TAT ATG CTT CAG-3‘	24bp	55.0	404 bp
Goat	mtD-loop	VPH-GF	Forward	5‘-TCC CAC TCC ACA AGC TTA CAG ACA-3‘	24bp	59.9	436 bp
		VPH-GR	Reverse	5‘-GCT TATATGCATGGG GGA GAT-3‘	21bp	54.5	-
Pig	mtD-loop	VPH-PF	Forward	5‘-AAT TTT TGGGGA TGCTTA GACT-3‘	22bp	52.5	712 bp
		VPH-PR	Reverse	5‘-TAT TTT GGG AGG TTA TTG TGT TGTA-3‘	25bp	52.6	-
Chiken	5-amino levulinate sythase gene	VPH-ChF	Forward	5‘-CCG GCA GCA GAG CAG ACT AACAAC-3‘	24bp	62.0	288 bp
		VPH-ChR	Reverse	5‘-AGG GGA TAC GCC GACTGCTGA G-3‘	22bp	63.1	-

* These primers are designed and validated under a DBT sponsored project at Deptt. Veterinary Public Health, College of Veterinary and Animal Sciences, G.B. Pant Universty of Agriculture and Technology, Pantanagar.

Source

For Species-Specific PCR

1. *Manual of methods of analysis of Foods (Meat and Meat Produts/Fish and Fish Products), LabManual-6, FSSAI, New Delhi.*
2. *Nagappa S. Karabasanavar, S. P. Singh, Umapathi V, Girish Patil, Sunil N.Shebannavar and Deepak Kumar.2011. Authentication of cara beef (water buffalo, Bubalus bublis) using highly specific polymerase chain reaction. European Food Research and Technology. 233: 985-989.*
3. *Nagappa S. Karabasanavar, S. P. Singh, Umapathi V, Deepak Kumar, GirishPatil and Sunil N. Shebannavar. 2011. A highly specific PCR assay for identification of raw and heat treated mutton (Ovis aries). Small Ruminant Research. 100: 153-158.*
4. *Nagappa S. Karabasanavar, S.P. Singh, Umapathi V., Deepak Kumar and Sunil N. Shebannavar. 2011. Identification of goat meat using highly species -specific polymerase chain reaction. Journal of Food Quality, 34 (2): 142-149.*
5. *Nagappa S. Karabasanavar. 2008. Molecular identification of animal speciesusing Polymerase Chain Reaction based techniques. Ph.D Thesis Submitted to the Govind Ballabh Pant University of Agriculture and Technology, P antnagar, Uttarakhand, India.*

For Multiplex PCR

1. *Manual of methods of analysis of Foods (Meat and Meat Produts/Fish and Fish Products), LabManual-6. FSSAI, New Delhi.*
2. *Sambrook, J. and Russell, D. W. 2001. Molecular cloning: A laboratory manual. 3rd edition.Cold Spring Harbour Laboratory Press, Cold Spring Harbour, New York.*
3. *Matsunaga, T., Chikuni, K., Tanabe, R., Muroya, S., Shibata, K., Yamada, J. and Shinmura, Y.1999a. A quick and simple method for the identification of meat species and meat products by PCR as say. Meat Science. 51: 143-148.*

2. Restriction Fragment Length Polymorphism (RFLP-Method2) [3]

Principle

PCR-RFLP is a rapid and versatile tool but often requires technical competence in interpretation of results. In PCR-RFLP, the conserved target gene is amplified and the resultant PCR product is digested with specific restriction endonuclease enzyme (s) to get a restriction pattern (fingerprint). These restriction enzymes splice the DNA at specific sites to yield species -specific pattern. The mutated nucleotides are recognized by the restriction enzymes leading to a specific recognition of the sequence. The mitochondria accumulate 10 times more mutations per unit time compared to the nuclear sequences. Hence, mt DNA has been the most targeted sequence. Also, PCR-RFLP has been widely used for species identification especially targeting the **mt cyt-b gene**.

Different restriction enzymes (AfiIII, AluI, ApaI, AvaII, BamHI, DraI EcoRI, HinfI, NcoI, NspI, PstI, RsaI, SfuI, StyI and TaqI) could be used depending on the

Different Steps	*PCR-RFLP [1,2,3]*	*FINS [1,2,3,4]*
Reagents	i-10X reaction buffer, ii-200μM of each dNTP, iii-2.5mM MgCl2, iv-20 pmoles each of forward and reverse primers, v-2U of Taq DNA polymerase, vi- 50ng of purified DNA and vii-Autoclaved milli Q water.	A-PCR amplification of a conserved sequence-For the identification and confirmation of the unknown species, the DNA is isolated and PCR is performed using universal primers as reported by Kocheretal. [1] against the flanking sequence of mt12SrRNA gene. Conserved sequnces in sufficient lenth carry information in the algorithm of ATGC that could establish origin of species thereby aiding in meat speciation. Reagents-PCR reaction mix- i- 10X Assay buffer [160 mM $(NH_4)_2SO_4$, 670 mM Tris-HCl, pH 8.8, 0.1 per cent tween-20, 25 mM $MgCl_2$],ii- 1 μl (200μM each) of dNTP mix [sodium salts of dATP, dCTP, dGTP and dTTP 10 mM each in water *i.e.*, 40 mM total pH 7.5],iii-1 μl or 20 Pico moles each of forward (5‘ -CAA ACT GGG ATT AGA TAC CCC ACT AT-3‘ 26 mer) and reverse (5‘- GAG GGT GAC GGG CGG TGT GT-3‘ 20 mer) primers,iv-1.66 U Taq DNA polymerase e. 50 ng of purified DNA,v- Autoclaved nuclease free water and v-Agarose gel (2 per cent in 1x TBE and 0.5 μg/ml ethidium bromide).
Apparatus	i-500 μl microtubes,ii-Thermocycler and iii-Pipettes,0.1 ml, 1 ml.	i-Thermal cycler,ii- Electrophoresis unit,iii-UV illumination and iv- Gel documenttion system.
Procedure	The method as described by Partis *et al.* [1] is followed for PCR-RFLP. PCR amplification reaction is set in a volume of 50 μl with 5 μl of 10X reaction buffer, 200μM of each dNTP, 2.5mM MgCl2, 20 pmoles each of forward and reverse primers, 2U of Taq DNA polymerase, 50ng of purified DNA and autoclaved milli Q water to make up the reaction volume. The reaction is performed within 500 μl microtubes and PCR amplification is carried out using a PCR System 9700. The reaction conditions used for PCR amplification consisted of 1 cycle of initial denaturation (at 94ºC for 5 minutes), 35 cycles each of denaturation (94ºC for 45 seconds), annealing (60ºC for 45 seconds), extension (72ºC for 45 seconds) and a final extension at 72ºC for 10 minutes. After completing the reaction, tubes with PCR products are held at 4ºC until further analysis.	PCR is performed in a 50 μl reaction volume; consisting of 5 μl of 10X Assay buffer [160 mM (NH4)2SO4, 670 mM Tris-HCl, pH 8.8, 0.1 per cent tween-20, 25 mM MgCl2] ; 1 μl (200μM each) of dNTP mix [sodium salts of dATP, dCTP, dGTP and dTTP 10 mM each in water *i.e.*, 40 mM total pH 7.5], 1μl or 20 Pico moles each of forward (5‘-CAA ACT GGG ATT AGA TAC CCC ACT AT -3‘ 26 mer) and reverse (5‘- GAG GGT GAC GGG CGG TGT GT-3‘ 20 mer) primers; 1.66 U Taq DNA polymerase, 50 ng of purified DNA and autoclaved NFW to make the final volume. The tubes flash spun and the PCR was performed using a Thermal cycler.

Contd.

Different Steps	PCR-RFLP [1,2,3]	FINS [1,2,3,4]
Purification of PCR amplified product	The PCR amplified products are subjected to purification using commercially available purification kit as per the protocol of the kit.	-
Analysis	The PCR-RFLP is performed as per the protocol described by Partis *et al.* [1].PCR products obtained from cytochrome b gene are subjected to digestion with suitable restriction enzymes (Alu I, Hinf I, Taq I and Rsa I). Reagents-a- Enzyme buffer mix- 2 µl of 10X buffer and 1 µl of restriction enzyme-i-. Alu I,ii- Hinf I and iii- Rsa I.b- Reaction mix- i-10 µl of purified PCR product and 3 µl of enzyme buffer mix and the volume is made upto 30 µl using autoclaved Milli Q water, c-3 µl of loading dye, d-100 bp ladder and e-2.5 per cent agarose gel. Apparatus-i-Water bath (37ºC),ii-Electrophoresis unit, iii-UV trans-illuminator and iv-Gel documentation system. Procedure-Enzyme buffer mix is prepared by mixing 2 µl of recommended 10X buffer and 1 µl of restriction enzyme.The reaction mix is prepared by mixing 10 µl of purified PCR product with 3 µl of enzyme buffer mix and the volume is made upto 30 µl using autoclaved Milli Q water.The reaction mixture is subsequ-ently incubated at 37ºC for Alu I, Hinf I and Rsa I and 65ºC for Taq I for 12 hours. Thereafter, the enzyme activity is arrested by adding 3 µl of loading dye to the above reaction mixture. The digested product is finally subjected to electrophoresis along with 100 bp ladder in 2.5 per cent agarose gel at 7 volts/cm for 2 hours. The bands are then visualized under UV trans-illuminator and analyzed under gel documentation system for bands of desired molecular weight.	PCRprogram-The cycling conditions involve an initial denaturation (94°C for 5 min) followed by 30 cycles of denaturation (94°C, 45s), primer annealing (60°C,45s,) and extension (72°C,1 min); after the final extension (72°C, 5 min), the PCR products are held at 4°C. Agarose gel (2 per cent in 1x TBE and 0.5 µg/ml ethidium bromide) electrophoresis (50 V, 2 hrs) is performed and the amplified products are visualized under UV illumination and the gel is documented.

Contd.

Different Steps	*PCR-RFLP [1,2,3]*	*FINS [1,2,3,4]*
		B-Sequencing of PCR products from mt12SrRNA gene-The PCR products are sequenced using automatic DNA sequencer at a DNA sequencing facility. The sequences obtained are aligned using Megalign and compared by Clustal' method.Any other equivalent software could also be used for this purpose. C-BLAST analysis of mt 12S rRNA gene sequence-i-Log in to the NCBI site (http://www.ncbi.nlm.nih.gov/blast/Blast.c gi),ii-Nucleotide blast (BlastN) is selected,iii-The unknown sequence (Accession number,gi,or FASTA) is entered,iv- Nucleotide collection and highly similar sequences (Megablast) option are selected,v- The sequence is subjected to Blast and vi-The data retrieval is in two modes, i.e., distance tree and alignments. a- Distance tree of results: Searched by sequence ID, taxonomic name, sequence title or the Blast name. Further, a focused sub -tree as well as the alignment report of the desired species (or accessions) was b rowsed. b- In Alignments' option, the sequence show significant alignments along with their accession numbers, title, score, E (bits) value, followed by the detailed information about the length of sequence, percent identity, gaps, *etc.* Based on the distance tree and alignments with the query sequence, the nearest species with the maximum score was found and the animal species is identified.

enzyme recognition and splicing sites. The PCR-RFLP has been one of the most commonly used techniques for the differentiation of meat species.

3. Forensically Informative Nucleotide Sequencing (FINS Method-3) [4]

For meat species identification, a forensically informative fragment of a particular gene is made use of to solve the legal problems. The conserved region of a gene specific to a species is amplified and the PCR product is analyzed. A species-specific PCR would be rather appreciable better if the samples are sub-optimal in the forensic analysis over the RFLP.

The details related to the Reagents, Apparatus, Procedure, PCR Programme and analysis of PCR products by electrophoresis *etc.* for both PCR-RFLP and FINS are given in Table given on previous pages.

Primers for PCR RELP

The following primers are used for PCR RELP-

Universal Primers Used for PCR-RFLP [2]

Sl.No.	*Sequence*	*Length*
1	5'-CCATCCAACATCTCAGCATGATGAAA-3'	26
2	5'-GCCCCTCAGAATGATATTTGTCCTCA-3'	26

Interpretation for PCR RELP

Based on the following table, meat species are identified:

Restriction Enzymes and their Digested Fragments (in bp) for PCR-RFLP

RE	*Cattle*	*Buffalo*	*Sheep*	*Goat*	*Pig*	*Chicken*	*Horse*
AluI	170,189	189,170	*	*	114,245	*	303
Hinf I	117,198	*	296	161,198	*	161,188	234
RsaI	*	325	*	*	*	148,211	*
TaqI	355	161,198	*	141,218	141,218	*	*

* indicates uncut products.

Source

For PCR RELP

1. *Partis, L., Croan, D., Guo, Z., Clark, R., Coldham, T. and Murby, J.2000.Evaluation of a DNA fingerprinting method for determining the species origin of meats.Meat Science.54: 369-376.*

2. *Carr, S.M. and Marshall, H.D.1991.Detection of intraspecific DNA sequence variation in the mitochondrial cytochromeb gene of Atlantic cod (Gadus morhua) by the polymerase chain reaction.Can.Bull.Fish. Aquat.Sci./Bull.Can.Sci.Halieut. Aquat. 48 (1): 48-52.*

3. *Manual of methods of analysis of Foods (Meat and Meat Produts/Fish and Fish Products), LabManual-6. FSSAI, New Delhi.*

For FINS

1. *Kocher, T. D., Thomas, W. K., Meyer, A., Edwards, S. V., Paabo, S., Villablanca, F.X. and Wilson, A.C.1989. Dynamics of mitochondrial DNA evolution in animals: amplification and sequencing with conserved primers. Proc. Natl. Acad. Sci.U.S. A. 86: 6196-6200.*

2. *Nagappa Karabasanavar, Suresh P. Singh, Sunil N. Shebannavar, Kotresh M. Angadi, Girish S.Patil and Umapathi.V.Pillai.2010. Identification of avian species using polymerase chain reaction and sequence analysis of mitochondrial 12s rRNA gene. Veterinarski Arhiv. 80 (5): 653-661.*

3. *Website: http://www.ncbi.nlm.nih.gov/b last.*

4. *Manual of methods of analysis of Foods (Meat and Meat Produts/Fish and Fish Products), Lab Manual-6, FSSAI, New Delhi.*

Index

B

C

F

L

M

Q

R

U

V

W

X